Telecommunication services for developing economies

Telecommunication services for developing economies

Proceedings of the ITC Specialist Seminar
Cracow, Poland, 22-27 April, 1991

Edited by
Janusz Filipiak

*Telecommunications Department
The University of Mining and Metallurgy
Cracow, Poland*

1991

ELSEVIER
AMSTERDAM • LONDON • NEW YORK • TOKYO

ELSEVIER SCIENCE PUBLISHERS B.V.
Sara Burgerhartstraat 25
P.O. Box 211, 1000 AE Amsterdam, The Netherlands

Distributors for the U.S.A. and Canada:

ELSEVIER SCIENCE PUBLISHING COMPANY, INC.
655 Avenue of the Americas
New York, N.Y. 10010, U.S.A.

Library of Congress Cataloging-in-Publication Data

```
ITC Specialist Seminar (1991 : Kraków, Poland)
   Telecommunication services for developing economies : proceedings
of the ITC Specialist Seminar, Cracow, Poland, 22-27 April, 1991 /
edited by Janusz Filipiak.
     p.   cm.
   Includes bibliographical references and index.
   ISBN 0-444-89332-6
   1. Telecommunication--Traffic--Congresses.  2. Telecommunication
systems--Congresses.  3. Integrated services digital networks-
-Congresses.  4. Computer networks--Congresses.  I. Filipiak,
Janusz, 1952-   .  II. Title.
TK5102.5.I85  1991
384--dc20                                                  91-38977
                                                               CIP
```

ISBN: 0 444 89332 6

©1991 ELSEVIER SCIENCE PUBLISHERS B.V., ALL RIGHTS RESERVED.

No part of this publication may be reproduced, stored in a retrieval system, or transmitted, in any form or by any means, electronic, mechanical, photocopying, recording, or otherwise, without the prior written permission of the publisher, Elsevier Science Publishers B.V., Permissions Department, P.O.Box 521, 1000 AM Amsterdam, The Netherlands.

Special regulations for readers in the U.S.A.-This publication has been registered with the Copyright Clearance Center Inc. (CCC), Salem, Massachusetts. Information can be obtained from the CCC about conditions under which photocopies of parts of this publication may be made in the U.S.A. All other copyright questions, including photocopying outside of the U.S.A., should be referred to the copyright owner, Elsevier Science Publishers B.V., unless otherwise specified.

No responsibility is assumed by the Publisher for any injury and/or damage to persons or property as a matter of products liability, negligence or otherwise, or from any use or operation of any methods, products, instructions or ideas contained in the material herein.

pp. 137-148, 329-341, 453-460: Copyright not transferred.

Printed in The Netherlands

Preface

This volume is the proceedings of the ITC'91 Specialist Seminar which was held in Cracow from 22 to 27 April 1991. The Cracow seminar follows previous meetings which were organized in Brussels (1986), Lake Como (1987), Beijing (1988), Adelaide (1989), and Morristown (1990). The seminars are coordinated by the International Advisory Council of the International Teletraffic Congresses. The First International Teletraffic Congress was held in 1955 in Denmark. Then the congresses were convened every three years.

The ITC events, both congresses and seminars, bring together people from telephone administrations, equipment manufacturers, and universities who are working in the area of network design, engineering, and administration in order to provide them with a forum for a free exchange of ideas, opinions and experience, as well as to facilitate personal contacts. The Seminar organized in Cracow had an additional dimension of including people from Eastern Europe in that exchange.

Papers contained in this volume are arranged in order of their presentation. 59 papers were given which includes three invited lectures and keynote papers delivered by K. Sriram, W. Henderson, and J. Filipiak. The papers can be broadly categorized into the following groups: Network services and evolution, Network structure and management, Traffic control in the Broadband ISDN, Performance and queueing models, High speed MAN and LAN, Packet and fast packet switching networks, Management and planning of ISDN and circuit-switching networks, Analysis and design of teletraffic systems.

As an editor of this volume and chairman of the Organizing Committee I am very indebted to Professor A. Pach, Doctors A. R. Pach, Z. Papir, M. Wroniewicz-Rams, E. Chlebus, and K. Wajda for their help and cooperation in bringing the Cracow Seminar into effect. Special thanks go to Irena Wojdylo for the work which she put to prepare this volume. Finally, I wish to acknowledge the support of the Polish Ministry of Communication.

Janusz Filipiak
Cracow, May 1991

Table of Contents

Invited Lectures

Wideband Packet Technology Modelling and Performance
K. SRIRAM ... 3

F-UNI: A Model of a Flexible User/Network Interface for B-ISDN
J. FILIPIAK ... 11

A Survey, and Some New Results, on Stochastic Petri Nets
W. HENDERSON ... 33

Network Services and Evolution

Rapid Service Delivery and Customization in a Developing Network Infrastructure
P.S. RICHARDS ... 49

Transition from Analogue to Digital Networks
Chr. ASGERSEN ... 61

Telecom Networks Evolution Towards Secure Dynamic Structures Routing Aspects of the Network Digitization
M. PIÓRO, M. de MIGUEL, I. PITA ... 73

Enhanced Telecommunications for Developing Regions
B. CRAIGNOU ... 85

Architecture and Control Aspects of Data Service Evolution to Broadband
M. WERNIK, R. KOSITPAIBOON, P. CARBONE ... 97

Network Management and Services

Traffic Management of Corporate Utility Networks
V.R. SAKSENA, T.J. SCHONFELD ... 111

On Telecommunication Networks Throughput: Analysis of the CCITT Recommendations and 2nd Study Group Questions
M.A. SCHNEPS-SCHNEPPE ... 123

Experiences and Expectations of Introducing PSTN New Services in Japan
J. MIZUSAWA, M. AKIYAMA ... 137

Telecommunication Services - User Requirements
G.A. HORNE ... 149

Network Structure and Management

A Statistical Study of Real-Time Telephone Traffic
Variations for Network Management
D. STERN — 163

Off- Vs. On-line Network Management - Implementation and Performance
Study of Circuit-Switched Traffic Control Methods
E. CHLEBUS — 169

Behavior Characterization of Alternate Routing in a Non-Hierarchical
Homogeneous Network
Y. ONOZATO, J. KANIYIL, S. NOGUCHI — 185

The Communication Spanning Tree Problem: An Heuristic Algorithm
F.J.M. SALZBORN — 199

Flexible Protection of Transmission Resources
J. LUBACZ, M. JAROCIŃSKI, A. TOMASZEWSKI, O.G. SOTO — 207

Information Transfer Techniques, B-ISDN and ATM

Traffic Control in the B-ISDN
J.W. ROBERTS — 221

An Efficient Approach to Analyze Discrete-Time Queues with Markovian
Arrivals and Services in BISDNs
Z. ZHANG — 233

A Queueing Model for Buffer Overflow in Multicast Communications
M.N. YUNUS — 245

Optimization of ATM Multi-Service Networks - Some Early Investigations
A.H. ROOSMA — 257

Consecutive Cell Loss and Buffer Level Fluctuation Rate:
A New Set of ATM Performance Parameters
F.M. BROCHIN — 269

Analytical Expressions for Blocking Probabilities in a B-ISDN
J.M. KARLSSON — 281

Buffer Dimensioning and Effective Bandwidth Allocation in ATM
Based Networks with Priorities
Z. DZIONG, K.-Q. LIAO, L. MASON — 293

Connection Admission Control in ATM Networks
E. DUTKIEWICZ, G. ANIDO — 305

A Two-Layer Optimization Structure for Access Control and Bandwidth
Sharing in High-Speed Integrated Networks
R. BOLLA, F. DAVOLI — 317

Performance Analysis of Resource Allocation and Routing Techniques
in B-ISDN
T. UHL, J. ULMER — 329

Generalised Karlsson Measurements for ATM-Networks
B. VEIRØ — 343

Performance and Queueing Models

Performance Models for Hybrid Broadband Networks
C. ROSENBERG, A. Le BON 353

Buffer Sizing in Bulk Service Systems
G. HÉBUTERNE, C. ROSENBERG 365

Performance Analysis of the $H_2/G/1$ System
I. EHRIEL 371

High Speed LAN and MAN

An Alternative Solution to the Electro-Optic and Service Bottleneck Problems in Multi Gbits/s LANs: The SUPERLAN Architecture
A. POPESCU, R.P. SINGH 381

A Study on Performance Improvement Algorithm in DQDB MAN
T. YOKOTANI, H. SATO, S. NAKATSUKA 397

The Adaptive Leaky Bucket: A New Approach for Bandwidth Enforcement in DQDB Networks
A. LOMBARDO, S. PALAZZO, D. PANNO, R. PIGNATELLI, L. SUSANNA 409

The Cambridge Backbone Network. An Overview and Preliminary Performance
K. ZIELIŃSKI, D.J. GREAVES 417

Simulation Study of Some Aspects of Fairness in DQDB
G. DALLOS 427

Switching Techniques

On Reliability and Throughput of Internal Interconnection Networks of ATM-Switching Nodes
H. DAHMS 439

Combinatorial Features of the Clos Type Switching Network
W. KROMOŁOWSKI, M. SZYMANOWSKI 453

Comparison of Some Combinatorial Properties of Direct and Indirect Binary N-Cube Interconnection Networks
G.G. VESELOVSKY, M.V. KUPRYANOVA 461

Multistage Interconnection Network as a Fast Packet Switch
Z. HULICKI 473

Packet and Fast Packet Switching Networks

Packet Network Structures Based on Multilink Interfaces
F.E. MARTÍN, E. GRANEL 489

Study of S-ALOHA Packet Radio Networks with a Split-Channel Configuration
J. WOŹNIAK 501

Adaptive Isarithmic Flow Control in Fast Packet Switching Networks - Heavy Traffic Case
M. COTTON, L.G. MASON 513

Performance Analysis of Multi-Layer Token Ring Local Area Networks
M. KWIATKOWSKI 525

Management and Planning of ISDN and Circuit Switched Networks

Routing Optimization and Dimensioning of Networks with Revenues: Numerical Results
A. GIRARD, B. LIAU, N. BOUMZEBRA 539

Traffic Modeling in Networks with Incomplete Data
J.A. SCHMITT 551

A General Purpose Model for Circuit-Switched Networks
M. LEBOURGES 563

A Dynamic Non-Hierarchical Routing with Incomplete Data
M. KONVIT, M. BORIK 575

End-to-End Blocking in an Integrated Services Network with Link Capacity Allocation Control
H. YOSHINO, Y. HOSHIAI 581

Dynamic Adaptive Routing Algorithm for Intercity Progressive Rate Telephone Network
M.J. BORIK, M. KONVIT 591

Performance of Hybrid Switching Networks with Priority: Movable Boundary Case
M.S. MOUSTAFA, M.I. MARIE 601

On Some Teletraffic Problems of RSU Control System Investigation in Telecommunications
B. GOLDSHTEIN, S. BRUSILOVSKY, R. RERLE 607

Digital Cross Connects Application for the Future Subscriber Network
N. SOKOLOV 617

Design and Control of Telephone Systems

On Priority Assignment Problems in SPC Systems
D. BURSZTYNOWSKI, W. BURAKOWSKI, W. SYSKI 623

Overload Control of SPC-Switches Using Optimal Alarming
T. RYDÉN, G. LINDGREN 635

Design of Two-Level PSTN
M. PIÓRO, A. TOMASZEWSKI, J. LUBACZ, M. JAROCIŃSKI 647

Agent Scheduling for ACD Switches
P. NOWIKOW, K. WAJDA 655

Analysis and Design of Teletraffic Systems

Study of Message Delays in the Presence of Long Messages and
Correlated Arrivals in Signalling System No.7 Networks
M. GHASSEMI, R.A. SKOOG   663

Accelerated Simulation of Rare Events Using RESTART Method with
Hysteresis
M. VILLEN-ALTAMIRANO, J. VILLEN-ALTAMIRANO   675

The New Criterion for the Comparison of Queueing Systems and
Systems with Losses
A. KUCHERJAVY   687

Author Index   691

Invited Lectures

Survey Paper

Wideband Packet Technology Modelling and Performance

K. Sriram

AT&T Bell Laboratories, Room 3H-607, Holmdel, New Jersey 07733, U.S.A.

EXTENDED ABSTRACT

Wideband packet technology (WPT) provides a means for efficient and flexible integration of voice, voice-band data, facsimile, digital data, image, and signaling traffic. An AT&T WPT product, known as the Integrated Access and Cross-Connect System (IACS), performs the integration and packet multiplexing functions, and also provides circuit and packet cross-connect capabilities. Appropriate coding, compression, packetization, and multiplexing methods are used in the IACS to maximize bandwidth efficiency while providing excellent quality of service. The IACS is currently in use in several countries for voice/voice-band data/facsimile compression and circuit-multiplication on domestic/international communications links. This paper will present an overview of the design concepts, modelling, and performance of the IACS.

Voice, Voiceband Data, and Facsimile - Coding and Compression:

Various options are available in the IACS for voice coding[9] [10], e.g., 64 kb/s PCM (G.711), 32 kb/s ADPCM (G.721), 32 kb/s embedded ADPCM (G.727), etc. Normally, the IACS codes voice sources using 32 kb/s embedded ADPCM. The voice samples are typically packetized over 16 ms intervals, and, hence, the packet size for voice is 64 bytes plus header. The IACS also employs digital speech interpolation (DSI), which can be described as follows. Voice packets are generated only during talkspurts and not during silence periods and, thereby, many more voice sources are statistically multiplexed on a given transmission link. The actual statistical gain would be somewhat less than the inverse of the speech activity factor, which ranges from 28% to 42% depending on the user population, cultural and language characteristics. Due to the combination of 32 kb/s ADPCM coding and DSI, the IACS provides better than 4:1 compression of voice. Normally, 24 and 30 voice calls (64 kb/s PCM) are transmitted on DS1 (1.544 Mb/s) and CEPT1 (2.048 Mb/s) links, respectively. The IACS increases the voice call capacity on these links to at least 96 and 120, respectively.

The IACS uses a signal classifier to quickly detect the speed of any modems used for voiceband data calls. It can classify the modem speed and assign an appropriate coding rate, e.g., 32 kb/s ADPCM for 1.2 or 2.4 kb/s modems, 40 kb/s for 4.8 kb/s modems, and 64 kb/s for modems with 7.2 kb/s and higher speeds.

The IACS compresses G3 facsimile calls through demodulation to extract the baseband signal, and packetizes and transmits the resulting 9.6 kb/s. The receiving IACS can reconstitute, from the arriving packets, a G3 signal for delivery to the receiving facsimile machine. Thus, the facsimile calls are compressed in the IACS by about 5:1 ratio.

Digital Data:

The IACS provides digital circuit emulation (DICE) service for interfacing with 64 kb/s digital data channels or DS0A subrate channels. It can be provisioned to remove DDS idle code, or the redundant copies of DS0A format. With the virtual data link capability (VDLC) for HDLC-based protocols, the IACS detects and eliminates (for compression purpose) the HDLC flags and any network idle codes. The IACS also supports LAPD frame relay services.

Congestion Control by Block Dropping on Voice:

For congestion control, the IACS takes advantage of the fact that voice is tolerant to loss of a small fraction of the less significant information in the digitized voice samples. The packet loss requirement for voice is on the order of .1% or less. However, voice is much more tolerant to loss of less significant bits (as compared to losing whole packets). A loss of even 10% of bits (the less significant ones) causes negligible degradation in perceived voice quality[11][12].

The voice packets are organized as described below to implement block dropping[3][4][15]. Each voice source is sampled at 8 kHz rate and encoded using an embedded ADPCM scheme at 32 kbps rate. Over a 16 ms interval, 128 samples are collected and organized into a packet containing four blocks. All the least significant bits from the 128 4-bit samples are contained in block #1 of the packet, the next significant bits are contained in block #2, and the two most significant bits are contained in blocks #3 and #4. A congestion measure is defined in term of the total number of packets of voice and data currently waiting for transmission (see [3][4] for details). When the congestion measure exceeds a first threshold, the block #1 containing the least significant bits is dropped from the packet about to enter transmission. If the congestion measure exceeds a second threshold, then both blocks #1, #2 are dropped. The packet header contains a field which indicates the number of blocks that are droppable in a voice packet. This field is set to zero in voiceband data and facsimile packets so that they are protected from block dropping[15].

Traffic Smoothing due to Block Dropping:

The superposition of packet sequences generated by packetized voice sources with speech detection exhibit high burstiness (relative to a Poisson process) due to inherent correlations between successive interarrival times in the superposition stream[1]. These correlations tend to cause significantly larger queueing delays and packet losses than would be predicted by a Poisson model[1][2]. It has been shown that block dropping on voice packets significantly smooths the burstiness in the superposition packet voice process by speeding up the packet service rate during critical periods of congestion in the queue[3]. In fact, in a packet voice multiplexer with block dropping, an accurate prediction of packet delay and queue length can be obtained by modelling the superposition packet arrival process by a Poisson process. Traffic smoothing as a result of block dropping on the voice packets has significant implications in terms of affecting delay and buffer overflow reductions in packet voice/data multiplexers[3]. Block dropping also increases the system capacity by about 20 to 25 percent[3] (in addition to the 4:1 compression already gained by 32 kb/s ADPCM and DSI).

Voice Quality Determination by Subjective Listening Tests:

Performance evaluations of packetized voice have been done to understand the subjective effects of block dropping on speech quality[11][12][13]. We developed methodologies for evaluating the effects of block dropping on voice under the following conditions: (1) at a fixed load when instantaneous fluctuations occur due to talker activity/inactivity, and (2) under variable load when variations occur due to call on/off. In the experiment, block dropping was introduced under software control[12]. This allowed us to evaluate speech quality under the influences of dynamic traffic fluctuations and the concomitant temporal changes in bit-rate. (The instantaneous voice bit-rate fluctuates due to block dropping.)

Results of subjective tests demonstrated that, in general, speech quality can be affected by variations in the bit-rate over time, especially in a heavily loaded network when the mean bits per sample over a call duration is low (e.g., $\bar{b} = 3.0$). However, with prudent traffic engineering of the network, the mean bits per sample, $\bar{b}$, over a call duration can be maintained fairly high (e.g., $\bar{b} = 3.7$), and then the voice quality remains robust to temporal variations in bits per sample resulting from nominal load fluctuations[12].

Dynamic Bandwidth Allocation for Voice and Data:

The IACS multiplexes three kinds of packets: signaling, voice (including voiceband data and facsimile), and digital data. The signaling packets are given highest priority, and are transmitted shortly after they are received in the signaling queue. This guarantees that the signaling packets experience negligible delay and zero packet loss. When there are no packets in the signaling queue, either voice or data packets are served according to the (T_1, T_2) - scheme[4]. First the waiting voice packets are served until a maximum of T_1 ms or until the voice queue is exhausted, whichever occurs first. Then the data packets are served likewise with a corresponding time slice allocation of T_2 ms. Both voice and data queues are subject to interruptions (upon completion of the packet in service) to serve any signaling packets that may have been received in the signaling queue while the service for either voice or data was in progress. During such excursions to the signaling queue, the T_1/T_2 timer for the voice/data queue is suspended. And the timer is resumed when the signaling queue is exhausted, and the service is returned to the interrupted voice or data queue.

Usually the signaling traffic intensity is very small compared to aggregate voice and data traffic intensities. Therefore, the (T_1, T_2) - scheme essentially allocates the bandwidth on the transmission link to the aggregate voice and data traffic in the ratio of T_1 to T_2. In other words, this scheme guarantees a minimum bandwidth of $\{T_1/(T_1+T_2)\}C$ for the aggregate voice traffic and $\{T_2/(T_1+T_2)\}C$ for the aggregate data traffic, where C is the overall transmission capacity of the link (signaling traffic is assumed to use only a negligible portion of C). Thus the priority scheme provides protection to each type of traffic so long as that traffic remains within its guaranteed bandwidth. The values of the transmission intervals T_1 and T_2 for selecting from the voice and data queues, respectively, can be selected to accommodate packet delay requirements for voice and data traffic. A duration of the voice transmission interval T_1 that is much larger than the data transmission interval T_2 will decrease delays for voice packets at the expense of increased delays for data packets, and vice versa. The values of the intervals T_1 and T_2 can be chosen either to reserve certain minimum bandwidth proportions for voice and data or to adjust delays for voice and data packets, as required. Detailed performance results on the (T_1, T_2) - scheme are reported in [4]. Extensions of the methods presented here to broadband ATM networks can be found in [46][47].

Acknowledgements:

The author wishes to acknowledge many colleagues at the AT&T Bell Laboratories with whom he has worked over the past several years on various issues related to the design, modelling, and performance of wideband packet technology and the IACS.

REFERENCES & BIBLIOGRAPHY

The following bibliography (by no means exhaustive) lists publications on wideband packet technology (WPT), the AT&T Integrated Access and Cross-connect System (IACS), and some related papers on speech coding, packetized voice, and voice support in broadband ATM networks.

MODELLING AND PERFORMANCE

1. K. Sriram and W. Whitt, "Characterizing superposition arrival processes in packet multiplexers for voice and data," *IEEE Journal on Selected Areas in Commun.*, vol. SAC-4, no. 6, September 1986, pp. 833-846.

2. H. Heffes and D. M. Lucantoni, "A Markov-modulated characterization of packetized voice and data traffic and related statistical multiplexer performance," *IEEE Journal on Selected Areas in Commun.*, vol. SAC-4, no. 6, September 1986, pp. 856-868.

3. K. Sriram and D. M. Lucantoni, "Traffic Smoothing Effects of Bit Dropping in a Packet Voice Multiplexer," the *IEEE Trans. on Commun.*, July 1989, pp. 703-712.

4. K. Sriram, "Dynamic Bandwidth Allocation and Congestion Control Schemes for Voice and Data Integration in Wideband Packet Technology," *Proc. of the IEEE Supercomm/ICC'90*, Atlanta, Georgia, April 1990, vol. 3, pp. 1003-1009.

5. K. Sriram, G. A. Mariano, D. O. Bowker, and W.J. Giguere, "An Integrated Access Terminal for Wideband Packet Networking: Design and Performance Overview," *Proc. of the International Switching Symposium*, Stockholm, Sweden, June 1990, vol. 6, pp.17-24.

6. S. Dravida and K. Sriram, "End-to-End Performance Models for Variable Bit Rate Voice over Tandem Links in Packet Networks," *IEEE Journal on Selected Areas in Communications*: Special Issue on Voice and Visual Communications over Packet Networks, pp. 718-728, June 1989.

7. S. Dravida and V. R. Saksena, "Analysis of a Packet Voice Multiplexer," *Proceedings of the 31st Midwest Symposium on Circuits and Systems*, St. Louis, August 1988.

8. S.G. Eick, "Access interface congestion controls for packetized voice transport in wideband networks," *Proc. of the IEEE GLOBECOM'88*, Hollywood, Florida, November 1988, pp. 226-230.

VOICE CODING, PERCEPTUAL QUALITY, PACKETIZATION AND PROTOCOL

9. M. H. Sherif, D. O. Bowker, G. Bertocci, B. A. Orford, and G. A. Mariano, "Overview of CCITT Embedded ADPCM Algorithms," *Proc. of IEEE Supercomm/ICC'90*, Atlanta, Georgia, April 1990, vol. 3, pp. 1014-1018.

10. M. H. Sherif, G. Bertocci, D. O. Bowker, B. A. Orford, and G. A. Mariano, "Overview and performance of CCITT/ANSI embedded ADPCM algorithms," accepted for publication in *IEEE Trans. Commun.*.

11. D. O. Bowker and C. A. Dvorak, "Speech Transmission Quality of Wideband Packet Technology," *Proc. of IEEE GLOBECOM'87*, Tokyo, Japan, November 1987, pp.1887-1889.

12. V. R. Karanam, K. Sriram, and D. O. Bowker, "Performance Evaluation of Variable Bit Rate Voice in Packet Networks," *AT&T Technical Journal*: Special Issue on Performance Modeling and Analysis, September-October 1988, pp. 41-56. Also in part in the *Proc. of GLOBECOM'88*, Hollywood, Florida, November 1988, pp.1617-1622.

13. D. O. Bowker and C. B. Armitage, "Performance Issues for Packetized Voice Communications", *Proc. National Comm. Forum* 41, (3), pp. 1087-1092 (1987).

14. J. M. Holtzman, "The interaction between queueing and voice quality in variable bit rate packet voice systems," *Proc. ITC 11*, Kyoto, Japan, Sept. 4-11, 1985, Paper 2.2A-4.

15. M.H. Sherif, R.J. Clark, and G.P. Forcina, "CCITT/ANSI Voice Packetization Protocol," *International Journal of Satellite Communications*, vol. 8, pp. 429-436 (1990).

16. "CCITT Recommendation G.727 - 5-, 4-, 3-, 2- Bits/Sample Embedded ADPCM", July 1990.

17. "CCITT Recommendation G.764 - Voice Packetization: Packetized Voice Protocol," July 1990.

18. "Facsimile Transport in Packet Circuit Multiplication Equipment," AT&T's contribution to CCITT SG XV, July 1990.

SYSTEM OVERVIEW AND APPLICATIONS

19. D. Sparrell, "Wideband Packet Technology," *Proc. of IEEE GLOBECOM'88*, Hollywood, Florida, November 1988, pp.1612-1616.

20. W.J. Giguere, "New applications of Wideband Technology," *Proc. of the IEEE Supercomm/ICC'90*, Atlanta, Georgia, April 1990, vol. 3, pp. 997-999.

21. M.H. Sherif, A.D. Malaret-Collazo, and M.C. Gruensfelder, "Wideband Packet Technology in the Integrated Access and Cross-Connect System (IACS)," *International Journal of Satellite Communications*, vol. 8, pp.437-444 (1990).

22. A. H. Daecher, "On the Road to Universal Information Services (UIS): Wideband Packet Technology", Entelec 89

23. M. H. Sherif, F. D. Fite and M. C. Gruensfelder, "On the Road to Universal Information Services: The Integrated Access Terminal", Entelec 89

24. S. B. Andrews, E. J. Messerli, and G. W. R. Luderer, "Faster Packet for Tomorrow's Telecommunications", AT&T Technology - Products, Systems and Services, Volume 3, Number 4, 1988, pp 24-33.

25. R. W. Muise, T. J. Schonfeld and G. H. Zimmerman, "Experiments in wideband packet technology," *Proc. Zurich Seminar on Digital Communication*, March 1986, pp. D4.1-D4.5.

26. G. W. R. Luderer, J. J. Mansell, E. J. Messerli, R. E. Staehler, A. K. Vaidya, "Wideband Packet Technology for Switching Systems", *Proc. of the International Switching Symposium*, Phoenix, Arizona, March 1987.

27. J. J. Kulzer and W. A. Montgomery, "Statistical Switching Architectures for Future Services", *Proc. of the International Switching Symp.*, Florence, Italy, May 1984.

OTHER RELATED PUBLICATIONS

28. T. Bially, B. Gold and S. Seneff, " A technique for adaptive voice flow control in integrated packet networks," *IEEE Trans. Commun.*, Vol. COM-28, March 1980, pp. 325-333.

29. T. Bially, A. J. McLaughlin and C. J. Weinstein, "Voice communication in integrated digital voice and data networks," *IEEE Trans. Commun.*, Vol. COM-28, September 1980, pp. 1478-1490.

30. R. V. Cox and R. E. Crochiere, "Multiple user variable rate coding for TASI and packet transmission systems," *IEEE Trans. Commun.*, Vol. COM-28, March 1980, pp. 334-337.

31. D. J. Goodman, "Embedded DPCM for variable bit rate transmission," *IEEE Tran. on Commun.*, COM-28, No. 7, July 1980.

32. Y. Yatsuzuka, "High-Gain Digital Speech Interpolation with ADPCM Encoding," *IEEE Tran. on Commun.*, COM-30, No. 7, April 1982, pp. 750-761.

33. M. Listani and F. Villani, "Voice communication handling in X.25 packet switching networks," *Proc. IEEE GLOBECOM'83*, San Diego, November 1983, Paper 2.4.

34. D. Grillo, F. Villani, M. Calabrese and R. Pietrojusti, "Impact of low and high bit rate coding in an integrated packet network," *Proc. International Switching Symposium,* Florence, Italy, May 1984, Paper 42B1.

35. D. Grillo, "Interactive voice application handling in wide area packet switched networks," *Proc. ITC 11*, Kyoto, Japan, Sept. 1985, Paper 5.4A-4.

36. C. M. Corbalis, "A design example of a T1-based fast packet voice switch," *Proc. of the IEEE International Conference on Commun.*, Seattle, June 1987, pp. 36-3.1-36.3.5.

37. H. Saito, "Optimal control of variable rate coding in integrated voice/data packet networks," *IEICEJ Technical Report*, January 1988, pp. 13-18.

38. N. Yin, S. Q. Li, and T. E. Stern, "Congestion Control for Packet Voice by Selective Packet Discarding," *Proc. of GLOBECOM'87*, Tokyo, Japan, November 1987, vol. 3, pp.1782-1786.

39. D. W. Petr, L. A. DaSilva, Jr., and V. S. Frost, "Priority discarding of speech in integrated packet networks," *IEEE J. Selected Areas in Commun.*, June 1989, pp. 644-656.

40. J. Suzuki and M. Taka, "Missing packet recovery techniques for low-bit-rate coded speech," *IEEE Journal on Selected Areas in Commun.*, June 1989, pp. 707-717.

41. J. G. Gruber, "Delay related issues in integrated voice and data networks," *IEEE Trans. on Comm.*, June 1981, pp. 787-800.

42. N.S. Jayant and S.W. Christensen, "Effect of packet losses in waveform coded speech and improvements due to an odd-even sample interpolation procedure," *IEEE Trans. Commun.*, pp. 101-109, February 1981.

POSSIBLE EXTENSIONS OF WPT CONCEPTS TO BROADBAND ATM NETWORKS

43. N. Kitawaki, H. Nagabuchi, M. Taka, and K. Takahashi, "Speech coding technology for ATM networks," *IEEE Communications Magazine*, pp. 21-27, January 1990.

44. H. Nakada and K.-I. Sato, "Variable Rate Speech Coding for Asynchronous Transfer Mode," *IEEE Trans. on Comm.*, March 1990, pp. 277-284.

45. K. Kondo and M. Ohno, "Variable Embedded ADPCM Coding Scheme for Packet Speech on ATM Networks," *Proc. of IEEE GLOBECOM'90*, San Diego, CA, Dec. 1990, pp. 523-527.

46. K. Sriram, "Methodologies for Bandwidth Allocation, Congestion Avoidance, and Transmission Scheduling in Broadband ATM Networks," to be published.

47. K. Sriram, R.S. McKinney, and M.H. Sherif, "Voice Packetization and Compression in Broadband ATM Networks," *IEEE Journal on Selected Areas in Commun.*, April 1991. (Also, appeared in part in *Proc. of the ITC Seminar*, Morristown, New Jersey, October 1990.)

F-UNI: A Model of a Flexible User/Network Interface for B-ISDN[1]

Janusz Filipiak

Telecommunications Department, The University of Mining and Metallurgy,
Al. Mickiewicza 30, Pl-30-059 Cracow, Poland

1 Introduction

This paper is aimed at introduction of a new model of the user/network interface in a broadband ISDN, which we shall call the Flexible User Network Interface (F-UNI). The F-UNI model is developed to encompass new features of ATM, ISDN, Open Network Architecture (ONA), Intelligent Network (IN), and Personal Communication Network (PCN). It is an abstraction of existing interfaces.

Basically, we view the UNI as a set of program units though some simple functions are usually executed by a dedicated logic. In Section 2 the intricacy of interactions between the program units is illustrated. We indicate that in order to efficiently design the interface software it must be structured. Working out a consistent structural model or in other words the UNI architecture is a necessary first step before the international standardization activities can be undertaken.

In order to structure the interface software we first develop *horizontal* and *vertical views* of the interface. This is done in Section 3. In fact, in that section we construct a three dimensional F-UNI space with dimensions determined by the decisionmaking level, service type, and functional distance to the core network. The vertical view of the interface is obtained by cutting cross-wise through the program units performing management and control functions.

The F-UNI architecture encompasses the existing standards for ISDN and ATM as well as the Open System Interconnection model and Signalling System No.7. The F-UNI picture of the OSI model is presented in Section 4. Remaining sections of the report present ATM protocols.

2 Passive and active program units

The UNI is activated by external events. While inactive, the interface can be treated as a set of passive program units kept in a database. One active program module (called

---

[1]This work was done under contract of the Teletraffic Research Center, The University of Adelaide, Australia, with Overseas Telecommunication Commission, Australia

MINITOR or SUPERVISER) is needed to monitor input ports. When a service request is detected some program units are activated. The active program units are called *decision units* (DEUs).

The DEUs communicate and invoke other DEUSs to provide the requested service. An actual method of activating and passing control between computer programs in time sharing system depends on the operating system in use. Moreover, the interface software can be implemented in the multiprocessor environment.

To illustrate different situations which may occur consider two examples.

EXAMPLE 1. Task T is broken down into several subtasks ST_i, $i = 1...3$, which are performed by the program units $P(ST_i)$. The $P(ST_i)'s$ are subroutines of the program $P(T)$, which invokes them one by one in a predetermined sequence using the subroutine call. After the subroutine $P(ST_i)$ is executed the control returns to the main program $P(T)$, see Fig. 1a.

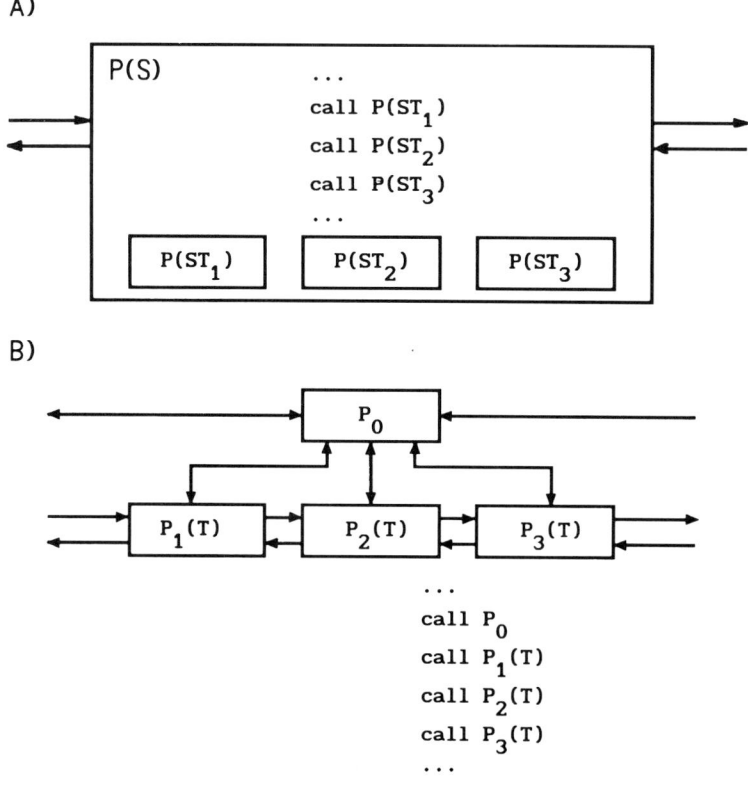

Figure 1

EXAMPLE 2. Consider task T which is processed by consecutive processors (active programs, DEUs) $P_1(T)$, $P_2(T)$, and $P_3(T)$. There is another active program unit P_0, which monitors operation of P_1, P_2, and P_3, setting up parameters of P_i modules and possibly changing their mode of operation, Fig. 1b. We shall call P_0 the manager (*or the management DEU*). Usually, it is assumed that the control and other information is exchanged between $P'_i s$ by means of primitives.

Note that in contrast to the management module P_0 in Example 2 the DEU $P(T)$ in Example 1 participates in execution of an actual task. DEUs such as $P(T)$ will be called the *control DEUs*.

Comment 1 Each DEU in Example 2 can have an internal structure similar to the one shown in Fig. 1a, and *vice versa*, the management DEU can be appended to the system considered in Example 1. The new constructs illustrate how new designs grow in complexity.

Comment 2 In Example 2 we have not specified how primitives are exchanged between DEUs. In what follows we shall concentrate on the functional system description. This means that we shall define a type and format of exchanged information rather than the detailed implementation dependent features.

Comment 3 The set of DEUs interacting to perform a given task (function) can be conceived as one aggregated DEU. By the same token, the aggregated DEU can be decomposed into several DEUs including management and control DEUs.

Comment 4 As regards an internal DEU structure we shall assume that the DEU has input and output ports, internal memory and algorithm. Executing an algorithm it can call external and internal programs and subroutines and in that way activate other DEUs. The algorithm written in Nonprocedural Language reacts to external events.

Comment 5 The DEU can be activated to perform subsequent tasks one by one. Alternatively, several tasks can be performed concurrently by one DEU.

Comment 6 The distinction between processors, control DEUs and management DEUs is relative and done according to the performed function. To give an example, from the viewpoint of a high level DEU the management DEU can be treated as a processor.

Comment 7 If the performed task (performed function, provided service) is relatively simple then usually the main program is broken down into several subroutines as has been illustrated in Example 1. If the task is complicated, then the separate subtasks are defined and implemented in a form of independent decision modules, Fig. 1b.

Comment 8 A given DEU may invoke different DEUs depending on some internal or external conditions.

DEU often requires that some additional tasks are accomplished (services are provided) for a successful job completion. Those additional tasks can be performed by embedded or external DEUs. If a subtask S is specific to a given job then the DEU is provided with a subordinate unit $P(S)$. Instead, if $P(S)$ is common for several applications it is *extracted* from several systems and kept in a database as a passive program unit. $P(S)$ is invoked when needed or, alternatively, installed as a permanently active DEU. If some functions are extracted from original DEUs they must contain submodules I interfacing new DEUs.

DEU_1 which activates (converts from passive into active unit) DEU_2 will be called the *higher level* DEU. DEU_1 modifies DEU_2 parameters and options. A given DEU may have its own knowledge database with passive program units. Alternatively, it may access a common database. DEU also stores in its memory and information about the system status or has an access to a database in which such information is recorded. Again, if some information is common for several DEUs then it is more convenient to *extract* it and store in a common database.

If a given function is extracted from several systems as an independent DEU then the new DEU requires that the separate control and management system is developed for it.

An algorithm which is executed by a given DEU to perform its function includes submodules A_{ij}. Submodule A_{ij} takes action j to respond to request i. A_{ij} consists of executing a piece of code, invoking a passive program unit, or using services of other DEUs. The DEU may include some other algorithms, e.g., an algorithm for negotiating its parameters with higher level DEUs.

3 F-UNI space

The ATM network can be viewed as a core of the B-ISDN network. The user/network interface transforms service specific data to the ATM format which is common for all services.

As eluded to before the structural evolution of complex systems is governed by the *extraction principle*.

If a given function is performed in several systems then in many cases it is profitable to extract it and implement in a separate new system.

> The ATM network has been devised to extract transmission, switching, and multiplexing functions from monoservice networks.

The extraction principle also determines the interface architecture. The interface software and hardware is organized into tiers, Fig. 2. Functions which are common to several applications are extracted and implemented in a tier closer to the network. Service specific functions are located close to the user.

The tier configuration of UNI corresponds to the standard ISDN architecture determined by access points R, S, T and U, which is illustrated in Fig. 2. Traditionally, points R, S, T and U separated different pieces of equipment. Within the F-UNI architecture we also use them to determine boundaries between functions performed by the software. In fact, in what follows we are more interested in what happens at the software level rather than in the hardware in which that software is imbedded. At this stage we do not decide how many tiers need to be distinguished at the particular interface. Points R, S, T and U in Fig. 2 define one tier each. However, other arrangements are possible (zero or more than one tier between two adjacent points).

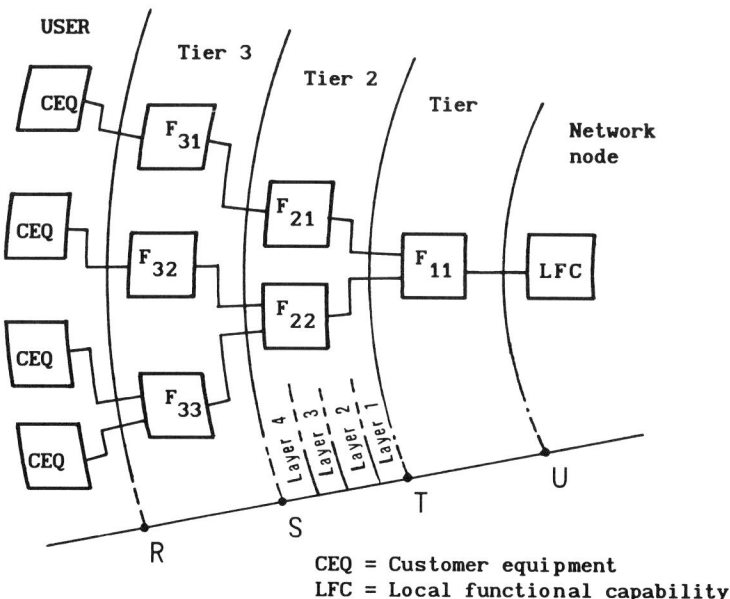

Figure 2

Functional blocks shown in Fig. 2 correspond to active program modules. They segment the data blocks into smaller units, send them to other modules, reassemble received blocks, perform sequencing, detect errors, retransmit mutilated blocks, and perform other functions related to handling of user data. In our model that type of program units are called *processors*. Processors are located in the horizontal plane at the transport level. They exchange the data blocks through the port (files) which are called the *Service Access Points* (SAPs).

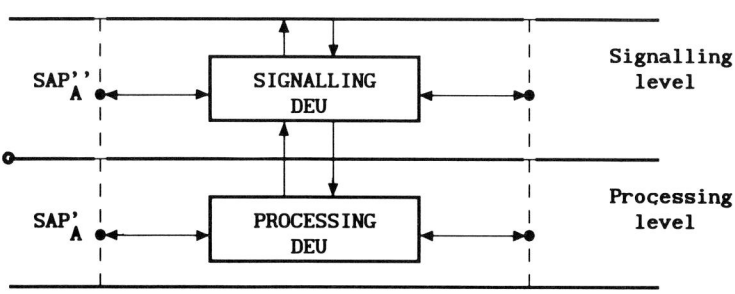

Figure 3

We shall assume that each processing DEU is associated with one or more signalling DEUs as is shown in Fig. 3. Signalling DEUs perform management and control actions. For instance, they allocate identifiers to data blocks and define SAPs through which those data blocks are exchanged. Signalling DEUs belong to the control or signalling plane which is situated above the transport plane.

One can envisage the third *management* plane. DEUs belonging to that plane allocate interface resources to different traffic types and groups of users. We stress that the tier structure is the same in each plane, Fig. 4.

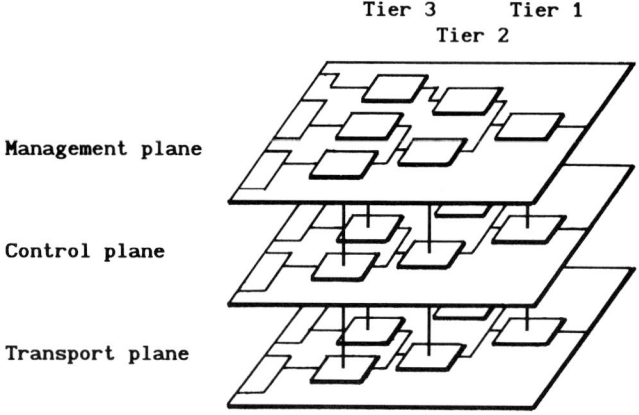

Figure 4

Note that all DEUs defined so far deal with servicing of traffic generated by users. Additionally, there is a group of DEUs to perform the operation, administration, and maintenance (OAM) functions. There may exist a separate OAM DEU for: (a) each DEU belonging to any of the three horizontal planes, (b) each tier, or alternatively, (c) each plane. We shall not consider the OAM DEUs for a while, assuming that they belong to the dual OAM space.

Cutting through the interface at a specific level (transport, signalling, management) we obtain one of the horizontal planes. We shall say that the horizontal planes give the *horizontal view* of the interface. The *vertical plane* is obtained by cutting cross-wise through the decisionmaking hierarchy.

Each vertical plane gives a picture of a protocol set for one of the services.

Fig. 5b presents the vertical plane corresponding to the crossection shown in Fig. 5a.

From what we have said so far it is easily seen that the interface is defined in the three dimensional F-UNI space which is shown in Fig. 6. The dimensions are:

Interface level Transport, Signalling, Management

Interface tier Service dependent and independent tiers and layers

Service Various service types and profiles

Each tier can be decomposed into layers, and layers into sublayers.

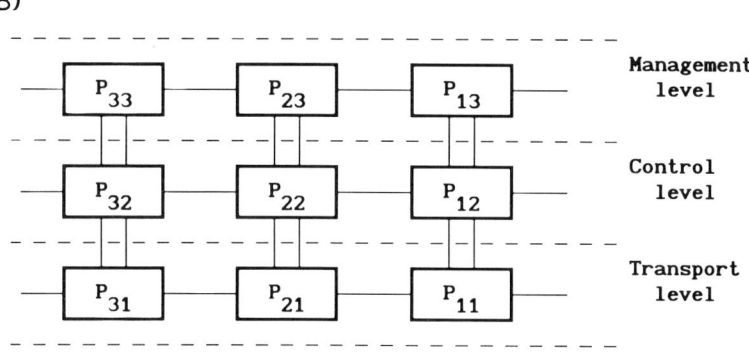

Figure 5

Before proceeding further we illustrate how the OSI architecture falls into the proposed framework.

4 F-UNI picture of the OSI architecture

The OSI architecture is usually depicted as is shown in Fig. 7. To fit it into our model we rotate the OSI hierarchy by 90 degrees to the left and separate the layer transport and management functions or, in other words, we break down the OSI layer entity into three DEUs: transport, control, and management. The new picture is shown in Fig. 8.

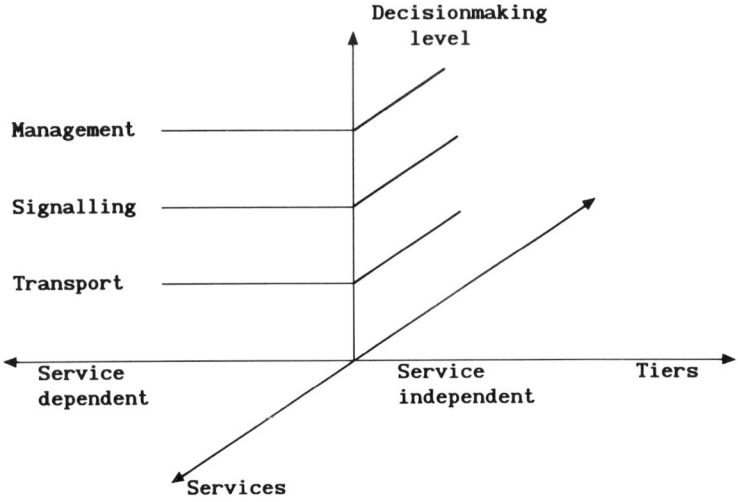

Figure 6

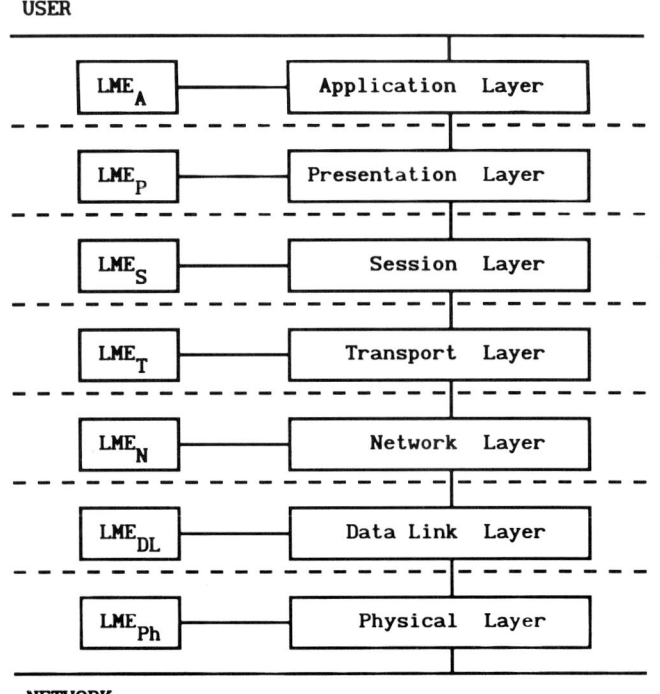

Figure 7

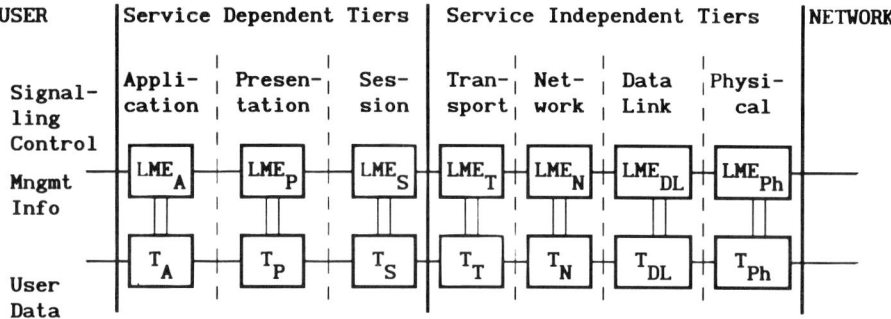

Figure 8

According to the F-UNI model the transport, control, and management modules of each layer interact with decisionmaking modules of the adjacent layers. In the OSI model this is done by exchanging primitives through the Service Access Point (SAP). The concept of SAP is retained in the F-Uni model with the only difference being that in F-UNI the clear distinction is made between flows of user, signalling, and management data.

As compared to early ISO documents, the OSI layer entity is currently being split into several subentities. One of them is the Layer Management Entity. The F-UNI architecture takes into account that trend in the OSI evolution.

The OSI layers can be grouped into two layers:

- *Service Independent Tier* comprising Physical, Data Link, Network and Transport layers, and

- *Service Dependent Tier* consisting of Session, Presentation, and Application layers.

Protocols of the Service Dependent Tier (SDT) can reside in separate DTE units and be situated to the left of S/T point. Even if it is not the case, it is still easily seen that rotating the OSI architecture by 90 degrees gives an option of situating the OSI interface with respect to the ISDN reference points.

5 ATM protocol stack

The currently developed protocol stack for an ATM network comprises:

- Physical layer
- ATM layer
- ATM Adaptation (AAL) Layer

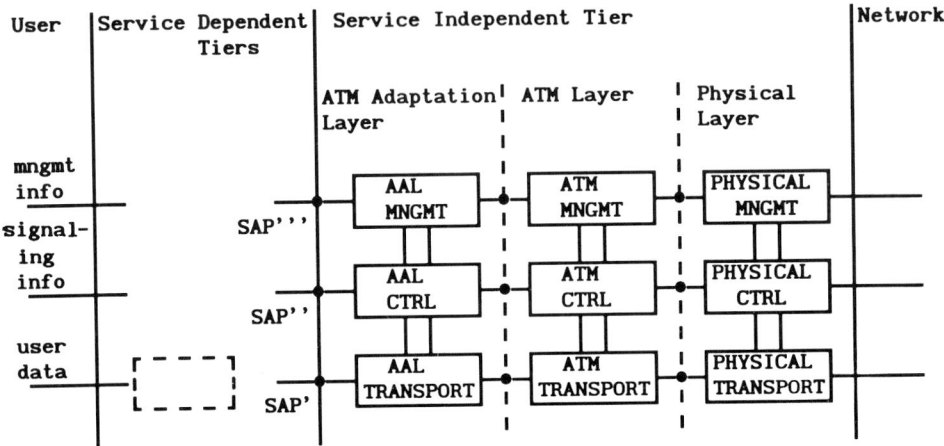

Figure 9

The AAL layer is subdivided into the Convergence (CS) and Segmentation and Reassembly (SAR) sublayers. Fig. 9 shows a vertical view of the corresponding F-UNI structure. We shall assume that Physical, ATM, and AAL protocols belong to the Service Independent Tier. The Service Dependent Tier protocols for B-ISDN has yet to be developed. There is a space for them in the F-UNI as indicated in Fig. 9.

In a remainder of this report we discuss functions of SIR DEUs.

6 ATM Physical layer

The Physical layer is divided into two sublayers:

- Transmission Convergence (TC) sublayer, and
- Physical Medium sublayer

The basic function of the Physical layer is to convert the flow of valid cells received from the ATM layer into a data flow matching the transmission system format. The valid cell is defined as a cell which has error free header.

An important feature of the ATM interface is self delineation, which means that cell boundaries can be restored independently of the transmission system used. To preserve that property the Physical layer merges the flow of cells with an appriopriate information for cell delineation.

Two primitives are exchanged across the boundary between the Physical and ATM layers:

PH-DATA-REQUEST: The ATM transport DEU requests an appriopriate Physical Layer transport module that the SDU associated with this primitive (valid cell) be transmitted.

PH-DATA-INDICATION: The ATM transport DEU is notified by the Physical Layer module that the SDU associated with the primitive coming from its peer is available.

Note that the control information carried by these primitives is exchanged at a relatively low decisionmaking level.

Decisionmaking and transport modules of the Transmission Convergence (TC) sublayer are shown in Fig. 10. Particular DEUs perform the following functions.

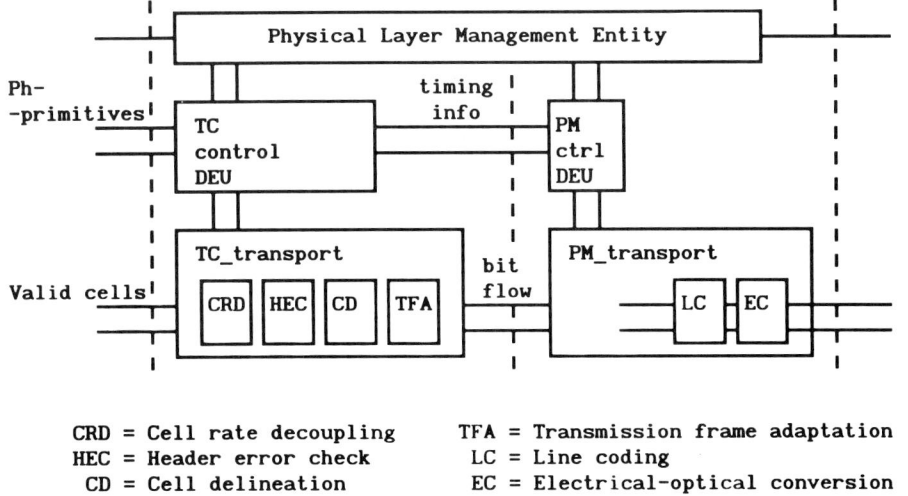

Figure 10

Cell rate decoupling To adapt an intensity of cell flow to payload of the transmission system the flow of valid cells is stuffed with idle cells. In a receive direction the idle cells are removed.

HEC sequence verification For cells sent in the transmit direction the HEC (Header Error Check) submodule generates the HEC sequence and inserts it into the cell header. Error check sequences of received cells are verified and if errors cannot be corrected cells with mutilated headers are discarded.

Cell delineation This module adds some information to the cell flow so that the receiving end can detect cell boundaries. Boundaries of received cells are identified.

Transmission frame adaptation Cells are put into transmission system specific envelopes (frames). Two basic types of envelopes are: (a) cell equivalent (no external envelope is added to the cell), and (b) synchronous transmission system (SDH) envelope.

The Physical Layer Management Entity receives an information from transmission OAM about a frame type, available payload, and others. It also exchanges and information with the ATM layer management entity.

Functions of the Physical Medium layer strongly depend on transmission system characteristics. They include line coding, electrical-optical transformation, and other functions needed to transmit continuous bit stream received from the Transmission Convergence layer.

7 ATM layer

As elluded to before, the ATM layer functions are associated with the cell header. The cell header fields are as follows: They include:

GFC Four bits in the cell header are allocated to flow control functions. GFC (Generic Flow Control) information can be used to discard some cells, determine precedence constraints, enforce access fairness, and others.

VPI The Virtual Path Identifier (VPI) is used to distinguish cells belonging to the same Virtual Path Connection (VPC). The specific VPI value assigned to the VPC at the UNI is service independent. However, some VPI values can be reserved for signalling channels.

VCI The Virtual Circuit Identifier (VCI) identifies cells belonging to the same Virtual Channel Conncetion (VCC). The specific VCI value assigned to the VCC at the UNI is service independent. However, some VCI values can be reserved for signalling channels.

PT Two bits are alocated to Payload Type field. If both bits are set to zero (PT = 00) the cell carries a user information.

RES One bit is not defined being reserved for future applications.

CLP Cell Loss Priority. If that bit is set to one the cell carrying it can be discarded from the network during a congestion period.

NOTE 1: The CLP value is not decided by the ATM layer. Also, the four types of AAL protocols do not have options for doing that. This is because only the service dependent module knows which cells can be discarded without adversly affecting the service quality. Therefore, it seems that using the CLP bit may cause implementation problems. Additional primitives will be needed to stamp information blocks which can be discarded by the network.

The control information and user data are exchanged between the AAL and ATM through the Service Access Points in a form of primitives. At each SAP the Service Data

Unit (SDU) is the cell information field. Two ATM/AAL primitives have been defined so far.

>ATM_DATA_REQUEST: One of AAL modules requests that the ATM Service Data Unit (ATM_SDU) associated with that primitive be transported to a distant system. The ATM_SDU is placed in the cell payload field without modifications.

>ATM_DATA_INDICATION: The ATM protocol notifies the AAL layer that the ATM_SDU has arrived from a distant system.

The ATM layer modules append headers to incoming data units forming cells. The cells are multiplexed. The multiplexing consists in forming non-continuous composite stream from cells arriving at the Service Access Points (SAPs). The SAPs are situated at the boundary between the ATM and AAL layers. Fig. 11 gives a horizontal view of the ATM transport plane.

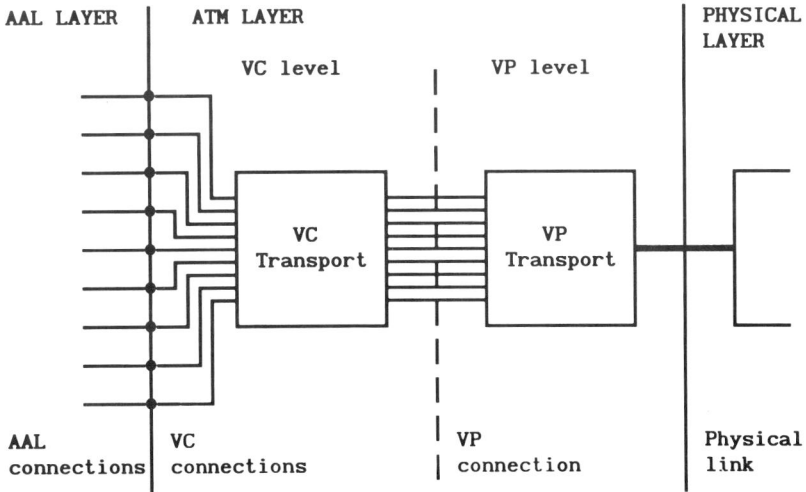

Figure 11

The ATM layer can be viewed as divided into two sublayers: Virtual Channel Connection sublayer, and Virtual Path Connection sublayer. A vertical view of the corresponding decisionmaking hierarchy is shown in Fig. 12. The VC_TRANSPORT unit appends headers to incoming SDUs. The VC_CONTROL module processes the control part of the ATM_DATA_REQUEST received from the AAL layer. Based on the SAP identification number it determines a VCI number for a particular cell.

In a similar way, the VP_TRANSPORT module writes the VP identifier to the VPI field of the cell header. The VPI number is determined from the VCI number.

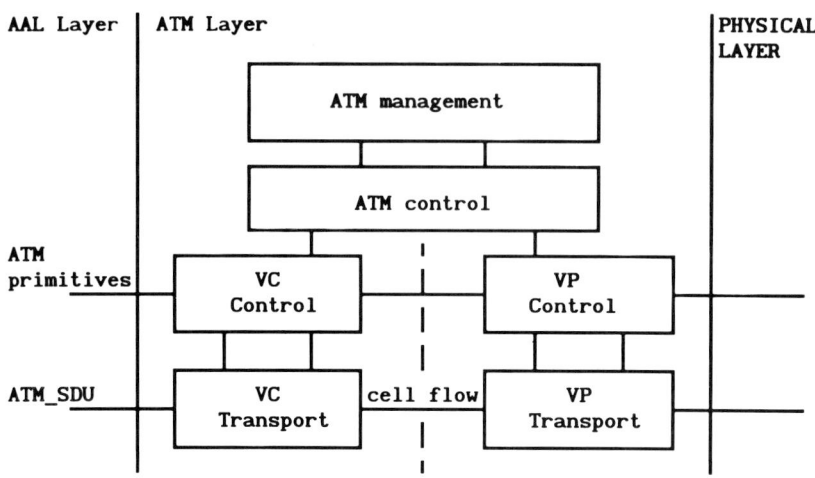

Figure 12

The VCI and VPI numbers are allocated to virtual path and channel connections by the ATM_CONTROL decisionmaking unit. The ATM_CONTROL DEU receives requests for VCC and VPC establishment. It negotiates VCC and VPC parameters with the network. After the connection is suives requests for VCC and VPC establishment. It negotiates VCC and VPC parameters with the network. After the connection is successfully set up, which means that it is confirmed by all interested parties, the VCC and VPC are activated.

The ATM_MNGMT (management) DEU keeps an information about VP and VC connections in progress and updates a pool of available VP and VC identifiers. It knows the status of adjacent layers as well as the network status. That latter information can be used during a connection set up. For instance, a connection request can be blocked or its parameter changed based on a local information kept by the ATM_MNGMT DEU.

ATM_MNGMT DEU updates its information according to changes which occur in the core network and interworking systems. The update cycle length is longer than interarrival time of VP and VC requests. The ATM_MNGMT DEU can perform complicated functions.

8 ATM Adaptation layer

The cell information field is carried transparently by the ATM layer. A part of that field is used by the AAL DEUs to carry the AAL layer control information to peer DEUs in distant systems.

The AAL layer performs the following functions:

- Segmenting and reassembly
- Handling of transmission errors
- Handling of quantization effects due to cell information field size
- Handling of lost and misinserted cell conditions
- Flow control and error control

It seems at present that various traffic types which are sent across the ATM network cannot be served by one AAL software module. Therefore, the traffic types were classified as is shown in Table 1.

TABLE 1

	Clas A	Class B	Class C	Class D
Timing relation between source and destination	Required	Required	Not required	Not required
Bit rate	Constant	Variable	Variable	Variable
Connection mode	Connection-oriented	Connection-oriented	Connection-oriented	Connectionless

CCITT Recommendation I.362 gives the following examples of services in the classes A, B, C, and D:

Class A: Circuit emulation: constant bit rate video
Class B: Variable bit rate video and audio
Class C: Connection-oriented data transfer
Class D: Connectionless data transfer

The AAL layer segments higher layer Protocol Data Units (PDUs) into blocks which can fit into an information field of the ATM cell. In the receiving direction it restores an original information structure. The ALL layer is divided into two sublayers:

- Convergence Sublayer (CS)
- Segmentation and Reassembly (SAR) sublayer

Four functionally different transport modules were standardized for the Convergence Sublayer. According to a current version of the standard there are no SAPs between the sublayers. For some traffic types the SAR and/or CS can be empty. To illustrate functions of the AAL layer let us briefly discuss the AAL type 1 protocol.

8.1 AAL type 1

The AAL type 1 has capabilities that enable the user/network interface to support the class A service. The functions performed by the AAL type 1 include:

- Segmentation and reassembly of user information
- Handling of cell delay variation
- Handling of lost and misinserted cells
- Source clock frequence recovery
- Monitoring of AAL Protocol Control Information (AAL PCI) for bit errors
- Monitoring of user information field for bit errors

Current CCITT documents do not specify sublayers to which particular functions belong.

The AAL detects transmission errors. Within UNI the errors are indicated to the AAL_CONTROL module. The AAL_CONTROL DEU can process the error information to take some action, or can simply pass that information to higher layers in the Service Dependent Tier.

To detect lost and misinserted cells the SAR_PDU carries the Sequence Number (SN). This number is protected by the four bit Sequence Number Protection (SNP) code.

The CCITT Draft Recommendation I.363 (B-ISDN ATM Adaptation Layer (AAL) Specification) considers additional functions which can possibly be provided by the AAL layer. They include:

- Monitoring of end-to-end QOS for circuit emulation which could possibly be achieved by calculating a CRC for CS-PDU payload, and and transmitting the result in the CS-PDU or by the use of a control cell.
- Forward error correction combined with bit interleaving for high quality audio and video to protect against bit errors.
- A time stamp pattern insertion in the CS-PDU to provide for an explicit time indication.

It seems, however, that such and other very specific functions could possibly be performed by the higher layer DEUs.

> QUESTION 1. It is not clear how the Constant Bit Rate (CBR) service can be provided by the AAL type 1 to the higher layers.

To elucidate an answer to this question we note that the CBR service provision requires that all links of the VP or VC connection have specific characteristics. This in turn implies that the AAL_CONTROL module negotiates those characteristics with the network DEUs. Because the VP and VC connections are established by the ATM layer decision modules the AAL_CONTROL DEU must do that indirectly via the ATM DEUs.

The forgoing comment indicates that DEUs residing in different layers exchange more control and signalling information than it is currently standardized by CCITT. In this context the three-dimensional F-UNI architecture provides a good framework for discussion of signalling flows.

9 Signalling procedures

In this and following sections we concentrate on information flows carrying management and signalling messages. Consider a decisionmaking unit DEU which is located in the mth layer and nth level of the F-UNI space, Fig. 13. In order to perform its function the considered DEU exchanges a control information with other DEUs.

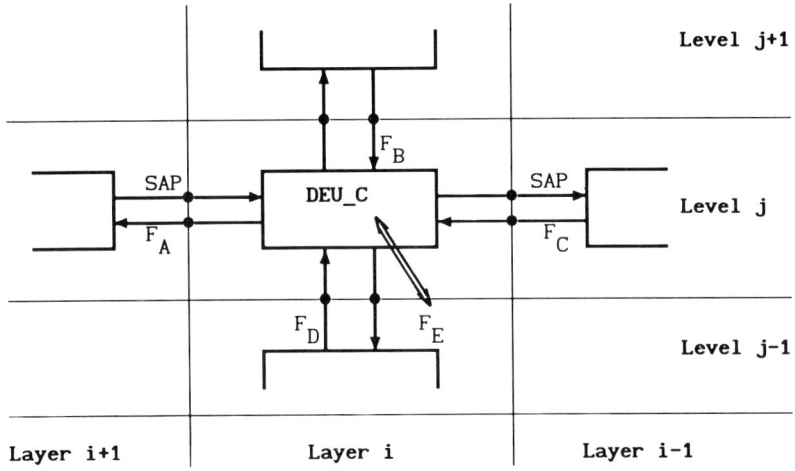

Figure 13

ASSUMPTION 1. The set of DEUs with which the F-UNI DEU exchanges control information comprises its immminent neighbours and *peer* distant systems. This gives the following information flows (see Fig. 13):

F_A: $DEU(i,j) \leftrightarrow DEU(i+1,j)$.
F_B: $DEU(i,j) \leftrightarrow DEU(i,j+1)$.
F_C: $DEU(i,j) \leftrightarrow DEU(i-1,j)$.
F_D: $DEU(i,j) \leftrightarrow DEU(i,j-1)$.
F_E: $DEU(i,j) \leftrightarrow$ distant *(peer)* DEUs.

The control information F_A and F_C is exchanged between layers across the Service Access Points in a form of primitives in the same way as in the OSI model. Instead, the flows F_B and F_D refer to an information transfer among the transport, control, and management planes. The information is exchanged through the common data blocks or files. The higher level DEU writes the control data (control parameters, addresses, options) to the output file. The lower level DEU reads that information. The communication in the opposite direction is organized in the same way.

In order to explain how F_E signals are sent consider an operation of ith layer transport module $DEU(i,1)$. It receives the Service Data Unit (SDU) from the higher layer transport entity and forms the Protocol Data Unit by appending the header and trailer to the SDU. The PDU is forwarded to the lower layer. Moreover, each layer can establish a virtual connection by allocating a common logical channel number to a sequence of PDUs. Taking this into account, the ith layer control and management DEUs have two options of sending a control information to a distant system.

- *In-slot signalling.* Write a control information in a header or trailer of a PDU which carries user data (SDU received from the higher layer).

- *Out-slot signalling.* Write a control information in the payload field of the PDU allocated to signalling.

In the first case some header and trailer fields are pre-allocated to carry the signalling information. In the second case, the PDU header values are pre-assigned to to identify the control PDUs.

NOTE 1. Two signalling modes have been named after inslot and outslot signalling systems applied in traditional circuit-switched networks.

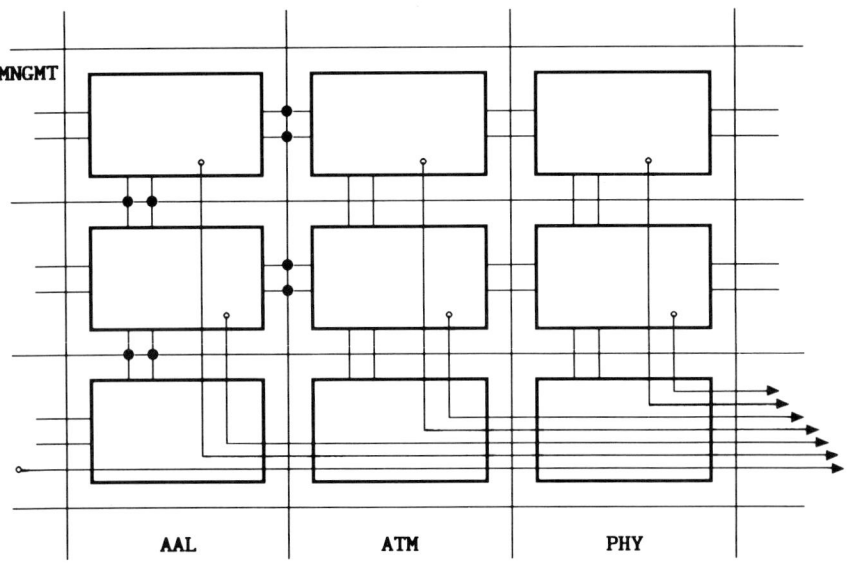

Figure 14

In the case of outslot signalling the ith layer connection can be established by the ith layer management or control DEU. The ith layer control connection starts in the transport module of that layer. It receives SDUs from the $DEU(i,2)$ and $DEU(i,3)$. Fig. 14 shows one user connection and six control connections which can possibly be established in the Service Independent Tier of the F-UNI architecture.

The CCITT has pre-assigned some cell header values at the ATM UNI interface to identify cells which are reserved for passing of management and control signals generated by DEUs in the ATM and Physical layers. The similar marking can be envisaged for higher layer PDUs.

The control information can also be exchanged in the message mode. Both in the case of inslot and outslot signalling one message can extend over several cells.

9.1 Meta-signalling

Meta-signalling is a procedure for establishing, checking and releasing signalling virtual channels. For each direction, the meta-signalling is performed in a dedicated permanent virtual channel connection having fixed standardized VCI and VPI values. The meta-signalling function comprises:

- Allocation of capacity to signalling channels

- Procedures needed to establish, release, and check the status of signalling channels

- Association of requests with service profiles

- Provision of means to distinguish between simultaneous requests

CCITT has preassigned fixed VCI and VPC values to metasignalling channels.

10 Call and connection control

A multicast and multipoint services require that several VC and VP connections be established during one call. Also, in case of some services a need arises to modify connection parameters during a call. To fulfil those requirements the separation of call and connection (bearer) control has been postulated. This feature can easily be seized in the developed model.

Consider the macroscopic view of the F-UNI interface, Fig. 15, which reveals the Service Independent (SIT) and Service Dependent (SDT) Tiers. In each tier we have transport, control, and management modules. The separation of call and connection control means that:

- *Call control* is performed by the SDT_C control module of the Service Dependent Tier.

- *Connection control* is performed by the SIT_C control module of the Service Independent Tier.

For some services the transport DEU of the Service Dependent Tier can be empty.

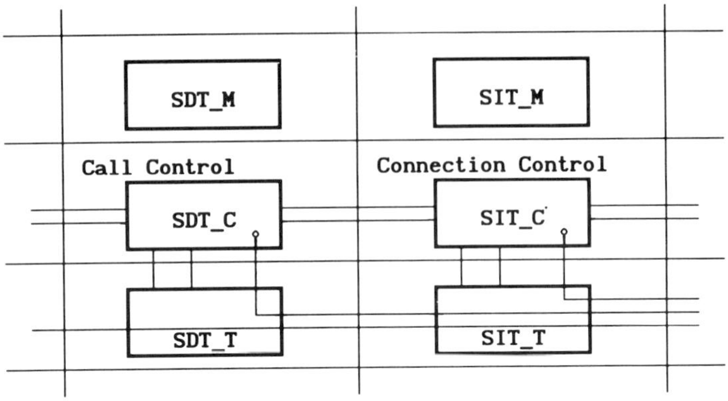

Figure 15

11 Application of SS7 to B-ISDN signalling

An application of the CCITT Signalling System No. 7 to call and connection control at the B-ISDN user/network interface is currently being discussed. Fig. 16 illustrates the corresponding F-UNI model. Fig. 16A shows the horizontal view of the transport plane. The user is connected to: (a) signalling SS7_DEU module, and (b) the SDT_T transport module which carries user data. Both modules are located in the Service Dependent Tier. Call control procedures reside in the CALL CONTROL DEU in the control plane of the Service Dependent Tier, Fig. 16A. The call connection messages are received by the SS7_DEU module. They are forwarded to the CALL CONTROL DEU, which communicates with:

- SDT management DEU (not visualised in Fig. 16B)
- AAL connection modules in the AAL layer
- VCC and VPC control modules in the ATM layer (indirectly via the AAL DEUs)
- Network call control DEUs
- Call control DEUs of other users

SDT management DEU can use the SS7_DEU to communicate with network and other users management systems. The Telecommunication Management Network protocols can possibly be used to exchange the management information.

The SS7_DEU uses services of the AAL type 3 protocol in the AAL layer to establish the SS7 SCCP connections. Figures 16C and 16D indicate that the CALL CONTROL DEU extends over the SDT_T transport module and controls the flow of user data.

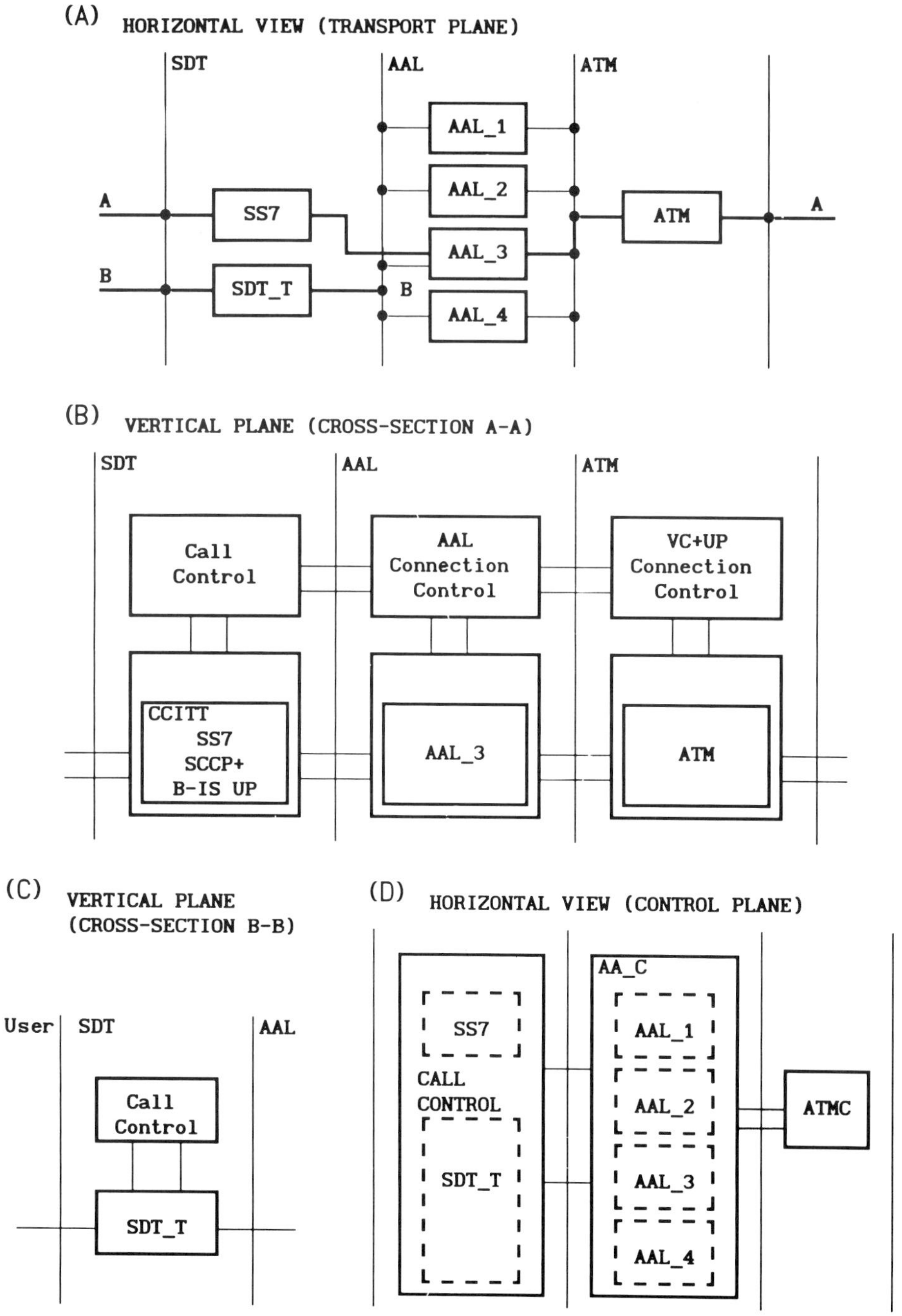
(A) HORIZONTAL VIEW (TRANSPORT PLANE)
SDT
AAL
ATM
AAL_1
AAL_2
AAL_3
AAL_4
A
SS7
ATM
A
B
SDT_T
B
(B) VERTICAL PLANE (CROSS-SECTION A-A)
SDT
AAL
ATM
Call Control
AAL Connection Control
VC+UP Connection Control
CCITT SS7 SCCP+ B-IS UP
AAL_3
ATM
(C) VERTICAL PLANE (CROSS-SECTION B-B)
User
SDT
AAL
Call Control
SDT_T
(D) HORIZONTAL VIEW (CONTROL PLANE)
SS7
CALL CONTROL
SDT_T
AA_C
AAL_1
AAL_2
AAL_3
AAL_4
ATMC

Figure 16

12 Concluding remarks

A new model of a user/network interface in a broadband ISDN has been presented. The model is called the Flexible User/Network Interface (F-UNI). The F-UNI architecture is developed to encompass new features of ATM, ISDN, Open Network Architecture (ONA), Intelligent Network (IN), and Personal Communication Network (PCN). The interface software is structured in a three dimensional F-UNI space with dimensions determined by the decisionmaking level, service type, and functional distance to the core network. The F-UNI architecture encompasses existing standards for ISDN and ATM as well as the Open System Interconnection model.

References

1 J. Filipiak, F-UNI: A model of a Flexible User Network Interface for B-ISDN, TRC Report 3/91, The University of Adelaide, January 1991

2 CCITT Draft Recommendation I.150: B-ISDN ATM functional characteristics

3 CCITT Draft Recommendation I.361: B-ISDN ATM layer specification

4 CCITT Draft Recommendation I.362: B-ISDN ATM Adaptation Layer (AAL) functional description

5 CCITT Draft Recommendation I.363: B-ISDN ATM Adaptation Layer (AAL) specification

A Survey, and Some New Results, on Stochastic Petri Nets

W. Henderson

Teletraffic Research Centre, Dept. of Applied Mathematics, The University of Adelaide, Adelaide, South Australia

Abstract

This paper is an introduction to the modelling tool of Petri nets (PNs) but concentrates mostly on its performance counterpart, stochastic Petri nets (SPNs). It is a short and personal survey with the majority of the paper dedicated to recent research from the Teletraffic Research Centre of Adelaide University.

1. Petri Nets

Petri nets (PNs) were introduced as a modelling tool in 1962 by Carl Petri [38], and designed to include the analysis of systems which exhibit concurrent and conflicting behaviour.

Formally, a PN is a bipartite directed graph, defined by the five-tuple,

$$P = (\mathcal{P}, \mathcal{T}, \mathbf{I}, \mathbf{O}, \mathbf{m}_0).$$

$\mathcal{P} = \{p_1, \ldots, p_r\}$, is a finite set of places, represented by circles, and $\mathcal{T} = \{t_1, \ldots, t_s\}$, is a finite set of transitions, represented by bars. The directed arcs connect places to transitions and transitions to places. The input function $\mathbf{I} : \mathcal{T} \to \mathbb{Z}_+^r$, maps a transition $t \in \mathcal{T}$ to its input bag $\mathbf{I}(t)$, which is an $r \times 1$ column vector, giving the input places (including multiplicity) of transition t. The output function, $\mathbf{O} : \mathcal{T} \to \mathbb{Z}_+^r$, maps a transition t to its output bag $\mathbf{O}(t)$, which is an $r \times 1$ column vector, giving the output places (including multiplicity) of t.

The state (or marking) of a PN is an $r \times 1$ column vector giving the number of tokens in each place in the PN. $\mathbf{m}_0$ is the initial marking of the PN. Transition t is "enabled" in marking $\mathbf{m}$ if $\mathbf{m} - \mathbf{I}(t) \geq 0$. In this case transition t can fire, absorbing the tokens in $\mathbf{I}(t)$, depositing the tokens in $\mathbf{O}(t)$, thus creating the marking $\mathbf{m} - \mathbf{I}(t) + \mathbf{O}(t)$.

Example 1

The Petri net of Figure 1 represents a resource allocation problem in which resources A, B and C are required to perform a task. The numbers of tokens present in places p_1, p_2 and p_3 denote the amounts of resources, A, B and C respectively, available for use. When transition t_1 fires it represents a customer arriving to take one of each of the three types of resource. The input bag for t_1 is therefore $\mathbf{I}(t_1) = (1,1,1,0,0)$. After a customer accesses resources, it uses A and B together by holding them in place p_4 whilst C is independently held in place p_5. Using a single token in place p_4 to represent a unit of resource A and a unit of resource B defines $\mathbf{O}(t_1)$ as $(0,0,0,1,1)$. Transition t_2 fires to signal a customer finishing with A and B, returning them together to places p_1 and p_2 respectively and t_3 fires to return a unit of resource C to place p_3. Consequently the marking of Figure 1 is the result of two customers having both finished with A and B but neither having yet finished with resource C. In this model there is no provision for customers to queue. Customers continue to arrive while the resources they require are being utilised but such customers are lost.

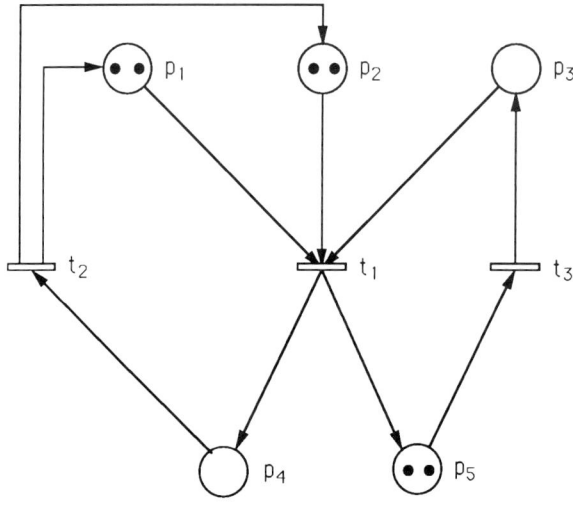

Figure 1

The above definition covers the structure and dynamic behaviour of a basic PN. Later refinements, introduced to attain greater modelling power, include
(i) inhibitor arcs (see, for example [37] and [41]) which prevent transitions from firing unless the input places, corresponding to the inhibitor arcs, are empty;
(ii) coloured and high level PN [22] in which transitions are enabled and fire under a variety of different conditions;
(iii) timing, where stochastic and deterministic times are added to places and/or transitions.

PNs without timing have been used for verifying and specifying telecommunication protocols and for computer systems ([32], [36]), however it is the timing refinement with which we are predominantly concerned in this paper.

Excellent surveys of PN theory can be found in Peterson [37], Reisig [41] and Murata [34].

2. Stochastic Petri Nets

The name Stochastic Petri net (SPN) is given to a PN in which random variables with given distribution functions are associated with timed events of the PN. Early work on SPNs used deterministic random variables defined on:

(i) the time between a transition becoming enabled and firing (the firing time) ([39], [31], [46]).

(ii) the time between a transition absorbing its input bag and depositing its output bag (the absorption time) ([40]).

(iii) the time a token is retained in a place ([44]).

Whenever the firing time random variable is defined on a discrete set of time points there is the possibility of groups of transitions ending their firing times concurrently. As these transitions may be in competition for the same tokens conflict can arise. A variety of procedures have been postulated to determine which of the conflicting transitions fire ([46], [40], [20], [45]).

SPNs with exponentially distributed firing times ([35],[33]), combine the simplicity and ease of representation of PNs with well-known techniques in Markov analysis and, being drawn from continuous distributions, do not have conflict problems. Consequently they are used frequently throughout the literature.

Generally distributed random variables have been incorporated into SPNs either by restricting consideration to semi-Markov processes ([35], [5], [9], [19]) or by replacing the general distribution with phase type distributions represented by SPNs ([6], [8]). Henderson and Lucic [15] generalised the former by relaxing the semi-Markov restriction to any SPNs which are insensitive (see Section 3) to their generally distributed random variables.

Further generalisations of the basic SPN structure include transitions with zero firing times ([29]), random selection of input and output bags ([13],[17] and [19]), high-level SPNs ([19], [28]) and regenerative simulation of SPNs ([12], [14]).

The major limitation of SPNs is that the graphical representation of systems becomes more and more difficult when system size and complexity increase. The number of states of the associated Markov Chain grows exponentially with the dimensions of the net and with the number of tokens in the initial marking. To overcome the large state space problem authors have:

(a) aggregated subnets ([3], [1], [2], [16], [19]);

(b) merged transitions ([9], [15]);

(c) amalgamated markings ([9], [15]);

(d) aimed for a particular performance measure ([43]);

(e) extended queueing results ([10], [17], [19]).

The following sections outline some recent work developed at the Teletraffic Research Centre of the University of Adelaide. Section 3 discusses exact aggregation and disaggregation techniques [16] and Section 4 explains how recent extensions of queueing network product form solutions to SPNs ([17], [19]) can be used to develop fixed point algorithms suitable for the approximate analysis of large telecommunications networks.

3. Aggregation and Disaggregation

Aggregation procedures must always be questionable unless they are supported by some measure of the error entailed. Before trying to estimate the error it is valuable to be aware of those procedures in which no error arises. That is the purpose of this section. If the aggregation procedure yields insensitive SPNs then an exact equilibrium distribution for the SPNs can be found.

A process is said to be "insensitive" to it's generally distributed lifetimes if the equilibrium distribution of the process depends upon these general distributions only through their means. Insensitivity properties appear regularly in the theory of networks of queues when certain queues with generally distributed service times are placed at the nodes of a Jackson network and a product form equilibrium solution obtained ([4], [23]).

Henderson and Lucic [16] define their aggregation procedure, allowing for the possibility that routing from an aggregated subnet may depend on the length of time spent in that subnet, as follows:

1. From the original net P, amalgamate sets of transitions and places and replace them with a box b and a transition t. Input bags of tokens enter box b and leave by being fired through transition t. Let $G_t(\cdot)$ be the distribution function of the time spent in the aggregated subnet b. Although the time in the subnet can be worked off at a state dependent speed and $G_t(\cdot)$ can depend on the type/colour of the tokens we shall simplify the explanation by ignoring these features. This process is continued until the desired aggregation has been completed. The aggregated SPN so created is $\overline{P}$. Assume that, after tokens have taken time y to be processed by t, the marking changes from $\overline{n}$ to $\overline{m}$, with age dependent probability $p_{\overline{n},\overline{m}}(t,y)$.

2. Create the averaged net Q from $\overline{P}$ so that, due to tokens being fired by t out of b, the marking changes from $\overline{n}$ to $\overline{m}$, with age independent probability
$$p_{\overline{n},\overline{m}}(t) = \int_0^\infty p_{\overline{n},\overline{m}}(t,y)\, dG_t(y).$$

3. Without altering the collection of routeing probabilities $p_{\overline{n},\overline{m}}(t)$, or the means of the distributions $G_t(\cdot)$, create the net M, by replacing all distributions $G_t(\cdot)$ with negative exponential distributions.

Note that $\overline{P}$ is a SPN with age dependent probabilistic routing so that, when t fires, an output bag is created according to a probability distribution. In general, the $\overline{P}$,

Q and M processes, will have different equilibrium distributions. However, using an extension of insensitivity theory proved by Rumsewicz and Henderson [42] and some standard results by Matthes [30], and König and Jansen [26] Henderson and Lucic [16] showed:

Result 1: If Q is insensitive to its generally distributed transitions, then the equilibrium distribution of Q is identical to the equilibrium distribution of $\overline{P}$ (and therefore a marginal equilibrium distribution of the original process P) and is given by the equilibrium distribution of M.

Result 2: If Q is insensitive to its generally distributed transitions the marginal distribution of P, given by the equilibrium distribution of $\overline{P}$, is independent of the equilibrium distribution of the aggregated subnets when considered in isolation. Consequently, if we wish to find the equilibrium distribution for the original SPN we need to find:
(a) The corresponding marginal equilibrium distribution of the aggregated net,
(b) The equilibrium distribution for the aggregated subnets considered in isolation.

To find the equilibrium distributions for M (and therefore $\overline{P}$) and the subnets we can use simulation, global balance equations, product form solutions or any other technique available. The big advantage is that each of the resultant processes has a state space considerably smaller than the original and therefore relatively easier to handle.

Example 2

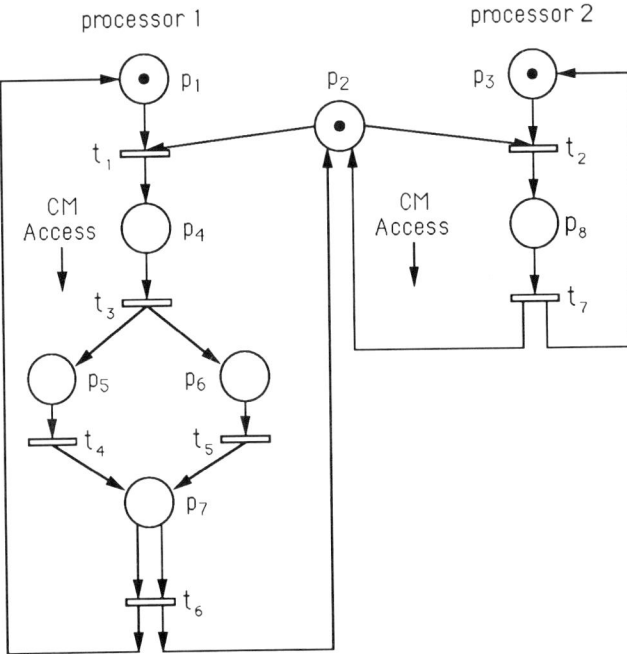

Figure 2

The SPN of Figure 2 is a model of a dual processor where processor 1 accesses two areas of memory concurrently, represented by the places p_5 and p_6. Let transitions t_i for $3 \leq i \leq 6$ have generally distributed firing times. Even if these transitions have negative exponentially distributed firing times the equilibrium distribution of the SPN does not appear to have a simple form.

Now aggregate places p_4, p_5, p_6 and p_7 and transitions t_3, t_4, t_5 and t_6 and represent the aggregation by a box b_1 and a transition T_1 to create the $\overline{P}$ SPN of Figure 3.

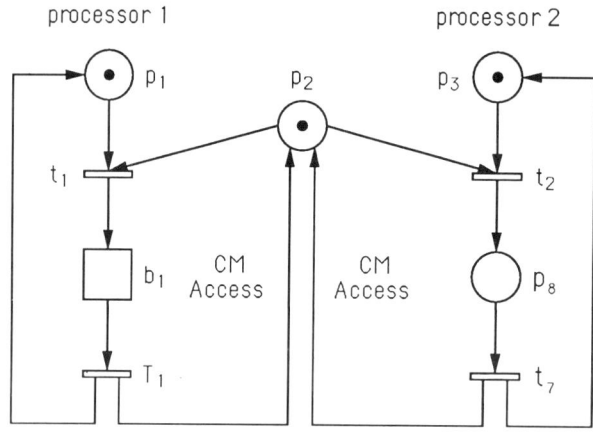

Figure 3

Step one of the aggregation procedure is now complete. In this case there is no age dependent routing, step two of the procedure is therefore redundant and $\overline{P}$ and Q are equivalent.

Since the firing time of transition T_1 cannot be assumed to be negative exponentially distributed, a simple solution can only be found if we prove insensitivity with respect to that firing time. Satisfying certain balance equations is a necessary and sufficient condition for insensitivity (see [30] and [26]). Using these balance equations or, in this case, observing that the process is semi Markov, Q can be seen to be insensitive to the transition firing time of T_1.

With the mean of the firing time distributions of t_1, t_2, t_7 and T_1 as μ_1, μ_2, μ_7 and $\overline{\mu_1}$ respectively and A a normalising constant the equilibrium distribution for the markings of M is readily observed to be:

$$\pi(1,1,1,0,0) = A\overline{\mu_1}\mu_7, \quad \pi(0,0,1,1,0) = A\mu_1\mu_7, \quad \pi(1,0,0,0,1) = A\mu_2\overline{\mu_1} \quad (1)$$

As a consequence of Result 1 a marginal equilibrium distribution of P is given by equation (1). The symmetry of the SPN also suggests insensitivity with respect to transition t_7. Further, when a pre-emptive resume protocol is assumed for transitions t_1 and t_2, the SPN of Figure 3 can also be shown to be insensitive to these transitions ([16] or [28]). Consequently t_1, t_2 and t_7 can also represent subnetworks as did transition T_1. Thus we have the power to further increase the complexity of the model with regard to the transfer of the processor to main memory and still derive exact equilibrium distributions.

Consider now the task of retrieving information apparently lost in the aggregation procedure. Now that the marginal equilibrium distribution for P has been found and insensitivity proved, Result 2 indicates that the remaining task is to derive the equilibrium distribution of the subnets in isolation.

Consider any marginal marking in which tokens are in b_1. In this example $(0,0,1,1,0)$ is the only relevant marking with one token somewhere inside b_1. Now isolate the subnet by looping the token back to place p_4 whenever it fires out of t_6.

Let (i,j,k,l) be the marking of the subnet representing i,j,k and l tokens in places p_4, p_5, p_6 and p_7 respectively. With negative exponentially distributed firing times throughout, an invariant measure for the subnet is,

$$\pi(1,0,0,0) = C, \quad \pi(0,1,1,0) = \frac{C\mu_3}{(\mu_4+\mu_5)}, \quad \pi(0,0,0,2) = \frac{C\mu_3}{\mu_6} \qquad (2)$$

$$\pi(0,1,0,1) = \frac{C\mu_3\mu_5}{\mu_4(\mu_4+\mu_5)}, \quad \pi(0,0,1,1) = \frac{C\mu_3\mu_4}{\mu_5(\mu_4+\mu_5)} \qquad (3)$$

Combining Equations (1),(2) and (3), with the aid of the independence property derived in Result 2, leads to the the exact equilibrium distribution of the original SPN. For example

$$\pi(1,1,1,0,0,0,0) = A\overline{\mu_1}\mu_7, \quad \pi(0,0,1,1,0,0,0) = AC\mu_1\mu_7$$

$$\pi(0,0,1,0,1,1,0,0) = AC\mu_1\mu_7 \frac{\mu_3\mu_5}{\mu_4(\mu_4+\mu_5)}$$

with A and C normalising constants from the appropriate distributions.

4. Fixed Point Approximations

Fixed point techniques are now commonly used to find approximate equilibrium distributions to stochastic systems which either do not have, or have complex, unusable closed form equilibrium distributions. One of the best examples of the latter is a loss network with fixed routing which is known to have a product form equilibrium distribution for the number of customers on each of the routes [7]. However the complex nature of the state space boundary and the large number of routes over which the product is taken makes the normalising constant impossible to calculate except in the simplest of cases. Consequently, although invariant measures are known, the equilibrium distribution, and therefore performance measures based on this distribution, cannot be calculated. When the network becomes more realistic, with trunk reservation and alternative routing included, even the product form equilibrium distribution is lost. In either case authors have preferred to find equilibrium distributions using fixed point algorithms [25] [24] [11].

In this section an example is presented in which resources are accessed and used by customers in a variety of ways. We postulate an approximation technique which breaks down the original SPN into a set of simpler SPN "modules" with interrelated parameters. An extreme version of this technique, applied to loss networks, becomes the Erlang fixed point approximation discussed by Kelly [24] [25]. In our example the accuracy of the result is improved significantly by making less stringent assumptions. More importantly, by looking at these problems from a SPN viewpoint gives a new perspective on the fixed point technique and naturally leads to new insights.

The essence of our approach is to ensure that, in breaking down the SPN, the resultant modules have a closed form solution. Consequently the product form results in the literature, especially those related to SPNs, such as Henderson, Lucic and Taylor [17] and Henderson and Taylor [19], provide crucial building blocks for our technique by identifying the form of modules allowed. With a general model of resource allocation the modules assume a particularly simple form. However the basic concepts of product form solutions postulated in [17] and [19] are not limited to these simple forms, suggesting that the same technique can be extended to much more complex SPN structures.

To illustrate the basic ideas involved in the approach consider the following example.

Example 3

As pointed out in Section 1 the Petri net of Figure 1 represents a loss network. Assume that customers arrive in a Poisson stream with intensity λ_1. Customers who arrive when the required resources are not available are lost. Because the stream is Poisson and therefore memoryless this is equivalent to transition t_1 having a negative exponentially distributed firing time (with mean $[\lambda_1]^{-1}$).

Assume that t_2 and t_3 fire at rates λ_2 and λ_3 respectively and, for simplicity, that initially there is only one resource available in each of p_1, p_2 and p_3.

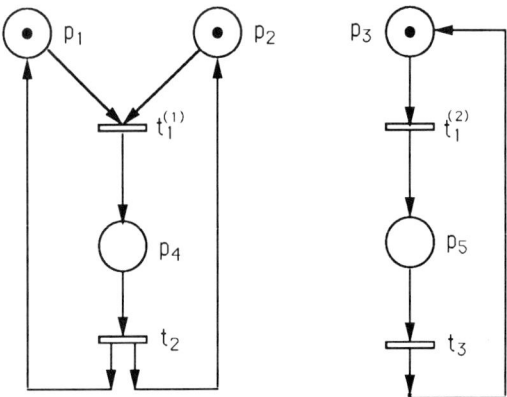

Figure 4

The procedure is to create modules by dividing transitions whilst keeping the throughputs of the new transitions approximately equal to that of the original. In this case transition t_1 of Figure 1 is split into transitions $t_1^{(1)}$ and $t_1^{(2)}$ with firing intensities $\mu_1^{(1)}$ and $\mu_1^{(2)}$ respectively.

Observe that the resultant SPN of Figure 4 is composed of two distinct SPNs each of which has a product form solution. We refer to these simpler SPN structures as "modules".

Let $\Pi(\mathbf{n})$ be the equilibrium probability that the SPN of Figure 1 has marking $\mathbf{n}$ and $\Pi_1(\mathbf{m}_1)$ and $\Pi_2(\mathbf{m}_2)$ be the equilibrium probabilities that the left and right hand modules of Figure 4 have markings $\mathbf{m}_1$ and $\mathbf{m}_2$ respectively. The aim is to choose $\mu_1^{(1)}$ and $\mu_1^{(2)}$ so that the throughputs of $t_1^{(1)}$ and $t_1^{(2)}$ are approximately the same as that of the Figure 1 transition t_1. The throughput of $t_1 = \lambda_1 \Pi(1,1,1,0,0)$, that of $t_1^{(1)}$ is $\mu_1^{(1)} \Pi_1(1,1,0)$ and of $t_1^{(2)}$ is $\mu_1^{(2)} \Pi_2(1,0)$ where the markings $(1,1,0)$ and $(1,0)$ represent the tokens in places p_1, p_2 and p_3 in the modules of Figure 4.

Now equate throughputs and make the assumption that $\Pi(1,1,1,0,0) = \Pi_1(1,1,0)\Pi_2(1,0)$ i.e. that the modules of Figure 4 can be used as an approximation of the SPN of Figure 1. This gives

$$\mu_1^{(1)} = \lambda_1 \Pi_2(1,0)$$

and

$$\mu_1^{(2)} = \lambda_1 \Pi_1(1,1,0).$$

Noting that

$$\Pi_2(1,0) = \frac{\lambda_3}{\mu_1^{(2)} + \lambda_3} \quad \text{and} \quad \Pi_1(1,1,0) = \frac{\lambda_2}{\mu_1^{(1)} + \lambda_2}$$

we obtain two equations in the unknowns $\mu_1^{(1)}$ and $\mu_1^{(2)}$ which, in this case, can be put together to give

$$\lambda_3 \left(\mu_1^{(1)}\right)^2 + (\lambda_1 + \lambda_2 \lambda_3 - \lambda_1 \lambda_3) \mu_1^{(1)} - \lambda_1 \lambda_2 \lambda_3 = 0.$$

This equation has a positive solution for any positive λ_1, λ_2 and λ_3. Thus it can be solved to give $\mu_1^{(1)}$ and hence $\Pi_1(1,1,0)$ and subsequently $\mu_1^{(2)}$ and $\Pi_2(1,0)$.

The exact result for the loss probability in this problem with $\lambda_1 = \lambda_2 = 2$, $\lambda_3 = 1$ is 0.7. The approximation gives us .7071, a surprisingly accurate result for such a small network.

An alternative approach to this example appears a little "over the top" but is nevertheless instructive. Consider the three modules of Figure 5, the result of splitting transition t_1 into three transitions $t_1^{(1)}$, $t_1^{(2)}$, and $t_1^{(3)}$, with the intention, once again, of setting the firing intensities $\mu_1^{(1)}$, $\mu_1^{(2)}$ and $\mu_1^{(3)}$ of the three new transitions to keep their throughputs equal to that of t_1.

With the firing intensities of $t_2^{(1)}$, $t_2^{(2)}$ and t_3 as λ_2, λ_2 and λ_3 respectively the assumption that the modules act separately gives

$$\mu_1^{(1)} = \frac{\lambda_1 P(T_1 \text{ is enabled})}{P(T_1^{(1)} \text{ is enabled})}$$

$$= \frac{\lambda_1 \lambda_2 \lambda_3}{\left(\mu_1^{(2)} + \lambda_2\right)\left(\mu_1^{(3)} + \lambda_3\right)}.$$

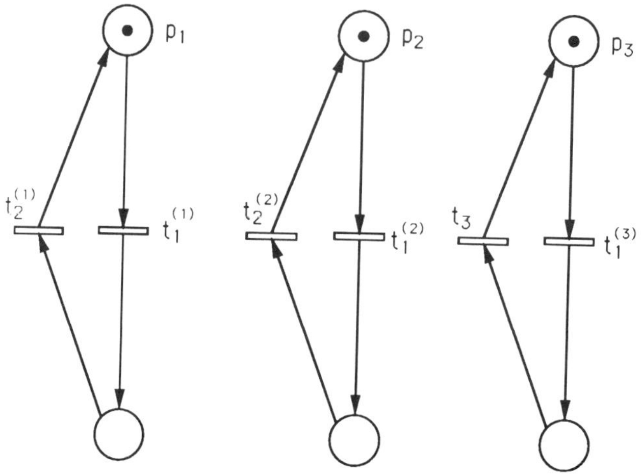

Figure 5

Similar expressions hold for $\mu_1^{(2)}$ and $\mu_1^{(3)}$. With $\lambda_1 = \lambda_2 = 2$, $\lambda_3 = 1$, as used in the previous method, we can establish that $\mu_1^{(1)}$ satisfies

$$\left(\mu_1^{(1)}\right)^3 + 4\left(\mu_1^{(1)}\right)^2 + 8\mu_1^{(1)} - 8 = 0$$

with solution $\mu_1^{(1)} = 0.7064$. Consequently $\mu_1^{(2)} = 0.7064$, $\mu_1^{(3)} = 1.0922$ and P (customer is lost) $= 0.739$.

Comparing 0.739 with the loss probability 0.7071 obtained earlier and the exact result 0.7 we note the obvious; the more extreme the approximation technique, the worse the estimate for the loss probability.

The latter approach is equivalent, for this example, to the Erlang fixed point approximation. The former method retains the flexibility of choosing the modules to suit the problem at hand.

In both methods the equilibrium probabilities for the markings of the original SPN are approximated from the modules, and the parameters for the modules are approximated using equilibrium probabilities from the other modules. The result is a system of equations linking parameters to probabilities to parameters. Looking at the fixed point technique from an SPN modular viewpoint identifies the steps in the analysis and immediately suggests ways of improving the approach.

Henderson and Taylor [18] follow this example with a generic Petri net for resource access problems and a general flexible procedure for creating the modules and deriving fixed point approximations. The essence of the approach is to create final modules with closed form equilibrium solutions and then use properties of the closed form solutions to solve the resultant equations.

The Erlang fixed point technique always has modules of the form of Figure 5. In many situations this extreme breakdown approach is necessary. On the other hand, when it is not necessary, better results can be achieved by finding less extreme approximations.

Conclusion

SPNs are valuable modelling tools particularly suited to many problems which arise in communications networks and protocols. As such they are worthy of attention. However SPNs have a broad definition, they therefore span a wide range of models and pose a wide range of problems. Consequently general solutions, whether they be analytic or algorithmic, are unlikely to cover more than a very small subset of SPNs. This paper has attempted to show that solutions can be found for classes of SPNs and that these solutions can apply as approximations to larger classes. There are many avenues to follow and I hope that this paper will encourage others to open up some of these research areas.

Acknowledgement

The authors would like to thank the Teletraffic Research Centre at the University of Adelaide and Telecom Australia for supporting the research presented in this paper.

References

[1] Ammar H.H and Islam S.M.R. (1989) - *On Bounds for Token Probabilities in a Class of Generalised Stochastic Petri Nets*, Proc. 3rd Int. Conf. on Petri Nets and Performance Models, Kyoto, Japan, pp 221-227, Dec. 1989.

[2] Ammar H.H and Islam S.M.R (1989) - *Time Scale Decomposition of a Class of Generalised Stochastic Petri Nets*, IEEE Trans on Software Engineering, **15**, no. 6, pp 809-820, 1989.

[3] Balbo G., Bruell S.C. and Ghanta S. (1988) - *Combining Queueing Networks and Generalized Stochastic Petri Nets for the Solution of Complex Models of System Behaviour*, IEEE Trans. on Computers, **37**, pp 1251-1268, 1988.

[4] Baskett F., Chandy K., Muntz R., and Palacios J.(1975) - *Open, Closed and Mixed Networks of Queues with Different Classes of Customers*, J. Assoc. Comp. Mach., **22**, pp 248-260, 1975.

[5] Bertoni A. and Torelli M., (1981) - *Probabilistic Petri Nets and Semi-Markov Processes*, Proc. 2nd European Workshop on Petri Nets, Bad Honnef, West Germany, Sept. 1981.

[6] Bobbio A. and Cumani A. (1984) - *Discrete State Stochastic Systems with Phase-Type Distributed Transition Times*, Proc. AMSE Intl. Conf. on Modelling and Simulation, Athens, pp 173-192, 1984.

[7] Burman D.Y., Lehoczky J.P. and Lim Y. (1984) - *Insensitivity of Blocking Probabilities in a Circuit Switching Network*, J. Appl. Prob., 21, pp. 850-859, 1984.

[8] Chen P., Bruell S.C. and Balbo G. (1989) - *Alternative Methods for Incorporating Non-Exponential Distributions into Stochastic Timed Petri Nets*, Proc. 3rd Intl. Workshop on Petri Nets and Performance Models, Kyoto, Japan, pp 187-197, Dec. 1989.

[9] Dugan J.B., Geist R.M., Nicola V.F. and Trivedi K.S. (1984) - *Extended Stochastic Petri Nets: Applications and Analysis*, Performance 84, Elsevier Sci. Publ. B.V. (N-Holland), pp 507-519, 1984.

[10] Florin G. and Natkin S. (1989) - *Matrix Product Form solution for Closed Synchronized Queueing Networks*, Proc. 3rd Int. Conf. on Petri Nets and Performance Models, Kyoto, Japan, pp 29-37, 1989.

[11] Girard A. (1990) - *Routing and Dimensioning in Circuit Switched Networks*, Addison Wesley, 1990.

[12] Haas P.J. and Shedler G.S. (1986) - *Regenerative SPNs*, Perform. Eval. 6, pp 189-204, 1986.

[13] Haas P.J. and Shedler G.S, (1986) - *Stochastic Petri Nets with Simultaneous Transition Firings*, Proc. Int. Workshop on Petri Nets and Performance Models, Madison, WI: IEEE Computer Society Press, pp 24-34, Aug. 1987.

[14] Haas P.J. and Shedler G.S. (1989) - *Stochastic Petri Net Representation of Discrete Event Simulations*, IEEE Trans on Software Engineering, 15, no. 4, pp 381-393, 1989.

[15] Henderson W. and Lucic D. (1988) - *Applications of Generalised Semi Markov Processes to Stochastic Petri Nets*, Performance of Distributed and Parallel Systems, Ed. T. Hasegawa, H. Takagi, Y. Takahashi, pp. 315-328, 1988.

[16] Henderson W. and Lucic D. (1991) - *Exact Results in the Aggregation and Disaggregation of Networks*, Submitted.

[17] Henderson W., Lucic D. and Taylor P.G. (1989) - *A Net Level Performance Analysis of Stochastic Petri Nets*, J. Aust. Math. Soc. Ser.B, 31, pp 176-187, 1989.

[18] Henderson W. and Taylor P.G. (1991) - *Fixed Point Approximations for Resource Access Using Stochastic Petri Nets*, Submitted.

[19] Henderson W. and Taylor P.G. (1991) - *Embedded Processes in Stochastic Petri Nets*, to be published in IEEE, Trans. on Software Engineering, 1991.

[20] Holliday M.A. and Vernon M.K. (1987) - *A Generalised Timed Petri Net Model for Performance Analysis*, IEEE Transactions on Software Engineering, SE-13, No. 12, pp 1297-1310, 1987.

[21] Jansen U. and König D. (1980) - *Insensitivity of Steady State Probabilities in Product Form for Queueing Networks*, Elektron, Informatsionsverarb. Kybernetik., **16**, pp 285-397, 1980.

[22] Jensen K. (1983) - *High-level Petri Nets*, Informatik-Fachberichte, **66**, pp. 166-180, 1983.

[23] Kelly F.P. (1976) - *Networks of Queues*, Adv. Appl. Prob., **8**, pp 416-432, 1976.

[24] Kelly F.P. (1991) - *Loss Networks*, Preprint.

[25] Kelly F.P. (1988) - *Routing in Circuit-Switched Networks: Optimization, Shadow Prices and Decentralization*, Adv. Appl. Prob., **20**, pp 112-144, 1988.

[26] König D. and Jansen U., (1974) - *Stochastic Processes and Properties of Invariance for Queueing Systems with Speeds and Temporary Interruptions*, Trans 7th Prague Conference Inf. Th., Stat. Dec. Fns. and Rand. Proc., pp 335-343, 1974.

[27] Lin C. and Marinescu D.C. (1988) - *Stochastic High-Level Petri Nets and Applications*, IEEE Trans. on Computers, **37**, pp 815-825, 1988.

[28] Lucic D. (1990) - *On Exact Equilibrium Distributions of Stochastic Petri Nets*, Ph. D. Thesis, Adelaide University, South Australia.

[29] Marsan M.A., Conte G. and Balbo G. (1984) - *A Class of Generalised Stochastic Petri Nets for the Performance Analysis of Multiprocessor Systems*, ACM Transactions on Computers, **2**, no. 2, pp 93-122, 1984.

[30] Matthes K. (1962) - *Zur Theorie der Bedienungsprozesse*, Trans. 3rd Prague Conference Inf. Th., Stat. Dec. Fns. and Rand. Proc., pp 513-528, 1962.

[31] Merlin P. (1974) - *A Study of the Recoverability of Computer Systems* Ph. D. Dissertation, University of California, Irvine 1974.

[32] Merlin P. (1979) - *Specification and Validation of Protocols*, IEEE Trans Comm, **COM-27**, No. 11, pp 1671-1680, Nov. 1979.

[33] Molloy M.K. (1981) - *On the Interpretation of Delay and Throughput Measures in Distributed Processing Models*, Ph. D. Dissertation, University of California, Los Angeles, 1981.

[34] Murata T. (1989) - *Petri Nets: Properties, Analysis and Applications*. Proc. of the IEEE, **77** No. 4, pp 541-580, 1989.

[35] Natkin S. (1980) - *Les Reseaux de Petri Stochastiques et leur Application a l'Evaluation des Systemes Informatiques*, These de Docteur Ingegneur, CNAM, Paris, France, June 1980.

[36] Noe J.D. (1971) - *A Petri Net Model of the CDC 6400*, Proc. ACM SIGOPS Workshop on System Performance, pp. 362-378, 1971.

[37] Peterson J.L. (1981) - *Petri Net Theory and the Modelling of Systems*. Englewood Cliffs, NJ: Prentice-Hall, Inc., 1981.

[38] Petri C.A. (1962) *Kommunikation mit Automaten*, Ph. D. dissertation, University of Bonn, West Germany, (In German);

[39] Ramchandani (1973) - *Analysis of Asynchronous Concurrent Systems by Petri Nets*, Ph. D. Dissertation, Dept. of Electrical Engineering, Massachusetts Institute of Technology, Cambridge, Massachusetts, July 1973.

[40] Razouk R.R. and Phelps C.V. (1984) - *Performance Analysis using Timed Petri Nets*, Proc. 1984 Intl. Conf. on Parallel Processing, pp 126-128, Aug. 1984.

[41] Reisig W. (1984) - *Petri Nets*, EATCS Monographs on Theoretical Computer Science, 4. New York: Springer-Verlag, 1985.

[42] Rumsewicz M. and Henderson W., (1989) - *Insensitivity with Age Dependent routeing*, Adv. Appl. Prob., **21**, pp 398-408, 1989.

[43] Sanders W. and Meyer J. (1989) - *Reduced Base Model Construction Methods for Stochastic Activity Networks*, Proc. Third Int. Workshop on Petri Nets and Performance Models, Kyoto, IEEE Computer Press.

[44] Sifakis J. (1977) - *Use of Petri Nets for Performance Evaluation*, Measuring, Modelling and Evaluating Computer Systems, Ed. H. Beilner, E. Gelenbe, North-Holland, pp 75-93, 1977.

[45] Woo D.L., Phelps C.V. and Sidwell R.D. (1986) - *Timed Petri Nets Probability Semantics*, Proc. 7th European Workshop on Applications and Theory of Petri Nets, Oxford, UK, pp 131-149, June 1986.

[46] Zuberek W.M. (1980) - *Timed Petri Nets and Preliminary Performance Evaluation*, 7th Annual Symposium on Computer Architecture, La Baule, France, pp 88-96, May 1980.

Network Services and Evolution

Rapid Service Delivery and Customization in a Developing Network Infrastructure

P.S. Richards

Bell-Northern Research, Ottawa, Canada

Abstract
The Intelligent Network Standards Recommendations, which are currently being developed by the CCITT, will open up unique opportunities for the modernization of a developing network infrastructure. For the first time, network providers will be able to invest in a set of standardized, service-independent functions, putting in place a prebuilt capability reusable across a broad spectrum of service definitions. This service-independent infrastructure will provide the basis for the low-risk prototyping and customization of advanced services, as well as full-scale deployment. Rapid service delivery and customization under the control of the service provider will be a major strategic advantage in a competitive, rapidly evolving market environment.

1. INTRODUCTION

This paper examines the potential role of the Intelligent Network (IN) in the modernization of a developing network infrastructure to provide advanced services.

During the past five years, the IN concept has been refined, and a number of specific architectural proposals have been made by various players within the telecommunications industry. This progress has been summarized in a number of recent papers, for example [1], [2], [3]. The concept is now sufficiently mature that a major international standardization effort has been launched within Study Group XI of the CCITT, [4]. The aim of this effort, which is known as IN Capability Set 1 (CS-1), is to achieve a usable set of standards recommendations for formal release by the end of 1992. Looking beyond 1992, further Capability Sets will be defined and standardized, but all will be supersets of CS-1. The purpose of this paper is to outline an initial set of IN capabilities, based on CS-1 standards, which would enable low-risk introduction of a wide range of advanced services together with rapid service delivery and customization capabilities.

The paper is organized as follows. Section 2 provides an overview of CS-1 as currently being defined in SG XI of CCITT. The CS-1 standards will comprise a set of functional and interface specifications, which will drive specific product implementations into the mid 1990s and beyond. Section 3 outlines some of the product implications. Finally, section 4 sets out a generic modernization scenario for a developing network environment, and uses it to illustrate:

- The deployment of a service-independent network infrastructure, for the low-risk introduction of advanced services,

- a process for rapid service delivery and customization by the service provider (or by the equipment manufacturer on behalf of the service provider), and

- implications for performance specification and management.

2. OVERVIEW OF IN CAPABILITY SET 1

IN Capability Set 1 (CS-1) standards will not be formally available until the end of 1992. This section therefore provides a brief synopsis of the intent as of late October 1990. This intent is captured in a CCITT Working Document entitled "IN CS-1 Guidelines and Work Plan" [5]. The remainder of this section is based on the October 1990 version of [5].

2.1 CS-1 Based Services

The value of CS-1 will ultimately be determined by the services which it will support. A candidate set of CS-1 based services is shown in Table 2.1, extracted from [5]. CS-1 is intended to provide the infrastructure for "single-ended, single point of control" services. This category, referred to as "Type A" in [5] covers services which are invoked on behalf of a single user (originating or terminating party) at the call set-up or tear-down phases of a call.

The Type A category leads to the following advantages in the context of CS-1 standardization:

- It includes a wide range of services, many of which have already proven their value in proprietary, service-specific implementations (e.g. Freephone).

- It depends on well-understood control relationships between network components, as outlined in section 2.2 below. Functional and interface specifications are therefore achievable within the required timeframe: technical agreements by the end of '91, leading to formal standards recommendations by the end of '92.

- It minimizes complexity, both for the service provider and equipment manufacturer, in the transition to a rapid service delivery process. This aspect is covered in more detail in section 2.2 below.

Table 2.1
Candidate Set of CS-1 Services from [5]

Freephone	Security Screening
Virtual Private Network (VPN)	Premium Rate
Universal Personal Telecommunications	Split Charging
User-defined Routing	Account Card Calling
Abbreviated Dialing	Credit Card Calling
Originating Call Screening	Automatic Alternate Billing
Terminating Call Screening	Televoting
Call Forwarding	Mass Calling
Call Distribution	Follow-Me-Diversion
Call Volume Distribution	Malicious Call Identification
Destination Call Routing	Completion of Call to Busy Subscriber
Selective Call Forward (Busy/Don't Answer)	Conference Calling (under discussion)

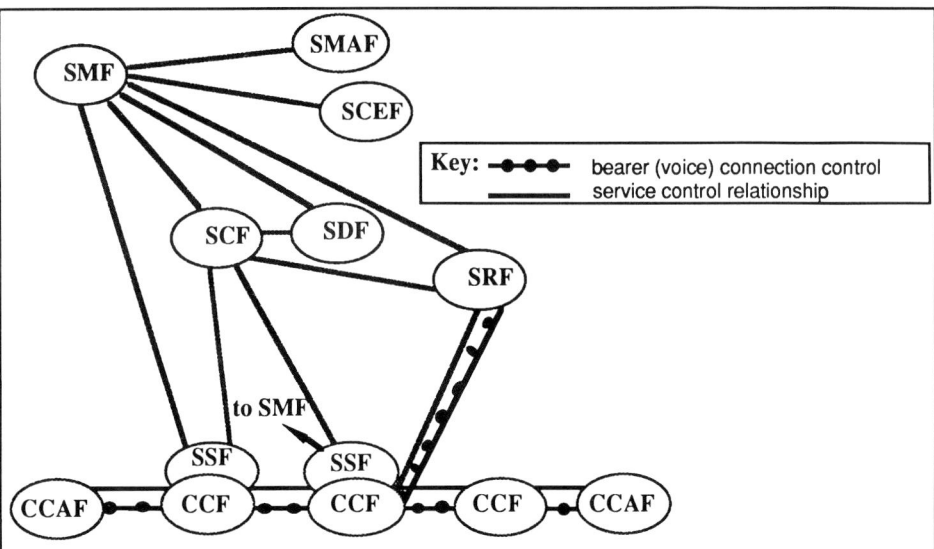

CCF -	Call Control Function: refers to call and connection handling in the classical sense (e.g., that of an exchange).
CCAF-	Call Control Agent Function: provides the user access to the network.
SSF -	Service Switching Function: interfaces with CCF and SCF. It allows CCF to be directed by the SCF.
SRF -	Specialized Resources Function: provides a category of resources for access by other network entities. Examples of resources include DTMF sending and receiving, protocol conversion, speech recognition, synthesized speech provision.
SCF -	Service Control Function: contains the IN service logic and handles service-related processing activity.
SDF -	Service Data Function: handles access to service-related data and network data and provides consistency checks on data. It hides from the SCF the real data implementation and provides a logical data view to the SCF.
SCEF -	Service Creation Environment Function: allows an Intelligent Network service to be defined, developed, tested and input to SMF. Output of this function involves service logic and service data templates.
SMAF -	Service Management Access Function: provides an interface (e.g., screen presentation) to the SMF.
SMF -	Service Management Function: involves service management control, service provision control and service deployment control.

Figure 2.1
Functions and Control Relationships for Capability Set 1 (extracted from [5])

2.2 CS-1 Functions and Interfaces

The CS-1 functions and interfaces are summarized in Figure 2.1, extracted from [5]. The following is intended to outline the purpose of the principal functions, their control relationships, and their interfaces.

2.2.1 CS-1 Functions

The Call Control Function (CCF) provides call processing for basic telephony services, and also for advanced, switched-based services. The Service Switching Function (SSF) is an augmentation of the CCF which provides a well-defined, service-independent interface to the Service Control Function (SCF).

A CS-1 call may be initiated in a number of ways. In a typical service scenario (e.g. Freephone) the originating party dials a sequence of digits (e.g. 800). This sequence is recognized by the CCF/SSF logic, which triggers an appropriate query to the SCF. The precise set of CS-1 service initiation modes and the corresponding set of CCF/SSF triggers is still under discussion in CCITT SG XI. The intent is to provide a sufficient set to satisfy the requirements of the candidate services listed in Table 2.1.

On receiving a query from the SSF, the SCF refers to its own service logic to determine how to proceed. In the Freephone example, a simple database look-up operation is sufficient to translate the Freephone number dialled by the originating party into the correct destination party address. However, a much wider range of service options and capabilities will be available under CS-1 including:

- Conditional database operations, conditioned on parameters such as:
    - time of day
    - day of week
    - calendar information (e.g. statutory holidays)
    - geographical area from which call originated
    - percentage or numerical call distribution criteria
    - etc.

- the ability for the SCF to query external Service Data Functions (SDF), such as a geographically distant database, and

- the ability for the SCF to launch a "prompt-and-collect" sequence with the calling or called party in order to obtain additional information, for example a Personal Identification Number (PIN) for authorization purposes. This is done with the help of the Special Resource Function (SRF), which contains the necessary voice prompts and the means to collect responses from the user.

When the SCF has executed the necessary service logic transactions, it formulates a response to the SSF containing the instructions for call completion. The SSF/CCF then proceeds with completion according to the instructions.

Most of the functionality which has just been described is *service-independent*, meaning that it is designed to be reused across all CS-1 based services. However, there is also some *service-specific* functionality, which defines the operation and external appearance of a specific service. *Service-specific* functionality in CS-1 comprises:

- Service logic and data within the SCF,

- customized prompts or announcements within the SRF, and

- activation/deactivation status of triggers in the SSF/CCF.

Service-specific functionality may be added, deleted or modified under the rapid service delivery and customization process. This process would typically be under the direct control of the service provider. The functions which enable this direct control are:

- the Service Creation Environment Function (SCEF) which allows service logic to be developed and tested,

- the Service Management Access Function (SMAF) which allows entry of customer-specific data in the context of service logic defined in the SCEF, and

- the Service Management Function (SMF), which encompasses the service data management functions required for service customers, as well as the means to update SCF locations with new versions of *customer-specific, service-specific* logic and data.

2.2.2   CS-1 Control Relationships
The principle intent of the CS-1 control architecture is summarized below:

- Call control responsibility remains with the CCF at all times. This includes control over switch-based resources, and control of CS-1 service interactions with switch-based services. In the event of SCF response time-out, the CCF/SSF will be capable of a default call processing sequence.

- Real-time execution of service logic on behalf of a calling or called party is the responsibility of the SCF. The SDF and SRF act as "assistants" to the SCF in this function.

- Responsibility for the integrity of *customer-specific, service-specific* logic and data rests with the SMF.

2.2.3   CS-1 Interfaces
The SSF-SCF interface is a primary focus of the CS-1 standardization effort. The CS-1 definition is intended to ensure that the following two conditions are met:

1) The CCF/SSF requires no "knowledge" of service-specific, customer-specific logic and data. This property is essential to the *service-independence* of the CCF/SSF.

2) Conversely, the SCF requires no "knowledge" of switch-based call processing and switch-based services. In particular, call-state information is not required by the SCF. This property is essential to achieving an agreed interface specification within the CS-1 standardization schedule.

Condition 1 is obvious, but condition 2 requires some elaboration. There exists a category of services (referred to in [5] as Type B) which would require transfer of call state information from SSF to SCF, and manipulation of call states by the SCF, in order to satisfy service-specific logic -- for example services involving tight service-specific control of simultaneous connections to multiple parties. Because of the complexity of the SSF-SCF interface specification required, Type B services will not be accommodated within the CS-1 specification, but will be left for further study in CS-2. Additional considerations in this decision include operational complexity of IN-supported Type B services for the service provider, and implementation complexity for equipment vendors.

Other interfaces that will be standardized under CS-1 include SCF-SDF and SCF-SRF. Interfaces involving SCEF, SMAF, and SMF will be defined but not standardized in CS-1. The role of the service provider in rapid service delivery and customization is new to the telecommunications industry, and it is preferable that commercially viable proprietary solutions to these interfaces be explored before standardization is attempted.

3. PRODUCT IMPLICATIONS

One of the stated aims of the CS-1 standards is that they should be implementable using today's product technologies. This applies particularly to the current generation of digital, stored-program switching machines.

This section provides a brief overview of the match between CS-1 functions and current generation products, and of the match between CS-1 interfaces and existing protocols.

3.1 Support of CS-1 Functions

The CS-1 CCF is available today, represented by the call processing function of modern digital switches. From the switch developer's viewpoint, the SSF would be developed as a set of call processing "features", to be implemented and delivered to the service provider as a new switch software release. Assuming that SS7 is already in place, no switch hardware upgrade would be necessary.

The SCF and SDF will likely be supported through evolved versions of today's Service Control Point (SCP) products. The main difference is that the service-specific software in today's SCPs would need to be augmented with the service-independent operations required in the CS-1 SCF. Given sufficient transaction-processing capacity, the existing software and the CS-1 SCF could co-exist on the same physical SCP.

The SRF could be implemented within the switching system, or as a separate physical unit referred to as an Intelligent Peripheral (IP). In its most basic form it would contain tone receivers and vendor-defined announcements. However, depending on customer and service requirements, versions with voice recognition and customizable announcements may be desirable.

Unlike the previous functions, the SMF, SCEF and SMAF do not have stringent real-time requirements. They will likely be implemented using available computing platforms, tools, and database environments.

3.2 Support of CS-1 Interfaces

The current view is that all the signalling interfaces targeted for CS-1 standardization would be supported using SS7 TCAP (see also Section 2.2.3). A SS7-based signalling infrastructure is therefore an essential prerequisite for deployment of CS-1 functions. The SRF to CCF interface, which must support voice channels as well as a control interface, will likely be Primary Rate ISDN.

4. NETWORK MODERNIZATION FOR ADVANCED SERVICES

4.1 Assumptions and Modernization Scenario

The CS-1 framework outlined in the previous sections provides for considerable flexibility in deployment of advanced service capabilities. This is particularly the case in a developing network infrastructure, where the network must be modernized not only to bring basic telephone service to high penetration levels, but also to provide for advanced services.

The network provider's options for deployment of a modern telecommunications infrastructure will be discussed assuming:

- An existing base of switching and transmission equipment, which can no longer be extended economically, and which therefore needs capping and replacing,

- a low to moderate penetration of basic telephone service, resulting in a large latent market demand which must be satisfied across all customer sectors,

- a desire to provide businesses with advanced telecommunications services, to improve their global competitiveness, and

- a specific annual capital budget for telecommunications network modernization.

Let us now contrast two possible approaches to network modernization:

1) A service-by-service approach, in which each service is individually justified, planned, developed and implemented.

2) The IN approach, in which a service-independent infrastructure is deployed. This provides the basis for rapid, low-risk service delivery and customization.

The service-by-service approach of scenario 1 is attractive in that it provides for straight-forward management and economic accountability. Each advanced-service project must be justified, usually on economic grounds, using indicators such as Return on Investment (ROI). Projects that pass the test proceed to completion, and those that fail are terminated. However, each go/no-go decision carries with it a risk. The "go" decision may be based on forecasts or assumptions that are optimistic, leading to wasted investment. Conversely, there is the risk that a "no-go" decision may be based on pessimistic or insufficient information, leading to a wasted opportunity.

These risks may be acceptable if the project decisions are being made in a reasonably stable environment -- for example, a monopoly provider serving a mature telecommunications marketplace. However, in a volatile environment, the market criteria used in the initial project justification may bear little resemblance to the market reality at project completion. This may be due to a market undergoing rapid development, or the actions of new competitors, or both. Under these circumstances, the "service-by-service" approach breaks down.

The IN approach to network modernization (scenario 2) provides a means of minimizing the "wasted-investment" and "wasted-opportunity" risks in the launch of IN-supported services. This is done by:

- Minimizing the service-specific portion of the total network modernization investment, by deploying a flexible, service-independent IN infrastructure, and

- making use of the increased velocity with which IN services may be developed, deployed and customized to meet volatile market needs.

These two aspects of IN are discussed in more detail in sections 4.2 and 4.3 below

4.2 Deployment of the Service-independent Infrastructure
In a developing network environment, the deployment of a new network infrastructure will be assumed to take place in two stages:

Stage 1: Deployment of digital switching and transmission, together with SS7, to drive the penetration of basic telephony services.*

Stage 2: Building on Stage 1, the selective deployment of service-independent IN capabilities.

In this approach, the first stage of modernization represents investment in an up-to-date technology base. The first services provided on this base will likely be basic telephony, and, in a situation of strong latent demand, these services will by themselves constitute sufficient economic justification for the modernization.

Since SS7-based signalling is now a mature technology, it makes economic sense to incorporate it for trunk signalling right away. (The intermediate signalling technology step based on multi-frequency tones was valid historically, but it has been superseded by SS7).

Referring to the CS-1 functions outlined in section 2, only the Call Control Function (CCF) is put in place through the first stage of modernization. The result of steady Stage 1 investment will therefore be an expanding network of CCFs, overlaying and superseding the old analog network.

Depending on modernization priorities, Stage 2 investment could begin as soon as the first CCF is in place, or, at the other extreme, it could wait until Stage 1 modernization has achieved widespread penetration. Let us therefore consider a likely intermediate scenario, in which modernization of several major urban switching centers has been completed, along with a long-distance network to interconnect them. In this scenario, it is assumed that the network provider now wishes to put in place a selection of IN-based services (e.g. from the list in Table 2.1), serving the communities which have access to the modernized switches. The Stage 2 modernization to achieve this will now be addressed.

Unlike Stage 1 modernization, which is driven by latent demand for an established service, the only purpose of Stage 2 is to put in place an infrastructure for new services. A typical low-risk approach would therefore be to put in a limited Stage 2 deployment for trial purposes, which can later be expanded as demand develops. This would consist of the following minimal set of functions:

- Two SSFs, located at two separate CCF switches giving access to the communities participating in the trial. These switches will be referred to as Service Switching Points, (SSP).

- One SCF, located in a separate piece of equipment referred to as a Service Control Point (SCP).

- A Service Management System (SMS), incorporating SMF, SMAF, and SCEF.

- Appropriate signalling and data connectivity: SS7 TCAP used on signalling network links between each SSP and the SCP, and a data link between the SCP and SMS.

---

* This stage also provides the infrastructure for data services, as well as a means to evolve to broadband capabilities.

The SSPs are implemented by means of a software upgrade to the existing CCF switches. The other functions would require appropriate hardware platforms together with software, as outlined in section 3.

Based on the trial results, decisions would be made on extending Stage 2 modernization to accommodate growth in service transactions, and increased geographical coverage. These two aspects can be managed largely independently, as follows:

The aggregate growth in service transactions (aggregate over all users and services) may be accommodated by adding modular capacity at the original SCP, and eventually by introducing new SCP platforms. Since the SCP platform is inherently service-independent, various growth strategies are possible. For example, multi-service SCPs could be dedicated to specific geographical areas or businesses, or, at the other extreme, a group of single-service specialized SCPs could serve a whole region or country.

Increased service reach and geographical coverage is handled simply by upgrading more switches with SSF software, and by providing appropriate SS7 TCAP connectivity to the SCP(s).

In summary, Stage 2 modernization puts in place the network infrastructure for market testing and expansion of IN services. Since this infrastructure is service-independent, the return on the investment is not dependent on the success or failure of any one particular service. The risk is shared over the aggregate of services which are eventually offered.

4.3 Rapid Service Delivery and Customization

The previous section outlined scenarios for the deployment of a service-independent infrastructure. This section illustrates how that infrastructure may be used for the rapid development, delivery and customization of IN Services. The three key requirements are:

- Reusable Service-Independent Building blocks (SIBs) implemented as software functions within the SCP and SSP,

- the means whereby the service provider can create service logic to act on the SIBs, as well as being able to enter service data in the context of that service logic, and

- a software environment within the SCP that can process the service logic and data.

At the time of writing this paper, the set of SIBs for CS-1 is still being defined. In the SCP, SIBs will typically encompass operations such as: data look-ups, time checks, calendar checks, launch query (e.g. to another database), launch a prompt-and-collect sequence (to get information from the end-user), or launch a response (to the SSP).

In the SSP, SIBs will typically encompass the service-independent triggering functions, the means to launch TCAP queries to the SCP and the means to act on responses.

The set of SIBs which is specified will determine the scope of CS-1 based services. The service list in Table 2.1 is a good indicator of this scope.

Although CS-1 SIBs are being defined to accommodate service creation by operating companies, they may also be applied by equipment vendors for the rapid configuration and delivery of turnkey services.

The CS-1 Recommendations will not attempt to standardize the service creation and service management process. Since this is a new area for the industry, it will be left to service providers and manufacturers to find the best approach. The following is therefore intended only as an illustrative scenario of how service creation might be used by a service provider to effect rapid service delivery and customization.

Service creation scenario...
Based on customer service requirements, a service designer develops the logic for a "new" CS-1-based service using the Service Creation Environment and its associated tools. This is done using a high-level, graphical representation (e.g. CCITT's Specification and Description Language (SDL)). This logic specifies the appropriate operations to be invoked in the SCP, and their correct order, to represent the intent of the service. While still within the SCE, the service logic is tested (using appropriate test data) to carry out functional and performance checks.

When the service designer is satisfied with the service logic, it is downloaded to the SMS. Here the logic is combined with individual customers' data and placed in a master database.

When the service is to be launched, the logic/data combinations are downloaded to the SCP. When activated by a service transaction, the logic/data is interpreted and processed by a set of service-independent generic programs. SIB functions are called up in the correct order to act on the data and execute the service.

Service customization...
Subsequently, another customer requests a special, "customized" version of this service. This request is satisfied by the service designer using the SCE, following the procedure outlined above to create the modified version. No changes are necessary within the SCP or SSP software.

4.4 Performance Specification and Management

The rapid service delivery and customization scenario of the previous section has fundamental implications on the way real-time performance will be specified and managed in an IN environment.

Traditionally, the dimensioning and provisioning of telecommunications networks has been based on the need to accommodate a particular forecast of service demand with a specified grade of service. However, in an IN environment, it will be necessary to provision the service-independent infrastructure without any forecasts of service mix, or penetration or traffic levels. At the same time, any particular IN-based service must be assured of access to sufficient resources to meet its own, specified performance requirement. An approach to this problem has been suggested in [6]. A brief summary follows:

The service-independent infrastructure may be modelled as pools of network resources which are shared by all users of IN-based services. Examples of CS-1-based resources would be computing modules for transaction processing in the SCP, and SS7 signalling links and terminating equipment. The utilization of these pools of resources is monitored, and equipment is added when the utilization rises above a predefined threshold level. The setting of the predefined threshold involves a trade-off between average equipment utilization and the probability of a resource shortage. For large pools of equipment shared across many services and service-users, this trade-off can be managed reliably and precisely.

Having specified acceptable "probability of resource shortage" levels in the service-independent infrastructure, it is possible to predict the traffic performance (delay, loss parameters) for any particular IN service. This would be done using performance analysis tools incorporated into the Service Creation Environment.

This approach is covered in more detail in [6].

5. SUMMARY

Intelligent Network Capability Set 1 Standards Recommendations are being developed in CCITT SG XI, aiming for formal availability at the end of 1992. They will be implementable using existing product technologies.

These standards and the associated network products will open up unique opportunities for the modernization of a developing network infrastructure. For the first time, network providers will be able to invest in a set of standardized, service-independent functions, putting in place a prebuilt capability reusable across a broad spectrum of service definitions.

The service-independent infrastructure will provide the basis for the low-risk prototyping and customization of advanced services, as well as full-scale deployment. Rapid service delivery and customization under the control of the service provider will be a major strategic advantage in a competitive, rapidly evolving market environment.

ACKNOWLEDGEMENT

The author would like to thank the many colleagues in BNR and CCITT SG XI whose ideas, comments and suggestions contributed to the content of this paper.

REFERENCES

1. "Application of a Service-Independent Architecture", I. Ebert, P. Richards, J. McGee, Bell Northern Research, Canada, ISS'90, Stockholm, May 1990.

2. "Evolution of Intelligence in Switched Networks", H. Bauer, J. Jacoby et al, AT&T Bell Labs, U.S.A., ISS'90, Stockholm, May 1990.

3. "Strategy for Implementation of the Intelligent Network", K. Schulz, G. Glaeser et al, Deutsche Bundespost Telecom, ISS'90, Stockholm, May 1990.

4. Section 2 of Meeting Report of CCITT WP XI/4, COMXI-R11-E, December 1989.

5. "Intelligent Network Capability Set Number 1 Guidelines and Work Plan", Section 2.2 of SWP XI/4-1 Meeting Report, CCITT TD439R1, October 1990.

6. "Traffic Performance Specification and Modelling in the Intelligent Network", P.Richards, R. Armolavicius, et al, Bell Northern Research, Canada. Paper accepted for publication, ITC-13, Copenhagen, June 1991.

TRANSITION FROM ANALOGUE TO DIGITAL NETWORKS

Chr. Asgersen

Copenhagen Telephone Company,
Teglholmsgade 1, P.O. Box 330, DK-1790 Copenhagen V, Denmark

ABSTRACT

The paper concerns strategies for transition from analogue to digital networks. Two main strategies are dealt with: The Replacement Strategy and the Overlay Strategy. The Overlay Strategy which is chosen by KTAS, The Copenhagen Telephone Company, is given a specially comprehensive description.

The strategies dealt with concern networks for telephone traffic, but handling of data traffic in case of implementation of the strategies mentioned is briefly treated. Also a short look at the broadband perspective is taken.

The strategies dealt with are looked upon basically from a traffic point of view. This is important as the traffic loads of the network determine the capacity needed and consequently to a high degree the costs of the network.

1. INTRODUCTION

The Copenhagen Telephone Company has chosen an overlay strategy by the introduction of digital exchanges in the network.

I think it could be useful to know of the choice of this strategy as it may be advantageous in countries with developing economies.

In the early eighties when the prices of digital switching equipment became lower than the prices of analogue equipment it was considered how the digital equipment could be utilized most profitably.

The new facilities offered by the digital technic seemed to be of interest primarily for business customers. The service provided by the existing analogue exchanges would still be acceptable for the residential customers.

A condition for the utilization of the facilities provided by the digital exchanges was that Common Channel Signalling could be provided. To fulfil this condition the digital network had to be coherent.

Another main point was that the access for the business customers to the digital network should be possible in a short time at any location.

2. CHOICE OF STRATEGY

Besides the main objectives the strategy should fulfil the following demands:

- To prevent bad transmission too many analogue/digital and digital/analogue transition points in series should be avoided.

- More transit equipment than strictly necessary should be avoided.

- More signalling transfer equipment than necessary should be avoided.

Obviously it was clear that establishing of new digital exchanges as incoherent spots all over the network where replacement of worn-out equipment was needed did not fulfil the objectives.

It was decided that a firm strategy should be elaborated.

Two main strategies which could possibly meet the requirements were considered: An overlay strategy and a replacement strategy.

2.1 The Overlay Strategy

The overlay network considered consisted of a separate layer of preferably small digital switches side by side with the analogue exchanges. (Fig. 1).

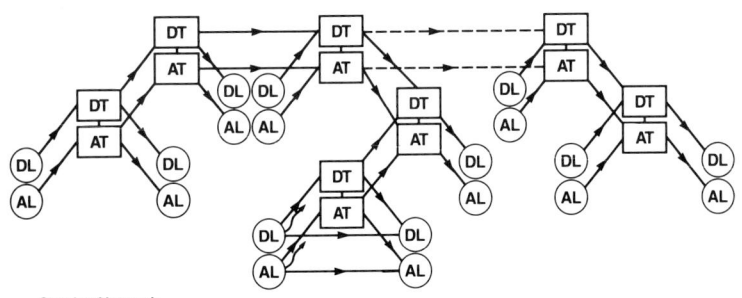

Overlay Network

Figure 1

In fig.1 the following denotations are used:

    AL   Analogue local exchange

    AT   Analogue transit exchange

    DL   Digital local exchange or remote subscriber stage

    DT   Digital transit exchange / homing switch for remote subscriber stages

The network structure is designed in such a way that the traffic between customers connected to the digital exchanges is routed by digital equipment exclusively.

The old analogue exchanges are retained untouched except for the gateways to the digital overlay network.

The residential traffic is assumed to be routed in the old analogue network.

The digital local switches could be remote subscriber-stages homing on the digital switch at the toll-centre location.

2.2 The Replacement Strategy

The idea behind the replacement strategy considered is to enable the existing network to absorb the digital equipment in a controlled way.

The principal analogue transit exchanges at the upper levels of the network hierarchy should be replaced as a first step. (Fig.2). The denotations in fig. 2 are the same as in fig. 1.

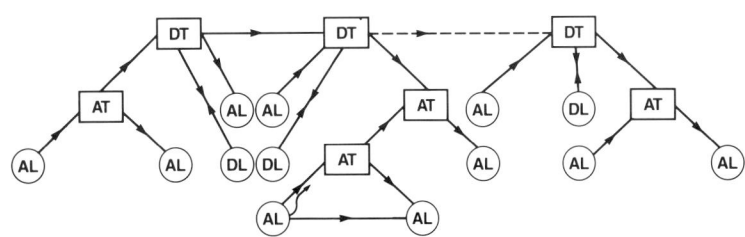

Hierarchical Network with the upper level replaced.

Figure 2

Having the upper part of the network digitalized and being coherent the common channel signalling could be used.

The important-long distance traffic could benefit from the digital technique.

An existing analogue local exchange subordinated a replaced analogue toll-exchange should be connected to the new digital switch replacing the old toll-exchange.

New digital local exchanges should be connected to the new digital switches at the upper levels of the hierarchy.

In this way the routing of the traffic through the network is not changed radically.

Connecting the digital local exchanges to the digitalized upper level of the network hierarchy limits the number of analogue/digital and digital/analogue transition points in series.

It was assumed that the exchanges replaced were not necessarily the oldest or most worn-out ones. The better parts of the replaced equipment might be used for extensions and renewals elsewhere in the network.

2.3 Comparison of the strategies

By the replacement strategy the transit exchanges carry a large amount of traffic between analogue exchanges. This is avoided by the overlay strategy retaining the analogue transit exchanges.

By the replacement strategy the connections between the local switches and the transit exchanges on which they home tend to be longer than by the overlay strategy.

These facts founded in traffic considerations supported by other considerations were in favour of the overlay strategy which was chosen.

3. SOLUTIONS OF TRAFFIC PROBLEMS IN THE OVERLAY NETWORK

A main problem concerning the traffic flow in an overlay network is that traffic which used to be internal in an exchange is now changed to be external; fig.3.

The loading of the transit routes tends to increase.

Especially the loading of the analogue toll exchanges having the transition points for analogue/digital and digital/analogue traffic tends to increase.

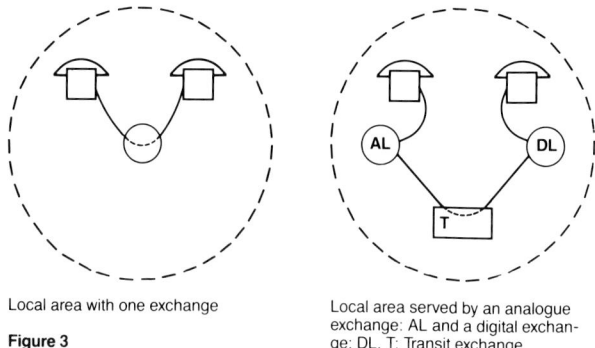

Local area with one exchange

Figure 3

Local area served by an analogue exchange: AL and a digital exchange: DL. T: Transit exchange.

This is bad as extensions with analogue equipment should be avoided, if possible.

A solution to this problem is to use a network pattern as shown in fig. 4.

Besides the denotations used in fig. 1 and fig. 2 the following denotations are used:

    HU Direct high usage groups with overflow to a final

    SP Service protection group with overflow to a final

    F   Final trunk group

    S   Simple trunk group without overflow

The structure shown in fig.4 is characterized by the use of alternative routing in such a way that traffic which tends to overload the old network is routed into the new network.

In this way unnecessary coupling in series of transit exchanges is avoided.

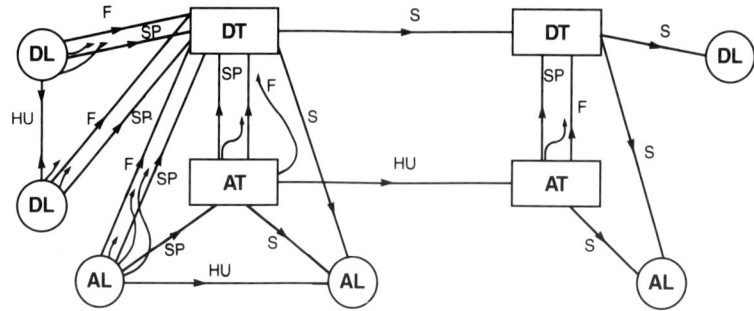

Details of Overlay Network Pattern utilizing the Service Protection Concept

Figure 4

The traffic routing in a network with a pattern as shown in fig. 4 is as follows:

Traffic between digital local switches (DL) is routed by direct trunk groups or via one or more digital switches (DT).

Traffic between analogue exchanges (AL) is routed by direct trunk groups or via one or more analogue transit exchanges (AT).

Simple groups without overflow from AL to AT are changed to service protection groups (SP) with overflow to finals (F) towards the digital switch (DT).

Traffic from analogue exchanges (AL) to digital exchanges is routed via a service protection high usage (SP) to DT with overflow to the final (F) towards DT.

Traffic from a digital switch (DL) to an analogue exchange (AL) is routed via DT by a simple group to AL.

Traffic overflowing outgoing trunk groups from AT is routed via the trunk group, F, between neighbouring AT and DT.

In this way the digital overlay network is used for the protection of the analogue network against overloading.

The overlay pattern described comprises a number of advantages:

The change of the trunk groups towards the old transit exchanges from simple groups to service protections makes it possible to control the loads of the analogue transit exchanges. Any increase of traffic will be absorbed by the finals towards the digital transit switches.

The network pattern enables effective utilization of the trunk groups as well as the switches.

It is possible to avoid extensions needing analogue equipment.

Another advantage is that uniform blocking conditions for the individual calls are obtained.

As I find it very important that a network has the property of providing uniform blocking conditions for the individual calls I will expand a little on that subject.

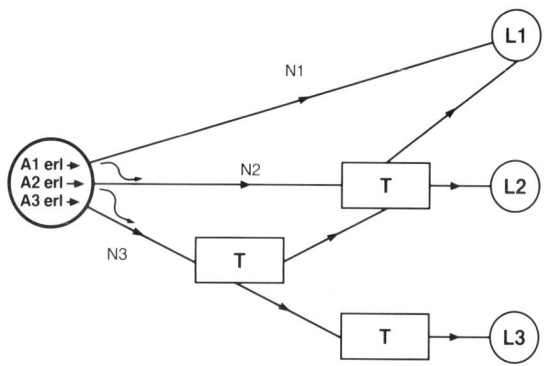

Figure 5

In fig. 5 is shown a part of a network. A1, A2 and A3 are traffic loads offered to the local exchanges L1, L2 and L3 respectively. T denotes transit exchanges.

A1 is routed by the direct high usage group N1 with overflow to another high usage group (N2) which again has alternative route via the final N3.

A2 is routed by the high usage N2 with alternative route via the final N3.

A3 has only one possible route, namely the final N3.

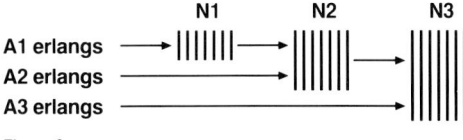

Figure 6

The blocking of the traffic loads A1, A2 and A3 is denoted B_{A1}, B_{A2} and B_{A3} respectively. B_{N1}, B_{N2} and B_{N3} are the blockings caused by N1, N2 and N3 respectively.

Referring to fig. 6 it will be seen that the blockings turn out to be:

$$B_{A1} = B_{N1} \times B_{N2} \times B_{N3}$$

$$B_{A2} = B_{N2} \times B_{N3}$$

$$B_{A3} = B_{N3}$$

The order of magnitude of the blockings experienced by A1, A2 and A3 will be different.

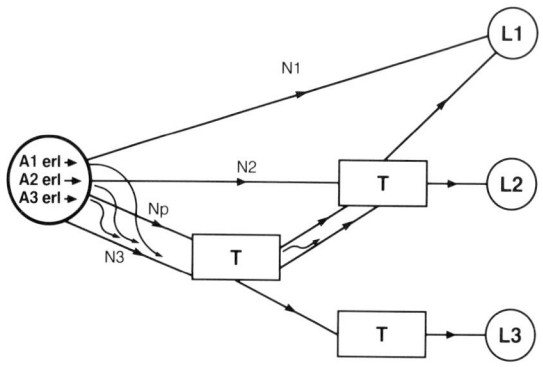

Figure 7

In fig. 7 is shown a network with the same exchanges as in fig. 5. The traffic loads offered (A1, A2 and A3) are the same too. The routing is altered in accordance with the Single-Stage Alternative Routing principle (1).

A1 is offered to the high usage N1 with overflow to the final N3. A2 is offered to the high usage (N2) with overflow to N3. A3 is offered to the service-protection group (Np) with overflow to N3.

The overflow pattern is shown in fig. 8. The denotations will be understood by referring to the previous figures.

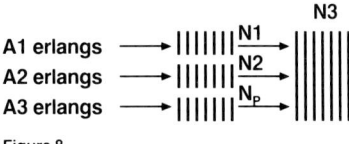

Figure 8

The blockings experienced by the offered loads A1, A2 and A3 turn out to be:

$$B_{A1} = B_{N1} \times B_{N3}$$

$$B_{A2} = B_{N2} \times B_{N3}$$

$$B_{A3} = B_{Np} \times B_{N3}$$

By this routing pattern uniform blocking conditions for the individual calls can be obtained.

A network feature I find extremely important is that it is possible to design the network basically from given goals for the blockings caused by the network.

The overlay network described here has this feature.

The pattern is so simple that it is possible to deduct the engineering values for the individual trunk groups according to given end-to-end blockings.

Considerations of space forbid me to go into details with this, but in (3) the principle of end-to-end dimensioning of networks is described.

4. DATA TRAFFIC AND BROADBAND SERVICES

As you will know telephone networks are to some extent able to handle data traffic by use of modems, but large amounts of data traffic need data networks if the data traffic should be handled fast enough and economically.

Besides data traffic a modern society also needs broadband services such as transmission of television and video-conferences.

How should the strategies dealt with above be judged in the light of the need for data transmission and broadband services?

If the data traffic and broadband traffic are served by dedicated networks isolated from the telephone network the strategies mentioned are not affected.

Tendencies of integration exist, however.

The possibility currently of interest in connection with the integration of telephone and data traffic is the ISDN concept.

Narrowband ISDN exchanges are to day commercially available.

Would it be wise to choose ISDN at an initial stage of digitalization?

I will not answer this question here. It is beyond the scope of the paper, but I will notice that the overlay strategy is convenient as well in case of digital exchanges without ISDN being chosen as in case of the tiger leap into the ISDN era being taken.

If a conventional digital network is chosen as a first step and it is planned to implement ISDN exchanges as a second step, then the overlay concept is also adequate.

The ISDN exchanges can simply be arranged in an additional overlay network.

So is it done with the digital network of the Copenhagen Telephone Company.

The telephone traffic between the conventional exchanges and the ISDN exchanges can be routed in accordance with the principles described above.

Data traffic can be routed directly on data-links connecting the ISDN exchanges.

The development of telecommunication systems integrating broadband services with telephone and data services is in rapid progress, but so far they are not ready for commercial operation.

Therefore I think it is too early to discuss how to connect broadband networks to narrowband networks, but again it seems likely that an overlay pattern which provides gateways for the telephone traffic in transition points selected with due regard to economic routing in the old network could be an adequate solution.

5. CONCLUSION

One of the main advantages of the overlay technique, using the pattern described is that the facilities provided by the digital technique, are rapidly widely available.

Another important advantage is that unnecessary traffic loadings of the transit exchanges, especially the old ones, are avoided.

A further important advantage is that the simple and firm network structure makes it easy to survey the blockings caused by the network and that design of the network from given grade-of-service objectives is possible.

Of course this brief paper does not so far elucidate all the problems of transition from analogue to digital networks, but I hope it is demonstrated that fundamental considerations concerning the traffic flow in the network are important for the economy and service.

REFERENCES

1. Chr. Asgersen, A NEW DANISH TRAFFIC ROUTING PLAN WITH SINGLE STAGE ALTERNATE ROUTING AND CONSISTENT USE OF SERVICE PROTECTION FINALS FOR FIRST ROUTED TRAFFIC TO TANDEM OFFICES, Proceeding of the 5th International Teletraffic Congress, New York 1967, page 407.

2. Roger Wilkinson, SIMPLIFIED ENGINEERING OF SINGLE STAGE ALTERNATE ROUTING SYSTEMS, Paper no. 75, Session 7, the 4th International Teletraffic Congress, London 1964.

3. Chr. Asgersen, END-TO-END DIMENSIONING OF TRUNK NETWORKS, A CONCEPT - BASED ON EXPERIENCE FROM THE DANISH NETWORKS. Proceeding of the 11th International Teletraffic Congress, Kyoto 1985, Paper no. 6.1-2.

TELECOM NETWORKS EVOLUTION TOWARDS SECURE DYNAMIC STRUCTURES
Routing Aspects of the Network Digitization

Michal Pióro[1], Manuel de Miguel and Isabel Pita

Alcatel Standard Electrica, Spain

Abstract
A natural scenario for the digitization of PSTN leads to a two-level network structure. The digital upper transit level will provide an excellent environment for some kind of dynamic routing. In the paper we discuss a way of introducing dynamic routing to such transit networks in order to achieve efficiency in traffic routing at minimum cost. The discussion is based on the results illustrating various aspects of the traffic performance of different alternative routing schemes.

1. INTRODUCTION

According to a natural scenario for the digitization of intercity and metropolitan PSTN, two network levels will be formed. The digital upper level (UL) will be composed of transit exchanges, serving the traffic between local exchanges of the lower level, delivered by the access network.

The transit UL network need not contain many exchanges. The intercity network of a European country would consist of perhaps 10 to 60 transit exchanges. For metropolitan networks this number would typically be smaller.

A number of examples illustrate this trend. A robust two-level network structure with 28 transit nodes has been proposed for the Spanish national intercity network [1,2]. For the British Telecom trunk network, 53 transit nodes are foreseen [3]. The Warsaw area metropolitan network will soon be equipped with the upper level of 8 digital exchanges interconnected by optical fibre cables [4].

The efficiency of the discussed solution has its source in the fact that the high concentration of traffic will be effectively carried via digital equipment.

Adequate attention should be paid to provide robustness at the physical layer, the junction (trunk group) layer, and the routing layer. The first two aspects are briefly described in Section 2.

The remaining sections are devoted to the impact of routing on traffic handling efficiency of the transit level and on its robustness. In section 3 we discuss some of our results in

---

1. On a leave from the Warsaw University of Technology

this area obtained for two reference networks (an 11-node network and a 6-node network, both with realistic offered traffic). In section 4 the conclusions and their impact on the scenario of implementing a routing system are presented.

2. SURVIVABILITY IN SWITCHING

Switching survivability may be solved according to [2]. At the UL, called the nodal network in [2], nodal transit exchanges will be duplicated, and put into two different geographical locations within the same area. Each of the two nodal exchanges will carry half of the traffic from/to their common area. The trunk groups of the access network will be doubled, and two-way trunk groups between nodal areas will be quadrupled (cf Fig.1). Of course, physical diversity at the transmission layer will further enforce the robustness of trunk groups by realizing them on disjoint transmission paths.

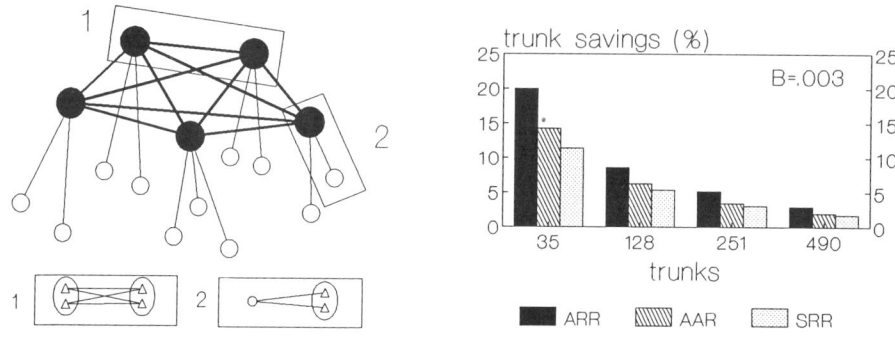

Fig.1. Network structure.   Fig.2. Trunk savings.

The duplication of trunk groups between nodal exchanges might seem an overexpansive solution in terms of trunking capacity. Since the number of input ports in UL exchanges is proportional to this capacity, the additional costs here will not be negligible even with the cheap transmission on optical fibre cables. Table 1 illustrates the order of magnitude of the amount of increased capacity. A secured "super" group organised with bidirectional trunks is considered (Fig.1(1)), with or without mutual overflow. The required capacity is compared with that required by an ordinary trunk group. The traffic offered from each side of the group is 200 Erl. (notice that at UL, the exchange-to-exchange offered traffic will rather be high since this is the traffic from one whole nodal area to another). The considered blocking levels for dimensioning are 0.3% and 5%.

Table 1
Additional trunk provision for secured trunk groups

system	no overflow		mut. overflow		common group	
blocking	0.3%	5%	0.3%	5%	0.3%	5%
trunks	492	420	452	400	439	393
difference	12%	6.8%	2.9%	1.8%	0%	0%

The case of 0.3% blocking corresponds to a network dimensioned for direct routing. With mutual overflow, only 3% more of capacity is required; at the same time a lot is gained in robustness: in the case of a node failure the network will still carry 50% of the traffic from/to the affected nodal area. In a network dimensioned under assumption of alternative routing, trunk groups admit much higher blocking (5% or more) and the amount of required additional capacity decreases noticeably.

The modularity of digital trunk groups is a potential source of further savings in trunking. With direct routing, modular trunk groups are, on the average, overdimensioned by half of the module size. With introducing alternative paths this kind of capacity overprovision can be alleviated [5].

For survivable networks the issue of robust design arises. The idea is to overdimension the network to ensure a minimum grade of service in the case of a major failure of an exchange or of a transmission link. Below we give two examples.

Case 1
Dimension the trunk groups in such a way that in the case of any of the exchanges breakdown, the network will still be able to carry x% of the traffic from/to the affected nodal area.

Case 2
Dimension the trunk groups in such a way that in the case of loosing y% of trunk capacity from/to any nodal area to a number of other nodal areas, the network will still be able to carry x% of the traffic from/to the affected area.

For high x, 80% say, the fulfilment of the requirement in Case 1 will, even with alternative routing, be expensive. In the second case the alternative routing (dynamic or static) will make the overdimensioning much lower with respect to direct routing. This will be further clarified in section 3.1.

3. COMPARISON OF ROUTING STRATEGIES

In this section we will summarise our results on the behaviour of alternative routing and the related capacity savings.

The studies have been performed with a set of procedures now available within the design and performance evaluation tool called Escorial [2]. Among others, we have considered the following routing strategies (all with overflow paths admitting only one transit node):

- FDR: fixed direct routing without overflow choices.
- FAR: alternative routing with fixed sequences of overflow choices; two versions have been considered:
  * FAR/ARR: FAR with ARR (automatic rerouting [6]), providing effective access to available overflow paths, based eg on crankback signalling
  * FAR/AAR: FAR with AAR (automatic alternative routing [6]), providing only limited access to available overflow paths (if the first leg of an overflow path is available and the second is not, the call is lost).
- SRR: sticky random routing of the DAR type [7].
- RCR: residual capacity routing of the DCR type [8].

Dynamic trunk reservation (DTR) has been used for protecting fresh traffic. Besides DTR (fixed reservation levels), also ATR (automatic trunk reservation) with the number of dynamically reserved trunks proportional to the traffic overflowing the direct path, has been considered.

3.1. An 11-node symmetric network

A fully connected symmetric junction network is characterised by the following parameters: M-number of nodes, A-traffic offered to each OD pair, n-trunk group size, s-DTR parameter and c-number of overflow choices.

With fixed n,s,A and c, blocking attributes depend on the value of M, yet they quickly stabilise. For the evaluations, M=11 has been chosen because the blocking attributes of symmetric networks are virtually the same for all M>=11, provided c<=9. Since 9 overflow choices are certainly sufficient to carry the traffic, M=11 appears to be a proper choice.

Dimensioning for nominal traffic: trunking savings

Fig.2 shows, for three alternative routing strategies and different trunk group sizes, the percentage of trunks that can be saved with respect to FDR. The design level for exchange-to-exchange loss is equal to 0.3% for nominal traffic matrix. In the case of FAR/ARR the most efficient trunk reservation level has been chosen. For simplicity we have not taken into account the secured substructure of trunk groups. As discussed in Section 2, this would further increase the savings from alternative routing.

With RCR the savings are virtually the same as with FAR/ARR. The reason for that is explained in Fig.3, showing average network blocking with RCR for several updating cycle lengths (180 sec. mean call holding time has been assumed). In all the cases of offered traffic the trunk groups have been dimensioned to assure 0.3% blocking under FAR/ARR (the same DTR parameter s has been applied for FAR/ARR and for RCR). It is clear that the difference in the average network blocking probability between FAR/ARR and RCR is negligible in the range of the updating cycle length between 10 and 30 sec.

Fig.2 exhibits a well known tendency: the greater the trunk group size the smaller the trunk savings from alternative routing. Savings are substantial for low and medium trunk group sizes (20% for n around 30, 10% for n around 120), but decrease noticeably for greater trunk groups (only 3% for n around 480).

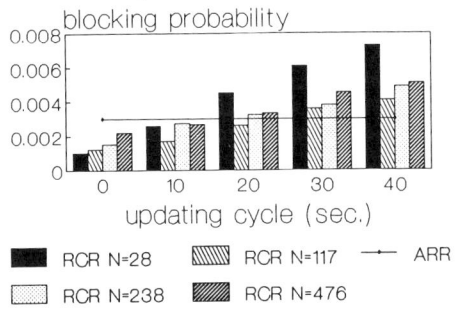

Fig.3. Blocking vs updating cycle length.

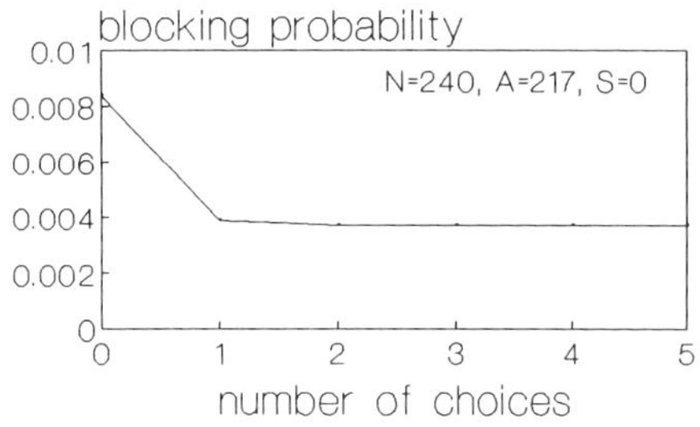

Fig.4. Blocking vs number of AAR choices.

The greatest savings are achieved with FAR/ARR using an optimised DTR parameter. For FAR/AAR (2 overflow choices and no DTR) and for SRR, the corresponding savings are smaller. Adding more overflow choices to FAR/AAR will not, as shown in Fig.4, decrease the cost any further. The difference in trunking between FAR/ARR and FAR/AAR is 6% for A=21.68 Erl., 3% for A=103.5 Erl., and only about 1% for greater traffic. SRR needs another 1% more trunks than FAR/AAR for greater traffic (3% for low traffic).

To sum up: for realistic networks (200 Erl. carried on the average from one area to another) the trunk capacity saving of 7% can be expected if alternative routing is applied. Out of this, 85% is gained by the application of the simple FAR/AAR. The gain would be greater if the secured substructure of trunk groups were assumed.

Robust dimensioning: trunk savings

We have also addressed the question of trunk savings related to the robust design criteria. To model the failure of one of the two nodes serving a nodal area, we have doubled the traffic to/from the remaining node. Then we have found the smallest n (number of trunks is still the same for all groups) with which the network still could carry 80% of the affected traffic. Savings in trunk provision are negligible. Alternative routing simply cannot help in such a situation, since all outgoing (incoming) trunk groups of the node are highly overloaded. What alternative routing can actually do is to allow using all trunks connected to the node as a common pool of resources for the whole outgoing/incoming traffic.

In a network dimensioned as above, in normal conditions FDR would ensure excellent grade of service. Still in the case of a less severe failure or overload, alternative routing would be advantageous. To illustrate this (and Case 2 of section 2) we have doubled the traffic only in two directions (to model the situation when two trunk groups lose half of their capacity). Now alternative routing is effective and this can be translated into trunk savings (cf Fig.5). To carry 80% of affected traffic, FDR requires 100 additional trunks per trunk group (almost 40% more). FAR/ARR with DTR, and also RCR, can carry the traffic without any extra capacity (n=251 is the capacity that assures 0.3% blocking in nominal traffic conditions, A=219 Erl., under FDR). Also FAR/AAR with multiple overflow paths and without DTR is able to carry the required traffic. FAR/AAR with one (optimised) overflow path requires 20 additional trunks per trunk group (8% more).

To sum up: a network designed to carry 80% of affected traffic in the case of partial capacity loss (anywhere in the network) exhibits great savings due to alternative routing, ranging from 30% (one overflow choice with reconfiguration) to 40% (multiple overflow choices).

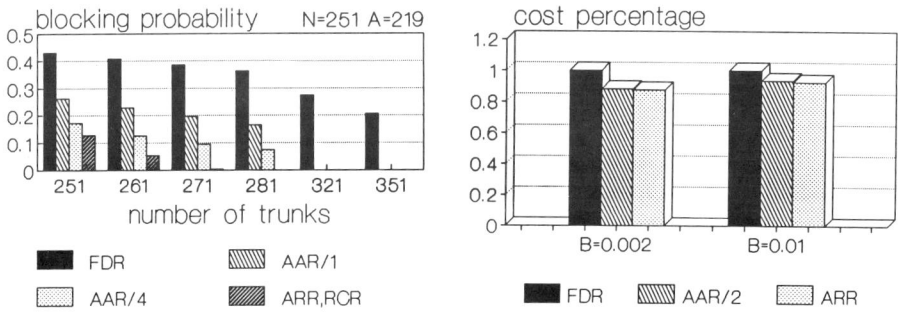

Fig.5. Failure of 2 links: greatest OD pair blocking.

Fig.6. Secure network cost.

3.2. A 6-node non-symmetric network

We have also studied a 6-node network with realistic offered traffic (6000 Erl. in total, 200 Erl. offered per one OD pair on the average) and with the secured structure of nodes and trunk groups, as described in Section 2. The entire distribution of exchange-to-exchange blocking probability has been examined, not only the average network blocking. Below we summarise the main observations.

* With alternative routing, savings in the transmission and switching cost, with respect to direct routing only, can reach about 12%-18% in the network dimensioned for nominal traffic, depending on the design loss level B (cf Fig.6).

12% saving in trunk provision means also that the amount of traffic of this order can additionally be carried on the same loss level in an existing junction network. The greatest gain is observed in the transition from FDR to FAR/AAR (up to 95% out of the 12% saving achievable with FAR/ARR). Any better access to available overflow paths does not yield much savings. The savings could be increased if the robust design criteria were obeyed.
* The traffic handling efficiency of a truly dynamic routing strategy (as RCR with updating cycle of 10 sec.) is comparable with FAR/ARR in the following situations:
  - non-nominal traffic conditions (cf Fig.7)
  - global overload
  - overload focused on a selected node (cf Fig.8)
  - failure of a transmission link.

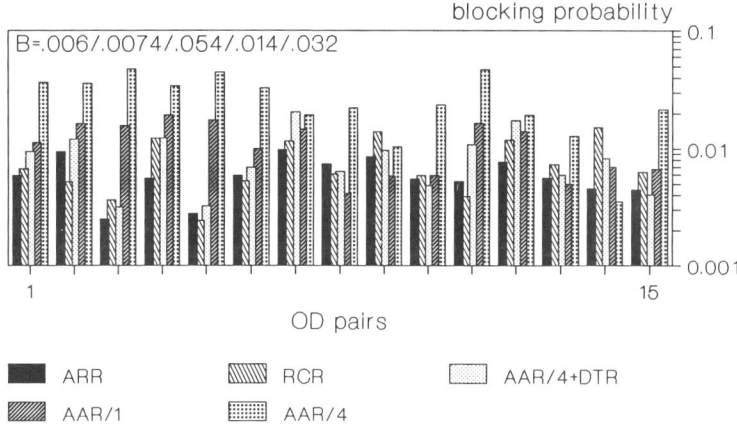

Fig.7. 5% shifts in traffic.

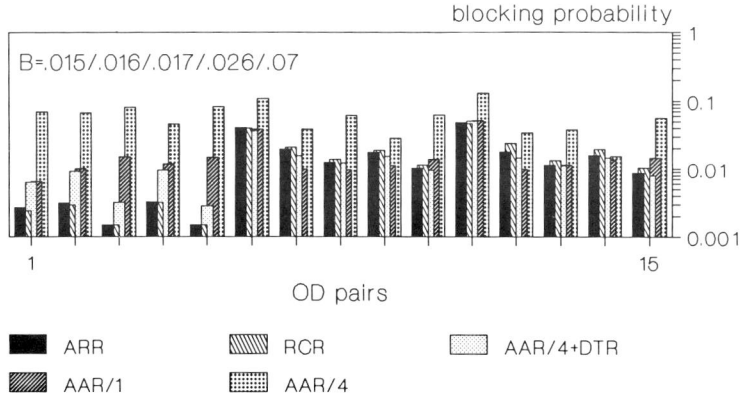

Fig. 8. 10% Focused overload.

In all the cases FAR/ARR uses DTR (fixed reservation levels), whilst RCR uses ATR. Three versions of FAR/AAR are considered: with 1 overflow choice and without DTR, with 4 overflow choices and without DTR, and with 4 overflow choices and with DTR. In all the cases routing sequences have been optimized for nominal traffic and then used in all other situations. It is clear that FAR/AAR with multiple choices and without DTR should not be used: it is apparently inferior to FAR/AAR with just one optimized overflow choice. FAR/AAR with multiple overflow choices and with DTR is superior to FAR/AAR with only one choice, although the difference is not significant.

* DTR mechanism (and ATR) seems to be too crude to cope with abnormal situations in the sense of fairness (ie equalisation of GOS for all traffic streams). The overflowing handicapped traffic streams should have priority over other transit streams in access to overflow paths.

4. INTRODUCING DYNAMIC ROUTING TO THE TRANSIT NETWORK

The most important conclusions from Section 3 are as follows.

a) In UL networks with realistic trunk group sizes designed for nominal traffic conditions, trunking savings (equivalently - increased carried traffic) due to the application of alternative routing, instead of direct routing, is significant. The savings are more substantial if the robust design criteria are obeyed.

b) In a properly dimensioned network trunking savings are the greatest in transition from FDR to FAR/AAR. This is true even with only one overflow choice for each traffic stream, provided the overflow recommendations can be changed in the case of a major traffic shift, failure or overload.

c) Giving effective access to many overflow choices through ARR combined with DTR, makes reaction to abnormal situations automatic.

d) FAR/ARR (with DTR) and RCR (with short updating time and ATR) are approximately equivalent in the day-to-day network operation (in [9] a similar conclusion is stated for robust networks as described in section 2).

Below we discuss a scenario of the deployment of dynamic routing, or more generally, alternative routing to a newly installed transit upper level of an intercity or metropolitan PSTN. We do not assume any particular switching system, only that exchanges are digital.

As to installing UL, we have to realise that in the first place digital transmission and switching equipment will be installed as quickly as possible. Thus we cannot expect an effort to implement any version of dynamic routing if this does not have a clear economical justification, unless such a system is provided within the switching system without any additional cost to the operator.

We propose two phases for the routing deployment. During the first phase a fast implementation of FAR/AAR is scheduled. In this way, with simple means, a considerable increase in the network traffic handling efficiency and robustness will be achieved in short time. During the second phase a more sophisticated routing, giving potentially full (but controlled) access to available network resources, will be implemented.

Phase 1
- Step1 After installing the upper level, a fixed alternative routing system is implemented with overflow paths via one transit node.
- Step2 Installation of an off-line routing patterns reconfiguration facility for asynchronously adjusting nodes' routing tables when such a necessity arises.

Comments to Step 1.
The use of non-hierarchical paths. Alternative routing in UL is non-hierarchical. Therefore the implementation of traffic overflow would require changes in the call handling process at switching nodes: calls should be routed not only by destination but also by origin. Such a differentiation is available within CCITT SS7. Still, we prefer not to assume the immediate implementation of SS7, rather to assume the in-band signalling. Then it would suffice to have a possibility to apply different routing rules to first routed calls (customer calls according to [10]) than to transit calls. In fact, with overflow paths restricted to one transit node, it is sufficient to forbid the transit calls to overflow.

The number of overflows paths. ARR could be difficult to implement even with SS7, and thus AAR, which is consistent with today's call set-up techniques, should be assumed. With FAR/AAR the question is if multiple overflow choices should be allowed, rather than 1 or 2 overflows. The answer depends on two factors: the design criteria and the DTR facility at switches.

Let us consider the network dimensioned to carry nominal traffic at low blocking (eg 0.3%). With AAR the use of more than 2 overflow paths in the network designed for nominal traffic does not improve performance (cf Fig.4 and also [11]). In conditions different from nominal, and with DTR, it is certainly advantageous to use multiple overflows in all cases, especially in traffic shifts, focused overloads and equipment failures.

The case of the absence of the DTR facility is of more interest since we should not insist on this possibility at switches. In global overload condition, focused overload and transmission link failure, FAR/AAR with a limited number of overflow choices performs well also without DTR. Allowing more alternative choices could degrade the network performance.

DTR combined with multiple overflow choices helps to cope automatically with traffic shifts, light focused overloads and minor failures.

In the case of the network dimensioned according to the robust design criteria, multiple choices are advantageous also without DTR (because now even in the situation of a major failure the network is not congested). On the other hand, FAR/AAR with only one or two overflow choices, and with the possibility of changing the currently used overflow paths in abnormal situations will perform sufficiently well (cf Fig.5).

In a network designed according to the robust design criteria, its robustness will gradually decrease, unless new capacity is not being continuously installed. Thus we should not assume this kind of design criteria. It is then recommended to use a limited set of overflow paths and have a possibility of on-demand adjusting routing patterns by the routing reconfiguration facility installed in step 2.

<u>Overflow paths with one transit.</u> UL nodes will be almost fully connected by trunk groups and there should always be enough paths composed of two trunk-groups (via one tandem) for overflow calls. Thus longer, more costly paths can be eliminated. With the one-transit paths, the looping problem can also be easily eliminated.

Comments to Step 2.

An off-line routing reconfiguration facility available within Network Management (NM) will anyhow be required to recompute routing patterns in longer cycles (weeks, months), so the facility could as well be used to reroute traffic in the unexpected situations of equipment failures, overloads and traffic shifts. The reconfiguration facility, according to our findings, does not have to be based on sophisticated network models and methods. It should rather be fast in order to be able to reoptimize routing in a short time on a workstation. Such a facility is currently under study at Alcatel SESA.

Proceeding beyond Phase 1.

The argumentation behind Step 1 also favours SRR. Furthermore, with SRR the reconfiguration facility is of less importance. However, the necessary changes in the switches software to implement SRR may be significant.

Another possible way to proceed would be to implement ARR by means of crankback signalling, and to introduce FAR/ARR with full access to available network resources. However, to do this the switches software would have to be enhanced at least in two major aspects: to implement handling of crankback signalling, and to implement the DTR facility. The latter is unavoidable since with effective access to overflow paths we have to protect fresh, first routed traffic. Otherwise, in overload the overflow calls could dominate the network causing a serious throughput degradation.

Another drawback of crankback is its intensive signalling when the network is already overloaded, and so is the signalling network.

The question is: is this worthwhile? It seems that we would invest much and gain little. The issue is that FAR/ARR would not significantly increase the efficiency over FAR/AAR (with reconfiguration) in the day-to-day operation. The sophistication of routing is worthwhile if it could help to implement and automate the congestion control functions of NM, aiming on throttling at source the call streams with low chance of completion. This position was presented in [12].

The integration of congestion control (CC) with DCR system was described in [13]. DCR is a centralised system and as such is well suited for taking over the CC functions.

Phase 2
Step1 Install a central routing processor (CRP) at the management center, and provide it with means to communicate with nodes's central data collectors, and with means to change the nodes routing tables through sending routing recommendations.
Step2 Implement the CC functions in switches (as call gapping) and control them automatically by the CRP.

Comments to Step 1.
A dynamic routing system does not have to be centralised. Distributed implementation is feasible in modern networks. However, the centralised solution has important advantages. By introducing the centralised routing system after Phase 1, an effective access to available overflow paths can be achieved (with a short updating cycle length) without major interference in the switches operating software. With updating cycle of the order of tens of seconds and with only one overflow recommendation, the centralised routing system would ensure effective access to available overflow paths, as compared with FAR/ARR. According to our studies, the updating cycle of about 30 sec. with 1-2 overflows (AAR) would be effective in the day-to-day routing operation.

Another advantage of the centralised solution is that DTR or ATR is not implemented in the switches software, but in the CRP. This gives a possibility of applying more sophisticated, and thus more effective protective controls, more difficult to implement in a decentralised solution. For instance, DTR protection parameters could be differentiated, depending on the particular stream of overflow calls. This could enhance network fairness.

5. FINAL REMARKS

It turns out that in PSTN, already fixed alternative routing with a reconfiguration facility can ensure many of the benefits of dynamic routing, in terms of handling the traffic admitted to the network. This includes abnormal situations. The greatest benefits are achieved by replacing the plain direct routing with an alternative routing scheme, using one or two fixed overflow choices for each traffic stream.

As the implementation of such a simple routing does not require major adjustments in the standard software of a switching system, we suggest this routing solution for the first years of a newly formed digital upper transport layer of PSTN.

An off-line routing reconfiguration facility placed at the NM center(s), communicating with switches through the MML (Man-Machine Language) commands, will enhance routing in failure and overload situations.

The simple routing with the reconfiguration facility, operating in a network with transmission and switching secured by duplicating transit exchanges, will ensure efficient traffic handling, even against the robust design criteria.

Further sophistication of the routing strategy should go in parallel with the automation of the on-line Network Management functions of congestion control and reconfiguration.

Acknowledgement

The authors wish to thank Mr. Oscar Gonzalez Soto, Alcatel Standard Electrica, for helpful discussions.

REFERENCES

1 J.Linares, Evolution strategy of the TELEFONICA Network, Proc. 5th International Network Planning Symposium, Mallorca, (1989).
2 M.de Miguel, A.Bartolome, F.Martin, Escorial, the tool for planning advanced national networks with security, paper accepted for ITC13, Copenhagen, (1991).
3 P.B.Key and G.A.Cope, Distributed dynamic routing schemes, IEEE Communications Magazine, **28**(10), (1990).
4 M.Pioro, A.Tomaszewski, J.Lubacz and M.Jarocinski, Design of two-level PSTN, paper 2.3, ITC Specialists' Seminar, Cracow, (1991).
5 M.Pioro and A.Tomaszewski, Modular engineering of telephone networks with dynamic routing, paper accepted for ITC13, Copenhagen, (1991).
6 CCITT Recommendation E.170.
7 R.J.Gibbens, F.P.Kelly and P.B.Key, Dynamic Alternative Routing - modelling and behavior, Proc.ITC-12, Torino, (1988)
8 W.H.Cameron and S.Hurtubise, Dynamically Controlled Routing, Telesis, Bell Northern Research ,June, (1986).
9 E.Granel et al., A comparative study of several dynamic routing algorithms with adaptive preselection and selection phases, submitted to ITC13, Copenhagen, (1991).
10 CCITT Recommendation E.412.
11 P.Chemouil, J.Filipiak and P.Gauthier, Analysis and control of traffic routing in circuit-switched networks, Computer Networks and ISDN Systems 11(1986).
12 F.Caron, Automating Network Management: A Service Oriented Approach, Proc.ITC Specialists' Seminar, 13.4, Adelaide, (1989).
13 J.Regnier and W.H.Cameron, State-dependent dynamic traffic management for telephone networks, IEEE Communications Magazine, **28(10)**, (1990).

Enhanced telecommunications for developing regions

Beatriz Craignou

FRANCE TELECOM, Centre National d'Etudes des Télécommunications
38, rue du Général Leclerc   92131 Issy-les-Moulineaux   FRANCE

Abstract

This paper considers the introduction of new telecommunication services in developing country local networks. After evoking development policy, digitization aspects and services, some ideas on demand analysis are given. Infrastructure requirements and new trends in network configuration are underlined. Particular aspects of new services for rural regions are addressed.

1. GENERAL NETWORK DEVELOPMENT POLICY

At present, most countries are driving telecommunication networks towards an Integrated Digital Network (IDN), as the first step towards the Integrated Service Digital Network (ISDN), universally accessible and capable of providing a very wide range of telecommunication services using a limited number of powerful capabilities. In the developing world, the speed to reach these objectives is lowered by financial constraints, provider dependance, deplorable estate of the existing network and because the priority aim is to provide, by alll means, at least traditional telephone service, which is desperately lacking in many regions of the planet.

In the last two decades, industrialized countries began the process of full digitization. Nevertheless, for most of these countries, the goal of a nationwide IDN will only be completed by the end of the century. The technical features of digital systems, the overall cost reductions of service provisioning and the demand for digital communication services are unquestionable reasons to support the trends towards digital networks.

Meanwhile, telematic applications entered into the business world and in some aspects are considered positive for quality of life. The convergence of computers and telecommunications was understood by policy makers : information technology undoubtedly plays an important role in the economy of a country.

For many developing countries a gradual network digitization has just begun. Digital technology deployment, for better telephony and new service offering is a fundamental part of their modernization plans. Not having an adequate support for telematic applications, not all the benefits of informatics can be reaped.

Telecommunications modernization includes effectual synchronization and signalling implementation. Network internal associated signalling channel is inadequate for modern requirements because it is slow, supports only a small signalling repertoire and is poorly adapted to control and digital exchanges. Most countries have therefore scheduled a progressive transition to CCITT Common Channel Signalling System (CCSS) N°7, which has not the above disadvantages and is capable of evolving to meet requirements such as those of Intelligent Networks and ISDN. If the latter is not an immediate objective and the CCSS N°7 is not chosen for network development, at least some signalling interfaces must be implemented at International Switching Centres in order to coordinate with ISDNs existing in other countries.

If the primary objective of an Administration is to provide the essential infrastructure for modern telecommunication facilities in the country, the national telecommunications policy, which is part of general development plans concernig feeding, housing, health, roads, education, should not discriminate against any region, sector of activities or subscriber location as far as equipment implementation is concerned. Telecommunication services must be offered with equity to all users, in farms, in villages, to city dwellers andbusiness sectors.

2. PREPARING NEW SERVICES INTRODUCTION

To prepare for the introduction of new services, besides their technical description, specific applications, basic network requirements, accesses, terminals and environment, the demand and the opportunity to offer them must be analysed in detail. Traffic models ought to be built. Other considerations deal with the choice of the most appropriate technical systems, the impact on network dimensioning, operation, maintenance and personal training, economic feasibility, tariffs, legal aspects, scope of future evolution.

Most of these tasks rely on data gathering, greatly facilitated if an information system exists. In any case, the administration policy concerning traditional and new services offerings is essential to derive the implementation strategy.

2.1. Services and applications

Telephony characteristics are known. The basic service was far enriched with SPC and digital technology, providing for instance, call transfer, call forwarding, call waiting, three party call, and so on. ISDN terminals allow much more sophisticated supplementary services.

The term *new i*is used if the service has been commercially introduced in the last decade, results from the latest technologies being deployed world-wide, or employs existing or emerging technical standards that facilitate international traffic and simplify operations. Thus, ISDN telephony is a new service.Some new services offer in fact voice and non-voice transfer. For example, audioconferencing and videoconferencing combine a number of voice and non-voice features.

Non-voice telecommunication services are defined as services which offer point-to-point or point-to-multipoint comunications other than voice, involving real time and store and forward communications. These non-voice services present new implementation aspects. Some services may involve an external data base provider and quality of service is a fundamental element of success, which depends on the service provider as well as on the network infrastructure. Administrations are led to active commercial roles. New services are introduced even though the basic service telephone demand is far from being satisfied in all countries.

As financial resources are limited and expected revenues will only represent for some years a low percentage of telephone revenues, new services could appear as a non priority for developing countries, which is an unfair conclusion. Telecommunications are a vehicle for economic growth. Their absence hinder the whole development.

The main services to be considered are : teletex, videotex, facsimile, electronic message handling, audiographic conference (teleconference), videoconference, still picture transmission, paging and mono-directional services, bi-directional dispatching services with mobile stations and bi-directional data exchange communications using VSAT.

Among several applications of these services, the best known deal with office communications, teleconferencing, computer to computer or terminal to computer communications, residential applications, encripted voice communications . Several services can be used for text and file transfer, data base access, renting and booking, tele-shopping, home banking, tele-health, tele-education, public services, tourism information, stock exchange, news, weather forecasts, messaging, and so on.

2.2. Demand forecasting or market segmentation

To study the relationship between socio-economic activities and needs for non-voice services, a meaningful measure of the demand should be taken. This is a difficult step when the market is limited and customers are not aware of new services potential.

How to overcome these difficulties ? Some times there is little historical data concerning other services that are going to be partially substituted by the new service. Answering a questionnaire, only a few customers express their intentions about a service from which they imagine with difficult its usefulness. Experiences of countries where new services have a significant penetration, or new technologies are being implemented, may guide the search for a pragmatic service demand evaluation, provided that the economic structures could be comparable.

It is neither feasible nor necessary to analyse user behavior or the impact of the different parameters related to a service over the whole population. Such analyses can be performed on a sample of users where the various population segments are represented.

As an example, traditional household segments include farmers, owners, liberal professions, managers, employees, labourers and service personnel and the non working population. Activity sectors for establishements include industry, banks and insurance, business, private services, public services and administrations. The number of employees is also an important parameter for establishment segmentation. [1] treats of the constitution of representative samples, advantages and limitations.The proportion in which the segments are represented may vary according to the studied service.

Demand forecasting of new services must take into consideration interaction with traditional services. Figure 1 illustrates the case of teletex and facsimile in relation to the traditional mail, telex and telephone services. Teletex is greatly used for document exchange between dedicated terminals or micro-computers. Telex growth has diminished but this service, enriched with modern terminals, remains a useful telecommunication means.

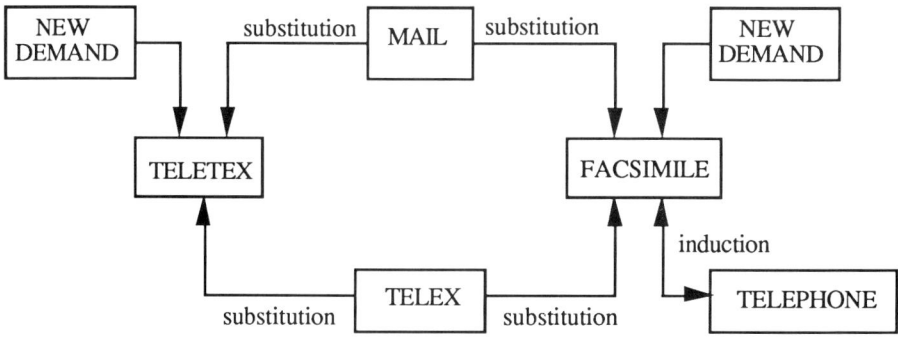

Figure 1. Relations between communication modes for business traffic

Indeed, the introduction of a service can not be strictly based upon fuzzy forecasts. Each country has to determine its own priorities and define its own strategy. The specific existing economy, the technical and operational environment and experiences in other countries may influence decisions and provide new service implementation guidelines. The case of facsimile is to be mentioned : its usefulness and success are not contested ; doubts about offering this service come rather from lack of lines and qualified exchanges than from lack of expressed demand.

Some services appear to have priority when : demand comes from a business segment for which these new services are essential, foreseen custumers are ready to invest in these new services, the services can, for a limited time, palliate the defficiency of basic telephony.

For the time being, ISDN service availability and penetration is relatively slow. Intended, first for business customers, their use by residential subscribers might be stimulated by broadband ISDN services such as video distribution.

During the present period, CCITT GAS 12 is studing the introduction of new non-voice services in developing countries. Guidelines on forecasting methology and other important aspects are give in [2] and [3].

3. NETWORK INFRASTRUCTURE

3.1. High performance networks for diversified services

Enhanced telecommunications require an effectual network infrastructure. Switching, transmission, distribution and access networks should be adapted to allow new service provisioning.

A modern switching system allows the offering of a great number of *services,* ranging from basic telephony to sophisticated services requiring digital capabilities, high speed, particular bandwidths, and so on.

The ISDN concept appears as a solution to provide all services in a more integrated manner. Each one of the current telecommunication networks provides a specific set of dedicated services (voice, circuit switched data, packett switched data, video).These networks are supported by analogue and digital equipment for subscriber access, switching and interexchange transmission. Without ISDN environment, a customer of a number of telecommunication services has separate access lines, user network interface requirements and terminals almost for each service.

The key to a successful ISDN implementation resides in the coherence of services, protocols and equipment. Usually, field trials go on for some time before decisions are made.

3.2. Network architecture evolution

The technological evolution and cost trends allow the provisioning of switching systems with a decentralized architecture. This leads to an optimal distribution of functions between the different processors, a higher processing capacity and great hardware and software modularity, thus making the system adaptable to different circumstances and extensible in favorable economic conditions.

The implementation of decentralized switching systems should satisfy user needs as well as planners and management people.

The user subjective notion of *quality of service* often refers to : traffic fluidity, short set up time, no congestion, satisfactory voice and data transmission quality, independent of the distance, no noise, no distortion, permanent availability.

Accesses to the network should be able to evolve. So, an analogue telephone line should be easily replaceable by an ISDN basic rate access without involving important network modifications.

For the planner, network design or optimization will be easier if he incorporates *flexible* switching and connection equipment. Figure 2 shows an example of local network organization.

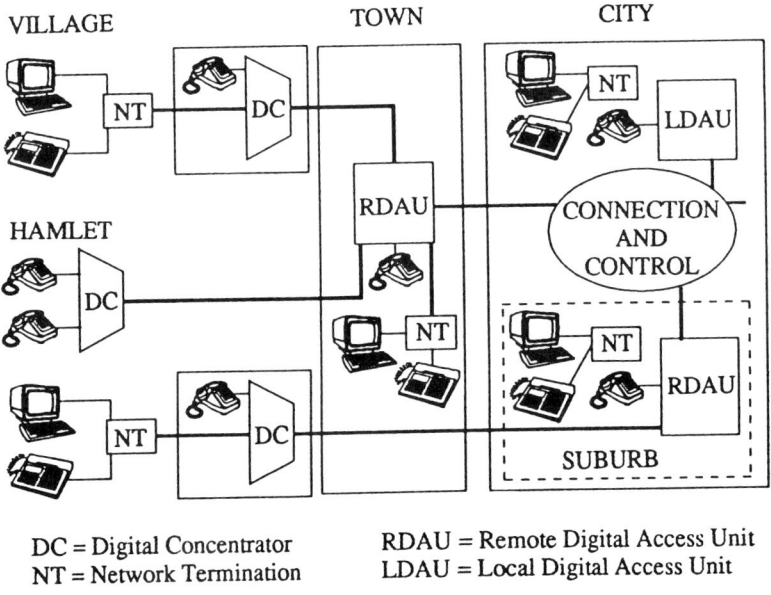

Figure 2. A high performance local network organization

Subscriber connection would be optimized if the units are located close to the subscribers. Then, the possibility of different decentralization levels is quite interesting, particularly in low populated areas. Connection units could be local to the host exchange or remote from it, when they are implemented several tenths of kilometres away, near the subscribers.

At present, switching technology provides the planner with exchanges, connection units and concentrators suitable to build a performant and cost-effective network. Besides, this equipment complemented with power groups and transmission equipment may be temporarily installed when a building is not yet available or in case of unpredictable events.

Connection units must be able to carry different amounts of traffic. The number of subscribers generating very high or very low traffic should be studied. Implementation in various topological conditions should also be studied.

The system should offer the possibility of undergoing long term functional and technological changes, evolving towards broadband ISDN and its associated Asynchronous Transfer Mode.

As far as subscriber connection is concerned, the possibility of different decentralization levels in a modern architecture is particularly interesting in areas where the expected number of subscribers is low. This could be the result of low population density, or of financial constraints which lead to a low number of lines, in spite of a very large population.

3.3. New trends

Several new technologies are emerging which will provide the network planner with a means to meet network modernization requirements, reach very high transmission quality, network intelligence and broadband capacity to satisfy expected customer communication needs and to improve network management and maintenance.

Conventional local networks are mostly based on a mixture of microwave and copper technologies, where generally the latter are dominant. Whereas this type of network can still handle the current telephone service, it represents an enormous financial expenditure in investment, operation and maintenance costs. Moreover, such networks will not fit the overall network of tomorrow, which will have to provide new services of a different nature, format and bit rate.

The use of optical fibres (OF) in the local network overcomes these barriers. However, systematic replacement of copper cables and microwave systems by OF is not cost-effective. Therefore, the opportunity of OF introduction, considering its advantages and the requirements of this technology leading to the implementation of Synchronous Digital Hierarchy, ought to be methodically examined.

Optical technology, applicable to switching, transmission and distribution networks opens up new ways for efficient network architecture and enhancement. This technology allows the evolution of systems, reliability, very high transmission quality, network flexibility, easy technical management, decreasing investment, operation and maintenance costs.

OF features overturn engineering and operational concepts on which existing networks are founded, at least in three aspects : transmission capacity, regeneration pace and immunity to electro-magnetic disturbances.

Beyond the tremendous improvements of its specific qualities, the main trends concern the definitive recognition of the single mode fibre (SMF) to be spread in all parts of the network and its advantages over the multimode fibre. With theoretical unlimited bandwith, due to the absence of intrinsic intermodal dispersion, the SMF is a major alternative to traditional systems because of its competitive cost, high reliability and potential for significant growth both in volume and type of services.

As far as equipment is concerned, the recently defined Synchronous Digital Hierarchy (SDH) standards, based on the SONET concept offer, for instance, digital transport structures with suitable adapted payloads over transmission networks. The main features of SDH are : flexibility of the frame structure, enhanced Operation, Administration, Maintenance and Provisioning (OAM & P) capabilities and standardized optical interfaces. These features can be used together with network management to provide a high capacity and flexible transmission network capable of supporting broadband services.

SDH provides for high bit rate channels to be transmitted on optical bearers. SDH based facilities offer significant advantages over earlier Plesiochronous Digital Hierarchy systems. One of its benefits is the ability to efficiently add and drop traffic without the necessity of demultiplexing and multiplexing again the entire payload. This capability allows the cost effective handling of traffic, as well as the establishment of bi-directional shared protection ring structures which provide full service survivability in the event of a single cable failure.

Digital transmission and switching networks will be deeply modified to be adapted to SDH. These networks, making use of optical fibres are more reliable than the old analogue networks and more recent digital networks.

For some time, specific equipment based on SDH will coexist with the present plesiochronous based equipment and later replace it. In the future, these systems will be able to support new broadband services that require end-to-end synchronization and quality assessment.

4. ENHANCED SERVICES FOR RURAL CUSTOMERS

4.1. Around rurality concepts

Rural areas are typically made up of a number of fragmented low income or low populatied communities spread over a very wide land. Because of their activity sectors, few people are concerned with new services. Investments to provide telecommunications to low populated areas are seldom recovered from the income they generate.

Very often, rural areas do not enjoy high priorities and lag far from big city development plans. These regions produce basic goods like food for the whole country. If they are forsaken by their residents, the whole country might suffer the consequences. In spite of that, not many things are done to improve quality of life in rural regions, perhaps because natural resources like clean air and fresh vegetables are supposed to satistfy rural dwellers.

Rural networks are part of the general network. The benefit of new technologies, initially introduced in the upper network structure, intended for big cities and industrialized areas, must be shared with rural regions to keep the same standards and quality of service in the whole network. The telecommunication policy, as part of a national policy, concerns all sectors. The socio-economic benefit brought by new services should be one of the most important factors to influence strategic decisions on rural development.

When the network architecture is such that any subscriber can be reached and provided with a line of sufficient high transmission quality, the concept of rural region is not longer that of a completely *isolated* and possible *forgotten* remote area.

4.2. Telecommunications and economic development

To lessen the decline of traditional industries and to bolster up new economic issues, like information based activities, a potential of new telecommunication services for rural business enterprises has to be offered. New rural economic development requires a reliable infrastructure for enhanced telecommunications. The increasing use of computers and telecommunications improves productivity and stimulates private investments. Then, it is important to identify types of industries and quality of life concerns where information links, between rural and urban areas and among rural residents, could contribute to their development.

Emerging telecommunications technologies contribute to the economic growth of rural regions which depends, among other factors, on the use of telecommunication services for rural business, residents and administrative entities.

Under the present circumstances ISDN success is not expected in rural areas. However, these regions can indirectly benefit from all the technical advances in telecommunication domains.

Modern exchanges, enriched signalling, qualified trunk groups, good state of distribution cables, high quality of service, maintenance, staff training, and so on, result in an effectual network shared by *all* subscribers.

Forecasting techniques can not be rigorously applied to rural regions, nor should the results of some surveys be considered alone in deciding to offer new services outside big cities.

The concept of socio-economic service to determine the benefit of traditional and new telecommunication services and the best distribution of total investments in rural areas is considered in [4].

A study [5] on the Economic Implications of Stimulating Applications of Information Technology and Telecommunications (IT&T) has been conducted on behalf of the Commission of the European Communities (DG XIII F), to support the planning of a Community action known as ORA, whose broad objectives are to stimulate the provision of technologies, services and infrastructures suited to business activities and public services in rural areas. A methodology for measuring the economic and employment impact of new IT&T investment has been developed. The "ability to benefit from IT&T" has been compared using a rurality index in six case studies performed in different countries. The relationship between rurality and employment as a per percentage of GDP spent on IT&T was also estimated. In all cases benefits were positive but the degree of benefit varies from one place to another.

If a region lacks general infrastructure such as roads or electricity, an improvement in the telecommunication systems may do little to improve local employment and economic conditions. But, when a minimum threshold of general development has been achieved, telecommunications investments may bring a great improvement to the local community.

As soon as telecommunications improve, local business enjoyt direct effects (lower expansion costs and time losses, better suppy and utilization of goods and services, and so on) and the benefit of a stimulated economy. However, telecommunication operators experience a negative cashflow for ten years or so, before utilization increases sufficiently to compensate for the initial capital investment. Therefore, it seems reasonable that local enterprises work in partnership with the operator to alleviate some of the initial investments.

To develop new information technologies and services, a "Programme Télématique" was launched in France two decades ago. The minitel, a simple computer terminal, now found in more than 5.7 million households, given to diffuse the computerized telephone directory, is used for a wide range of interactive consumer and business services. In households, it is used identically by urban and rural residents. It is considered helpful to virtually diminish distance barriers, save time and improve quality of life.

Among telematic applications, those dealing with agriculture, health, manufacturing, handicraft, transportation, training, marketing, stock exchange, tourism, for instance, are attractive for rural users.

With enhanced telecommunications, many enterprises, particularly those whose activities are based on information transfer, could be encouraged to move to the countryside.

4.3. Strategic locations for attractive services

Some potential users have difficulties to communicate, due to physical or psychological handicaps. Some people just ignore the new communication possibilities or do not feel any need of them, even assuming that they could manage to use the terminals and could afford to pay the price, which is not always evident. Then, in spite of theoretical possibilities, telecommunication services are not accessible to everybody everywhere.

The existance of a modern network infrastructure does not solve all communication problems in rural areas. The community, local authorities and concerned entities play a crucial role, as partners of the Administration, for services development. If tariffs are affordable and innovation is encouraged, new uses of telecommunications will be stimulated. Service providers have to offer information that really interests rural users.

As the low use of telecommunication services in low populated areas discourages providers from extending the network to these regions, it is understandable that some particular points have to be designed to concentrate the traffic, if services are going to be offered and intended to be used by the community.

As for the basic telephone service, Public Call Offices or other public places are best suited for telecommunication provisioning. Then, at least in post offices or some places accessible to everybody, some services can be offered. This has the avantage that some qualified people could help the user to manipulate a terminal and make use of the service. In that case, someone is available to receive, store and distribute messages, if the persons to whom they are addressed are not present, and to take care of the terminals. A disadvantage of this solution is that the service is available only during the office opening hours.

Facsimile penetration increases quickly. This service must be offered to villages, so that people in the community could use it without buying a fax terminal. Public services, small industries, merchants, shops, could benefit from it.

To provide new services for villages, the implementation of community teleservice centres could be one of the adopted means. This idea started to be materialized in Scandinavian countries, extended to other European countries, and is now being spreaded to developing countries in Asia and Latin America. In [6] interesting experiences and realizations of these centres are described.

A community teleservice centre may be defined as a centre with dataprocessing and telecommunication facilities, placed in a local community in a geographically or socially remote region, so that these facilities can be used by all people in the community.

5. CONCLUSION

A strong interdependance between development policy and service implementation does exist. Every country should consider the pros and cons of particular service offerings, hopefully having in mind that new telecommunication technologies are a strategic resource to subside distance, lack of education, poverty, physical handicaps.

It is not evident to apply classical forecasting techniques for new services in a lack of reliable data. Consequently, methods have to be adapted to the prepare some forecasts, suitable for network resources dimensioning and to guide decisions concerning enhanced **telecommunication services.**

A modern network configuration is basic for advanced service offer. Besides diminishing implementation, running and maintenance costs, such a network provides a high quality of service and varied means to satisfy communication needs, which result in higher income for telecommunication companies.

Even where ISDN is more a philosophical concept than a target network, enhanced telecommunications services can open new horizons for development activities in cities and rural regions. Efficient telecommunications are vital for social welfare and economic growth.

ACKNOWLEDGEMENTS

The author benefitted from fruitful discussions on these subjects with many colleagues around the world, particularly in CCITT GAS 7 and GAS 12 Working Groups.

REFERENCES

[1] Chabrol J.L. and Craignou B.- An information system applied to telecommunication service demand and traffic forecasting. ITC-13, Copenhague, 1991.

[2] CCITT GAS 10 Handbook. Planning Data and Forecasting methods. ITU, Geneva, 1987.

[3] CCITT GAS 11 Handbook. Strategy for the Introduction of a Public Data Network in Developing Countries. ITU, Geneva, 1987.

[4] Carrier C., Craignou B, Nugroho A. and Sugondo K.- A socio-economic rural communication service concept. ITC-13, Copenhague, 1991.

[5] Opportunities for Applications of Information and Communication Technologies in Rural Areas. Economic development and employment implications. Vol II. Commission of the European Communities. DGXIIIF, Brussels, 1990.

[6] Proceedings of the Community Tele-Service Centre Symposium Improving Teleservices in Rural Areas. Warsaw, 1990.

Architecture and Control Aspects of Data Service Evolution to Broadband

Marek Wernik, Rungroj Kositpaiboon, Peter Carbone

Bell-Northern Research
P.O. Box 3511 Station C
Ottawa, Ontario, Canada K1Y 4H7

Abstract

This paper addresses Virtual Data Networking (VDN) service and technology evolution to broadband. After identifying VDN attributes, a Frame Relay standard which satisfies wideband VDN requirements is described and high-speed trunking requirements for Frame Relay wide-area networking are defined. The Broadband ISDN protocol enabling Frame Relay service evolution to broadband VDN is then proposed with emphasis on the use of the Asynchronous Transfer Mode (ATM) network for Frame Relay trunking and on interworking aspects. The traffic management and control requirements enabling evolution from Frame Relay service to broadband VDN are further discussed.

1. INTRODUCTION

Fiber transmission and increasing demand for high-performance data communications are the main factors which will drive evolution of telecommunications networking and services in the industrially developed countries in the nineties.

Several technologies, standards, and services are currently being researched, experimented and deployed to replace or complement conventional low-speed data networks which can no longer meet competitive corporate market requirements.

Developing countries which are quickly modernizing their telecommunications infrastructure will benefit by selecting evolvable technology and services instead of installing soon-to-be obsolete systems or investing in interim solutions.

This paper describes high-speed data communications requirements driving the introduction of VDN and provides guidance for selecting preferred services and technologies.

Data Applications Evolution

The increasing demand for high-performance data communications in recent years can be attributed to several factors. Data applications have been evolving from master-slave to peer-peer and client-server paradigms appropriate for distributed computing environments. This created the need for Local Area Networks (LANs) which are currently penetrating every business sector and the installed base is doubling every two years in North America [1].

In many large corporate networks, a single 10 Mbit/s Ethernet LAN does not have sufficient capacity to carry the total traffic. Parallel and hierarchical LAN architectures are therefore employed with the tendency towards high-speed LANs such as Fiber Distributed Data Interface (FDDI) operating as 100 Mbit/s backbone networks.

At the same time, the characteristics of information carried on LANs is changing. Today it is predominantly alphanumeric with traffic bimodally distributed with peaks around 50 bytes and 1500 bytes [2]. Larger bursts resulting from database management and image transfer which can reach the size of 100 K bytes or larger are becoming more and more frequent. This forces LAN managers to install separate LANs for this type of traffic in order to maintain acceptable performance. It is expected that FDDI LANs will provide capability to support these situations. However, other applications, where transfer of larger data objects at high speed is required, such as high-resolution (interactive) image transfer, CAD/CAM (Computer Aided Design/Computer Aided Manufacturing) graphics, and scientific visualization will be creating the need for even higher throughput to meet performance requirements. Similarly, supercomputer access requires speeds in the several hundred Mbit/s range. Standards such as High Performance Parallel Interface (HPPI) are being developed today to enable short-distance interconnect in these cases. In addition to higher data traffic volumes and more stringent performance requirements, emerging multimedia applications will be driving the introduction of integrated voice, data, image, and video workstations interconnected locally using multimedia LANs or PBXs with capacities reaching Gbit/s. Experimental multimedia campus networks are already operational or planned for the near future [3].

Data applications are also becoming more distributed, involving multiple sites, and exploiting shared access to common computing and storage resources. This can be illustrated by using the example of a large corporation increasing its CAD/CAM operations. Today CAD/CAM networking is limited primarily to local traffic involving file transfer between workstations and hosts located either on the same LAN or on the high-speed backbone LAN at the same location. Traffic between locations is very sporadic and involves primarily off-line transfer of archive files. Interactive communications between sites (for example database access) is minimized and the preferable solution is rather to keep databases in each location synchronized and updated occasionally. The need for more interactive communication between locations will arise due to the penetration of distributed processing such as special purpose CAD engines, and when distributed databases become more cost effective than today's practice of database duplication. That will create new requirements for better reliability and instantaneous recovery from communications facility failure, which will involve very high speed transfer of database contents.

Corporate Data Networking Requirements

All these evolutionary factors: increased traffic volumes, ubiquity of LANs, new traffic patterns with more stringent performance requirements, and more distributed applications are imposing new requirements for the corporate data networking in Metropolitan and Wide Areas.

These requirements are primarily :

- Increased speed or bandwidth and improved performance;
- Flexible connectivity (e.g. point-to-multipoint connectivity); and
- Usage-dependent cost.

Today, users have three alternative solutions for data networking: packet switched networks such as X.25 operating at 9.6 kbits/s to 56 kbit/s speeds, leased lines at px64 kbit/s, nxT1 and T3[*] rates, and switched circuits predominantly at 64 kbit/s and recently also at px64 kbit/s up to 2 Mbit/s.

As shown in Figure 1, the capabilities of X.25 networks and leased circuits are orthogonal. Private lines offer bandwidth but do it very inflexibly and for wideband rates at cost too high for many small users because of the fixed and distance sensitive pricing. Low speed public packet network in turn has been designed for low performance transmission facilities and therefore error control protocol overhead is responsible for its inadequate performance for high-speed data applications. Emergence of public wideband circuit switched network [4] provides significant advantages over private circuits in terms of provisioning, survivability and also allows for usage time-dependent billing options to the users. However, it still has some limitations in terms of connectivity since the number of simultaneous active connections on an access interface is limited by its multiplexing hierarchy (e.g. maximum 24 connections of 64kbit/s on a 1.5Mbit/s access).

The fourth option which is gaining increasing acceptance among CPE (customer premises equipment) vendors and service providers, is Virtual Data Networking (VDN) which is based on the introduction of high-speed public packet networks. The VDN service should provide simple data communications capabilities to the user equivalent in terms of performance to Private Line service, and combining connectivity and cost advantages of other networking alternatives shown in Figure 1. The VDN service is also a more flexible alternative to provide usage dependent cost based on the actual traffic volume transported by the network and the offered performance.

Furthermore, in order to be successful in competing with other alternatives, a VDN service must satisfy the following requirements :

• End-to-end standard connectivity enabling wide area networking;
• Evolvability from wideband to broadband capability; and
• Minimum impact on existing CPE networking and protocols.

Attributes	X.25 network	Leased circuits	Switched circuits	Virtual data network
High speed or bandwidth	No	Yes	Yes	Yes
Flexible connectivity	Yes	No	Limited	Yes
Usage-dependent cost	Yes	No	Time-dependent	Volume & performance-dependent

Figure 1 Data networking options.

---

[*] T1 (1.5Mbit/s) and T3 (45Mbit/s) are elements of North American digital multiplexing hierarchy.

2. NARROWBAND/WIDEBAND VDN

Service considerations

Frame Relay service (FRS) is a defined and standardized service that is currently available and satisfies VDN requirements,

The high-speed operation of Frame Relay service is achieved by the elimination of the several error control and flow control functions from the network (Figure 2). The remaining LAPD core functions provide end-to-end transport of variable length frames with no acknowledgement (non-assured delivery) nor flow control capabilities. It however ensures that frames, if not lost, are delivered with no errors and in the same sequence as they were transmitted.

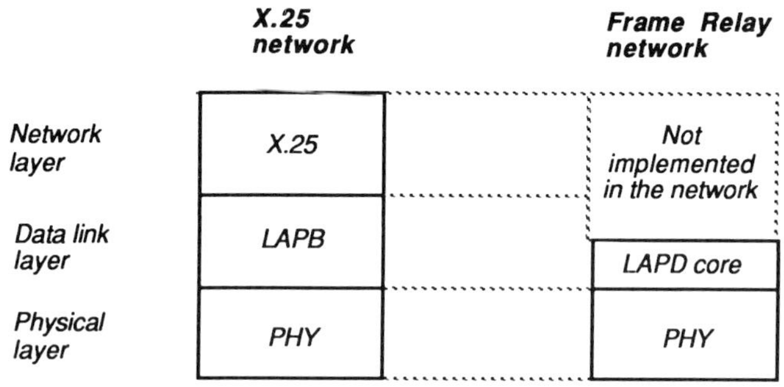

	X.25 network	Frame Relay network
Network layer	X.25	Not implemented in the network
Data link layer	LAPB	LAPD core
Physical layer	PHY	PHY

Figure 2 Protocol comparison between X.25 and Frame Relay networks.

FRS is a connection-oriented service enabling point-to-multipoint permanent or switched virtual channel connectivity between users. FRS can be introduced with minimum impact on existing CPE by implementing LAPD core functions in software on top of the existing physical interface of a router. Since FRS is a connection-oriented service, it is also directly compatible with network architectures born upon private line attributes such as SNA. Furthermore, connection-oriented operation simplifies security control. The physical access interface is currently defined by CCITT and ANSI for rates up to 2 Mbit/s.

When FRS is used with a permanent connection option, it provides equivalent virtual leased line service, where service rate can be anything within the limit of physical interface speed. Several other services can be implemented on top of such a FR transport once it is installed in the network as shown on Figure 3. These services can be implemented within the CPE or provided by a carrier.

One of the services which can possibly be supported by FR transport is the Switched Multi-megabit Data Service (SMDS) which is a connectionless service currently being defined by Bellcore [5]. The additional functions which have to be implemented on top of LAPD core is the layer 3 SMDS Interface Protocol (SIP) and some address management features such as address screening for security and multicasting/broadcasting. Implementation of these functions at high-speed is currently a design challenge.

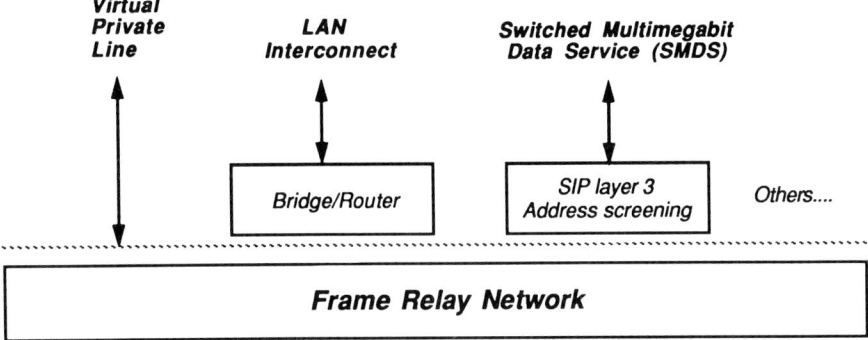

Figure 3   Services offered by a Frame Relay network.

Architecture considerations

In order to provide ubiquitous wide area VDN service, the development of standards for access (user-network interfaces) and trunking (interswitch interfaces) is mandatory. With peak service rate being 2 Mbit/s, the trunk capacity must be at least 45 Mbit/s to exploit statistical advantage of packet mode traffic. Although frame transport can in principle be extended to operate at these rates, cell-based ATM technology is favoured. This is due to its extendability to even higher rates (150 and 600 Mbit/s) and its compatibility with B-ISDN. Using the ATM Virtual Path (VP) concept, the virtual trunks can be established permanently between FRS nodes as shown in Figure 4, thus creating a second level of VDN connectivity. ATM trunks should be able to carry LAPD core frames transparently in the same manner as they would be carried through a physical point-to-point trunk facility. The trunk interface definition enabling such an operation is discussed in the next section.

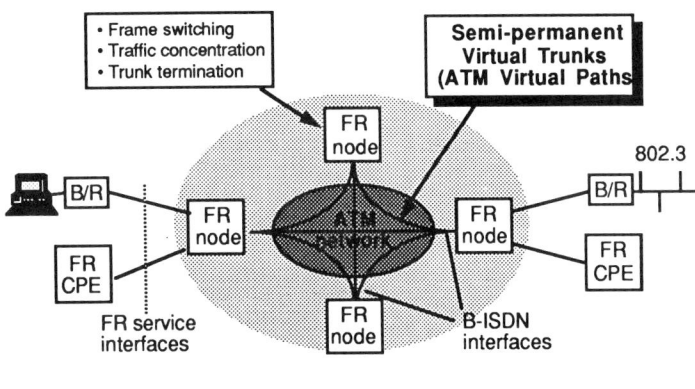

Figure 4   Frame Relay networking using ATM.

3. BROADBAND VDN

Service considerations

For service continuity, and ease of interworking, the evolution of VDN should retain key service attributes provided to the users, and should modify only transport capability, if necessary, to support broadband rates. With FRS-based VDN, these attributes which should be directly extendable to broadband are error-free delivery, sequence preservation, no retransmission nor flow control.

The broadband services with these attributes are now being defined in T1S1 and CCITT for B-ISDN services standardization as part of the ATM Adaptation Layer (AAL) definition. In particular, the AAL for variable bit rate services operating in non-assured mode (no retransmission), with no flow control but with sequenced delivery of frames to ensure reliable frame loss detection will satisfy broadband VPN requirements [6].

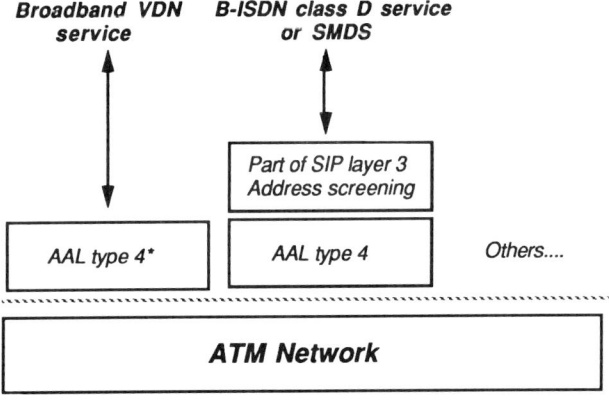

Figure 5 Support of broadband VDN service and SMDS on an ATM network.

Architecture considerations

When ATM is used as a transport mechanism, the broadband VDN service can be implemented using a connection-oriented AAL. Connectionless service, such as SMDS, can be implemented on top of the ATM broadband transport as shown in Figure 5. The service provided by the connection-oriented AAL can also be used to provide Frame Relay trunking as discussed earlier. In this case Frame Relay frames can simply be encapsulated in AAL protocol data units (PDUs) and then segmented into cells and routed using ATM virtual path connectivity (see Figure 6).

The architecture employing common ATM transport for VDN, FRS trunking, and future B-ISDN services has advantage of simple evolution and ease of interworking. The interworking between wideband and broadband VDN services can be illustrated using example of a bank, which in order to accommodate an increasing number of transactions, upgrades its headquarters data processing capabilities to a more powerful mainframe that can be accessed via a broadband interface. All branches using existing narrowband FRS service should be able to communicate with the headquarters without being forced to upgrade their interfaces. Interworking is simple when broadband host and FRS terminal communicate using the same format of frames (LAPD core). These frames can then be encapsulated in AAL PDUs and carried through ATM network similarly to FRS frames carried through virtual trunks. However, if different frame format or functions is used by the terminal and broadband host (e.g. one of the high speed protocols such as discussed in [7]), then translation of the LAPD core frame format and functions is necessary.

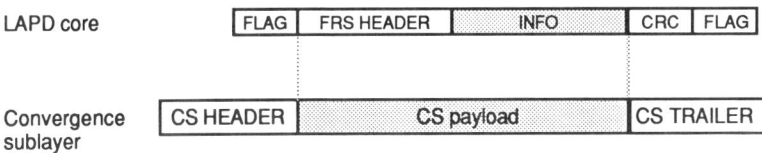

Figure 6   Encapsulation of LAPD core frames into B-ISDN CS-PDUs.

4. PERFORMANCE AND CONTROLS ASPECTS

For its initial deployment, a VDN network may be engineered or dimensioned to support a certain level of traffic with no specific traffic control functions required in the network. Based on careful and continuous traffic monitoring, a reasonably accurate provisioning of near-term demand increase (e.g. monthly) may be achieved. These demand increase forecasts can be used to engineer or dimension the network to provide acceptable levels of service availability and quality. The engineering may not allow for an efficient utilization of resources since the network always has to be over-dimensioned to a certain "safe" level, in particular if a stringent quality of service is required.

Even with very careful network engineering, if no traffic management per connection is provided, statistical traffic arrival may create short-term congestion, in particular at the egress from the network (demultiplexing points). Consequently, mechanisms must be put in place to control its effect without affecting network stability.

Frame Relay, for example, adopts a congestion avoidance mechanism, which is based on monitoring traffic levels in the network and notifying users when the levels indicate approaching overload. Users are then requested to decrease their traffic by activating appropriate rate or window flow control protocols at higher layers, which currently exist in most data equipment.

In broadband VDN, traffic management and control techniques must be modified to be effective in a very different environment with diverse performance requirements of applications, higher unpredictability of user traffic, and higher speed of the network.

The diversity of applications' bandwidth and performance requirements implies that an accurate provisioning would be difficult to achieve and the network efficiency would be even lower than in the unimedia network case. Additional control mechanisms are therefore required to achieve economical, more efficient utilization of network resources. The definition of the minimum set of simple control functions which allow for exploitation of statistical nature of broadband data traffic is currently subject of discussion in standards bodies.

These control mechanisms required for ATM networks to provide a guaranteed[**] ATM quality of service (QOS) should satisfy the following requirements:

• ATM traffic controls should not rely on AAL protocols which are B-ISDN service specific, nor on higher layer protocols which are application specific;
• ATM traffic controls should maximize network efficiency while minimizing network complexity; and
• ATM traffic controls should handle different congestion conditions (e.g. temporary congestion due to short-term traffic fluctuation and long-term congestion due to network overload or failure) while maintaining acceptable performance to the users.

Current ATM traffic control proposals

Currently two approaches to broadband traffic management and control are being investigated, the first reflecting network provider preference to control access to the network for guaranteed performance, and the second which reflects data user preference for LAN-like networking.

The first approach is preventive and is based on the Connection Admission Control (CAC) and the Usage Parameter Control (UPC) functions provided by the network [8,9]. The second approach is reactive and based on cooperation of users with network which explicitly or implicitly notifies users about the load increase [10,11]. Both approaches have advantages and disadvantages and may coexist in the case of broadband VDN.

With the first approach, at connection set-up time, a user provides some traffic or service description to the CAC which in turn will allocate the necessary bandwidth for the connection throughout the network. After connection establishment, the UPC will monitor the connection's traffic and compare with the specified traffic or service descriptors. Some action will be taken by the UPC if a violation to the traffic or service description is detected, such as discarding excess traffic. In this case, the ATM QOS does not rely on user cooperation. However, if there is a need to negotiate some traffic descriptors at connection set-up time, it is unclear how the user can accurately characterize his application's traffic. Furthermore, it will be impossible for the network to develop a usage parameter control algorithm which can quickly detect and react to a violation of the traffic descriptors (i.e. be effective), and at the same time be completely transparent for compliant users (i.e. be fair), especially if the long-term average rate is one of the traffic descriptors to be negotiated, as shown in [12].

A promising solution to this problem was proposed in [12], which consists of standardizing a simple UPC algorithm and incorporating these UPC parameters in the B-ISDN service definition. Bandwidth should be allocated based on the worst case traffic admissible by UPC in order to provide a guaranteed ATM QOS for the traffic admitted by the UPC. However, it still implies that the user may have to estimate his application's traffic characteristics in order to select an appropriate service for his application. Alternatively, he may subscribe for a service and try it until he is not satisfied with his end-to-end performance and then subscribe for higher service rate. In addition, the user may implement the standard UPC algorithm in his terminal as a traffic shaper to avoid cell

---

[**] By guaranteed ATM QOS we mean, a finite probability that cells are lost or not delivered to their destination in time because of congestion, is guaranteed for most (e.g. 99%) of the time.

discard by the UPC and make the best use of the service provided by the network. Although this new scheme for preventive controls meets all the requirements mentioned above, its relative complexity must be judged against expected efficiency gains. Also, if large bursts constitute a significant portion of the traffic, the efficiency gains of this scheme may diminish.

With the second approach, the data users are responsible for maximizing their end-to-end performance by reacting to different traffic load levels in the network, similar to the case of today LAN environments. This reaction can be activated by a detection of packet loss or excessive delay (i.e. implicit notification), or by explicit messages (i.e. explicit notification). There are existing protocols that operate in both cases in today LAN environments. Therefore, the only function implemented in the network is the generation of messages such as the Explicit Congestion Notification (ECN) as adopted in the FR standard. The use of ECN in a high speed wide area packet network has been investigated in [13]. It is shown that at high speed (i.e. 45Mbit/s), the network performance is very sensitive to the buffer threshold set to generate ECN messages, as well as to the reaction to these messages taken by the users (i.e. traffic reduction rate). This will be even more critical in B-ISDN operating at higher speeds (i.e 150Mbit/s and higher). In order to be effective, the buffer threshold for ECN generation must be set very low and the users should reduce their traffic rates significantly. With a low buffer threshold, ECN messages may be generated when there is no congestion but only because of statistical fluctuation of the traffic. This may lead to unnecessary throttling of the users, and consequently to high end-to-end delay under light load conditions. To avoid this problem, either the network has to be operated at sufficiently low utilization or large buffers will be required to allow for a higher ECN threshold.

It is clear that the reactive control approach based on ECN as currently defined for Frame Relay service minimizes network complexity and is appropriate for wideband data networks. However, in broadband networks, more sophisticated traffic controls would also be required in order to maximize network efficiency and better cope with temporary congestion due to statistical traffic fluctuation of the variety of services, without affecting the performance perceived by the users.

Likely solution

The most likely solution would be a combination of preventive controls that do not rely on user cooperation, and reactive controls which should be activated only when severe, continuous congestions occur due to network overload or equipment failure.

Under normal conditions (i.e. light to medium load), occasional temporary congestions may occur due to statistical traffic fluctuation. These congestion events should be controlled within an acceptable limit specified by the offered performance (i.e. cell delay and loss rate), and should be perceived by the users and the network as normal events that do not require any dramatic action. Preventive controls should ensure that the network is operating under these normal conditions by limiting the total traffic generated into the network to a "safe" level. The schemes proposed in [12] and [14] are possible candidates for such situation.

On the other hand, under network overload conditions (e.g. when preventive controls fail to limit the input traffic, or in the case of equipment failure), severe and continuous congestions may occur which will lead to unacceptable performance perceived by the users if no actions are taken. The network may reduce the total traffic of the congested link by rerouting some connections or by terminating certain connections in order to ensure that connections that are not terminated will still perceive acceptable performance. In order to ensure that reactive controls are activated only under network overload conditions, a reliable traffic and performance measurement scheme is required. Ideally, a network overload condition should be detected before network performance is degraded to an unacceptable level. In addition, it may be desirable to have a mechanism that can notify the end users of this potential network overload, similar to, but not necessarily identical to, ECN defined in Frame Relay. This additional feature would also simplify the interworking of broadband VDN service based B-ISDN/ATM with narrowband/wideband VDN service based on Frame Relay.

5. SUMMARY

Virtual Private Data Network is a data service which, when offered by a public network, becomes a cost effective alternative to leased line solution. The Frame Relay standard has been defined by ANSI and CCITT to enable VDN supporting up to 2 Mbit/s frame-based switched access. Frame Relay switches will be deployed in Telecommunication Operating Company offices and will be networked using high-speed (45 Mbit/s and higher) trunks.

Carriers can ensure the evolution from Frame Relay-based VDN to support future broadband applications by adopting ATM technology and B-ISDN services having similar attributes to Frame Relay Service. A B-ISDN connection-oriented service such as provided by an ATM Adaptation Layer which does not support flow control and retransmissions but provides sequenced delivery of frames, is a good candidate to become basis for broadband VDN. The same service is also well suited to support Frame Relay virtual trunking based on ATM technology. Furthermore, this ATM platform can also be used to support a broadband connectionless data service similar to SMDS, and future multimedia applications.

In order to be competitive with Leased Line services, simple and effective traffic management and control must be put in place to allow for high performance (low cell loss and delay) with utilizations which render VDN economical. It is likely that a combination of preventive controls and reactive controls will be employed, which reflects the grade of service requirements dictated by the end users.

6. REFERENCES

[1]   Dataquest Inc., Telecommunication Industry Service, October 1989.
[2]   R. Gusella, "A Measurement Study of Diskless Workstation Traffic on an Ethernet", IEEE Transactions on Communications, Vol. 38, No. 9, September 1990.
[3]   Special Report on Gigabit Network Testbeds, IEEE Computer, September 1990.
[4]   R. Das, P. Carbone, "Dialable Services for FDS1 and DS1 Rates - The Services Network of the Early 1990's", ICC'91.
[5]   Bellcore TA-TSY-000772, Generic System Requirements in Support of Switched Multi-megabit Data Service, Issue 3, October 1989.
[6]   R. Breault, "AAL type 4 bearer service for frame relay support", T1S1.5 contribution no. 91-031, January 1991, Houston, Texas.
[7]   W. Doeringer, et al., "A Servey of Light-Weight Transport Protocols for High-Speed Networks", IEEE Transactions on Communications, Vol. 38, No. 11, November 1990.
[8]   G. Woodruff, R. Kositpaiboon, "Multimedia Traffic Management Principles for Guaranteed ATM Network Performance", IEEE JSAC, vol.8, April 1990.
[9]   A. Eckberg, D. Luan, D. Lucantoni, "Meeting the challenge: Congestion and Flow Control Strategies for Broadband Information Transport", Globecom'89.

[10] K. Ramakrishnan, R. Jain, "An Explicit Binary Feedback Scheme for Congestion Avoidance in Computer Networks with Connectionless Network Layer", Proc. ACM SIGCOMM'88, August 1988.
[11] C. Cooper, K. Park, "Toward a Broadband Congestion Control Strategy" IEEE Network Magazine, vol.4, May 1990.
[12] R. Kositpaiboon, V. Phung, "Usage Parameter Control and Bandwidth Allocation for B-ISDN/ATM Variable Bit Rate Services", Multimedia'90, November 1990, Bordeaux, France.
[13] O. Aboul-Magd, H. Gilbert, M. Wernik, "Flow and Congestion Control for Broadband Packet Networks", ITC'13, Copenhagen, June 1991.
[14] P. Boyer, J.-R. Louvion, D. Trancher, "Intelligent Multiplexing in an ATM Network", Multimedia'90, November 1990, Bordeaux, France.

Network Management and Services

Traffic management of corporate utility networks

V. R. Saksena and T. J. Schonfeld

AT&T Bell Laboratories, Holmdel, NJ 07733, USA

Abstract

This paper describes the traffic management function for corporate utility networks. As users move towards consolidating multiple applications and protocols on a common backbone network, the task of network management and control becomes quite complex. The role of traffic management is to proactively manage network resources from a traffic and performance point of view to ensure that the network delivers high quality of service while operating at peak efficiency. An overall framework for traffic management is proposed and key issues central to an effective functional architecture are emphasized.

1. INTRODUCTION

The proliferation of multi-protocol multi-service networks is driving the need for a networking solution that can consolidate disparate networks into a single backbone network while retaining the logical transparency from an applications perspective [1]. Such a "utility" network will not only reduce capital costs, but will also streamline network operations and management tasks tremendously. Most corporations are recognizing this need and are aggressively deploying enterprise networks that can integrate the commonly used asynchronous and synchronous protocols as well as many of the peer-to-peer protocols, e.g., DECnet, TCP/IP, and XNS that are dominant in local area network (LAN) and environments.

The dynamic nature of the present day corporation, which requires timely access to information to maintain its own competitive edge, has led to stringent performance requirements on the network. With distributed computing applications now on the horizon, requiring thousands of transactions per second, networks are now being stressed as never before, necessitating the need for traffic management systems that can monitor the pulse of the network and maintain acceptable performance levels at all times.

Traffic management represents a series of interrelated tasks whose collective goal is to engineer and administer the network on an ongoing basis so as to meet the network's performance objectives in a cost-effective manner. Measuring, reporting, analyzing and troubleshooting network performance - the key elements of traffic management - are gaining popularity with the network managers since they allow them to respond quickly and proactively to their business needs. Monitoring traffic trends in the network also allows the network managers to more effectively plan their network evolution and be better prepared for new workloads and changes in end-user demands. A traffic management system should, therefore, add value beyond simple monitoring and reporting capabilities by providing cost-effective recommendations in the form of dynamic network performance improvements. The payoff is in allowing the networks to run at high efficiencies without having to go out and acquire more capacity than what is needed.

In this paper, we address the design of a traffic management system for utility data networks. We describe the service requirements imposed by diverse applications and an approach for engineering network components to meet these requirements. We then propose a multistep process that utilizes the vital statistics collected from network components to produce reports on the state of the network and recommendations for tuning network performance. The process is mapped into a system architecture that will provide the network manager with the needed intelligence to proactively manage network resources.

2. THE UTILITY NETWORKING SCENARIO

Traditionally corporate networking has taken the approach of building separate backbone networks for separate applications. Asynchronous terminal-to-host applications have used low-speed dial configurations for a long period of time. Payroll, inventory, and other business applications have used synchronous networks primarily based on IBM's Systems Network Architecture (SNA). The emerging distributed, peer-to-peer applications that are LAN-based use bridge/router technology for wide-area networking. Since most of these applications tend to be highly bursty, i.e., possess a high peak-to-average ratio, the dedicated networking facilities are inefficiently utilized. Furthermore, as these networks grow in size, the cost of network equipment and facilities, as well as the cost of managing separate backbones becomes prohibitively high. To minimize these costs, corporations are deploying utility wide-area networks that can efficiently integrate multiple protocols on a common backbone.

Figure 1 shows the generic architecture of a corporate utility network. The network is accessed by asynchronous terminals and hosts, synchronous devices (cluster controllers, front-end processors) running SDLC or similar protocols, and LAN interconnect bridges and routers running TCP/IP, DECNet, XNS or other internetworking protocols. The backbone nodes can be either time-division multiplexers (TDMs) or packet switches. Time-division multiplexers provide channelized integration of communication links over high speed network facilities. This backbone network solution lowers network cost to some extent while providing fully transparent, non-interfering integration. A more cost-effective solution is obtained by using packet switching technology to statistically share high bandwidth facilities among different applications. For bursty data traffic, the packet-based architecture also provides performance improvements over the TDM architecture because data bursts have access to the full link bandwidth on demand. The packet network, however, requires careful engineering to ensure that the applications do not unduly interfere with each other and cause user-perceived performance degradation.

Early on, most backbone networks used TDM technology for simplicity as well as to permit the integration of voice traffic. As packet technology matured and data traffic grew to a point where a separate data network became cost justifiable, users began to evolve to packet-based backbones which are ideally suited to the characteristics of data traffic. To date, the CCITT standard X.25 protocol has provided the common underlying glue for integrating multiple native protocols on a utility backbone at speeds below 1 mbps [2]. Recent advances in fast packet switching technologies, e.g., frame relay and cell relay, will allow the evolution to speeds beyond 1 mbps [3], [4]. This paper focuses on the packet-based utility network architecture (as opposed to the TDM-based architecture) since this is perceived to become the dominant architecture over time.

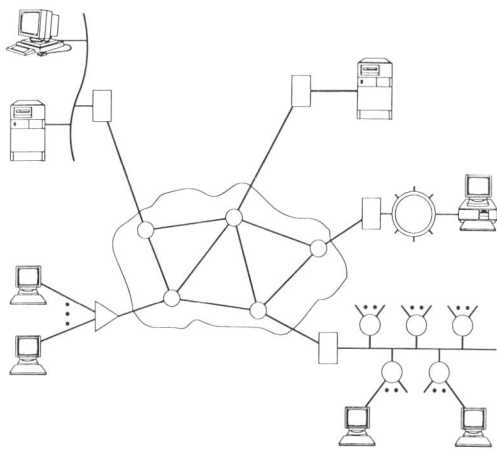

FIGURE 1 UTILITY NETWORK ARCHITECTURE

3. FRAMEWORK FOR TRAFFIC MANAGEMENT

An overall network control plan is comprised of distributed network controls and centralized network management controls. The distributed network controls are those that are exercized by the network nodes to control the traffic flow in real-time. These include routing and call set-up controls, flow controls, and buffer management controls. The purpose of routing and call set-up controls is to minimize the occurence of congestion by spreading the traffic flows over multiple paths across the network. However, due to the statistical multiplexing of bursty traffic sources, the probability of short-term congestion is not zero. The objective of flow controls, and buffer management and scheduling controls is to minimize performance degradation under such circumstances and enforce fair and efficient use of network resources.

Since no distributed control scheme can deal adequately with every unanticipated event, centralized network management capabilities are invoked to control the network operation based on global network status. These centralized management capabilities operate at different time scales as shown in Figure 2. At the slowest time-scale (minutes to hours) are the congestion management capabilities. These capabilities serve to optimize performance within the constraints of available network resources. They allow the network manager to keep the network operating near maximum efficiency during periods of unexpected traffic surges, short-term overload conditions or a sudden reduction in network capacity as in the case of failures. In the absence of these capabilities, the carried load would drop below the engineered capacity of the network accompanied by severe performance degradation.

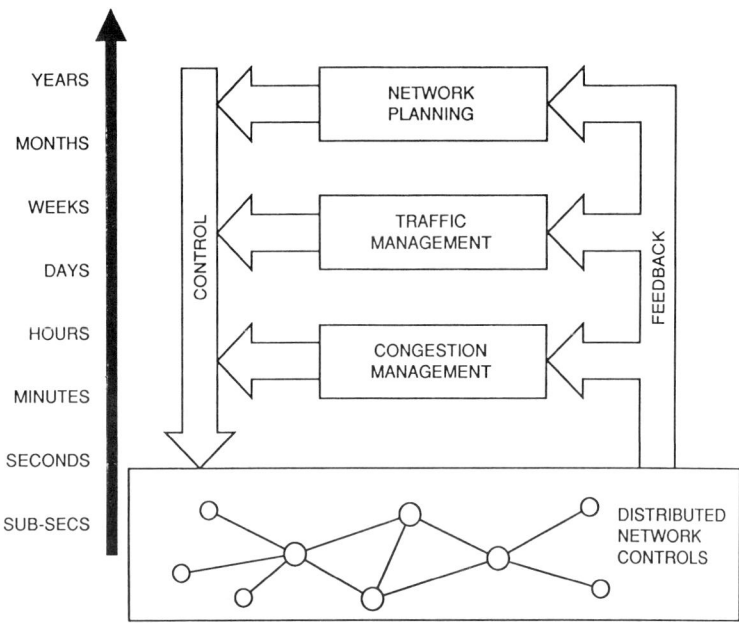

FIGURE 2 TIMESCALES IN NETWORK CONTROL

On a weekly or monthly basis, the traffic management function computes statistical measures of performance to detect problems resulting from consistent patterns of network demand. The objective of traffic management is to proactively identify and relieve traffic related problems before they become service affecting. If the performance on some network connections appears to exceed the pre-established performance objectives, actions are taken to alter the steady-state pattern of flows via routing changes or to add capacity at appropriate places in the network. Such incremental tuning is aimed at alleviating performance bottlenecks without attempting a complete network redesign. These actions ensure that the network adapts to steady-state changes in the offered load while continuing to maintain acceptable performance levels.

On a longer term (semi-annually or annually), the network planning function estimates traffic growth trends and redesigns the network to accomodate the expected growth in traffic. Network planning allows for an orderly network evolution under various scenarios of budget and technological constraints. It allows the planners to determine cost-effective strategies for deploying new technologies and vendor services to support emerging end-user applications in a manner that leads to a graceful evolution from the embedded base.

The rest of the paper focuses on the concepts, issues, and implementation related to the traffic management function.

4. APPLICATION PERFORMANCE ENGINEERING

It is the primary goal of the traffic management function to ensure that a packet-switched backbone network supporting multiple services and applications is engineered to satisfy multiple performance requirements. Basically, performance requirements are established for the call set-up and data transfer phases.

At the call set-up level, there are two key performance objectives: call set-up delay and call blocking. These objectives are particularly important in a switched asynchronous terminal-to-host environment and must be kept sufficiently low for acceptable service. Typical values include 1-in-1000 for call blocking and 500 msec for call set-up delay.

During the data transfer phase, two key performance objectives apply: response time and throughput. The response time objective is critical for real-time interactive applications such as character echoplexing, telnet or remote login, and on-line transaction processing. Typical values for response time objective vary between 100 and 400 msec. The throughput objective applies to bulk data transfer applications such as ftp and uucp. Typical values vary between 10 and 300 kbps.

In order to ensure that the utility network is properly engineered for multiple applications, the application level performance objectives have to be mapped into component objectives. By monitoring the load on network components, the traffic management fucntion will then be able to ensure compliance with service objectives. Table 1 shows how the user-perceived performance is influenced by the characteristics of network components.

Table 1
Performance objectives and their dependencies on network characteristics

Performance Attribute	Network Dependencies
Response Time	Queueing delays at nodes, trunks, access modules; propagation delay.
Throughput	Queueing and propagation delays; packet loss and error rates; virtual-circuit flow control.
Call Set-up Delay	Call controller delay; queueing and propagation delays.
Call Blocking	Virtual-circuit limits on nodes, trunks; alternate routing.

The first step in deriving component objectives is to determine the load-service characteristics of network components. These characteristics provide a simple way to map the component load into service levels such as delay or blocking. Figure 3 shows an example of a load-service curve for a network trunk supporting a mix of echoplex and SDLC applications. The load-service curve is influenced by several key parameters. These include, application specific parameters such as data unit size and the application mix, the nature of the packet generation process, the processing capacity of the component, and the queueing/scheduling discipline used by the component.

In some cases, the data unit size is fairly universal, while in some other cases it is not. For instance, echoplexing almost always involves single characters while SDLC message sizes vary anywhere from 200 bytes to 2000 bytes. In LAN interconnect applications such as ftp, the block sizes are typically large and are constrained only by the LAN maximum packet size, e.g., 1500 bytes on Ethernet. Both the application data unit size and the traffic mix are obtained after a careful study of the end-user applications profile.

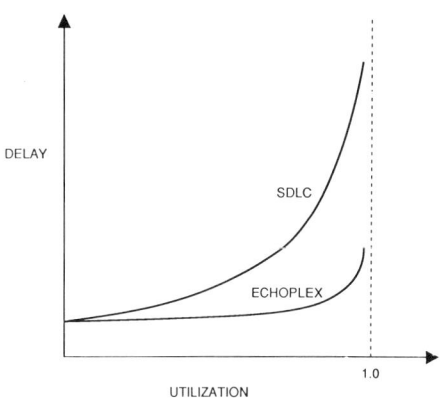

FIGURE 3 LOAD-SERVICE CHARACTERISTICS FOR DELAY

The nature of the packet generation process is quite critical for estimating queueing delays at network components. File transfer sources generate traffic in large amounts, e.g., 20 kbytes to 1 mbyte. However, packets from a single file source are spaced out due to the end-to-end window flow control mechanism. Due to the nature of round-trip delays, the packet interarrival times are highly correlated and so are the service times. These correlations manifest themselves in large queueing delays that significantly effect user-perceived performance [5]. Alternatively, interactive telnet traffic tends to be a lot smoother and can be reasonably approximated by a Poisson process. Before the traffic management function is implemented, a sound model of individual traffic sources as well as their superposition needs to be developed.

The queueing and service discipline used by network components has a direct impact on issues such as fairness and buffer utilization. FIFO service disciplines with shared buffers are most commonly used due to their inherent simplicity and efficient buffer usage. However, they have poor fairness properties when it comes to protecting light users from heavy users. Round-Robin service disciplines with dedicated buffers tend to be fair albeit with somewhat higher buffer requirements. Priority queueing is also used in many instances to support delay-sensitive applications. The queueing delays and the load-service characteristics are very much dependent upon the service discipline used which must be well understood.

Once the load-service characteristics are obtained, the next step is to use them and the end-to-end objectives to derive component objectives. Figure 4 shows an example of how this can be done for the delay objective. A typical "reference connection" is selected and the end-to-end delay objective is allocated to individual components on the basis of component costs and efficiency. The component delay objective is then inversely mapped into a utilization objective (which can be directly measured) on the basis of the load-service curve. The component objectives ensure that if each network component is operating below this objective, then the end-to-end objective will always be satisfied. This simple approach has been used extensively [6], [7].

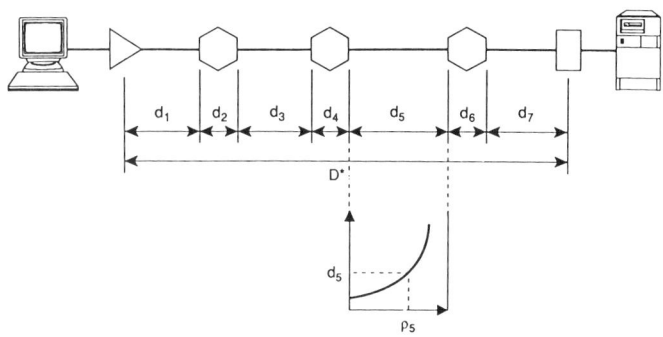

FIGURE 4 DETERMINATION OF COMPONENT OBJECTIVES

5. TRAFFIC MANAGEMENT ARCHITECTURE

Traffic management of a utility network is a multi-step process that begins with the collection of vital statistics such as packet counts, utilizations, virtual call attempts, blocked attempts, and lost/errored packets. This can be a daunting task due to the sheer complexity of interfacing with multiple network elements. These include multiplexers, access lines, trunks, switch processors, call controllers etc. The frequency of data collection is an important parameter. The data collection interval should not be too large, because then it will not be possible to capture the burstiness in the offered load. On the other hand, it cannot be too small either because of the excessive overhead in processing and storing large volumes of data. Typical intervals vary from 10 to 30 minutes. The collected data is processed to detect outliers and to eliminate inconsistent and errored measurements. The validated data can then be used by application programs responsible for carrying out the traffic management functions.

The most efficient scheme for data gathering and storage is to have a centralized system with an open interface that will allow any element management system (EMS) to connect to it and transmit management data using standardized protocol and message sets. The measurement data is stored in a centralized database that also serves as the repository for other configuration information such as network topology and routing tables. The centralized database thus acts as the focal point for storing critical management data that constitutes a basic input to all traffic management activities.

Figure 5 shows a conceptual view of how traffic management functions can be architected. There are three key application modules: the routing administration (RA) module, the performance evaluation (PE) module, and the administrator's workbench (AWB) module.

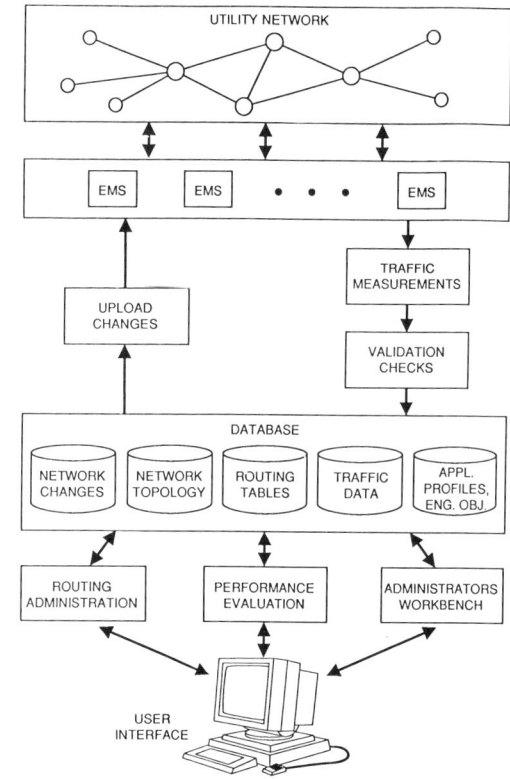

FIGURE 5 TRAFFIC MANAGEMENT ARCHITECTURE

The RA module is responsible for generating new routing tables as well as for validating existing routing tables so as to detect erroneous conditions such as loops, dead-ends, and circuitous paths that, if allowed to persist, will lead to service degradation and inefficient use of network resources. In the analysis mode, RA provides an exhaustive summary of routing problems along with the conditions under which they occur. This allows the administrator to analyze the severity of these problems and to formulate a plan for corrective actions. RA reporting processes should be appropriately structured to effectively calibrate and track the quality of routing tables. The impact of corrective actions should be clearly quantifiable in terms of properly chosen metrics. In the synthesis mode, RA incrementally generates new routing tables as the network topology changes. These changes may be due to network growth or due to configuration adjustments effected to relieve service problems.

The PE module processes the raw measurements collected from the network and provides statistics on service quality parameters such as response time, throughput, and blocking. Typically these statistics include mean, 95th percentile, and peak data. Thresholding capabilities based on application performance engineering objectives are provided to identify observed or potential performance problems and exception conditions. Service levels should be reported for individual network components as well as on an end-to-end basis. The latter reports will allow the network service levels to be compared against the desired service objectives, while the former reports will allow for bottleneck detection. Network administrators should have the flexibility to view performance reports on a daily, weekly, or monthly basis to sectionalize persistent performance problems from transient exceptions. Historical trend reports should also be produced by the PE module to allow the network administrators to characterize traffic growth patterns critical for network planning.

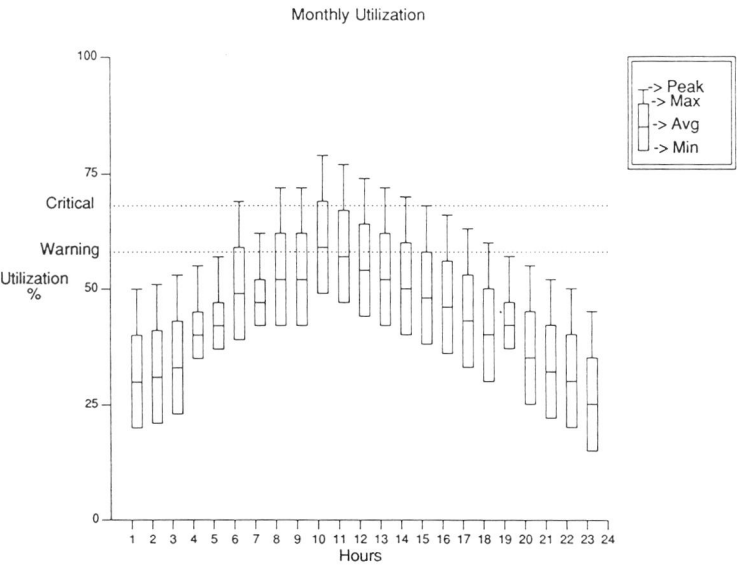

FIGURE 6 MONTHLY UTILIZATION STATISTICS FOR A COMPONENT

Figures 6 and 7 show examples of component utilization reports that provide insightful information on the load patterns. Figure 6 summarizes the monthly utilization statistics across time-consistent hours. The "min" value denotes the smallest utilization observed for that hour, the "avg" value denotes the average value observed for that hour, and the "max" value denotes the highest utilization observed for the hour. In addition to the hourly statistics, the report also shows the *peak* utilization observed during shorter intervals of 5 or 10 minutes during a monthly span of that hour. The report allows the traffic engineer to estimate traffic burstiness by computing the peak-to-average ratio. The report also shows two thresholds, "warning" and "critical," that allow for proactive determination of component exhausts. Figure 7 shows the complete distribution of hourly and peak utilizations over the month. If a component exceeds the engineered threshold, this report indicates the *frequency* of threshold crossing which is an important parameter for taking engineering decisions.

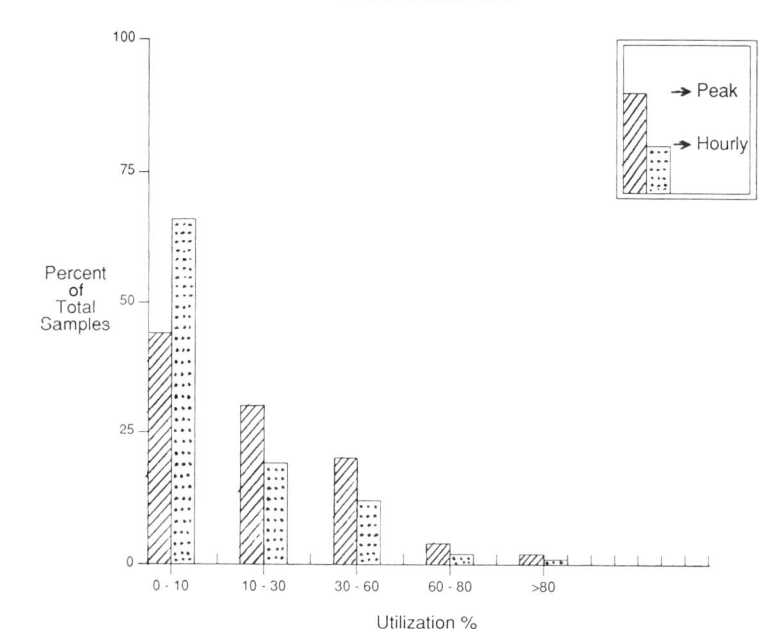

FIGURE 7 MONTHLY UTILIZATION DISTRIBUTION FOR A COMPONENT

The AWB module provides the value-added intelligence needed to diagnose performance problems, to identify cost-effective relief alternatives, and to analyze "what-if" scenarios for network configuration changes. Performance bottlenecks flagged by PE are passed to the AWB module where algorithmic procedures are used to determine least-cost solutions. The relief actions typically involve a combination of parameter tuning (flow control, virtual-circuit limits, etc.), load balancing via routing changes, and capacity augmentations. "What-if" analysis capabilities provide a powerful tool for the administrators to proactively plan for incremental changes. It allows them to control changes more effectively rather than function in a reactive mode when adhoc changes cause unnecessary problems. The recommended changes can be stored in the centralized database and downloaded automatically into the network according to predetermined schedules.

The user-interface is a key component of any traffic management system. Through effective use of menus, screens, and graphical displays, one can effectively hide the underlying complexities of the system and allow the administrators to derive full benefits from its capabilities. Collectively, the three modules provide the network administrator with a comprehensive set of traffic management tools and decision-support capabilities needed to maintain acceptable network performance at a minimum cost.

6. REFERENCES

[1] N. Lippis, "Data comm buys a multiprotocol network," *Data Communications*, December 1990, p80.

[2] E. Mier, "New signs of life for packet switching," *Data Communications*, December 1989, p90.

[3] J. McQuillan, "Broadband networks: the end of distance," *Data Communications*, June 1990, p76.

[4] N. Lippis, "Frame relay redraws the map for wide-area networks," *Data Communications*, July 1990, p80.

[5] K. Fendick, V. Saksena, and W. Whitt, "Dependence in packet queues," *IEEE Transactions on Communications*, Vol. 37, No. 11, November 1989, p1173.

[6] K. Fendick and V. Saksena, "Traffic engineering for the Accunet Packet Service network," *Proc. IEEE Globecom*, Florida, December 1988.

[7] S. Jidarian, D. Shapiro, D. Sheng, and J. Gottfrid, "Traffic administration of a corporate wide area network," *IEEE Network Operations and Management Symp.*, San Diego, February 1990.

ON TELECOMMUNICATION NETWORKS THROUGHPUT: ANALYSIS OF THE
CCITT RECOMMENDATIONS AND 2nd STUDY GROUP QUESTIONS *

M.A.SCHNEPS-SCHNEPPE

University of Latvia, Raina blwd 19, Riga, Latvia, USSR

Abstract

The critical analysis of the CCITT Recommendations is done by pointing at the unfavourable gap between teletraffic theory and Recommendations, on the one hand, and between Recommendations and newer telecommunication technology, especially of computer expansion, on the other hand. New traffic model, explaining the knee-shaped carried traffic curves versus offered traffic, is built.

1. CHALLENGE TO TELETRAFFICERS FROM COMPUTER EXPANSION

Perhaps the most important factor in the progress of telecommunication during recent years has been the integration of computers and communication systems (C&C). As for me as a specialist in telecommunications I confess with regret that in this C&C marriage the computer party is the more aggressive part. Do we really have to put up with it?

Let us turn to the history of teletraffic theory. In 20's and later a lot of devices for telephone loading and blocking simulation were built. The Monte Carlo method as we call it now was used. But the birth of this method is not today connected with teletraffic. During World War II the Monte Carlo method was extensively used in studying solutions of differential equations arising from work with the atomic bomb. The name Monte Carlo for this method is from that period.

"Money talks,"- we might say for comfort. And really the Los Alamos project was like Gulliver compared with telephone exchange Lilliputian.

This is not the case of the C&C marriage. Both parties are mo-

---

* This paper was written during author's visiting professorship at the Institute of Mathematical Statistics and Operations Research, Technical University of Denmark.

neyed. The investments in telecommunications are fully comparable with those in computers. Why to step back? Why not meet challenge if we are conscious of these new problems?

In this context it is time to recall Arne Jensen's words in 'Foreword' in the special teletraffic engineering issue of 'Telecommunications Journal'from 1984.
"New problems arise in the future with new integrated networks carrying different types of services. The modular construction of the routes, and their modular connection to the exchange require to a greater extent co-ordinated decisions to dimensioning of the different groups in the various traffic streams"[1].

This is for the future. But what do we have today?   Let me turn to one significant case.

The CCITT Recommendation Q.543 contains a caustic reproach to teletraffic researchers, especially to the 2nd Study Group engaged in traffic engineering and network management. This recommendation entitled "Digital exchange performance design objectives" states the up-to-date overload control strategy not yet implemented in the Series E Recommendations.

According to Rec.Q.543, an effective overload control strategy will prevent the rapid decrease in processed call attempts with increased overload (see Curve B in Fig.1). The relatively gradual decrease with overload control (Curve C) is due to the increasing processing overhead in exercising the overload control. Overload is defined as the level of call attempts offered to the exchange in excess of the exchange engineered capacity.

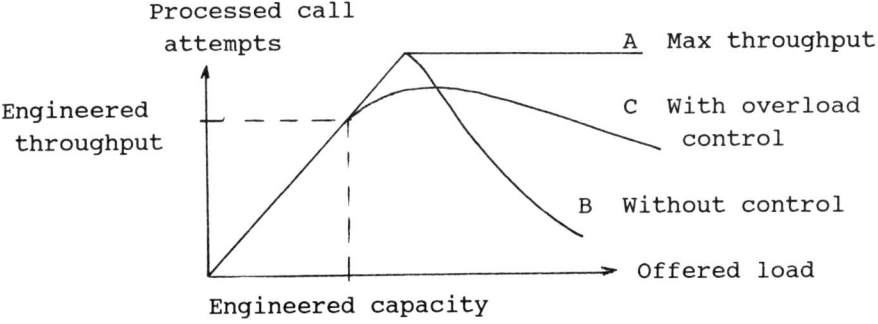

Fig.1. Exchange throughput at an overload

The extremal shape of carried call attempts or, in other words of carried load versus offered load is the highly important feature of modern telecommunication systems, not yet sufficiently taken into account by teletraffic engineers.

During the 12th ITC Geza Gosztony, the chairman of the CCITT 2nd Study Group, pointed out the slow process of standardizing efforts in CCITT and stated that the preparation of recommendations will have to be dramatically accelerated [2]. This is without doubt. But the success is to a great extent up to us. The 2nd Study Group has to be more active itself.

Let me stop to discuss two cases by G.Gosztony. At first, he refers to Recommendation Q.543 and to Fig.1, respectively, as a basic model for processor capacity allocation. At the same time he illustrates himself by figure contained only the increasing part of curves in Fig.1 thereby omitting the very essence of the problem regarding the overload process of switching system. I should think that the teletraffic models to be developed in nearest future are to take into account the knee-shaped carried load curves without fail. As a consequence, in the next sections I shall discuss the phenomenon of feedback in telecommunication as a basis for overload study and, therefore, as an approach for carried load study in a whole range of offered load values.

The second case illustrates the unfavourable gap between up-to-date teletraffic theory and implementation in the CCITT recommendations. In [2] G.Gosztony has mentioned the Rec.E.524 "Overload approximations for non random traffic". It contains the comparison of blocking calculation methods for individual traffic streams sharing the same overflow group (Fig.2). Three methods are discussed: 1) interrupted Poisson process method taking into account three or rather four moments of input loads, 2) equivalent capacity method and 3) approximative Wilkinson Wallstrom method. The accuracy of methods is tested by calculating the examples.

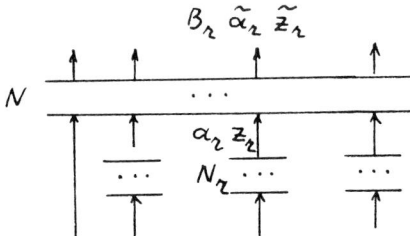

Fig.2. Traffic streams shared the same overflow group

In my opinion, this approach is too complicated for international standardizing level. Moreover, the extremely simple and accurate Hayward's approximation for overflow traffic [3] known since 1980 exists but it is not even mentioned in references to Rec.E.524. Let me survey it.

If we are given N channels (Fig.3) and the input stream with mean α and peakedness z, where z is defined as the variance-to-mean ratio of the number of busy channels on an infinite trunk group offered this traffic, then the resulting

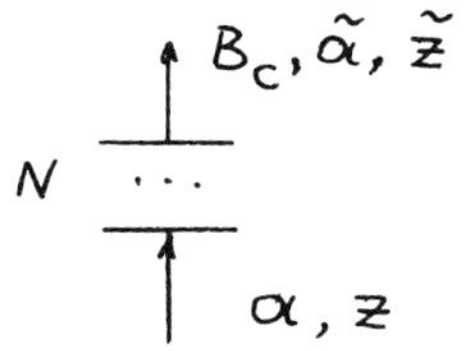

$$\begin{array}{c} \uparrow B_c, \tilde{\alpha}, \tilde{z} \\ N \xrightarrow{\cdots} \\ \uparrow \alpha, z \end{array}$$

Fig.3. Simple overflow case

blocking probability (call congestion) B_c according to Hayward's approximation is obtained via Erlang loss function

$$B(N,\alpha) = \frac{\alpha^N/N!}{\sum_{i=0}^{N} \alpha^i/i!} \equiv \left[\alpha \int_0^\infty e^{-\alpha t}(t+1)^N dt\right]^{-1} \quad (1.1)$$

in the following way:

$$B_c = B\left(\frac{N}{z}, \frac{\alpha}{z}\right) \quad (1.2)$$

For overflow stream we have mean call rate

$$\tilde{\alpha} = \alpha B_c \quad (1.3)$$

and peakedness

$$\tilde{z} = z - \tilde{\alpha} + \frac{\alpha z}{N + z + \tilde{\alpha} - \alpha} \quad (1.4)$$

If we have n input streams with parameters α_i, z_i, $i=\overline{1,n}$ there exist cited in Rec.E.524 Lindberger's formulas [4] for individual overflow stream blocking probability

$$B_i = \left(1 + \frac{1-B}{1+B}\left(\frac{z_i}{z} - 1\right)\right) B \quad (1.5)$$

and for moments

$$\tilde{\alpha}_i = \alpha_i B_i \quad (1.6)$$

$$\tilde{z}_i = 1 + (\tilde{z} - 1)\frac{\tilde{\alpha}_i}{\tilde{\alpha}} \quad (1.7)$$

where β is calculated by (1.2) at $a = \sum_{i=1}^{n} a_i$ and $z = (1/a)\sum_{i=1}^{n} V_i$. Formulas (1.1) - (1.7) are much simpler than those proposed by Rec.E.524. Thus they may supply a revision of this Recommendation.

It is hard to explain why these widely known results were not implemented in the CCITT Blue Book. Moreover, how to explain the call of the 2nd Study Group for current period 1988-1992 for more contributions concerning the accuracy of methods named in Rec.E.524. Such call might be reasonable for a scientific not for a routine and widely usable standardization process.

As a consequence of this, a proposal concerning interaction between CCITT and International Teletraffic Congresses may be stated. The aim of ITC activity, in my opinion, is to prepare proper background for standardization on international level. The results of congresses and seminars might be extracted for use by CCITT study groups. A similar idea has already been discussed by A.Lewis during the last ITC:"Perhaps extracts from some ITC seminars would be included in CCITT handbooks giving wider dissemination to the valuable but little known achievements of these seminars"[5].

The critical analysis of Rec.Q.543 and Rec.E.524 points out the unfavourable state of the CCITT 2nd Study Group. Mr.Geza Gosztony does his best, I think, but we, the teletrafficers, are really to blame. It is time to increase our activity, to be more busting in comparison with our colleagues from computer side.

Let us turn now to the overload study. The problem may be split in two: 1)for control computer of SPC-switch and 2)for channel groups and network as a whole.

2.OVERLOAD OF SPC-SWITCH AS A POSITIVE FEEDBACK

Shortly after the introduction in the 1960's of SPC-telephone switches it was noticed that their performance sometimes degraded drastically under overload. Analysis of measurements taken during the decrease of the performance revealed two causes. First of all, as more call generate tasks for the same processor, each task has to wait longer, thus increasing the time necessary to set up a call. As a result, more customers hang up prematurely. In the second place, the time, which the processor has spent working on tasks for this call request which does not get completed, represents wasted capacity at the very time when capacity is most needed. This leads to a positive feedback in the form of repeated call attempts.

As a model that describes the behaviour of the switch at overload we have chosen one recently proposed by R.K.Boel [6] and shown in Fig.4.

Call requests flow into the system at a rate of λ. Denote by X_t the number of call requests waiting to be set up and use it as the system state. Each of these X_t call requests with arrival rate ν generates tasks to be executed by the control processor. Denote the average duration of one task by $1/\mu$ and

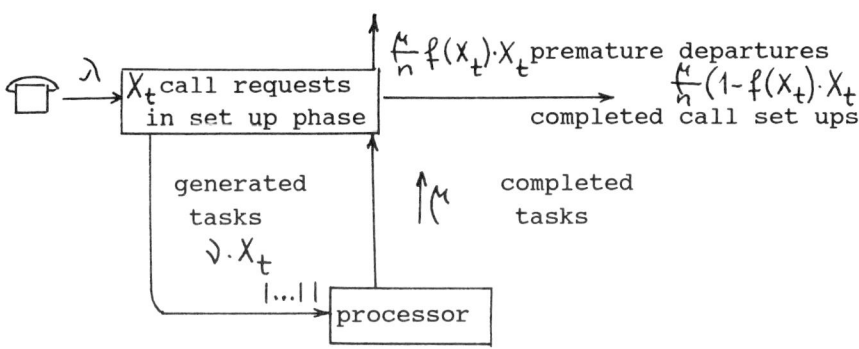

Fig.4. Model of a telephone switch

assume the average number of tasks per call requests is n. Thus, the load per call request is at least n/μ and for stability the inequality $\lambda n/\mu < 1$ is to be fulfilled.

Instead of unobserved quantity λ we may observe current state X_t and use X_t for control. It is clear that for stability of processor the input rate νX_t has to be less than output rate μ. Thus, in Boel's opinion, the inequality $X_t < \mu/\nu$ has to be managed. Such requirement, I think, is too strong. The quantity X_t is a random variable and its values may theoretically go up to infinity. Thus, from the cases $X_t > \mu/\nu$ for short time intervals do not follow the unstability with certainty.

Furthermore, Boel makes two proposals. Firstly. He supposes that set up time consists of two phases. The first phase equals $\frac{n}{\mu}$ in the average. After the generation of the last task the call request still has to wait some extra time $T(X_t)$ depending on current load of system X_t and increasing with growth of X_t. Hence the total time between initiation of the call request and completion is

$$\frac{n}{\mu} + T(X_t)$$

Secondly. The probability, denoted by $f(X_t)$, exists that the caller experiencing the corresponding average delay will hung up prematurely. Thus, we have the rate of succesful call completions

$$\frac{M}{n}(1-f(X_t))X_t$$

depending on current load of exchange and the rate of premature departures

$$\frac{M}{n}X_t \cdot f(X_t)$$

Therefore, for implementation of the Boel's model two unknown functions $T(X_t)$ and $f(X_t)$ are to be built.

This proposal is rather unusual but nevertheless it seems more realistic than the extraordinary complicated procedure for exchange performance analysis recommended by Rec.E.502 "Traffic measurement requirements for SPC (especially digital) telecommunication exchanges". According to Rec.E.502, 20 traffic flows through exchange are to be measured (Fig.5) and 23 types of da-

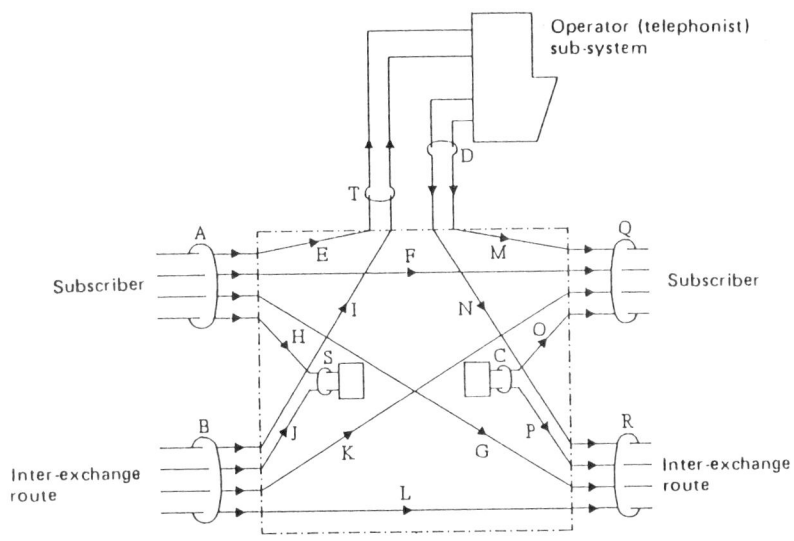

A — Originating traffic
B — Incoming traffic
Q — Terminating traffic
R — Outgoing traffic
F — Internal traffic
G — Originating outgoing traffic
H — Originating system terminating traffic
J — Incoming system terminating traffic
K — Incoming terminating traffic
L — Transit traffic

O — System originating terminating traffic
P — System originating outgoing traffic
S — System terminating traffic
C — System originating traffic
T — Operator terminating traffic
D — Operator originating traffic
E — Originating operator terminating traffic
I — Incoming operator terminating traffic
M — Operator originating terminating traffic
N — Operator originating outgoing traffic

Fig.5. Exchange traffic flow diagram

ta are to be gathered for many objects: subscriber line groups, circuit groups, common control units, auxiliary devices, destinations etc. It seems hardly implementable for performance monitoring and network managemewnt purposes.

If this is not the case "not seeing the wood for the trees"? Rec.E.502, in the opinion of CCITT, will be the basis for measurements in the ISDN networks. Why so complicated? Moreover, the Boel-type model may be discussed as a contribution to the 2nd Study Group for the SPC-switch control.

To work out the admission procedures for overload protection in cases of distributive controlled SPC-switch it is reasonable to choose multiqueue case instead of one task queue. In this connection we may refer to the earlier paper by P.J.Kuehn [7] and some recent papers (B.Wallstrom and H.Voigt [8], P.Tran-Gia[9]) containing the proper survey.

3. TRAFFIC MODELS

Now we turn to the routine teletraffic engineering problems concerning circuit groups of network. Conventional traffic models consider the trunk networks exchanges separately so that congestion in the network is assumed not to affect the exchange traffic and vice versa. Moreover, such models do not take into account at least two aspects: 1) that subscriber call attempt generation rate increases as congestion increases because of the repeated attempts and 2) that subscriber call generation and holding time habits are unstable, varying with the service the subscriber perceives.

This effect is equivalent to introducing a feedback into the subscriber traffic patterns. The more the network is overloaded the more interaction exists between subscribers, exchanges and network.

Let me start with Question 15 for the 2nd Study Group entitled "Traffic models and measurements required for traffic offered to the network and grade of service". Despite a lot of works, an acceptable solution to the offered traffic problem has not yet been found. According to Rec.E.501, the traffic offered A may be expressed via the traffic carried A_c in the following way:

$$A = A_c \frac{1-WB}{1-B} \quad (3.1)$$

Fig.6 borrowed from Rec.E.501 explains the meaning of constants B and W :

$$B = \frac{N_L}{N} = \text{measured blocking probability on the circuit group}$$

$$W = \frac{N_{LR}}{N_L} = \text{proportion of blocked call attempts that re-attempt}$$

Unfortunately, the formula (3.1) cannot explain the knee-shaped curves B and C (see Fig.1). Thus, the formula is based upon inadequate traffic model. What to do?

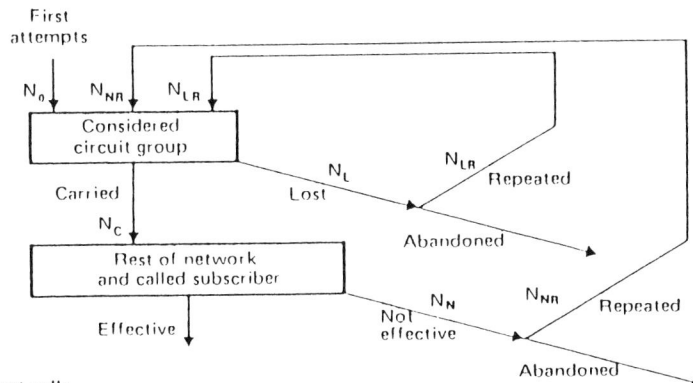

N_0 First attempt calls
N_C Carried calls
N_l Lost calls
N_{LR} Lost calls repeated
N_N Non-effective calls
N_{NR} Non-effective calls repeated

Fig.6. The CCITT traffic model

Question 15 considers that the traffic offered cannot be observed directly and that the repeated attempts are unavoidable.

To bridge the gap between theory and practice, one must first of all quote the CCITT terms and definitions (Rec.E.600):
1) traffic carried - the traffic served by a pool of resources,
2) traffic offered - the traffic that would be carried by an infinitely large pool of resources. It follows from these definitions that traffic offered A is greater than traffic carried A_c. In reality, this is not always the case. The inequality

$$A(\infty) > A_c(\infty)$$

is true for an imaginary infinitely large circuit group. But for finite N circuit group the inequality

$$A(\infty) < A_c(N)$$

may be fulfilled. Why?

The problem is that the traffic observed $A_c(N)$ is frequently equated with its largest component only i.e. with effective traffic, or chargeable traffic A_e being disregarded its second component - setting up traffic A_s, constituted by the call set up times (Fig.7). We mean

$$A_c(N) = A_e(N) + A_s(N)$$

As offered traffic $A(N)$ increases, effective traffic $A_e(N)$ first of all also increases, but then, having reached its peak it begins to fall and, finally, tends to zero when values of offered traffic become extremely large. At the same time, due to repeated attempts, the setting up traffic A_s increases monotony with increase of A. Thus, at high subscriber perseverance the carried traffic $A_c(N)$ may exceed theoretically defined offered traffic $A(\infty)$.

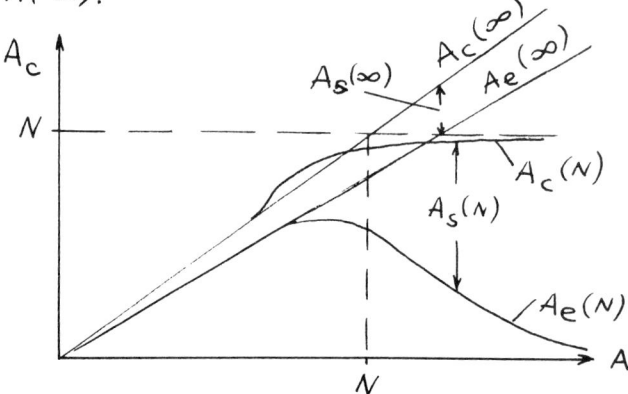

Fig.7. Relationship between offered and carried traffic

To explain the curves shown in Fig.7 we propose the extended traffic model (Fig.8). What is new in this figure in comparison with Fig.6? Besides the holding time T we introduce set up times:

t - set up time for considered circuit group

t_a - set up time for rest of network including called subscriber. In Fig.6 and formula (3.1) respectively only holding time T have been counted.

For a rough analysis we propose the independence between states of the circuit group, the rest of the network and the called subscriber whereas the dependence between subsequent call attempts and the changes of set up times with load increase will be counted. The simple carried traffic formulas gain by the named proposal of independence.

Each call attempt (the first one as well repeated ones) gives rise to carried traffic equal

t with probability β
$t + t_a$ with probability $(1-\beta)\bar{\pi}_a$
$t + t_a + T$ with probability $(1-\beta)(1-\bar{\pi}_a)$

Therefore, the total carried traffic per attempt is as follows

$$y = t\beta + (t+t_a)(1-\beta)\bar{\pi}_a + (t+t_a+T)(1-\beta)(1-\bar{\pi}_a) = \quad (3.2)$$

$$= t + (1-\beta)t_a + (1-\beta)(1-\bar{\pi}_a)T$$

(there $\sum_{i=1}^{\infty} i x^{i-1} = \frac{d}{dx} \sum_{i=1}^{\infty} x^i = \frac{d}{dx} \frac{x}{1-x} = \frac{1}{(1-x)^2}$, $x < 1$)

The blocking probability of attempt equals

$$p = 1 - (1-\beta)(1-\bar{\pi}_a) \qquad (3.3)$$

The carried traffic per call depends on the average number of attempts per call, M, determined by

$$M = 1\cdot[(1-p)+p(1-W)] + 2(pW)(1-pW) + 3(pW)^2(1-pW) = \quad (3.4)$$

$$= \sum_{i=1}^{\infty} i(pW)^{i-1}(1-pW) = (1-pW)\sum_{i=1}^{\infty} i(pW)^{i-1} = \frac{1}{1-pW}$$

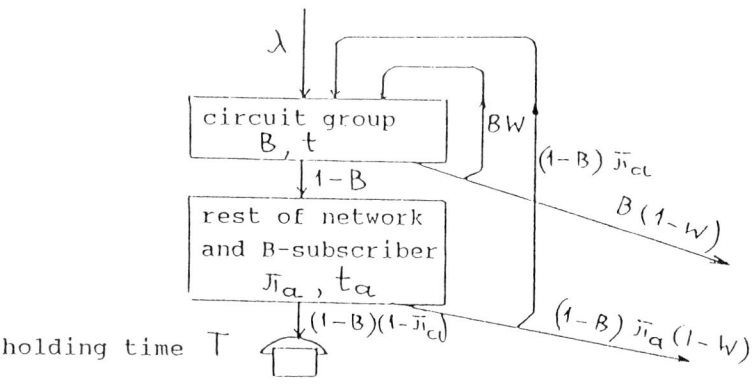

Fig.8. Extension of the CCITT traffic model

Probability of call demand loss

$$P_c = p(1-W) + (pW)p(1-W) + (pW)^2 p(1-W) + \ldots = \\ = p(1-W) \sum_{i=0}^{\infty} (pW)^i = \frac{p(1-W)}{1-pW} \qquad (3.5)$$

The supplementary (up to one) probability indicates the proportion of call attempt strings ending with successful attempt and equals

$$1 - P_c = 1 - \frac{p(1-W)}{1-pW} = \frac{1-p}{1-pW} \qquad (3.6)$$

Therefore, when traffic offered

$$A = \lambda T \qquad (3.7)$$

we have from (3.2) - (3.4) the carried traffic

$$A_c = \lambda M y = \lambda \frac{t + (1-\beta)t_\alpha + (1-\beta)(1-\bar{\pi}_\alpha)T}{1 - W(1-(1-\beta)(1-\bar{\pi}_\alpha))} \qquad (3.8)$$

and from (3.6) the effective (charged) traffic

$$A_e = \lambda (1 - P_c) T = \frac{\lambda T (1-\beta)(1-\bar{\pi}_\alpha)}{1 - W(1-(1-\beta)(1-\bar{\pi}_\alpha))} \qquad (3.9)$$

Discussion. At the numerical analysis of formulas A_c and A_e an important assumption was taken. Namely, if circuit number N is given then obviously with the increase of load λ the blocking probability β increases also going up to one and A_c increases going up to N. But as to A_e, for achieving a strongly marked tendency of knee-shaped effect we have used formula

$$t = t_o / (1 - \beta) \qquad (3.10)$$

determining set up time through considered circuit group.

This proposal is similar to Boel's proposal regarding extra waiting time in control processor which depends on current load of system. As to dependence between t and λ then instead of (3.10) another increasing function may be used. (It needs to carry out measurements of set up time versus offered load.)

For practical use of proposed approach the tables of repeated call system are needed (for survey cf.[10]) to calculate the blocking probability B as a function of circuit number N and offered load A. Usable approximations with satisfactory accuracy have been received also by queueing theory [11].

Numerical analysis has shown that above mentioned inequality

$$A(\infty) < A_c(N)$$

for some region of offered load is only fulfilled if persistent probability W is near to one, say W =0.95, but at W =0.8 the inequality

$$A(\infty) \geq A_c(N)$$

has fulfilled for all values of λ.

References

[1] A.Jensen, Telecomm.J., 51 (1984) p.312
[2] G.Gosztony, ITC 12 Torino (1988)
[3] A.A.Fredericks, BSTJ, 59 (1980) No1
[4] K.Lindberger, ITC 10 (1983)
[5] A.Lewis, ITC 12 Torino (1988)
[6] R.K.Boel, Comp Netw ISDN Syst, 20 (1990) p.179
[7] P.J.Kuehn, BSTJ, 58 (1979) p.671
[8] B.Wallstrom and H.Voigt, ITC 12 (1988)
[9] P.Tran-Gia, ITC 12 (1988)
[10] S.N.Stepanov, ITC 12 (1988)
[11] M.A.Schneps-Schneppe, ITC 6 Munich (1970)

Experiences and Expectations of Introducing PSTN New Services in Japan

Jun-ichi MIZUSAWA and Minoru AKIYAMA

Department of Electrical Engineering, The University of Tokyo, 7-3-1, Hongo, Bunkyo-ku, Tokyo, 113 Japan

Abstract
Telecommunication services have been considered as major concerns of carrier's business since mid 1970s after having fulfilled the telephony demands in Japan. Furthermore, in 1985, the competition rule is introduced into Japanese telecommunication market which accelerated the development and the offer of new telecommunication services. In this paper, taking typical new service examples, the experiences learned from new network service introduction are described. Moreover, future scenarios of service introduction are explained.

1. INTRODUCTION

Planning the introduction of new network services, there are two basic choices listed below.
1) Construct a new public switched network for a new network services,
2) A new service should be realized within a framework of existing PSTN by adding new service functions.

New public network construction(1) is a better solution to realise public network evolution, but it has a lot of difficulties to overcome, especially financial backup and a long-term incomings and outgoings balance. Method (2) requires to add new service functions to a PSTN for providing a new network service. Adding new functions can be classified into two following categories.
2a) Existing PSTN is utilized as an access network for a new network service.
2b) By adding new service functions such as database, a PSTN itself gets new capabilities to provide new services.

An access network capability (2a) for a new service network is commonly utilized in case of following network services.
-PC communications (Personal computers to a host computer data transfer)
-Facsimile network service (Store and forward type facsimile network services)
-NCC(New Common Carrier) toll telephone services
-Value Added Network services (Example: Voice-mail service)
Adding new functions (2b) is the method to enhance PSTN capabilities by introducing new service functions. The new functions are mostly computer based database technologies, CCS

(Common Channel Signalling) technology, and switching node technology which can be summarized as Intelligent Network Technologies.

In this paper, compling with these categories, three typical network service experiences are explained.
(1) New public network construction:DDX packet network
(2a) An access network:DDX-TP service
(2b) Adding new service functions:Freedial service
After that, narrow band ISDN network design principles and expectations of new network services are explained based on these experiences.

2. NEW PUBLIC NETWORK CONSTRUCTION:DDX-PACKET

First example is the new network introduction called DDX packet. It is X.25 type public data network. Presently it became the second largest sales public network in Japan next to the telephone network.
In 1980, DDX packet started its service in the form of independent public data network. At the designing period prior to the start of service, the discussion items for principal network design were as follows.
Q1) Which of the public data network should be selected, DDX-C(X.21) public network or DDX-P(X.25) public network?
Q2) How a customer terminal should be connected to DDX-P network, via PSTN network or via an independent subscriber line?
Q3) What type of a terminal will be the major network terminals, an existing start-stop type slow speed terminal or a new X.25 high speed terminal?

In the first design phase, the decision and the reason were as follows.

A1) Both DDX-P and DDX-C were selected. The reason was both network have particular applications. DDX-P is for conversational communication, and DDX-C is for files transfer.
A2) The particularity of the project was that the basic design concept has the independence of a new network service. Therefore an independent subscriber line was selected. And an access from the PSTN was not allowed at the initial DDX-P service phase.
A3) It was expected that the most of terminals would be a slow speed existing terminal. Consequently, PAD (Packet Assembly and Disassembly) functions were developed which provide X.28 and X.29 protocols together with other various kinds of protocol sequences.

After ten years exercises, following results are shown.
R1) DDX-P has now great advantages over DDX-C in terms of subscriber numbers. The reason is the capability of logical link multiplexing which gives merits to subscriber computers,

even though it requires sophisticated communication software. This shows the basic characteristics of a new network service is quite important to judge an advantages of a new network service.

R2) A new network independence is an ideal concept. In case of DDX-P, after the service was started, everybody asked DDX-P project group why DDX-P can't get a number of customers as it had been prospected before. It continued for five years. One of the reasons why it was difficult to get new customers with independent network configuration is that an independent access line asks costly monthly charge for a customer. This was a barrier for a customers to be familiar with a new network service. This fact shows that a new network service should, in some way, be easy to access to let our customers to be familiar with. In case of DDX-P, an access from the PSTN for a slow speed terminal were provided from 1985,which is called DDX-TP service. DDX-TP contributed very much to the traffic increase of DDX-P, because the number of DDX-TP subscribers increased a lot by virtue of the spread of PC communications.

R3) Concerning DDX-P customer terminals, now 95% of them are X.25 type terminals. Therefore, the estimation *(A3)* went wrong. Slow speed terminals do the access from PSTN (via DDX-TP).A customer of a subscriber line prefer to utilize an advantage of logical link multiplexing. This resulted PAD functions of DDX-P is effective only for an access route from PSTN. PAD functions for subscriber lines are now installed within customers premises. This implies a direct subscriber line are for professional use, such as service providers or a private network designer, and they try to bring out maximum advantage of a new network service.

Concerning a new terminal (X.25 terminal in this case), the fact that it takes at least five years to get a variety of terminals for a new network service should also be explained. Because those who are responsible for business networks are always very careful about the quality of a new network service, and terminal manufactures are very steady to invest their budget for new terminal developments.

DDX-P now has approximately 59 thousands direct subscriber line customers after ten years operation.

3. AN ACCESS NETWORK:DDX-TP SERVICE

PSTN was originally designed for telephone service. But recently, with the progress of new network technologies and also with the introduction of free competition for new network services, basic requirements for PSTN is changing. The very basic requirement is the access capability for a new service network, which is the category *(2a)* mentioned in section 1. New network examples are, a facsimile store and forward type network, a new common carrier's network which provides toll telephone services, and other value added networks. In this section, taking DDX-TP service as an access network example, the role that an access network plays is explained.

Considering the DDX-P design principle, the network independence mentioned above seems half successful but half unsuccessful. The successful aspect is customer inducement to an introduction of a X.25 cluster terminal which has an advantage of X.25 logical link multiplexing capabilities. The unsuccessful aspect was explained with increased number of DDX-TP subscribers, who do access to PC communication host computers located in DDX-P network. From a network designer's viewpoint, following two contradictory results can be obtained.

R1) PSTN access capability has a limitation of bit rates and qualities of service, therefore a direct subscriber line should be provided to invite real evolution of a new network service.

R2) Contrary, an access network capability of PSTN is always very attractive for encouraging customers to touch with a new network service and let them be a supporter of a new network service.

The reason why these contradictory results should be both accepted can be explained with following data. At present, the number of PSTN access customers via DDX-TP service is five times as large as that of direct access. This is the advantage for item *(R2)*. But the actual traffic of DDX-P network are mostly produced by direct access customers (Fifteen times more). This is the data which supports item *(R1)*.

The consideration could be summarized as follows. Both concepts *(R1)* and *(R2)* help together the evolution of our customers to learn new network services and qualities. And finally it allows a smooth change from an old technology based network to a new one. Furthermore, the role of PSTN as an access network will be very important to invent a lot of new network services. This concludes that access capabilities to plural network should be well designed within PSTN. Actually, access capability enforcement is now requested, and POI (Point of Interface) capabilities are important for new network service providers, particularly the numbering capability for the access is indispensable for encouraging new carriers' business.

4. NEW SERVICE INTRODUCTION:FREE-DIAL

The third category *(2b)* is the new service actualization by means of taking new service functions onto an existing PSTN. One of a typical network service is Freedial (freephone). The network structure of Freedial is based on Database and CCS (Common Channel Signalling No.7) technologies. At the early stage of designing Freedial, there are several discussion items which relate to a network structural design.

Q1) Whether Freedial service capabilities should be realized within a switching node software, or it should be implemented in the outside of a switching node such as a database.

This is an architectural discussion of PSTN design. After the decision of applying a database and CCS technologies, there were next questions.

Q2) Logical number translating capabilities are allowed to be implemented in a database. Then, call control functions such as routing function, called party busy-idle management and originating area screening etc., should be included within a database software or it should be left for a switching node software.

Q3) Which node should have database inquiry capabilities via CCS signalling network, a local switch, a toll switches or a designated toll switch?

With the successful operation of Freedial service, following conclusions are obtained.

A1) There are not any difficulties to introduce call control functions into a database, that is other than a switching node. An existing PSTN is realizing so called distributed processing for a call control. Together with the distributed call control, an advantage of introducing centralized call control functions was made clear in following means.

-Nation wide call rerouting is possible with rather simplified algorithm.

-The initial investment to introduce a new service is rather small compared with installing new software functions in many switching nodes.

A2) As for a database inquiry node, a dedicated toll switching node is selected because a local switch did not have a CCS capability. In case the traffic handling amount is rather small, a restricted number of database inquiry nodes is advantageous. When a lot of service handling information inquiries are required, then inquiries from all local switching nodes are preferable because of a short response time.

Freedial started its service in 1985. The service is now very popular in Japan having more than 200 thousands of subscribers. The reason why it is so popular is considered as follows compared with other type of services mentioned before.

R1) This service requires no newly developed subscriber terminals. Therefore, all PSTN subscribers are potential new service customers.

R2) Its service function is the very basic feature of PSTN, i.e., charging. The variation of PSTN basic features are quite important for producing new popular network services.

As for the conclusion, Freedial type service, which has very basic simple PSTN feature, has great advantage compared with other services which require a new network installation. The experience of Freedial service introduction resulted in enforcing the development of other new services which employ Database and CCS technologies. Furthermore, it provoked network architectural discussion, i.e., Intelligent Network. Now IN technologies are strategically important for carriers from a view point of network businesses.

5. ISDN NETWORK DESIGN

The experiences described above tell us fundamental conditions for the design and development of new network services.

I1) PSTN basic telephone service is always weighty network service. Therefore, if a new network service is based on the basic telephone service, you can expects the benefits of 50 million existing customers.

I2) Having entered into network service competition, an access capability to a new network become more and more requested as a basic PSTN service function. The access capability can be considered as a gate function for other new service carriers. PSTN should have flexible numbering, routing and charging facilities which can fulfill access design requirements of new network providers.

In Japan, ISDN service started in 1988. Prior to this, the fundamental network structure design principles are discussed. ISDN basic design concept was concluded as that item *(I1)* should be respected. That is, PSTN basic telephone services should be replaced with ISDN network service in the near future. ISDN design principles were as follows.

D1) Overlay network structure: ISDN network is totally duplicated with existing PSTN network. And it is not a separate independent network.

D2) Numbering plan of the ISDN network is designed with the rule of the existing PSTN. Therefore, there are no distinct difference between PSTN telephone dialling and ISDN telephone dialling.

D3) ISDN charging principle is approximately equal to that of PSTN telephone service. In case of ISDN telephone, 2B+D is the basic service. Therefore monthly charge of an ISDN line is approximately twice of that of a telephone. Call by call charge is exactly the same with PSTN telephone.

D4) D70 local switch has both analog module and ISDN module. Therefore, analog subscriber telephone lines can form a key number group service together with ISDN subscriber lines. Furthermore, an ISDN telephone can handle a variety of telephone services as a PSTN telephone can do.

Even in the 21st century, PSTN service will still be the very basic and major network service. Therefore, it is indispensable to change over from PSTN to ISDN network in order to let our customers enjoy the benefits of a new digital network. To achieve this object, the most important strategy is to replace PSTN network with ISDN network while our customers do not intend to or while they do not aware of.

At present, having started 2B+D basic service three years ago, there are over 15 thousands customers. Most of their utilization are for G4 facsimile terminals and for leased lines backup of a private computer network. Required conditions of new ISDN telephone terminals are both simple and kind human interface, and merits of having ISDN terminals in their home. Now NTT (Nippon Telephone & Telegraph Corp.) started selling an economical ISDN telephone set which costs only ¥30,000. NTT is also replacing an analog pay-telephone with an ISDN digital pay-telephone in order to let the public be familiar with a digital telephone.

With the introduction of narrow band ISDN network which is based on digital network technologies, there are four principal network structure reforms which are going on or should be proceeded from now on.

T1) PSTN network will be replaced with a digital ISDN network. This will be successful if ISDN network can provide a basic telephone service without any tariff increase.

T2) Network integration is now going on between ISDN packet and DDX-P service. Now both network has a common toll stage packet network.

T3) PSTN access network capability (a gateway node) is now requested. This means PSTN (ISDN) should have enough flexibilities to accept numbering, routing and charging services requests from new network service providers.

T4) Network service competitions are now going on. Therefore there are so many new services which are under development, such as VPN (Virtual Private Network), Masscalling (Telephone broadcasting) service, etc., enhancing the technologies of database and CCS which have become very popular after the development Freedial.

These are the movements of PSTN evolution in Japan. Still there are questions the network designers should face and study.

S1) Hardware technology : Broadband ISDN and fiber optics subscriber lines.

S2) Software technology:Intelligent network technology.

S3) Network service:Personal telecommunication services.

These are key factors for the evolution of PSTN. And there are several scenarios which can be illustrated.

6. ACTUAL TRAFFIC STATUS OF JAPANESE PUBLIC NETWORK

In this section, examples of traffic characteristics are explained; firstly PSTN traffic characteristics, and secondly ISDN traffic characteristics. These figures are illustrated from the measured traffic data. It is estimated that they have the accuracy of 5% errors, and also the sampling numbers of them are limited. Therefore, one can not examine these data in detail, but the figures would be useful for getting general ideas about the public network traffic tendencies.

6.1. TOLL TRAFFIC

Now the major PSTN toll traffic is the telephone traffic. With the popularization of PC(Personal Computer) communications and facsimile communications, the traffic characteristics of PSTN are gradually changing. Figure 1 shows the share of toll traffic types between Tokyo and Osaka. Now, the share of facsimile traffic is over 15% in business hours. Japanese has Kanji characters which has the cause to promote facsimile terminal introduction rather than PC keyboard communications.

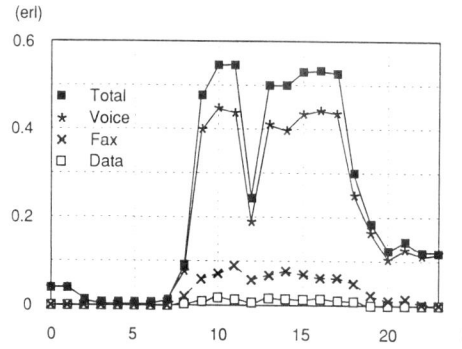

Figure 2 shows the holding time distribution of the toll traffic between Tokyo and Hakata. Here, the mean holding time of total traffic is 117 seconds and that of voice traffic is 152 seconds. The facsimile traffic has shorter mean holding time of 94 seconds.

Figure 1 Hourly variation of traffic intensity per circuit (PSTN:Tokyo ←→ Osaka)

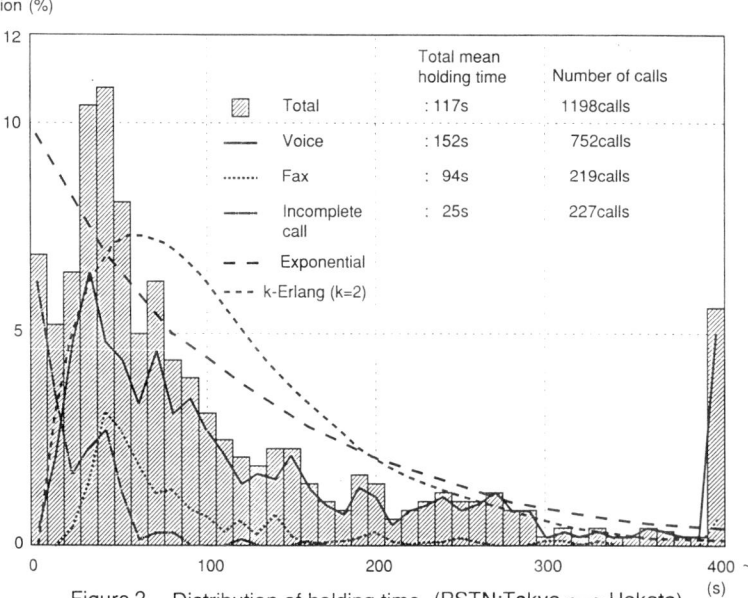

Figure 2 Distribution of holding time (PSTN:Tokyo ←→ Hakata)

6.2. ISDN TRAFFIC

Digital local switch D70 has both an analog module and an ISDN module. ISDN module is named ISM and it provides basic and primary ISDN interfaces. Figure 3 traffic data is one of the measured data of ISM traffic handling which is located in a business district.

Figure 3 shows the daily distribution of the number of calls. About 90% is the 64kbps traffic which carries G4 facsimile

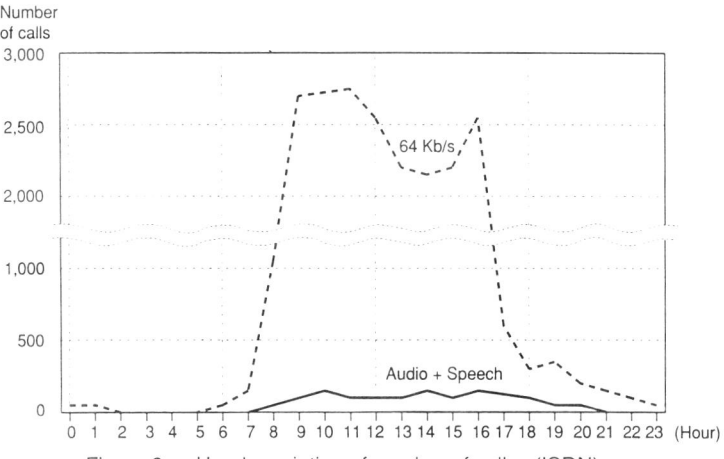

Figure 3 Hourly variation of number of calls (ISDN)

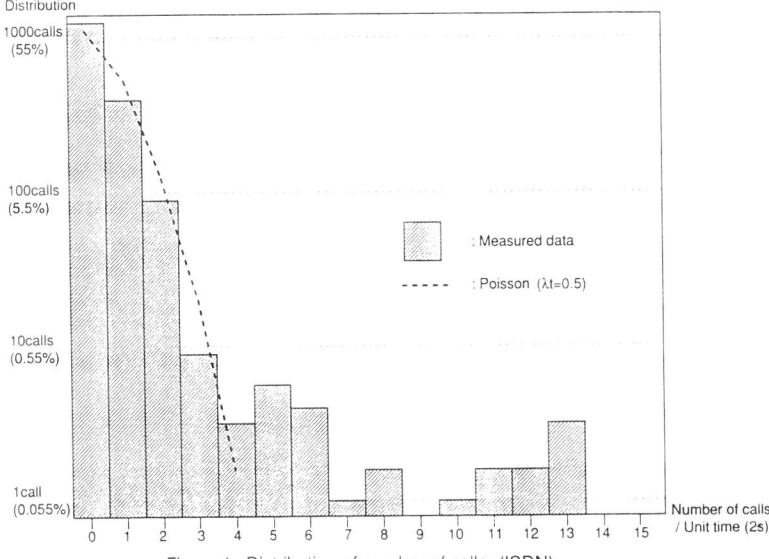

Figure 4 Distribution of number of calls (ISDN)

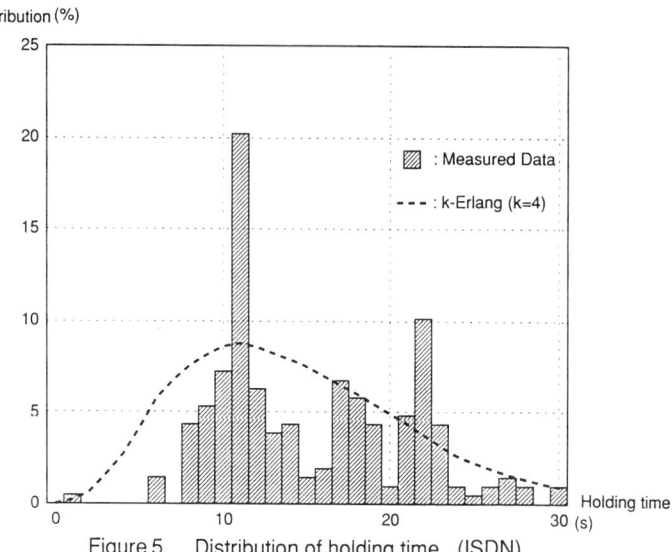

Figure 5 Distribution of holding time (ISDN)

signals. The voice traffic occupies only 10%.

The unit time distribution of call occurrence is illustrated in Figure 4. The data is obtained measuring all calls from 10:00 a.m. to 11:00 a.m.. In the module, all ISM customers have a basic I interface. Some companies have plural basic interfaces, and they operate them together for G4 facsimile transmission. This is the reason of peak numbers over 5 on the X axis, which indicate the number of simultaneous calls. It is easily expected that with an introduction of primary rate interface, the number of simultaneous call occurrence will increase.

The call holding time distribution of ISM is illustrated in Figure 5. As it is explained, most of terminals are G4 facsimile, therefore the call holding time is rather short and the graph has plural mountaintops which correspond to a number of pages.

7. PUBLIC NETWORK DESIGN SCENARIOS

7.1 FROM ISDN TO VIPA(VIRTUALLY INTEGRATED PUBLIC ACCESS)

Having experienced above mentioned movements of PSTN, it is certain that the fundamental role of PSTN is now changing. The first changing is explained with the phrase "from plain telephone service to integrated services". This changing is now in progress and another new causes of PSTN changing could be indicated as follows.
C1) New network construction.
  A typical example shown is DDX packet network constructions.

Now, there are so many movements to construct new networks based on recent social and technological tendencies.

C2) New network services introduction.
Freedial is explained as an example. Carriers are doing competition in this field, and it creates new telecommunication markets.

C3) New network capabilities will be provided for customers.
Narrowband-ISDN and Broadband ISDN are examples. In other words, multi-media services will bloom in 21st century.

C4) Multi-carrier services should be provided.
Examples are New Common Carriers, and Value Added Network providers. In Japan, there are approximately one thousand Type II carriers now.

C5) Multi-service-providers should be connected.
There will be many network service providers and information providers.

There may be many ways to judge these technical and social movements and identifies principal PSTN directions. Here, present authors judged three major points. These three items can be summarized as the development of Virtually Integrated Public Access technology.

N1) Logical subscriber lines, which realises a variety of access for all types of new network services, should be prepared. It should also be reliable access.

N2) Flexible gateway functions, which solves numbering, routing and charging problems, should be implemented for all types of network interconnection.

N3) Virtual network definition and virtual call control technology are required in order to provide simple new network introduction and to realize sophisticated call control for coming new services such as broadcasting.

PSTN designers were tends to be slow to change the fundamental design concepts of PSTN because the system is too large. PSTN fundamental architectural concept in Japan is fixed approximately 30 years ago. Through the experiences of introducing PSTN new network services, it is clearly shown that PSTN principal architectural design should be dynamic in 21st century. It is possible to say that the social circumstances, i.e., recent political and economical movements, are asking more dynamic PSTN technologies mentioned above.

7.2 FROM UNIQUE TO DIVERSE TRAFFIC DESIGN

This paper presented the fundamental PSTN design concepts and experiences. As the summary of this paper, the personal view of PSTN traffic design aspect is explained as follows.

The telephone traffic was a major traffic in PSTN last 30 years. It has not been required to consider the new service traffic characteristics because it occupied only a part of PSTN traffic. Now new traffic such as facsimile is getting larger percentages in PSTN. Therefore, from now on, it is necessary to do the research concerning these new types of traffic. In order to clarify the traffic characteristics of a new network service, the classification of PSTN structural element would be helpful. Present authors propose following 5 categories for new PSTN traffic discussion.

P1) Media type :
 Voice, Video, Data, Facsimile.
P2) Switching and transmission technology Type :
 Circuit switching, Packet switching, ATM, MAN.
P3) Communicator's type and its combination :
 Man, Woman, Host computer, Terminal.
P4) Topological pattern of path connection :
 1 to N, N to M, One way transfer, Both way transfer.
P5) Network type and its combination :
 Local, Access, Toll, Gateway.

The network type and its combination will be a new discussion item. Especially, the traffic design of an access network will be very important because PSTN traffic pattern changes from unique to diverse.

8. CONCLUSION

The experiences of introducing new network services are explained. New network design is classified into three categories, New public network construction, Access network construction and Adding new service function. The examples are DDX packet network, DDX Telephone to Packet access service and Freedial service. Following them, ISDN network design concepts are explained, together with the present traffic characteristic examples. Finally, the new direction of PSTN fundamental design concept is indicated in terms of VIPA (Virtually Integrated Public Access) by summarizing new requirements and experiences.

REFERENCES

1 NTT DATA BOOK '90, NTT, 1991
2 NTT International Symposium 90, NTT, 1990
3 ISDN System, Review of the Electrical Communications Laboratories, NTT, Vol.35 No.5 1987

TELECOMMUNICATION SERVICES - USER REQUIREMENTS

G. Alan Thorne
Telecommunication Managers' Association (TMA) at the
European Telecommunications Standards Institute (ETSI)

1. INTRODUCTION

Telecommunications services and standards have largely been set by the network providers, Users have not been involved.

Suppliers have been technology driven and due to lack of formal standards and the need to be competitive, ad hoc terminal standards have evolved. Again, users have not influenced this process.

The result is that in Europe we have an inconsistent range of telecommunication services, equipment and cost structures. This inconsistency results in additional costs to the user not only directly in the cost of terminals and the service but also in the indirect costs arising out of users spending a large amount of time attempting to overcome the inconsistencies. For example large customers with sites in many different locations have to undergo analysis and selection of equipment and services in each country when in practice they would like to carry out only one exercise.

2. USER REQUIREMENTS

2.1 General

Practical experience sugests that in respect of telecommunications equipment and services, users would ideally like to have:

- o tariffing related to cost of provision and where there is a monopoly, there should be an independent regulatory control mechanism

- o a range of telecommunications services and equipment which are interoperable and designed to match residential and business requirements

- o a choice of local, National and International suppliers of telecommunications network services and equipment

- <u>numbering as a common European/International resource</u> administered independently of any PNO. The administering body should be required to implement progressive plans to meet such future needs such as personal numbering

- suppliers of services to be licensed to meet <u>defined quality standards</u>. Such standards to include:
  - adherence to International/European technical standards
  - charging
  - availability
  - liability
  - error rates
  - common tones
  - common terminology

- the ability of users to <u>maintain their own equipment</u> and services after satisfactorily being registered by a recognised institution eg. BSI **The current UK position of not allowing users to maintain their own equipment is believed to be anti competitive and to contravene European law**

- an <u>independent appeals procedure</u> in case of dispute with the supplier of the telecommunication service

- a licence requirement which imposes <u>cross guarantee of service</u> in the event that if a PNO fails. eg. other PNOs to provide service along the lines of that in the banking and travel sectors

- 'social' services to be provided and jointly funded by all licence suppliers

- an access charge on all operations to cover the cost of rural services

- complex services which demand complete co-operation to functions such as 999, Directory Enquiry, Personal Numbering, Location Registers, should be provided through common funding

2.2 Market Sectors

Before any services can be defined and standards written it is essential to understand the basic applications to be addressed by the technology. These will vary according to market sector.

Residential

The residential non business category comprises the largest sector in any country with low utilisation of expensive local line; the UK uses less than 10 minutes per day, the US approximately 25 minutes and Japan approximately 40 minutes. This indicates that there is potential for growth in traffic even where, in Europe, there is a high penetration of lines per head of population.

The residential market buys the lowest cost product. After obtaining basic switched telephony services a significant number of residential users will be attracted by telecommunications services with an entertainment value. These value added services have been found in the US and UK to include racing information, placing bets and the ability to participate in TV game shows from the home.

Experience to date with some of the information services is not all encouraging. Home banking, and home shopping have not gained the widespread acceptance that was anticipated. This non acceptance has not been caused by any fundamental lack in technology but rather that people's requirements are largely satisfied by other competing ways of conducting their business.

Mixed Business and Residential

In Western Europe there is a growing number of people working from home who require a business line for various purposes such as appointment setting, call outs, receiving orders and/or a fax line. These options provide the user with increased security, and mobility whilst keeping them fully integrated with their company.

There is an opportunity for the telecommunication operators to work with terminal equipment suppliers to package a range of business, security and information service products for this group.

Business - Single Site

The business single-site category mainly comprises companies involved in selling to larger companies and to the local community. They require good communication links to larger companies, for example Fords will only deal with a supplier if they invest in a data link and software to Ford's specification. In this category there are opportunities for the Telecommunications Operator to offer more effective support in a growing geographical market place. It is unfortunate that this sector is often low on the network providers' priority list and they are therefore not well served. Again, it is difficult to see broadband applications other than video-conferencing being required by this sector.

Multi-Site Business

The final category - multi-site business users is the biggest revenue earner for the Telecommunications Operator and is small enough to address cost effectively.

The sector wants a cost effective service which may not necessarily be the cheapest, together with 100% availability and absolute reliability, combining such facilities as centralised operator, call transfer, telephone conferencing, call back, all across the network and a provision to contact mobile staff.

A multi site company wishes to appear as a single site company, independent of country boundaries. Internal voice and non-voice communications and customer contact should be simple.

The market place is not only Europe but the World. The widening geographical market increases the value of effective telecommunication systems.

A study carried out by ICC for the European Commission confirmed:

    o   Reliability of communications between sites is more important than with customers

    o   The telephone, fax and pc are the three most important terminals

2.3 Standards

Traditionally standards for Telecommunication systems have been set by the network operator often in the purchasing specification for products required to build the telecommunications network.
Similarly terminal equipment standards have often only been documented in purchasing contracts. Operators in each country have developed their purchasing specifications independently of each other. Technical requirements were often influenced by the equipment suppliers eg ITT and Siemens which operated across more than one country.

Additionally CEPT developed standards which could be followed by the Telecommunication operators. Throughout Europe today we therefore have telecommunication networks which, although basically follow the same technical criteria, differ in important elements. This in the long term is an unacceptable situation.

Suppliers have developed their own proprietary standards and in many instances satisfied customers' immediate needs. The question arises: can we, as an industry, develop harmonised standards suitable to business requirements in a speedy manner?

The harmonised standards for Telecommunications are now being developed by ETSI.

It is the European Telecommunications Standards Institute (ETSI) which must develop the standards to enable the EC objective of an open harmonised market to be created.

However, the user is not well represented at ETSI. Such representatives as there are, are mainly from major User Associations and can make a significant contribution.

Many User Associations have debated the Users requirements for standards. Annex A provides a list of the requirements for standards from the users point of view.

It is important that all members of ETSI and the European Commission understand the requirements of the user when developing regulations and standards.

It is the user which will dictate the success of the standards making process in their take up and use of the resulting products and services.

3. PRIORITIES

The user priorities for telecommunication services and standards do vary according to the business and its location within Europe. However the underlying priorities are:

3.1 Public Network Harmonisation

Users need a powerful and effective telecommunications network independent of national boundaries. The type of information carried over the network should fall outside the responsibility of the network provider.

The users focus on applications and not on the technology. It is therefore of a priority to work on the harmonisation of the network infrastructure where interconnection and inter-operability should really exist and enable applications to be developed suited to differing customer requirements.

The debate therefore centres round the most effective way of achieving European wide networks. Currently the majority of the effort is being placed into producing standards for the new services eg ISDN and GSM.

There are many arguments which could be tabled supporting a move towards a closer harmonisation of the existing analogue networks.

3.2 Private circuits

Major multi site businesses view a private network as extremely important to their businesses. Standards for private networks should be compatible with those for the public networks. As with public networks, country boundaries should be transparent.

3.3 Choice of terminal equipment

Choice is required in the equipment to be connected to the network. Therefore the network interface should be well defined and documented.
The same terminal interface should be available both on PBXs and the network.

3.4 Value Added Services

Users wish to be able to develop their own services for their own use, shared use and third party use. Shared and third party use of services requires common standards to be followed by system suppliers.

Users and Service providers should have the same access to the network as the network operators in respect to the utilisation of the network as a communication tool.

3.5 Alternative technologies

Users require a choice of service to provide a competitive element and true diversification. Choice can be satisfied by competitive technologies eg fixed versus mobile communications. However, real choice lies in being able to purchase the same services from more than one network provider.

In reality no one technology can satisfy all requirements. An economic solution is realised in a combination of fixed, mobile and satellite services. The different services should adhere to common standards and provide full inter-operability.

The users' attention is focussed on applications not on technology and it is therefore a priority to work on the harmonisation of the network infrastructure across Europe enabling applications to be developed, suited to differing customer requirements.

The debate therefore centres around the most cost effective way of achieving European wide networks. Currently the majority of effort is being placed in providing standards for narrow band ISDN and Pan-European mobile services (GSM).

In the area of optical fibres we appear to be technology limited. UK Cable operators are finding a combination of optical fibre, coax and twisted pair to be the most economic solution to delivering multi services to the home.

In the last few years terminals, storage and processor speeds have seen impressive advances and trends would indicate that future services could revolve around local storage, mass delivery of information (broadcast or fibre), local search, local processing, and low speed transaction processing over the network for orders.

4. USER CONCERNS

4.1 User involvement

Today the majority of businesses do not see the need to have people involved with influencing regulations or standards. Indeed there are many examples where taking part in the standards making activity has contributed to Telecommunications managers losing their jobs.

It is often argued by businesses that if they input effort to developing the right standards for their business their competitors benefit.

Conversely today a telecommunications supplier needs to develop products which match standards and have them available as soon as possible after a standard has been ratified. Failure by a manufacturer to adhere to a standard can seriously effect the marketability of a product.

4.2 Regulatory Constraints

The ability for network operators to prevent users having choice in the fixed network has been allowed through the EC Services Directive making switched telephony services a 'Reserved Service'. Any member state can now, if they wish, have a monopoly service provider for telephony

Users should consider mounting a campaign to remove the anti competitive practice of reserved services.

4.3 Mandatory standards

Mandatory Standards must be limited to the lowest Common Denominator addressing only the essential requirements for apparatus approval.

Voluntary Standards should be written to the highest common factor.

It would appear that many PTO representatives try to increase the complexity of the mandatory standards.

4.4 Opportunities

There are major investments in existing technologies. Any change has to be clearly justified and beneficial to customers and suppliers alike. The rate of change is dictated by demand and by financial and human resources. Priorities should be based on social and economic factors; significant research is required to ascertain just what should be our strategy.

To achieve these advances co-operation is vital. Competition is healthy and a fundamental requirement of our society. However only through co-operation between PTO's, terminal manufacturers and information providers will we succeed in building the total size of the market and hence the increased use of networks ensuring increased revenues and hence more funds for research and development.

To achieve users requirements, the industry quickly needs standards with the minimum amount of mandatory standards and most being voluntary. Users require common billing and documentation, one stop shopping and a code of practice on quality and availability of service irrespective of PTO or technology.

Education of the user is critical. The technologists might develop the perfect solution but unless the industry presents their particular technology in a way users can relate to and understand its value, it will not succeed. We need to establish a strategy in this area.

5. ACTION BY USERS

There are three main routes through which individual companies can have direct influence on services and standards:

o Direct negotiation with equipment and service provider

In any purchasing specification European standards can be specified. Users can gain an understanding of these from their trade association or the National Standards Organisation. To assist suppliers in developing the right products and in supporting the right standards, users and suppliers must learn to work together to identify business applications for emerging technologies.

o   Direct negotiation with Official Bodies

In the UK OFTEL is responsible for co-ordinating input on UK related standards and regulatory matters. The Department of Trade and Industry (DTI) is responsible for co-ordinating views on European Standards and Regulatory matters. It is these European rules which are progressively taking precedence over the UK laws.

The European Commission (DGXIII) in Brussels welcomes direct input on regulatory matters, and ETSI direct input on standards.

o   User Associations

There are a number of User Associations playing an active role in lobbying as well as dissemination of information relating to the many new regulations and standards being developed. The Telecommunications Managers Association (TMA), The Telecommunications User Association (TUA), ECTUA, INTUG and ITUSA in particular are promoting the users' interest. However these associations require greater support in order to provide effective lobbying at both a UK and European Level.

If users have an interest in ensuring the right products and services are made available to business enabling greater business efficiency then we would encourage them to make contact with one or more of the bodies mentioned.

6. CONCLUSIONS

European Telecommunication Standards are required in order to create European Wide networks and a competitive environment. As the Green Paper stated, the Commission aims through standards and regulations:

" To provide the European Users with a broad variety of telecommunications services on the most favourable terms"

However we are in danger of failing to achieve this goal. Users must recognise that their interests are not entirely safeguarded by the Administrations, PTOs and Manufacturers. The User Associations are using their best resources to support user interests at ETSI and in the Commission but significant extra effort is required.

Ultimately only closer co-operation between users, suppliers and regulators will enable us to wisely invest in the right mix of technology, enabling our business to prosper in the world markets and improve our own quality of life.

ANNEX A

STANDARDS THE USERS' REQUIREMENTS

The European Council for Telecommunication User Associations (ECTUA) has debated these issues at length. The consolidated view is:

o   Users require International voluntary market driven standards agreed by the largest number of interested parties in order to communicate with the largest number of other users in an open environment

o   Standardisation should result in functional performance and cost benefits

o   Standards should be justified by technical or economic factors and be timely available and open to all

o   Standards must be simple to apply and use and be well documented

o   Standards must not inhibit evolution of technology or constraints on applications

o   European standards must be developed only in the absence of International Standards

o   Standards should be voluntary and only applied when they are creating opportunities for better function, better performance, or cost reduction.

o   Standards should become mandatory only after consultation with all interest groups.

o   Technical Regulations (NETs) for assessing terminal equipment should be constrained to the essential requirements of protecting the network and user from harm and safeguarding network integrity

Network Structure and Management

A STATISTICAL STUDY OF REAL-TIME TELEPHONE TRAFFIC VARIATIONS FOR NETWORK MANAGEMENT

Daniel STERN

FRANCE TELECOM CNET/PAA/ATR
38-40 rue du General Leclerc 92131 Issy les Moulineaux FRANCE

Abstract

In this contribution, we present a method allowing the network managers to detect an abnormal situation in the network. The elaboration of a diagnosis is based on a global view of the network entities which are monitored. To this purpose, we apply mutivariate statistical analysis to data collected periodically. The paper illustrates the proposed methodology on data related to node state.

1. INTRODUCTION

Several studies are currently carried out at CNET on telephone Network Management. Among them, we are working on the problem of making diagnosis : how to detect any disturbance existing within the network, through the analysis of traffic data which are periodically collected.

Making diagnosis consists first of seaking if any monitored object : node, trunks group or traffic relation is far enough from the nominal status, second of finding out the cause of the trouble.

This paper is a contribution limited to the nodes data. It shows that this data are sufficient for making diagnosis for several types of network disturbances.

2. METHODOLOGY

The approach is based on a global view of the state of all the switches. In fact, we do not analyze the data for each switching machine separately, but rather for the whole network at once. Therefore we obtain an information which is more appropriate and more synthetic.

To this purpose, it is necessary to apply mutivariate statistical methods such as the principal component analysis ([1] and [2]), rather than univariate.

The data related to the nodes at instant t may be considered as a table where the "individuals" are the nodes and the variables are the NM indicators which are retained.

The principal component analysis (PCA) approach is particularly adapted to such a type of information because it analyses the correlations between the variables. This method allows to represent the individuals and the variables in an axis system, the principal components, which are the eigenvectors of the correlation matrix. Moreover, once the analysis is made, it is possible to project, *a posteriori*, new entities or new variables as additional elements. The analyzed data are then used as a reference and new data allow one to detect an eventual variation from the reference.

3. CASE STUDY

The data used in this study are simulated data generated by the SuperMac software [3], developed at CNET, which is a Network Management station simulator.

Among its functionalities, SuperMac includes an event-by-event network simulator which simulates calls in the network under study, according to traffic parameters and situations described in a scenario. This allows to evaluate the performances of the network under various abnormal conditions such as :

- distortion with respect to nominal traffic load,
- focused or overall overload,
- mass calling ;
- trunk group failure,
- node breakdown,
- terminal equipment failure.

The software also elaborates periodical raw measurements and CCITT parameters from all the network entities :

- node status : efficiency, occupancy, loss ;
- trunk group status : overflow, bids/circuit/hour, seizures/circuit/hour, answer bid ratio, answer seizure ratio, occupancy ;
- traffic stream status : efficiency, loss, answer seizure ratio.

We have considered the French long distance network which is a part of the whole intercity network shown in figure 1.

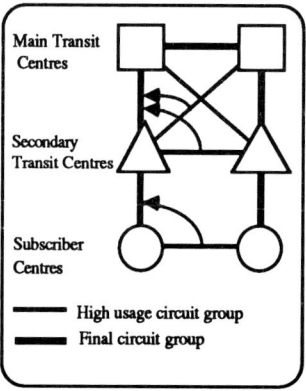

Figure 1 : French intercity network

The simulated network is a model of this network and was designed according to actual data extracted from the France Telecom databases. It has been sized using a software tool which is currently used in operational services of France Telecom for dimensioning the telephone network.

This network consists of 45 transit centres including 6 main transit centres and 39 secondary transit centres, 3 of which having also main transit functions ; they are connected by 1458 trunk groups which amount to 99000 circuits.

Four different situations have been analyzed :

* the nominal conditions which are used as the reference ;
* a local overload which amounts to 400 % : traffic grows from 47 to 235 erlangs;
* an overall overload : the whole traffic is increased by 50 % ;
* a main-transit centre breakdown.

We focused our analysis on the nodes. Data of the first scenario are used as reference: only these data are analysed ; data from the three other scenarios are projected as additional data. This allows to position on the one hand each centre regarding the position of the others nodes under nominal conditions ; on the other hand, it allows to visualize, in abnormal conditions, the deviation of each node regarding its position under nominal conditions.

As mentioned above, there are 3 parameters related to the nodes : efficiency, occupancy and loss. But, since the loss is always null in nominal conditions, this parameter cannot be retained as a variable of the analyzed data even though it is non-null in abnormal conditions.

Finally, for each scenario and every 5 minutes, we have a table composed of 45 rows : the nodes, and of 2 columns : efficiency and occupancy. In the nominal case, six subsequent tables (which amounts half an hour) were averaged, in order to smoothe stochastic fluctuations. In each disturbance scenario, we use a table containing the "raw" data, i.e. not averaged, of the first 5 minutes after disturbance beginning.

4. RESULTS

4.1. Nominal Conditions

The two variables (occupancy and efficiency) are 61 %-correlated. Since two variables are analyzed, the matrix has only 2 eigen vectors, therefore two axes amount 100 % of the variance.

Figure 2 represents the projection of data on the axes 1 (horizontal) and 2 (vertical) obtained from the PCA. The variables as well as the individuals (nodes) are projected.

The two variables determine two diagonals along which the most representative nodes are projected, these nodes are the main transit centres.

The diagonals correspond to reverse scales of occupancy and efficiency. The centres located in upper left corner are far from the variable occupancy because their values of occupancy are lower than the mean ; the centres located in lower left corner are far from the variable efficiency because their values of efficiency are lower than the mean (0.77). So MA74 MTC has a 0.42 occupancy which is the lowest value, etc... Also LL09 MTC is the only centre with an efficiency less than 1.0.

The other nodes are all projected within the axis origin : they have mean occupancy and efficiency.

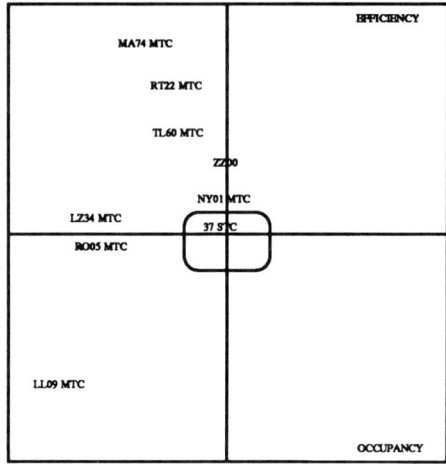

Figure 2. - Representation of Nominal Conditions -

4.2. Local Overload

Figure 3 displays the projection of data collected during the overload, as additional data, on the graph resulting from the analysis of the reference data.

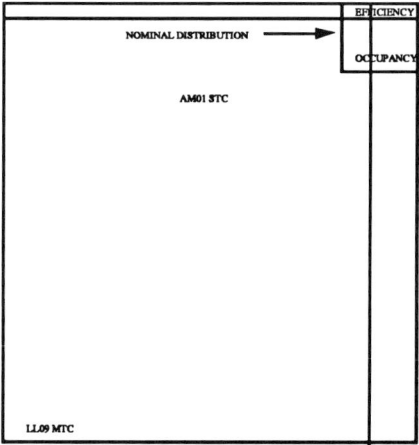

Figure 3. - Projection of Data concerning Local Overload -

The representation of the nominal situation (in the bold square) is smaller due to the variation of the position of two nodes. The positions of the other nodes remain within the nominal region, this means that the overload does not cause them any significant variations.

The two moved nodes are AM01 and LL09. As in the previous case, the positions in the lower left corner are explained by low values of efficiency : respectively 0.60, which is the minimum value and 0.94.

An important fact is that AM01 is the switching centre where the overload is originating and LL09 is its hierarchical transit centre. The graph illustrates therefore consistently the real situation.

4.3. Overall Overload

As in the previous case, Figure 4 represents the projection of data obtained during the overall overload, as additional data on the graph resulting from the analysis of the reference data. The result is an explosion of all the nodes, including the secondary transit centres, out of their reference frame.

The main transit centres are the most affected by the overload. The increase of traffic moves all the nodes diagonally towards a region of low occupancy level. The lowest centre is RT22 MTC which occupancy amounts 0.19.

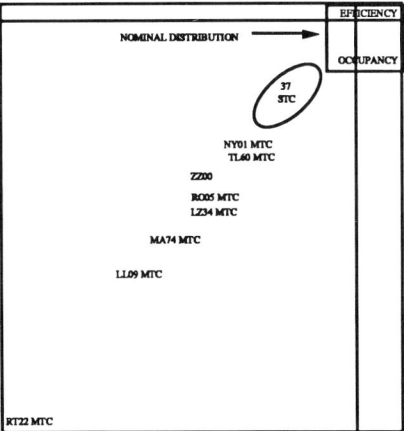

Figure 4. - Projection of Data concerning Global Overload -

4.4. Node Breakdown

Figure 5 displays the effect of the failure of a main transit centre (LL09). As in the previous cases, the nominal situation is quite reduced. The position of the affected node corresponds to the fact that its efficiency is null which is consistent with the reality. Four other main transit centres have gone out of the reference frame. This is due to the traffic lost by LL09 MTC which causes call reattempts in the network and therefore overload on these centres.

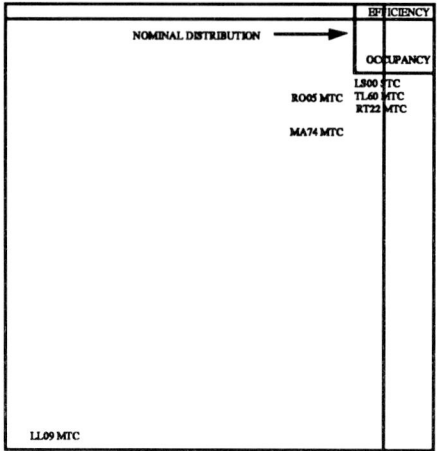

Figure 5. - Projection of Data concerning Node Failure -

5. CONCLUSION

The results presented in this contribution show that the PCA approach allows one to detect the 3 abnormal situations used as examples.

It is easy to associate a projection shape with a disturbance. First rule, the farthest centres from their reference frame are the most affected centres. Second rule, the more important is the number of centres out of the reference frame, the more severe is the disturbance. Third rule, the graphical proximities between centres correspond to their relationship in the network.

It is emphasized that few data are needed to have a correct diagnosis ; this feature is interesting since a large amount of measurements is collected for network management purposes. Applications of the PCA approach concerning other entities like trunk groups or destinations should illustrate the usefulness of such an approach.

REFERENCES

[1] K. Pearson, "On lines and Planes of Closest Fit to Systems of Points in Space", on Phil. Mag., 1901, vol. 2, n° 11, pp 559-572.

[2] H. Hotelling, "Analysis of a Complex of Statistical Variables into Principal Components" J. Educ. Psy., 1933, vol. 24, pp 417-441, pp 498-520.

[3] F. Herrmann, D. Stern, P. Chemouil, "SUPERMAC : A Software Tool For The Performance Evaluation Of Network Management", ICCC Symposium '89, On Computer Communication, Beijing, China, 1989, pp 66-69.

OFF- VS. ON-LINE NETWORK MANAGEMENT - IMPLEMENTATION AND PERFORMANCE STUDY OF CIRCUIT-SWITCHED TRAFFIC CONTROL METHODS

Edward CHLEBUS

Telecommunications Department, The University of Mining and Metallurgy, al.Mickiewicza 30, 30-059 Kraków, Poland

Abstract

The paper presents a comparative study of various traffic routing strategies aiming at overall network blocking minimization. Circuit-switched traffic is modelled by means of static and dynamic flows as well as Markov decision processes. The corresponding optimization problems are formulated and solved by methods of control theory. Various implementations of the optimal control are proposed. They are evaluated by means of simulation for the BNR-Toronto network.

1. INTRODUCTION

It is obvious that alternate routing schemes introduced in 1950's, consisting in hierarchical overflow on alternate paths according to the predetermined fixed sequence, do not suit the capabilities offered by the common channel interoffice signalling (CCIS) and the stored - program common control switching (SPC) technique. In order to fully exploit in the network management practice the flexibility provided by the technological innovations, the systematic analysis of network operation is needed.

The telecommunication network constitutes a field for the application of methods typical for control theory. However, establishing unified methodology in that domain is very difficult. It is due to the following obstacles:
- great geographical dispersion of network elements;
- discrete and stochastic activity of subscribers;
- great dimension of the object under study.

Modern trends in control theory suggest multilayer control of complex systems. That approach suits well the analysis of telecommunication network performance [4,11,13,24]. The very basic aspect of network management is the observation time scale, with respect to which we can distinguish two various levels of traffic control:
- flow control corresponding to the long-term analysis;
- short-term operation which relates to processing individual calls at network nodes.

The first control level is based on off-line statistics collected over a long period of time. The gathered data are then exploited to determine the overflow sequence which is changed several times a day. The method exploits the phase shifts between peak hours in different time zones. The necessary condition of its successful application is geographical distribution of network nodes. This is the DNHR - type approach [2,3].

The second routing concept is a consequence of more dynamic tuning of network control to changing load conditions. The right choice of instantaneous state-dependent route congestion measure determines the overflow sequence which is updated periodically with fixed frequency. The order of the update cycle length is several seconds. Such idea has been successfully adopted in the BNR-Toronto trial [5,23].

From the foregoing considerations we conclude that treatment of the optimal routing problem is not uniquely determined. Various approaches providing the common framework for the system analysis are possible. The aim of this study is to examine their usefulness for establishing routing mechanisms.

2. ROUTING OPTIMIZATION

2.1. Static model of traffic flows

Let us focus our attention on a mesh-structure network (cf. Fig 2.1).

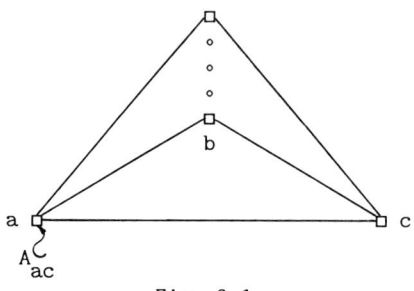

Fig. 2.1.

Traffic A_{ac} from source a to destination c can be routed over the direct link (a,c) or one of possible transit paths [(a,b),(b,c)]. To satisfy the requirements of stability transit paths composed of more than two links are not attempted.

Our aim is to minimize the total traffic loss in the whole network. That yields the quality criterion

$$J = \sum_F \sum_b \alpha_b A_{ac} B_b \qquad (2.1a)$$

subject to constraints:

$$\alpha_b \in <0,1> \qquad (2.1b)$$

$$\sum_b \alpha_b = 1 \qquad (2.1c)$$

where
- F — the set of all source-destination flows {a,c} distinguished within the network
- α_b — traffic control variable
- B_b — blocking probability of a given path between source a and destination c given by

$$B_b = \begin{cases} P_{ac} & \text{for } b = 0 \quad (2.2a) \\ P_{ab} + P_{bc} - P_{ab}P_{bc} & \text{for } b \neq 0 \quad (2.2b) \end{cases}$$

where b = 0 holds for a direct link. Links blocking probabilities P_{qr} are determined by applying the Erlang B formula.

Consider an optimum solution of the nonlinear problem

$$P_\alpha^* : J(\underline{\alpha}^*) = \min_{\underline{\alpha}} J(\underline{\alpha}) \quad (2.3)$$

where $\underline{\alpha}$ is the vector of control variables. The Kuhn-Tucker necessary conditions [25] state that for each traffic flow {a,c} there exists constant γ_{ac} such that

$$\frac{\partial J}{\partial \alpha_b^*} = \gamma_{ac} \quad \text{if } \alpha_b^* > 0 \quad (2.4a)$$

$$\frac{\partial J}{\partial \alpha_b^*} \geq \gamma_{ac} \quad \text{if } \alpha_b^* = 0 \quad (2.4b)$$

The condition (2.4) are interpreted as follows. For the optimum solution, the partial derivatives of the total network traffic loss with respect to the control variables of a given source-destination load, are equal for all routes which are used for that load and are less than or equal to the analogous derivatives calculated for the unused routes.

The Frank - Wolfe multicommodity optimization method [12,18] appears to be suitable to approach the conditions (2.4). It has been applied to improve the performance of the BNR-Toronto network operated under various load conditions [8]. Optimization runs for two cases, i.e.
- 30% total overload;
- perturbated load (i.e. offered traffic is assumed to be different from the engineered one whereas its global intensity remains unchanged and is equal to 177.5 Erlangs)

are visualized in Fig. 2.2.

In order to fix the initial flow each source-destination load A_{ac} has

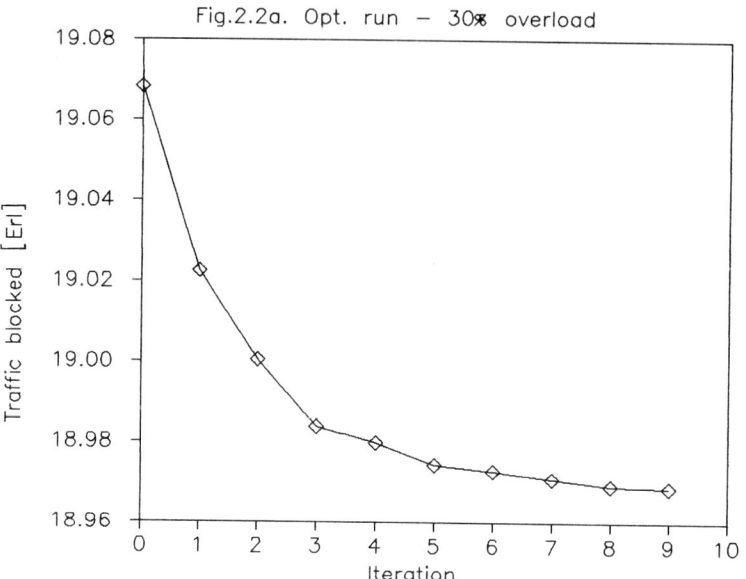

Fig.2.2a. Opt. run — 30% overload

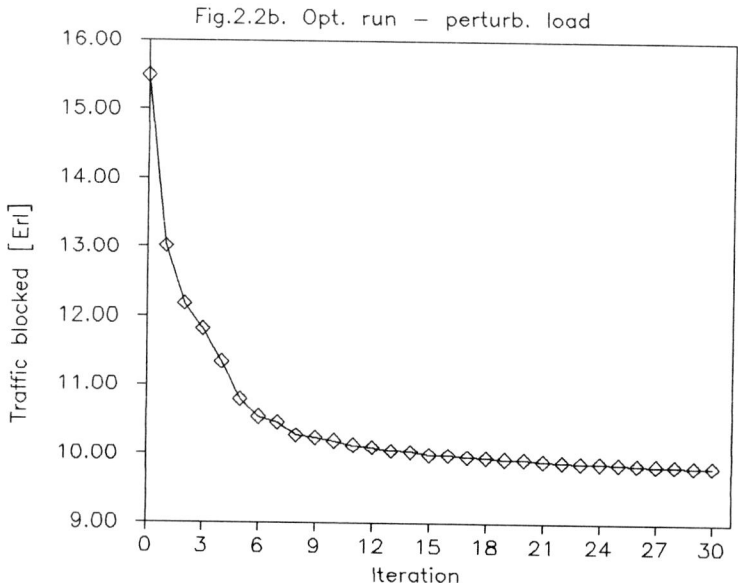

Fig.2.2b. Opt. run — perturb. load

been assigned to a direct link if it exists, otherwise traffic distribution is provided proportionally to "paths capacities" defined as

$$m_b = \min(m_{ab}, m_{bc}) \tag{2.5}$$

where m_{pq} denotes trunk group size of a link (p,q).

Optimization runs have been terminated when a difference between values of the objective function obtained in two subsequent iteration steps decreased under 0.001. An exception to that rule was made only for slow convergence of the algorithm occuring for perturbated load.

Observe, that in the first considered case an optimization gain is very small (cf. Fig. 2.2a). This can be explained as follows.

When comparing nominal traffic loads with the relevant trunk groups sizes we conclude that the Toronto network is dimensioned for high grade of service. Thus, we can not expect a significant decrease in overall network blocking by partial traffic distribution over transit paths [8].

Our case is analogous. The studied overload is uniformly dispersed and results in similar increase in traffic loss both on direct links and permissible alternative paths. That precludes possibility of transit routing on a large scale.

Partial load sharing over transit routes appears to be an efficient method of counteracting congestion when real load differs much from the one for which the network was engineered, provided that the total traffic intensity remains the same. In this case a recomputation of traffic control variables significantly improves quality of service without changing network configuration and dimensionality (cf. Fig. 2.2b).

The analysed cases have proved usefulness of the applied Frank - Wolfe algorithm as a method for network programming. The obtained results are the most spectacular for perturbated load. Note, however, that in other studied conditions considered in [8], i.e. nominal load and 100% overload of one node, a gain is also achieved despite circumstances which do not suggest traffic redistribution as a way of decreasing blocking.

2.2. Dynamic model of traffic flows

The time-dependent state evolution of the finite capacity M/M/m system is given by the nonlinear differential equation

$$\overset{\circ}{x}(t) = -G[x(t)] + A \tag{2.6}$$

where $G(\circ)$ denotes the total outflow of the system.

Very accurate approximation of $G(\circ)$ based on Newton's method of iteration has been proposed in [16]. The approach developed in [11] is much more convenient for engineering purposes. The approximation function is chosen to be

$$G(x) = \begin{cases} x & \text{for } x < m/2 \tag{2.7a} \\ m/2 - \hat{a} \ln\left(2\dfrac{m-x}{m}\right) & \text{for } m/2 \leq x < m \tag{2.7b} \end{cases}$$

where

$\hat{a}$ - parameter fixed as: $\hat{a} = 3.25 + 0.2708$ m

In order to provide the suitable description for a mesh-structure network the problem of accurate modelling of links in series appears. To tackle this problem let us consider a transit path [(a,b),(b,c)] (see Fig. 2.1)

Strictly applying Eq.(2.6) the dynamic flow model of the link (a,b) has the form [11]

$$\overset{\circ}{x}_{ab}(t) = - G_{ab}[x_{ab}(t)] + \alpha_b A_{ac} \qquad (2.8a)$$

For the tandem link (b,c) we have

$$\overset{\circ}{x}_{bc}(t) = - G_{bc}[x_{bc}(t)] + x_{ab}(t) \qquad (2.8b)$$

Observe, that the model (2.8) takes into account that traffic offered to any link of a transit path is a consequence of blocking occuring only on the preceding links in the series [13,22].

Another approach which appears frequently in the telecommunications literature [9,14,17,20,21] incorporates blocking both on up- and downward links before being offered to a given link. Under the above assumption the transit path state is modelled by a following nonlinear state equation [7]

$$\overset{\circ}{x}_b(t) = - \{ G_{ab}[x_b(t)] + G_{bc}[x_b(t)] - x_b(t) \} + \alpha_b A_{ac} \qquad (2.9)$$

By applying the Pontryagin minimum principle to the models (2.8) and (2.9) it is shown in [7,11] that the number of blocked calls is minimized if

$$\alpha_b > 0 \quad \text{for} \quad \gamma_b = \hat{\gamma} \qquad (2.10a)$$

$$\alpha_b = 0 \quad \text{for} \quad \gamma_b \geq \hat{\gamma} \qquad (2.10b)$$

where γ_b denotes a costate variable corresponding to the path [(a,b),(b,c)] and $\hat{\gamma}$ is given by

$$\hat{\gamma} = \min_b \{ \gamma_b \} \qquad (2.11)$$

When the stationary state is reached, a value of γ_b, provided that the model (2.8) is applied, is uniquely determined by (cf. [11])

$$\Omega^{(1)}(x_{ab}, x_{bc}) = 1 - \left(\frac{dG_{ab}(x_{ab})}{dx_{ab}} \frac{dG_{bc}(x_{bc})}{dx_{bc}} \right)^{-1} \qquad (2.12)$$

Analogously for the model (2.9) we obtain [7]

$$\Omega^{(2)}(x_{ab}, x_{bc}) = \frac{dG_{ab}(x_{ab})}{dx_{ab}} + \frac{dG_{bc}(x_{bc})}{dx_{bc}} \quad (2.13)$$

The approximation (2.7) for congested links, i.e. $x_{pq} \geq m_{pq}/2$, gives

$$\Omega^{(1)}(x_{ab}, x_{bc}) = 1 - \frac{m_{ab} - x_{ab}}{\hat{a}_{ab}} \cdot \frac{m_{bc} - x_{bc}}{\hat{a}_{bc}} \quad (2.14)$$

and

$$\Omega^{(2)}(x_{ab}, x_{bc}) = \frac{\hat{a}_{ab}}{m_{ab} - x_{ab}} + \frac{\hat{a}_{bc}}{m_{bc} - x_{bc}} \quad (2.15)$$

2.3. Markovian model

In order to minimize the number of rejected calls we minimize carried traffic taken with the minus sign. The cost corresponding to the stationary policy $\underline{\alpha}(x)$ is given by

$$J[\underline{\alpha}(x)] = \pi(x) \, k(x) \quad (2.16)$$

where

$\mathbf{x} = [x_1, x_2, \ldots, x_l, \ldots, x_L]$ is a state of the network determined by the occupancy of all paths distinguished between origin-destination pairs; $\pi(x)$ denotes probability to be in state x

All elements of the column vector $\mathbf{k}$ are determined

$$k(x) = -\sum_l x_l \quad (2.17)$$

The unknown optimal control vector which we look for

$$\underline{\alpha}^*(x) = [\alpha_1^*(x), \alpha_2^*(x), \ldots, \alpha_l^*(x), \ldots, \alpha_L^*(x)] \quad (2.18)$$

can be obtained by solving the following functional equation [15]

$$J[\underline{\alpha}^*(i)] = \min_{\underline{\alpha}(i)} H_i[\underline{\alpha}(i), v] \quad \bigwedge_{i=1, \ldots, n} \quad (2.19)$$

H_i denotes the Hamiltonian functional given by

$$H_i[\underline{\alpha}(i), v] = k(i) + \sum_{j=1}^{n} Q_{ij}[\underline{\alpha}(i)] v_j \qquad (2.20)$$

where
- n - size of the state space
- v - the optimal dual vector
- Q - the transition matrix

Introducing various approximations to the Hamiltonian functional (cf. [10]) we obtain that the following path "length" measures

$$\Omega^{(3)}(x_{ab}, x_{bc}) = y_{ab} \left[\frac{E(m_{ab}, y_{ab})}{E(x_{ab}, y_{ab})} - 1 \right] +$$

$$+ y_{bc} \left[\frac{E(m_{bc}, y_{bc})}{E(x_{bc}, y_{bc})} - 1 \right] \qquad (2.21)$$

and

$$\Omega^{(4)}(x_{ab}, x_{bc}) = \frac{E(m_{ab}, y_{ab})}{E(x_{ab}, y_{ab})} + \frac{E(m_{bc}, y_{bc})}{E(x_{bc}, y_{bc})} \qquad (2.22)$$

where
- $E(\circ, \circ)$ - the Erlang B formula
- y_{qr} - traffic offered to a link (q,r)

must be minimized. Note, that the congestion index (2.22) was in a different way derived in [19].

3. IMPLEMENTATIONS

The main interest of a network supervisor centers on finding rules which apply to routing of single calls according to a predefined quality criterion. In the following we discuss how to implement solutions provided by the presented theoretical considerations.

3.1. Off-line control

In order to solve the optimization problem (2.3) the traffic specification is reqiured, which can be obtained from measurements performed over a long period of time. It is obvious that traffic intensity fluctuates during a day. To find a set of optimal controls α^* related to periodically changing load conditions the multicommodity optimization method needs to be applied many times to each one of the traffic matrices.

3.1.1. Random Load Sharing (RLS)

Assume, that multiple overflows are allowed. Consider the k-th switching attempt. The probability that an entering call will be directed to the b-th route is given by

$$\alpha_b^{(1)} = \alpha_b^* \tag{3.1a}$$

when we take $k = 1$ or generally

$$\alpha_b^{(k)} = \frac{\alpha_b^*}{1 - \sum_{n=1}^{k-1} \alpha(n)} \tag{3.1b}$$

where $\alpha(n) = \alpha_r^*$ means that the r-th route tried in the n-th attempt was busy. This idea is implemented using a random number generator with samples uniformly distributed over the interval (0,1).

3.1.2. Sequential Routing (SR)

The other implementation is a simple consequence of the following reasoning arguments. It is obvious that the greater the control variable α_b^* the greater the probability of successful call completion on the b-th route. That suggests the idea of Sequential Routing (SR) which is well known from the previously cited DNHR implementation.

All routes are ordered according to the decreasing values of α_b^*. They are tried by a call starting with that one which leads via a transit node

$$\hat{b} = \arg [\sup_{b} \{ \alpha_b^* \}] \tag{3.2}$$

Note, that the foregoing class of routing schemes does not take into account an instantaneous network state. The policies established in the sequel avoid that weakness.

3.2. On-line control

Consider the state-dependent congestion measures (2.12) and (2.13). Observe, that these formulas can be used to determine a succesion of routes linking a given source-destination pair. An overflow sequence, updated periodically, provides the optimal routing adaptation to changing load conditions. In order to avoid significant redistribution of traffic, a direct link, if exists, is always favoured independently of its current congestion. Transit paths are attempted when all trunks of a direct link are busy. They are tried starting with the shortest one in the sense of "length" defined by (2.12) and (2.13).

That algorithm is similar in nature to the policy used in the BNR-Toronto trial [5,23] except that the formula for a route quality (assuming no prediction mechanism and no protection level) was then chosen to be

$$\Omega(x_{ab}, x_{bc}) = \min(m_{ab} - x_{ab},\ m_{bc} - x_{bc}) \qquad (3.3)$$

Note, that the proposed procedure is convenient for real-time control since prior traffic specification is not required.

3.3. Two-layer network control

Observe, that approaches to traffic control presented so far follow various information patterns and are rather complementary than opposite in nature.

A scheme that successfully brings together their benefits has been proposed in [1]. It works as follows. The first choice path is determined after the formula (3.2) by the higher-layer optimization algorithm operating with respect to long-term time horizon. If the first path is busy, the successive paths are selected from the list of other paths on the basis of dynamic congestion index being valid in the lower-layer. Hence, such advanced approach to traffic routing is a hybrid of previously presented off- and on-line control methods.

Now we investigate two-layer network management in more detail. The results of Section 2.3 are very suitable for implementation of this concept.

Note, that the optimal control variables $\underline{\alpha}^*$ uniquely determine the following vectors:

$\underline{x}^*$ - optimal network state

and

$\underline{y}^*$ - optimal fictitious traffic offered to each link

They are reported in [8].

Let us try to use these optimal parameters for routing purposes. The idea of keeping an instantaneous network state $\underline{x}$ in close proximity to the optimal one $\underline{x}^*$ suggests itself. When transit paths are considered, the following heuristic measure estimates a deviation from the required state

$$\Omega^{(5)}(x_{ab}, x_{bc}) = x_{ab}^* - x_{ab} + x_{bc}^* - x_{bc} \qquad (3.4)$$

Application of the optimal vector y^* for traffic routing is not straightforward. In order to exploit it we adopt the formulas (2.21) and (2.22). Observe, that to calculate their numerical values information on traffic y_{qr} offered to each link (q,r) must be known in advance. Unfortunately, this parameter is not provided straightway by the load specification matrix. That is so, because the traffic corresponding to each source-destination pair is split among several paths. Thus, the load of a given trunk group is composed of contributions from different flows that use it. In order to omit that difficulty we make the following heuristic assumption [19]

The traffic allocation which is determined by the solution of the nonlinear multicommodity flow problem provides a suitable approximation to an allocation which is realized under optimum state-dependent routing.

Hence, we propose to use Eqs. (2.21) and (2.22) taking

$$y_{qr} = y_{qr}^* \qquad (3.5)$$

Now we can apply these formulas to real-time control.

3.4. Routing algorithm

In order to use the path congestion measures (2.12), (2.13), (2.21), (2.22), (3.3) and (3.4) for routing purposes we pursue the idea of the previously mentioned BNR-Toronto experiment. Given a vector of a current network state consider the following routing algorithm:

Step 1: Select one of the following path "length" measures: $\Omega^{(i)}$, $i=1,2,3,4$ (Ω, $\Omega^{(5)}$)

Step 2: Every update period T for each traffic flow {a,c} calculate the "length" of all possible transit paths [(a,b),(b,c)].

Step 3: Order them according to the increasing (decreasing) "length".

Step 4: A call overflowing from a direct link (a,c) forward to the destination node c in succession determined in Steps 1-3 starting with the "shortest" ("longest") transit path.

The order of the update cycle length T is assumed to be several seconds.

4. SIMULATION EXPERIMENT

In order to investigate efficiency of all developed concepts they have been implemented for the previously mentioned BNR - Toronto network

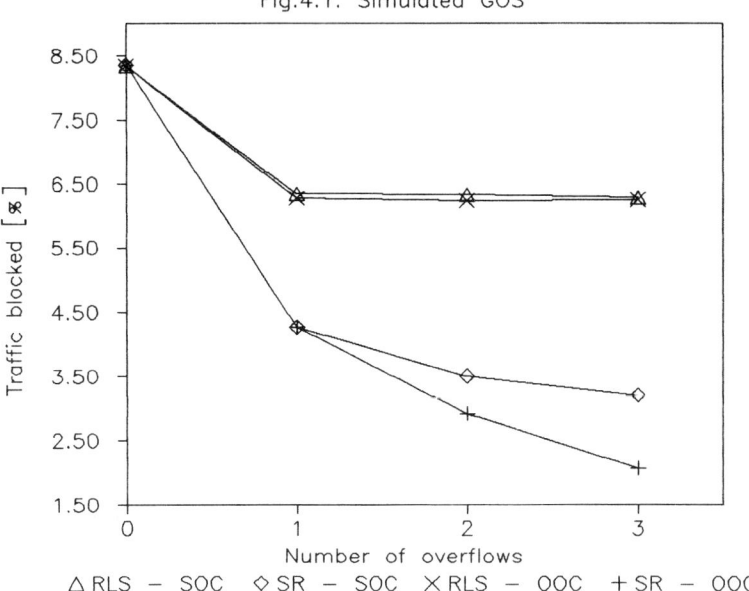

Fig.4.1. Simulated GOS

△ RLS — SOC   ◇ SR — SOC   × RLS — OOC   + SR — OOC

operated under traffic 30% higher than nominal. The network specification is reported in [5,6].

Performance of the schemes has been investigated under the following assumptions. 60 000 seconds of network operation have been simulated. For the first 10 000 seconds the network was reaching the steady state. The statistics were collected in the time period (10 000, 60 000) s.

For each simulation run we evaluated network grade of service (GOS) as a function of the following factors:

- signalling technique (both Sequential Office Control (SOC) and Originating Office Control (OOC) systems have been implemented);

- number of admissible overflows;

- update cycle length T (only for the schemes which use periodically modified routing tables);

The results of simulation are depicted in Fig. 4.1 - 4.3.

5. CONCLUSIONS

The systematic analysis of the simulation results suggests the following remarks:

■ Random Load Sharing (RLS) is the optimal first choice policy. Note, however, that increasing the number of admissible overflows we can not obtain a significant improvement of service quality.

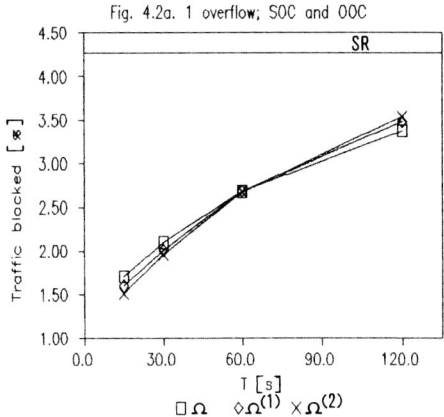

Fig. 4.2a. 1 overflow; SOC and OOC

□ Ω ◇$\Omega^{(1)}$ ×$\Omega^{(2)}$

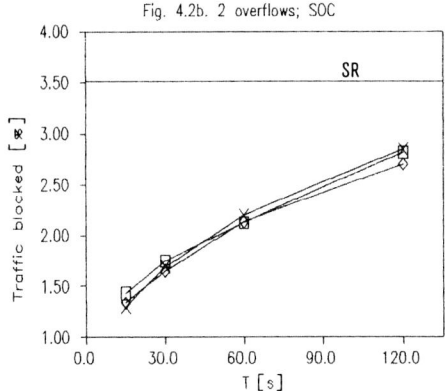

Fig. 4.2b. 2 overflows; SOC

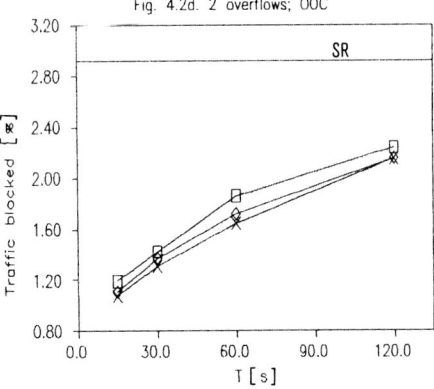

Fig. 4.2d. 2 overflows; OOC

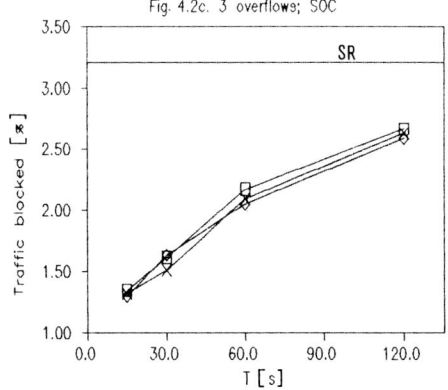

Fig. 4.2c. 3 overflows; SOC

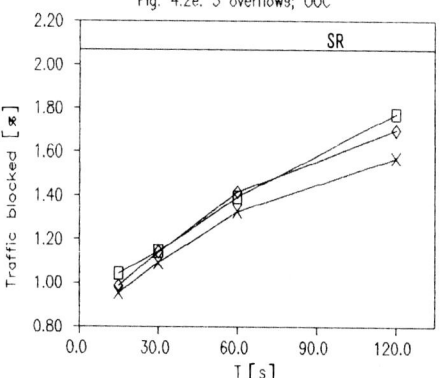

Fig. 4.2e. 3 overflows; OOC

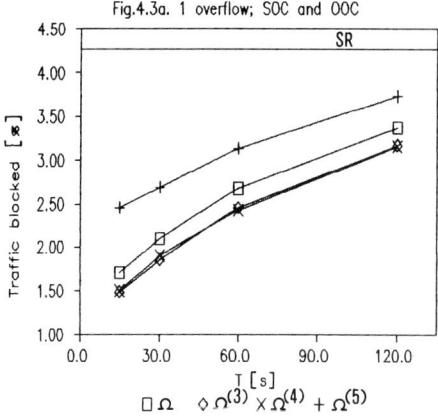

Fig.4.3a. 1 overflow; SOC and OOC

□ Ω ◊ $\Omega^{(3)}$ × $\Omega^{(4)}$ + $\Omega^{(5)}$

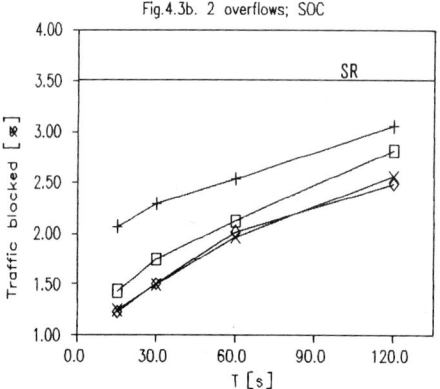

Fig.4.3b. 2 overflows; SOC

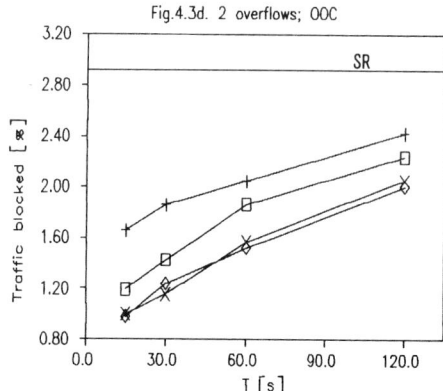

Fig.4.3d. 2 overflows; OOC

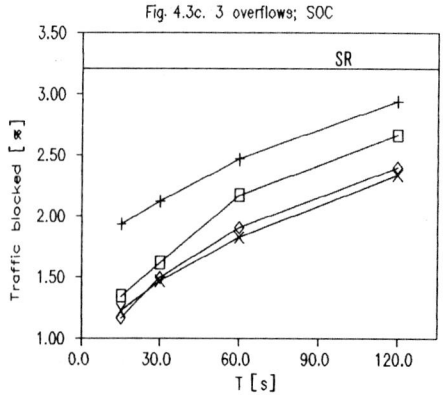

Fig. 4.3c. 3 overflows; SOC

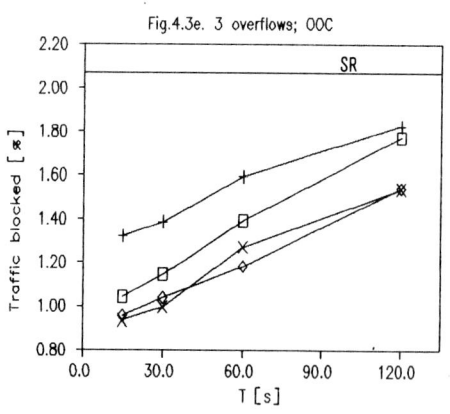

Fig.4.3e. 3 overflows; OOC

- Sequential Routing (SR) gives satisfactory results. The greater the number of admissible overflows the better the routing efficiency.

- All the policies using periodically modified routing tables work better than the off-line algorithms - RLS and SR. They fully reveal their advantages for short reconfiguration times and many admissible overflows. This is consistent with the intuition. When the update cycle length is of the order of the mean holding time τ, performance of those schemes is comparable with SR.

- The metrics $\Omega^{(3)}$, $\Omega^{(4)}$ are the best of all investigated congestion measures. They result in very efficient dynamic routing schemes.

- SOC and OOC signalling techniques give the same results in the following cases:
  - no overflow is assumed;
  - one overflow is admitted under an assumption that the direct link is the first choice route.

  Simulation confirms that conjecture. In other cases, OOC signalling provides more efficient utilization of network resources.

Summarizing the results of the study we must lay stress on the importance of traffic modelling problems and following implementation issues. In the paper we have evaluated various implementation proposals. Observe, that optimal control aiming at network blocking minimization can be implemented in many ways resulting in quite different network performances.

REFERENCES

[1] Ash G.R.: Use of a trunk status map for real-time DNHR, Proc. of Int. Teletraffic Congress, ITC-11, Kyoto, 1985
[2] Ash G.R., Cardwell R.H., Murray R.P.: Design and optimization of networks with dynamic routing, BSTJ, 60 (8), 1981, 1787-1820
[3] Ash G.R., Kafker A.H., Krishnan K.R.: Servicing and real-time control of networks with dynamic routing, BSTJ, 60 (8), 1981, 1821-1846
[4] Ash G.R., Chemouil P., Kashper A.N., Katz S.S., Yamazaki K., Watanabe Y.: Robust design and planning of a worldwide intelligent network, IEEE JSAC, 7 (8), 1989, 1219-1230
[5] Cameron W.H., Galloy P., Graham W.J.: Report on the Toronto advanced routing concept trial, Proc. of "Networks '80" Conf., Paris, 1980
[6] Chemouil P., Filipiak J., Gauthier P.: Analysis and control of traffic routing in circuit-switched networks, Computer Networks and ISDN Systems, 11 (3), 1986, 203-217
[7] Chlebus E.: A dynamic flow model of a transit path and its application to adaptive routing in telephone networks, to appear in Information and Decision Technologies
[8] Chlebus E.: Application of optimization techniques to the design of routing control systems in circuit-switched networks, Ph.D. dissertation, Telecommunications Department, The University of Mining and Metallurgy, Cracow, 1990

[9] Chlebus E.: Validation of models for a telephone network employing routing over transit paths, to appear in Applied Mathematical Modelling
[10] Chlebus E., Bernussou J.: State-space reduced optimal routing in circuit-switched networks, accepted for presentation at Int. Teletraffic Congress, ITC-13, Copenhagen, 1991
[11] Filipiak J.: Modelling and Control of Dynamic Flows in Communication Networks, Springer Verlag, New York, 1988
[12] Frank M., Wolfe P.: An algorithm for quadratic programming, Nav. Res. Log. Quart., 1956, 95-110
[13] Garcia J.M., Hennet J.C., Titli A.: Optimization of routing in interurban telephone networks, Large Scale Systems, 2, 1981, 257-267
[14] Girard A., Côté Y.: Sequential routing optimization for circuit-switched networks, IEEE Trans. Commun., 32 (12), 1984, 1234-1242
[15] Howard R.A.: Dynamic Programming and Markov Processes, The M.I.T. Press, Cambridge, Massachusetts, 1960
[16] Jagerman D.L.: Methods in traffic calculations, AT&T BLTJ, 63 (7), 1984, 1283-1310
[17] Kelly F. P.: Routing in circuit-switched networks: optimization, shadow prices and decentralization, Adv. Appl. Prob., 20, 1988, 112-144
[18] Kennington J.L., Helgason R.V.: Algorithms for Network Programming, John Wiley & Sons, New York - Chichester - Brisbane - Toronto, 1980
[19] Krishnan K.R., Ott T.J.: State-dependent routing for telephone traffic: theory and results, Proc. of Conf. on Decision and Control, CDC-25, Athens, 1986
[20] Le Gall F., Bernussou J.: An analytical formulation for grade of service determination in telephone networks, IEEE Trans. Commun., 31 (3), 1983, 420-424
[21] Lin P.M., Leon B.J., Steward C.R.: Analysis of circuit-switched networks employing originating office control with spill-forward, IEEE Trans. Commun., 26 (6), 1978, 754-765
[22] Schneider K.S., Minoli D.: An algorithm for computing average loss probability in a circuit-switched communication network, IEEE Trans. Commun. 28 (1), 1980, 27-32
[23] Szybicki E., Bean A.E.: Advanced routing in local telephone networks; performance of proposed call routing algorithms, Proc. of Int. Teletraffic Congress, ITC-9, Torremolinos 1979
[24] Winnicki A., Paczyński J.: An approach to design of three-layer controlled telephone networks, Large Scale Systems, 1, 1980, 245-256
[25] Zangwill W.I.: Nonlinear Programming: A Unified Approach, Prentice - Hall, Englewood Cliffs - New Jersey, 1969

BEHAVIOR CHARACTERIZATION OF ALTERNATE ROUTING IN A NON-HIERARCHICAL HOMOGENEOUS NETWORK

Yoshikuni Onozato[+], Jaidev Kaniyil[+] and Shoichi Noguchi[++]

[+]Department of Communications and Systems, University of Electro-Communications, Tokyo 182, Japan.

[++]Research Center for Applied Information sciences, Tohoku University, Sendai 980, Japan.

ABSTRACT

It is well known that the alternate routing schemes in non-hierarchical networks exhibit a sudden behavioral change under increasing loads. This change in the behavior had been earlier characterized by the fold catastrophe. This paper extends the results reported in reference [1]. It is shown that at higher loads there exist two distinct regions in each of which Liapunov functions exist. The Liapunov functions are identified with respect to the dynamic flow balance functions. Two equilibrium states can be identified one of which corresponds to high blocking while the other corresponds to low blocking. The dynamics of the system around the fold point is studied by simulations. When the load is increased beyond the fold point, the system undergoes oscillations between the two equilibrium states. At higher input rates, the system finds itself placed in the equilibrium state with higher blocking. This provides another qualitative interpretation of the behavioral change in the alternate routing schemes in homogeneous non-hierarchical networks.

I INTRODUCTION

Alternate routing schemes are incorporated in both hierarchical and non-hierarchical networks to increase the network efficiency and utilization. When the desired route is blocked, alternate routing schemes select the next desirable (alternate) route. In hierarchical networks, alternate routes through the links higher in the hierarchy are selected only if those through links lower in the hierarchy are blocked. Alternate routing schemes in the non-hierarchical networks are more complex as there exist too many candidate routes. In practice, only a few of the possible alternate routes would be attempted.

Nakagome and Mori [2] and Mori and Nakagome [3] predicted the behavioral properties of such a scheme by quantitative analyses. It was found that, under light loads, the blocking probability approaches zero. This situation was identified as the 'normal state' of the system. Under heavy loads, the blocking probability increases drastically. This situation was identified as the 'congestion state'. While the load increases, at a certain load, there is an abrupt transition from the normal state to the congestion state. Krupp [4] and Akinpelu [5,6] further studied the congestion aspects by simulation.

Most of the approaches towards illustrating the presence of congestion have been by taking recourse to the quantitative performance estimates. The initiation of the congestion is said to occur when the direct routed calls suddenly decrease. An effective characterization of the interplay between the structural properties and the control parameters had been forbidding with the conventional mathematical tools like queuing theory and Markov process modeling. Since, most often, a high degree of idealization is involved in the modeling for the quantitative estimation of parameters, the accuracy of prediction of congestion directly depends on how close the idealization is with respect to the actual system. Approximating the blocking by the Erlang B formula, references [2 and 3] probed for the performance issues of non-hierarchical alternate routing networks. These results provide the motivation to probe further for the details on the internal mechanism of congestion that would afford a better understanding of the conditions leading to it.

In this paper, it is demonstrated that the manifestation of congestion is considered to be a structural property of the system. The behavioral characteristics of non-linear systems are often estimated by inspecting the dynamics of the equilibrium state of the system under perturbation. 'Summarizing Functions' [7] indicate whether the system is stable or not. One such summarizing function is the Liapunov function [7]. Alternatively, the concept of 'Potential Function' has helped in the characterization of structural stability issues of a wide class of interdisciplinary dynamic systems [8,9]. The behavior pattern of these dynamic systems can be conceived to be the property of the potential function possessed by the system. We identify a suitable potential function of the alternate routing scheme under consideration. Once a potential function is obtained, the results from catastrophe theory can be applied on it. In general, a system whose potential function is a Liapunov function is a stable one.

The study reported in this paper is an extension of the work reported in reference [1]. From the dynamic flow balance equation, we identify the Liapunov functions. These functions suggest the existence of the potential function. The results presented in this paper can be enlisted as follows:

   i)   A model, under the concept of 'effective input rate' and 'effective throughput rate', of alternate routing scheme in a homogeneous symmetrical network is formulated with respect to the state equation. The blocking probability is considered as the state variable.

ii) The integral Liapunov function [7] of the system, which specifies the dynamics of the perturbed system, is identified from the dynamic flow equation. The integral Liapunov function is extended so as to obtain the potential function throughout the state space. Upon disturbance, the internal mechanism governing the system state trajectories between equilibrium states can be considered to be related to the potential function of the system.

iii) The results from catastrophe theory [8-11] can be applied on the model to estimate the value of the control parameter exceeding which congestion springs up. The model affords a better understanding of the phenomenon of congestion.

iv) The previously reported results can be interpreted in the light of structural changes of the system.

In section II, the formulation of the state equation is discussed. Thereafter, in section III, the analysis is carried out with respect to the stability aspect of the system. Initially the Liapunov function is found out. A potential function is thereupon identified after which, as reported in ref. [1], the results from catastrophe theory is employed to characterize the behavior. In section IV, a numerical example is discussed. Here, the system behavior is simulated. Our conclusions are noted in section V.

II STATE EQUATION

A. State dependent carried traffic

We consider a homogeneous fully connected network with N nodes. At any given node i, generation of new messages directed towards each of the remaining N-1 nodes is assumed to be a Poisson process with mean rate λ; thus, the total traffic generated at any one node constitutes a Poisson process with the aggregate mean rate $(N-1)\lambda$. A link services messages according to exponential distribution with mean μ. The node attempts to transmit a newly incoming message over that single link which connects the origination and the destination nodes. If this direct link is blocked, the message is attempted to be sent over any of the N-2 alternate paths consisting of two links. Alternate paths through more than two links are not considered. We also assume that all of the links have the same blocking probability b. The model of the links outgoing from an arbitrary node i is illustrated in Fig. 1 as per the following details.

The messages originating at the node would be blocked over the direct route with a probability b. These blocked messages would be alternate-routed. Let the alternate routing be successful with a probability P_s. Then, P_s can be calculated as:

$$P_s = 1-(1-s^2)^{N-2} \qquad \ldots\ldots (1)$$

So, the traffic originated from the node i which is denoted by λ_{io} is:

$$\lambda_{io} = \lambda(N-1)(1-b) + \lambda(N-1)bP_s \qquad \ldots (2)$$

There will be alternate routed messages from nodes m, m≠i, through the node i. The amount of the alternate routed traffic through the node i, denoted as λ_{ia}, can be calculated as:

$$\lambda_{ia} = \lambda(N-1)bP_s \qquad \ldots (3)$$

Hence, from eqs. (2) and (3) the effective traffic rate, denoted as $\tilde{\lambda}$, accessing the link group successfully from node i is:

$$\tilde{\lambda} = \lambda_{io} + \lambda_{ia}$$
$$= \lambda(N-1)(1-b+2bP_s) \qquad \ldots (4)$$

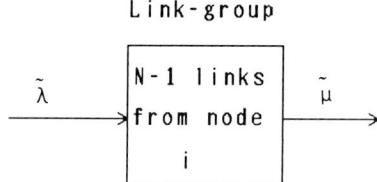

Fig. 1 Flow through the link-group associated with the node i

Our model should be valid at higher traffic intensities at which the abrupt changes in the behavior of the system has been found by the analysis of refs. [2-6]. At these higher traffic intensities, all of the links would be busy. The effective service rate of the (N-1) links of the link-group, denoted by $\tilde{\mu}$, is:

$$\tilde{\mu} = (N-1)\mu \qquad \ldots (5)$$

Note that the above equations have been employed in the analysis reported in ref. [1]. We would now model the flow of the traffic at the link group by the dynamic flow principles [12].

$$\begin{Bmatrix} \text{Rate of growth} \\ \text{of average number} \\ \text{of entities in} \\ \text{the link group} \\ \text{at time t} \end{Bmatrix} = \begin{Bmatrix} \text{Average rate} \\ \text{of traffic} \\ \text{arrivals at} \\ \text{the link group} \\ \text{at time t} \end{Bmatrix} - \begin{Bmatrix} \text{Average rate} \\ \text{of traffic} \\ \text{departures} \\ \text{from the link-} \\ \text{group at time t} \end{Bmatrix} \quad \ldots (6)$$

Let $\bar{x}$ denote the average amount of the entities (carried traffic) within the link-group. Then, the state equation of the model can be formulated as:

$$\frac{d\bar{x}}{dt} = \tilde{\lambda} - \tilde{\mu} \qquad \ldots (7)$$

From eqs. (4), (5) and (7),

$$\frac{d\bar{x}}{dt} = F(b) \qquad \ldots (8)$$

where

$$F(b) = \lambda(N-1)(1-b+2bP_s) - (N-1)\mu \qquad \ldots (9)$$

Fig. 1 shows the model for the dynamic flow at the node i. Fig. 2 indicates the values of F(b) for various values of ρ, where ρ is $\lambda/\tilde{\mu}$. In view of eqs. 4, 5 and 7, we consider that b, instead of $\bar{x}$, designates the state of the system. We will now proceed to formulate the state equation with respect to the state variable b. We may represent db/dt as follows:

$$db/dt = (d\bar{x}/dt) / (d\bar{x}/db) \qquad \ldots (10)$$

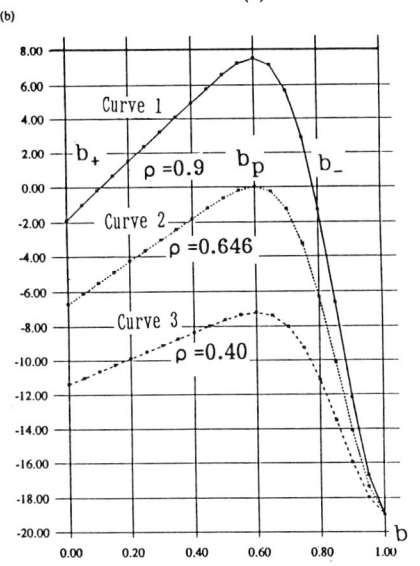

Fig. 2 (a) F(b) for various values of ρ (N=20)

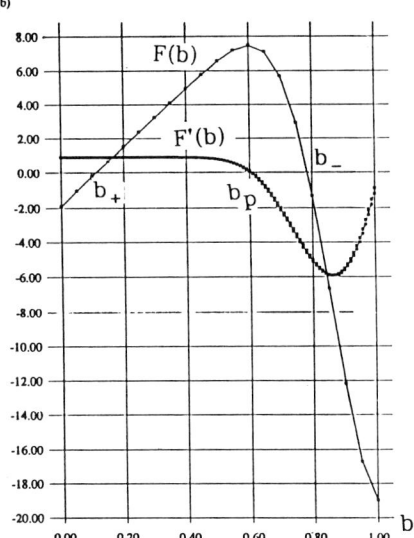

Fig. 2 (b) F(b) and F'(b): (ρ=0.9)

Fig. 2 Shape of F(b): (N=20)

From Little's result (see, [12, 13]),

$$\bar{x} = \tilde{\lambda}/\tilde{\mu} \qquad \ldots (11)$$

Since $\tilde{\mu}$ is independent of b, we may express $d\bar{x}/db$ as:

$$d\bar{x}/db = F(b)/db = F'(b) \quad \ldots\ldots (12)$$

Fig. 2 (b) shows the relation between $F(b)$ and $F'(b)$. Then the state equation of the model is formulated as:

$$db/dt = f(b), \quad \ldots\ldots (13)$$

where,

$$f(b) = F(b)/F'(b) \quad \ldots\ldots (14)$$

The above equation represents a time invariant equation (see [7] for details on time invariant equations).

III ANALYSIS FOR STABILITY

A. Identification of a Potential Function

It can be noted from eq. (9) that $F(b)$ is a single-valued and continuous function of b, $0 \leq b \leq 1$ and that $\partial F/\partial b$ is positive at b=0 and negative at b=1. Since $F(b)$ is expressed in various powers of b, maximum of which is N, the states at which $F(b)=0$ cannot be analytically found. The shape of $F(b)$ is important for our analysis. In the low blocking region, the shape of the curve is concave. The shape changes from concave to convex at $b=\{2(N-1)-\sqrt{2(N-1)}\}/(2N-3)$ onwards. Also, it is found that $F(b)$ would have only one peak. In Fig. 2, $F(b)$ at various values of ρ, for a 20 node network, is shown. It can be noticed that for high values of ρ, $F(b)$ has positive values for some range of b (see, curve 1); but at lower ρ's, the situation is different. When ρ is reduced, at $\rho=0.646$, the maximum value of $F(b)$ is zero (see, curve 2). When ρ is still reduced, the value of $F(b)$ is always negative (see, curve 3). Since we are interested in the high traffic situations, firstly we will consider the curve 1 and then generalize our results to be applicable to curves 2 and 3. Let b_p denote the value of b at which the maximum of $F(b)$ occurs. The equilibrium states of the system are those states at which $F(b)=0$. Thus, b_+ and b_- represent the only two equilibrium states of the system; the slope of $F(b)$ is positive at b_+ and negative at b_-. Obviously, when the system finds its equilibrium state at b_-, there exists a higher degree of blocking in the system. In this state, as has been observed in refs. [2-6], the alternate routed calls increase at the cost of direct routed calls. Thus most of the successful calls would engage two links thereby decreasing the throughput. The following lemma predicts the stability aspects with respect to the equilibrium states:

Lemma: An Integral Liapunov Function $V_L(b)$ can be constructed in the region $b_p \leq b \leq 1$ as follows:

for $b_p \leq b \leq 1$,

$$V_L(b) = -\int_{B=b_p}^{b} F(B)dB + C, \quad \ldots(15)$$

where C is a constant given by the following relation:

$$C = -\int_{B=0}^{b_p} F(B)dB. \quad \ldots(16)$$

The above function V_L completely describes the system dynamics in the region $b_p \leq b \leq 1$. The purpose of adding the constant C to the integration is to facilitate the potential function of the system to be discussed later.

The evolution of the system at states other than the equilibrium states would be to locally minimize the function V_L. In the region $0 \leq b < b_p$, it can be proved that $\hat{V}_L$, as given by the following equation,

$$\hat{V}_L = -\int_{B=0}^{b} F(B)dB, \quad 0 \leq b \leq b_p, \quad \ldots(17)$$

is an integral Liapunov function; the function has a local minimum at b=0.

The above two functions enable us to predict the evolution of the system perturbed to any state in the entire state space: the system evolves by reducing V_L (or $\hat{V}_L$) as if it were traversing down a 'potential hill' to locally minimize it. If the system is perturbed to any state in the region $b_p < b \leq 1$, the final equilibrium state would be at b_-, because V_L is locally minimized at this state; similarly, if it is perturbed to anywhere in the region $0 \leq b < b_p$, the evolution would be towards the state at b=0, because V_L has the least value at this state. The system in equilibrium corresponding to any of these equilibrium states continues to remain in the same equilibrium state till the occurrence of a perturbation. Based on these global properties, we select the following function V(b) as the potential function (see [8,11]) of the system:

$$V(b) = -\int_{B=0}^{b} F(B)dB, \quad 0 \leq b \leq 1, \quad \ldots(18)$$

Note that the constant C in eq. (16) has been defined to facilitate the above global summarizing function. It can be seen that at b_-, the potential function has local minimum; at b=0, the value of the function is the least in the region $0 \leq b \leq b_p$. If the system is in equilibrium at

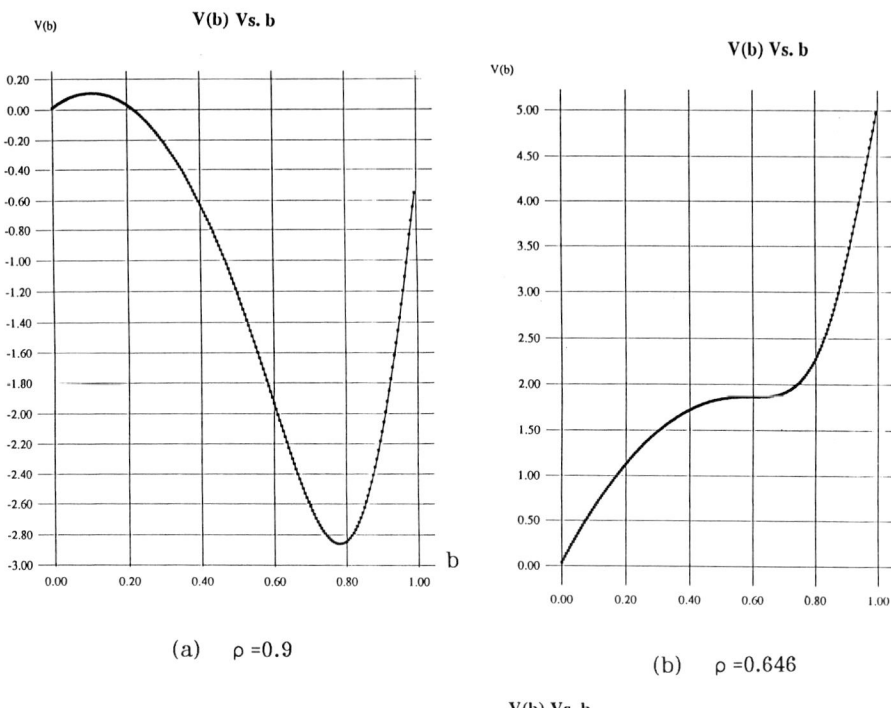

(a) ρ =0.9

(b) ρ =0.646

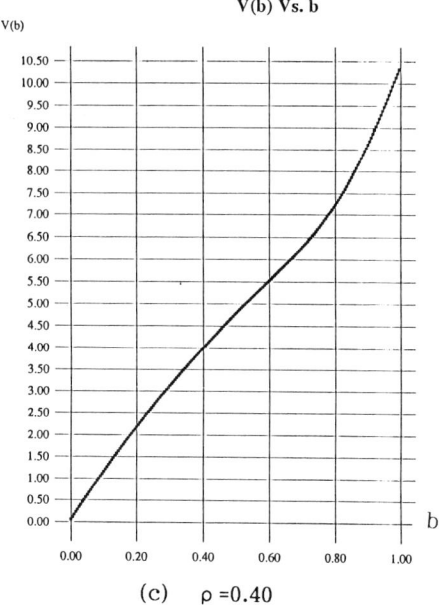

(c) ρ =0.40

Fig. 3 Potential Function Vs. b: (N=20)

b_-, small perturbations will not cause a change in the equilibrium state, because the system evolves towards b_- itself from the perturbed states. Hence, this state corresponds to stable equilibrium states (recall that b_- was already found to be a stable equilibrium state under the considerations of integral Liapunov function). Fig. 3 indicates the potential function associated with each of the plots in Fig. 2 (a). Note that the destination state towards which the perturbed system evolves is the same whether interpreted in terms of the Liapunov function or the potential function.

B. Application of Catastrophe Theory [1]

The details on the stability behavior of the alternate routing scheme under consideration is comprehensively represented by the potential function found above; the shape of the function suggests the nature of the behavior. Once the potential function is known, we may apply the results from catastrophe theory as discussed in ref [1]:

B.1 Identification of catastrophe

We examine how $V(b)$ determines the system's stability behavior, and the circumstances in which small perturbations would result in an abrupt jump of the equilibrium state. The following theorem specifies the nature of the behavior:

Theorem [1]: There exists a fold catastrophe in the model; other higher order elementary catastrophes do not exist.

The above theorem establishes the stability properties of the equilibrium states in accordance with the classification of elementary catastrophes (see [8-11]). The potential function of the perturbed system around the equilibrium state would correspond to the canonical function V_c specified by the following equation:

$$V_c = (b^3/3) - ab \quad \ldots(19)$$

where 'a' corresponds to the control parameter of the system. The plots in Fig. 3 corresponds to various values of 'a'. In Fig. 3 (a) it corresponds to a>0, in Fig. 3(b) to a=0 and in Fig. 3(c) to a<0 (see, ref. [8]).

B.2 Fold point [1]

The fold point, designated by (b_0, ρ_0), can be found by solving the following relation

$$\partial V/\partial b = \partial^2 V/\partial b^2 = 0, \quad \ldots (20)$$

(see, ref. [8]).

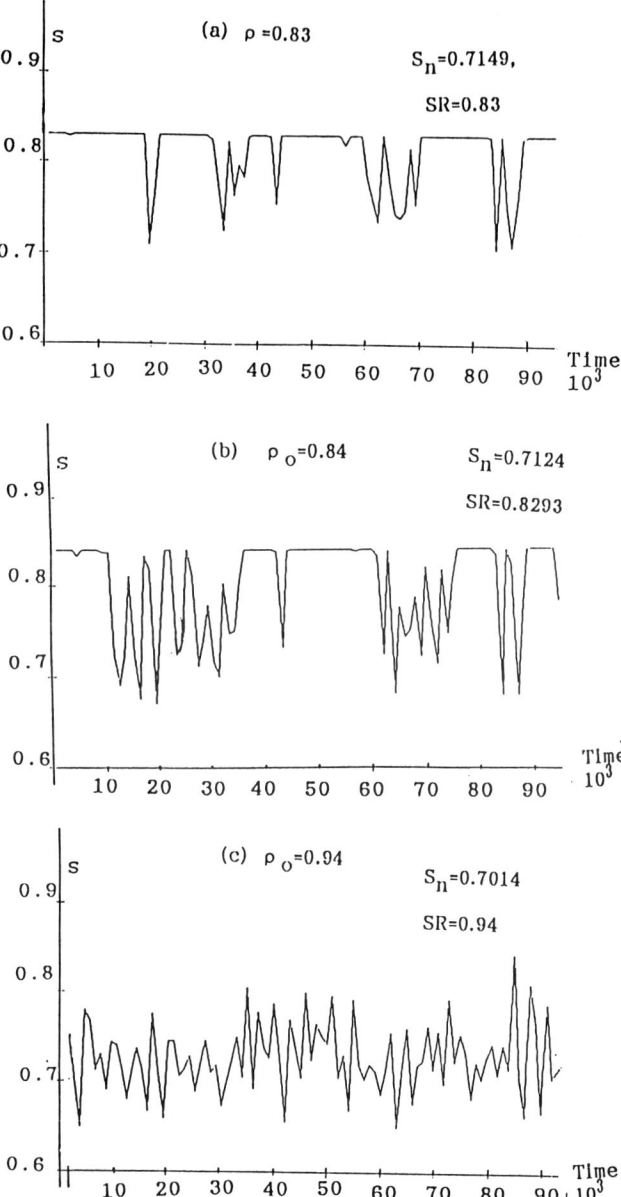

Fig. 4 Transient throughput (N=6, Trunk Group=100)

IV NUMERICAL EXAMPLE

It can be considered that the messages on the direct links would not be blocked with a higher level of probability as long as the load intensity corresponds to levels less than that at the fold point. Hence, most of the messages can be considered to be transferred over the direct links without detour. We would regard that the throughput corresponding to $b=0$, denoted by S_o, closely approximates with ρ, i.e.,

$$S_o = \rho, \quad \text{for} \quad \rho < \rho_o \quad \ldots\ldots(21)$$

But, beyond the fold point there exist two equilibrium states: the one at b_- is stable while the other at b_+ is unstable. Correspondingly there would exist two rates of throughput: at the rate S_n if the system is in equilibrium at b_- and at the rate S_p if it is in equilibrium at b_+. By numerically solving for b in eq. (9), the throughput S_n and S_p can be calculated as:

$$S_n \cong \rho(1-b+bP_s)\Big|_{b=b_-}, \quad \text{for} \quad \rho > \rho_o \quad \ldots(22)$$

and

$$S_p \cong \rho(1-b+bP_s)\Big|_{b=b_+}, \quad \text{for} \quad \rho > \rho_o \quad \ldots\ldots(23)$$

The folding nature of the throughput beyond the fold point was presented in ref. [1]. It was also established in [1] that this method agrees with that employed by Nakagome and Mori [3]. Further, S_p approaches S_o at higher values of ρ. Thus, the equations (21)-(23) provide with both an easy and analytically agreeable method of throughput estimation.

Fig. 4 illustrates the simulation result on the transient throughput, denoted by S, of a 6 node non-hierarchical network under stochastic load of mean ρ. For this network, it can be numerically found that $\rho_o = 0.79$. At loads greater than ρ_o, the output is distorted. From the structural point of view, this can be interpreted as the result of excursions of the system between the two equilibrium states: b_+ and b_-. Moreover, it is to be noticed that at higher loads, the amplitude of fluctuations decreases.

As seen in the above figures, the throughput oscillations depend on the proximity of the equilibrium states to the fold point: oscillations are less if the load intensity is in the close vicinity of the fold point. Suitable performance indices are to be specified to characterize the system sojourn in between the two equilibrium states. Let the transient throughput be in the vicinity of that corresponding to the stable equilibrium state for N_p units of time. Similarly, let the transient throughput be in the vicinity of that corresponding to the stable equilibrium state for N_n units of time. Then, we define the ratio SR, which we call the Sojourn Ratio, as follows:

$$SR = (N_p S_p + N_n S_n)/(N_p + N_n) \quad \ldots\ldots(24)$$

If the long term average is more than SR, the system can be considered to stay at the stable equilibrium state more number of times than it does at the unstable one. On the other hand, if SR is more than the long term average, the system would be visiting the unstable equilibrium state more frequently than it does the stable equilibrium state.

As a crude method of approach for calculating SR, we divided the region between S_p and S_n into horizontal strips with equal width. If the transient throughput falls in the top-most strip, we consider that the throughput approximates with S_p; if it falls in the bottom most strip, we consider that the throughput approximates with S_n. In Fig. 4 (a), the load is 0.83; the throughput corresponding to S_n seldom occurs and it is more or less stable at S_p. When the load is slightly increased to 0.84, we see that the transient throughput is more distorted. S_n corresponding to the load 0.84 is 0.7124. The ratio of the period for which the system throughput is S_p to that for which the throughput is 10.95:1. Then, as per eq. (24).,

$$SR = \{(10.95 \times 0.84)(1 \times 0.7124)\}/(10.95+1) = 0.8293.$$

When the load is further increased to 0.94, we see that the oscillations are around only S_n and that the transient output never falls in the top most region. Thus, SR is the same as S_n which is found to be 0.7014.

V CONCLUSION

A model of non-hierarchical alternate routing scheme in a homogeneous fully connected network is formulated. The state variable of the system is considered to be the blocking probability. From the dynamic flow consideration principles, a Liapunov function which dictates the dynamics of a perturbed system is found out. This enables to identify a potential function of the system. The dynamics of equilibrium states beyond the fold point is studied through simulation. When the input increases past the fold point but remains in the neighborhood of it, the system exhibits oscillations between two equilibrium states: one stable and the other unstable. The blocking probability corresponding to the stable equilibrium state is higher than that corresponding to the unstable one. The more the increase in the input rate, the more the system stays at the stable equilibrium state. Finally, at much higher input rates than that corresponding to the fold point, the system stays always at the stable equilibrium state.

ACKNOWLEDGEMENT

We are grateful to Professor T. Goto, Professor K. Ishizaka and Professor H. Shimizu of the University of Electro-communications for their encouragement and support during the course of this study. We would like to thank Mr. K. Mizuno for providing us with the simulation results.

REFERENCES:

[1] Onozato, Y., Suzuki, M. and Noguchi, S. 1987. "An approximate analysis of the alternate routing in homogeneous Networks", in IEEE GLOBECOM'87, Tokyo, Nov., pp.6.4.1-6.4.5.

[2] Nakagome, Y. and Mori, H. 1973. "Flexible routing in the global communication network", in Seventh International Teletraffic Congress, No.426.

[3] Mori, H. and Nakagome, Y. 1975. "Traffic characteristics of the communication networks with alternate routing - analysis of polyhedron-type network nodes", Trans. IECE Japan (in Japanese), Vol. J58-A, No.6, June, pp.331-338.

[4] Krupp, S. 1982. "Stabilization of alternate routing networks", IEEE Int. Commun. Conf., Philadelphia, June, pp.3I.2.1-3I.2.5.

[5] Akinpelu, J.M. 1983. "The overload performance of engineered networks with hierarchical and non-hierarchical routing", in Tenth International Teletraffic Congress, Montreal, June, Session 3.2, paper 4.

[6] Akinpelu, J.M. 1984. "The overload performance of engineered networks with non-hierarchical routing", AT&T Bell Labs.Technical Journal., Vol. 63, No.7, Sept., pp.1261-1281.

[7] Luenberger, D.G. 1979. Introduction to Dynamic Systems, John Wiley & Sons, New York.

[8] Fararo, T.J. 1978. "An Introduction to Catastrophes", Behavioral Science, Vol. 2, pp.291-153.

[9] Haken, H. 1979. Synergetics: An Introduction to Non-equilibrium Phase Transition & Self-organisation, Springer Verlag, Berlin Heidelberg.

[10] Nelson, R. 1987. "Stochastic Catastrophe Theory in Computer Performance Modelling", Journal of the Association for Computing Machinery, Vol. 34, No.3, July, pp. 661-687.

[11] Onozato, Y. and Noguchi, S. 1985. "On the thrashing cusp in slotted ALOHA systems", IEEE Trans. Commun., Vol. COM-33. No. 11, pp. 1171-1182.

[12] Filipiak, J. 1988. Modelling and control of Dynamic Flows in communication Networks, Springer Verlag, Berlin Heidelberg.

[13] L. Kleinrock, "Queueing Systems", Vol.1, John Wiley & Sons, Inc., Toronto.

The communication spanning tree problem: an heuristic algorithm

F.J.M. Salzborn

Teletraffic Research Centre, The University of Adelaide, Adelaide, South Australia

Abstract

Suppose for a collection of nodes the two-way traffic between each O-D-pair is given and one wants to design a tree network with the links dimensioned such that all the offered traffic can be carried. Such a tree network is called a communication spanning tree. The problem considered here is that of finding minimum cost communication spanning trees, assuming that the cost of the tree is equal to the sum of the cost of its links and that the cost of each link is proportional to the traffic carried on that link. In this research the cost of a link is actually taken to be the length of the link multiplied by the traffic carried. This problem is known to be NP-complete, so one cannot expect to find efficient algorithms for solving it. Here we will discuss an heuristic algorithm that appears to give good results. It is well-known that there is a polynomial algorithm for the special case that the distances are ignored and the link cost depend on the carried traffic only.

1. COMMUNICATION SPANNING TREES

This paper is concerned with a particular problem in topological network design. Suppose we are given a set of nodes N which are numbered from 1 to n. Let $r(p,q)$ denote the traffic between nodes p and q in both directions; so the traffic matrix $R = (r(p,q)|p,q \in N)$ is symmetric. A spanning tree T connecting these nodes, such that all communications take place by using the links of the tree, is called a communication spanning tree. Suppose that for each link (i,j) a cost per call $d(i,j)$ is given. Let $l_T(p,q)$ denote the cost per call of the path from p to q in the tree T, in other words the sum of the costs of the links of that path; so this is just the cost of a call from p to q if the communication tree T is used. The total cost of communications tree T is therefore

$$c(T) = \sum_{p<q} l_T(p,q) r(p,q). \tag{1}$$

The problem to be considered here is that of finding a spanning tree T for which the cost $c(T)$ is minimized.

There is an alternative way of formulating the problem.

A spanning tree T determines a collection of $n-1$ cut–sets, the *fundamental cut–set collection* $\mathcal{F}_T$ *of* T, as follows. Let (i,j) be a link of T; removing (i,j) from T determines a partitioning of N into the sets $X_i^{(i,j)}$ of nodes still connected to i and $X_j^{(i,j)}$ of nodes still connected to j. $\mathcal{F}_T$ consists of the $n-1$ cut–sets $(X_i^{(i,j)}, X_j^{(i,j)})$, for $(i,j) \in T$.

Let $\mathcal{F}_T$ be the fundamental collection of cut–sets of T, and let $(X, \overline{X})$ be the fundamental cut–set corresponding to the link (i,j) of T. Then the amount of traffic that has to be carried by this link, if the links of the tree T are the only ones available, is equal to the traffic across this cut–set:

$$\text{cap}(i,j) = r(X, \overline{X}) = \sum_{i \in X} \sum_{j \in \overline{X}} r(i,j). \tag{2}$$

So $\text{cap}(i,j)$ is the capacity to which link (i,j) has to be dimensioned to enable it to carry all the traffic assigned to it. Clearly, adding up the traffic carried by each link (i,j) of T multiplied by the cost, will also give the total cost of T:

$$c(T) = \sum_{(i,j) \in T} d(i,j) \text{cap}(i,j). \tag{3}$$

The problem of finding the spanning tree T with minimum cost $c(T)$ is called the *communication spanning tree problem*. It has been shown to be NP-hard in [3], so for problems of any reasonable size one has to use heuristic methods to find reasonable solutions in reasonable time. In section we will discuss a method that appears to give good results.

For the case that the costs for all links are identical, in which case we can take $d(i,j) = 1$ for all links (i,j), there does exist a polynomial algorithm. The cost $c(T)$ of a tree T is then equal to the total number of hops for all the calls. As shown by Hu [2] the spanning tree that minimizes the total number of hops is the so called Gomory–Hu tree, described in [1]. This problem is also investigated in [5] and [6].

2. OPTIMAL TWO-HUB NETWORKS

In this section the following problem is considered. Given a set of nodes N, numbered $i = 1, \ldots, n$, a symmetric traffic matrix $R = (r(p,q) \mid p, q \in N)$, its entries indicating the capacity that has to be available for OD–pair p, q, and a cost matrix for the links $D = (d(i,j) \mid i, j \in N)$, find a two–hub communication spanning tree of minimum cost.

A two–hub network, with hubs h_1, h_2 is a spanning tree of N such that each of its links has as one of its endpoints either node h_1 or node h_2 or both. Clearly (h_1, h_2) must be a link of the network, otherwise it would not be connected.

Let S_1 be the set of nodes linked to hub h_1, excluding the other hub h_2, and similarly, S_2 the set of nodes linked to hub h_2. So S_1 and S_2 form a partitioning of the nodes that are not hubs, the set $N \setminus \{h_1, h_2\}$. We define $X_1 = S_1 \cup \{h_1\}$ and $X_2 = S_2 \cup \{h_2\}$, so (X_1, X_2) is a cut–set of the network. The corresponding two–hub network will be denoted by $T(h_1, h_2, S_1, S_2)$.

Its cost is

$$c(T(h_1, h_2, S_1, S_2)) = \sum_{i \in S_1} d(i, h_1) tr(i) + \sum_{j \in S_2} d(j, h_2) tr(j) + d(h_1, h_2) r(X_1, X_2), \quad (4)$$

where

$$tr(i) = \sum_{j=1}^{n} r(i, j) \quad (5)$$

is the total traffic at node i, and

$$r(X_1, X_2) = \sum_{i \in X_1} \sum_{j \in X_2} r(i, j) \quad (6)$$

is the total traffic (or capacity) crossing the cut–set (X_1, X_2).

This cost depends not only on the hubs h_1 and h_2, but also on which of the remaining nodes are linked to h_1 and which to h_2. However for fixed hubs there exists a polynomial algorithm to determine the optimal selection of these sets S_1 and S_2, as we will now show.

Consider the complete *directed* graph DG for the node set N, i.e. this is the graph which contains every directed arc (i, j), for $i, j \in N$; note that arcs (i, j) and (j, i) are now distinguished. For each arc (i, j) of DG we define an arc–number $u(i, j)$ as follows.

$$\text{for } i \neq h_1, j \neq h_2: \quad u(i, j) = d(h_1, h_2) r(i, j), \quad (7)$$
$$\text{for } j \neq h_2: \quad u(h_1, j) = d(h_1, h_2) r(h_1, j) + d(j, h_2) tr(j), \quad (8)$$
$$\text{for } i \neq h_1: \quad u(i, h_2) = d(h_1, h_2) r(i, h_2) + d(i, h_1) tr(i), \quad (9)$$
$$\text{for } i = h_1 \text{ and } j = h_2: \quad u(h_1, h_2) = d(h_1, h_2) r(h_1, h_2). \quad (10)$$

For the cut–set value of a cut–set (X_1, X_2) with $h_1 \in X_1$ and $h_2 \in X_2$ we get:

$$
\begin{aligned}
u(X_1, X_2) &= \sum_{i \in X_1} \sum_{j \in X_2} u(i, j) \quad (11)\\
&= \sum_{i \in S_1} \sum_{j \in S_2} d(h_1, h_2) r(i, j) + \sum_{j \in S_2} [d(h_1, h_2) r(h_1, j) + d(j, h_2) tr(j)] + \\
&\quad \sum_{i \in S_1} [d(h_1, h_2) r(i, h_2) + d(i, h_1) tr(i)] + d(h_1, h_2) r(h_1, h_2) \\
&= d(h_1, h_2) r(X_1, X_2) + \sum_{i \in S_1} d(i, h_1) tr(i) + \sum_{j \in S_2} d(j, h_2) tr(j) \quad (12)\\
&= c(T(h_1, h_2, S_1, S_2)). \quad (13)
\end{aligned}
$$

So the value of this cut set is equal to the cost of the corresponding two–hub network.

This means that, for given hubs h_1 and h_2, to find the optimal way of linking the remaining nodes to these hubs, all one has to do is to find the cut–set (X_1, X_2) of the directed graph DG with minimum value. From the well–known 'maximum flow–minimum cut' theorem of Ford and Fulkerson we know that this is equivalent to finding a maximum flow from h_1 to h_2 in the graph DG, where the values $u(i, j)$ are regarded as upper capacities and the lower capacities are 0. For this task algorithms are available of complexity $O(n^3)$. The maximum flow value will actually be the minimum cost of a two–hub network with these hubs, but the algorithm also provides the minimum cut–set (X_1, X_2). Then $S_1 = X_1 \setminus \{h_1\}$ and $S_2 = X_2 \setminus \{h_2\}$ are the sets of nodes linked to the two hubs.

To find the minimum two–hub network one has to repeat this process for each pair of possible hubs, that means $1/2 n(n-1)$ times. So the whole procedure has then complexity $O(n^5)$.

3. SPLITTING STAR HEURISTIC

One of the possible outcomes of the procedure described in the previous section is, that for the optimum two-hub network one of the sets S_i is empty, say $S_2 = \emptyset$, which means that hub h_2 is only linked to the other hub but no other nodes are linked to it, so it is not really a hub: the network is a star, with h_1 as centre node. This implies that h_1 must be the node with maximum traffic $tr(h_1) = \max_{i \in N} tr(i)$, as the star is the least cost star.

On the other hand we can start off with a star network and apply the two–hub procedure of the previous section to test whether the star network is the minimum cost tree. More generally, we can formulate a necessary condition for optimality of a spanning tree T as follows.

Consider a node h with more then one neighbour. We define

$$N(h) = \{j \in N \mid (h, j) \in T\}, \tag{14}$$

and the component connected to node $j \in N(h)$, when node h is removed from T, is indicated by X_j^h. This component will be called a *pseudonode* of h. For the set consisting of the hub h and its pseudonodes we define the traffic matrix $\hat{R}$ and distance matrix $\hat{D}$ as follows:

$$\hat{r}(h, X_j^h) = \sum_{i \in X_j^h} r(h, i), \tag{15}$$

$$\hat{d}(h, X_j^h) = d(h, j), \tag{16}$$

$$\hat{r}(X_{j1}^h, X_{j2}^h) = \sum_{i \in X_{j1}^h} \sum_{j \in X_{j2}^h} r(i, j), \tag{17}$$

$$\hat{d}(X_{j1}^h, X_{j2}^h) = d(j1, j2). \tag{18}$$

The link costs for the links between pseudonodes are just the costs for the links between the corresponding nodes in the original network. For instance, if the link cost is just the Euclidian distance, then pseudonode X_j^h is supposed to be located at the same position as the corresponding neighbour node j of h.

If there exists a two–hub network for the communications tree problem defined by the traffic matrix $\hat{R}$ and distance matrix $\hat{D}$ that is cheaper than the star network with hub h then the spanning tree T is clearly also not optimal for the original problem; all one has to do is to replace the links connecting node h with its neighbours by the links suggested by the optimal two-hub network of the sub-problem and one gets a cheaper tree. This action we will call *splitting the star* centered at node h. So a necessary condition for optimality of a spanning tree T is that none of its stars can be split.

This suggests the heuristic procedure that is proposed here for the communication spanning tree problem: choose a star centered at some node and try to split it; if no star can be split, stop. The procedure starts with the optimal two-hub network for complete set of nodes. If this is already a star, the procedure stops immediately.

We will illustrate the heuristic method by means of an example. There are 15 nodes, the positions of the nodes are shown in figure 1, the link cost per call is the length of the link and the corresponding cost matrix D is shown in table 1. The traffic matrix R is displayed in table 2. Figure 1 shows the best tree designed by the algorithm; figure 2 shows intermediate trees.

The graph on the top lefthand side of figure 2 shows the optimal two-hub network for the problem. The nodes centred around node 8 constitute a pseudonode of node 12, together with the other nodes linked to node 12 (singleton pseudonodes), this determines the star centred on node 12 and one can check whether it is profitable to split this star. The result is shown in the top righthand tree of figure 2: the star is split, the two hubs are nodes 12 and 14. This 'splitting'-test can be applied to any node. The bottom lefthand tree shows how the star with node 8 as center is split into a two-hub tree with hubs at nodes 8 and 5, then the star around node 5 is split once more to get the last tree of this picture. Finally the star with center node 8 is split to get the tree shown in figure 1. Now no more stars can be split and the procedure stops.

Table 1
Distance matrix

	1	2	3	4	5	6	7	8	9	10	11	12	13	14	15
1	0.0	6.0	3.0	5.2	3.9	7.8	8.1	5.8	9.2	12.0	9.8	10.2	11.7	13.3	13.9
2	6.0	0.0	3.0	1.4	4.3	4.5	13.5	10.1	13.6	16.4	13.1	10.3	9.5	12.9	12.5
3	3.0	3.0	0.0	2.5	3.0	5.8	10.8	7.8	11.4	14.2	11.4	10.0	10.4	13.0	13.2
4	5.2	1.4	2.5	0.0	2.9	3.6	12.3	8.7	12.3	15.0	11.7	9.0	8.6	11.7	11.5
5	3.9	4.3	3.0	2.9	0.0	4.0	9.6	5.9	9.4	12.1	8.8	7.0	7.8	10.0	10.3
6	7.8	4.5	5.8	3.6	4.0	0.0	13.0	9.0	12.2	14.6	10.8	6.4	5.0	8.5	8.1
7	8.1	13.5	10.8	12.3	9.6	13.0	0.0	4.1	3.2	5.1	5.8	10.8	14.3	13.5	15.0
8	5.8	10.1	7.8	8.7	5.9	9.0	4.1	0.0	3.6	6.4	4.1	7.1	10.3	10.0	11.3
9	9.2	13.6	11.4	12.3	9.4	12.2	3.2	3.6	0.0	2.8	2.8	8.5	12.4	10.8	12.5
10	12.0	16.4	14.2	15.0	12.1	14.6	5.1	6.4	2.8	0.0	4.0	10.0	14.0	11.7	13.6
11	9.8	13.1	11.4	11.7	8.8	10.8	5.8	4.1	2.8	4.0	0.0	6.1	10.0	8.1	9.8
12	10.2	10.3	10.0	9.0	7.0	6.4	10.8	7.1	8.5	10.0	6.1	0.0	4.0	3.2	4.2
13	11.7	9.5	10.4	8.6	7.8	5.0	14.3	10.3	12.4	14.0	10.0	4.0	0.0	4.2	3.2
14	13.3	12.9	13.0	11.7	10.0	8.5	13.5	10.0	10.8	11.7	8.1	3.2	4.2	0.0	2.0
15	13.9	12.5	13.2	11.5	10.3	8.1	15.0	11.3	12.5	13.6	9.8	4.2	3.2	2.0	0.0

Table 2
Traffic matrix

	1	2	3	4	5	6	7	8	9	10	11	12	13	14	15
1	0	19	0	33	42	0	68	30	65	0	0	46	0	94	0
2	19	0	39	24	0	49	0	0	0	0	90	0	11	41	38
3	0	39	0	0	0	0	0	0	0	0	75	0	31	12	86
4	33	24	0	0	41	0	81	0	32	0	0	0	0	0	0
5	42	0	0	41	0	0	0	98	14	68	31	68	0	69	0
6	0	49	0	0	0	0	0	0	0	0	91	0	74	53	0
7	68	0	0	81	0	0	0	0	95	88	0	0	0	0	7
8	30	0	0	0	98	0	0	0	1	45	94	0	78	35	64
9	65	0	0	32	14	0	95	1	0	0	77	100	0	60	0
10	0	0	0	0	68	0	88	45	0	0	31	49	0	69	14
11	0	90	75	0	31	91	0	94	77	31	0	0	6	0	0
12	46	0	0	0	68	0	0	0	100	49	0	0	0	49	0
13	0	11	31	0	0	74	0	78	0	0	6	0	0	0	68
14	94	41	12	0	69	53	0	35	60	69	0	49	0	0	96
15	0	38	86	0	0	0	7	64	0	14	0	0	68	96	0

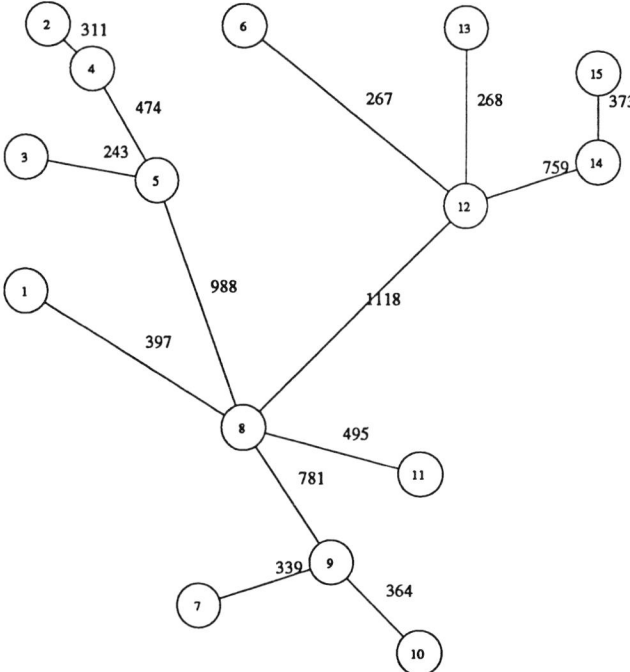

Figure 1. Best communication spanning tree found.

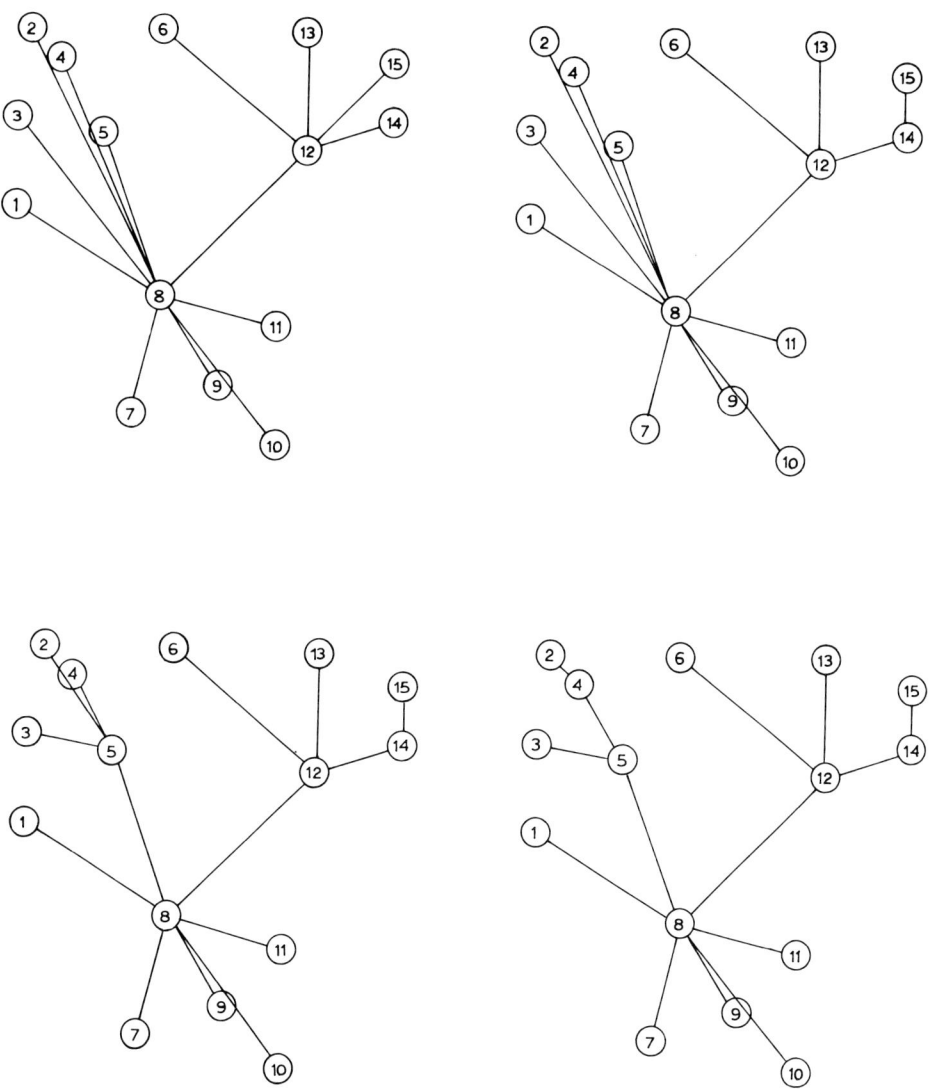

Figure 2. Intermediate networks.

4. REFERENCES

1. R.E. Gomory and T.C. Hu, Multiterminal network flows, SIAM Journal of Appl. Math., 9 (1961) 551–570.

2. T.C. Hu, Optimum communication spanning trees, SIAM Journal of Computing, 3 (1974) 188–195.

3. D.S. Johnson, J.K. Lenstra and A.H. Rinnooy Kan, The complexity of the network design problem, Networks, 8 (1978) 279–286.

4. V.M. Malhotra, M.P. Kumar and S.N. Maheshwari, An $O(|V|^3)$ algorithm for finding maximum flows in networks, Information Processing Letters, 7 (1978) 277–278.

5. F.J.M. Salzborn, A program for finding the communication spanning tree with minimum number of hops, TRC Report No.2, (1990).

6. F.J.M. Salzborn, Gomory–Hu trees and the problem of finding a spanning tree with minimum number of hops, TRC Report No.12, (1990).

Acknowledgement

This research was funded under contract with Telecom Australia.

FLEXIBLE PROTECTION OF TRANSMISSION RESOURCES *

J.Lubacz, M.Jarociński, A.Tomaszewski

Institute of Telecommunications, Warsaw University of Technology
Nowowiejska 15/19, 00-665 Warszawa, Poland

O.G.Soto

ALCATEL Standard Electrica, S.A.
Ramirez de Prado 5, 28045 Madrid, Spain

Abstract

The paper concerns rearrangement of network transmission resources with the use of digital cross connects. Problems concerning modelling network configuration and rearrangement are discussed. An approach to modelling is proposed and its performance for a sample network of practical size is illustrated.

1. INTRODUCTION

The foreseen extensive use of digital cross connects (DCCs) together with the SDH based multiplexing principle open new, powerful options in network configuration and rearrangement. The new options come together with more rigorous requirements concerning reliability and trafficability that are associated with the utilisation of high capacity optical transmission systems and the ongoing evolution towards integrated, multiservice traffic environments. Thus the development of flexible, dynamic network configuration and rearrangement options are not only an interesting challenge but also a practical necessity.

In this paper we consider the problem of modelling networks in the context of the new options. The eventual result of the discussion is a network configuration and rearrangement algorithm proposed in Sec.4. The performance of the algorithm was tested for a network of practical size; numerical results are presented in Sec.6.

In the presentation we shall use the generic, layered architecture of network transmission resources proposed by the authors in [1]; we recall it in Fig.1. Our main object of interest here is the Synchronous Network (SN) layer, subject to configuration and rearrangement with DCCs. Configuration and rearrangement concern SN paths

---

* *This research has been carried out within ALCATEL's project in the network analysis area*

(chains of SN links) which are built upon the Physical Network (PN) paths by means of synchronous subdivision of PN paths' capacity. PN paths terminate at DCC ports; endpoints of SN paths are interfaces between DCCs and switch (exchange) ports.

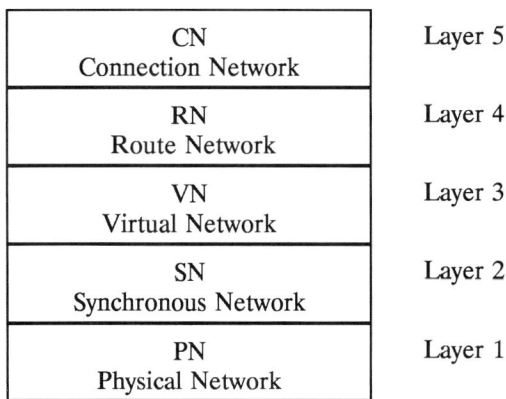

Fig.1. Network transmission resources generic structure

SN rearrangement is triggered by PN paths failures, or by events such as e.g. traffic pattern changes. Such events result in new demands concerning Route Network (RN) and/or Virtual Network (VN) configuration; these demands can be translated into requirements concerning SN capacity/configuration. In general, the principle of such a translation depends on the type of network considered - IDN, ISDN or B-ISDN. In the following we shall assume that such a translation is done, i.e. our discussion will be independent on the network type.

In the classical network protection principle, usually referred to as 1:1, the object of protection is the set of SN links that are created on one PN path. If the PN path fails, all the SN links are switched *en bloc* to a redundant PN path (standby). The protected path and the standby are associated in fixed pairs; a standby from one pair cannot be used in the case of a failure of a PN path from another pair.

In the flexible network protection principle considered in this paper - Path Protection (PP), no such fixed association is assumed; all redundant PN paths are treated as a common, shared pool of protection resources. Moreover, in contrast to the 1:1 protection, every SN link is treated individually; the object of protection is each individual SN path. If a PN paths fails, then a separate protection path is created for every SN path that goes through the PN path that failed. A protection path is functionally an SN path; it is created on redundant PN paths from the common pool of redundant resources. In effect, SN paths affected by a PN path failure are switched to protection paths, each of which can be potentially created on a different set of redundant PN paths. The creation of protection paths can be formulated even more universally. In general, a PN path failure results in that some active SN paths become idle - these could be potentially also used in creating protection paths, i.e. appended to the pool of redundant resources. In our research we shall assume such generalization.

2. THE BASIC MODEL

The SN layer planning comprises the design of a basic network configuration and a rearrangement procedure. A <u>basic network configuration</u> (BNC) is a network configuration (NC) that meets nominal capacity demands at no-defect conditions. Potential events that can make BNC fail to satisfy the capacity demands should be identified. These are, apart from PN paths defects, e.g. events related to temporary traffic volume growth. The latter are observed as an increase of capacity requirements in a number of RN layer links. Whatever the case - PN path failure or traffic growth - the BNC becomes unsatisfactory since the capacity provided on SN paths is not sufficient any more. The rearrangement procedure is triggered by each of the events that leads to shortage of resources in the SN layer; as a result the BNC is transformed into a satisfactory NC (SNC).

An SN layer network configuration is described with a set of SN paths and their capacities. The NC is satisfactory if for each RN link the total capacity of SN paths connecting its origin and destination nodes, matches the capacity demand of the link. The SN paths are directed attributed paths in the graph of PN. The objective cost function allowing for comparison of different SNCs expresses the total cost of PN paths usage. The cost of a PN path depends on the path length and the capacity of the transmission system provided. This capacity constrains the total capacity of SN links allocated on the PN path.

Thus the design of an SNC for each event that triggers the rearrangement procedure - <u>triggering event</u> - forms a multi-commodity problem with potentially different sets of the graph edges and capacity demand matrices. The multi-commodity flow problems associated with different triggering events are interdependent since the final capacity assigned to a PN path must be high enough to accommodate the greatest flow; thus capacity constraints are common to all the problems.

We shall consider the two following cases:

(1) <u>restricted rearrangement</u> - ongoing connections cannot be switched to different RN paths during SN paths rearrangement;
(2) <u>unrestricted rearrangement</u> - the constrained from (1) is not assumed.

Let $G = (V,E)$ be a graph modelling the SN layer; V is the set of nodes which correspond to DCCs, and E is the set of arcs (directed edges), which correspond to PN paths; $e = <v_1, v_2> \in VxV$.
The following notation will be used:

- $R \leq VxV$ denotes the set of RN layer links; r denotes a link from R;

- $b(r)$ and $t(r)$ denote the beginning and the termination node of link r, respectively;
- S is a set of possible triggering events; $s=0$ denotes nominal capacity demands and no-defect conditions;
- function $d:S \rightarrow 2^E$ associates the set of faulty PN paths with each of the possible triggering events;
- $a(r,s)$, $r \in R$, $s \in S$, is a capacity demand in r when triggering event s occurs, expressed in multiples of m_0;
- $c(e)$, $e \in E$, is a cost of a capacity unit on PN path e; PN path cost function is assumed to be linear with respect to the capacity of the transmission system used;
- m_2, being a multiple of m_0, describes the modularity of the transmission systems used on PN paths;
- $i(e)$, $e \in E$, is an initial capacity of the transmission system installed on PN path e, expressed in multiples of m_2;
- $q(e)$, $e \in E$, denotes the final optimized capacity of the transmission system installed on PN path e, expressed in multiples;
- x is the restoration factor describing what percentage of RN link's capacity should be restored in the presence of PN paths defects.

3. LINEAR PROGRAMMING APPROACH

Each of the multi-commodity problems and their interdependence can be expressed straightforwardly if SN layer design is formulated in terms of a linear program. If such a formulation is adopted, the nature of the constraints are such that the solution to the problem has to be achieved with mixed integer/linear programming methods. The capacity of PN paths is a multiple of module m_2, capacities of SN paths are modular also.

Unrestricted rearrangement.

Let $f(r,e,s) \geq 0$ denote the capacity on arc e, used by SN paths of link r when triggering event s occurs. From the conservation of flow requirement in a graph node the following should be fulfilled for every fixed triple $s \in S$, $v \in V$, $r \in R$:

$$\sum_{\substack{e=<\cdot,v> \\ e \notin d(s)}} f(r,e,s) - \sum_{\substack{e=<v,\cdot> \\ e \notin d(s)}} f(r,e,s) = \begin{cases} a(r,s) & \text{iff} \quad t(r)=v,\ s=0 \\ -a(r,s) & \text{iff} \quad b(r)=v,\ s=0 \\ a(r,s) \cdot x & \text{iff} \quad t(r)=v,\ d(s) \neq \varnothing \\ -a(r,s) \cdot x & \text{iff} \quad b(r)=v,\ d(s) \neq \varnothing \\ 0 & \text{otherwise} \end{cases} \quad (1)$$

The amount of capacity actually used in at least one SNC resulting from transformation of BNC is given by $u(e)$, $e \in E$, which satisfies the following condition:

$$u(e) \geq \sum_{r \in R} f(r,e,s) \quad \text{for each } s \in S, \quad (2)$$

while the constraints expressing the required capacity of PN paths are as follows:

$$q(e) \cdot m_2 \geq u(e), \quad q(e) \geq i(e) . \tag{3}$$

The objective of the optimization is to minimize the cost function which takes into account the total cost of the required augmentations of PN paths capacity and the amount of capacity actually used (defined by (2)):

$$\sum_{e \in E} c(e) \cdot \{A \cdot m_2 \cdot [q(e) - i(e)] + B \cdot u(e)\} \tag{4}$$

where A and B are arbitrary non-negative constants, chosen to penalize the augmentation of existing resources and/or ineffective use of allocated capacity, respectively.

Equations (1) - (3) and cost function (4) define a discrete optimization program. All variables are assumed to be non-negative integers.

Both, the number of variables and the number of equations, are roughly equal to $dim(S) \cdot dim(E) \cdot dim(R) \approx dim(V)^4$.

Restricted rearrangement

Let P(r) be the set of paths in graph G along which the capacity required by link r can be allocated in BNC. $P(r)$ can be constructed by evaluating the prescribed number of paths with lowest unit cost, providing required degree of spatial diversity.

Let $h(r,p,s) \geq 0$ denote the capacity of the SN path routed along $p \in P(r)$ when triggering event s occurs. For BNC, i.e. nominal capacity demands and no-defect conditions, the following should be satisfied:

$$\sum_{p \in P(r)} h(r,p,0) = a(r,0) \quad \text{for all } r \in R. \tag{5}$$

If triggering event s occurs new allocation of capacity is required. Let $g(r,e,s) \geq 0$ denote additional capacity on arc e traversed by SN paths of link r when s occurs. The conservation of flow requires that for every fixed triple $s \in S$, $v \in V$, $r \in R$:

$$\sum_{\substack{e=\langle\cdot,v\rangle \\ e \notin d(s)}} g(r,e,s) - \sum_{\substack{e=\langle v,\cdot\rangle \\ e \notin d(s)}} g(r,e,s) = \begin{cases} [a(r,s) - \sum_{p \sim d(s)=\varnothing} h(r,p,s)] \cdot v & \text{iff } t(r)=v \\ -[a(r,s) - \sum_{p \sim d(s)=\varnothing} h(r,p,s)] \cdot v & \text{iff } b(r)=v \\ 0 & \text{otherwise} \end{cases} \tag{6}$$

where v equals unity if $d(s)$ is an empty set and x otherwise.

An additional set of constraints is needed:

$$h(r,p,s) \geq h(r,p,0) \quad \text{for all } r \in R, s \in S, \text{ and } p \in P(r) \text{ such that } p \cap d(s) = \varnothing . \tag{7}$$

The capacity analogous to the one appearing in (2) is expressed with the following conditions:

$$u(e) \geq \sum_{\substack{r \in R, \\ p \in P(r): e \in p}} h(r,p,s) + \sum_{r \in R} g(r,e,s) \quad \text{for all } e \in E, s \in S. \tag{8}$$

Formulae (5) - (8) together with (3) and (4) constitute the modified formulation of the problem.

The complexity of the linear program formulated above is enormous; we can somewhat reduce it if predefined paths in G are used only. This can be achieved by extending formula (5) for all values of s instead of introducing the set of variables $g(r,e,s)$ and expression (6). Expression (8) should be modified accordingly.

With such a modification the complexity of the problem can be reduced by a factor $dim(V)$, both with respect to the number of variables and the number of equations. However even in this case the discrete nature of the optimization and the huge space of feasible solutions make the problem intractable even for network sizes of practical interest.

In [2] the problem of transmission network planning with reliability issues taken into account is considered. The work presents a mixed integer-linear programming formulation with predefined sets of paths for each origin-destination node pair. The problem is approached through continuous relaxation. The resulting fractional values achieved by means of linear programming methods are cut off with a myopically directed adjustment method. This requires dual manipulation of the linear programming solution. The proposed bounding procedures allow for a branch-and-bound approach. The method was tested for relatively small networks only.

Our general conclusion from this discussion is that a heuristic approach is needed if networks of practical size are to be considered.

4. HEURISTIC APPROACHES

It is common for most heuristic approaches to separate the design of BNC from the rearrangement procedure planning. As far as BNC is concerned, the capacity for each RN link is usually provided by the set of SN paths with the lowest possible unit costs, while providing the required degree of spatial diversity. A set of SN paths from one RN link is constructed by first identifying in the graph the required number of disjoint shortest paths and then assigning capacity to them.

Heuristic methods known to the authors concentrate on planning the transformation of a BNC, given *a priori*, into an SNC. The methods aim at evaluating the required augmentation of transmission resources of BNC that allows for rerouting of faulty SN paths whatever defect occurs.

In general this is an NP-complete problem. The practical solution is to decompose the problem and predefine the order of SN paths restoration. This allows for replacement of the multi-commodity flow problem with a sequence of single-commodity problems.

The most common approach is to neglect the interdependence of network configurations related to different defects [3]. The rearrangement procedure is not designed for all potential defects simultaneously; each defect is considered separately so transmission capacity is adjusted to one failure at a time. Faulty SN paths are restored one by one, each time triggering only the necessary augmentation of transmission resources. Usually only one module of capacity is restored. The order of restoration is predefined, or the path to be restored is selected dynamically based on the resulting augmentation cost. The problem of optimal path routing is approached through shortest path labelling algorithms.

The problem can be simplified even further if single-link defects are assumed. Transformation into a single-commodity problem is possible if only these parts of the faulty SN that traverse a defected link paths are rerouted. This may however result in inefficient utilisation of capacity.

If each defect is analyzed separately, the restoration of the capacity for one defect does not take into account the augmentations required for restoration of SN paths affected by other defects. The performed augmentations are not correlated so the provided resources cannot be fully utilised.

In [4] a method is proposed to overcome the above difficulties. All defects are considered simultaneously. The lists of the RN links affected by each defect are ordered according to the amount of capacity still unrestored. At each iteration of the algorithm one module of the capacity is restored for the first RN link on every list. The capacity is allocated along the shortest paths in the graph of unassigned resources. If no path can be found, capacity restoration is impossible and the augmentation of PN paths is necessary. Restoration is retried after a PN path that blocks the maximum number of restorations has been augmented. This PN path is identified in the bounding cuts of all unrestored RN links of the current iteration. The augmentations are repeated until all restorations required at the current iteration are performed.

There are two objectives while restoring the faulty capacity. The RN layer links affected by the same defect should effectively share network resources that have already been allocated. At the same time, capacity augmentations required for different defects should be correlated so that different defects exploit the same redundant capacity.

The method from [4] aims at supporting the second requirement, but all methods known to the authors fail to take into account the first one. The use of excessively long paths that require no augmentation of resources is not prohibited. This may result in the increase of necessary capacity augmentations when consecutive restorations associated with the same defect are examined.

5. THE PROPOSED ALGORITHM

In the algorithm proposed in the following we try to overcome some of the above mentioned difficulties associated with heuristic methods.

The algorithm employs the shortest path labelling method with a modified cost function. SNCs for all triggering events are constructed concurrently. This is achieved by introducing an iterative process of path restoration. In each iteration all triggering events are scanned. For each event all RN links that require restoration are analyzed, however only one module of a single RN link is actually restored. Iterations are repeated until the construction of the SNCs associated with all triggering events is completed. The choice of the RN link, the capacity module of which is restored, is made upon the RN link's cost; lowest cost RN link is selected in every iteration. The cost function in the shortest path algorithm used to make the choice takes into account capacity augmentation cost and the total amount of capacity used for a module restoration.

The specific feature of the algorithm is that it aims at allocating the capacity in small portions and spreading it cost effectively over the whole network. It also introduces correlation between consecutive capacity allocations; in this way, although indirectly, the correlation between BNCs associated with different triggering events is introduced.

The algorithm is as follows. We keep the notation introduced above.

Step 1

The BNC is designed. For each RN link $r \in R$, the set $P(r)$ of cheapest disjoint paths is constructed; a shortest path labelling algorithm is used for this purpose. The number of paths in $P(r)$ is equal to a predefined small positive integer value. Let the cost of path $p \in P(r)$ be denoted by $c(p)$. The amount of capacity allocated along each of the paths depends on the distribution of their costs. A part $s(p)$ of RN link r, $r \in R$, capacity demand that is to be routed along path p, $p \in P(r)$, equals:

$$s(p) = \frac{c(p)^{-1}}{\sum_{q \in P(r)} c(q)^{-1}}, \quad \text{where} \quad c(p) = \sum_{e \in E:\ e \in p} c(e) . \tag{9}$$

Let $f(e,0)$ be the total capacity of SN paths of the BNC used on PN path e:

$$f(e,0) = \sum_{\substack{r \in R \\ p \in P(r):\ e \in p}} s(p) \tag{10}$$

Step 2

Let $l(r,s)$ denote the capacity that should be allocated for RN link r, $r \in R$, when triggering event s occurs:

$$l(r,s) = a(r,s) - \sum_{\substack{p \in P(r): \\ p \wedge d(s) = \emptyset}} s(p) \ . \tag{11}$$

Let $Q(s)$, $Q(s) = \{(r,q)\}$, be the set of required restorations necessary to construct SNC for triggering event s. Capacity q that should be allocated for RN link r equals:

$$q = l(r,s) - (1-v) \cdot a(r,s) \ . \tag{12}$$

Initial capacity of PN paths for triggering event s NC is expressed as in (10) but faulty SN paths are excluded from the summation.

<u>Step 3</u>
The set S is scanned in a random order.

For each triggering event s, if set $Q(s)$ is not empty, one capacity module of size $min(m_1, q)$ of request $(r,q) \in Q(d)$ is restored along path p. The request is the one for which the cost of its restoration path is lowest.

The routing in graph G and the cost of restoration path of capacity module Δ is evaluated with a shortest path labelling algorithm; the cost $x(e)$ of PN path e is defined as follows (a_1, a_2, a_3 are arbitrary positive constants ($a_1 \leq a_2 \leq a_3$)):

$$x(e) = \begin{cases} a_1 \cdot c(e) \cdot \Delta & \text{iff } f(e,s) + \Delta \leq z(e) \\ a_2 \cdot c(e) \cdot (f(e,s) + \Delta - z(e)) & \text{if } f(e,s) + \Delta \leq q(e) \cdot m_2 \\ a_3 \cdot c(e) \cdot (f(e,s) + \Delta - q(e) \cdot m_2) & \text{otherwise} \end{cases} \tag{13}$$

where $z(e) = \max_s f(e,s)$.

After one capacity module has been restored, the set $Q(s)$ is updated and the values of $q(e)$ and $f(e,s)$ are redefined.

Step 3 is repeated until all sets $Q(s)$ are empty.

6. NUMERICAL EVALUATION

The performance of the algorithm described in the previous section was evaluated for a 28 node, 45 span, intercity network of a grid/mesh type. Node-to-node capacity demand ranged from 1 to almost 1000 primary PCM groups. The evaluation was split into partial tasks described below; the final results are summarized in Fig.2.

<u>Task 1</u>: 1:1 protection

First, the network was designed and analyzed assuming the standard 1:1 protection mechanism for a nominal traffic pattern and capacity demand. The capacity demand was next increased by 25, 50 and 100 percent of the nominal volume, and the design was repeated for each new capacity demand. For each of the resulting network configurations the total cost of active and redundant PN paths capacity was computed. The cost of a PN path was defined as the sum of the cost of PN links that form the path; a PN link cost was assumed proportional to the number of its capacity modules and its length.

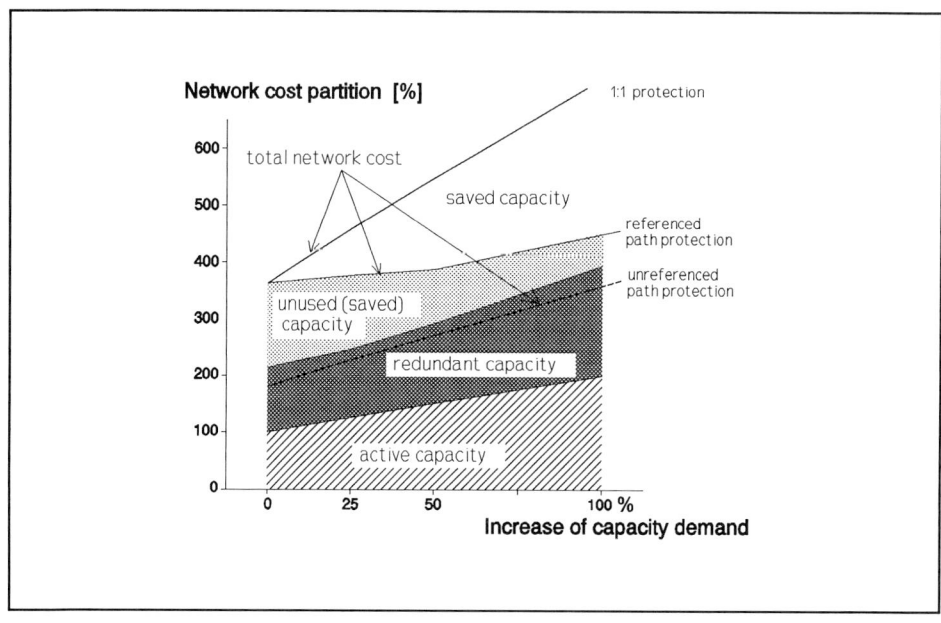

Fig.2. Network cost partition

Task 2: referenced path protection

The network configurations from task 1 were used as a starting point for network design assuming path protection and the algorithm described in the previous section; new redundant PN paths were designed (for each capacity demand) and their total cost was computed. Protection paths were established on redundant PN paths from task 1 whenever it was cost effective (the network configuration from task 1 was thus treated as a reference point); otherwise the paths were augmented. As the overall effectiveness of path protection is greater than the 1:1 protection then some capacity of the redundant PN paths from task 1 remains unused - is saved.

Task 3: unreferenced path protection

Here only the active PN paths configuration and capacity were taken from task 1 - the redundant PN paths were designed from scratch (without any reference to the redundant paths from task 1) assuming path protection and the algorithm from the previous section; the total cost of redundant paths was computed.

As can be seen from the figure, path protection shows considerable superiority over the classical 1:1 protection; the price to be paid for the more efficient and flexible capacity utilization is the cost of DCCs hardware and software. Note that the capacity that is unused when substituting the network from task 1 with the network from task 2 can be utilized effectively to absorb growing capacity demand. In particular, if the demand grows by 50%, the total network capacity has to be increased only by 5%, while maintaining 1:1 protection would require 50% capacity increase (comp. Fig.2). Note also that the referenced version of path protection (task 2) still leaves some unused capacity (it decreases as the global capacity demand increases); this could be used to carry additional traffic resulting from e.g. local traffic pattern changes and overloads.

7. REFERENCES

[1] Lubacz J., Jarociński M., Pióro M., Soto O.: On the ATM network architecture and routing principles; *7th ITC Specialists' Seminar*, Morristown, 1990

[2] Gavish B., (et al.): Fiberoptic Circuit Network Design Under Reliability Constraints; *IEEE JSAC*, Vol.7, No.8, Oct.1989, pp.1181÷1187

[3] Sallai Gy., Kollath G.: Management of Metropolitan Transmission Networks Under Transmission Failure Conditions; *Budavox Rev.*, pp.17÷25

[4] Lindberg P.: Optimization of a standby protection network; *NETWORKS'80*, Paris 1980, paper VIII.3 (pp.192÷196)

Information Transfer Techniques, B-ISDN and ATM

Traffic control in the B-ISDN

J. W. Roberts

France Telecom, CNET/PAA/ATR, 38 rue du Général Leclerc,
92131 Issy les Moulineaux, France.

Abstract

We consider the critical problem of traffic control for variable bit rate sources in the ATM based B-ISDN. It is shown how specific controls must be designed for distinct service categories. Statistical multiplexing with low delays can be achieved if source peak rate is restricted to a small fraction of the multiplex bandwidth. It is argued that admission control at burst level is necessary for a large class of on-off sources. The integration of high speed LAN-LAN interconnections remains a serious traffic control problem.

1 Introduction

The design of the B-ISDN, which we assume will be based on ATM technology, depends critically on the definition of an effective traffic control capable of guaranteeing required quality of service for a wide variety of connection types. In the present context, the term traffic control includes the actions of routing and resource allocation, necessary for setting up virtual connections, as well as the protective measures required to maintain throughput in overload situations. It has a wider meaning than flow control which term we would reserve for actions affecting the flow of information on established connections. While experience gained in present circuit switched and packet switched networks is invaluable, their traffic control techniques cannot easily be adapted to the B-ISDN. The high transmission speeds, the widely differing service requirements and the need to operate in a public (and therefore non-cooperative) environment impose the definition of novel procedures having an impact both on the nature of the services which can be offered and on the way the network is constituted.

Despite the recent publication of a large number of papers on this subject (of which the references cited in this paper only constitute a sample), it remains difficult to extract a consensus point of view: some solutions depend on mechanisms which other authors maintain are not feasible; reactive controls are ruled out from the start by some but are seen to be essential by others ... The limited aim of the present paper is to clarify the issues involved and hopefully to throw some light on the nature of the stumbling blocks preventing agreement.

An important source of misunderstanding results from different ideas about the nature of broadband services. We therefore begin, in Section 2, by a discussion of service characteristics, identifying different categories requiring different traffic control actions. In Section 3, we then discuss the particular constraints and requirements of the B-ISDN influencing the choice of appropriate traffic control actions. An important objective of traffic control being to allow a high level of resource utilization, we examine, in Section 4, how statistical multiplexing efficiency depends on source characteristics. This efficiency can only be realized if there is a strict control on user access; we discuss admission control and necessary bandwidth enforcement at both call and burst levels in Section 5. Conclusions derived from the foregoing discussion are summarized in Section 6. It must be stressed here that these conclusions engage the responsibility of the author alone and should not be taken for the point of view of his employer.

2 B-ISDN service characteristics

While all agree that much uncertainty surrounds the nature of future demand for broadband connections, it is possible to distinguish a number of service categories according to the requirements they may place on network traffic control.

2.1 Service classification

Various classifications of broadband services are possible: e.g. according to whether they are distributive or interactive [WRI 90] or whether they provide access to people, to information or to information processing [PAT 90]. We identify the following representative service classes which are distinct by the type of broadband connection they require:

- *video communication*: videophone, videoconference, shared viewing of a visual object;

- *image retrieval*: numerous professional and residential applications — medecine, publishing, home shopping, ...;

- *LAN-LAN interconnect*: an extension to a wide area of LAN capabilities is seen as an early broadband requirement [AMI 90];

- *multimedia communication*: sound, video and data can be used simultaneously in numerous applications based on multimedia workstations.

An essential characteristic of these services is their required bit rate. The rate of a video codec depends on required image quality; a mean rate of a few megabits/s is probably sufficient for most communication applications. Image retrieval bit rate should be high enough to provide a rapid response time; a typical photograph is coded in around 100 Kbytes allowing display within one second at megabit/s rates. LAN-LAN interconnection would ideally be at the LAN rate of some tens of Mbit/s or more although only a limited number of end applications actually depend on these very high speeds (distributed computing [LIN 90], animated image sequences, ...). A data communication service involving file transfers can be loosely assimilated to the image retrieval service.

2.2 Traffic characteristics

With the significant exception of video codecs, whose frame to frame bit rate varies in a continuous range [VER 89], most VBR sources are of the on/off type: periods of activity or bursts with cell emissions at constant rate alternate with periods of inactivity or silence. In the following, to simplify the discussion, we consider that multimedia communications are simply a superposition of such sources; we do not study specific multimedia traffic controls.

For purposes of traffic control, it is useful to distinguish between sources whose rate variation characteristics (e.g. mean rate, expected burst length, ...) can be predicted in advance and sources whose activity pattern depends on factors which only become apparent during the call. Among the former we would include packetized speech terminals and standard video codecs while the latter would include terminals and hosts involved in most interactive data communications and image retrieval sessions.

It also seems important to distinguish services according to whether they are more or less adapted to sharing bandwidth. We have in mind an extension of circuit switching where portions of bandwidth are attributed to individual calls; this is an appropriate operating mode when services are delay sensitive and have a small bandwidth requirement with respect to the multiplex rate. Bandwidth is not usually shared in data networks (including LANs) where packets are emitted at the multiplex rate and multiplexing efficiency is achieved at the expense of considerable buffering; services multiplexed in this way must be tolerant of potentially large and varying delays.

2.3 Source categories for traffic control

With respect to the above characteristics, we identify three VBR source categories for which different traffic control strategies would appear to be necessary:

- category 1: sources whose statistical characteristics are sufficiently well known when the call begins and whose peak rate is low compared to the multiplex rate (e.g. standard communication video codecs);
- category 2: on/off sources whose statistical variations are not known in advance and whose bit rate when active is low compared to the multiplex rate (e.g. image retrieval);
- category 3: on/off sources whose bit rate when active is high compared to the multiplex rate (e.g. LAN-LAN interconnection).

These categories do not cover all conceivable VBR sources but are sufficiently representative for the present discussion.

3 Traffic control design principles

The principal difficulties in designing an effective traffic control for the B-ISDN are due to the high transmission speeds, the heterogeneous nature of offered traffic and the need to operate in a public environment with users who cannot be assumed to behave for the good of the network.

3.1 Traffic routing and resource allocation

The primary function of traffic control is to allocate network resources (bandwidth and memory) to satisfy a new user connection request without affecting the quality of service provided to existing connections. Resources may be allocated at several levels: quasi-permanently for LAN-LAN interconnection, on a call by call basis or at the start of each burst in an on/off connection. The following design principles for a resource allocation procedure are adapted from [WOO 90]:

- it should be possible to allocate the resources of individual network links without consideration of other network elements or particular node architectures;
- requested connections should be described by a small number of traffic descriptors and network performance should be insensitive to other source characteristics;
- it should be possible to operate the network at a sufficiently high load.

When insufficient resources are available to set up a new request it should be denied access. This is the essential role of *admission control* which may act at any or all of the levels identified above [FIL 89, HUI 89].

3.2 Congestion control

For packet switched data networks, the main function of traffic control from the network point of view is the prevention of throughput degradation and loss of efficiency due to overload [GER 80]. While this, of course, remains an objective in the B-ISDN, it may be argued that congestion would be avoided by an effective admission control. This is to extend the routing principles of circuit switched networks and, for certain broadband services, would be a more appropriate congestion control than to restrict the data flow of established connections, however 'fairly' this might be done.

If admission control is only implemented at call level, there may occur overloads due to the simultaneous transmission of too many bursts. Excess cells belonging to all active communications will then be rejected resulting in a general service degradation. This unfortunate situation would be avoided if admission control were also performed at burst level: an on/off connection entering an activity period would only be allowed access if the current overall arrival rate were sufficiently low.

One particularity of a public network is the need to protect service quality against the acts of users who, for one reason or another, might use their potential to emit cells at up to the access rate (of around 150 Mbit/s) to provoke network congestion. It is necessary to ensure that users conform to the traffic characteristics assumed when making an admission control decision and which determine bandwidth or buffer requirements.

3.3 Impact of high transmission speeds

The fact that propagation time in a high speed long distance network becomes a significant part of response time has a strong impact on the operation of window based data transmission protocols. Window size corresponds to the maximum number of transmitted data units which can remain unacknowledged. In the present context, extremely large windows are necessary to maintain a continuous flow of data: the time between emission and acknowledgement of a 1 Kbyte packet transmitted across the USA at 45 Mb/s is 264 times the transmission time [MIT 90]. If the window size is the only flow control, multiplex buffers must be dimensioned in consequence: even though considerably less than a full window size must be reserved in each node [MIT 90], the required memory for all active virtual channels is likely to be huge.

The efficiency of burst admission control depends on the relative values of the time required to reserve bandwidth and the burst transmission time which can be unfavourable for high speed connections and short bursts [DOS 90b,BOY 90a].

3.4 Impact of heterogeneous traffic

To meet the different quality of service requirements of different service classes, we distinguish the following possibilities:

- service classes are not differentiated and the network is designed to meet the most stringent constraints;
- separate resources are reserved for distinct service classes allowing service specific traffic controls [COO 90];
- resources are shared but access is controlled to optimize utilization while meeting the different quality of service constraints [HYM 90].

While the first possibility would correspond to an ideal for an integrated network, it is difficult to see how the conflicting requirements of categories 1 and 2, on the one hand, and category 3, on the other hand, can be met without dedicating resources. In the interests of simplicity, we believe the differential treatment of different service classes considered in the latter possibility should probably be limited to the use of two cell loss priority classes, as considered in [KRO 91].

4 Statistical multiplexing of VBR sources

It is important to determine the extent to which statistical multiplexing is advantageous for different service categories. For illustration purposes, we consider an isolated multiplexing stage.

4.1 Burst scale and cell scale congestion

It is useful to distinguish congestion phenomena occuring in a time scale typical of an on/off source activity period or a video frame duration (burst scale) and phenomena occuring within the time scale of an inter-cell interval (cell scale) [NOR 91]. The former type of congestion occurs when the cell arrival rate exceeds multiplex capacity and leads to a more or less constantly growing queue as long as the arrival rate excess lasts. Cell scale congestion is due to the simultaneous arrival of cells from independent sources and occurs even though the cell arrival rate is less than capacity. We then distinguish two alternatives for defining an acceptable multiplex load:

- burst scale delay – a large buffer is provided to absorb burst scale congestion with the buffer overflow probability limited to 10^{-9}, say;

- burst scale loss – a (much smaller) buffer is provided to absorb cell scale congestion only with an overflow probability of 10^{-9}, given that the arrival rate is less than capacity ; the acceptable multiplex load is then such that the arrival rate only exceeds capacity with a very small probability (10^{-9}, say).

Higher link occupancy is achieved in the first alternative at the expense of longer delays. If burst scale admission control is used, the acceptable probability of arrival rate exceeding capacity in the second alternative may be higher (10^{-4}, say).

4.2 Burst scale delay

The problem with absorbing burst scale delays is that performance depends significantly on the detailed characteristics of multiplexed sources such as the distribution of burst lengths for on/off sources [BEN 90,WOO 89] or the autocorrelation function for VBR video sources [ROB 91]. Without knowing these characteristics, it is impossible to dimension buffers to meet a prescribed cell loss rate. Burst scale delays may also be considered too long for sources like VBR video codecs [VER 89].

A network carrying category 3 services only, however, would have to be dimensioned to absorb burst scale delay since this is the only way of achieving high multiplex utilization for such services. Buffers would need to be largely dimensioned, with memory reserved for each VC depending on end to end window size [DOS 90b,MIT 90,RAM 90]. The cell loss probability would not then be a buffer dimensioning criterion.

4.3 Burst scale loss

If the objective is to maintain the cell arrival rate below link capacity (with probability 10^{-9}), the allowed traffic mix for a given link can be determined simply from the arrival rate stationary distribution [APP 90,UOS 90, ...]. If λ_t is the arrival rate process and

C the link capacity, the cell loss rate may be approximated by the 'freeze-out fraction':
$\phi = E\{(\lambda_t - C)^+\}/E\{\lambda_t\}$. Note that this parameter is much less sensitive to individual source characteristics than any delay based criterion and, for on/off sources in particular, it is sufficient to know the source mean rate and peak rate.

Given link capacity and a number of known traffic classes (i.e. sources of category 1), it is possible to determine the set of traffic mixes compatible with the imposed grade of service. The surface defining this set turns out to be approximately planar (e.g. [APP 90]) allowing the definition of acceptable mixes by a linear function: a source of given type requires an 'equivalent bandwidth' (somewhere between mean and peak rates) and the sum of the equivalent rates of all accepted calls must be within capacity. Buffer size is here determined by cell scale models to ensure minimal loss as long as the input rate is within multiplex capacity and is typically around 100 cell places.

4.4 Multiplexing efficiency

Statistical multiplexing efficiency may be gauged by the load realized when meeting quality of service constraints. The following results from [VIR 90] show how this efficiency for on/off sources depends on source parameters.

Consider a superposition of homogeneous on/off sources with on probability α and peak rate h. Let the link capacity be Ch. The maximum number of sources N compatible with a freeze-out fraction of 10^{-9} is determined for different values of C from the binomial distribution of λ_t. The realized multiplex load is then $(N\alpha/C)$.

For $C = 100$ and $\alpha < 0.2$, realized load is approximately constant and greater than 70%. For $C = 10$, on the other hand, the realized load is no greater than 31% for $\alpha < 0.2$. These figures would suggest that to achieve efficient statistical multiplexing with burst scale loss, the source peak rate should barely exceed one hundredth of multiplex capacity. A more comprehensive discussion of multiplexing efficiency may be found in [LOU 90].

5 Admission control and bandwidth enforcement

For service category 1, traffic control could be based on call level admission control alone while for category 2, it is also necessary to implement admission control at burst level. Admission control is based on required bandwidth which must be enforced by a policing mechanism. The specific requirements of category 3 services are briefly discussed

5.1 Call level admission control only

Assuming call statistical characteristics are known in advance, it is possible to determine an equivalent bandwidth requirement and to perform admission control by comparing this with available link capacity determined from the declared characteristics of calls in progress. If we do not aim to absorb burst scale delay, the necessary characteristics for an on/off source are simply the mean and peak bit rates (cf. Section 4).

In practice, it would be necessary to deduce an equivalent bandwidth assuming the worst case parameters compatible with the policing mechanism employed. For the mean rate in particular, this requirement can introduce considerable discrepancies between the bandwidth allocated and that actually used by a VBR source.

5.2 Mean rate policing

Among the various devices proposed to enforce a given mean rate while allowing a measure of burstiness, the leaky bucket appears as the most efficient [RAT 90b]: a counter is incremented on each cell emission and decremented at constant rate a ; any cell arriving when the counter has attained its maximum M is discarded (or 'tagged' as suitable for discarding in case of congestion [ECK 90,GAL 89]). Broadly speaking, a determines the long term mean rate while M limits the maximum duration of an activity period.

To fix a and M requires a difficult compromise between three factors [ECK 90]:

- the margin between the mean rate authorized by a and the real source mean rate;
- the responsiveness to excessively long bursts which is determined by M;
- the proportion of cells discarded or tagged.

In general, for randomly varying burst and silence lengths, to achieve a low discard probability, either a must be set considerably higher than the source mean rate or M must be very much greater than the expected burst length. Conversely, if a and M are tightly dimensioned, the user must accept to lose more than a negligible proportion of cells at the network interface. The difficulty of achieving a satisfactory compromise severely limits the field of application of a traffic control based on admission control at call level only [RAT 90a,KOS 90].

The device of not discarding offending cells but only tagging them has the advantage of restricting user traffic only in case of network congestion. The drawback is that no service quality standards can be guaranteed for the tagged traffic [COO 90] and the quality of service of non-tagged cells can suffer significantly from an overload of tagged cells [GRA 91].

5.3 Burst level admission control

For services of category 2 and some services of category 1, it is necessary to implement admission control at burst level, as discussed for example in [FIL 89,HUI 89, ...]. This is described as 'in-call negociation' in [DOS 90b] and an implementation known as the Fast Reservation Protocol (FRP) is presented in [BOY 90a].

As previously noted, the time required to perform admission control is especially critical in a high speed environment. A variant of the FRP supposes that link capacity can be reserved 'on the fly' by the first cells of a burst as and when it arrives in each switching node [BOY 90b]: on each link of its path, a VC would either be enabled to transmit cells or not; its status would be changed to enabled, if capacity allowed, following an analysis performed in the time necessary to transmit a burst preamble; this status would be removed at the end of the burst, notified by an appropriately identified trailer; cells arriving on a non-enabled VC would simply not be taken into account; special procedures would be necessary to inform the user in case of blocking to allow him to make a new attempt.

If the admission decision depends only on the required burst bit rate being less than available capacity, burst blocking probability will increase substantially with this rate especially at times of momentary overload. This leads us to suggest that a minimum capacity should be available before a burst of *any* rate can be accepted. This free capacity limit effectively corresponds to the maximum source peak rate which can be multiplexed in this way and would need to be fixed with regard both to user requirements and achievable multiplexing efficiency.

Call level admission control can only be based on the source peak rate and, since this is the only information available for existing calls, available capacity must be estimated by real time traffic measurements. The type of measurements to be made and the manner of calculating residual capacity remain to be defined. The same limitation on maximum peak bit rate would apply here as for burst level admission control.

5.4 Peak rate policing

If burst admission control is employed, it remains necessary to verify that a user conforms to a declared peak rate. Theoretically, it would be sufficient to check that the inter-cell interval is not smaller than the inverse of the peak rate; in practice, it is necessary to account for cell jitter introduced by delay stages situated between the source and the observation point. These may be in the user's premises (e.g. a LAN) or in an an adjacent network when policing is performed at a network–network interface. Because of jitter it is necessary to estimate the peak rate by observing a number of successive inter-cell intervals [BOY 90a,NIE 90].

As highlighted in [BOY 90a], the required observation period (manifested by the value of the leaky bucket threshold, for example) can be quite long, thus limiting the responsiveness of the policing mechanism. Since short sharp high rate bursts can have a very negative effect on multiplex congestion, to provide the level of protection required in a public network, it appears necessary to enforce the peak rate by actively spacing out cells arriving too close together. This is the function of the 'spacer–controller' described in [BOY 90a].

5.5 LAN-LAN interconnection

Very high rate bursts are emitted on LAN-LAN connections in an extremely variable fashion [LEL 90]. We believe an adapted traffic control for this type of traffic would be based on both end to end windowing coupled with some form of rate control (e.g. as in [RAM 90]).

A leaky bucket based rate control (proposed in [AMI 90]) would presumably need to resolve the difficult compromise mentioned in Section 5.2 by allowing a high cell discard probability, effectively 'shaping' the offered traffic to avoid network congestion. The drawback is that the end user would experience very poor end to end performance [LEL 89].

This service category clearly poses very serious traffic control problems which seem far from finding a solution. An important question is whether the B-ISDN should, in fact, be designed to support this type of traffic (which, however, corresponds to the only existing commercially viable application [TIM 90]), or whether users should be encouraged to establish individual, identifiable calls whose characteristics (notably, peak bit rate) allow the application of call and burst admission control as described above. An alternative is to provide LAN-LAN interconnection via gateways having the role of reducing the peak bit rate.

6 Conclusions

Traffic control must be designed with specific service categories in mind. We have identified three distinct categories of VBR services typified, respectively, by video communication, image retrieval and LAN-LAN interconnection.

Services of the first category have a relatively low bit rate with respect to the multiplex rate and their statistical variations can be well characterized in advance. A significant statistical multiplexing advantage can be achieved assuming burst scale loss operation. Admission control at call level only is sufficient provided basic source characteristics such as mean bit rate can be enforced. We note, however, that reliance on policing mechanisms like the leaky bucket may require considerable bandwidth overallocation for on/off sources with variable burst and silence lengths.

The second category of services is characterized by a sequence of bursts and silences which is unknown before the call begins. The burst bit rate is small enough to allow a significant multiplexing advantage with burst scale loss. It is necessary for such services to perform admission control at burst level as well as at call level. To maintain efficiency for short burst lengths, it would be preferable to implement a resource reservation scheme acting 'on the fly' as each burst arrives. To equalize blocking probabilities for services of different bit rates, we suggest that a minimum amount of resources be available before admitting any new call or burst. This effectively defines a maximum bit rate (probably of only several Mbit/s) for which service quality is guaranteed.

Multiplex buffer requirements for these first two categories would be determined to meet cell loss rate requirements, assuming the input rate were less than link capacity. A buffer of some one hundred cell places seems sufficient.

The third category of services has a burst bit rate which is high with respect to the multiplex rate. This means that statistical multiplexing efficiency can only be achieved with large buffers absorbing burst scale delay. Memory would be allocated for each virtual connection and required buffer dimensions would depend on end to end window sizes rather than on a required cell loss rate; indeed, the latter cannot otherwise be guaranteed since the buffer overflow probability depends significantly on many uncontrollable source characteristics including the distributions of burst and silence lengths.

There is a clear incompatibility between the first two categories and category 3 and it would appear that the B-ISDN should therefore contain at least two dedicated sets of resources. Alternatively, it may be decided that the B-ISDN should only offer connections of up to a certain maximum peak bit rate (of several Mbit/s) with LAN-LAN interconnection rate reduced by appropriately designed gateways.

Acknowledgement

The author has greatly benefitted from his discussions on the subject matter of this paper with his colleagues P. Boyer and F. Guillemin.

References

[AMI 90]   Amin-Salehi B., Flinchbaugh G.D., Pate L.R. Implications of new network services on B-ISDN capabilities. Infocom '90, 1990.

[APP 90]   Appleton J. Modelling a connection acceptance strategy for asynchronous transfer mode networks. ITC Seminar, Morristown, 1990.

[BEN 90]   Bensaou B., Guibert J., Roberts J. Fluid queueing models for a superposition on on/off sources. ITC Seminar, Morristown, 1990.

[BOY 90a] Boyer P., A congestion control for the ATM. ITC Seminar, Morristown, 1990.

[BOY 90b] Boyer P., Louvion J.R., Tranchier D. Intelligent multiplexing in an ATM network. Third IEEE COMSOC Multimedia, Bordeaux, November 1990.

[COO 90]   Cooper C.A., Park K.I. Toward a broadband congestion control strategy. IEEE Network Magazine, May 1990.

[DOS 90b] Doshi B.T., Dravida S. Congestion control for bursty data in high speed wide area packet networks: In-call parameter negociations. ITC Seminar, Morristown, 1990.

[ECK 90]   Eckberg A.E., Luan D.T., Lucantoni D.M. An approach to controlling congestion in ATM networks. Intl J. of Digital and Analogue Comm Systems. Vol 3, pp 127-135, 1990.

[FIL 89]   Filipiak J. Structured systems analysis methodology for design of an ATM network architecture. IEEE JSAC, Vol 7, N 8, October 1989.

[GAL 89]   Gallassi G., Rigolio G., Fratta L. ATM: bandwidth assignment and enforcement policies. Globecom 89, Dallas 1989.

[GER 80]   Gerla M., Kleinrock L. Flow control: a comparative survey. IEEE Trans Commun. Vol COM-28, No 4, April 1980.

[GRA 91]   Gravey A., Boyer P., Hébuterne G. Tagging versus strict rate enforcement. COST 224 document TD(91)004.

[HUI 89]   Hui J.Y. Resource allocation for broadband networks. IEEE JSAC, Vol 6, No 9, December 1988.

[HYM 90]   Hyman J., Lazar A., Pacifici G. Real time scheduling of switching nodes based on asynchronous time sharing. ITC Seminar, Morristown, 1990.

[KOS 90]   Kostutufor. On admission control and policing in an ATM based network. ITC Seminar, Morristown, 1990.

[KRO 90]   Kroener H., Theimer T.H., Briem U. Queueing models for ATM systems – a comparison. ITC Seminar, Morristown, 1990.
[KRO 91]   Kroener H., Hbuterne G., Boyer P., Gravey A. Priority management in ATM switching nodes. To appear IEEE JSAC, 1991.
[LEL 89]   Leland W.E. Window based congestion management in broadband ATM networks: the performance of three access control policies. Globecom '89, Dallas, 1989.
[LEL 90]   Leland W.E. LAN traffic behavior from milliseconds to days. ITC Seminar, Morristown 1990.
[LIN 90]   Lindisky W.P. Data communications needs. IEEE Networks Magazine. March 1990.
[LOU 90]   Louvion J.R., Boyer J., Gravereaux J.B. Statistical multiplexing of variable bit rate sources in ATM networks. Third IEEE CAMAD, Turin; September 1990.
[MIT 90]   Mitra D., Mitrani I., Ramakrishnan K.G., Seery J.B., Weiss A. A unified set of proposals for control and design of high speed data networks. ITC Seminar, Morristown, 1990.
[NIE 90]   Niestegge G. The 'leaky bucket' policing method in the ATM network. Intl J. of Digital and Analogue Comm Systems. Vol 3, 1990.
[NOR 91]   Norros I., Roberts J., Simonian A. Virtamo J. The superposition of VBR sources in an ATM multiplexer. To appear IEEE JSAC, 1991.
[PAT 90]   Patterson J.F., Egido C. Three keys to the broadband future: a view of applications. IEEE Network Magazine, March 1990.
[RAM 90]   Ramamurthy G., Dighe R.S. Distributed source control: a network access control for integrated broadband packet networks. Infocom '90. 1990.
[RAT 90a] Rathgeb E.P., Theimer T.H. The policing function in ATM networks. ISS'90, Stockholm, 1990.
[RAT 90b] Rathgeb E.P. Policing mechanisms for ATM networks - modelling and performance comparison. ITC Seminar, Morristown, 1990.
[ROB 91]   Roberts J., Guigert J., Simonian A. Network performance considerations in the design of a VBR codec. To appear, ITC 13, Copenhagen 1991.
[TIM 90]   Timms S., Broadband communications in the 1990s: a market model. ICC'90, 1990.
[UOS 90]   Uose H., Shioda S., Mase K. Fast cell loss rate evaluation methods and their application to ATM network control. ITC Seminar, Morristown, 1990.
[VER 89]   Verbiest W. Pinnoo L., A variable bit rate video codec for ATM networks. IEEE JSAC, Vol 7, N 5, June 1989.
[VIR 90]   Virtamo J. Statistical multiplexing of on-off sources. COST 224 document TD(90)034, 1990.
[WOO 89]   Woodruff G., Kositpaiboon R. Fitzpatrick G., Richards R. Control of ATM statistical multiplexing performance. ITC Seminar, Adelaide, 1989.
[WOO 90]   Woodruff G.M., Kositpaiboon R. Multimedia traffic management principles for guaranteed ATM network performance. IEEE JSAC, Vol 8, No 3, April 1990.
[WRI 90]   Wright D.J., To M. Telecommunication applications of the 1990s and their transport requirements. IEEE Network Magazine, March 1990.

An Efficient Approach to Analyze Discrete-Time Queues with Markovian Arrivals and Services in BISDNs

Zhensheng Zhang

1220 SW Mudd Bldg., Center for Telecommunications Research
Columbia University, New York, NY 10027, USA

Abstract

This paper presents a new, efficient computational procedure to calculate the queue size distribution in a queueing system with a number of independent sources. Each source is characterized by a finite state discrete-time Markov chain. The service process can be Markovian as well. Explicit expression for the queue size distribution is obtained directly. The method is based on computing the modified spectral expansion of the state distribution of the system. This representation requires evaluating the roots inside the unit cycle of the entire system characteristic function and solving a set of linear system equations. If the arrival process is composed of multiple two-state Markov chains, by using a decomposition technique the problem of finding the roots becomes simple. The complexity of the algorithm is compared with that of the standard generating function approach. We observe that our method is more efficient than the generating function approach in most cases.

1 Introduction

In this paper, we consider a queueing system with multiple independent types of traffic. Each input may represent data, voice, or video, etc. The arrival process is composed of a number of independent finite state discrete-time Markov chains. The service process can also be a Markov chain. The buffer capacity of the system is assumed to be infinite. Since Broadband ISDN will support many bursty sources and the burstiness of the traffic can be modeled as Markov chains, the practical importance of the model is growing today. A better understanding of the performance of the BISDNs in such an environment becomes an essential requirement for network design and management. The queueing systems with integrated traffic have been analyzed extensively through various approaches [1-6]. In [7], the generating function of the queue size distribution is obtained for the model mentioned above. Later on that result is extended to the case in which the number of arrivals is a random variable governed by the state of the underlying Markov chain [8]. No attempt is made for the queue size distribution in either [7] or [8].

This model is considered again in a recent paper by Li [9] using the generating function approach. The queue size distribution is obtained. In the paper, the problem of obtaining the generating function for the queue size distribution requires the evaluation of the roots inside the unit cycle and solving a set of linear system equations resulting from the roots for the boundary probabilities. Due to the analytical property of the generating function inside the unit cycle, all the roots inside the unit cycle are cancelled out in the expression of the generating function. Li therefore refers to these roots as vanishing roots. The roots outside the cycle, called non-vanishing roots, are then needed to construct the queue size distribution. Each non-vanishing root contributes a geometric term to the queue size distribution. The coefficients of the geometric terms are computed from the generating function, which in turns requires the boundary probabilities.

The vanishing roots contribute *nothing* to the structure of the queue size distribution. Then, do we have to compute them? In other words, can we construct the queue size distribution using the non-vanishing roots directly? In this paper, we demonstrate that one can actually obtain the queue size distribution explicitly without using the vanishing roots at all. To this end, we first simply assume that the solution is in the form:

$$P(i) = \begin{cases} b & i = 0 \\ \sum_k a_k z_k^i G_k, & i \geq 1 \end{cases}$$

and then determine the elements $\{b, a_k, G_k, z_k\}$, where $1/z_k$ is the so called non-vanishing root. That is the *key idea* of this paper. It turns out that the approach is nothing but a modified spectral expansion for the state distribution of the system.

To evaluate all the elements $\{b, a_k, z_k, G_k\}$ for a large system is by no means an easy task. However, by using a decomposition technique, we show that the problem of finding the pairs (z_k, G_k) for the entire system can be decomposed into many small eigenvalue and root finding problems. A set of linear system equations resulting from the boundary conditions is obtained for the unknowns $\{b, a_k\}$. When the arrival process is composed of multiple two-state Markov chains, finding the pairs is no longer a problem. The main work in this approach hence becomes to find the solution of the linear system equations. The size of the system equations can be reduced significantly since there are many identical terms. Therefore all the elements can be computed efficiently.

The method proposed here is compared with the standard generating function approach in terms of the computational complexity. We observe that our method is more efficient than the generating function approach in most cases, since it does not require the computation of the vanishing roots.

Our method is even more promising when it is used to analyze finite buffer queueing systems. Using the same approach, the queue size distribution can be obtained at the expense of slightly increasing the size the linear equations [10]. This is somewhat surprising since finite queues rarely have elegant solutions. It is well-known that, however, the generating function approach cannot apply in most of finite buffer queueing systems.

The powerful spectral expansion method has been established in the works [11-14], in which the stochastic fluid models are considered. In [11-14], the Kronecker product form to the eigenvalues and eigenvectors is also given. A multiple birth-death queueing model is considered in [15].

2 Single Markov Chain Case

2.1 System Model

In integrated networks, information to be transmitted is segmented into small, fixed length cells. Time is divided into fixed length slots. The transmission time of a cell is one slot. Cell arrivals can occur only at slot boundaries. We first assume that the cell arrival process is a discrete-time Markov chain and there are M (deterministic) servers (channels) in the system. We then extend the results to the case in which both the arrival and service processes are composed of multiple independent Markov chains. The buffer capacity is assumed to be infinite.

Denote the number of cell arrivals in the n-th slot by $a(n)$. We assume that $\{a(n), n \geq 1\}$ is a homogeneous, discrete-time, irreducible and positive recurrent Markov chain with transition matrix given by $Q = \{Q(i,j), 0 \leq i, j \leq N\}$, where

$$Q(i,j) = P(a(n+1) = j | a(n) = i), \ 0 \leq i, j \leq N. \tag{1}$$

Let $v_j, j = 0, ..., N$, denote the steady-state probability that the Markov chain is in state j. It is well-known that $v = (v_0, ..., v_N)^T$ satisfies the following linear equations

$$v = vQ, \quad \sum_{j=0}^{N} v_j = 1. \tag{2}$$

The average number of arrivals during a slot is equal to $\bar{a} = \sum_{i=0}^{N} i v_i$.

2.2 General Structure of the Solution

Let $q(n)$ be the queue size at the end of the n-th slot. The state of the system at the end of n-th slot is defined by $(q(n), a(n))$. The process is a two-dimensional Markov chain on the stat space $\{(i,k), i \geq 0, 0 \leq k \leq N\}$. The evolution of the queue size is governed by the following equation

$$q(n+1) = (q(n) + a(n) - M)^+ \tag{3}$$

where $(x)^+ = max(0, x)$.
Let

$$p(n,i,k) = P(q(n) = i, a(n) = k).$$

A necessary and sufficient condition for the existence of the equilibrium probabilities $\{p(i,k)\}$ is $\bar{a} < M$. We henceforth assume that this condition holds in this section, in which case

$$p(i,k) = \lim_{n \to \infty} P(q(n) = i, a(n) = k).$$

Let $P(i) = [p(i,0), p(i,1), ..., p(i,N)]$ be a vector of dimension $N+1$. The balance equations that the joint probabilities must satisfy are given by

$$p(0,k) = \sum_{m=0}^{M} [\sum_{l=0}^{M-m} p(l,m)] Q_{m,k} \tag{4}$$

and for $i \geq 1$,

$$p(i,k) = \sum_{m=0}^{N} p(i-m+M, m) Q_{m,k} \tag{5}$$

where $p(i,k) = 0$ if $i < 0$. Writing these equations in matrix form, we have

$$P(0) = P(M)Q_0 + P(M-1)(Q_0 + Q_1) + ... + P(0)(Q_0 + ... + Q_M) \tag{6}$$

and for $i \geq 1$,

$$P(i) = \sum_{m=0}^{N} P(i-m+M)Q_m \tag{7}$$

where

$$Q_m = E^{(m)}Q$$

with

$$E^{(m)} = (E^{(m)}_{i,j}),\ E^{(m)}_{m,m} = 1,\ E^{(m)}_{i,j} = 0,\quad \text{for } i \neq m \text{ or } j \neq n. \tag{8}$$

The general **structure of the solution** of (6) and (7) is given by

$$P(i) = \begin{cases} b & i = 0 \\ \sum_{k,l} a_{k,l} z_{k,l}^i G_{k,l} & i \geq 1 \end{cases}, \tag{9}$$

where b and $G_{k,l} = G_k(z_{k,l}) = [g_{k,0}(z_{k,l}), ..., g_{k,N}(z_{k,l})]$ are vectors of dimension $N+1$. $z_{k,l}$ and $a_{k,l}$ are scales. $z_{k,l}$ are the roots of

$$det(D(z)) = 0$$

in $0 < |z| < 1$ and the pair $(z_{k,l}, G_{k,l})$ satisfies

$$G_{k,l} D(z_{k,l}) = 0 \tag{10}$$

where

$$D(z) = z^{N-M} I - T(z)Q$$

with

$$T(z) = diag(z^N, z^{N-1}, ..., 1).$$

It is easy to see that

$$b = v - \sum_{k,l} \frac{a_{k,l} z_{k,l} G_{k,l}}{1 - z_{k,l}} \tag{11}$$

since $v = \sum_{i=0}^{\infty} P(i)$, where v is given in (2). The modified spectral expansion method proposed here requires the computation of $\{a_{k,l}, z_{k,l}, G_{k,l}\}$. If we can calculate the elements $\{a_{k,l}, z_{k,l}, G_{k,l}\}$, then the queue size distribution can be obtained explicitly. As we will see later, decomposition results for our model allow the pair $(z_{k,l}, G_{k,l})$ to be accurately computed. The coefficients $\{a_{k,l}\}$ are obtained from the boundary equations (6), (7) ($1 \leq i \leq N - M$).

2.3 Eigenvalues and Eigenvectors

In this subsection, we present some key relations among the $\{z_{k,l}\}$, the $\{G_{k,l}\}$ and the system parameters. These relations will be used in the decomposition technique described in the next section. To find these relations, we first consider the eigenvalues and eigenvectors of $T(z)Q$. Let $\lambda_k(z)$ be the eigenvalue of $T(z)Q$ and $G_k(z)$ be the corresponding left eigenvector, that is

$$G_k(z) T(z) Q = \lambda_k(z) G_k(z).$$

For simplicity, we assume that rank(Q)=N+1. For the case when rank(Q) < N+1, the reader is referred to [8].

Lemma 1. Under the stability condition $M - \bar{a} > 0$, the eigenvalues $\{\lambda_i(z)\}$ can be numbered so that

$$\lambda_l(z) = z^l \alpha_l(z), \quad \alpha_l(0) \neq 0, \quad 0 \leq l \leq N$$

Proof See [8].

Since

$$det(D(z)) = \prod_{k=0}^{N} [z^{N-M} - \lambda_k(z)],$$

the roots of $det(D(z))$ can be solved separately for each k from

$$z^{N-M} - z^k \alpha_k(z) = 0.$$

It was proven in [8] that, for each k, the equation

$$z^{N-M} - z^k \alpha_k(z) = 0$$

has $(N - M - k)^+$ roots in $0 < |z| < 1$. Denote these roots by $z_{k,j}$, $j = 1, ..., N - M - k$, $k = 0, ..., N - M - 1$. There are total of $\sum_{k=0}^{N-M-1}(N - M - k) = \frac{(N-M)(N-M+1)}{2}$ roots in $0 < |z| < 1$. Once the roots are obtained, the corresponding eigenvectors are given by $G_{k,l} = G_k(z_{k,l})$.

2.4 Boundary Conditions

Using the boundary equations (6), (7) ($1 \leq i \leq N - M$) and (11), we obtain the following linear equations for the unknown coefficients $\{a_{k,l}\}$. The reader is referred to [16] for detailed derivations.

$$\sum_{k=0}^{N-M-1} \sum_{l=1}^{N-M-k} a_{k,l} z_{k,l}^{-(N-M-i)} G_{k,l} C_i(z_{k,l}) - (v - \sum_{k=0}^{N-M-1} \sum_{l=1}^{N-M-k} \frac{a_{k,l} G_{k,l}}{1 - z_{k,l}}) E^{(M+i)} = 0 \quad (12)$$

where

$$C_i(z) = \sum_{k=i+M+1}^{N} z^{N-k} E^{(k)} = diag(0, 0, ..., 0, z^{N-i-M-1}, ..., z, 1) \quad (13)$$

and v is given in (2). From (13), we notice that there are $N - M - i + 1$ equations in (12) for each i, $i = 1, 2, ..., N - M$. The number of equations in (12) is equal to $\frac{(N-M)(N-M+1)}{2}$. The coefficients $\{a_{k,l}\}$ can thus be determined uniquely from (12).

Up to now, the algorithms for computing all the elements $(a_{k,l}, z_{k,l}, G_{k,l})$ in the modified spectral expansion for the state distribution have all been described. The solution of (6) and (7) is now given by

$$P(i) = \begin{cases} v - \sum_{k=0}^{N-M-1} \sum_{l=1}^{N-M-k} \frac{a_{k,l} z_{k,l} G_{k,l}}{1 - z_{k,l}} & i = 0 \\ \sum_{k=0}^{N-M-1} \sum_{l=1}^{N-M-k} a_{k,l} z_{k,l}^i G_{k,l} & i \geq 1 \end{cases}. \quad (14)$$

From (14), one may obtain various system performance measures. Two performance parameters of interest are the **queue size distribution** and the **queue size tail distribution**. These two distributions are given by

$$q(i) = \lim_{n \to \infty} P(q(n) = i) = \sum_{k=0}^{N-M-1} \sum_{l=1}^{N-M-k} a_{k,l} z_{k,l}^i G_{k,l} e, \quad i \geq 1 \tag{15}$$

and

$$t(i) = \lim_{n \to \infty} P(q(n) \geq i) = \sum_{k=0}^{N-M-1} \sum_{l=1}^{N-M-k} \frac{a_{k,l} z_{k,l}^i G_{k,l} e}{1 - z_{k,l}}, \quad i \geq 1 \tag{16}$$

respectively, where $e = (1, ..., 1)^T$ is a column vector of dimension $N+1$.

2.5 Discussion

In this section, we compare the complexity of the method presented in this paper with that of the generating function approach. Generally, using a generating function approach, one must first obtain the generating function of the state distribution, then derive the state distribution from the generating function. The later involves the inversion of a z-transform, which is by no means an easy task. In some situations it is even impossible to derive the distribution from its generating function.

In [9], an excellent method deriving the queue size distribution from its generating function is given, which involves six major steps:

a. Find all the eigenvalues $\tilde{\lambda}_i(z)$ and all the left and right eigenvectors of $diag(1, z, ..., z^N)Q$.

b. Compute all the vanishing roots of $z^M - \tilde{\lambda}_i(z)$ in $0 < |z| \leq 1$.

c. Obtain the boundary probabilities by solving a set of linear system equations of order $\frac{M(M+1)}{2}$.

d. Compute all the non-vanishing roots of $z^M - \tilde{\lambda}_i(z)$ in $|z| > 1$.

e. Compute the parameters c_{ki}^o given in equation (2.12) in [9], which involves the boundary probabilities obtained in step c.

f. Compute the queue size distribution using the parameters obtained in step e.

The approach presented in this paper consists of the following four major steps:

1. Find all the eigenvalues $\lambda_i(z)$ and all the left eigenvectors $G_i(z)$ of $diag(z^N, ..., z, 1)Q$.

2. Compute the roots of $z^{N-M} - \lambda_i(z)$ in $0 < |z| < 1$ for each i, $i = 0, 1, ..., N$.

3. Solve a set of linear system equations to obtain the coefficients, the number of equations is equal to $\frac{(N-M)(N-M+1)}{2}$.

4. Calculate the queue size distribution from (15).

As one can see, steps 1, 2, and 4 in the approach proposed here are equivalent to the generating function approach steps a, d and f (except finding all the right eigenvectors). The reciprocal of the root computed in step 2 is the very non-vanishing root computed in step d, noting the differences between the two matrices in step 1 and step a, and the two root equations in step 2 and step d. The computational requirement in steps b and e is extensive, especially for larger M. In contrast, the approach presented here does not require these two steps at all. The number of linear equations in both step 3 and step c depends on the value of M. (The coefficients c_{kl}^o computed in step e are, essentially, the coefficients $a_{k,l}$).

From the above observations, we conclude that if M is small (e.g., close to 1) and if the vanishing roots are available (or at least not difficult to obtain), the generating function approach is preferred. On the other hand, if M is small but finding the vanishing roots requires extensive computation, or if M is greater than $N/2$, the algorithm proposed here is much more efficient.

For example, consider the case when $M = N - 1$. Using the method presented here, one notices that $D(z)$ has only one root in $0 < |z| < 1$. The state distribution is given by

$$P(i) = \begin{cases} v - a_1 z_1 G_1/(1 - z_1) & i = 0 \\ a_1 z_1^i G_1 & i \geq 1 \end{cases},$$

where (z_1, G_1), $0 < |z_1| < 1$, satisfies $G_1 D(z_1) = 0$. The computation required is simply to compute the root z_1, G_1 and v (finding a_1 is straightforward). However, using the generating function approach, in addition to finding the root z_1 and G_1, one must find the $(N-1)N/2$ vanishing roots. Then, one has to solve the resultant $(N-1)N/2$ linear equations. The advantage of the current approach becomes apparent.

More significantly, the method presented here can be readily used to analyze finite buffer queueing systems. The complexity of computing the queue size distribution for the finite buffer case is about the same as or slightly more than that for the infinity case [10]. On the contrary, the generating function approach does fail in most cases in dealing with finite buffer queueing systems.

3 Multiple Markov Chains

The results presented in the previous section are in theory straightforward. A direct application of the results is not appreciated when the system size is large. However, large systems often consist of a number of independent smaller systems and the computation for each smaller system is not difficult. In this section, we use the decomposition technique to show that the bulk of the computation can be performed at each smaller system. Thus the computation requirement can be reduced significantly.

Let the arrival process consist of K independent discrete-time Markov chains and

$$A(n) = \sum_{k=1}^{K} a_k(n)$$

be the total arrivals during a slot, where $0 \leq a_k(n) \leq N_k, k = 1, ..., K$. Let $Q^{(k)}$ be the transition matrix for the k-th Markov chain. The transition matrix for the K-dimensional Markov chain $\{(a_1(n), ..., a_K(n)), n \geq 1\}$ is given by

$$Q = \otimes_K Q^{(k)} \tag{17}$$

where $\otimes_K$ denotes the K-fold matrix Kronecker production operation. Let $P(i) = \{p(i, j_1, ..., j_K)\}$ be the equilibrium vector probability of dimension $\prod_{k=1}^{K}(N_k + 1)$.

The system stability condition is $\overline{A} < M$, where $\overline{A}$ is the average number of cells arriving during one slot.

Let $\underline{j} = [j(1), j(2), ..., j(K)]$, $|\underline{j}| = \sum_{k=1}^{K} j(k)$ and

$$E_i = \sum_{|\underline{j}|=i} E^{(j(1))} \otimes ... \otimes E^{(j(K))} \tag{18}$$

where $E^{(m)}$ is given in (8). The balance equations are given by

$$P(0) = P(M)E_0Q + P(M-1)(E_0 + E_1)Q + ... + P(0)(E_0 + ... + E_M)Q \tag{19}$$

and for $i \geq 1$,

$$P(i) = \sum_{l=0}^{T} P(i + M - l)E_l Q \tag{20}$$

where $T = \sum_{k=1}^{K} N_k$. The **solution** of (19) and (20) is given by

$$P(i) = \begin{cases} b & i = 0 \\ \sum_{\underline{j},l} a_{\underline{j},l} z_{\underline{j},l}^i G_{\underline{j},l} & i \geq 1 \end{cases}, \tag{21}$$

where b and $G_{\underline{j},l}$ are row vectors of dimension $\prod_{l=1}^{K}(N_l + 1)$, and the pair $(z_{\underline{j},l}, G_{\underline{j},l})$ satisfies

$$G_{\underline{j},l} D(z_{\underline{j},l}) = 0 \tag{22}$$

with

$$D(z) = z^{N-M} I - T_1(z) \otimes ... \otimes T_K(z) Q$$

The dimension of the matrix $D(z)$ is $\prod_{l=1}^{K}(N_l + 1) \times \prod_{l=1}^{K}(N_l + 1)$. Finding the solutions of $det(D(z)) = 0$ directly leads to formidable computational problems because of its huge size. However, by using the decomposition technique, the solutions can be obtained easily as shown below.

Let $\lambda_{j(k)}(z)$ and $G_{j(k)}(z)$ be the eigen-pair of the matrix $T_k(z)Q^{(k)}, 0 \leq j(k) \leq N_k, k = 1, 2, ..., K$. Using the Kronecker product property [17], we have

$$\lambda_{\underline{j}}(z) = \prod_{k=1}^{K} \lambda_{j(k)}(z) \tag{23}$$

$$G_{\underline{j}}(z) = \otimes_K G_{j(k)}(z). \tag{24}$$

Thus the problem of finding all the eigen-pairs of $T(z)Q$, whose dimension is $\prod_{l=1}^{K}(N_l + 1)$, is reduced to find all the eigen-pairs of a much smaller size matrix $T_k(z)Q^{(k)}$. For $N_k \leq 4$, explicit expressions for the eigen-pairs of $T_k(z)Q^{(k)}$ can be obtained.

Once the eigenvalues and eigenvectors of $T(z)Q$ are obtained, the roots $\{z_{\underline{j},l}\}$ can be computed from the equation

$$z^{N-M} - \lambda_{\underline{j}}(z) = 0 \tag{25}$$

separately for each j using an iterative substitution algorithm. The corresponding eigenvector $G_{j,l}$ is then given by $G_{j,l} = G_j(z_{j,l})$.

The coefficients $\{a_{j,l}\}$ can be determined from (12) with $k, E^{(m)}$ being replaced by j, E_m correspondingly.

From the above analysis, we notice that if all the eigen-pairs of $T_k(z)Q^{(k)}$ can be explicitly obtained, finding the pairs $(z_{j,k}, G_{j,k})$ is not difficult any more. The main computational requirement becomes to find the solution of a set of linear equations. However, by combining many identical terms, the size of the linear system equations can be reduced significantly, as we will see later. In the next section, we consider the case in which explicit expressions for the eigen-pairs are available and carry out the analysis in more details.

4 Markovian Service Process

When the service process is also a Markov chain, independent of the arrival process, the previous analysis can be easily applied by using a simple service conversion [9], [14] as described in the following. Let $c(n)$ denote the number of cells transmitted during a slot. We assume that $c(n)$ is a discrete-time Markov chain in the state space $\{0, 1, .., C\}$ with transition matrix $Q^c = (Q^c_{i,j})$.

The queue size follows the evolution equation

$$q(n+1) = (q(n) + A(n) - c(n))^+ \tag{26}$$

Let $a_{K+1}(n) = C - c(n)$, then

$$q(n+1) = (q(n) + A(n) + a_{K+1}(n) - C)^+. \tag{27}$$

The model reduces to the one studied in the previous subsection. The transition matrix for $\{a_{K+1}(n)\}$ is given by $Q^{(K+1)} = (Q^{(K+1)}_{i,j}) = (Q^c_{C-i,C-j})$.

5 Multiple i.i.d. Two-State Markov Chains

In this section, we consider the case in which the arrival process is composed of K independent identical two-state Markov chains. The transition matrix for each two-state Markov chain is given by

$$Q = \begin{pmatrix} p_f & 1 - p_f \\ 1 - p_0 & p_0 \end{pmatrix}. \tag{28}$$

The average number of arrivals per slot is equal to $K\frac{1-p_f}{2-p_f-p_0}$. The system stability condition is

$$K\frac{1-p_f}{2-p_f-p_0} < M$$

Let $\underline{i} = [i(1), ..., i(K)]$, $i(k) = 0, 1$ and $|\underline{i}| = \sum_{k=1}^{K} i(k)$. The eigenvalues and left eigenvectors of the matrix $diag(z, 1)Q$, for each k, are given by

$$\lambda_{i(k)}(z) = \frac{p_0 + p_f z}{2} \pm \sqrt{(\frac{p_0 + p_f z}{2})^2 - z(p_0 + p_f - 1)}, \quad i(k) = 0, 1 \tag{29}$$

$$g_{i(k)}(z) = (1, \frac{\lambda_{i(k)}(z) - p_0}{z(1 - p_f)}), \quad i(k) = 0, 1. \tag{30}$$

From (23) and (24), we have

$$\lambda_{\underline{i}}(z) = \lambda_0^{K-|\underline{i}|}(z)\lambda_1^{|\underline{i}|}(z) \tag{31}$$

and

$$G_{\underline{i}}(z) = g_{i(1)}(z) \otimes g_{i(2)}(z) \otimes \ldots \otimes g_{i(K)}(z). \tag{32}$$

Since all the two-state Markov chains are i.i.d., we have that the state $\underline{i}$ is equivalent to state $\underline{i}'$ if $|\underline{i}| = |\underline{i}'|$. That is

$$\lambda_{\underline{i}}(z) = \lambda_{\underline{i}'}(z), \quad \text{if } |\underline{i}| = |\underline{i}'|.$$

Let us denote $\lambda_{\underline{i}}(z)$ by $\lambda_{|\underline{i}|}(z)$. We can also group all the eigenvalues corresponding $\lambda_{|\underline{i}|}(z)$ into one equivalent single vector, denoted by $G_{|\underline{i}|}(z)$, which is given by

$$G_{|\underline{i}|}(z) = \sum_{|\underline{s}|=|\underline{i}|} G_{\underline{s}}(z)$$

An explicit expression for $G_{|\underline{i}|}(z)$ has been presented in [9].

The root equations then become

$$z^{K-M} - \lambda_{|\underline{i}|}(z) = 0 \tag{33}$$

In [9], an efficient algorithm to solve the root equations is given. It can be shown that there are $(K - M - |\underline{i}|)^+$ roots in $0 < |z| < 1$ in (33) for each $|\underline{i}|$. Denote these roots by $z_{|\underline{i}|,k}$, $k = 1, \ldots, K - M - |\underline{i}|$, $|\underline{i}| = 0, 1, \ldots, K - M - 1$.

Let $G_{|\underline{j}|,\underline{s}}(z)$ denote the s-th element of $G_{|\underline{j}|}(z)$. It can be proven that if $|\underline{s}| = |\underline{s}'|$, then $G_{|\underline{j}|,\underline{s}}(z)$ and $G_{|\underline{j}|,\underline{s}'}(z)$ are equal to each other. Let us denote $G_{|\underline{j}|,\underline{s}}(z)$ by $G_{|\underline{j}|,l}(z)$ in this case. That is

$$G_{|\underline{j}|,l}(z) = G_{|\underline{j}|,\underline{s}}(z), \quad \text{for } |\underline{s}| = l. \tag{34}$$

Note that $G_{|\underline{j}|,l}(z)$ is a scale function.

In order to simplify the boundary equations for the coefficients, we rewrite the eigenvector $G_{|\underline{j}|}(z)$ by eliminating all (except one) the identical terms in $G_{|\underline{j}|}(z)$ and denote the resultant new vector by $H_{|\underline{j}|}(z)$. The vector $H_{|\underline{j}|}(z)$, whose dimension is $K+1$, is given by

$$H_{|\underline{j}|}(z) = [G_{|\underline{j}|,0}(z), \ldots, G_{|\underline{j}|,K}(z)]$$

where $G_{|\underline{j}|,l}(z)$ is given by (34).

From (34), we observe that

$$p(i, \underline{s}) = p(i, \underline{s}'), \quad \text{if } |\underline{s}| = |\underline{s}'|$$

If we denote $p(i,\underline{s})$ by $p(i,|\underline{s}|)$ and let $P_c(i) = (p(i,|\underline{s}|))$, a vector of dimension $K+1$, then the solution is given by

$$P_c(i) = \begin{cases} v - \sum_{|\underline{j}|=0}^{K-M-1} \sum_{k=1}^{K-M-|\underline{j}|} \dfrac{a_{|\underline{j}|,k} z_{|\underline{j}|,k} H_{|\underline{j}|}(z_{|\underline{j}|,l})}{1 - z_{|\underline{j}|,l}}, & i = 0 \\ \sum_{|\underline{j}|=0}^{K-M-1} \sum_{k=1}^{K-M-|\underline{j}|} a_{|\underline{j}|,k} z_{|\underline{j}|,k}^i H_{|\underline{j}|}(z_{|\underline{j}|,l}), & i \geq 1 \end{cases} \tag{35}$$

The boundary equations for the coefficients $\{a_{|\underline{j}|,l}\}$ are given by [16], for $1 \leq i \leq K - M$,

$$\sum_{|\underline{j}|=0}^{K-M-1} \sum_{l=1}^{K-M-|\underline{j}|} a_{|\underline{j}|,l} z_{|\underline{j}|,l}^{-(K-M-i)} H_{|\underline{j}|}(z_{|\underline{j}|,l}) C_i(z_{|\underline{j}|,l})$$
$$-[v - \sum_{|\underline{j}|=0}^{K-M-1} \sum_{l=1}^{K-M-|\underline{j}|} \frac{a_{|\underline{j}|,l} H_{|\underline{j}|}(z_{|\underline{j}|,l})}{1 - z_{|\underline{j}|,l}}] E^{(M+i)} = 0. \tag{36}$$

There are $\frac{(K-M)(K-M+1)}{2}$ equations in (36).

The **queue size distribution** is given by

$$q(i) = \begin{cases} 1 - \sum_{|\underline{j}|=0}^{K-M-1} \sum_{k=1}^{K-M-|\underline{j}|} \frac{a_{|\underline{j}|,k} z_{|\underline{j}|,k} H_{|\underline{j}|}(z_{|\underline{j}|,l}) u}{1 - z_{|\underline{j}|,l}} & i = 0 \\ \sum_{|\underline{j}|=0}^{K-M-1} \sum_{k=1}^{K-M-|\underline{j}|} a_{|\underline{j}|,k} z_{|\underline{j}|,k}^{i} H_{|\underline{j}|}(z_{|\underline{j}|,l}) u, & i \geq 1 \end{cases}, \tag{37}$$

where $u = (u_i)$, with $u_i = \binom{K}{i}$, $i = 0, 1, ..., K$.

The discussion presented in section 2.5 for both methods applies here with $N = K$. Numerical results will be reported soon regarding the performance of the queueing system and the efficiency of the algorithm.

6 Conclusion

In this paper, we have analyzed the queueing system with multiple Markovian arrivals and services. An efficient computational procedure to calculate the state distribution has been presented. We have showed that if the arrival process is composed of multiple two-state Markov chains, considerable reduction in computational complexity can be achieved using the decomposition technique. The advantage of the current method over the standard generating function has been discussed. If the number of servers is small (e.g., close to 1) and if the vanishing roots are available (or at least not difficult to obtain), the generating function approach is preferred. On the other hand, if the number of servers is small but finding the vanishing roots requires extensive computation, or if the number of servers is large, the algorithm proposed here is much more efficient.

The approach can be used to analyze finite buffer queueing systems as demonstrated in [10]. The result can also be extended to the case in which the number of arrivals is a *bounded* random variable, rather than being fixed as assumed here, and the random variable is governed by the state of the Markov chain.

Acknowledgement The author thanks Thomas E. Stern for many help discussions and for making the reference [15] available.

References

[1] J. Daigle and J. D. Langford "Queueing analysis of a packet voice communications system " *IEEE Infocom'85*, pp. 18-26.

[2] H. Heffes and D. M. Lucantoni, "A Markov modulated characterization of packetized voice and data traffic and related statistical multiplexer performance", *IEEE JSAC* SAC-4, No. 6, 1986, pp. 855-868.

[3] A. Leon-Garcia, R. Kwong and G. Williams, "Performance evaluation methods for an integrated voice/data link" *IEEE Trans. on Commu.* COM-30, pp.1848-1857, 1982.

[4] San-qi Li and J. Mark, "Performance tradeoffs in an integrated voice/data services TDM system", *Journal of Performance Evaluation*, Feb. 1988, pp.51-64.

[5] K. Sriram, P. K. Varshney and J. G. Shanthikumar, "Discrete time analysis of integrated voice/data multiplexers with and without speech activity detectors" *IEEE JSAC-1*, pp.114-1132, 1983.

[6] K.W. Fendick, V. R. Saksena and W. Whitt, "Dependence in Packet Queues", *IEEE Trans. Comm.* Vol. 37, N0. 11, 1989, pp. 1173-1183.

[7] San-qi Li "A General Solution Technique for Discrete Queueing Analysis of Integrated Traffic on ATM", *IEEE INFOCOM'90*, (also to appear in IEEE Trans. Comm.).

[8] Z. Zhang "Analysis of a Discrete-time Queue with Integrated Bursty Inputs in ATM Networks", *1990 Conference on Information Science and Systems*, Princeton University, March 21-23.

[9] San-qi Li "Generating Function Approach for Discrete Queueing Analysis with Multiple Markovian arrival and Markovian Processes", *Proceedings of 7th International Teletraffic Congress Seminar*, Morristown, NJ, October 9-11, 1990, (also to appear in *IEEE Trans. Comm.*).

[10] Z. Zhang "Finite Queues with Multiple Markovian Arrivals and Services", submitted for publication.

[11] D. Anick, D. Mitra and M. M. Sondhi, "Stochastic theory of a data-handling system with multiple sources", *Bell System Tech. J.* 61, 1982.

[12] L. Kosten "Stochastic theory of data-handling systems with groups of multiple sources", *Performance of Computer-Communication Systems*, eds. H. Rudin and W. Bux, Elsevier, Amsterdam, pp. 321-331, 1984.

[13] D. Mitra "Stochastic theory of a fluid model of products and consumers coupled by a buffer", *Adv. Appl. Probability*, 20, 1988, pp.646-676.

[14] T. Stern and A. I. Elwalid "Analysis of Separable Markov-Modulated Rate Models for Information-Handling Systems" *To appear in Adv. Appl. Prob.*

[15] A.I. Elwalid, D. Mitra, T.E. Stern "Statistical multiplexing of Markov-modulated sources: theory and algorithms" 1st International Workshop on the numerical solution of Markov chains, Raleigh, NC, Jan. 1990.

[16] Z. Zhang "An Efficient Approach to Analyze Discrete-time Queues with Multiple Markovian Arrival and Service Processes", Columbia University, CTR-Technical Report, 1991.

[17] A. Graham *Kronecker Products and Matrix Calculus: with Applications*, John Wiley & Sons, 1970.

A queueing model for buffer overflow in multicast communications

M. Naim Yunus

TELEKOM MALAYSIA, Research and Development Division, P.O. Box 11812, 50758 Kuala Lumpur, Malaysia

Abstract

We consider a protocol in multicast (point-to-multipoint) communications where a sender sends copies of similar messages to a number of recipients and waits for acknowledgments from each of these recipients. The sender has a finite number of buffers to store arriving acknowledgment packets before processing them. A queueing model for buffer occupancy is described and calculations are made for the expected number of lost packets. The relationship between the number of buffers and lost traffic is discussed.

1. INTRODUCTION

In multicast communications, a source node will transmit copies of messages to a selected number of possible sources. This mode of communications is gaining importance in modern information technology such as in new telecommunications services, distributed systems, and parallel processing [3,4]. This includes applications in teleconferencing, computer conferencing, electronic mail and distributed data processing.

In the case of single multicast, a single source controls information flow. Destinations may, in this case, communicate via the source. However when members of a communication group are allowed to multicast to fellow group members, then each node has the same status as a source. This activity may be referred to as a group multicast. Members of a group need not remain static. Instead members can dynamically change according to situations that arise.

Proposals for packet-switching fabrics which support point-to-multipoint communications have already been made [1,5,6]. A multicast packet switch should be able to replicate packets and also to perform switching functions. For example, the broadcast Banyan network has been investigated for this purpose [1,5]. So has the ring network [7]. For self-routing, stage-type architectures, it is vital that the ability of the network to cope with congestion with multicasting-intensive applications be investigated.

In this work we consider a particular protocol [2] in which a source node, which we call the sender, multicasts to a group of nodes, called the recipients. Each recipient would then acknowledge receipt of the message. This is accomplished by sending an acknowledgment packet to the sender. We assume that the sender has N buffers, each being able to store one acknowledgment packet. In the context of a queueing system this would correspond to N waiting spaces. We investigate the expected number of acknowledgment packets lost for a given number of buffers at the sender, assuming that the time taken for the recipients to respond is distributed negative exponentially and that the service time at the buffers follows a similar distribution, but with a different parameter. This is done by calculating the steady-state probabilities of being in each possible state. Knowledge of these probabilities can lead to the calculation of other measures of interest.

2. NETWORK MODEL

A sender node has a stream of waiting messages to be transmitted sequentially, each of them to be transmitted simultaneously to N recipient nodes. This model is slightly different to the one described by Danzig [2] where the sender has only one message to multicast. Once this message is multicast the system remains idle. In our model we assume that the awaiting messages are queueing at the sender and the sender would accept one at a time for multicasting. When the sender accepts a message, it makes N copies and multicasts them to the recipients. Once this is done, the sender then waits for the recipients to respond, in the form of acknowledgment packets.

We assume that messages do not collide in the transmission media and that they require negligible transmission times. These assumptions will make our analysis easier but may overestimate the number of lost packets in real-life cases [2]. We also assume that all messages (either multicast or responses) are successfully transmitted to their destinations, that is we are assuming that the network is very reliable. The case where the network is not assumed reliable calls for a protocol whereby timeouts are employed by the sender in receiving responses [2].

Arriving responses will be stored in the buffers at the sender before it processes them. We are interested in the calculation of expected number of lost responses, assuming one response occupies one buffer, given the arrival and service discipline and distribution, and when the number of buffers is known. This measure will enable network designers to optimize the number of buffers to be used and also for implementing timeouts when the sender knows that congestion at its buffers is hindering successful responses from being processed.

The whole cycle then repeats itself when all N recipients have responded, regardless of whether the responses were successful or otherwise. In this respect we are assuming that the sender knows when all recipients have responded. As the cycle goes on infinitely, we are able to calculate the expected amount of time the system spends in each possible state.

3. FORMULATION OF QUEUEING MODEL

We begin by assuming that there are N recipients. We assume that all recipients are homogeneous, that is, upon receiving a multicast message, the time taken for each recipient to respond is τ, where τ is distributed negative exponentially with parameter λ. Once a recipient has responded it will remain idle until the next message is received, when it will respond according to the same distribution as before. As mentioned in the previous section, all messages are assumed to be successfully transmitted.

Let the number of buffers be b and let the arriving responses be buffered according to the FIFO discipline. The sender will then process the awaiting responses using the same discipline. A response is assumed lost when it finds all b buffers being full upon arrival.

We assume that the sender serves the buffers according to the negative exponential distribution with parameter μ. We define the state (i,j) where i ($i=0,1,...,N$) is the sum of the number of responses processed and the number of responses lost, and j ($j=0,1,...,b$) is the number of occupied buffers. For each state (i,j) we define p_{ij} as the steady-state probability of being in that state.

For the queueing model we are considering, we can construct the state transition diagram as shown in Fig. 1a below, for the case of $N=4$ and $b=2$. Notice that p_{40} is not shown, as in our model the state (4,0) is identical to state (0,0). This is because once all

Fig. 1a : State Transition Diagram for N=4, b=2

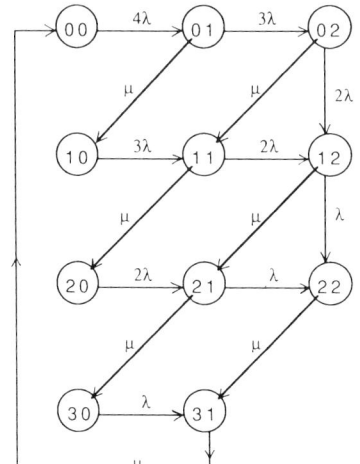

recipients are known to have responded (successfully received by the sender or not), the sender immediately multicasts the next message. Even though this assumption may not be satisfied, it justifies the analysis so as to compute certain characteristics per cycle of multicast, as given below by equations (6) and (7). Based on a similar diagram for the general case, one can readily derive the following balance equations:

Row 0:

$$(N-j+1)\lambda p_{0,j-1} = (\mu+(N-j)\lambda)p_{oj}, \qquad j=1,2,...,b \qquad (1)$$

Rows $i = 1,2,...,N\text{-}b$:

$$\mu p_{i-1,1} = (N-i)\lambda p_{i0} \qquad (2a)$$

$$\mu p_{i-1,j+1} + (N-i-j+1)\lambda p_{i,j-1} = ((N-i-j)\lambda + \mu)p_{ij}, \qquad j=1,2,...,b-1 \qquad (2b)$$

$$(N-i-b+1)\lambda(p_{i-1,b}+p_{i,b-1}) = ((N-i-b)\lambda+\mu)p_{ib} \qquad (2c)$$

Rows $i = N\text{-}b+k$, for $k = 1,2,...,b\text{-}1$:

$$\mu p_{i-1,1} = (N-i)\lambda p_{i0} \qquad (3a)$$

$$\mu p_{i-1,j+1} + (N-i-j+1)\lambda p_{i,j-1} = ((N-i-j)\lambda + \mu)p_{ij}, \qquad j=1,2,...,b-k-1 \qquad (3b)$$

$$\mu p_{i-1,b-k+1} + \lambda p_{i,b-k-1} = \mu p_{i,b-k} \qquad (3c)$$

Using these equations it is possible to obtain closed-form expressions for every p_{ij}, but it is a very tedious exercise. Alternatively, for given N, b, λ and μ, one can iterate to obtain numerical solutions. To do this, we initially set $p_{00} = 1$. Using the recursive relations in Row 0, we can compute $p_{01}, p_{02}, ... , p_{0b}$ sequentially. Then we repeat the same process for subsequent rows to obtain numerical solutions to all p_{ij}. This scheme can be represented by the diagram in Fig. 1b (on the next page), which shows this left-right, up-down direction of solving the balance equations.

Of course the sum of all these probabilities will exceed one. Actual probabilities are obtained by normalization. If $S = \sum_{i,j} p_{ij}$, then the actual probabilities, π_{ij}, ($i = 0,1,...,N$; $j = 0,1,...,b$) are given by

$$\pi_{ij} = \frac{p_{ij}}{S} \qquad (4)$$

The π_{ij} computed in this manner is the steady state probability of being in state (i,j).

4. CALCULATION OF BUFFER OVERFLOWS

Calculation of buffer overflows or lost packets can be undertaken by noting that losses can only occur when all buffers are full. This situation occurs for all states with $j=b$, i.e. for states (i,b), $i = 0,1,...N\text{-}b\text{-}1$. When the system is in any one of these states a loss occurs when there is an arrival. For example, a loss will occur in state $(2,b)$ if an arrival,

Fig. 1b : General Direction of Solving Balance Equations

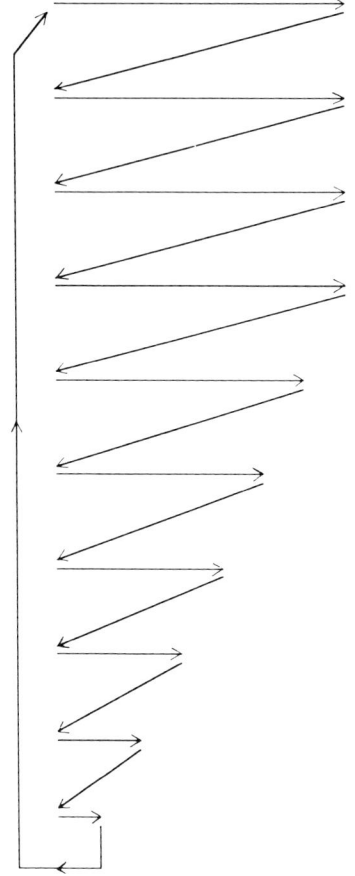

which arrives at rate $(N-b-2)\lambda$, takes place. Hence, summing all possible cases, we can write an expression for expected number of lost packets per unit time as

$$L_T = \sum_{i=0}^{N-b-1} (N-b-i)\lambda \pi_{ib} \tag{5}$$

Recalling the model, we find that we are actually considering a system which we assume achieves steady-state by going through an infinite number of cycles. We define a cycle as the flow, in the transition rate diagram, from state $(0,0)$ to state $(N-1,1)$, and back to state $(0,0)$. For the case of $N=4$ and $b=2$, please refer to Fig. 1a. If we are interested to compute the expected number of lost packets per cycle then we need to find an expression for the cycle time, which is the time taken for one full cycle to take place.

An expression for the cycle time, T_C, can be obtained by noting that the mean time the system spends in state (0,0) per cycle, say, is equivalent to the probability of being in state (0,0), π_{00}, multiplied by T_C. Since the mean time in state (0,0) is $1/N\lambda$, we therefore have

$$T_C = \frac{1}{\pi_{00}} \cdot \frac{1}{N\lambda} \tag{6}$$

Thus we can write the expected number of lost packets per cycle as

$$L_C = L_T \cdot T_C$$
$$= L_T \cdot \frac{1}{N\lambda\pi_{00}} \tag{7}$$

In the next section we apply equations (5), (6) and (7), to obtain the expected loss per unit time, the cycle time and the expected loss per cycle for several examples.

5. NUMERICAL EXAMPLES

We apply equations (5), (6) and (7) on two systems with $N=30$ and $N=200$, to represent a small and a large system. For $N=30$ we let $\lambda=20$ and vary μ. For $N=200$ we let $\lambda=100$ and again with varying μ. The results are shown in Figs. 2, 3 and 4, for $N=30$, and in Figs. 5, 6 and 7, for $N=200$. From these figures we can identify the following features:

(a) As we increase the number of buffers, the expected lost traffic per time, L_T, decreases monotonically (Figs. 2, 5). This is expected as an increase in buffer size would mean more packets are buffered, thus reducing the number of losses. For λ the decrease seems to be exponential. However the decrease becomes more linear as $\mu >> \lambda$. Of course we expect that $L_T \to \infty$ as $b \to \infty$.

(b) The cycle time, T_C, increases almost linearly as b increases (Figs. 3, 6). However for small values of b, the increase is not linear at all. This can be seen from the numerical results, but not so clear from the graphs. However this increase is monotonic. Increase in buffer size means more packets can wait for service. This in turn would result in more packets being served. Thus the system stays in certain states longer and this contributes to the cycle time.

(c) The expected lost traffic per cycle, L_C, decreases almost linearly as we increase b (Figs. 4, 7), but again for small values of b the decrease is not linear at all. We do expect this behaviour. As a check we calculate this same measure

Fig. 2 : N=30 with (λ,μ) - lost traffic/time

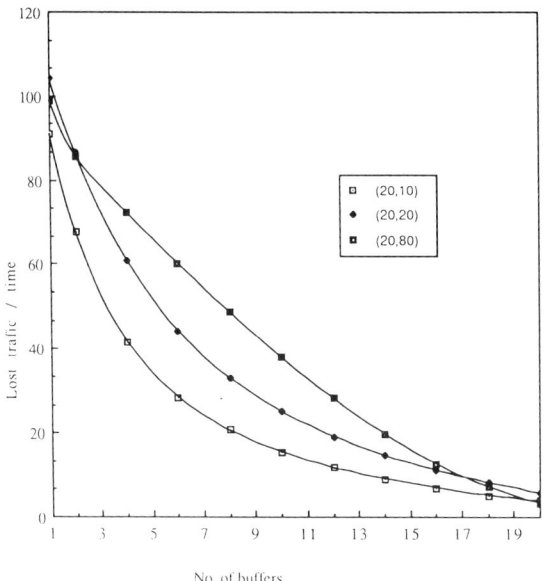

Fig. 3 : N=30 with (λ,μ) - cycle time

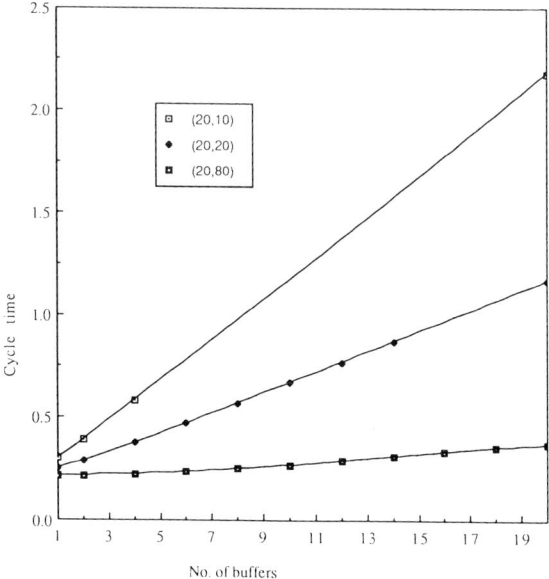

Fig. 4 : N=30 with (λ,μ) - lost traffic/cycle

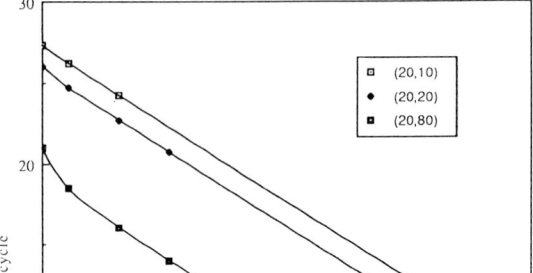

Fig. 5 : N=200 with (λ,μ) - lost traffic/time

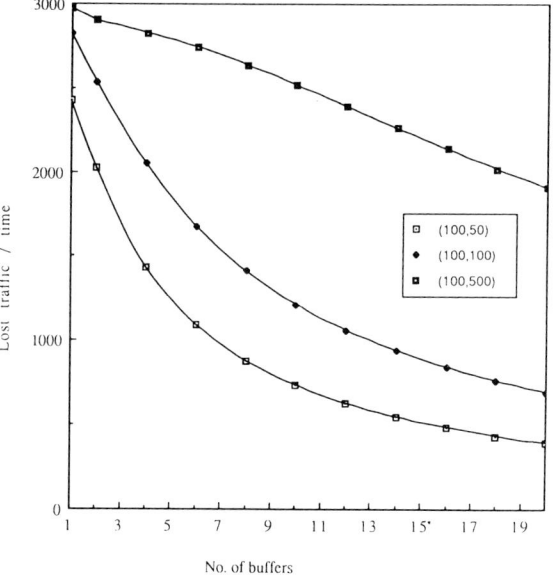

Fig. 6 : N=200 with (λ,μ) - cycle time

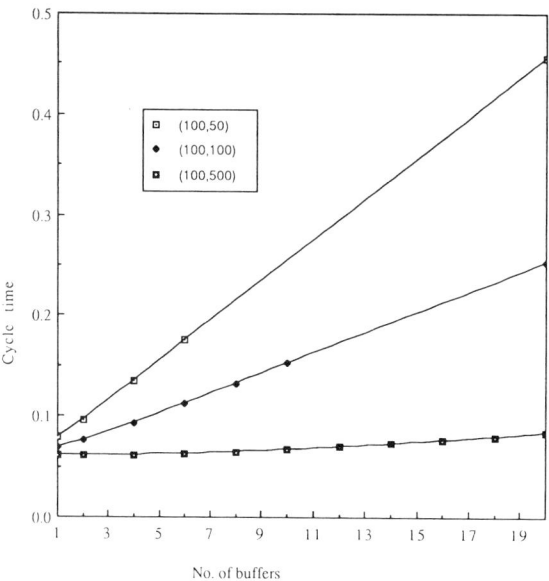

Fig. 7 : N=200 with (λ,μ) - lost traffic/cycle

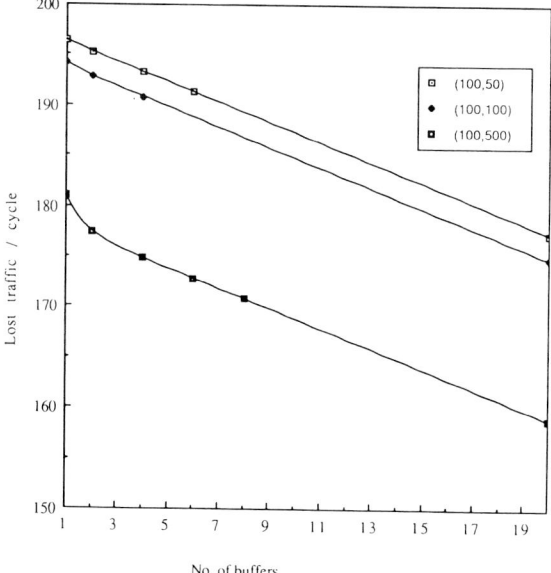

using the dynamic programme from [2]. The dynamic programme is outlined in (8) below. The results obtained via this dynamic programme agree with the values we obtained using (7). However Danzig's model [2] only enables one to compute the expected lost traffic per cycle.

From numerous numerical experiments, we also found that L_C depends on the ratio λ/μ. Thus we have the same results for $\lambda=5, \mu=10$ and for $\lambda=200, \mu=400$. Another interesting relation is that if both λ and μ are multiplied by x, then T_C would be divided by x, but L_T would be multiplied by x too. One can immediately see that the product of L_T and T_C, which is L_C, remains unchanged.

As mentioned in (c) above, we also used the dynamic programme from [2] to compute L_C. Defining $L_N(s)$ as the expected number of losses when n recipients respond with s buffers begin occupied, and the first of n responses arrive immediately, we can write the following fast dynamic programme [2], for $s=1,2,...,b-1$, and $n=1,2,...,N$:

$$L_n(0) = L_{n-1}(1)$$
$$L_n(s) = P_{ns} L_{n-1}(s+1) + (1-P_{ns}) L_n(s-1)$$
$$L_n(b) = P_{nb} (1+L_{n-1}(b)) + (1-P_{nb}) L_n(b-1)$$
$$L_0(b) = L_0(s) = 0 \qquad (8)$$

where

$$P_{ns} = \frac{n\lambda}{n\lambda + u_0(s)\mu}$$

and $u_0(s)$ is the unit step function.

6. DISCUSSION AND FURTHER RESEARCH

We have presented the balance equations for a queueing model relevant in multicast communications. These equations enable us to numerically compute the steady state probabilities. Although Danzig [2] had also considered a similar model, he did not use this approach. Thus he was only able to present a dynamic programme to compute the expected lost traffic per cycle. Using our approach, we can investigate two more interesting performance measures, namely the cycle time and the expected lost traffic per unit time. We can also treat cases when λ is time-dependent and investigate this using the state-transition diagram. Another area of research would be to consider cases where there is relationship between the recipients in responding. This would result is some form of coeficients for the λ's in the state-transition diagram. The same thing happens for the case when the service rate is dependent on the number of responses.

Research described above also shows that one can easily solve this particular multicast model by means of recursive relations based on the state-transition diagram. The numerical procedure used so far seems to be stable. Furthermore it is very quick and uses minimal space.

The various measures described above when plotted against the buffer size will assist the system designer in optimizing the number of buffers required in order to satisfy a required grade of service. One can also investigate the performance against changes in the values of the parameters N, λ and μ.

ACKNOWLEDGMENTS

This research was supported by the Australian Research Council and by Telecom Australia. I gratefully acknowledge the hospitality extended by Les Berry and his staff at the Centre for Teletraffic Research, School of Information and Computing Sciences, Bond University (Gold Coast, Queensland), where this research was undertaken. I am also indebted to him for making the visit possible and also to Nico M. van Dijk (Vrije Universiteit, Amsterdam) for fruitful discussions on some aspects of this work.

REFERENCES

[1] R.G. Bubenik, J.S. Turner, "Performance of a broadcast packet switch", *IEEE Trans. Commun.*, 37(1), 1989, 60-69.
[2] P.B. Danzig, "Finite buffers and fast multicast", *Perf. Eval. Rev.*, 17(1), May 1989, 108-117.
[3] A.J. Frank, L.D. Wittie, A.J. Bernstein, "Multicast communication on network computers", *IEEE Software*, May 1985, 49- 61.
[4] L. Hughes, "An introduction to multicast communication", in *Computer Communication for Developing Countries - ICCC 1988* (ed. S. Ramani, A. Garg), Elsevier, Amsterdam, 1988, 193-213.
[5] T.T. Lee, "Nonblocking copy networks for multicast packet switching", *IEEE J. SAC*, 6(9), 1988, 1455-1467.
[6] G. Nathan, P. Holdaway, G. Anido, "A multipath multicast switch architecture", *4th. Aust. Fast Packet Switch. Workshop*, OTC, Sydney, July 1989.
[7] M. Rahnema, "The fast packet ring switch: a high performance efficient architecture with multicast capability", *ICC 1989*, Boston, 884-891.

Optimization of ATM Multi-Service Networks - Some Early Investigations

André H. Roosma

PTT Research, P.O.Box 421, 2260 AK Leidschendam, Netherlands

Abstract

This paper explores the optimal dimensioning of future ATM multi-service networks. As a first step, it investigates the important characteristics of these networks, as far as they are already standardized. Existing models for optimization of circuit-switched networks are analysed. Also, some indications are given towards the adaptation of such models to deal with the specific characteristics of ATM networks. Some interesting mathematical problems arise, requiring further investigation.

1. INTRODUCTION

Evolving ATM networks will facilitate a variety of telecommunication services with different, possibly varying, bitrates, using a fast packet (cell) switching technique.
Regarding the optimal end-to-end dimensioning and structure optimization of such networks at the switched level, only a very limited number of papers have been published so far (e.g. [23], [10], and some references therein).
This paper aims at providing some early investigations in this field. In a top-down approach, it will focus on deriving general methodology and principles, more than specific formulae.
Similarly, numbers and quantities given are only meant as rough indications.

A complicating factor in the design of ATM network optimization is the fact that many operational details of ATM networks have not been settled as yet (cf. also [18]). Complications of so called statistical multiplexing or new STM developments may result in hybrid solutions that can be treated at a network planning level as multi-bitrate circuit switching. This implies that network dimensioning methods should be flexible with respect to the details of technological implementations. Such flexibility is most easily obtained using a layered or modular approach.
It is likely that at a high level a lot of characteristics of todays circuit-switched networks will persist in future networks, such as the (virtual) calls, with all information routed along a fixed path. Therefore, special attention will be given in this paper to the possibilities of adapting available models for dimensioning and optimization of circuit-switched

single-bitrate networks (such as in [21]). It will build further on some basic material on the subject of ATM network or packet network design, such as [9], [10], [29].

2. ATM NETWORKS

According to the latest CCITT draft recommendation on ATM [1], an ATM transport network is structured as two layers: the ATM Layer and the Physical Layer. The transport functions of the ATM layer are subdivided into the Virtual Channel level and the Virtual Path level. Virtual Channel Connections (VCCs) are concatenated from Virtual Channel Links (VCLs) between ATM switches. Many VCLs can use capacity of Virtual Path Connections (VPCs) between these switches. Reallocation of existing physical capacity to these VPCs can be performed at a network management level. We will assume that VPCs will have a fixed capacity. (For more details on the use of VPCs refer to [26].)
In the rest of this, and the next section we will examine the specifics of ATM networks and their consequences for network optimization.

2.1. An ATM link concept

In the context of this paper, we could use the term 'link' for any network element – either physical or functional – involved in carrying traffic, having a non-marginal cost per unit capacity, and hence influencing the costs and performance of the network substantially.
Here, we confine ourselves to the link concept that is most in line with the classical concept of the (functional / switched network) trunk group: the concept of a virtual path connection (VPC) – a functional connection of specified fixed capacity between two exchanges. This capacity may be implemented through ports of these switches and (shared) transmission equipment (including digital cross-connects / VP switches) in between.
(In practice, establishing a VPC between two exchanges will not always be necessary. Especially if there are no VP switches in between, VCLs may also be set up directly over the Physical Layer. Yet, capacity – here modelled as a 'link' – needs to be planned for these VCLs in that case as well.)

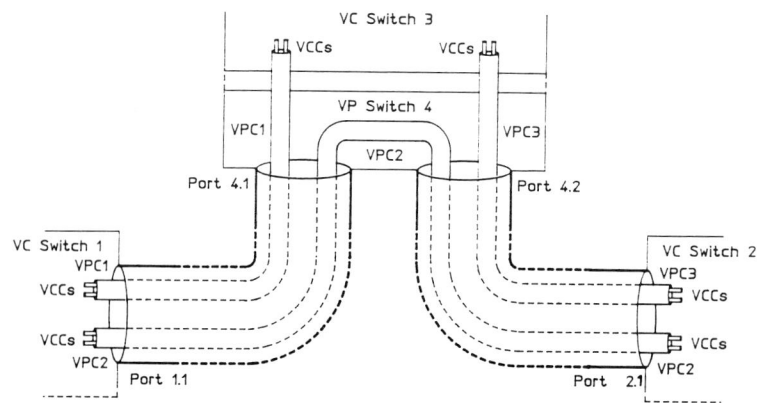

Figure 1. Illustration of VCCs, VPCs and physical ports. VP Switch 4 may be incorporated in VC Switch 3, just colocated or completely detached

Current developments on ATM switches focus on 150 Mbit/s physical ports. More than one VPC may use one port. Using more than one port for a VPC will not be possible, due to induced cell sequence integrity problems. So, the speed of a VPC is limited to this 150 Mbit/s, and all VPCs to/from an exchange must be allocated to the 150 Mbit/s ports, in such a way that all VPCs reach their destination. This does not necessarily mean that only VPCs destined for one other exchange may be allocated to one port, since it is possible to do cross-connecting on VP level in between exchanges, or make a through-connection of a VPC through a third exchange, as illustrated in Figure 1. (in [11] these possibilities are explicitly exploited, cf. also [2]). All this makes the allocation problem a challenging optimization problem in itself.

2.2. The costs of ATM networks: what to be optimized?

Something needs to be said about what parts will contribute most significantly to the costs of ATM networks if we want to create useful models for ATM network optimization. Network costs have always been divided into a fixed part and a capacity-dependent part, and in transmission and switching costs. In the process of digitalization the fixed switching (exchange) costs have increased due to increased complexity of operations and management (software!). In fixed transmission costs we see some greater one-time investments, associated with deployment of fiber, but they are quite stable otherwise. The capacity-dependent transmission costs have decreased dramatically and will likely continue to decrease in the coming decade (the costs of a 600 Mbit/s and a 150 Mbit/s transmission system making use of the same fiber, may differ only 50%).
The effect of ATM-ization is hard to predict. However, some first ideas are that the capacity-dependent switching costs will play an important role, as complexity of most switch architectures increases supralinearly with the number of ports. This may well imply that these capacity costs will be the most dominating ones in the dimensioning process. However, capacity-dependence itself will depend strongly on the switch architecture used.
In summary, the costs to be included in an object function for the optimal dimensioning problem will probably be dominated by the costs per port of the switches. Furthermore, this cost function will be a rather coarse step function as equipment is built up modularly. Details will depend on switch architecture, buffer sizes, etc.
All these arguments regard the costs at the physical level. Costs of physical ports will need to be attributed to the VPCs using them.

2.3. Network structure developments

In network structure optimization, the high fixed costs of exchanges and transmission media must well be taken care of. This may yield a tendency towards fewer, larger exchanges and transmission routes. However, predefined customer-oriented goals with respect to network performability (availability) will limit this tendency strongly (cf remarks and references on this issue in [22]). Supra-linear increase in switch complexity with size might also lead to smaller exchanges. Since VP switching will be less complex than VC switching, the economic threshold for creating a dedicated VPC (cf. 'high-usage link') between two exchanges will be quite low.

2.4. Services, call control and connection admission control

Future networks will be characterized by the fact that they carry a large variety of services. By services, we will understand: bearer services – the concept introduced to model the functional transport requirements from services to the network.

Call control and routing remain roughly the same for ATM networks as for classical circuit-switched networks, since ATM networks use the virtual call concept. However, there is one important change.

At the call-setup phase the characteristics of the required bearer service(s) are given. On this basis the network will try to route the call to the destination(s). This will involve establishing a (number of) VCC(s) on VPCs throughout the network. (The fact that calls may require any number of connections, each requiring a specific GOS/QOS, implies that traffic is best described in terms of connections in stead of calls.)

Thus, an essential element of the call control procedure is the connection admission control (CAC): the rules deciding whether a certain connection (VCC) can be allowed on a given link (VPC). Different CAC rules have been studied in literature. One of the most simple rules is Peakrate Allocation (PA): allow a new VCC on a VPC if the sum of all peakrates of established VCCs together with the peakrate of this new VCC does not exceed the speed of the VPC. PA leads to a lightly loaded network in case of bursty services. It has the advantage that burstiness of sources need not be taken into account in traffic management nor in network dimensioning, etc. In this case, the network can be regarded as a multi-rate circuit switched network (for which traffic models are available, such as [8]). The same applies if it is possible to derive a generally applicable virtual bandwidth requirement (in between peak and average rate) for each VCC, as suggested in [12]. First results of yet unpublished research at PTT Research indicate that statistical multiplexing gain is highly dependent on the number of VCCs to be multiplexed (or, on the VPC bandwidth if a certain mix of specified VCCs is multiplexed). So, this virtual bandwidth will at least depend on some parameter characterizing the ratio of peak or average VCC rate to total link bandwidth or the number of VCCs to be multiplexed.

2.5. Stochastics of traffic, and performance

As pointed out clearly in [24] and [17], traffic in an ATM network is characterized by stochasticity at three levels (or time scales): connection (or: call), burst (or: action) and cell (or: transmission) level. Hence, we also have performance issues at all of these three levels. The network may or may not have enough capacity to carry a new connection at the required lower level grades of service, a burst of information within an already established virtual connection may cause many cells to be lost or tremendously delayed, and a single cell may be lost due to more temporary buffer overflow or delayed beyond the required maximum delay due to congestion at cell level.

3. DIMENSIONING OF ATM MULTI-SERVICE NETWORKS

For the dimensioning of ATM networks, different approaches are possible. The performance requirements at different levels can all be integrated into one model, which may become rather complex, but overall optimality may be guaranteed best this way.

Alternatively, a hierarchical approach may be chosen, where end-to-end requirements at burst- and cell level are first distributed over the links and translated into reasonable buffer sizes and CAC rules. In a subsequent phase, the link speeds are determined such that requirements at connection level are met. Iteration between these phases might be used to circumvent the introduced sub-optimality.

In both cases, methods for circuit-switched networks could be used as a basis for designing methods for the ATM case. Therefore, dimensioning methods for classical circuit-switched networks will be discussed in the sequel. They will be used as a starting point to derive outlines for dimensioning methods for ATM networks.

3.1. Dimensioning of classical circuit-switched networks

For the dimensioning or optimization of 'classical' single-bitrate circuit-switched networks, a large variety of approaches has been used, ranging from brute-force heuristics severely constraining network structure, to mathematically justified methods using gradients or other hill-climbing optimization principles. Many algorithms have been investigated. Here, we mention just a few aspects of the most representative ones, focussing on attributes that might label them as interesting for extension to ATM multi-service networks. A more detailed discussion of basic principles as well as an indication of performance is given in Appendix A.

All methods make use – implicitly or explicitly – of the following models:

- a link model: a traffic / capacity / performance model for one link; e.g. the one moment Erlang B or the two moments ERT model;

- a network traffic model: a model relating end-to-end offered traffic to link offered traffic; e.g. cascade model (offered traffic = carried traffic on previous link) or reduced load approximation (cf.[28]), with link-by-link overflow or route overflow;

- a network performance model: a model relating end-to-end performance requirements to link performance requirements and link performance to end-to-end performance;

- an optimization model: an optimality principle (e.g. equal marginal costs for traffic on subsequent routes) together with a network calculation method (e.g. all links in a prescribed order), or an optimization algorithm (e.g. gradient projection, using Lagrange multipliers for constraints).

A method in which all of these models are easily adaptable to the ATM multi-service network case will be most favourable. If the models are explicitly (as opposed to implicitly) available in the algorithm, they are more easily replaced by another model, e.g. an ATM oriented one. So, this will be an attractive attribute. Some representative groups of algorithms are:

1. Heuristics based on marginal efficiency formulae for high-usage links; e.g. the one by Pratt [19];

2. Mathematical programming formulations; e.g. the ones by Berry [4], [5], [6] [7], Harris [15];

3. Recent combinations of the two approaches, as in [21], [25].

We have compared them on the basis of the above mentioned models used, evaluating also some additional criteria that will be relevant in ATM networks: accuracy (are requirements exactly satisfied), optimality, network structures that can be treated (strictly hierarchical, semi-hierachical, non-hierarchical), routing rules (single or multiple homing, Sequential or Originating Office Control), only uni- or also bidirectional trunk groups, performance criteria (final link or end-to-end blocking), determining parameters of service protection mechanisms (discerning different service levels for different streams on one link), etc.

The conclusion is that the approaches under 2 and 3 are most promising. The approach under 3 (esp. the one from [21]) has the advantage that all models are explicitly available in the algorithm, and cpu-time is shortest because of the strongly reduced dimensionality of the free parameter space. So, this approach looks promising as a basis for developing dimensioning methods for ATM networks. A disadvantage could be that derivatives of the link model must be available; in that respect methods under 2 are more easily adaptable.

3.2. Necessary changes for dealing with ATM networks

To adapt the circuit-switched network dimensioning methods to deal with ATM multi-service networks, a number of changes will be necessary. Where applicable, we will especially focus on the changes needed in the third group of methods (as from [21]).

The link model will need the most drastic revision. The link problem in the ATM case can be formulated as follows: Given link offered traffic streams and their characteristics, given either a marginal capacity or maximum blocking value at connection level and requirements at burst and cell levels, three parameters for that link should be determined (in stead of just one): buffer size, speed and CAC rule. The extra parameters for service protection can be translated into different CAC rules for different streams. An alternative may be to provide two or more VPCs with different performance, and route traffic streams (VCCs) over appropriate VPCs. Still another approach is found in extending circuit-switched selective trunk reservation to selective bandwidth reservation in the ATM case: in considering connection acceptance, a part of the bandwidth (speed) of the link is reserved for specific traffic streams.

Modularity can be taken into account, either on a link by link basis (allocating one VPC to each physical port), or on a multiple link basis (allocating VPCs towards one or more switches in a particular direction to one port). Alternatively, the allocation problem can be completely separated from the VPC dimensioning.

The network traffic and network performance models can be derived straightforwardly. The mere difference is in the description of traffic and performance at three levels: connection, burst and cell level. The performance requirements at burst and cell levels will be given on an end-to-end basis per route. In a similar way as in [21], it should be possible to derive approximative formulae to divide these over links in a route. At cell level, a cascade traffic model will theoretically be more accurate than, e.g., a reduced load approximation, but with cell losses in the order of 10, this is only an academic issue. A full load approximation (not counting blocking elsewhere on a route) may be the most practical here.

The optimization model needs little change if an external optimization algorithm is used. If the concept of marginal traffic is used (at connection level), the problem is to derive new formulae for the marginal quantities in an ATM multi-service environment from the link model. If statistical multiplexing is used, this will influence the efficiency of different size links differently, thus complicating both categories of optimization models.

An example

Adapting the approach from [21] to the ATM case, we could arrive at the following successive approximation outline:

1. initialize link performance parameters (all levels)

2. calculate traffic flows through the network

3. for all non-final links:

    - calculate link performance requirements (burst- and cell-level), and call-level marginal efficiency parameter(s)

    - determine optimal link connection allocation control, required buffer size and link speed from link offered traffic, link performance requirements and marginal efficiency parameter(s)

    - calculate resulting link performances

4. recalculate traffic flows through the network

5. for all final links:

    - calculate link performance requirements (all levels) from end-to-end requirements (using already obtained GOS at connection level on previous routes)

    - determine optimal link connection allocation control, required buffer size and link speed from link offered traffic and link performance requirements

    - calculate resulting link performances

6. recalculate traffic flows through the network

7. if solution sufficiently stable then stop else go to 3.

3.3. Performance and dimensioning of one ATM link

As derived above, a good link model is essential for developing network optimization models. Here we will show its feasibility. A single ATM link can generally be modelled as a single- or multi-server system with a finite waiting room (buffer). The capacity of a link is constituted by three parameters: link speed (bandwidth), buffer size and CAC rule.

Considering just one link, one can easily derive the following observations. Cell loss (rate) can be decreased by increasing link speed or buffer size. Cell delay by increasing speed or decreasing buffer size. Both, and the performance at burst level, are highly influenced by the CAC. Allowing less connections on the link will improve the performance at cell and burst level, but it will deteriorate the performance at connection level.

Thus, it is theoretically possible to derive the optimal CAC strategy, and the optimal speed and buffer size for each load level, given:

- (marginal) costs of buffer space and bandwidth,
- traffic load and quantified characteristics of the stochastics of the traffic at connection, burst and cell level,
- performance requirements at each level.

Further research will be needed to derive useful models.

4. CONCLUSIONS AND TOPICS FOR FURTHER STUDY

Optimization of evolving ATM networks forms a challenging Operations Research problem. The existence of both functional and physical sub-problems, each having its own constraints, creates a complex allocation problem. Methods for circuit-switched network optimization can be used as a basis to develop optimization methods for the functional subproblem. The convergence of resulting algorithms should be investigated. Reduced gradient techniques may also be used, as in [16]. Further research may demonstrate whether decomposing the functional problem in ones at connection and cell/burst levels will yield adequate results. Much depends on the feasibility of simple CAC rules. Further standardization and technological research are needed to specify a number of constraints and elements of the object function in more detail. An example of this is the fact that it may not be possible to provide VPCs with different QOS values over one physical port. More insight in the stochastic behaviour of traffic will facilitate the derivation of adequate link models (incl. marginal quantities). Through its concept of VP switching, ATM will have a large impact on optimal network topologies/structures. Possibly small VPCs may reduce the amount of VC switching, whereas large VPCs may yield considerable statistical multiplexing gains. It seems interesting to develop heuristic approaches along the lines of those in [20], based on mathematical programming formulations of this problem.

Acknowledgements

My colleagues, dr. J.L.(Hans) van den Berg, dr. Dick Brandt, Rein F.J. de Vries, Jos Wage, and J.C. (Kees) van der Wal are gratefully acknowledged for many discussions that were helpful and stimulating in conceiving this material. My colleagues Jan M. van Noortwijk (now with Delft Univ. of Technol.) and Theo A. Smit are acknowledged for providing helpful numerical material on burstiness of different traffic sources and on statistical multiplexing, respectively.

References

[1] CCITT SG XVIII WP 8, *Draft RECOMMENDATION I.311 - B-ISDN general network aspects*, Report R 34, Geneva, May 1990.

[2] R.G. Addie, R.E. Warfield, *Bandwidth Switching and New Network Architectures*, Austral.Telecomm.Research 22 (2), 1988, pp.45-51.

[3] A.K. Blaauw, *Optimal Design of Hierarchical Networks with Alternate Routing based on an Overall Grade of Service Criterion*, 1-st Internat. Telecomm. NETWORKS Planning Symp., Paris, 1980, paper V.3.4.

[4] L.T.M. Berry, *An Application of Mathematical Programming to Alternate Routing*, Australian Telecomm. Research 4 (2), Nov. 1970, pp.20-27.

[5] L.T.M. Berry, *A Method for Determining Optimal Integer Numbers of Circuits in a Telephone Network*, Austral.Telecomm.Research 11 (1), Jan. 1977, pp.133-137.

[6] L.T.M. Berry, *Optimal Dimensioning of Circuit Switched Digital Networks*, 10-th Internat. Teletraffic Congress, Montreal, 1983, paper 2.1.6.

[7] L.T.M. Berry, *Dimensioning Methods for Hierarchical and Non-Hierarchical Digital Telecommunications Networks*, ITC/CIC Intermat. Seminar on Teletraffic and Network, Beijing, Sept. 1988.

[8] S.-P. Chung, K.W. Ross, *Reduced Load Approximations for Multirate Loss Networks*, 7-th ITC Specialist Seminar, Morristown NJ, USA, Oct. 1990, paper 11.5.

[9] J. Filipiak, *Structured Systems Analysis Methodology for Design of an ATM Network Architecture*, IEEE Jl.Sel.Areas in Communic. 7 (8), Oct. 1989, pp.1263-1273.

[10] G. Gallassi, G. Rigolio, L. Verri, *Resource Management and Dimensioning in ATM Networks*, IEEE Network Magaz. May 1990, pp.8-17.

[11] M. Gerla, J.A.S. Monteiro, R. Pazos, *Topology Design and Bandwidth Allocation in ATM Nets*, IEEE Jl.Sel.Areas in Communic. 7 (8), Oct. 1989, pp.1253-1262.

[12] T.R. Griffiths, *Analysis of a Connection Acceptance Strategy for Asynchronous Transfer Mode Networks*, IEEE Global Telecomm. Conf. GLOBECOM '90, San Diego CA, USA, Dec. 1990, paper 505.4, pp.0862- 0868.

[13] R.J. Harris, *Concepts of Optimality in Alternate Routing Networks*, 7-th Internat. Teletraffic Congress, Stockholm, 1973, pp.427/1-6.

[14] R.J. Harris, *Comparison of Network Dimensioning Models*, 10-th Internat.Teletraffic Congress, Montreal, 1983, paper 4.3B.3. Austral.Telecomm.Research 18 (2), 1984, pp.59-69.

[15] R.J. Harris, *MINDER: An Interactive Planning Tool for Network Planners*, 12-th Internat. Teletraffic Congress, Torino, June 1988, paper 3.2B.4.

[16] R.J. Harris, M.H. Rossiter, *A Multi-class, Multi-commodity Flow Model of a Fast Packet Switched Network*, ITC Specialist Seminar, Adelaide, Oct. 1989, paper 17.3.

[17] J.Y. Hui, *Resource Allocation for Broadband Networks*, IEEE Jl. Sel.Areas in Communic. 6 (9), Dec. 1988, pp.1598-1608.

[18] P. O'Reilly, *Integrated Switching A Fresh Look*, 7-th ITC Specialist Seminar, Morristown NJ, USA, Oct.1990, paper 3.3.

[19] C.W. Pratt, *The Concept of Marginal Overflow in Alternate Routing*, 5-th Internat. Teletraffic Congress, New York, 1967, pp.51-58.

[20] A.H. Roosma, *Optimization of Digital Network Structures*, 11-th Internat. Teletraffic Congress, Kyoto, Sept. 1985, paper 2.4A.4.

[21] A.H. Roosma, *Dimensioning and Evaluation of Switched Networks with Service Protection, taking Transmission Breakdowns into Account*, 12-th Internat. Teletraffic Congress, Torino, June 1988, paper 2.1B.2.

[22] A.H. Roosma, *Planning and Modelling of Digital Networks: A Survey*, ITC/CIC Internat.Seminar on Teletraffic and Network, Beijing, Sept. 1988.

[23] H. Saito, *New Dimensioning Concept for ATM Networks*, 7-th ITC Specialist Seminar, Morristown NJ, USA, Oct.1990, paper 15.3.

[24] F.C. Schoute, *Simple Decision Rules for Acceptance of Mixed Traffic Streams*, 12-th Internat. Teletraffic Congress, Torino, June 1988, paper 4.2A.5.

[25] M. Shinohara, *Generalization of Pratt's Formula for Dimensioning Circuit-Switched Networks*, ITC Specialist Seminar, Adelaide, Oct. 1989, paper no. 11.2.

[26] I. Tokizawa, K-I. Sato, *Broadband Transport Techniques based on Virtual Paths*, IEEE Global Telecomm. Conf. GLOBECOM '90, San Diego CA, USA, Dec. 1990, paper 705B.4, pp.1269-1273.

[27] M.F.L. Van Nielen, A.H. Roosma, *Economics of Multi-Service Networks: New Marginal Costs Formulae and Approximations*, 12-th Internat.Teletraffic Congress, Torino, June 1988, paper 3.3iiA.2.

[28] W. Whitt, *Blocking When Service Is Required From Several Facilities Simultaneously*, AT&T Techn.Jl. 64 (8), Oct. 1985, pp.1807-1856.

[29] M. Yokoyama, K. Miyake, I. Nakajima, *Packet-Switched Network Planning System Architecture and Network Optimization Algorithms*, Rev. El.Comm.Labs 36 (3), 1988, pp.307-311.

Appendix A. Dimensioning methods for circuit switched networks

1. Efficiency formulae for high-usage links

Models by Pratt [19], and many others divide the links in an orderable hierarchical unidirectional network into high-usage and final links. (A network is said to be orderable if the links can be ordered in such a way that overflow traffic or carried traffic of a link is never offered to links with lower order.) Marginal costs per Erlang on high-usage and final links are equated, yielding optimal link sizes for high-usage links. Final links are dimensioned on a fixed final link blocking criterion. Many variants use just a single 'iteration' through the network, resulting in a robust and simple procedure that finds application on a large scale in network planning tools. The procedure yields a suboptimal network (due to shortcomings in the optimality principles, often used single-moment traffic characterization in marginal quantities, etc. Cf. also [14]). Blaauw [3] updated the method to include dimensioning on end-to-end constraints, deriving final link constraints from end-to-end constraints by an extensive optimization procedure.

2. Mathematical programming formulations

Berry [4], [5], [6] [7], Harris [15] and others have formulated the network dimensioning problem as a mathematical program, which is solved by standard techniques (e.g. Langrange multipliers / gradient projection). A central set of variables is the set of chain flows: a two-moment representation of traffic on different routes. Simple approximations are used to describe relations with numbers of circuits on links, etc.
This approach yields very good solutions. End-to-end constraints can be taken into account, as can more general network structures and routing rules. The other side is the huge mathematical program (with enormous numbers of variables and constraints): solving it costs a lot of computer time. Furthermore, due to the fact that different streams (with different characteristics) on a link cannot be assigned arbitrary individual blocking values, the blocking constraints may not be met exactly.

3. Recent combinations of the two approaches

More recently different authors have tried to combine advantages of the afore mentioned approaches, e.g. [21], [25]. The latter gives an extension of [19] to mathematical programming.

The basic approach in [21] (see also [27]) is based on the observation that the dimensionality of the problem of approach 2 can be significantly reduced by using explicit approximative formulae for the marginal overflow from one route to another and for the marginal variations in carried traffic on links in a route. Due to characteristics of these marginal quantities, the optimal solution can easily be found by a simple successive approximation procedure. Simple approximations were derived to distribute final route service requirements (calculated from end-to-end service requirements and service obtained on earlier routes) over links in final routes. Different service requirements on each final link are used to determine service protection levels for specific streams on these links.

Thus, a generalization of approach 1 arises, that can also handle non-orderable quasi-hierarchical bidirectional networks (final links, i.e. appearing in any last-choice route, should be identifiable), and end-to-end constraints. Though iterative in nature (as opposed to approach 1.), it yields very good solutions in a reasonable cpu-time.

A further comparison of the methods under 2. and 3.

Though at first sight the approaches under 2. and 3. seem quite different, it can be shown they are actually very close. Both obtain a sub-optimal solution satisfying the same equation ((41) in [4]). The claim in [14] that the model of [5] should yield cheaper networks than the approach of [19] must apparently be described to the fact that [19] assumed fixed final link blockings in stead of end-to-end constraints. This is circumvented in [21].

However, the approach in [21], like the others, is theoretically sub-optimal, since it does not take into account the combinatorial effects of routes of different traffic sources. Many practical examples have shown that this is only a theoretical point, especially since service protection was taken into the model. Service protection has the advantage of being able to discriminate service of different traffic streams on one link, thus reducing link costs if different streams have very different blocking constraints.

Furthermore, it can be derived that this approach is apparently better than a family of approaches yielding solutions denoted in [13] as 'Game theoretically optimized'. In the example treated in [13] (table 1) the solutions of those approaches are within .2% from the 'System optimized' (= total network costs optimized) solution. In both approaches, it appears optimal to treat high-usage groups first (cf. [5], section on Results). In fact, the only basic difference is the use of a gradient (models 2.) or determining in each iteration the optimal stepsize in each dimension of a reduced dimensionality problem (reduced to h-u links only, with final link sizes then following from the constraints) (models 3.).

Consecutive Cell Loss and Buffer Level Fluctuation Rate : a New Set of ATM Performance Parameters

Frank M. Brochin

NTT Communication Switching Laboratories,
3-9-11 Midoricho, Musashino-shi,
Tokyo 180, Japan

Abstract

We propose here to consider the probability of consecutive loss as well as the occupancy fluctuation rate when evaluating the performance of an ATM statistical multiplexer. After showing the influence of these parameters on the quality of transmission, we present a joint analysis of both the tagged arrival process and the buffer level fluctuations. We use this analysis to derive a method of computation of the probability of a tagged individual source losing an arbitrary number of consecutive cells in the output buffer of a multiplexer fed by a superposed mixed traffic. In our study, the arrival stream for the individual source is modeled after Neuts' versatile Markovian point process while the inter-arrival time distribution for the rest of the aggregate arrival traffic as well as the service distribution are both assumed to be of Phase-type.

1 Introduction

The Asynchronous Transfer Mode technique [4] has been recognized as the most effective mode of implementation of Broadband-Integrated Services Digital Networks. Its major advantages are its simplicity, its flexibility to support various types of service with a wide range of traffic characteristics, and its efficiency with respect to bandwidth, obtained by statistically multiplexing bursty sources. Recent research has focused on performance evaluation [10], the study of efficient control schemes [2, 12] and the analysis of teletraffic issues in ATM systems [6]. Of particular interest are problems related to the communication of voice and video in packet form. Indeed, because of the possibility of delay and congestion in the output buffers of switching nodes, cells from different sources may get discarded when competing to access a shared medium of transmission. This results in contention loss, an important parameter for the determination of the quality of service [5, 11].

Estimating the average probability of loss, however, whether for the aggregate traffic or for the individual lines, is not sufficient to give an adequate measure of performance in the context of packet voice and video communications [5, 8, 9, 11, 13]. Indeed, the loss rate for such systems changes slowly and may remain high for very long periods of time. This results in sources losing a large number of consecutive cells that arrive to buffers during periods of high loss rate. In certain cases, the loss of a string of successive cells may be more desirable than the loss of the same number of non-successive cells. This is true, for example, when we consider data transmission with an HDLC retransmission protocol, according to which a string of successive cells has to be retransmitted after each individual loss. However, for ATM voice and video services, the loss of consecutive cells introduces gaps in the transmission and is extremely damaging to the quality of service. For this reason, there is a real need to evaluate the lengths of consecutive periods of high and low loss rate, as well as devise methods of computation of the probability of an individual source losing an arbitrary number of consecutive cells when an aggregate traffic is transmitted through a statistical multiplexer.

This problem was approached in [5] where a formal relationship was derived by using Palm probabilities, between the average number of consecutively lost cells and the conditional probability of consecutive loss (the probability of a source losing a cell once we know that the previous one has already been discarded). This relationship was applied to an $M/M/1/\infty$ model and the distribution of the consecutive loss due to excessive delays in the queue was computed. An extension to a non-Markovian model, however, was not presented because of the difficulty of computing the conditional probability of consecutive loss.

In fact, the main problem is the lack of knowledge concerning the joint characteristics of both the tagged individual arrival stream and the buffer level fluctuations, the latter being determined by both the aggregate arrival process and the service process. In a recent paper [1], we studied these fluctuations, focusing on the transient loss dynamics of a queueing system. We introduced the concept of lossy and lossless periods which are the intervals of time during which the buffer level is above or below a certain value. We show here how to analyze these periods jointly with the tagged arrival process and derive a method of computation of the probability of an individual source losing consecutive cells in a superposed traffic. Our analysis also provides some information concerning the consecutive loss process and helps us comment on some results obtained in [5] and [8] by using Markovian property assumptions.

This paper is organized as follows. In Section 2, we present our model. We consider a finite-size queueing system and model the arrival stream for the individual source after Neuts' versatile Markovian point process. The inter-arrival time distribution for the rest of the aggregate arrival traffic as well as the service distribution are both assumed to be of Phase-type. In Sections 3 and 4, we study the buffer level fluctuations and define possible patterns of consecutive loss with respect to these fluctuations. We present a method of computation of the probability of each of these patterns in Section 5 and finally illustrate our method by numerical examples in Section 6. Section 7 is the conclusion.

2 The packet communication system

We model the output buffer of the switching node as a single server queue and assume that the total size of the system, including the server, is equal to K. We assume that the arrival and service processes are independent and distinguish between two components in the arrival process, the stream of cells from the tagged individual source and the rest of the superposed arrival traffic. We further assume that these two arrival streams are also independent.

The inter-arrival time distribution for the aggregate arrival process, excluding the tagged individual source, is modeled by a Phase-type distribution PH_1 [7]. We denote its parameters by $(M_1, M_1^0, \overline{\alpha}_1)$. The same assumption is made with respect to the service distribution and the corresponding parameters of PH_2 are denoted by $(M_2, M_2^0, \overline{\alpha}_2)$. The matrices M_i, $i = 1, 2$, are of size m_i and satisfy $(M_i)_{jj} \leq 0$ for $1 \leq j \leq m_i$ and $(M_i)_{jk} \geq 0$ for $j \neq k$. M_i^o is such that $M_i \overline{e}_i + M_i^o = \overline{0}$ where $\overline{e}_i$ is the transpose of the $1 \times m_i$ vector $(1, \ldots, 1)$. $\overline{\alpha}_i$ denotes the vector $(\alpha_i(1), \ldots, \alpha_i(m_i))$ and satisfies $\overline{\alpha}_i \overline{e}_i = 1$. For $i = 1, 2$, we define the transition matrices $Q_i = M_i + M_i^o \overline{\alpha}_i$.

We model the arrival process on the tagged individual line after Neuts' versatile Markovian point process [7]. We refer to this process as N_3 and denote its underlying Phase-type distribution parameters by $(M_3, M_3^0, \overline{\alpha}_3)$. Q_3 is defined in the same way as above. During any sojourn time of the Markov process Q_3 in state i, $1 \leq i \leq m_3$, there are compound Poisson arrivals with rate λ_i and group size density $(s_i(k), k \geq 0)$. At (i, j)-renewal transitions, there are group arrivals with probability density $(r_{ij}(k), k \geq 0)$ while at (i, j)-transitions, there are group arrivals with probability density $(q_{ij}(k), k \geq 0)$. The independence assumptions are the same as in [7].

For an arbitrary time instant t, our system is described by the number of customers present, denoted by $I(t)$, as well as by the phase states of the arrival and service processes, $J_1(t)$, $J_2(t)$ and $J_3(t)$. We also consider the Cartesian product set

$$J = [(1,1,1), (1,1,2), \ldots, (1,1,m_3), (1,2,1), (1,2,2), \ldots, (1,2,m_3), \ldots, (m_1, m_2, m_3)]$$

and assign a number $(J_1(t)-1)m_{23}+(J_2(t)-1)m_3+J_3(t)$ to each triplet $(J_1(t), J_2(t), J_3(t))$, where m_{ij} denotes the product $m_i m_j$. The equilibrium probabilities for the system at an arbitrary instant are represented by the vector $\overline{q}$ with entries $q_{j_1 j_2 j_3}^N$'s, where j_1, j_2 and j_3 refer to the different possible phase states of the system and N is the number of customers present. We use the matricial algorithm presented in [14] for the computation of $\overline{q}$.

Finally, we consider the three counting processes $\{N_i(t), t \geq 0\}$, $i = 1, 2, 3$, and define the matrices $P^i(n, t) = \{P_{kl}^i(n, t)\}$ by

$$P_{kl}^i(n, t) = P\{N_i(t) = n, J_i(t) = l \mid N_i(0) = 0, J_i(0) = k\} \qquad (1)$$

where $1 \leq k, l \leq m_i$.

3 Patterns of consecutive cell loss

A possible representation of the system occupancy as a function of time is presented in Figure 1. Note that the values taken are integers in the range $[0, K]$. We denote by x_i the instant at which the system becomes full for the ith time and by y_i the first instant thereafter at which cells may enter the system again. For $i \geq 1$, we define L_i and l_i as $L_i = [x_i, y_i)$ and $l_i = [y_i, x_{i+1})$ so that these random variables represent the periods of time during which cells arriving to the system are bound to be discarded (L_i) or sure to be succesfully transmitted (l_i). We shall refer to them here as the lossy and lossless periods [1]. We also define the ith loss cycle C_i as $C_i = L_i + l_i$ and $C_i = [x_i, x_{i+1})$. Note that the random variables C_i's as well as the L_i's and l_i's are not independent.

Let us consider an instant t_1 at which the tagged individual source loses a cell and take it as the time origin for our model. Note that the source may actually lose several cells at the same time because of the possibility of batch arrivals. We denote by n_1 the number of cells lost at t_1. The length of the lossy period during which this loss(es) occur will be refered to as L_1 and its remainder as seen at time t_1 by L_1^*. Notice that if the size of the batch arriving at time t_1 is larger than n_1, then L_1^* is equal to L_1. We will also denote the following successive periods by l_1, L_2, l_2 etc.

Now, the instant t_1 corresponds to the beginning of a series of at least N losses if $N \leq n_1$ or if $N > n_1$ but all of the $N - n_1$ cells (and possibly more) generated after t_1 by the tagged individual source are also discarded. This will occur if and only if all of these cells arrive during the lossy periods $L_1^*, L_2, L_3, \ldots$ and none arrives during any of the l_i's. Note that there is no requirement concerning what happens after the Nth loss since we consider possibly more than N consecutive losses.

To compute the corresponding probability, we distinguish between different patterns of loss according to the number n of loss cycles involved and also the way in which the $N - n_1$ cells get discarded in these cycles. To do so, we consider the n-uplets $(N_1, \ldots, N_n)$ which correspond to the occurence of N_1 arrivals in L_1^*, N_2 in $L_2 \ldots$ and N_n in L_n as well as no arrival during l_1, nor $l_2 \ldots$ nor l_{n-1}, conditioned on the fact that a loss occurs at t_1. To avoid taking into account the same events twice for different values of n, we restrict ourselves to the sequences $(N_1, \ldots, N_n)$ such that the Nth loss does not occur before the nth lossy period; that is, we consider all the n-uplets $(N_1, \ldots, N_n)$ in the set

$$S_{N-n_1}^n = \{(N_1, \ldots, N_n) / N_n \geq 1, \sum_{i=1}^n N_i = N - n_1\} \qquad (2)$$

The probability $P(N^+)$ of our individual source losing a string of at least N consecutive cells starting with the loss at t_1 (and conditioned on the occurence of that loss) is the sum of both the probabilities $P(n_1)$ for all $n_1 \geq N$ and the probabilities of the sequences of $S_{N-n_1}^n$ for all the possible values of n and n_1, with $n_1 < N$.

We define $P_{j_1 j_n}(N_1, \ldots, N_n)$ to be the probability that N_i cell(s) are lost during the lossy period L_i, $i = 1, \ldots, n$, that no cell is transmitted during any of the lossless periods

l_i, $i = 1, \ldots, (n-1)$, and that the phase of the system is j_n at the end of L_n conditioned on the fact that it was j_1 at t_1. The corresponding $m_{123} \times m_{123}$ matrix is denoted by $P(N_1, \ldots, N_n)$, where $m_{123} = m_1 m_2 m_3$. We also define $\pi(n_1, j_1)$ to be the joint probability that at the instant t_1, the tagged individual source loses n_1 cells at the same time and that the state of the system is j_1 just after. We obtain

$$P(N^+) = \sum_{n_1=N}^{\infty} P(n_1) + \sum_{n_1=1}^{N-1} \sum_{n=1}^{\infty} \sum_{(N_1,\ldots,N_n) \in S_{N-n_1}^n} \overline{\pi}(n_1) P(N_1, \ldots, N_n) \overline{e}_{123} \qquad (3)$$

where $\overline{\pi}(n_1)$ is a $1 \times m_{123}$ vector with components $\pi(n_1, j_1)$. Note that $S_{N-n_1}^n$ is of size $card(S_{N-n_1}^n)$ given by

$$card(S_{N-n_1}^n) = f(N-n_1, n) - f(N-n_1, n-1) \qquad (4)$$

where the $f(k,n)$ represent the number of different ways of losing exactly k successive cells in n successive cycles and may be efficiently computed by using the following set of recursive equations

$$\begin{aligned} f(k,n) &= f(k, n-1) + f(k-1, n-1) + \ldots + f(0, n-1) \\ f(k,1) &= 1 \end{aligned} \qquad (5)$$

for $k \geq 0$ and $n \geq 2$. Note that $f(0,n)$ is equal to one for $n \geq 2$.

In order to compute the consecutive loss distribution, we therefore only need to study the probabilities $\pi(n_1, j_1)$ as well as the matrices $P(N_1, \ldots, N_n)$. To do so, we first need to determine the joint distribution of the variables L_i's and l_i's.

4 Buffer level fluctuations

From a general point of view, the dynamics of the system between two arbitrary instants in time correspond to those of the $PH_1 + N_3/PH_2/1/K$ queueing system. However, because of the assumption of independence between the processes involved, during periods when no cell from the tagged source arrives to the system, the transient behavior of the queue is the same as that of the $PH_1/PH_2/1/K$ queue.

Now, when N is strictly larger than n_1, we consider a case where cells from the tagged individual source are discarded consecutively after t_1. Since these cells do not enter the system, they do not interfere with it. For this reason, throughout all the period starting at t_1 and ending with the first successful transmission of a tagged cell, the lengths of the lossy periods L_i's and those of the lossless periods l_i's are determined by the dynamics of the $PH_1/PH_2/1/K$ queueing system. However, note that the starting point for the analysis of this transient behavior is t_1 and therefore, that the phase states at that instant are obtained by considering the probability of loss for the individual source in the $PH_1 + N_3/PH_2/1/K$ queue.

Considering the PH_1 and PH_2 processes, we denote by $f_{i_1 j_n}(L_1^*, l_1, \ldots, l_{n-1}, L_n)$ the joint probability density function of the lengths of an arbitrary number n of successive lossy and lossless periods following t_1 and the phase $j_n = (J_1(y_n) - 1)m_2 + J_2(y_n)$ at the end of L_n, conditioned on the fact that L_1^* starts in phase $i_1 = (J_1(t_1) - 1)m_2 + J_2(t_1)$. Note that only the remainder of L_1 is considered. The corresponding $m_{12} \times m_{12}$ matrix is denoted by $f(L_1^*, l_1, \ldots, l_{n-1}, L_n)$. We also consider the instants of transition between the periods L_i's and l_i's and define

$$i_k = (J_1(x_k) - 1)m_2 + J_2(x_k) \tag{6}$$

$$j_k = (J_1(y_k) - 1)m_2 + J_2(y_k) \tag{7}$$

for $k \geq 1$. We make the exception that for $k = 1$, i_1 refers to the aggregate phase of the PH_1 and PH_2 processes at t_1 rather than at x_1. We have

$$f_{i_1 j_n}(L_1^*, l_1, \ldots, l_{n-1}, L_n) = \sum_{j_1=1}^{m_{12}} \sum_{i_2=1}^{m_{12}} \cdots \sum_{j_{n-1}=1}^{m_{12}} \sum_{i_n=1}^{m_{12}} f_{i_1 j_n}(L_1^*, j_1, l_1, i_2, \ldots, j_{n-1}, l_{n-1}, i_n, L_n) \tag{8}$$

Now, it was shown by the author in [1] that for a $PH/PH/1/K$ model, conditioned on the knowledge of the phase of the system at the instants of transition between successive lossy and lossless periods, the lengths of the L_i's and l_i's become independent random variables. Further, it was also shown that because of the properties of the PH densities, the distribution of the length of a lossy period that starts in a certain phase is the same as that of its remainder that starts in the same phase. For these reasons, we replace L_1^* by L_1 and write

$$f_{i_1 j_n}(L_1, l_1, \ldots, l_{n-1}, L_n) = \tag{9}$$
$$\sum_{j_1=1}^{m_{12}} \sum_{i_2=1}^{m_{12}} \cdots \sum_{j_{n-1}=1}^{m_{12}} \sum_{i_n=1}^{m_{12}} \phi(L_1, j_1 \mid i_1)\psi(l_1, i_2 \mid j_1) \ldots \psi(l_{n-1}, i_n \mid j_{n-1})\phi(L_n, j_n \mid i_n)$$

where $\phi(L_k, i_k \mid j_{k-1})$ denotes the probability that the lossy period lasts for a duration L_k and ends with the system stepping in state i_k, when it is observed in state j_{k-1}. $\psi(l_k, j_k \mid i_k)$ is defined in the same way with respect to the lossless period. We rewrite these equations in matrix form and obtain

$$f(L_1, l_1, \ldots, l_{n-1} L_n) = \prod_{i=1}^{n-1}[\Phi(L_i)\Psi(l_i)]\,\Phi(L_n) \tag{10}$$

where as above, we replaced L_1^* by L_1. A method of computation of the $m_{12} \times m_{12}$ matrices $\Phi(L)$ and $\Psi(l)$ for a more general definition of the lossy and lossless periods than that used here was presented by the author in [1]. Note that the entries $\phi_{ij}(L)$ and $\psi_{ij}(l)$ are defined as

$$\phi_{ij}(L) = \phi(L, j \mid i)$$
$$\psi_{ij}(l) = \psi(l, j \mid i)$$

5 The probability of consecutive loss

We may now complete our computation of the probability of consecutive loss. Let us first denote by u_k and v_k, $k \geq 1$, the phase states for the tagged individual arrival process at the instants of transition between lossy and lossless periods; that is $u_k = J_3(x_k)$ and $v_k = J_3(y_k)$. As before, we actually use $u_1 = J_3(t_1)$. Conditioned on the knowledge of the phase of the N_3-process at the instants of transition that separate the lossy and lossless periods, arrivals from the tagged individual source occur independently during the L_i's and l_i's. Further, since conditioned on the knowledge of the phase at the instants of transition that separate them, the successive L_i and l_i have independent lengths, we write

$$P_{(i_1-1)m_3+u_1(j_n-1)m_3+v_n}(N_1,\ldots,N_n) = \tag{11}$$

$$\int_0^\infty \int_0^\infty \cdots \int_0^\infty \int_0^\infty \sum_{j_1=1}^{m_{12}} \sum_{v_1=1}^{m_3} \sum_{i_2=1}^{m_{12}} \sum_{u_2=1}^{m_3} \cdots \sum_{j_{n-1}=1}^{m_{12}} \sum_{v_{n-1}=1}^{m_3} \sum_{i_n=1}^{m_{12}} \sum_{u_n=1}^{m_3}$$

$$\phi(L_1, j_1 \mid i_1)\, P^3_{u_1 v_1}(N_1, L_1)\, \psi(l_1, i_2 \mid j_1)\, P^3_{v_1 u_2}(0, l_1) \ldots \psi(l_{n-1}, i_n \mid j_{n-1})\, P^3_{v_{n-1} u_n}(0, l_{n-1})$$

$$\phi(L_n, j_n \mid i_n)\, P^3_{u_n v_n}(N_n, L_n)\, dL_n\, dl_1 \ldots dl_{n-1}\, dL_n$$

where for reasons explained earlier, we use the notation L_1 instead of L_1^*. This equation may be written in matrix form. We obtain

$$P(N_1,\ldots,N_n) = \prod_{i=1}^{n-1} \left[\int_0^\infty \Phi(L_i) \otimes P^3(N_i, L_i)\, dL_i \int_0^\infty \Psi(l_i) \otimes P^3(0, l_i)\, dl_i\right]$$

$$\left(\int_0^\infty \Phi(L_n) \otimes P^3(N_n, L_n)\, dL_n\right) \tag{12}$$

where the symbol $\otimes$ denotes the Kronecker product of two matrices. Note that using this product allows us to consider the evolution of the system in terms of its general phase, as it was defined in Section 2. For n equal to one, our formula is replaced by

$$P(N-n_1) = \int_0^\infty \Phi(L_1) \otimes P^3(N-n_1, L_1)\, dL_1 \tag{13}$$

To apply Equation (3), we now need to compute the probabilities $Pr(n_1)$ for $n_1 \geq N$ as well as the vectors $\overline{\pi}(n_1)$ for $n_1 = 1,\ldots,N-1$. In fact, both can be obtained from the computation of the latter for all values of n_1. Now, $\pi(n_1, j_1)$ for $j_1 = 1,\ldots,m_{123}$ and $n_1 \geq 1$, is the probability that conditioned on the fact that we have an instant of loss for the tagged source, n_1 cells are lost and the phase is j_1 just after that instant. This can be rewritten as

$$\pi(n_1, j_1) = \lim_{t \to \infty} Pr(n_1 \text{ cells are lost at } t,\, J(t^+) = j_1 \mid t \text{ is a loss instant for } N_3) \tag{14}$$

where t is a random instant of observation. Using a decomposition on all possible states for the system at equilibrium, we obtain

$$\pi(n_1, i, j, k) = \frac{1}{\beta} \sum_{n=0}^{K} \sum_{k_3=1}^{m_3} q^n_{i,j,k_3}[1_{k=k_3} \lambda_{k_3}\, s_{k_3}(n_1 + K - n) \tag{15}$$

$$+ M_3^\circ(k_3)\, \alpha_3(k)\, r_{k_3 k}(n_1 + K - n) + M_3(k_3, k)\, q_{k_3 k}(n_1 + K - n)]$$

where β is a normalizing constant. Note that the probability obtained is conditioned on the occurence of a loss for the tagged individual source.

6 Practical computation and numerical results

We apply our method to systems characterized by the mean arrival rates and the squared coefficients of variation of the inter-arrival time distributions of both the individual process and the superposed arrival process, as well as the mean service rate and squared coefficient of variation of the service time distribution. A detailed description of the models chosen for the processes PH_1, PH_2 and N_3, a practical method of computation of the right-hand term of Equation (12) as well as additional numerical results are presented in [3].

We choose to model the superposition of a finite number $n_{vl} + 1$ of sources and study the loss of consecutive cells for one of them. The mean arrival rate for each arrival process is chosen to be roughly 1320 cells per minute. The mean arrival rate for the aggregate arrival process (excluding the tagged source) is $\lambda_1 = 1320\, n_{vl}$ cells per minute. We choose a slotted transmission channel model with 3000 slots per second. The corresponding capacity is 136 voice lines, yielding nearly 100% utilization. The service distribution is assumed to be Erlangian with 5 states. We apply our method to different system sizes, ranging from 10 to 40, and to different numbers of superposed sources, ranging from 80 to 135. We also consider various squared coefficients of variation ca_1^2 and ca_3^2 for both the tagged and superposed arrival processes so as to measure the influence of the burstiness of the arrival traffic.

We first study the buffer level fluctuations. In Figures 2 and 3, we present numerical results for the mean and squared coefficient of variation of L_1^*. These results are obtained for $ca_1^2 = 0.5$ and $ca_3^2 = 0.2$ respectively. Note that the mean of L_1^* is normalized with respect to the inter-arrival time for the tagged source so as to give better information in terms of consecutive loss. Similar curves are shown in Figures 4, 5 and 6 for l_1 and results obtained for the coefficient of correlation between L_1^* and l_1 are presented in Figure 7.

In Figures 8 and 9, we present the results obtained for the probability of consecutive loss. We consider a system of size 30 and focus on the conditional probability of consecutive loss for different tagged arrival processes with squared coefficients of variation ranging from 0.2 to 50. Our results show that the conditional probability P_{cons} of the tagged source losing at least a second consecutive cell once it just lost one increases sharply with the burstiness of the arrival process. However, for arrival processes with squared coefficients of variation larger than 25, it becomes practically constant.

We denote the probability of loss for the tagged source by P_{loss} and define the ratio $r = P_{cons}/P_{loss}$. In Figure 10, we present numerical results for r. The different plots correspond to $K = 10, 20$ and 30, while the parameters are those chosen above for the study of the dynamics of the queue. These results show that the conditional probability of consecutive loss for our tagged source can be much larger than the probability of loss itself. This implies that the probability of losses occuring in strings of consecutive cells is much larger than what the probability of loss by itself could suggest. Such a conclusion was first reported in [8] and [9] after simulations. Our results also show how fast the ratio

r increases with the size of the buffer for a fixed load of the system (determined by n_{vl}) and decreases with the number of voice sources for a fixed buffer size.

It is important to note that the conditional probability of consecutive loss is nearly not affected by the buffer size, provided that the latter is large enough. Results show that for K larger than 20, the conditional probability of consecutive loss obtained is nearly the same. This confirms some findings presented in [11] and [13]. It can be explained by the fact that once the loss of a cell occurs, the loss rate remains high for a non-negligible amount of time and the buffer level fluctuates around high values only. This phenomenon justifies the high probability of consecutive losses as well as the fact that the dynamics of loss do not depend anymore on the maximum possible length of the queue.

It is also interesting to consider what happens when we decrease the proportion of traffic that corresponds to the tagged source (by adding more lines for the rest of the aggregate traffic) without altering the total load for the system. This can be done by considering a transmission channel with larger capacity and better illustrates the influence of the proportion of the tagged traffic on the conditional probability of consecutive loss. We apply our method to 5, 12, 20, 30, 50 and 100 voice lines (excluding the tagged one). The computations show that, once the total system load is fixed, the influence of the size of the buffer on the conditional probability of consecutive loss is negligible. This agrees with the comments made above. However, since the probability of loss decreases when the size of the buffer is increased, the ratio r varies greatly for the same number of voice lines involved and can become extremely high when the total load for the system is low.

7 Conclusion

In this paper, we presented a method of computation of the probability of a tagged individual source losing an arbitrary number of consecutive cells in the output buffer of a multiplexer fed by a superposed mixed traffic. Our method was based on a study of the transient dynamics of the buffer level fluctuations in an ATM environment, generalized so as to analyze the periods of loss and transmission jointly with the tagged arrival process. We modeled the arrival stream for the individual source after Neuts' versatile Markovian point process while the inter-arrival time distribution for the rest of the aggregate arrival traffic as well as the service distribution were both assumed to be of Phase-type. We showed that when the buffer is large enough (say K larger than 20), the conditional probability of successive loss remains the same for all values of K but increases with the burstiness of the arrival process. More importantly, we showed that the conditional probability of consecutive loss can become extremely large compared to the probability of loss for a particular tagged source. This phenomenon makes the occurence of losses of strings of consecutive cells from the same source more likely than what is suggested by the probability of loss. Because it introduces gaps in communications, it emphasizes the need to consider both the probability of consecutive loss as well as the buffer level fluctuation rate as important parameters of performance evaluation for ATM systems.

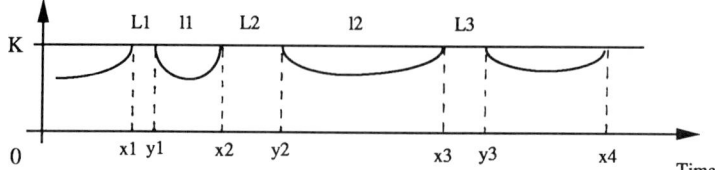

Figure 1

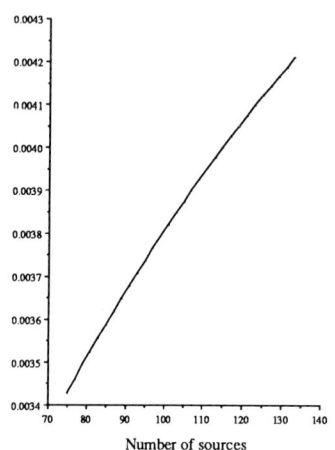

Figure 2

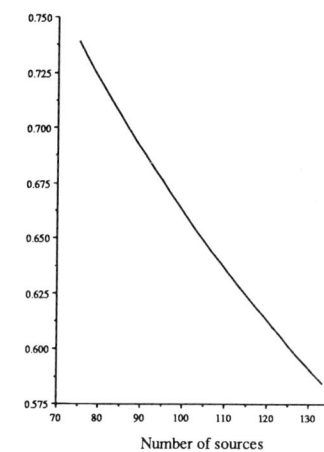

Figure 3

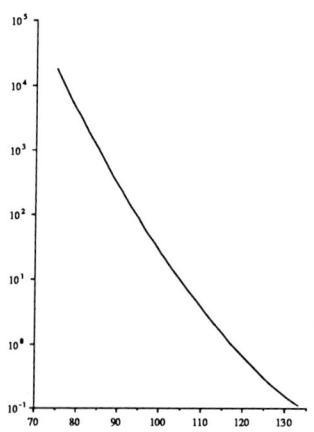

Figure 4

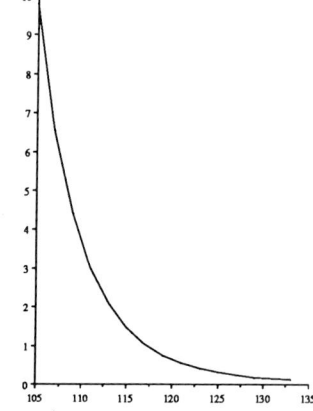

Figure 5

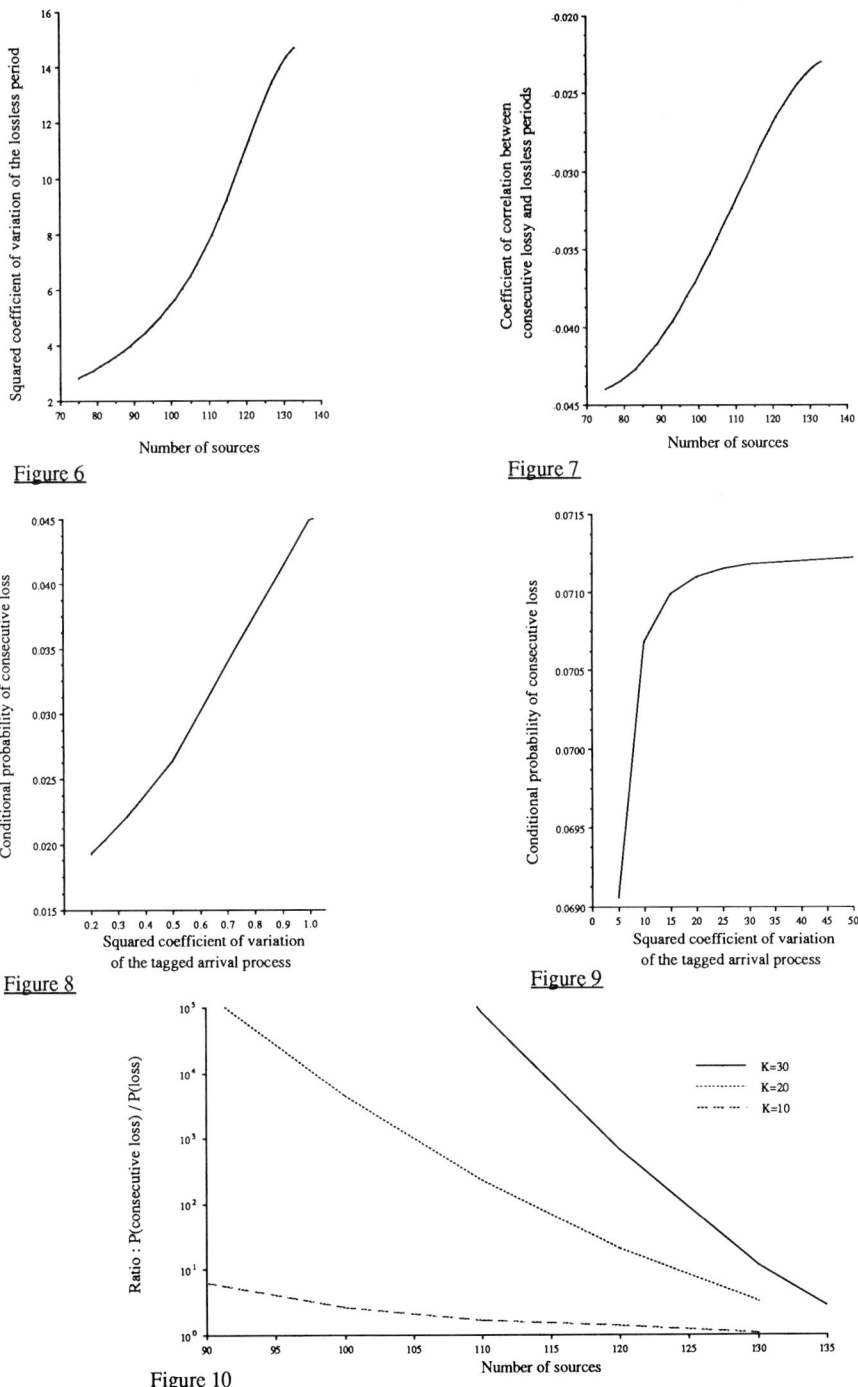

Figure 6

Figure 7

Figure 8

Figure 9

Figure 10

References

[1] F.M. Brochin, "Buffer level fluctuations in a bursty environment", to appear.

[2] F.M. Brochin, "The output process of a traffic shaper", in the *Proceedings of the Spring National Convention of the IEICE*, Japan, March 1991.

[3] F.M. Brochin, "A method of computation of the consecutive cell loss probability for an individual source in superposed traffic", to appear in the *Proceedings of the International Conference on the Performance of Distributed Systems and Integrated Communication Networks*, Kyoto, Japan, September 1991.

[4] CCITT, "Rec. I.121", Blue Book, Melbourne, 1988.

[5] J.M. Ferrandiz and A. Lazar, "Consecutive packet loss in real-time packet traffic", in the *Proceedings of the Fourth Int. Conf. on Data Comm. Syst. and their Perf.*, Barcelona, Spain, June 1990, pp. 306–324.

[6] K. Kawashima and H. Saito, "Teletraffic issues in ATM networks", in *Computer Networks and ISDN Systems*, Vol. 20, No. 1–5, December 1990, pp. 369–375.

[7] Marcel F. Neuts, "A versatile Markovian point process", in *Journal of Appl. Probab.*, Vol. 16, 1981, pp. 764--779.

[8] H. Ohta and T. Kitami, "Simulation study of the cell discard process and the effect of cell loss compensation in ATM networks" in *Transactions of the IEICE*, Vol. 73, No. 10, October 1990.

[9] V. Ramaswami and W. Willinger, "Efficient traffic performance strategies for packet multiplexers" in *Computer Networks and ISDN Systems*, Vol. 20, No. 1–5, December 1990, pp. 401–407.

[10] H. Saito and K. Kawashima, "Performance modeling of integrated voice and data communication networks" in *Stochastic Analysis of Computer and Communication Systems*, North Holland, New York, 1990.

[11] S.Q. Li, "Study of information loss in packet voice systems", in *IEEE Transactions on Communications*, Vol. 37, No. 11, November 1989, pp. 1192–1202.

[12] S. Sumita, "Synthesis of an output buffer management scheme in a switching system for multimedia communications", in *Proc. of the IEEE Infocom '90*, San Francisco, June 1990.

[13] G. Woodruff, R. Kositpaiboon, G. Fitzpatrick and P. Richards, "Control of ATM statistical multiplexing performance", in *Computer Networks and ISDN Systems*, Vol. 20, No. 1–5, December 1990, pp. 351–360.

[14] H. Yamada and F. Machihara, "Performance analysis of a statistical multiplexer with control on input and/or service process", *Technical Report of the IEICE*, December 1989.

Analytical Expressions for Blocking Probabilities in a B-ISDN

Johan M Karlsson

Department of Communication Systems

Lund Institute of Technology

P.O. Box 118, S-221 00 LUND

Sweden

Abstract

The rapid changes in technology and service requirements make it increasingly important to find an efficient way to evaluate different performance issues for the B-ISDN, when the offered traffic changes. The analytical approach in this paper gives a good estimate of the blocking probabilities for an outward trunk group in an ATM like environment. A recursive formula gives the steady state probabilities in a lost calls cleared service facility for heterogeneous traffic streams. These values are then used to obtain the blocking probabilities for different resource sharing policies (bandwidth allocation schemes) perceived by the different traffic demands from various sources. Further, the exact values for this splinted group is also compared to the values for a common group. These sources could be any application that is integrated in the network, for example telephony, low speed data, Hi-Fi sound and picture telephony. The analytical values obtained by these formulae follow very well the values of a simulation carried out to validate the accuracy of the results.

1 Introduction

The evolution of telecommunications is towards a multi-service network fulfilling all user needs for voice, data and video communications in an integrated way. Up to now, new networks were developed whenever a new service became relevant. This hardly seems an efficient and cost effective way to meet emerging communication needs. Some of these applications are computer related (e.g. communication among remote supercomputers performing jointly a task, others involve the transmission of images and video signals). While fiber optic technology provides the necessary bandwidth for transmission purposes, the creation of a network that can provide high bandwidth services to the users remains a challenge. The problems that evolve are mainly in switching and resource sharing. Since such a network will carry all applications in an integrated fashion it is an important task to be able to switch according to the bandwidth allocation algorithm. This is to be done in a packet switching mode which offers greater flexibility than circuit switching in handling the wide diversity of data rates and latency requirements resulting from the integration of services. Today, the world is thinking about this network as an ATM (Asynchronous Transfer Mode) network. It is shortly also to be defined by CCITT. One of the major goals is the realization of a single and worldwide integrated broadband communication network (IBCN).

2 Model description

The basic concepts are almost settled, but it still remains some time before a final international design is provided. This depends on the many different opinions regarding which switching technique to be used, and the complexity of the system. The following models are based on the basic conditions mentioned above, a further, more detailed, discussion could be found in [1, 2].
Owing to the high bandwidth that is required along with the low bit error rate on tomorrows transmission media (e.g. fiber optics) it is more efficient to look upon each transmission of a cell as potentially correct, instead of wrong (compare HDLC). Necessary error detection and retransmission are handled at higher layers and mainly end-to-end or edge-to-edge. In this way we avoid complex window flow type control on a link-to-link basis. The network consists of a number of nodes connected by trunk

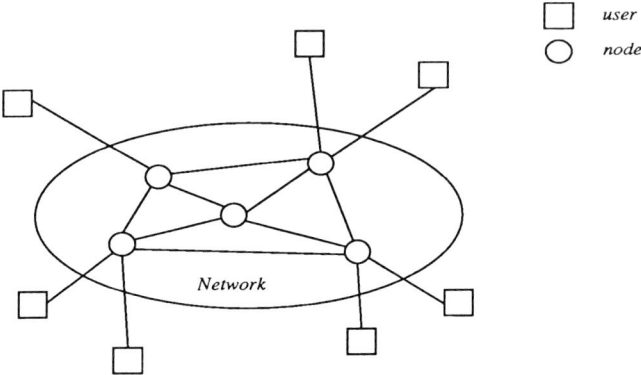

Figure 1: *A brief outline of the network analyzed, with access and transit nodes.*

groups as shown in Figure 1. There are two different types of nodes, access nodes and transit nodes. The distinction is that the access node has connection to subscribers and receives fresh traffic destinated for the network whereas the transit node only deals with traffic inside the network structure. The middle node in Figure 1 is an example of a transit node. Connecting the nodes we have trunk groups which have direct routes to other nodes in the network. Each trunk group consists of a number of links, which on their part consists of units of 64 kbps, further on called a cell, which is the smallest unit that could be allocated.

2.1 Traffic Characteristics

To describe the traffic characteristics for a not existing network is one of the major problems that we have to deal with at the moment. It could be discussed which new services that are going to evolve, to what extent they and already existing services will be used. In spite of the fact that it theoretical is possible to allocate any bandwidth to a user, it is undoubtedly some "standardized" sizes that would be predominant. In our analysis we have focused on some of these sizes, which would correspond to services like HDTV, picture telephony, Hi-Fi sound, lower speed data, ordinary telephony and some other service related to office based communication devices. To get a more objective view of the performance of the analytical work presented in this paper, different mixtures have been adopted. The different traffic types used, as well as the traffic mixtures, are shown in Table 1, further details could be found in [3, 4, 5].

Bandwidth (kbps)	Arrival- intensity	Departure- intensity	Traffic mix A	B	C
64	0.2	1	1		
128	0.24	3	2	1	
256	0.05	1	3	2	1
512	0.7	5	4	3	2
1024	2.7	30		4	3
2048	0.03	15			4

Table 1: *The different traffic types, their bandwidth and the different traffic mixtures.*

2.2 Trunk Group Model

The analysis in this paper is considering an outward trunk group emanating from one of the transit nodes, described earlier in this section. The trunk group consists of a number of links (N), where each link on their part consists of M cells, as shown in Figure 2. Traffic arrivals are all Poisson and are made up of a mixture of narrowband and broadband traffic types, as discussed in section 2.1. Since the links compose some restrictions on the calls position a product form solution is, unfortunately, not applicable, which would be expanded on in section 3.1. Some recursive formulae for blocking probabilities in a common group, i.e. without the link limitation, shared by different users have been known quit a long time [6].

3 Blocking Probabilities

We distinguish different traffic types according to the number of cells required per call. Calls of type i require d_i cells ($1 \leq d_i \leq 1792$) and the traffic offered by this source(s) is denoted by A_i. The probability that, at an arbitrary instant, the trunk group is

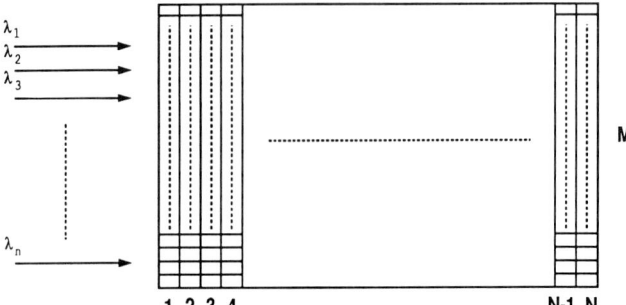

Figure 2: *Generic model of an outward trunk group from a transit node.*

occupied by k_i calls of type i ($i = 1, 2, ..., l$) is then

$$p(k_1, k_2, ..., k_l) = \frac{1}{G} \frac{A_1^{k_1}}{k_1} \frac{A_2^{k_2}}{k_2} \cdots \frac{A_l^{k_l}}{k_l} \quad (1)$$

where G is a normalization constant such that the sum of the $p(\mathbf{k})$ over all feasible states (i.e. $\sum_{i=1}^{l} k_i d_i \leq M \cdot N$) is equal to one. It follows directly from this definition that

$$Pr(n) = Pr(n \text{ channels occupied}) = \sum_{\sum_{i=1}^{l} k_i d_i = n} p(k_1, k_2, ..., k_l). \quad (2)$$

Further we use the convention to equate the offered traffic by the number of busy channels on an infinite trunk group, the mean and variance of such an operation with the definition made above are,

$$m = \sum_{i=1}^{l} A_i d_i \qquad v = \sum_{i=1}^{l} A_i d_i^2. \quad (3)$$

There will be a great number of different traffic streams satisfying these requirements. We have noticed empirically that, for complete sharing, the distribution $Pr(n)$ is approximately invariant for different offered traffics with given mean and variance, these being identified with the mean and variance of the total number of occupied cells in an infinite system. Noticing this, we could use a simpler traffic model to be able to analyze the behaviour of the system. We substitute the Poisson arrivals of traffic with different bandwidths by a bursty single cell traffic with negative binomial arrivals, producing the same mean and variance. Since these formulae are independent of the holding time distribution we only have to know the mean holding time. We define

$$\alpha = \frac{m^2}{v}, \qquad \beta = 1 - \frac{m}{v} \qquad and \qquad \sum_{n=0}^{N \cdot M} Pr(n) = 1$$

and are then able to calculate $Pr(n)$ by the following recurrence relation [8]

$$nPr(n) = [\alpha + (n-1)\beta] Pr(n-1). \quad (4)$$

3.1 Common versus Splinted Group

The model described above could be looked upon as a common group of $N \cdot M$ cells, divided into N **dependent** links of M cells each. Consequently, d_i-calls could be blocked in spite of the fact that the total system is in a state with $\geq d_i$ free cells, since a call has to be allocated entirely on one link. However, any new arrival could be served by any of the links and an unsuccessful attempt on one link implies a new try on another link. A call is never blocked before all links are accurately searched. To analyze the difference in the behaviour of a common group and the dependent links of this model a Markov chain is derived. As discussed, it is necessary to keep record of each trunk group separately as well as if a links state that defines that the link is occupied by calls adding up to, for example 1024 kbps in reality consists of 16 calls of 64 kbps each, 8 calls of 128 kbps, 8 calls of 64 kbps and 4 calls of 128 kbps or any any combination adding up to, in this case 1024 kbps. This problem was solved by symbolically derive the linear equations that arise from the Markov chain. Finally, the blocking states for different traffic types are added into separate blocking probabilities for that specific traffic type. The analyze is then accomplished by multiplying one of the traffic streams both characteristics (λ_i, μ_i) by the same value, which should not imply any difference if the product form solution is valid, i.e. there is no real difference between our model and a common group.

3.2 Analytical Approximations

The routing strategy must keep a call within one link. This implies that the blocking probability for an i-call is that none of the links has d_i or more free cells. In the following the requirements of an i-call in terms of cells is denoted by d_i. Since the trunk group is divided into links we have to make some special arrangements dealing with these properties. We number the links from 1 to N (see Figure 2). Then the blocking probability for a d_i-call could be expressed as

$$B_{d_i} = \sum_{k=0}^{N \cdot M + 1 - d_i} r_{d_i}(N, k) Pr(k) \qquad (5)$$

where we introduce conditional probabilities,

$$r_{d_i}(n, k) = Pr(\text{a } d_i\text{-call call cannot be carried in any of the first } n \text{ links} \mid k \text{ of the first } M \cdot n \text{ cells are busy}) \qquad (6)$$

for $1 \leq n \leq N$, $0 \leq k \leq M \cdot n$.

To express each of those probabilities could be rather cumbersome. However, they could be determined iteratively as below

$$r_{d_i}(n, k) = \sum_{j=k-M}^{k-(M-1)+d_i} r_{d_i}(n-1, j) \Upsilon(n, j, k) \qquad (7)$$

where we once again introduce conditional probabilities,

$$\Upsilon(n,j,k) = \Pr(j \text{ busy cells in link 1 to } n-1 \mid k \text{ busy cells in link 1 to } n). \qquad (8)$$

The iterative formula (7) expresses the states in which a d_i-call is blocked due to occupancy in link n. Then it remains to evaluate an expression for the last probabilities $\Upsilon(n,j,k)$. They would strongly depend upon the routing strategy, and consequently we proceed by discussing the different solutions to that problem. For a more detailed description of various allocation algorithms the reader is referred to [7, 9].

3.2.1 Random Search

In this algorithm the links in the trunk group form a logical ring so that there is a continuous succession of links $(1, 2, \cdots, N-1, N, 1, 2, \cdots)$. On arrival the algorithm searches forward from an haphazard place in the logical ring until a link with sufficient capacity is found or the search is completed, that is that the search is back to the original link.

Since this algorithm does not achieve our goal to fill up the links before allocating an empty link, the behaviour of this algorithm is the worse in respect of overall blocking and utilization of the trunk group [9]. We are, consequently, not going to show any result for this algorithm. However, a possible way of calculating the blocking probabilities is described below. If we look upon a single cell we mean by random selection that the probability, that the cell is allocated on a specific link, is proportional to the number of free cells on that link. This will ensure that the probability of k cells occupied in a link is equal for all links. We further assume that the same occupancy symmetry is maintained for a d_i-call, when $d_i \geq 1$. Following this assumptions we get,

$$\Upsilon(r,j,k) = h_{M(r-1),M}(j,k) \qquad (9)$$

where M is the capacity of the link (in cells) and $h_{m,n}$ is the hypergeometric distribution,

$$h_{m,n}(i,j) = \frac{\binom{m}{i}\binom{n}{j}}{\binom{m+n}{i+j}}. \qquad (10)$$

It should also be noted here that we are obliged to recalculate the values for the parameters obtain as we add new links to our iterative method. In this case it means that the values of $h_{m,n}(i, k-i)$ has to be recomputed as the m-value increases. Finally, we get

$$B_{d_i} = \sum_{k=0}^{N \cdot M + 1 - d_i} \sum_{j=k-M}^{k-(M-1)+d_i} r_{d_i}(N-1,j) \frac{\binom{M(N-1)}{j}\binom{M}{k}}{\binom{M \cdot N}{j+k}} \Pr(k). \qquad (11)$$

3.2.2 Sequential Search

This group of algorithms could be diveded into two major parts. The first one is when the algorithm always starts, irrespective of the calls bandwidth, at the first link in the group and searches sequential forward until a link with sufficient capacity is found or the search is completed. The second group is when the search is carried out from different places in the trunk group, dependent on the calls bandwidth demand. In this algorithm the links also form a logical ring as described in section 3.2.1.

For this algorithm we could model the system as an overflow system, see Figure 3, with a first choice group of m ($=M(r-1)$) channels and an overflow group of n ($=M$) channels and with an arrival rate that is negative binomial. With this model we could

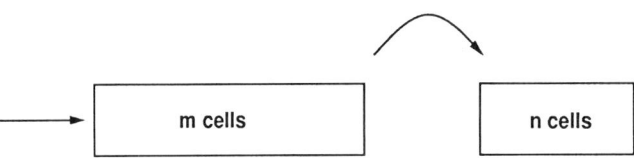

Figure 3: *Overflow model.*

define the state space and draw the system using a state diagram. By making cuts on this state diagram, we are able to equate the flows in either directions. We start with making cuts around each state on row n (i.e. states with the overflow group filled), which gives us equations on the form:

$$p(j,n) = (n + (\gamma_{j-1,n} + j - 1)p(j-1,n) \qquad (12)$$
$$- \gamma_{j-2,n} p(j-2,n))/j$$

where

$\gamma_{i,j}$ = the negative binomial arrival rate from state (i,j) $\qquad (13)$
to state (i+1,j) for $i = 1, 2, \cdots, m-1$ and
from state (m,j) to state (m,j+1).

This makes it possible to express $p(j,n)$ in terms of $p(0,n)$, for $j = 1, 2, \cdots, m$. $p(0,n)$ could then be assigned any value. We just have to keep record of the calculated values which will give us the normalizing constant at the end of the algorithm. Knowing this, we continue to make cuts between rows $j = n$ and $j = n-1$ which yield an equation of the form:

$$p(m, n-1) = \frac{n}{\gamma_{m,n-1}} \sum_{\forall i} p(i,n). \qquad (14)$$

Knowing $p(m, n-1)$, we progressively compute $p(i, n-1)$ from relations of the form:

$$p(i, n-1) = \alpha_{i,n-1} p(i+1, n-1) + \beta_{i,n-1} \qquad (15)$$

where the α and β values could be expressed in terms of the transition rates and previously calculated state probabilities.

$$\alpha_{i,n-1} = \frac{(i+n+\gamma_{i+1,n-1})p(i+1,n-1)}{\gamma_{i,n-1}} \tag{16}$$

$$\beta_{i,n-1} = -\frac{(i+2)p(i+2,n-1)+np(i+1,n)}{\gamma_{i,n-1}} \tag{17}$$

The same procedure is continued row by row to the end of the last row. State probabilities are then derived by dividing $p(i,j)$ for all i and j by the running total. This operation is relatively stable as long as some renormalizations are carried out through the calculations. We are now capable of expressing the last probabilities in the previous iterative scheme (7) as,

$$\Upsilon(r,j,k) = \frac{p(j,k-j)}{\sum_{\forall i} p(i,k-i)} \tag{18}$$

and finally,

$$B_{d_i} = \sum_{k=0}^{N \cdot M+1-d_i} \sum_{j=k-M}^{k-(M-1)+d_i} r_{d_i}(N-1,j) \frac{p(j,k-j)}{\sum_{\forall d_i} p(d_i,k-d_i)} Pr(k). \tag{19}$$

3.2.3 Best Fit

This algorithm tries to "optimize" to which link the call should be directed. As intimated by the name, this algorithms main objective is to gather the calls as much as possible on the same link, to prevent gaps on the links to occur. The algorithm searches all the links and allocates the call on the link with the highest utilization and with capacity enough to transfer the call.

This algorithm has shown to has the best behaviour among the algorithms on a full scale simulation of the system [9]. Because of that, this is the algorithm that we want to get some analytical expressions for the blocking probabilities carried out for. As described above, the algorithm tries to allocate the call on the link that is closest to full capacity, but still able to take care of the arriving call's cell requirement. To be able to detect this link, the algorithm has to go through all the links before deciding upon which link the call should be directed. Yet, another way to describe the same function is to have the links organized in a memory in order of used capacity. In this scenario the algorithm only have to search sequentially (!) until the first link capable of transmitting the call is found. Knowing this, we assume that the links are ordered in respect of used capacity in a memory, we are able to use our newly achieved knowledge of sequential selection, in section 3.2.2, to calculate the blocking probabilities of the arriving calls. It should be noticed however, that since the algorithm described is iterative, we need some starting values for $r_{d_i}(1,k)$ for different d_i:s and k:s. These could be set up and devoted to their values in the beginning of a computation. Further, we should also keep in mind that the $Pr(n)$ has to be recalculated each time we add a link to our iterative procedure. This is going to show up as an increase of capacity in the first choice group of m ($= M(r-1)$) cells in Figure 3.

4 Results

To be able to see the accuracy of the analytical models described above, the results for the *Best Fit* algorithm have been compared to simulation results carried out. The choice of the *Best Fit* algorithm is due to the fact that it in previous investigations [7, 9] has shown, without exceptions, the best overall performance. Because of the iterative way of solving the equations and the rapidly growing state space, the comparison is performed for a small model. The different traffic types involved are divided by the least common denominator in respect of size. The trunk group is in the same manner diminished according to the sizes of the arriving calls. The analytical models show a small discrepancy, as could be seen in Figure 4-6, due to the fact that the arrivals are no real batch arrivals. However, the approximation made gives us an adequate method of

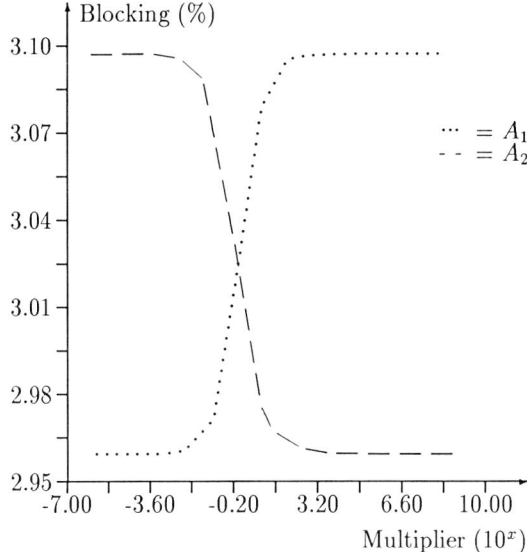

Figure 4: *Blocking probability for traffic type 1 versus the number that A_1 and A_2 are multiplied by, using traffic mixture A.*

getting the blocking probabilities in a fast and reliable way. The results are shown for different traffic mixtures, as explained in Table 1, with the *Best Fit* algorithm adopted. In Figure 4 the discrepancy between the blocking probabilities for traffic type 1 and 2 in traffic mixture A could be seen. The blocking probabilities show an S-shaped form when the offered traffic (A) is multiplied by a "constant". For a common group this

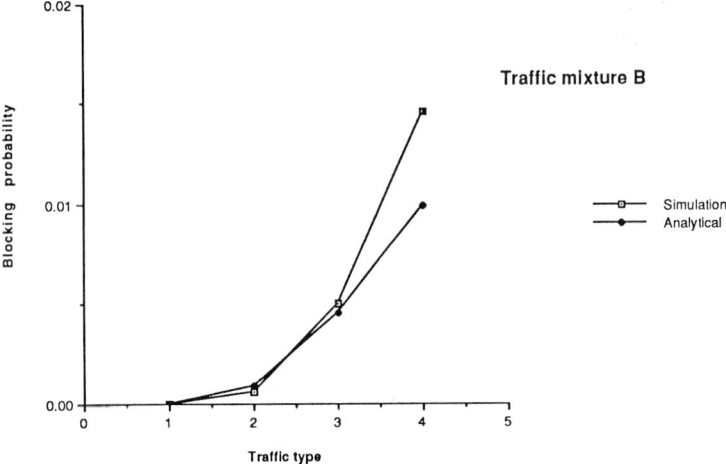

Figure 5: *The blocking probabilities for the different traffic types, with traffic mixture B adopted.*

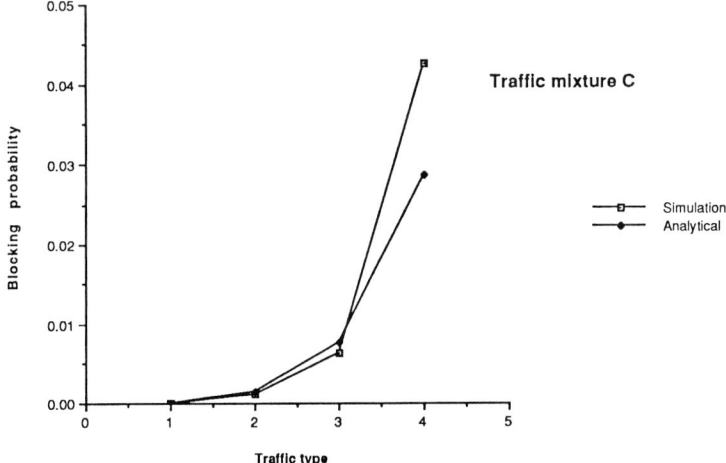

Figure 6: *The blocking probabilities for the different traffic types, with traffic mixture C adopted.*

would not imply any changes in the blocking probabilities. The difference is, however, not to alarming.

If we look at the comparison between our analytical approaches and simulations in Figure 5-6, it could be seen that the differences are bigger the broader the bandwidth is, this yields also relatively. This could be explained by the fact that the bursty **one** slot arrival process we used, does not capture the real behaviour of the bulk arrivals in the simulation. As the bulk sizes grow (broader traffic types) this inaccuracy gets worse, but is still in a reasonable range comparing to the simulation results. Both parameter settings show a similar performance. For narrowband calls the analytical approach shows a slightly higher value for the blocking probabilities, while showing a lower value for broadband traffic. This seems rather logical, since the bursty arrival process levels out the true bulk arrivals and consequently gets a more linear behaviour than the simulation.

5 Conclusions

A way to analytically calculate the blocking probabilities for different traffic sizes merged together in an outward trunk group has been proposed for an ATM like environment. The results are produced by an iterative method, which is built using probability theory, teletraffic theory and overflow theory. As the results of the comparison between the analytical approach and simulation results show, the analytical approach gives quite acceptable values for the blocking probabilities. The results are for a rather small model, but we should have in mind, that this model represents a full size model in respect of relative differences in the traffic demands and the size of the trunk group. An investigation of the difference of this model, i.e. a splinted group, compared to a common group is carried out as well. The approach given in this paper gives us a quick, accurate, estimation of the blocking probabilities for the outward trunk group. This could be to great use designing this part of the network, or to get a feeling of what signification different changes in the traffic mixture would have.

References

[1] M. De Prycker and J. Bauwens, "The ATD concept: one universal bearer service", *CEPT/BSLB Seminar on Broadband Switching, Albufeira, January, 1987.*

[2] P. Gonet, M. Servel and J. P. Coudreuse, "Packet-Oriented Techniques for Integrated Broadband Networks", *GSLB Seminar, Albufeira, 1986.*

[3] H. Armbrüster and B. Schaffer, "Realization Aspects of the Future Telecommunications Infrastructure", *Telecom Report 10 No. 5, 1987.*

[4] S.E. Minzer and D. R. Spears, "New Directions in Signalling for Broadband ISDN", *IEEE Communications Magazine, Vol 27, No. 2, February, 1989.*

[5] H. Suzuki, T. Takeuchi, F. Akashi, T. Yamaguchi, "Very High-Speed and High-Capacity Packet Switching for Broadband ISDN", *IEEE Journal on Selected Areas, Vol 6, No. 9, December, 1988.*

[6] J. S. Kaufman, "Blocking in a Shared Resource Environment", *IEEE Transactions on Communications, Vol 29, No. 10, October 1981*

[7] J. M. Karlsson, "Loss Performance in Trunk Groups with Different Capacity Demands", *To be published.*

[8] J. W. Roberts, "Teletraffic Models for the Telecom I Integrated Services Network", *Tenth International Teletraffic Congress, Montreal, 1983.*

[9] J. M. Karlsson, "Some Aspects on Link Allocation in an IBCN", *International Conference on Networks, Singapore, July, 1989.*

Telecommunication Services for Developing Economies
J. Filipiak (Editor)
© 1991 Elsevier Science Publishers B.V. All rights reserved

Buffer Dimensioning and Effective Bandwidth Allocation in ATM Based Networks With Priorities[1]

Zbigniew Dziong[2], Ke-Qiang Liao, Lorne Mason

INRS-Telecommunications, 3 place du Commerce, Verdun, Quebec H3E 1H6, Canada

Abstract

We consider an ATM multiplexer with several FIFO buffers operating in non preemptive priority regime. We assume that each connection or type of traffic (e.g. connectionless services) is allocated to one priority characterized by its grade of service constraints. Under very general assumptions we derive a model for buffer dimensioning and effective bandwidth allocation. It is based on upper bounds for grade of service measures (cell blocking, average cell delay). The model consists of one independent multiplexer for each priority. Each of these multiplexers has only one FIFO queue. The approach assumes that a model for buffer dimensioning and effective bandwidth allocation in a multiplexer with one FIFO queue is available. The advantage of the proposed approach is its simplicity and generality.

1 Introduction

It is well recognised that in case of ATM based networks preventive congestion control methods are of great importance (see e.g. [1,2]). The two main components of the preventive control are call admission control and source policing. In general the call admission control can be performed in several ways but from bandwidth management and implementation viewpoints the approach based on a logical bandwidth allocation to each call is most attractive. In this case a call is admitted if the bandwidth required is smaller or equal to the residual bandwidth available on all links constituting the chosen path. The bandwidth allocated to a call should provide that the grade of service constraints on the cell level are met under the condition that all calls do not exceed their specified parameters. To enforce this condition source policing can be implemented.

Up to now several models for bandwidth allocation and related buffer dimensioning have been proposed (e.g. [3,4,5,6]). Most cases reported are limited to an ATM multiplexer with one FIFO buffer and common GOS constraints for all call classes. But since introducing priorities for calls with different GOS requirements enables more efficient use of the bandwidth resources there is a need to investigate such schemes. Some results on bandwidth allocation in priority system with one buffer and overwriting mechanism are reported in [7]. In the paper we propose an approach for effective bandwidth allocation and buffer dimensioning in an ATM multiplexer with several FIFO buffers operating with non preemptive priority. It is assumed that each connection or type of traffic (e.g. connectionless services) is allocated to appropriate priority

---
[1] The work was supported by the NSERC Strategic Grant No. STR 0045121.
[2] On sabbatical leave from Warsaw University of Technology, Poland.

which is characterized by its grade of service constraint. The model is based on upper bounds for cell loss probability and average cell delay for each priority traffic. The theoretical background for these bounds is given in [8]. The central idea of the approach is the replacement of the considered priority scheme by a set of multiplexers serving each priority traffic. The resulting system is readily analysed by employing the method previously developed for traffic with a single grade of service constraint.

The paper is organized as follows. In Section 2 we recall some basic concepts connected with bandwidth allocation and buffer dimensioning for a multiplexer with one FIFO buffer using as an example the approach presented in [4]. Some new results illustrating the accuracy of this approach are presented. In Section 3 the bounds for the priority system are described and then the algorithm for bandwidth allocation and buffer dimensioning is presented. Some numerical examples and a discussion of the efficiency of the proposed scheme are given in Section 4.

2 No priority system

Up to now three practical approaches to call bandwidth allocation in a multiplexer with one FIFO queue and a common GOS have been discussed in the literature, namely: peak rate allocation, average rate allocation with link utilization threshold and effective bandwidth allocation. Any of these approaches can be used in the model for the multi-priority system described in the subsequent sections. In the following we illustrate some basic concepts of the effective bandwidth allocation.

Let the vector $\mathbf{x} = \{x_j\}$ denotes the multiplexer state where x_j denotes the number of j-th class calls carried on the multiplexer. For the sake of presentation we assume that x_j is a real variable. Each call class is characterized by a "limit" call bit rate process and its parameters: the peak rate, P_j, the average rate, A_j, etc. (in general we allow any process which is stationary). In particular it may be a process described by the alternating idle and burst states (e.g. voice calls, graphic stations) or a process with continuously changing bit rate (e.g. compressed video). Concerning connectionless services, we treat a data flow carried on a path as a semi-permanent connection. Thus all subsequent considerations based on the connection concept apply equally to connectionless services. By the "limit" process we mean a process whose parameters cannot be exceeded by a call from this class. This process can be defined by the parameters of a policing mechanism or by the type of source. It is also assumed that each connection process is not correlated with the others. Now we can define an admissible region as a function of call class normalized throughputs $\mathbf{U} = \{U_j\}$ where $U_j = \frac{x_j A_j}{L}$ and L is the line speed. An example of the admissible region for two call classes is presented in Fig.1 (convex line). The interpretation of admissible region is that for the states below the boundary, the GOS constraints are met for all call classes while above the boundary the GOS constraints are not met for at least one call class.

The important feature of the exact admissible region is its convexity previously indicated in several papers (e.g. [9,4,6]). This feature implies that any effective bandwidth allocation that provides maximum link utilization is state dependent. This follows from the fact that for each point on the boundary we should have $\sum_j x_j V_j(\mathbf{x}) = L$. It is obvious that this feature is not attractive from an implementation point of view. Observe that effective bandwidth allocation can be non-state dependent only in case the admissible region is linear. In fact this feature is possessed by one of the first models for the effective bandwidth allocation described in [3]. In this approach the effective bandwidth, V_j, allocated to the j-th class call is given by the link capacity divided by the maximum number of the j-th class calls, x_j^{max}, which the link can support while meeting the grade of service requirements. Henceforth we refer to this model as a linear approximation. The graphical interpretation of this approach in case of two call classes is given in Fig.1 (upper line). The weak side of this approach is that in some states (the area between exact and linear boundary) the GOS constraints are not met. An alternative solution was proposed in [4]. It is based on the construction of a linear surface tangent to the admissible region represented in the multidimensional link state space. This surface is treated as a boundary of a new admissible region. Then the steady effective bandwidth allocation for each call class can be easily evaluated from the maximum number of calls in the new admissible region. Henceforth this approach is called the modified linear approximation. One issue left to be clarified is the choice of tangent point. In general the objective is to maximize the average throughput. While the exact solution would require knowledge of the state distribution and complex analytical tools, we propose a simplified approach in which we maximize throughput for the mixture of calls proportional to the predicted or measured average state (operating point). This approach is illustrated in Fig.1.

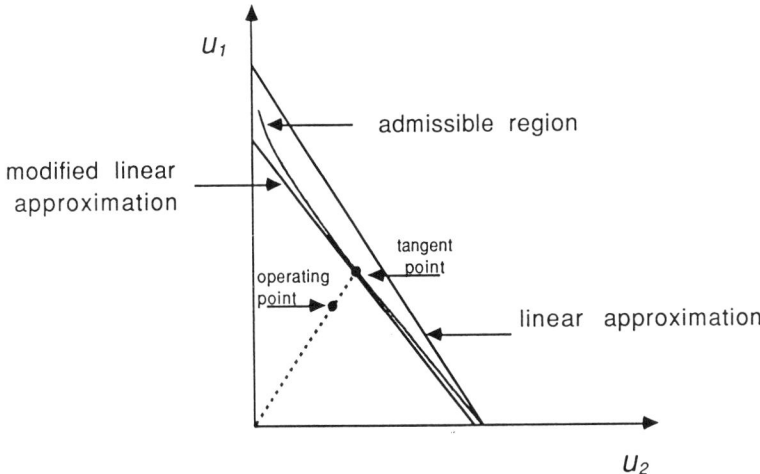

Fig. 1. Admissible region and its approximations.

Observe that the modified linear approximation can be used with any model providing an accurate admissible region. Up to now several models were proposed to evaluate such a region (e.g.[9,4,6]). In the remainder of this section we summarize the approach presented in [4] (henceforth called the non-linear approximation) and present some new results illustrating its accuracy.

The non-linear approximation is based on a two parameter (background softness, S_j, and background sparseness, H_j) characterization of the statistical properties of a call background process. The call background process is defined as a superposition of all calls carried on the link except the one under consideration. The background softness is defined as the difference between the sum of peak rates over all background calls and the sum of effective bandwidth allocated for these calls in the ATM link under consideration. The background sparseness is defined as the difference between the effective bandwidth required to carry all background calls in the ATM link under consideration and the average rate of these calls. The main idea behind this two parameter description is that it is related to the performance in the system under consideration (by means of the sum of effective bandwidth). This is an important feature since as indicated in [11] the first two or three input process moments provide only weak information relative to GOS constraints especially in the case of non-renewal processes.

The following considerations are applicable to the multiplexer in a state for which the GOS constraints are tight (admissible region boundary). It was shown [4] that by varying the multiplexer capacity in the homogenous case (all calls belongs to the same class) one obtains, for each call class, unique relations between the effective bandwidth and background softness, $V_j = f_j^S(S_j)$, and between the effective bandwidth and background sparseness, $V_j = f_j^H(H_j)$. These functions can be evaluated from an analytical model or by real measurement in a test bed multiplexer.

The key assumption in the non-linear approximation is that, for each point on the boundary of the admissible region, the background softness and sparseness of each call (characterized by V_j) should be at least as large as the background softness and sparseness (respectively) in the homogeneous case for class j resulting in the same bandwidth allocation, V_j. Additionally we assume that effective bandwidth evaluated from the linear approximation, V_j', is a lower bound for V_j. This can be stated as follows:

$$V_j = \max\{f_j^S(S_j), f_j^H(H_j), V_j'\} \quad ; j \in J \quad (1)$$

This set of equations can be used to determine the effective bandwidth allocation for each type of call by means of repeated substitution with initial values evaluated from the linear approximation (for details see [4]).

Table 1
Accuracy of linear and non-linear approximations ($D^c = 10$)

case					linear		non-linear	
j	P	B	S	L	x	D	x	D
1	10	200	2000	50	1	75.30 ($\pm$ 12.8)	1	10.61 ($\pm$ 3.18)
2	1	22	44		124	22.06 ($\pm$ 3.84)	105	2.34 ($\pm$ 0.49)
1	10	200	2000	50	12	28.74 ($\pm$ 3.72)	12	6.21 ($\pm$ 1.55)
2	1	22	44		62	12.55 ($\pm$ 1.65)	43	2.45 ($\pm$ 0.60)
1	10	100	300	50	1	21.01 ($\pm$ 1.05)	1	11.93 ($\pm$ 0.54)
2	10	10	30		14	12.17 ($\pm$ 0.46)	13	7.13 ($\pm$ 0.20)
1	10	100	300	50	6	14.14 ($\pm$ 1.13)	6	8.11 ($\pm$ 0.78)
2	10	10	30		7	8.83 ($\pm$ 0.66)	6	4.86 ($\pm$ 0.41)
1	10	200	2000	100	1	60.28 ($\pm$ 10.3)	1	10.44 ($\pm$ 2.58)
2	1	22	44		271	24.44 ($\pm$ 3.79)	251	2.82 ($\pm$ 0.43)

The preliminary assessment of accuracy of the non-linear approximation presented in [4] was based on the MMPP model used as the input (homogenous case) and as the reference (heterogenous case). The study showed that the non-linear approximation gives results close to the heterogenous model of MMPP. In the following we present the results of an additional study where the input to the non-linear approximation is based on a modified fluid approximation [10] with average cell delay, D, as the GOS measure and infinite buffer. In [10] it was shown that the model has good accuracy over a wide range of source parameters (modelled as a burst source). A sample of the results is presented in Table 1. Each case represents a mixture of two call classes characterized by different parameters (peak rate, P_j, average burst length, B_j, average silence length, S_j). The number of sources of the first call class, x_1, is fixed while the number of the second class sources, x_2, is evaluated by the non-linear approximation to provide average delay equal to the constraint, $D^c = 10$ (expressed in cell service time unit). Simulation is used to check the actual GOS. The linear approximation is also evaluated. The results confirm that the non-linear approach provides a GOS close to the constraint and is significantly more accurate than the linear approximation.

3 Priority system

For the sake of presentation we begin with a two priority system. In this case the multiplexer has two FIFO buffers Q_l, Q_h where Q_h has non preemptive priority. Each call can be allocated to one of the two priorities characterized by different grade of service constraints: blocking on the cell level, B_l^c, B_h^c and/or average cell delay, D_l^c, D_h^c. In other words all cells from a given call are entering one buffer (Q_l or Q_h) and the objective of the call admission control is to provide that the grade of service constraints are met for each call in progress. We consider two types of server: synchronous and asynchronous. In the synchronous case time is slotted with cell service time period and a cell to be served has to wait for the beginning of the slot. In the asynchronous case the cell can be served immediately on condition the server is free. The general structure of the considered system is presented in Fig.2a. Throughout this section we assume that a model for buffer dimensioning and effective bandwidth allocation in the multiplexer with one FIFO buffer is available. To simplify notation the waiting time is expressed in cell service time units.

3.1 High priority traffic

Consider a modified system where the low priority traffic is neglected and the buffer capacity Q_h is given by $K'_h = K_h - \delta$ where $\delta = 0$ in the case of synchronous server and $\delta = 1$ in the case of asynchronous server (see Fig.2b). Let $B_{j,h}$, $D_{j,h}$ denote the grade of service for the j-th class high priority call in the original system. Assuming that the state of high priority calls in the modified system is the same as in the original system, it can be shown [8] that the following relations hold:

$$B'_{j,h} \geq B_{j,h} \qquad (2)$$
$$D'_{j,h} \geq D_{j,h} - \delta \qquad (3)$$

where the prime indicates the parameters in the modified system. In fact for $\delta = 0$ these relations become equalities since in the synchronous system the low priority cells are not influencing the high priority cells. These results show that the GOS values in the non-priority system can be used to evaluate upper bounds for the GOS values for high priority calls in the original system.

Let us apply the model for the buffer dimensioning and bandwidth allocation in the modified, non-priority system under the following GOS constraints

$$B_h^{c'} = B_h^c \qquad (4)$$
$$D_h^{c'} = D_h^c - \delta \qquad (5)$$

where it is assumed that $D_h^c > 1$ for $\delta = 1$. Based on (2, 3) it is easy to show that if the evaluated buffer capacity $K_h = K_h' + \delta$ and effective bandwidth allocations are used in the original system for the high priority calls, the grade of service constraints are met for all classes of high priority calls

$$B_{j,h} \leq B_h^c \qquad (6)$$
$$D_{j,h} \leq D_h^c \qquad (7)$$

Observe that this result is valid regardless of the level of low priority traffic. In other words the evaluated buffer dimension and bandwidth allocation for high priority calls are independent of analogous values for low priority calls.

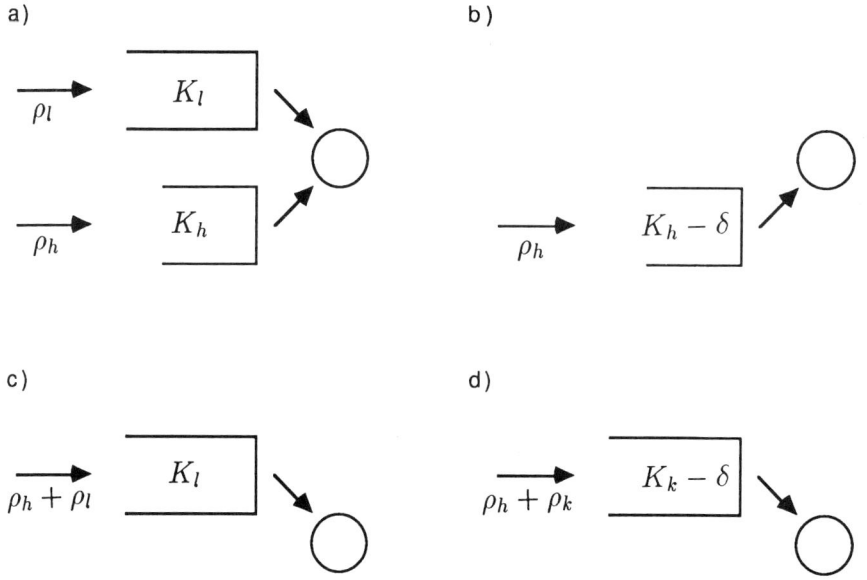

Fig.2. Structures of the original and modified systems.

3.2 Low priority traffic

Consider a modified system where both low and high priority calls are offered to the buffer Q_l (Q_h is not used) as illustrated in Fig.2c. Let B_l, D_l denote the cell loss probability and the average cell delay for low priority cells, respectively, and B, D denote the average cell loss probability and average delay for both high and low priority cells in the original system. Let us also define the total link throughput ρ, and throughputs of high and low priority traffic, ρ_h, ρ_l as

$$\rho = \frac{\sum_j x_j A_j}{L} \quad ; \quad \rho_h = \frac{\sum_j x_j^h A_j}{L} \quad ; \quad \rho_l = \frac{\sum_j x_j^l A_j}{L} \qquad (8)$$

where x_j^h, x_j^l denote numbers of the j-th class calls allocated to high and low priority, respectively. It is clear that $\rho = \rho_l + \rho_h$. In the following we assume (**A1**) that the difference between the average delay measures in the considered systems with finite buffers and the same systems with infinite buffers, for the states within the admissible region, is negligible. This assumption is resonable in the ATM environment since the blocking constraints are very stringent.

Assuming that the state of calls of both types in the modified system is the same as in the original system, it can be shown [8] that

$$B'' \geq B_l \qquad (9)$$

and under **A1**

$$D'' = D \qquad (10)$$

where the double prime indicates parameters in the modified system. Note that in the original system we have

$$\rho_l D_l + \rho_h D_h = \rho D \qquad (11)$$

Thus by using (10) we arrive at the following relation for average delay of the low priority cells

$$D_l = \frac{\rho}{\rho_l} D'' - \frac{\rho_h}{\rho_l} D_h \qquad (12)$$

Based on this relation and (9) the following GOS constraint for the modified system can be constructed

$$B^{c''} = B_l^c \qquad (13)$$
$$D^{c''} = \min_{\mathbf{x} \in X^b} \{ \frac{\rho_l}{\rho} D_l^c + \frac{\rho_h}{\rho} D_h \} \qquad (14)$$

where X^b denotes the set of states for which both the state of high priority traffic and the state of low priority traffic are located on the boundaries of the corresponding admissible regions. Let us assume for the moment that the constraint (14) can be evaluated. If the proposed constraints are used in the model for the buffer dimensioning and bandwidth allocation in the modified non-priority system and the result is applied in the original system it is easy to show that the following relations are valid

$$B_l \leq B_l^c \qquad (15)$$
$$D_l \leq D_l^c \qquad (16)$$

Up to now we assumed that the value of $D^{c''}$ is known, but in fact it is a function of the final result thus making the algorithm implicit. To find the solution one can apply an iterative algorithm with an arbitrary initial value $D^{c''} > D_h^c$, e.g.

$$D^{c''} = D_l^c \qquad (17)$$

Note that in many practical applications the GOS constraints for high priority traffic are several orders of magnitude smaller than the ones for low priority traffic. Thus the upper bound

$$D_l \leq \frac{\rho}{\rho_l} D'' \qquad (18)$$

achieved by neglecting the second term in (12) can be used to construct a simpler constraint for the modified system

$$D^{c''} = [\frac{\rho_l}{\rho}]_{min} D_l^c \qquad (19)$$

where $[.]_{min}$ denotes the minimum value over $\mathbf{x} \in X^b$.

An alternative upper bound for average delay of low priority calls can be developed under assumption (**A2**) that the average delays in the original system can be approximated by an $M/D/1$ system with infinite buffers and asynchronous server. In other words we assume that the cell superposition process is close to Poissonian for each priority. In this case the following relation can be developed [8] from the solution of $M/D/1$ system

$$D_l'' = D_l(1 - \rho_h) \qquad (20)$$

Thus to evaluate effective bandwidth allocation in the modified system we can use the following constraint for the average delay

$$D^{c''} = D_l^c(1 - [\rho_h]_{max}) \qquad (21)$$

where $[.]_{max}$ denotes maximum value over $\mathbf{x} \in X^b$. The advantage of this constraint is that all its input parameters can be evaluated apriori.

The proposed algorithm provides that the GOS constraints are met for the superposition of low priority calls. But if we assume (**A3**) that for any link state on the admissible boundaries, $\mathbf{x} \in X^b$, the proportion of cell losses for different call classes with low priority is similar in both original and modified systems

$$\frac{B_{j,l}}{B_{i,l}} \cong \frac{B_{j,l}''}{B_{i,l}''} \quad ; \quad i,j \in J \qquad (22)$$

the following relations for each call class are also valid:

$$B_{j,l} \underset{\approx}{\leq} B_l^c \qquad (23)$$

Under an analogous assumption for average delays we have

$$D_{j,l} \underset{\approx}{\leq} D_l^c \qquad (24)$$

3.3 General case

The approach for the system with two priorities can be easily extended to the general case with N priorities. Let $i = 1, ..., N$ denotes the priority index where $i = 1$ and $i = N$ correspond to the lowest and highest priority, respectively. To evaluate buffer dimension and bandwidth allocation for particular class k it is convenient to introduce an aggregated description of all calls with higher priority, ρ_h^k, B_h^k, D_h^k, defined as follows:

$$\rho_h^k = \sum_{i>k} \rho_i \qquad (25)$$

$$B_h^k = \frac{\sum_{i>k} B_i \rho_i}{\rho_h^k} \qquad (26)$$

$$D_h^k = \frac{\sum_{i>k} D_i \rho_i}{\rho_h^k} \qquad (27)$$

Consider a modified system where all calls with priority $i \geq k$ are offered to the buffer Q_k (the rest of the calls is neglected) and the buffer capacity is given by: $K_k''' = K_k - \delta$ as illustrated in Fig.2d. Observe that from the k-th priority traffic viewpoint the system is analogous to the one for low priority calls described in the previous section with the exception that the buffer capacity is reduced in the case of asynchronous server to take into account the influence of the calls with the lower priority as was done in the case of bounds evaluation for high priority calls. Thus based on (2, 3, 9, 12), the following bounds hold:

$$B_k \leq B''' \tag{28}$$

$$D_k - \delta \leq \frac{(\rho_h^k + \rho_k)}{\rho_k} D''' - \frac{\rho_h^k}{\rho_k}(D_h^k - \delta) \tag{29}$$

In case the original system can be approximated by an $M/D/1$ system with infinite buffers and asynchronous server (analog of **A2**) it follows from (3, 20) that

$$D''' \geq (D_k - \delta)(1 - \rho_h^k) \tag{30}$$

Thus to evaluate the buffer capacity and effective bandwidth allocation for the k-th priority calls we can solve the modified system with the constraints given by

$$B^{c'''} = B_k^c \tag{31}$$

$$D^{c'''} = \min_{x \in X^b} \{ \frac{\rho_k}{\rho_h^k + \rho_k}(D_k^c - \delta) + \frac{\rho_h}{\rho_h^k + \rho_k}(D_h^k - \delta) \} \tag{32}$$

or alternatively

$$D^{c'''} = (D_k^c - \delta)(1 - [\rho_h^k]_{max}) \tag{33}$$

It is clear that for the system with asynchronous server in the case $k = 1$ we should put $\delta = 0$ since there is no traffic with lower priority.

4 Discussion and numerical examples

In the following a discussion of the efficiency of the proposed approach for buffer dimensioning and bandwidth allocation for a two priority system is presented. To illustrate the discussion we evaluated admissible regions for high and low priority traffic in case of two call classes. The evaluation is based on the linear approximation for the non-priority model and average delay constraint only (infinite buffers). The calls are modelled as on-off sources characterized by peak rate, P_j, burst length, B_j, and silence length, S_j. The modified fluid approximation presented in [10] is used as an input to the linear approximation for both synchronous and asynchronous servers. The examples of admissible regions for high priority calls, H, and for superposition of high and low priority calls, T, are shown in Fig.3,4. The indexes s, a corresponds to synchronous and asynchronous servers, respectively. Different admissible regions for total state of calls correspond to different bounds for GOS constraints: T_1 - basic approach (14), T_2 - high priority factor neglected (19), T_3 - $M/D/1$ approximation (21), T_0 - initial region in the iterative procedure (17). The interpretation of the boundaries is that the state of high priority calls carried by the multiplexer cannot exceed the boundary H and the state of all calls carried by the multiplexer cannot exceed the boundary T. These boundaries can be translated into effective bandwidth allocations for high priority calls, V_j^h, and for all calls, V_j, respectively. It is obvious that while for a new low priority call the call admission control has to consider one condition, $V_j \leq L - \sum_i x_i V_i$, in the case of a new high priority call two conditions, $V_j^h \leq L - \sum_i x_i^h V_i^h$, $V_j \leq L - \sum_i x_i V_i$, have to be checked.

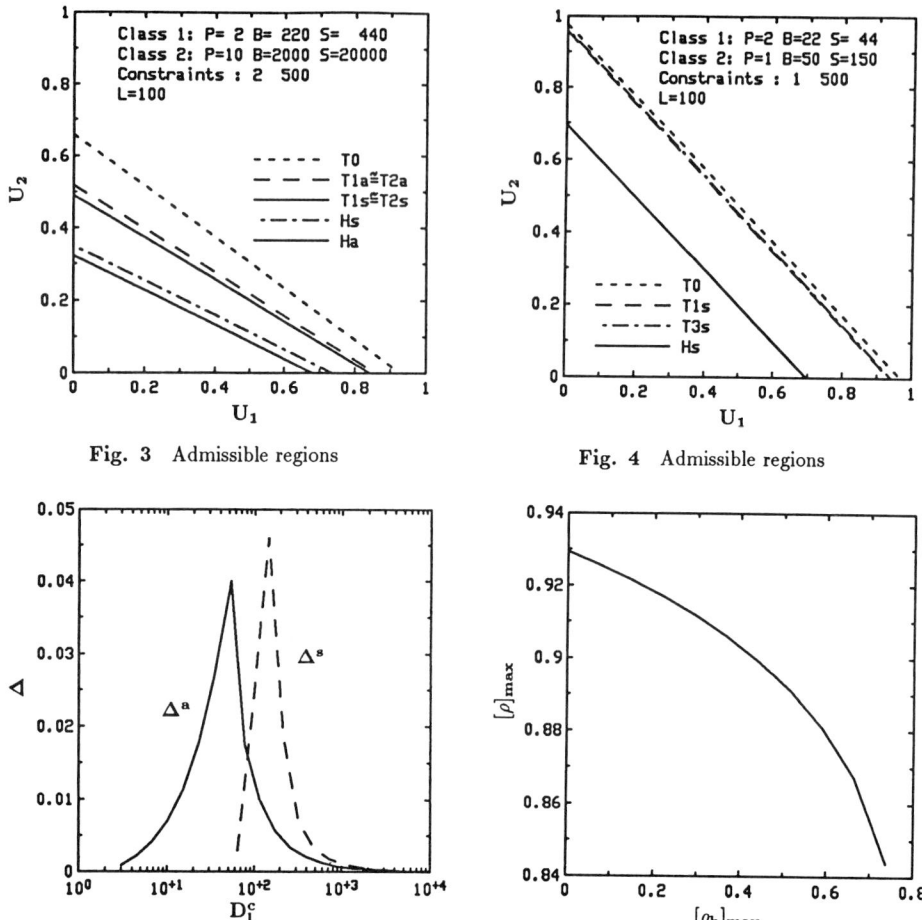

Fig. 3 Admissible regions

Fig. 4 Admissible regions

Fig. 5 Gap between T_1 and T_2 versus constraint for low priority calls (system parameters from Fig. 3)

Fig. 6 Maximum link utilization versus throughput of high priority calls (system parameters from Fig. 3)

Observe that in the case of synchronous server the modified system (Fig.2b) used to evaluate the boundary H^s is exact since we have equality in (2, 3) for $\delta = 0$. To consider the tightness of bounds (2, 3) in the case of asynchronous server ($\delta = 1$) let us define two factors

$$r_b = \frac{1}{D_h^c} \tag{34}$$

$$r_d = \frac{1}{K_h} \tag{35}$$

It can be shown that for $r_b \to 0$ and $r_d \to 0$ the used bounds provide the maximum throughput of high priority traffic. In other words the larger the buffer capacity and the larger the delay constraint the tighter the bounds. In case these values are small there is a gap between the bounds and real values causing some reduction of the admissible region for high priority calls. Some indication about the size of this gap is given by the difference between the boundaries H^s and H^a since the former can be treated as an upper bound for the exact solution, H_e^a. It is important to stress that from a practical point of view the gap between H^a and H_e^a can be treated as a positive factor. This follows from the fact that the smaller the admissible region for high priority traffic the larger the total admissible region (compare T_1^s and T_1^a in Fig.3). In fact as indicated below in some cases it might be reasonable to artificially reduce the admissible region for high priority traffic in order to increase link utilization.

Concerning the tightness of bounds for low priority traffic we begin with an analysis of the upper bound for blocking probability B_l (9). Let r denotes the ratio of average delays

$$r = \frac{D_h}{D_l} \qquad (36)$$

It can be shown that for $r \to 0$ we have

$$B'' = B_l \qquad (37)$$

Note that in practical cases it is most likely that $D_h^c \ll D_l^c$. This indicates that the admissible region evaluated by the proposed approach should be close to the maximum one. Concerning the average delay of low priority cells, the basic approach is based on the equality (12) so the constraint $D^{c''}$ is optimal. In case the approach is based on the bound where the high priority factor is neglected (18) it is clear that the smaller the ratio $r^c = \frac{D_h^c}{D_l^c}$ the closer the result is to the optimal one. This feature is illustrated in Fig.5 where the gap between T_1 and T_2 (expressed as the average difference in link utilization, Δ) is given as a function of D_l^c for the example from Fig.3. Observe that the behaviour of Δ is opposite to the predicted one for $D_l^c < 50$ in the case of asynchronous server and for $D_l^c < 144$ in the case of synchronous server. The reason is that in this range $D^{c''} < D_h^c$ so H is used as a lower bound for T_2. Moreover $T_1^s = H^s$ for $D_l^c < 64$ so $\Delta^s = 0$ in this range. The efficiency of the approach based on the $M/D/1$ model is illustrated in Fig.4. Since in this case the cell superposition process can be approximated by the Poissonian distribution, T_1 and T_3 are very close to each other.

Observe that the boundary T_0 can be interpreted as a limiting case for $[\rho_h]_{max} \to 0$. This suggests that by imposing additional restriction on the admissible region for the high priority calls we can increase the maximum link utilization, $[\rho]_{max}$. The relation between $[\rho]_{max}$ and $[\rho_h]_{max}$ is given in Fig.6 for the case of synchronous server. Application of this mechanism is reasonable in case the high priority calls do not constitute a majority of the traffic. In other words the maximum bandwidth allocated to high priority calls should provide sufficient GOS on the call level. Another possible application of the function presented in Fig.6 is to adapt the bandwidth allocation to the current value of ρ_h by using it as $[\rho_h]_{max}$. In general the adaptation should be based on the full algorithm for bandwidth allocation. But if we assume that for the new value $[\rho_h]_{max} = \rho_h$ the new boundary T_1 is approximately parallel to the original one, the effective bandwidth allocations can be approximated as follows

$$V_j = \frac{[\rho]_{max}^o}{f(\rho_h)} V_j^o \qquad (38)$$

where f(.) denotes the function presented in Fig.6 and ^{o} indicates the values from the original solution.

5 Conclusions

In the paper we have proposed an approach for buffer dimensioning and bandwidth allocation in an ATM multiplexer with priorities. The approach is based on performance bounds which can be expressed in terms of performance for a non-priority system. The main advantage of the proposed approach is its simplicity and generality. In particular the only performance model needed is the one for the multiplexer without priorities and with homogenous input. Moreover no restrictions on the call cell process are imposed. Numerical examples indicate that the approach is flexible and provides a mechanism for tradeoff between the throughput of high and low priority traffic. The accuracy of the approach depends mainly on the GOS constraints. Assuming that in ATM systems the constraints for high priority traffic are two or more orders of magnitude smaller than the ones for low priority traffic, the approach is close to optimal. The final assessment of the accuracy can be done once the standards for GOS constraints are set.

References

[1] Tutufor K., "On Admission Control and Policing in an ADM based Network", Proc. of ITC Specialist Seminar, Morristown, USA, October 1990.

[2] Mason L., Dziong Z., Liao K.-Q., Tetreault N. "Control Architectures and Procedures for B-ISDN", Proc. of ITC Specialist Seminar, Morristown, USA, October 1990.

[3] Akhtar S., "Congestion Control in Fast Packet Switching Network", Master Thesis, Washington University, 1987 (advisor: Prof. J.S. Turner).

[4] Dziong Z., Choquette J., Liao K.-Q., Mason L., "Admission Control and Routing in ATM Networks", Proc. of ITC Specialist Seminar, Adelaide, September 1989. Also published in "Computer Networks and ISDN-Systems" Vol. 20, December 1990, 189-196.

[5] Gallassi G., Rigolio G., Fratta L., "Bandwidth Assignment and Bandwidth Enforcement Policies", Proc. of Globecom'89, Dallas 1989.

[6] Appleton J., "Modelling a Connection Acceptance Strategy for Asynchronous Transfer Mode Networks", Proc. of ITC Specialist Seminar, Morristown, USA, October 1990.

[7] Gallassi G., Rigolio G., Fratta L., "Bandwidth Assignment in Prioritized ATM Networks", Proc. of Globecom'90, San Diego, California, December 1990.

[8] Dziong Z., Liao K.-Q., Mason L., "Performance Bounds in ATM Multiplexer with Priorities", in preparation.

[9] Hui J.Y., "Resource Allocation for Broadband Networks", IEEE Trans. Commun., vol.6, No.9, December 1988.

[10] Liao K.-Q., Mason L., "A heuristic approach for performance analysis of ATM systems", Proc. of Globecom'90, San Diego, California, December 1990.

[11] Holtzman J.M., "Coping with Broadband Traffic Uncertainties: Statistical Uncertainty, Fuzziness, Neural Networks", Proc. of Globecom'90, San Diego, California, December 1990.

Connection Admission Control in ATM Networks

E. Dutkiewicz and G. Anido

Development Unit, OTC Limited, GPO Box 7000, Sydney NSW 2001, Australia

Abstract
A three level management architecture forms a basis for managing and controlling network resources in ATM networks. It utilises the concepts of virtual channels, virtual paths, and virtual networks. Connection admission model using a 2-state Markov modulated Poisson process approximation for traffic streams is studied in order to obtain simple approximation algorithms.

1. INTRODUCTION

Broadband networks will be required to support a variety of services displaying a wide range of characteristics and demanding from the network different levels of performance. Of particular interest, at least in the early stages of the existence of broadband networks, will be the problem of supporting existing services which often require deterministic channels with stringent temporal and spatial requirements. Provision will also have to be made for other future services which may place different kinds of demand on the network. In order to deal with these various requirements traffic control and traffic management capabilities will have to be provided in such networks.

Admission control will form an important aspect of the control part of future broadband networks. A number of models have been proposed in the literature (see, for example [4]) in which source models are developed and studied in this context. In this paper a three level management architecture is presented for controlling and managing traffic in broadband networks. The problem of admission control is then studied under this framework. A queuing model for the connection admission control is developed and studied in order to obtain performance measures which can be used to describe the quality of service for connections.

In Section 2 of the paper we present a three level management architecture which allows a wide range of services to be accommodated and by means of which efficient use of the available resources can be made. In particular, use is made of the concepts of virtual channels and virtual paths which provide communication channels in ATM networks. The concept of virtual networks is defined wherein resource management of virtual paths is used as a means for providing guaranteed qualities of service in the network.

Using the above framework the problem of connection admission control is studied in Section 3. The basic function of the connection admission control is to ensure that connections are accepted provided there is enough resources to support them with guaranteed quality without degrading the quality of existing connections.

Results of studies based on the model presented in Section 3 are discussed in Section 4. A simple admission algorithm is proposed in Section 5.

2. MANAGEMENT ARCHITECTURE

The motivation for developing a management architecture for ATM networks is the need for a flexible framework under which network resources will be utilised in an efficient manner and which will guarantee to users required performance levels.

The management architecture which was first proposed in [1] and discussed further in [5] consists of three levels: ATM Resource Management, ATM Traffic Management, and ATM Traffic Control.

At the ATM Resource Management level virtual networks are set up and managed according to network demand. A virtual network is a group of virtual paths set up in the network to provide a specified quality of service. In particular, at this level use of resources in virtual networks is monitored and allocated resources are adjusted as demand for them changes.

At the ATM Traffic Management level resources allocated to individual virtual paths within virtual networks are managed. Connection admission control also take place at this level. This is based on the quality of service demanded by a connection, expected volume of traffic in the connection, and the current state of the virtual path into which the connection is to be admitted. Grade of service (in terms of call blocking probability) may be also taken into account in the admission process.

At the ATM Traffic Control level actions are taken to ensure that traffic volumes in individual virtual channels and virtual paths do not exceed allowed thresholds. This can be carried out by source policing traffic entering virtual channels and virtual paths. In virtual paths level traffic control is required in order to provide separation between different virtual paths and in this manner to guarantee quality of services provided by those virtual paths.

The management architecture showing the above levels is depicted in Figure 1.

3. CONNECTION ADMISSION CONTROL

Quality of service and grade of service are two criteria on which connection admission into ATM networks may be based. In order to distinguish these two terms it may be convenient to associate quality of service with cell level performance and grade of service with call level performance. As a result quality of service is often specified in terms of cell loss probability and cell delay and grade of service in terms of call blocking probability.

Approaches taken in the literature concentrate on the quality of service aspects of the connection admission control. A common approach relies on the knowledge of connection interarrival time distributions. The statistical parameters used to describe such distributions include average bit rate, bit rate variance, and peak bit rate. In [8] average bit rate and bit rate variance of individual connections is used in the admission algorithm. It is assumed that when a sufficient number of such connections is multiplexed together the resulting bit rate distribution will approach a normal distribution. This will then allow the determination of the maximum allowable load which can be permitted for a given cell loss performance.

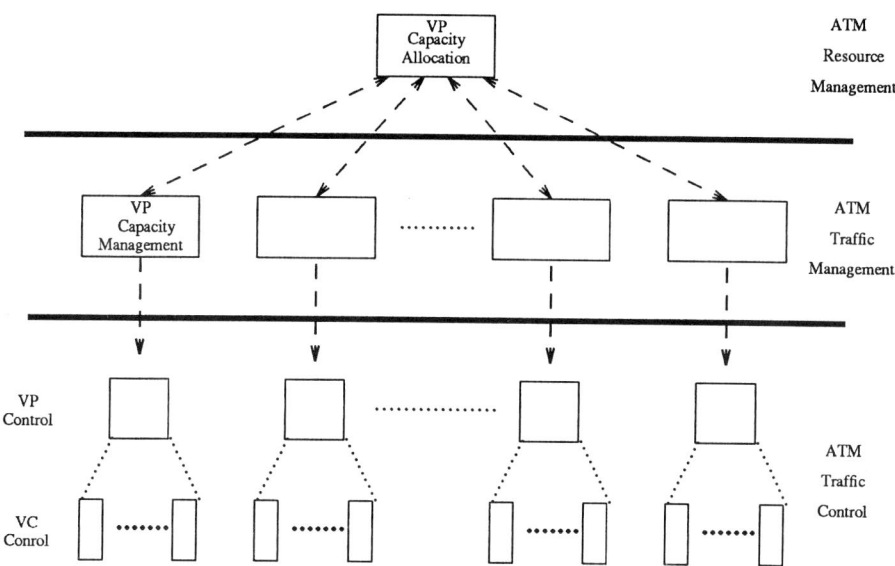

Figure 1: Traffic Management and Control Architecture for ATM

A similar approach is taken in [16]. In this case the acceptance decision is based on the mean bit rates and peak bit reates of the new and existing connections without the need for the bit rate variance.

The approaches based on the knowledge of bit rate distributions do not take into account the fluctuations of bit rates in time. Thus a model which incorporates a measure of burstiness might be more useful.

3.1 Connection Admission Model

In this model the criterion for connection admission is the quality of service requested by the connection. The grade of service requested by all connections is assumed identical.

Quality of service measure is measured in terms of cell loss probability. Cell delay performance, which depends on the size of the network and the route through the network, is regarded to be of less importance. This is particularly the case given that ATM buffers are expected to be small [8].

Using the idea of virtual networks, each virtual path belonging to the same virtual network will have to support the same quality of service. Connections which require different qualities of service are assigned into virtual paths within distinct virtual networks corresponding to the required qualities of service. In order to ensure that distinct qualities of service are provided by virtual network, traffic streams belonging to different virtual networks will have to be separated. One method which can be used to carry out such separation is through peak limiting the capacity assigned to virtual paths within different virtual networks. Capacity utilised by each virtual path will then be policed using a policing mechanism placed at the origin of each virtual path. As a result, the quality of service provided by a virtual network will be determined by the policing mechanisms associated with all virtual paths belonging to the virtual network and by the connection admission mechanism.

3.2 Queuing Model for Connection Admission Control

The policing mechanism at the origin of a virtual path will ensure that the peak bandwidth of the traffic carried in the virtual path does not exceed the allowed threshold. One method of implementing such a policing mechanism can be achieved by means of a leaky bucket [2].

A leaky bucket for policing virtual paths can be modelled as a short queue with a deterministic server. The service rate of the queue sets the allowed peak bit rate of the traffic carried by the virtual path and the length of the queue should be set so as to ensure separation between traffic carried by different virtual paths. However, setting the length of the queue to $N = 1$ so as to achieve complete separation would not allow short bursts of cells on the transmission line (due to jitter introduced by multiplexing equipment) to be accommodated. As a result the exact setting of the queue length will have to be a compromise between the above requirements.

The queue modelling the leaky bucket as the policing mechanism of the virtual path can be represented as G/D/1/N queue using Kendall notation where G represents a general input process, D a deterministic server, and N the length of the queue including the server.

The choice of the arrival process into the virtual path representing individual connections has been influenced by need for a process which can be used to model a wide range of different traffic streams and at the same time one which will allow tractable analysis to be carried out when applied to a queuing system. An arrival process which has attracted a lot of attention on account of satisfying the above criteria is the 2-state Markov modulated Poisson process.

A 2-state Markov modulated Poisson process (2-state MMPP) is a special case of the more general Markov modulated Poisson process (MMPP). An MMPP is a Poisson process whose instantaneous rate is a stationary random process which evolves according to an irreducible m state Markov chain. It can be characterised by an infinitesimal generator for the underlying Markov process **R** and an arrival matrix Λ given by:

$$\mathbf{R} = \begin{bmatrix} r_{1,1} & r_{1,2} & r_{1,3} & r_{1,4} & \cdots & r_{1,m} \\ r_{2,1} & r_{2,2} & r_{2,3} & r_{2,4} & \cdots & r_{2,m} \\ r_{3,1} & r_{3,2} & r_{3,3} & r_{3,4} & \cdots & r_{3,m} \\ r_{4,1} & r_{4,2} & r_{4,3} & r_{4,4} & \cdots & r_{4,m} \\ \vdots & \vdots & \vdots & \vdots & & \vdots \\ r_{m,1} & r_{m,2} & r_{m,3} & r_{m,4} & \cdots & r_{m,m} \end{bmatrix}$$

$$\Lambda = diag(\lambda_1,..,\lambda_m)$$

In the case of the 2-state MMPP the above reduce to:

$$R = \begin{bmatrix} -r_1 & r_1 \\ r_2 & -r_2 \end{bmatrix}$$

$$\Lambda = \begin{bmatrix} \lambda_1 & 0 \\ 0 & \lambda_2 \end{bmatrix}$$

The parameters λ_1, λ_2 required to specify Λ correspond to the intensity of the Poisson processes in states 1 and 2 of the underlying Markov process respectively, whereas r_1, and r_2 correspond to the transition rates of the Markov process.

The solution of the 2-state MMPP/D/1/K can be obtained by following analysis for more general processes as presented in the literature such as for the N/G/1 queue found in [13] and equivalent BMAP/G/1 queue found in [10]. An outline of an algorithm which has been implemented and used to solve the 2-state MMPP/D/1/K queue has been given in the Appendix.

Solution of the above queuing system allows modelling of the policing mechanism given an aggregate traffic stream consisting of all traffic streams entering a given virtual path. Connection admission will, however, have to deal with individual connections and as a result a method of obtaining a superposition of individual connections which can then be applied to the queue is needed. Two such superposition methods were investigated. In the first method for obtaining a superposition of 2-state MMPPs developed in [7] (Method I) the following characterising parameters have been used: m - mean arrival rate, v - variance of arrival rate, μ_3 - third moment of arrival rate, τ - time constant of the arrival rate.

When n 2-state MMPPs are superimposed together characterising parameters of the superimposed stream are given by:

$$m = \sum_{i=1}^{n} m_i$$

$$v = \sum_{i=1}^{n} v_i$$

$$\mu_3^* = \sum_{i=1}^{n} \mu_{3i}^*$$

$$\tau = \sum_{i=1}^{n} \frac{v_i}{v} \tau_i$$

where $\mu_3^* = \mu_3 - 3mv - m^3$

The defining parameters for the 2-state MMPP representing the superimposed stream can be obtained by inversion of the above equations as given in [7].

In the second method proposed in [14] (Method II) the arrival process is characterised by the following parameters: λ - mean arrival rate, c^2 - squared coefficient of variation of interarrival times, $Z(\infty)$ - asymptotic variance to mean ratio of the number of arrivals, $C(\infty)$ - asymptotic covariance of the number of arrivals.

When n 2-state MMPP streams are superimposed together the set of the characterising parameters for the superimposed stream is given by:

$$\lambda = \lambda_1 + ... + \lambda_n$$

$$Z(\infty) = \sum_{i=1}^{n} \frac{\lambda_i}{\lambda} Z_i(\infty)$$

$$C(\infty) = \sum_{i=1}^{n} C_i(\infty)$$

$$c^2 = E(T) 2\lambda - 1$$

where E(T) is the mean of the forward recurrence time and is given by: $E(T) = \frac{1+c^2}{2\lambda}$

The set of the parameters defining the 2-state MMPP can be obtained from the characterising parameters after some algebraic manipulation as given in [14].

In order to determine their suitability for the connection admission algorithm the two superposition methods were applied to the problem of superposition of n voice streams with the resulting aggregate stream used as an input to the queuing system described above. A voice stream was modelled by an interrupted Poisson process (IPP) based on the results presented in [15] and [14]. The following parameters were used to define each stream: $\lambda_1 = 54.44$ *cells/sec*, $\lambda_2 = 0$, $r_1 = 2.268$ sec^{-1}, and $r_2 = 1.532$ sec^{-1}.

Figures 2 and 3 show the defining parameters of the superimposed stream obtained by using the two methods. In both figures Method II used stream superposition in groups of 10 streams. The effect of stream grouping used in Method II is particularly evident in Figure 3. Method I is insensitive to such groupings. This means that superimposing an additional stream to an existing aggregate will produce the same result as the superposition of all individual streams. This becomes an important issue when the implementation of the algorithm is considered. In Method II the computationally expensive step is associated with the calculation of the coefficient of variation c^2 where the cost increases as 2^n with the number of different streams n.

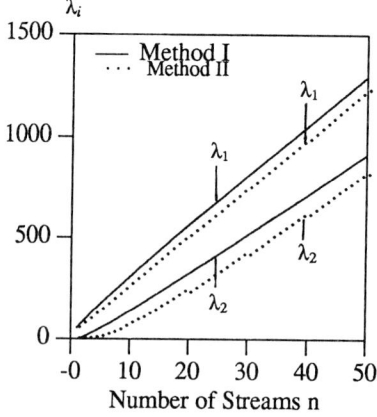

Figure 2: Component Rates λ_i for Superposition

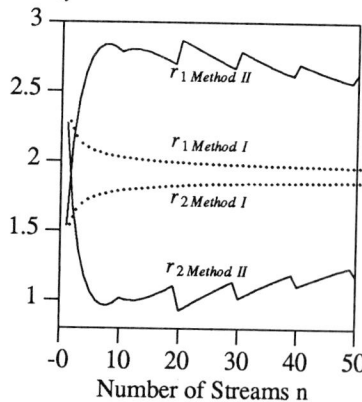

Figure 3: Phase Rates r_i for Superposition

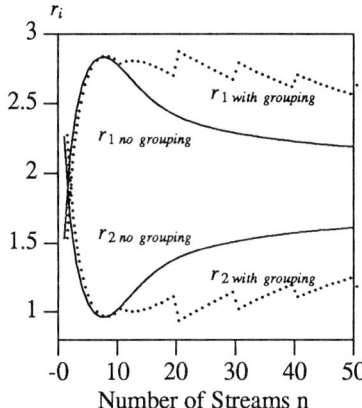

Figure 4: Effect of Grouping on Phase Rates in Method II

The effect of grouping streams in groups of 10 and applying Method II to carry out the superposition as opposed to using Method II without grouping is shown in Figure 4 for the resultant transition rates r_1 and r_2 of the superimposed stream.

The two superposition methods were also used to produce a superimposed stream which was subsequently applied to the 2-state MMPP/D/1/K queue. Figures 5 and 6 show the loss probability produced for two different queue length K and a number of different utilisation levels. Separate curves were obtained for Method II with stream grouping in groups of 10 and without stream grouping. The resultant curves indicate that stream grouping in Method II underestimates the resultant loss probability as compared to Method II used without stream grouping. Note also that Method I gives the most conservative results, however, as the number of streams increases the results obtained using the two methods converge. Queueing performance criteria obtained for the two superposition methods must be treated with caution. They show relative performance as there is no baseline against which they could both be judged. It should also be noted that traffic characterisation which will allow for accurate queueing performance prediction is a topic of some debate. As a result, the choice of the superposition method for subsequent studies was based on the ease of implementation. In this respect Method I was found more efficient.

4. SUPERPOSITION OF HETEROGENEOUS STREAMS

A virtual path which is associated with quality of service rather than being service specific can be expected to carry a number of dissimilar traffic streams. Algorithms for connection admission control will have to accommodate such wide range of services. The algorithm for solution of the 2-state MMPP/D/1/K queue was used to investigate performance of such queues when two types of different traffic streams were used. The two traffic types representing voice and video services have been chosen. The voice stream model has been presented earlier. The video stream was modelled as an IPP based on the studies in [11] with the resultant 2-state MMPP parameters as follows: $\lambda_1 = 11.1 * 10^3 \ cells/sec$, $\lambda_2 = 0$, $r_1 = 3.25 \ sec^{-1}$, and $r_2 = 0.64 \ sec^{-1}$.

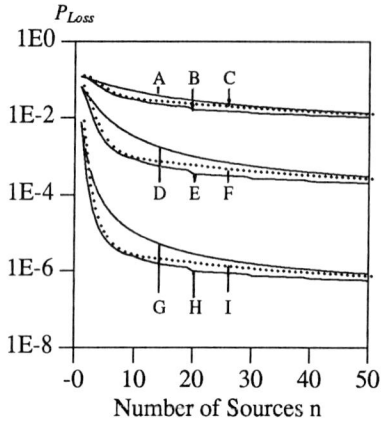
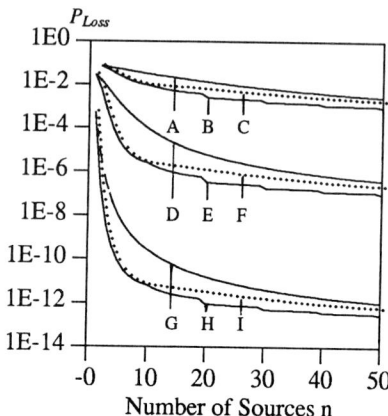

Figure 5: Loss probability for K=10 using Methods I and II.

Figure 6: Loss probability for K=20 using Methods I and II.

Legend for Figures 5 and 6:

A : Method I, $\rho = 0.8$ D : Method I, $\rho = 0.6$ G : Method I, $\rho = 0.4$

B : Method II with grouping, $\rho = 0.8$ E : Method II with grouping, $\rho = 0.6$ H : Method II with grouping, $\rho = 0.4$

C : Method II without grouping, $\rho = 0.8$ F : Method II without grouping, $\rho = 0.6$ I : Method II without grouping, $\rho = 0.4$

The two stream types were applied to the 2-state MMPP/D/1/K queue and boundaries for the resultant loss probability of 10^{-8} were obtained for different queue sizes K as shown in Figure 7. Also, having fixed the queue size to K=10, the boundaries for the same loss probability have been obtained for different queue service rates c (in Mbps, which correspond to different output link capacities). These are shown in Figure 8.

The acceptance boundaries shown in Figures 7 and 8 display near linearity, the consequence of which could be exploited in producing approximate acceptance algorithms. Simple approximations based just on the end points of the acceptance boundaries may not be possible as the boundaries display convexity. Further studies will be carried out to determine the amount of convexity for different configurations. Similar results concerning the linearity of acceptance boundaries have been noted in [6] where a bufferless network was assumed with traffic modelled as interrupted deterministic streams and the statistical behaviour was studied using a large deviation approximation.

The noted linearity of the acceptance boundaries could be used in developing simple connection admission algorithms utilising the concept of effective bandwidth. (For a survey of such methods see [12]).

Another important consideration in developing connection admission algorithms is the sensitivity of bandwidth requirement for a connection as the quality of service changes. In order to study this problem voice streams as defined previously were applied to a deterministic queue. The effective bandwidth defined as the available capacity over the number of streams carried, was obtained for such systems against loss probability. Figure 9 shows the resultant curves obtained for different values of queue length K.

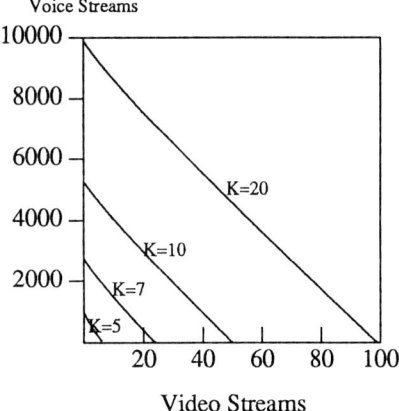
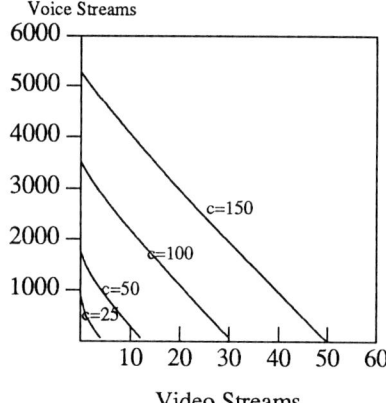

Figure 7: Acceptance Regions for Different Queue Sizes K

Figure 8: Acceptance Regions for Different Queue Service Rates c

Curves obtained in Figure 9 suggest that as the queue length increases changes in many orders of magnitude to cell loss result in very small changes to the effective bandwidth requirement. On the other hand, it should be noted that only a small change in effective bandwidth may result in changes to cell loss probability of many magnitudes. This could have important consequences on the sensitivity of the quality of service provided in a virtual path.

5. CONNECTION ADMISSION CONTROL ALGORITHM

A simple connection admission algorithm based on the models presented earlier has been developed. Connections requesting a particular quality of service can be admitted into a virtual path according to the algorithm steps depicted in Figure 10. Call parameters are used to determine the set of characterising parameters as well as connection type. It is assumed that there is a finite number of connection types possible and look-up tables can be used to match connection parameters to the required connection type if the connection type is not specified explicitly. In the next step in the algorithm the new connection is superimposed with the existing aggregate of connections carried by the virtual path. The resultant stream is then applied to the queuing system which models the virtual path and the queuing system is solved for cell loss probability. If the resultant cell loss probability exceeds that allowed for the virtual path the new connection is rejected, otherwise the new connection is accepted and the characterising parameters of the new aggregate connection in the virtual path are stored.

The process of removing connections from the virtual path is depicted in Figure 10. It involves only the calculation of the new aggregate stream without the need f .r solving the queuing system. The first step in the algorithm involves determination of the characterising parameters of the connection to be removed. This information can be obtained from a look-up table if the type of the connection is known. Next the new aggregate for the virtual path with the required connection removed is obtained and the characterising parameters for the new aggregate are saved.

Removal of an individual stream from a superimposed stream can be carried out in a simple fashion if Method I for stream superposition is used. Let m, v, μ_3, τ and $m_i, v_i, \mu_{3_i}, \tau_i$

denote the characterising parameters of the existing aggregate and the connection to be removed respectively. Then the characterising parameters of the new aggregate with the connection removed can be obtained from:

$$m_n = m - m_i,$$

$$v_n = v - v_i,$$

$$\mu_{3n}^* = \mu_3^* - \mu_{3i}^*,$$

$$\tau_n = \frac{1}{v - v_i}(v\,\tau - v_i\,\tau_i).$$

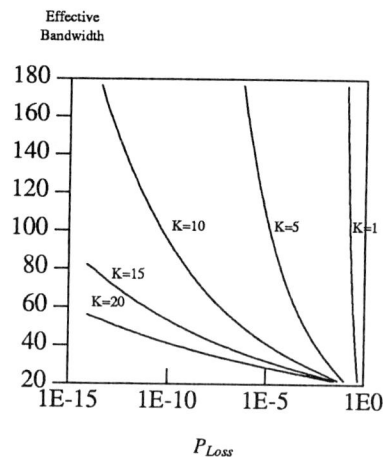

Figure 9: Effect of Effective Bandwidth on Loss Probability

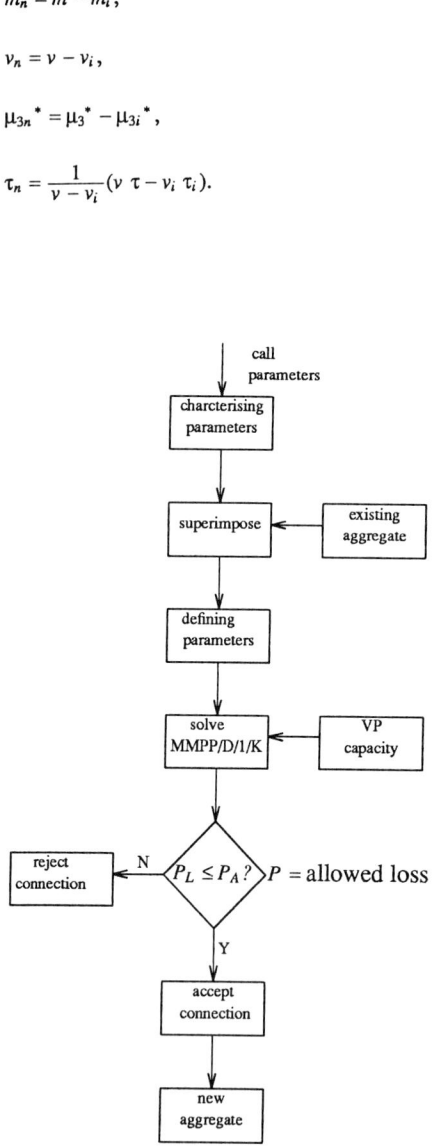

Figure 10: Algorithm Steps for Adding New Connections

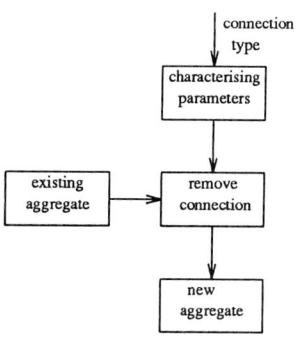

Figure 11: Algorithm Steps for Removing Old Connections

The main purpose of the above algorithm is to provide a basis for development and comparison of simple connection admission algorithms such as those using the idea of the effective bandwidth. It should also be noted that the above steps can form the basis of an algorithm which could be used in practice. However, the step involving the solution of the queueing system may impair the performance of the algorithm and as a result simplifications of this procedure may be required.

6. CONCLUSION AND FURTHER WORK

A versatile three level management architecture for controlling and managing traffic in future broadband networks has been proposed.

A queuing model for connection admission control has been presented and a connection admission control algorithm based on the model has been proposed.

The main purpose of the above algorithm is to act as a vehicle for further studies of simple connection admission algorithms. In particular, studies of acceptance regions with a wide variety of different traffic types for development of algorithms which use the concept of the effective bandwidth will be carried out.

7. APPENDIX

A matrix algorithmic solution of the 2-state MMPP/D/1/K queue from which queue length distribution and loss probability can be obtained can proceed according to the following steps. Let $\tilde{P}(x)$ denote the transition probability matrix for a queue with an infinite number of spaces. This matrix can be expressed as

$$\tilde{P}(x) = \begin{bmatrix} \tilde{B}_0(x) & \tilde{B}_1(x) & \tilde{B}_2(x) & . & . \\ \tilde{A}_0(x) & \tilde{A}_1(x) & \tilde{A}_2(x) & . & . \\ 0 & \tilde{A}_0(x) & \tilde{A}_1(x) & . & . \\ 0 & 0 & \tilde{A}_0(x) & . & . \\ . & . & . & . & . \\ . & . & . & . & . \end{bmatrix}$$

where $\tilde{A}_n(x)$ and $\tilde{B}_n(x)$ are 2×2 matrices. The stationary transition probability matrix $P = \tilde{P}(\infty)$ can be used to obtain the stationary queue-length density x at departure epochs. The defining system of equations is:

$$x \tilde{P}(\infty) = x, \quad x e = 1.$$

The effect of a finite queue is the truncation of the state space. An efficient algorithm for the computation of matrices A_n has been presented in [9]. Computation of the stationary probability vector x at departure epochs can then be carried out following simple matrix analysis [3]. Queue length distribution y at an arbitrary time t can be obtained next by application of the key renewal theorem [13]. The blocking probability for the queue can be then obtained as:

$$P_{block} = 1 - \sum_{n=0}^{N-1} y_n e.$$

8. ACKNOWLEDGEMENT

The permission of the Managing Director of OTC Limited to publish this work is gratefully acknowledged. The views and opinions expressed in this work are those of the authors and do not necessarily imply OTC Limited policy or future service offerings.

9. REFERENCES

1. Anido G., "Traffic Control and Management Mechanism for a Broadband Packet Network", ITC Specialist Seminar, Adelaide, 1989.
2. Aumann G., "Source Policing in the Broadband-ISDN", 4th ATERB FPS Workshop, Sydney, July 1989.
3. Blondia C., "The N/G/1 Finite Capacity Queue", Commun. Statist.-Stochastic Models, 5(2), 273-294 (1989).
4. Dittmann L., and Jacobsen S.B., "Statistical Multiplexing of Identical Bursty Sources in an ATM Network", IEEE GLOBECOM'88.
5. Dutkiewicz E and Anido G.J, "Traffic Management and Control in Broadband Networks", 5th ATERB FPS Workshop, Melbourne, July 1990.
6. Griffiths T.R., "Analysis of Connection Acceptance in Asynchronous Transfer Mode Networks", British Telecom Research Laboratories, 1990.
7. Heffes H., "A Class of Data Traffic Processes - Covariance Function Characterization and Related Queueing Results", BSTJ, vol. 59, no. 6, July-August 1979.
8. Joos P., Verbiest W., "A Statistical Bandwidth Allocation and Usage Monitoring Algorithm for ATM Networks", ICC/89, Boston, June 1989.
9. Lucantoni D.M., "Efficient Algorithms for Solving the Non-linear Matrix Equations Arising in Phase Type Queues", Commun. Statist-Stochastic Models, 1(1), 29-51 (1985).
10. Lucantoni D.M., "New Results on the Single Server Queue with a Batch Markovian Arrival Process", Stochastic Models, vol. 7, no. 1, 1991.
11. Malgris B., et al, "Performance Models of Statistical Multiplexing in Packet Video Communications", IEEE Transactions on Communications, vol. 36, no. 7, July 1988.
12. Mase K., and Shioda S., "Real-Time Network Management for ATM Networks", to be presented at the 13th ITC, Copenhagen 1991.
13. Ramaswami V., "The N/G/1 Queue and its Detailed Analysis", Adv. Appl. Prob., vol 12, pp. 222-261, Mar 80.
14. Rossiter M.H., "Sojourn Time Theory and the Switched Poisson Process", Telecom Australia Research Laboratories Report 7835, 1986.
15. Sriram K., Whitt W., "Characterizing Superposition Arrival Processes in Packet Multiplexers for Voice and Data", IEEE Journal on Selected Areas in Communications, vol. SAC-4, no. 6, September 1986.
16. Wallmeier E., "A Connection Acceptance Algorithm for ATM Network Based on Mean and Peak Bitrates", submitted to Int. J. Digital and Analogue Cabled Systems.

A two-layer optimization structure for access control and bandwidth sharing in high-speed integrated networks

R. Bolla, F. Davoli

Department of Communications, Computer and Systems Science (DIST), University of Genoa, Via Opera Pia, 11/A - 16145 Genova, Italy

Abstract
The access to a multiservice broad-band synchronous TDM network is considered, where hybrid frames are used to carry two basic traffic types (a circuit-switched isochronous and a packet-switched asynchronous one), generated by several users. Each user is assigned a portion of the total available bandwidth, in terms of slots/frame, which is dynamically allocated between the two traffic types at the user premises, by means of a local randomized decision rule, whose structure is determined by a local parameter and by the assigned bandwidth share. A two-layer optimization problem is defined, through which the users independently adjust their local parameter values and a central controller reassigns the bandwidth partitions, that play the role of coordination variables. Simulation results are provided that also compare the present scheme with a similar, though totally centralized, one.

1. INTRODUCTION

The advent of high-speed integrated networks, where a multitude of different traffic types with different grade of service requirements should be carried, is greatly increasing the importance of access control and bandwidth allocation in order to avoid congestion and achieve an efficient resource utilization. A great deal of work is being done to examine and find solutions to the above problems in the context of asynchronous multiplexing (ATM) networks, where all traffic is packetized and statistically multiplexed (see, for instance, [1-2]). ATM introduces a multi-layer environment [3] (one can distinguish, typically, cell, burst, and call layers), where admission and bandwidth controls are required, with very different time scales, and interaction among layers. On the other hand, even if ATM is very likely to become the universal traffic carrying integrated technique of the future, more "traditional" modes, like synchronous multiplexing (STM) and circuit switching, will still be in use for some time to come. These techniques are certainly less flexible in terms of traffic integration and bandwidth allocation; however, just because of their more structured nature, the call allocation control problems are more easily handled than with ATM.

Thus, there is a certain interest, in our opinion, in finding ways of increasing the flexibility and efficiency of synchronous transmission modes, while retaining their inherent capability to satisfy grade of service requirements for time-sensitive traffic. In this respect, as regards circuit-switched traffic with varying bandwidth requirements, access control policies have been determined, for instance, in [4].

In other papers [5, 6] as well as in the present one, we are interested in hybrid structures, where the slots of a synchronous TDM frame are used to carry both isochronous and asynchronous traffic types. Moreover, we consider a multiuser environment, and attempt to dynamically share the available bandwidth among users and their traffic types in order to satisfy a certain optimality criterion.

More specifically, we follow a parametric optimization approach, where each user independently decides on the scheduling of two traffic types (isochronous, circuit-switched, and asynchronous, packet-switched, respectively) by means of a fixed randomized local admission rule that is a function of the number of the user's calls in progress. Periodically, the users are reassigned shares of the global bandwidth (capacities), on the basis of the parametric minimization of a cost function, which may be evaluated over a finite [5] or even infinite [6] future time horizon. The cost function is averaged with respect to the probability distribution defined by the Markov chain of the number of isochronous calls in progress; the instantaneous value of this variable also affects inequality constraints at the intervention times. Unlike the situation considered in [5, 6], where the only parameter affecting a user's admission control function was the assigned capacity, and the optimization problem was carried out only on the capacity variables by a central decision maker, in this paper we suppose that each of these functions can be shaped by a local parameter. Thus, a two-layer optimization problem is defined, through which the users independently adjust their local parameter values and a central controller reassigns the bandwidth partitions, that play the role of coordination variables.

The paper is organized as follows. In the next Section, we define the traffic model and the decisional structure. A possible mathematical programming procedure is outlined in Section 3. In Section 4, we report simulation results and comparisons with those corresponding to the previous centralized schemes. The conclusions are drawn in Section 5.

2. TRAFFIC MODEL AND DECISIONAL STRUCTURE

Consider a TDM multiplexer that carries the traffic of M user sites on a link with total capacity C_T [slots/frame]. As in [5, 6], we suppose each user i to be characterized by two traffic types, a circuit-switched traffic with average call arrival rate $\lambda_1^{(i)}$ calls/s and average holding time $1/\mu^{(i)}$ s, and a packet-switched traffic with average arrival rate $\lambda_2^{(i)}$ packets/s, with fixed length packets (one slot). The circuit-switched traffic requires continuation of service (one slot/frame per call) until the end of the call. Packet arrivals follow a Poisson distribution, whereas the circuit-switched calls are supposed to form a birth-death process with respect to the frame.

The decisional structure, shown in Fig. 1, is made by M local decision makers (DM) and a central one. The latter assigns each user i a capacity $c_{t_k}^{(i)}$ in slots/frame at decision instants t_k, k=0,1,..., and this assignment lasts over the period $[t_k, t_{k+1} -1]$. At the same instants, the local DM's decide on a parameter value that will be specified in following; these parameters influence the (fixed) local admission control rules that each DM exerts at the beginning of each frame on the incoming call requests of the circuit-switched traffic.

We shall use a discrete time model, with the frame duration of b seconds representing a time unit. Moreover, in the decisional problem to be specified in the following, capacities $c_{t_k}^{(i)}$ will be treated as continuous variables (see [5] for a discussion of this point). By taking this fact into account, in order to make the bandwidth allocation problem meaningful, even in the worst case, it is assumed that

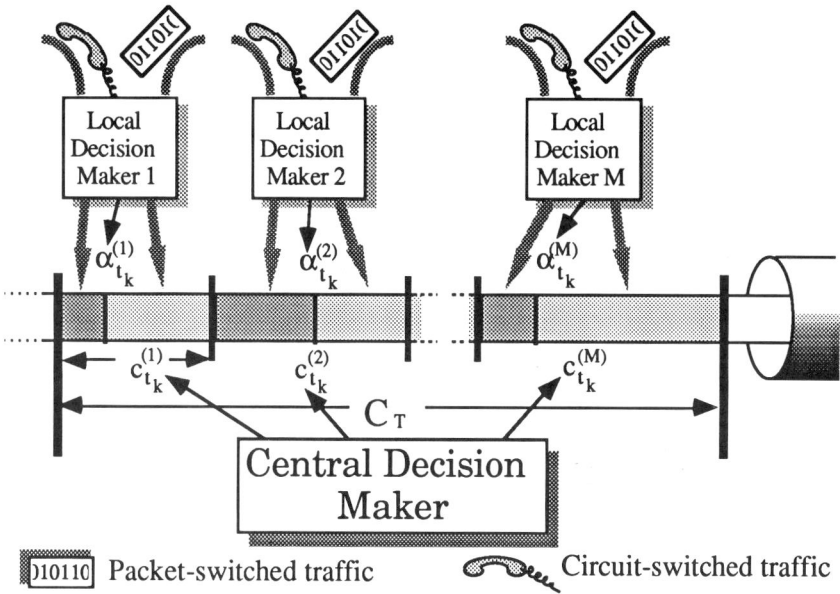

Fig. 1. System's and decisional structure

$$\sum_{i=1}^{M} (\lambda_1^{(i)}/\mu^{(i)} + \lambda_2^{(i)}b) < C_T - M+1 \qquad (1)$$

As regards the decisional structure, we have already mentioned that each DM has a local access controller; we suppose it to act according to a fixed randomized strategy. More specifically, in each frame t, DM i either accepts an incoming call with probability $\beta_t^{(i)}$ or blocks it with probability $1-\beta_t^{(i)}$. Let $r_t^{(i)}$ be the number of calls that require continuation of service in the next frame; then, we choose the following form for $\beta_t^{(i)}$

$$\beta_t^{(i)} = \begin{cases} 1 - \left[\dfrac{3(r_t^{(i)})^2}{(c_{t_k}^{(i)}-1)^2} - \dfrac{2(r_t^{(i)})^3}{(c_{t_k}^{(i)}-1)^3} \right]^{\alpha_{t_k}^{(i)}} & \text{if } c_{t_k}^{(i)} > r_t^{(i)}+1 \\ 0 & \text{otherwise} \end{cases}$$

$$t = t_k, t_k+1, \ldots, t_{k+1}-1 \qquad (2)$$

where $ß_t^{(i)}$ is continuously differentiable and $0 \leq ß_t^{(i)} \leq 1$, for every $c_{t_k}^{(i)} > 0$; $\alpha_{t_k}^{(i)} > 0$ is the other parameter to be chosen at time t_k, according to the criterion that will be specified below.

Fig. 2 shows a plot of the function for different values of $\alpha_{t_k}^{(i)}$. In particular, one may observe that $ß_t^{(i)}$ tends to a step function for $\alpha_{t_k}^{(i)} \to \infty$, whereas it flattens on the horizontal axis for $\alpha_{t_k}^{(i)} \to 0$.

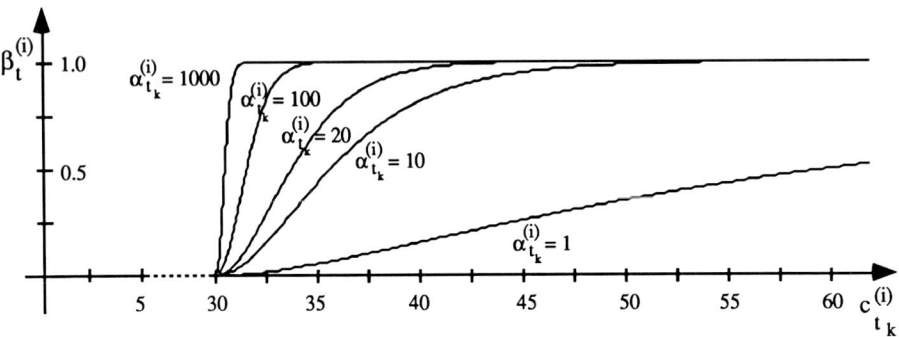

Fig. 2. A plot of $ß_t^{(i)}$ for various values of $\alpha_{t_k}^{(i)}$ and $r_t^{(i)} + 1 = 30$.

This form of $ß_t^{(i)}$ has been chosen in order to have a probability of acceptance that is decreasing for decreasing "available space" $c_{t_k}^{(i)} - r_t^{(i)}$ and everywhere differentiable with respect to $c_{t_k}^{(i)}$; the parametrization by $\alpha_{t_k}^{(i)}$, that gives rise to a family of curves with different slopes the proximity of $r_t^{(i)} + 1$, also allows a greater flexibility in comparison with the function used in [5, 6] that depended only on $c_{t_k}^{(i)}$. Note that $ß_t^{(i)}$ is always 0 when $r_t^{(i)} = \lfloor c_{t_k}^{(i)} \rfloor$, $\forall c_{t_k}^{(i)}$ (where $\lfloor x \rfloor$ denotes the first integer less than or equal to x), and it is still 0, with 0 derivative, for $r_t^{(i)} = \lfloor c_{t_k}^{(i)} - 1 \rfloor$, if $c_{t_k}^{(i)} - 1$ is an integer.

Each process $r_t^{(i)}$, $i = 1, ..., M$, is a Markov chain with transition probabilities (due to the birth-death assumption)

$$p_{jk}^{(i)} \equiv P(r_{t+1}^{(i)} = k | r_t^{(i)} = j) = \begin{cases} \mu^{(i)} bj & k = j-1 \\ 1 - \mu^{(i)} bj - \lambda_1^{(i)} b ß_t^{(i)} & k = j \\ \lambda_1^{(i)} b ß_t^{(i)} & k = j+1 \\ 0 & \text{otherwise} \end{cases} \quad (3)$$

Let $P_{t_k}^{(i)}$ be the transition probability matrix of $r_t^{(i)}$, $t \in [t_k, t_{k+1}-1]$, and let

$$\pi_t^{(i)}(s) = \Pr\{r_t^{(i)} = s\} \quad (4)$$

Clearly, $0 \leq r_t^{(i)} \leq \lfloor c_{t_k}^{(i)} \rfloor$, $\forall\, t \in [t_k, t_{k+1}-1]$. However, as $c_{t_k}^{(i)}$ can vary between 0 and C_T (at least in principle), we can take $\{0, 1, \ldots, C_T\}$ as the state space, and define

$$\pi_t^{(i)} \equiv \left[\pi_t^{(i)}(0), \ldots, \pi_t^{(i)}(C_T)\right]^T \tag{5}$$

where T denotes transpose.

Then, $P_{t_k}^{(i)}$ is a $(C_T+1) \times (C_T+1)$ - dimensional transition probability matrix (where all terms $p_{j,j+1}^{(i)}, j \geq \lfloor c_{t_k}^{(i)} \rfloor$ are null), and we can write

$$\pi_{t+1}^{(i)} = \left[P_{t_k}^{(i)}\right]^T \pi_t^{(i)}, \quad t \in [t_k, t_{k+1}-1] \tag{6}$$

We can now define the following parametric optimization problem: at each time instant t_k, $k = 0, 1, \ldots$, on the basis of the knowledge of all a priori information and of $r_{t_k}^{(i)}$, $i=1,\ldots,M$, find the capacity vector $c_{t_k} \equiv \text{col}[c_{t_k}^{(1)}, \ldots, c_{t_k}^{(M)}]$ and the parameter vector $\alpha_{t_k} \equiv \text{col}[\alpha_{t_k}^{(1)}, \ldots, \alpha_{t_k}^{(M)}]$ that minimize

$$\overline{J}_{t_k}(X_{t_k}; c_{t_k}, \alpha_{t_k}) = \sum_{i=1}^{M} \overline{J}_{t_k}^{(i)}(x_{t_k}^{(i)}; c_{t_k}^{(i)}, \alpha_{t_k}^{(i)}) \tag{7}$$

where

$$\overline{J}_{t_k}^{(i)}(x_{t_k}^{(i)}; c_{t_k}^{(i)}, \alpha_{t_k}^{(i)}) = \sum_{t=t_k}^{t_{k+1}-1} \left\{ \sum_{s=0}^{C_T} \left[(1-\beta^{(i)}(s))\lambda_1^{(i)} b\right] \cdot \pi_t^{(i)}(s) \right\} +$$

$$+ \sigma^{(i)} \sum_{t=t_k}^{t_{k+1}-1} \max\left\{0;\; \frac{L_{t_k}^{(i)}}{G} + \lambda_2^{(i)} b - \sum_{s=0}^{C_T}\left[c_{t_k}^{(i)} - s - \lambda_1^{(i)} b \beta^{(i)}(s)\right]\cdot \pi_t^{(i)}(s)\right\} \tag{8}$$

$\sigma^{(i)}$ is a weighting coefficient, and we have defined
- $L_{t_k}^{(i)}$ = number of packets in the packet queue of user i at time t_k;
- $X_{t_k} \equiv \text{col}[x_{t_k}^{(1)}, \ldots, x_{t_k}^{(M)}]$; $x_{t_k}^{(i)} \equiv \text{col}[r_{t_k}^{(i)}, L_{t_k}^{(i)}]$;
- $\beta^{(i)}(s) = \beta_t^{(i)}$ for $r_t^{(i)} = s$;
- $G \equiv t_{k+1} - t_k \equiv T/b$

The first sum over time in (8) contains terms that represent the average number of blocked calls (type 1 traffic) per frame at user i.

In the second sum, the nonzero argument of max{0 ; . } represents an approximation of the difference between the average input flow into and the average output flow out of the packet queue at user i. Actually, we consider the number of packets in the queue at time t_k, $L_{t_k}^{(i)}$, as uniformly spread over the interval $[t_k, t_{k+1}-1]$, and approximate the output flow with the average space available for packets in each frame (which is close to reality whenever the packet queue is saturated).

Of course, the capacity assignment must take into account the following constraints (beside the state equation (6))

$$\sum_{i=1}^{M} c_{t_k}^{(i)} = C_T \qquad k = 0, 1, \ldots \qquad (9)$$

$$c_{t_k}^{(i)} \geq r_{t_k}^{(i)} \qquad \begin{array}{l} i = 1, \ldots, M \\ k = 0, 1, \ldots \end{array} \qquad (10)$$

$$\alpha_{t_k}^{(i)} > 0 \qquad \begin{array}{l} i = 1, \ldots, M \\ k = 0, 1, \ldots \end{array} \qquad (11)$$

where (10) imposes continuation of the outstanding calls.

In the next Section, we shall describe a possible two-level optimization procedure that can be applied to the above stated problem.

3. A TWO-LEVEL OPTIMIZATION PROCEDURE

The structure of the above stated optimization problem is such to suggest a possible multilevel decomposition, where the parameters $c_{t_k}^{(i)}$ would play the role of coordination variables. In fact, local optimization problems for each DM i can be defined as

$$\tilde{J}_{t_k}^{(i)}(x_{t_k}^{(i)}; c_{t_k}^{(i)}) = \min_{\alpha_{t_k}^{(i)}} \overline{J}_{t_k}^{(i)}(x_{t_k}^{(i)}; c_{t_k}^{(i)}, \alpha_{t_k}^{(i)}) \qquad (12)$$

with the constraint

$$\alpha_{t_k}^{(i)} > 0 \qquad (13)$$

whereas the minimization of the global cost (7) is given by

$$\min_{c_{t_k}} \tilde{J}_{t_k}(X_{t_k}; c_{t_k}) = \min_{c_{t_k}} \sum_{i=1}^{M} \tilde{J}_{t_k}^{(i)}(x_{t_k}^{(i)}; c_{t_k}^{(i)}) \qquad (14)$$

under constraints (9) and (10).

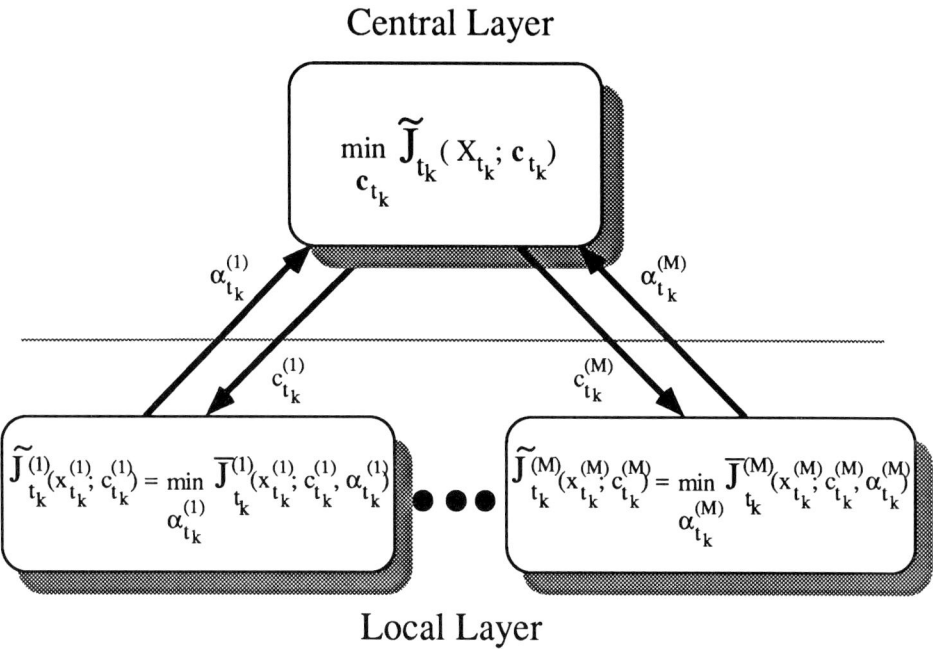

Fig. 3. Decomposition and coordination structure

A possible numerical approach to the above outlined optimization problem is given by the application of the so-called "model coordination method" [7]. A reason for its choice in this case is the feasibility of the intermediate results during the iteration procedure that will be shortly summarized hereafter.

Referring to Fig. 3, the following steps should be performed:

i) for a fixed value of variables $c_{t_k}^{(i)}$, all DM i's independently solve their local problems (12), (13);

ii) the resulting vector α_{t_k} is passed to the central DM, that finds the corresponding solution of problem (14), with constraints (9), (10), by applying a suitable descent procedure; the ensuing values of the components of vector c_{t_k} are passed to the local DM's, and the procedure is repeated until convergence (or for a suitable number of steps).

As regards the local optimization problems, their numerical solution is straightforward, simply reducing to a unidimensional search. The central optimization can be effected by means of the gradient projection method, as already done in [5, 6]; we briefly report here the essential points.

First of all, note that the terms $\pi_t^{(i)}(s)$, $i = 1, \ldots, M$, $s = 0, 1, \ldots, C_T$, $t = t_k, t_k+1, \ldots, t_{k+1}-1$, appearing in (8) and also in the gradient components $\partial \widetilde{J}_{t_k}/\partial c_{t_k}^{(i)}$, can be computed through recursion (6), initialized with

$$\pi_{t_k}^{(i)}(s) = \begin{cases} 1 & , \; s = r_{t_k}^{(i)} \\ 0 & , \; \text{otherwise} \end{cases} \tag{15}$$

The derivatives of $\pi_t^{(i)}(s)$ can also be computed recursively, by differentiating both sides of (6), which yields

$$\frac{d\pi_{t+1}^{(i)}(s)}{dc_{t_k}^{(i)}} = \sum_{j=0}^{\lfloor c_{t_k}^{(i)} \rfloor} \left[p_{js}^{(i)} \cdot \frac{d\pi_t^{(i)}(j)}{dc_{t_k}^{(i)}} + \frac{dp_{js}^{(i)}}{dc_{t_k}^{(i)}} \cdot \pi_t^{(i)}(j) \right] \qquad t \in [t_k, t_{k+1}-1] \tag{16}$$

initialized with $d\pi_{t_k}^{(i)}(j)/dc_{t_k}^{(i)} = 0$, $\forall j$. The expression of derivatives involving terms of the transition matrix $P_{t_k}^{(i)}$ is straightforwardly derived by (3) and (2). Obviously, some care must taken in differentiating the term $\max\{0; . \}$ in (8) (left and right partial derivatives have to be considered in the possible points of non-differentiability).

In order to apply the gradient projection method, we can reduce the problem to one involving only inequality constraints of the form (10), by finding, at each iteration of the method, the index m of the minimum gradient component, and by computing the derivatives $\partial \widetilde{J}_{t_k}/\partial c_{t_k}^{(i)} - \partial \widetilde{J}_{t_k}/\partial c_{t_k}^{(m)}$, $\forall i \neq m$; this corresponds to eliminating $c_{t_k}^{(m)}$ in (8) through

$$c_{t_k}^{(m)} = C_T - \sum_{i \neq m} c_{t_k}^{(i)} \tag{17}$$

In this way, gradient projections may take place only on the constraint set defined by $c_{t_k}^{(i)} \geq r_{t_k}^{(i)}$, $\forall i \neq m$. Denoting by w the iteration number of this optimization procedure, and defining $c_{t_k}^w \equiv \text{col}[c_{t_k}^{(1),w}, \ldots, c_{t_k}^{(M),w}]$, the descent step is given by

$$c_{t_k}^{(i),w+1} = \max\left\{r_{t_k}^{(i)}; c_{t_k}^{(i),w} - \gamma^w \left[\frac{\partial \tilde{J}_{t_k}}{\partial c_{t_k}^{(i)}} - \frac{\partial \tilde{J}_{t_k}}{\partial c_{t_k}^{(m)}}\right]\bigg|_{c_{t_k}^w}\right\}, \quad \forall i \neq m$$

(18)

where γ^w represents the stepsize.

4. SIMULATION RESULTS

In this Section, we show some simulation results referring to the application of the above described scheme, and also compare its performance with that of the previous centralized methods treated in [5, 6]. The example we consider has the following data:

$M = 3$ $C_T = 128$ slots/frame $T = 4$ s $b = 0.0025$ s
$\lambda_1^{(1)} = 18.29$ $\lambda_1^{(2)} = 9.14$ $\lambda_1^{(3)} = 4.57$ calls/s
$\lambda_2^{(1)} = 21485$ $\lambda_2^{(2)} = 10752$ $\lambda_2^{(3)} = 5371$ packets/s
$1/\mu^{(1)} = 1/\mu^{(2)} = 1/\mu^{(3)} = 1$ s $\sigma^{(1)} = \sigma^{(2)} = \sigma^{(3)} = 0.01$
$\alpha^{(1)} = \alpha^{(2)} = \alpha^{(3)} = 2$ (only in the case of constant $\alpha_{t_k}^{(i)}$)

(the indicated values of $\lambda_1^{(i)}, \lambda_2^{(i)}$ i = 1,...,3 correspond to a total normalized input traffic of 1.0 [calls+packets/slot]; the values corresponding to 0.4, ..., 0.9 are obtained through a proportional scaling; the total duration of every simulation was of 32 seconds).

The results corresponding to three different cases are shown in each figure, namely, the present case, labeled as "optimum α_{t_k}", and two others where α_{t_k} is kept constant. Of these, the one labeled "constant α_{t_k}" corresponds to the situation where the minimization with respect to c_{t_k} is carried on until convergence at each intervention time instant t_k. Actually, this should always be done, as function (7) and the position and value of the minimum obviously depend on X_{t_k}. However, especially if the distance between t_k and t_{k+1} is sufficiently large to allow a certain setting of initial transients, it would be reasonable to expect not too large variations with respect to X_{t_k}, as in fact the numerical results reported in [5] tend to confirm. This fact suggests the possibility of performing a few descent steps at each instant t_k, resuming the procedure from the previously reached values of the variables to be optimized. Even if constraints (10) are dynamically variable, a little thought (see [6] for a more detailed discussion) ensures that the starting point $c_{t_k}^w$ is always feasible with respect to the value of the new constraints.

The above observation is exploited in the results labeled "constant α_{t_k}, two steps", where only two descent steps were performed at each instant t_k. A similar procedure was applied also in the present case, with respect to the iterations between the two layers (whereas the search within a layer was always carried on until convergence).

The figures show the percentage of refused calls and the average queue length, respectively. The hierarchical optimization where α_{t_k} is included loses a little with respect to the calls, but shows a relevant gain in the average packet queue. It is worth noting that in the high load case the maximum queue lengths were 13,156 packet slots with constant α_{t_k}, and 5,753 with optimal α_{t_k}, whereas the refused isochronous traffic (expressed as percentage of slots per lost call with respect to the total load in slots) corresponded to 1.16% and 0.83%, respectively.

We do not explicitly report individual results regarding the performance of the single users; however, in the example considered, the allocation was always fair in this respect.

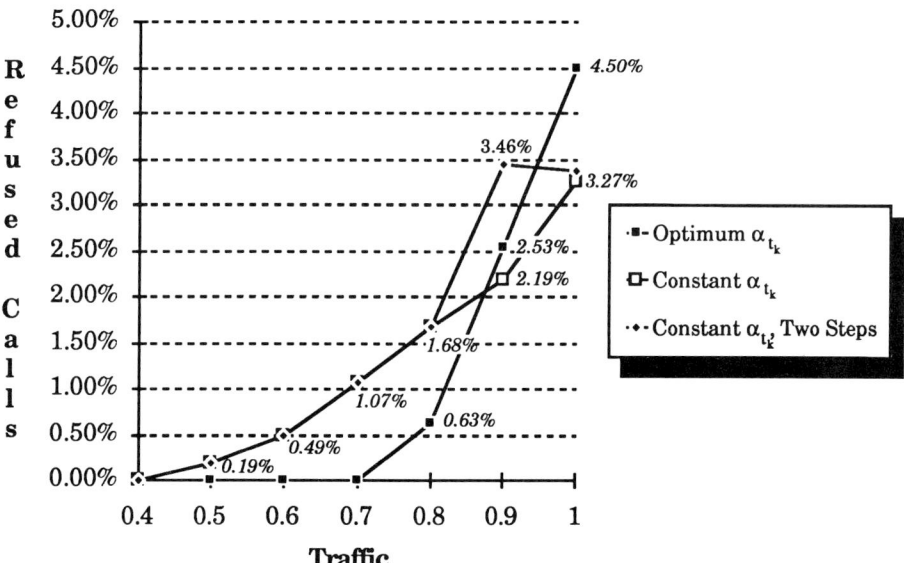

Fig. 4. Percentage of refused calls versus total normalized input traffic.

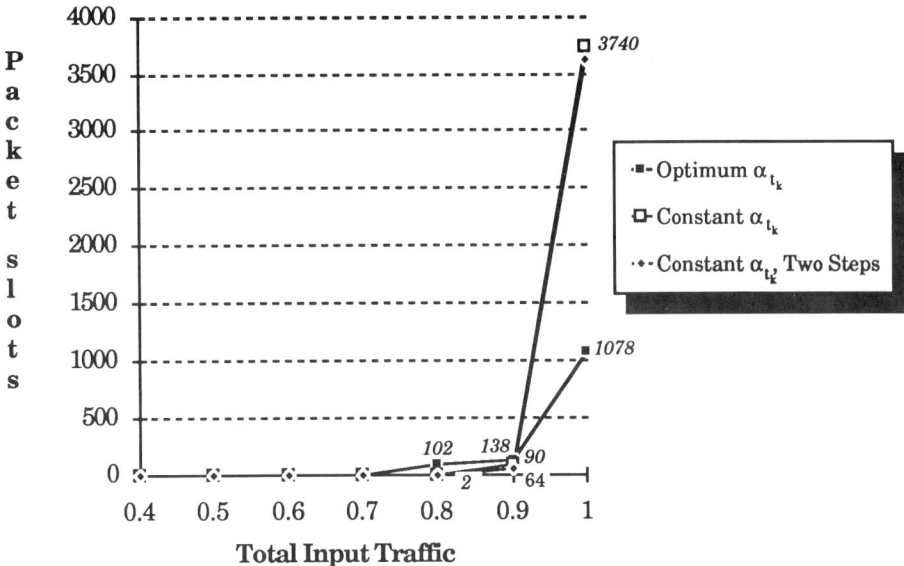

Fig. 5. Overall average queue length versus total normalized input traffic.

5. CONCLUSIONS

This paper has addressed the admission control and bandwidth allocation problem in the access area to a TDM network carrying two basic traffic types (circuit- and packet-switched, respectively) as a two-level parametric optimization. A control structure has been proposed, where the users independently decide on the admission of each isochronous call, on the basis of a fixed randomized decision rule, and also cooperate with a central decision maker in the minimization of an overall cost function that takes into account blocking probability as well as packet queue lengths. Each user tries to optimize a parameter that determines the shape of the acceptance function, while the central agent assigns bandwidth partitions to the users, by applying a gradient projection method in its minimization. Feedback on the queue lengths and the calls in progress is explicitly utilized. The performance of the method was tested by simulation, showing a certain capability of traffic balance and fairness in bandwidth allocation.

ACKNOWLEDGMENT

This work was supported by the National Council of Research of Italy (C.N.R.) under the Special Program on Telecommunications and by the Ministry of University and Scientific Research.

REFERENCES

1　G. Gallassi, G. Rigolio, L. Fratta, in Proc. Globecom '89, Dallas, TX (1989) 1788.
2　Z. Dziong, J. Choquette, K.Q. Liao, L. Mason, in Proc. ITC Specialist Seminar, Adelaide, Australia (1989) 15.3.
3　J. Filipiak, IEEE J. Select. Areas Commun., SAC-7 (1989) 1263.
4　K.W. Ross, D.H.K. Tsang, IEEE Trans. Commun., COM-37 (1989) 934.
5　M. Aicardi, R. Bolla, F. Davoli, R. Minciardi, in Proc. Internat. Conf. on Commun., Atlanta, GA (1990) 302.
6　M. Aicardi, R. Bolla, F. Davoli, R. Minciardi, in Proc. Globecom '90, San Diego, CA (1990) 41.
7　D.A. Wismer (ed.), Optimization Methods for Large Scale Systems, McGraw-Hill, New York, 1971.

Performance analysis of resource allocation and routing techniques in B-ISDN

T. Uhl and J. Ulmer

Institute for Electronic Systems and Switching
University of Dortmund
Dortmund, Federal Republic of Germany

Abstract

With respect to future broadband ISDN networks (B-ISDN) the ATM technique gains more and more importance. This paper provides a simulative analysis of resource allocation and routing in the B-ISDN. The main performance values, i.e. average establishment time and the blocking probability in this network are presented as functions of the offered traffic and different services. Furthermore, the obtained results comprise optimum values which are of particular interest. The present investigations also can be applied as a basis for the design of routing protocol in B-ISDN.

1. INTRODUCTION

With the publication of the CCITT Recommendation I.121 [1] the ATM switching concept gains more and more importance. These ATM networks allow to handle a large amount of data from different sources at high transmission speed.

One of the fundamental problems of control in the ATM networks is that of resource allocation and routing of the transmitted data blocks, the so-called cells. At the moment several different resource allocation methods and routing rules are existing, but only a few of them are relevant for practical use in ATM networks. Despite of several investigations [2-6] the problem of choosing the optimal control mechanism in such a network has not been solved.

The problem of resource allocation in this communication systems can be decomposed into the following tasks:

1) Choice of suitable routes for different services - establishment of virtual channels. An adequate routing strategy has to fulfil this task.

2) Allocation of bandwidth for the virtual connection according to limits of bandwidth used by different services that have been fixed in advance. Hence an appropriate link allocation scheme is required.

The factors being described in item 1) and 2) are having a strong influence onto the main performance values, i.e. the average establishment time and the blocking probability in the network. In this paper the effects on the performance values of known control mechanisms are investigated, if they are applied to ATM-networks.

In the following chapters the services of B-ISDN, the routing rules and the link allocation scheme will be described in detail. After a presentation of the model under study the simulation parameters will be explained. Then the results will be presented and discussed. The paper closes with a summary and an outlook to future studies.

2. SERVICES IN B-ISDN

The different services in B-ISDN can be split up into the following groups [1]:

Interactive Services
- Conversational services (voice)
- Messaging service (mail-box)
- Retrieval services (database)

Distribution Services
- without user individual presentation control (TV)
- with user individual presentation control (Videotext)

With regard to the traffic characteristics, the classes of services shown above represent two different groups: a) services with a steady bitstream (CBO traffic) and b) services without a steady bitstream (bursty traffic). The following figure shows some estimated data rates and burst factors for different services in B-ISDN [7].

Table 1
Data rates and burst factors for services in B-ISDN

service	data rate [Mbit/s]	burst factor
voice	≤ 0.064	2 - 3
data	≤ 0.064	> 10
pictures (steady)	≤ 1	1 - 10
video conference	≤ 2	5
TV	≤ 140	2 - 3

The traffic in B-ISDN can be described in several (e.g. three) levels (as shown in figure 1) [8].

The first level represents the activity of single sources. The most important parameters here are the mean intermessage time t_a and the mean holding time t_h of a service. These parameters usually are rather different for the various services (as depicted in table 1).

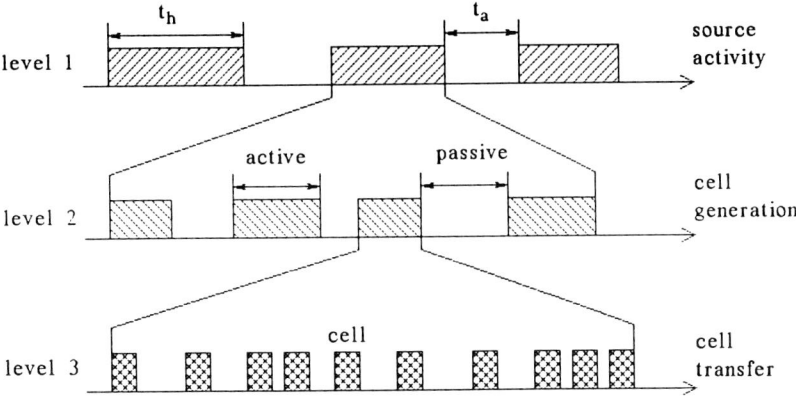

Figure 1. Three level model for the traffic in B-ISDN.

The second level describes the flow of character for an active service. The active and the passive phases of a service are presented here. The distribution of intermessage times for a bitstream can be described by the so-called burst factor (as shown in table 1). The bitstream generated by a source is divided into cells of constant length by a packetizer.

In the third level generated cells are considered that are transferred to the transport network. The transmitted cells of all sources represent the traffic in the B-ISDN. The statistical distribution of this traffic is the subject of intensive investigations today [9,10].

In this paper the signalling network of the ATM systems will be investigated. Therefore we just need to observe the traffic at the first level. It is possible then to determine the main performance value, i.e. the average establishment time and the blocking probability in the network.

3. ROUTING RULES

The new fast packet switching technique in ATM networks requires special properties for the implemented protocols. These properties have large influence onto the network efficiency. One of the most important protocols in B-ISDN is the routing protocol. Being responsible for the establishment of a new virtual connection it is responsible for the distribution of arriving cells.

Several routing rules are implemented in different existing computer networks [11]. However, they cannot be applied to ATM networks, because ATM networks differ too much from classical data networks (no storage of cells in the communication nodes is possible, extremely different character of services).

A routing protocol of high performance in B-ISDN has to fulfil the following features:

1) Different qualities of service have to be considered
2) maximum availability of the network between any pair of users has to be guarantied

3) flexible reaction on changes of the network topology must be possible
4) loops have to be avoided
5) decentralized realisation of the routing protocol functions is required

Taking into account these five items and other already known criteria for routing protocols in computer networks [12], we can derive the following characterisation of a routing rule in B-ISDN. The desired rule has to be adaptive, alternate and distributed with either periodic or aperiodic update interval.

The most important parts of the routing protocol are the so-called routing tables. The elements of these tables are containing the node numbers for each route from the source node to the sink node (different routes are possible). The routing tables can be determined by different algorithms. Some suitable of them can be found in [12,13]. Two methods for a determination of routing tables will be described in the sequel.

The first algorithm (it is called shortest-path-first (SPF) [14] here) generates trees of minimal distance to every destination node in the network. On the assumption that the hop distance is the criterion for this algorithm figure 2b shows an example of these trees. With these trees (obviously loop free) the routing tables can easily be created (fig. 2c).

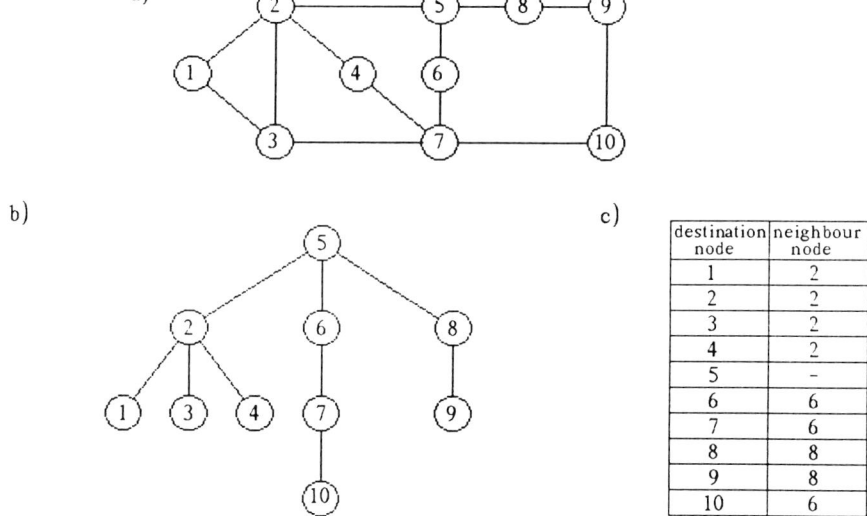

Figure 2. Illustration of the SPF algorithm. a) network topology, b) tree of minimal distance in node 5, c) routing table of node 5.

The second method, an iterative optimization algorithm [13], generates at least two alternate loop free routes for each destination node. Since the algorithm is unable to avoid ping-pong loops, the routing decision has to take care about that: arriving cell shall not be sent back to its native node. An example for a routing table created by the iterative optimization algorithm is given in figure 3.

destination node	1st route via node	2nd route via node	3rd route via node
1	2	6	-
2	2	6	-
3	2	6	8
4	2	6	8
5	-	-	-
6	6	2	8
7	6	2	-
8	8	6	-
9	8	6	-
10	6	8	-

Figure 3. The routing table for node 5 of the network topology of fig. 2 created by the iterative optimization algorithm.

Routing tables are necessary if a new virtual connection has to be established. In this case one specific route to the destination node has to be selected out of the set of possible routes. In ATM networks an additional parameter has to be taken into account, the so-called available bandwidth. A new connection will only be established if enough bandwidth for that service is available. The required bandwidth can be assigned by different methods depending on the considered service. These methods are usually called link allocation strategies and will be investigated in the next point.

4. LINK ALLOCATION STRATEGY

Each route between source and sink consists of several hops. In B-ISDN these hops will be generally described as trunks, which are consisting of several links. Every link comprises a fixed limit of bandwidth causing the second problem of control in B-ISDN, i.e. the link allocation problem. Up to now several different strategies are known for a solution of this problem. Some of the most important ones are: a) complete sharing (CS), b) complete partitioning (CP) and c) the movable boundary strategy (MB) [2].
Complete sharing means that all traffic types are allowed to occupy the whole trunk bandwidth on a FIFO basis. Under use of a complete partitioning strategy the aggregate trunk bandwidth is divided into distinct link groups (one for each traffic). Finally in case of the movable boundary method a traffic type may exclusively use as well dedicated links as links that is shared with other traffic types. In case of two services the methods above explained are illustrated in figure 4.
The permitted area Ω in fig. 4 for the i-th service (i = 1, 2) can be determined as follows

$$\Omega_1 = [\underline{b} \mid \sum b_{si} \leq B] \tag{1}$$

$$\Omega_2 = [\underline{b} \mid b_{si} \leq B/2; \sum \beta_i = B] \tag{2}$$

$$\Omega_3 = [\underline{b} \mid b_{si} \leq \beta_i'' \leq B - \sum_{j \neq i} \beta_j'] \tag{3}$$

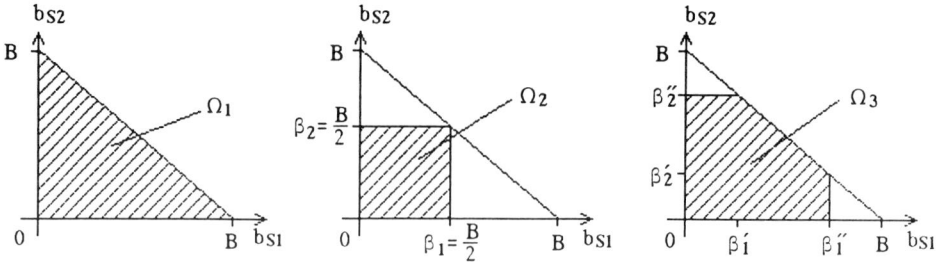

Figure 4. Graphs for different link allocation methods: a) CS, b) CP, c) MB. B = maximum trunk bandwidth, Ω = permitted area, b = limits of trunk bandwidth for services S_1 and S_2.

The different link allocation methods with respect to different routing rules (see section 3) will be compared in a special model which is described in the next chapter in detail.

5. MODEL UNDER STUDY

This study bases on the DATEX-P topology (see fig. 5) [13].

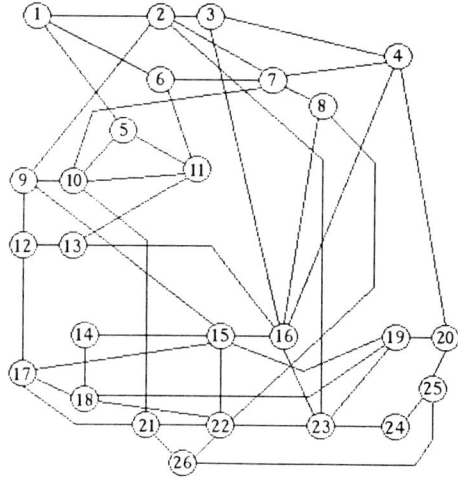

Figure 5. DATEX-P topology from 1982.

The topology of figure 5 shows identical appearance for the transport and signalling network. Figure 6 presents the assumed node model.

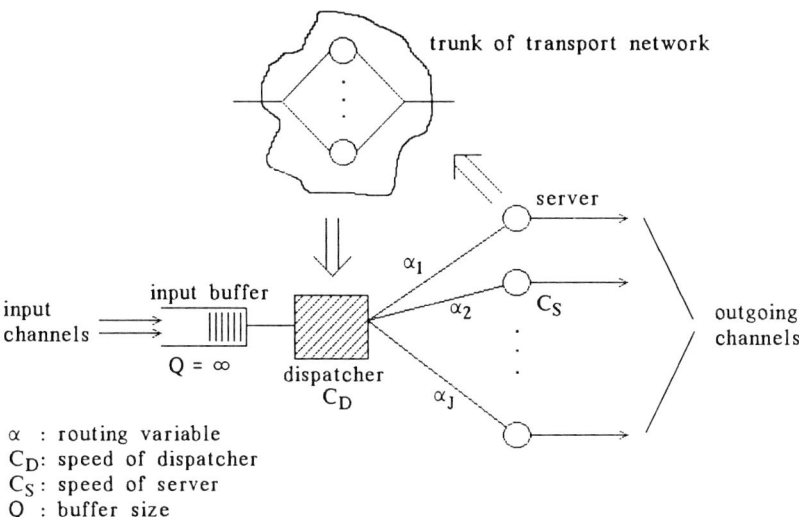

Figure 6. Model of the node under study.

Figure 6 depicts that there is a feedback between signaling and transport network. This leads to a dependence between the admittance for new services to the network and the state of the transport network.

The offered traffic can be divided into parts A_1 (voice), A_2 (data), A_3 (video) and A_4 (TV). The traffics A_1 to A_4 are supposed to be of Poisson type with mean arrival rates λ_i ($i=1,2,3,4$) and mean service time t_{hi} ($i=1,2,3,4$) being negative exponential distributed. The factors SB_i and BF_i ($i=1,2,3,4$) are describing the maximal bandwidth and the burst factor of the i-th service respectively (see table 1). The service cells are having the constant length $1/\mu_s$ bits/cell.

The termination rate of a call be denoted by ϵ_i, the total mean arrival rate by λ and the total offered traffic by A. Then the following equations holds true

$$A_i = \lambda_i t_{hi} \qquad (4)$$
$$\epsilon_i = 1/t_{hi} \qquad (5)$$
$$A = \sum A_i \qquad (6)$$
$$\lambda = \sum \lambda_i \qquad (7)$$

Up to a certain point of system complexity networks can be calculated exactly. In case of more complex network models exact calculations are often no longer possible or take great effort for finding an adequate solution. This applies to the model in this study as well. In such cases simulative techniques are a reasonable method for investigations.

For the comparative study in this paper the event-to-event simulation method has been used [15]. In each simulation the number of events has been limited to 100.000 successful calls, because in this case an acceptable computing time could be obtained (using a MICROVAX computer system) and most of the 95 % confidence intervals are less than 10 % of the mean performance values.

The investigations are based on the models shown in figure 6 and 7. The system parameters used for the simulative studies are as follows

t_{h1} = 3 min   B = 600 Mbit/s SB_1 = 64 Kbit/s BF_1 = 2
t_{h2} = 1 min C_D = 300 Mbit/s SB_2 = 1 Mbit/s BF_2 = 10
t_{h3} = 60 min C_S = 150 Mbit/s SB_3 = 2 Mbit/s BF_3 = 2
t_{h4} = 120 min $1/\mu_s$ = 424 bits/cell SB_4 = 32 Mbit/s BF_4 = 1

A = 1.8 - 1800 Erl $A_1 : A_2 : A_3 : A_4$ = 0.6 : 0.2 : 0.1 : 0.1

β_i' and β_i'', i = 1,2,3,4: see descriptions in fig. 10 and 11

Out of the long holding times for the video and TV services each holding time t_{hi} (i = 1,2,3,4) has been divided by the factor 1000. The offered traffic A_i is not allowed to be change. Therefore new modified arrival rates λ_i have been calculated according equation (4). These new parameters λ_i have been used in the simulation. A comparison with simulations made under neglect of the dividing by the factor 1000 has confirmed this way of reduction in computing time.

The obtained simulation results in case of different routing rules and different link allocation methods are presented in the next section.

6. OBTAINED RESULTS

The obtained simulation results will be presented into two distinctive parts. The first part considers the SPF routing rules and different link allocation methods.

The curves in figure 7 demonstrate that the CS policy produces good results under light traffic load. But if the offered traffic exceeded a certain threshold the losses grow rather fast.

The CP policy (see figure 8) has been proved to be a good link allocation scheme for all narrowband services. The broadband services suffer excessive blocking except in cases of light traffic load.

From figure 9 we can see that the MB policy is combining the advantages of the CS and CP methods. It is quite suitable method for B-ISDN. However, it should be pointed out that an appropriate determination of the parameter β leads to an optimization problem. The diagram in figure 10 demonstrate the appearance of the difficulty that just been mentioned.

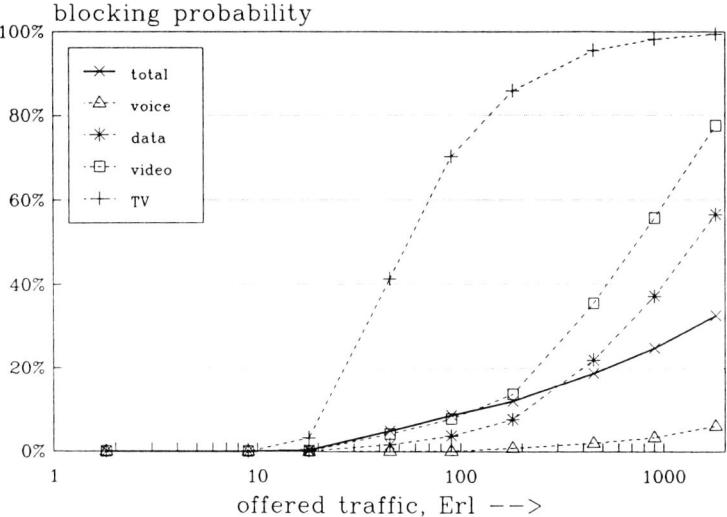

Figure 7. Blocking probability for different services versus total offered traffic for the CS police and SPF routing rule.

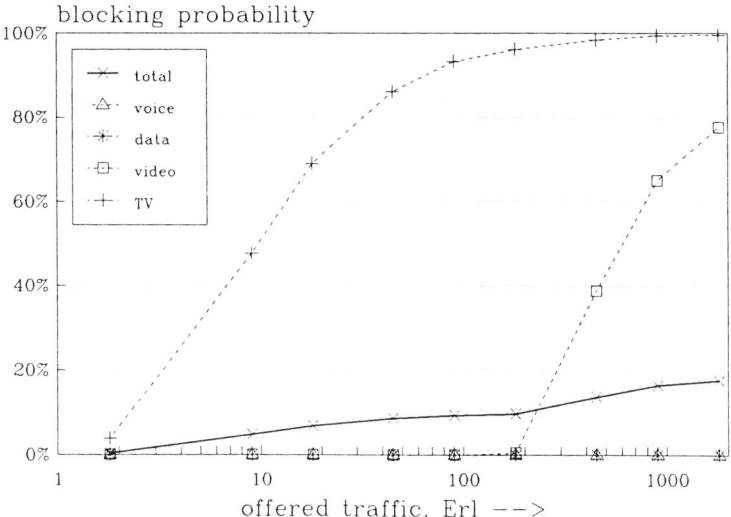

Figure 8. Blocking probability for different services versus total offered traffic for CP policy and SPF routing rule.

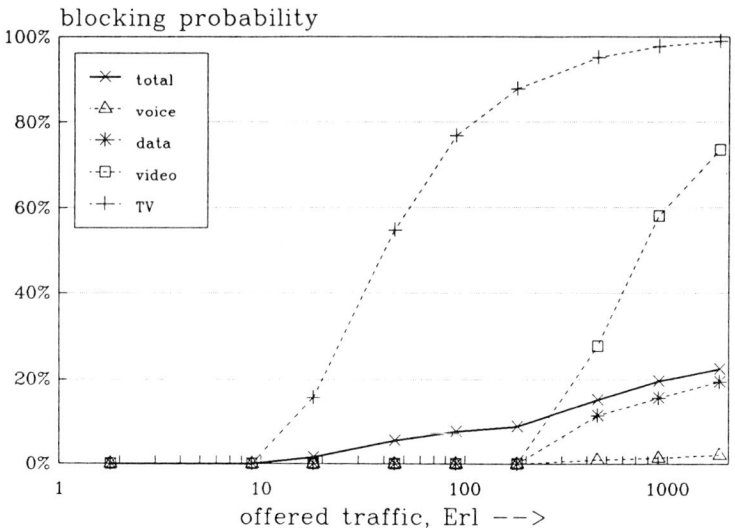

Figure 9. Blocking probability for different services versus total offered traffic for the MB police and SPF routing rule ($\beta_i'=0$, $\beta_1''=0.05$, $\beta_2''=0.1$, $\beta_3''=0.3$, $\beta_4''=0.7$).

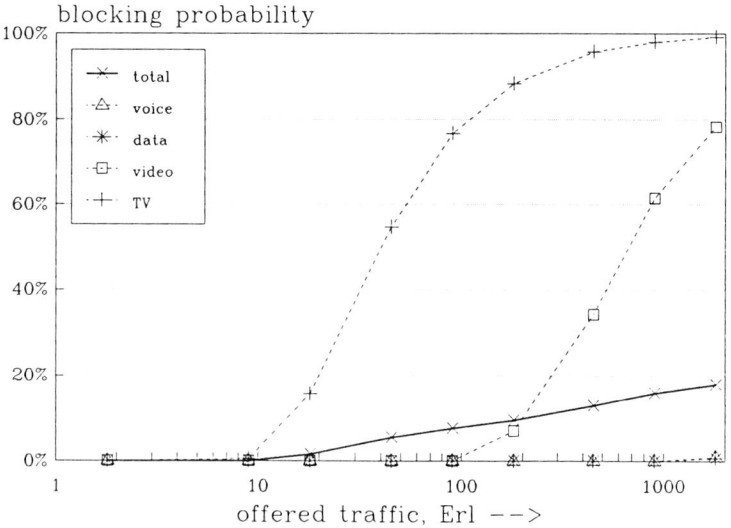

Figure 10. Blocking probability for different services versus total offered traffic for the MB policy and SPF routing rule ($\beta_i' = \beta_i''/2$, $\beta_1''=0.05$, $\beta_2''=0.1$, $\beta_3''=0.3$, $\beta_4''=0.7$).

It has to be emphasized that the simulation tool which has been developed for this study is suitable for a determination of the parameter β in case of different network topologies.

The second performance value, the average establishment time, has not changed in case of the SPF routing algorithm for different link allocation schemes. Its value results as 21.4 +/- 0.3 μs.

The second part of results deals with the alternate routing rules. First the pure alternate routing scheme (PAR) will be considered. Here the routing decision is based on the routing variables (elements of routing tables). This aim will be achieved by use of a random number generator [17]. Second, an alternate routing scheme with overflow (OAR) has been implemented which tries to use the shortest way first. If this is not possible (because the bandwidth is not sufficient) alternate routes are used. The routes are scanned in ascending order sorted by their length. In this part the speed of the channels to every local sink will be increased to 1.8 Gbit/s.

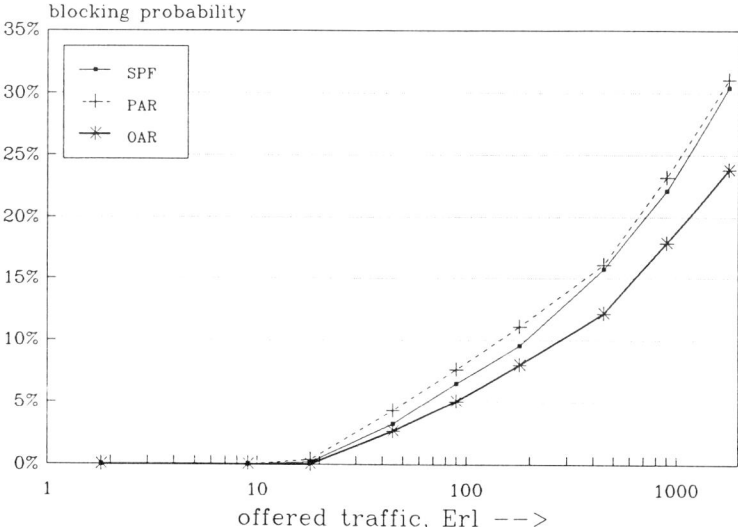

Figure 11. Total blocking probability versus offered traffic for the CS policy and different routing rules.

Figure 11 shows the total blocking probability for the CS link allocation scheme as a function of different routing rules. The PAR rule produces the worst results. The reason for these results is that the PAR scheme occupies more network resources for the same number of connections in comparison to the SPF scheme. Especially in case of low traffic the total blocking probability can be expected to be extremely worse, whereas in the overload situation the results are close to those that have be obtained by the SPF routing rule. The best results for each of the investigated network loads are obtained for the OAR rule. Using the latter scheme all possible routes are used optimal with just sharing of the available bandwidth.

Some interesting curves are shown in figure 12. It comes out that the combination of OAR and MB schemes is rather efficient. Once more it becomes clear that the MB method has a quite positive influence onto the performance of a network (comparison of the values from fig. 11 and fig. 12). Addionally the advantages of the OAR rule can be seen here.

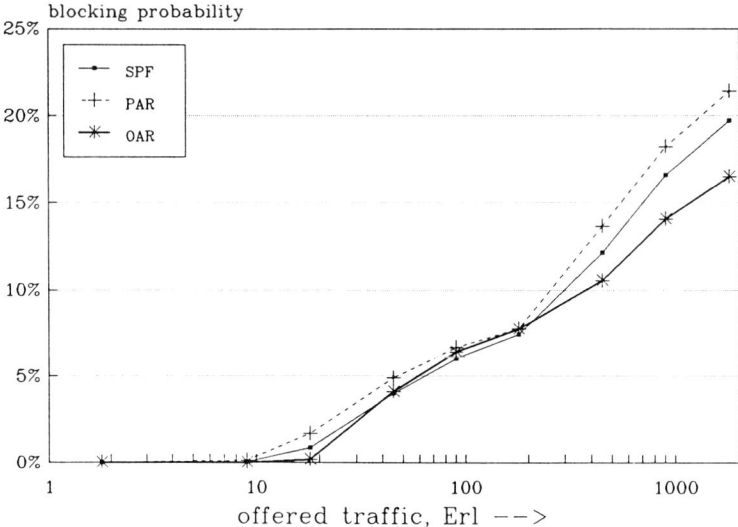

Figure 12. Total blocking probability versus offered traffic for the MB policy and different routing rules.

The establishment time is constant for SPF (21.4 +/- 0.3 μs) and PAR (32.1 +/- 0.3 μs) scheme, but is increasing for the OAR method with rising offered traffic (from 21.7 μs for $A = 1.8$ Erl to 23.1 μs for $A = 1800$ Erl).

7. CONCLUSION

In this paper the link allocation methods and routing techniques in B-ISDN have been investigated. For this purpose a new network planning tool based on simulations has been developed. The tool is applicable in case of various network topologies. The network parameters (capacity of node and channels, routing tables, etc.) can be supplied easily because of the dialogue-orientated operation of the program.

First the different link allocation policies have been investigated. The obtained results have shown that the movable boundary rule (MB) is the best strategy. For this method, however, it is necessary to choose the optimal control parameters depending on the network topology and the type of the desired services.

The impact of routing onto service call blocking has also been considered. A combination of alternate routing with overflow (OAR) and a suitably designed MB scheme guarantees a significant reduction of the total blocking probability in the network. Such combinations seem to be quite suitable for practical applications within B-ISDN.

The future studies are intended for an examination of reliability aspects (failure of nodes and/or channels).

8. REFERENCES

1. CCITT SG XVIII, Draft Recommendation I.121: On the Broadband Aspects of ISDN. Blue Book, Geneva, 1989.
2. B. Kraimeche and M. Schwarz, Bandwidth Allocation Strategies in Wide-Band Integrated Networks. IEEE J. on Selected Areas in Commun., Vol. SAC-4, No. 6 (1989) pp. 869-878.
3. B. Kraimeche and M. Schwarz, Analysis of Traffic Access Control Strategies in Integrated Service Networks. IEEE Trans. on Commun., Vol. COM-33, No. 10 (1985) pp. 1085-1093.
4. M. Gerla, Routing and Flow Control in ISDN's. Proc. of 8th Conference ICCC'86, Munich, 1986, pp. 643-647.
5. J. Hui, Resource Allocation for Broadband Networks. IEEE J. on Selected Areas in Commun., Vol. SAC-6, No. 9 (1988) pp. 1598-1608.
6. M. Gerla, J. Monteiro and R. Pazos, Topology Design and Bandwidth Allocation in ATM Nets. IEEE J. on Selected Areas in Commun., Vol. SAC-7, No. 8 (1989) pp. 1253-1261.
7. A. Bellman, Switching Architectures Towards the Nineties. Proc. of GSLB Seminar on Broadband Switching, (1987) pp. 75-84.
8. M. Krüger and T. Uhl, Asynchronous Transfer Mode (ATM) - a Concept for Broadband ISDN. Proc. of 2nd Intern. Seminar on Flow Control, Moscow, 1988.
9. K. Lindberger, Traffical Analysis of Statistical Multiplexing in ATM Networks. Proc. of 8th Nordic Teletraffic Seminar, Otnas, 1989.
10. L. San-Qi and J. Mark, Traffic Characterization for Integrated Services Networks. IEEE Trans. on Commun., Vol. COM-38, No. 8 (1990) pp. 1231-1243.
11. M. Schwarz and T. Stern, Routing Techniques Used in Computer Networks, IEEE Trans. on Commun., Vol. COM-28, No. 4 (1980) pp. 539-551.
12. J. Seidler, Principles of Computer Communication Network Design, Ellis Horwood Limited, Chichester, 1983
13. N. Mersch and T. Uhl, Algorithms for the Determination of Loop Free Alternate Routes in Communication Networks, to be published shortly, AEÜ, October 1991 (published in German language).
14. J. McQuillan, I. Richer and E. Rosen, The New Routing Algorithm for the ARPANET, IEEE Trans. on Commun., Vol. COM-28, No. 5 (1980) pp. 711-719.
15. R. Shannon, Systems Simulation: The Art and Science, Prentice-Hall, 1975.
16. T. Uhl and J. Ulmer, A Simulation Tool for Optimal Resource Allocation and Routing in B-ISDN, internal report, Institute for Electronic Systems and Switching, University of Dortmund, 1991.
17. T. Uhl, An Analysis of Delay in Packet Switching Networks in Case of a Specific Implementation for an Alternate Routing Scheme, Proc. of 2nd Intern. Seminar on Flow Control, Moscow, 1988.

Generalised Karlsson Measurements for ATM-Networks

Biørn Veirø

Copenhagen Telephone Company

Abstract

The paper first sets the context in which the generalized Karlsson measurements should be seen and continues with a statement of the requirements to the method. Then a brief review of the procedure is given and it turns out, that a modification of the procedure as proposed originally is needed. A proof of the revised method is outlined and it is shown that this new version is valid for quite general traffic distributions. After that some source models for real coded video data are given, but it turns out, that only simulation will be accurate enough to assess the generalized Karlsson measurements and a simulation study then shows, that the procedure works well.

1. INTRODUCTION AND BACKGROUND

In the old telephone networks the call admission worked as follows: The network had to find a route with a free channel all the way from the caller, A, to the receiver, B. If such a route was found the call was admitted, otherwise it was lost. In ATM-networks, however, the situation is different, because not only is it allowed to have on-going calls demanding different bandwidths, but also calls that do not demand the same bandwidth for the whole call duration.

In the part of the network, where the services carried have a fixed bandwidth assignment per call the situation is similar to the one in telephony: The network checks if the resources needed for the new call are available, and if that is the case, the call will be allowed, otherwise it will be lost. In this paper we only deal with the part of the network where variable bit rate (VBR) calls are allowed and in that part it is a problem to decide when a new call can be admitted. In the literature the call admission is usually based on a contract set up between the user and the network. The idea then is that the user will specify her/his needs to the network and based on those parameters the network has to decide if the call can be carried or not considering the degradation in grade of service for the calls already admitted to the network as a result of allowing the new call. Of course it is a problem for the user to know in advance what resources (s)he is going to need during the call, and therefore such parameters have either to be very coarse or to be easy to identify (e. g.: I am a videophone call, I am a file server call or the like).

Once a VBR call has been admitted, it will be necessary for the network to check if the user really abides by the contract (s)he negotiated. This is necessary partly to protect itself and partly to protect other well behaved users. A way to achieve this, which has gained a lot of interest in the literature is the so-called policing function, which either discards user information violating the contract immediately or marks it for later deletion if network congestion occurs further downstream. Several such mechanisms have been proposed the first one being the leaky bucket [Tur86], but also the moving window, the jumping window, the exponentially weighted moving average and ticket policing [Ive89]. (The marking procedure [ELL89] would work with any of the methods mentioned.). The user information is sent as 48 octets of payload and 5 octets header, called a cell [CCI90]. It has also been proposed to shape the source, i.e. once the calling terminal knows how the policing function works it might as well choose to transmit when it knows that the information will be accepted. It should be noted, that shaping has the nice side effect, that the network can be dimensioned as shown in [RSK91] subject to the problems that arise because shaping is done at the source and policing at the network access point. But as pointed out in [Kos90] it would be much more effective if we just shaped the user out of existence!

How nice such ideas may look on paper there is one minor problem: They do not work in practice because it turns out, that is impossible to police the mean value correctly as Rathgeb and Theimer [RTh90] point out. One way out of this would be to avoid the mean value policing altogether [Kos90]. The idea is, that the network a posteriori decides whether the call belonged to the class it declared at call set up time using some kind statistical test on the call, based on measurements performed during the call. If the network decides that the user violated the contract a penalty is paid via the charging system. This fine can be made dependent on the number of contract-violations for the user in question in order to avoid users who deliberately violate the contract every time they are connected.

The paper continues with a statement of the requirements to a measurement method as outlined above. Then a brief review of the approach presented in [Vei89] is given with a revision such that the proposed procedure will work for general distributions of the bandwidths we measure. A proof of the new version of the measurement procedure is then given and after that some source models for real coded video data are considered in order to get an idea what kinds of results one can expect with the measurement procedure such as the one proposed. Finally a simulation experiment is carried through as a (sort of) practical verification of the method.

2. REQUIREMENTS TO MEASUREMENTS

It seems reasonable to require from any measurement routine used in connection with customer charging, that it can be implemented in one single chip. It is therefore needed to get an idea how many cells a customer would generate. Before starting to assess various traffic scenarios it is worth while to find out what the maximum count that could possibly be required would be. So let us consider a 622Mb/s access line running continuously for one year. The number of cells generated is: $622*10^6/(53*8)*60*60*24*365 = 46.3*10^{12}$ corresponding to 46 bits. If we consider a VBR scenario where we have equal probabilities to measure 8 bit rate levels distributed uniformly between 0 and 622Mb/s, then the average bit rate would be 311Mb/s and if one measurement consists of 128 cells we would need 35 bits in each of the 8 counters. (The actual number of cells proposed in [Vei89] for one measurement was 160, but for the sake of the argument a power of two was selected here). A scenario producing a traffic as described would correspond to an extremely industrious user indeed.

Scenarios studied in the European research programme RACE [RAC89] suggest much lower values even for big business customers. The total of cells generated in one year is only 31.9Mcells/year based on "busy hour" durations corresponding to one full working day and this implies a need of 25 bits in a counter. If we consider measurements taken as averages over 160 cells and 8 bit rates then an assignment of 32 bits pr. counter (an assignment which is very practical using today's technology) would be very generous and 8 32-bit counters fit easily in one chip (it is not even needed to use VLSI-technology!).

Obviously there is no need for a measurement method which is based on the assumption, that it will be needed to save space in the chips for counters unless it is assumed, that one chip should be enough to measure all the existing connections in a PABX at a large enterprise in which case it will be difficult to say how many counters are really needed. The investigation in the sequel can then be seen as a proposal covering the case, where the aim is to fit as many counters as possible to one chip.

3. GENERALIZED KARLSSON MEASUREMENTS

The generalized Karlsson measurements work as follows: Consider a call given as a function of the bandwidth b(t) (see fig. 1). The call arrives at $t = T_1 = 0$ and ends at $t = T_2$. When the call has arrived it is coupled to a scanning procedure that generates a sample of b(t) each h seconds (see below how the sample value of b(t) is generated). Let us assume that the scanning procedure is independent of the call arrival process. Then if the first sample arrives at time t_x we can assume also, that t_x will be a random variable distributed uniformly over [0,h]. If we have N scans during the call it is a well known result, that the expectation of the actual duration $T_2 - T_1 = T_2$ is the same as the expectation of N· h.

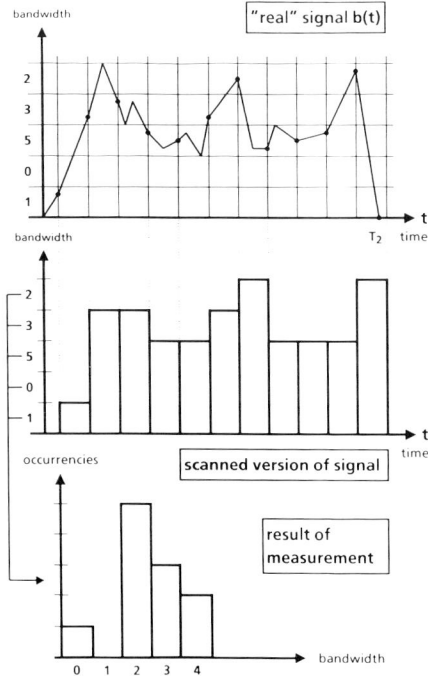

Figure 1. Principle for Generalized Karlsson Measurements.

We now return to the problem of how to obtain the individual measurement values. If we restrict ourselves to the ATM layer of the communication protocol then the only way to perform bandwidth measurements is to count cells. Hence a sample of b(t) consists of counting how many of the cells passing the measurement device in a predetermined time interval, that belong to our signal b(t). Let us assume that the maximum bandwidth allocated to a given call can be controlled effectively (e. g. by the access line itself). Let us also assume, that we want to distinguish exactly L measurement levels. Then we need to count for at least a period corresponding to L cells, but as proposed in [Vei89] we shall use k· L worth of time, where k is an integer constant > 1. It was furthermore argued, that k = 20 would be an appropriate value to choose and we will continue to use that value here. If we assume that c cells have been counted in the measurement duration, then in the original version, the value that the measurement took was the rounded value of c/k.

The proof given was an approximation proof, which in broad terms said that if we make the samples as close to each other as possible, then we will get the correct result. This result is rather obvious, as it simply means that we count all the cells in this case.

Trying to generalize the old proof we run into problems, because if we do not go to the limit where all cells are counted, we can not be sure to get the correct value. To see this let us consider the function $b(t) \in \mathbf{b}(t)$, where $\mathbf{b}(t)$ is the ensemble of possible outcomes, and let $t \in [0,T_2]$. By scanning we replace the function b(t) by its value at equidistant instants $U+i, i \in [1, ..., N]$ as we without loss of generality use the scanning interval as a unit of time. Since the scanning is assumed independent of the call, U is distributed uniformly over [0,1]. Then we have:

$$E\{\sum_{i=1}^{N} b(U+i)\} = \sum_{i=1}^{N} E\{b(U+i)\} = \sum_{i=1}^{N} \int_{i}^{i+1} b(t)dt = \int_{0}^{T_2} b(t)dt$$

This means that we will only get the correct expected value of the sum, when we let the measured value of b in the point t_i be represented by $b(t_i)$, which means that the measurement only can be guaranteed to be correct in the case, where $b(t_i)$ happens to be one of the approximated values.

Hence we propose a slight modification of the method as proposed originally: Let us assume (again) that the number of cells counted in the k· L time slots is c. Then we take a random number, x, whose distribution is rectangular between 0 and 1. We now represent the measurement by the value $\lfloor x + c/k \rfloor$, where $\lfloor \ \rfloor$ denotes the integer part. The advantage with adding such a random number is, that we now have a case similar to the one for the measurement, on the time axis and we are able to repeat the same proof for the individual sample values as the one used in the literature (see [Vei89] for references) for the duration. Such a method is also known to provide the correct expected value and to be valid for any distribution of the sample values of b(t).

The disadvantage is, of course, that for each measurement we need to generate a random number, but fortunately good generators for that exist for a wide range of processors (see e. g. [MZT90]). To be fast enough the it will have to be implemented in hardware, so the number of bits calculated in section 2 is challenged by this change. Whether such a generator could be shared for a number of applications needs further study. But if just one generator can be used for one whole switching centre, it would not matter.

4. APPLICATION TO SOURCES

4.1 Modelling

In [Vei89] an example was used where VBR voice data was considered using the source model provided in [SrW86]. As it is probably more cost effective to keep the voice traffic in the telephone network we have to look for other kinds of services that one could realistically imagine in an ATM network and for this purpose coded video data making the evaluation of source models possible is provided in [VPV88] and [MAS88], but in [MAS88] only 10 seconds worth of data is provided while [VPV88] offer data corresponding to 30 minutes of video and therefore their data is used (and because data corresponding to [VPV88] was benevolently put at my disposal). For the source modelling [VPV88] and later [VeP89] models are considered based directly by measurements, so that their representation of the probability density function is by a (dense) histogram scaled to have the area 1. [MAS88] study a fluid flow model which is refined in [SNR89]. Also in the European research project COST 224 fluid flow models have been considered by several authors [NRS91]. [NFO89] propose an auto-regressive model, but the fit they show to experimental data does not seem convincing. Such a model is also studied in [Hua88]. [Has72] subsamples the data he has collected in order to avoid to formulate a reliable mathematical model.

I studied data from a video conference scene from [VPV88] (fig. 2) and a video telephony scene from [VeP89] (fig. 3) and tried fit a lognormal, a gamma and a Weibull distribution to it all by finding distributions which had the same mean and variance as the data. The results (fig. 2 and 3), where even the best fit (the lognormal distribution) were so bad that more refined fitting methods (like fractile fitting) were not even considered. The parameters calculated are given in table 1.

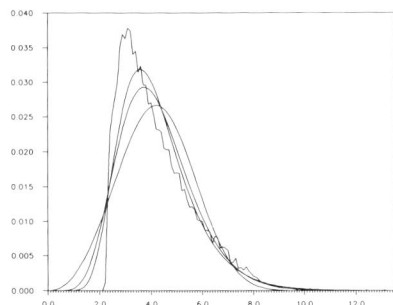
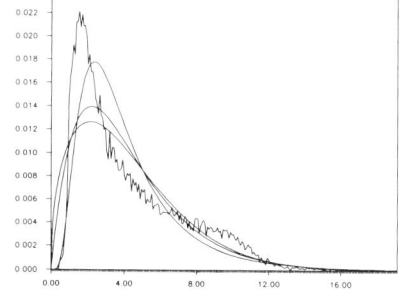

Figure 2. Videoconference density function (relative frequency of bandwidth in Mbit/s)

Figure 3. Videotelephony density function (relative frequency of bandwidth in Mbit/s)

	Videoconf.	Videotelf.
$E(x) = \mu$ for original data $V(x) = \sigma^2$ for original data	4.218 Mb 2.061 (Mb)^2	4.245 Mb 8.706 (Mb)^2
Weibull: density $(k/\beta) * (x/\beta)^{k-1} * \exp(-(x/\beta)^k)$	$k = 3.228$ $\beta = 4.708$ Mb	$k = 1.462$ $\beta = 4.687$ Mb
Gamma: density $(\Gamma(k)\beta^k)^{-1} x^{k-1} * \exp(-x/\beta)$	$k = 8.632$ $\beta = 0.4887$ Mb	$k = 2.070$ $\beta = 2.051$ Mb
Lognormal: density $(x\beta\sqrt{2\pi})^{-1} * \exp(-0.5 * ((\log x - \alpha)/\beta)^2)$	$\beta = 0.3311$ $\alpha = 15.20$	$\beta = 0.6278$ $\alpha = 15.06$

Table 1: Parameters calculated for experimental data.

4.2 Simulated Measurement Experiment.

In order to get an impression of how the proposed measurement procedure would work in practice, a simulated measurement experiment was carried out. As a genuine cell stream was not available, the original data was reconstructed to yield something that looked like a possible cell stream from the experiments under consideration. This was done by rescaling the measured frequency functions so that number of occurrences of the various bit rates were expressed rather than their relative frequencies. As each of the original measurements corresponds to the amount of information contained in one video frame, we know that it also corresponds to 40ms worth of data. Each frame was then offered to an access line of 34 Mb/s in the form of ATM cells with 48 bytes of payload information and 5 bytes of overhead information. This gives an information rate of 48/53 = 0.906, which is somewhat higher than the 0.875 Verbiest et al. had in [VPV89] and [VeP89]. This is because they did not use standard ATM cell size and this has to be taken into account when data from our two experiments are compared.

For a given 40ms data rate it can then be calculated how many cells this would correspond to when offered to a 34Mb/s line. On such a line 3208 cells pass in the 40ms frame and now let the number of cells we need to achieve the data rate in the original experiment be j. Then we distribute the j cells randomly among the 3208 cells available in the frame by picking j distinct random cell numbers such that they have a uniform distribution over [1, 3208]. The result of this is then used as our image of the original cell stream.

It turns out that when the data is reconstructed in the way described, it is relatively seldom to have bursts of consecutive cells. In the videoconference scene we get the average burst length 1.18 cells and 1.33 cells for the videotelephony scene. I have been informed that this is not realistic for real video codecs where bursts of 4 cells should be quite common but still the statistical behaviour of the data generated in the way described above is sufficiently close to the real data to be useful for conclusions.

On the data thus generated the measurement method described in section 3 is applied for both the videoconference and for the videotelephony scenes. We take the number of levels L to be 8 and the constant factor k to be 20 as proposed originally. Then the duration of the individual sample is 2.00ms. The measurement is repeated 3 times for each scene with different spacing between consecutive sample values. The times chosen were 2.5ms, 100ms and 10s, where the time is to be understood as the time between the start of one sample to the start of the next one. Obviously this means, that the smallest spacing we can select is 2ms.

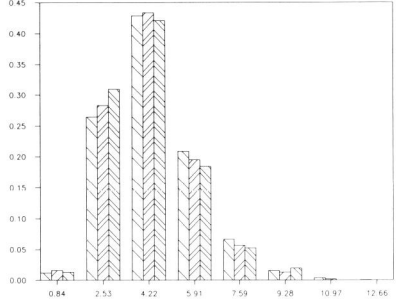 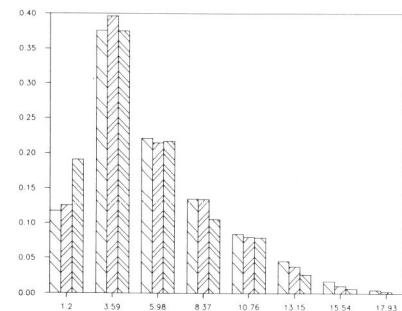

Figure 4. Measurement on Videoconference scene (the leftmost column is for 2.5ms spacing, the middle for 100ms and the last for 10s)

Figure 5. Measurement on Videotelephony scene (the leftmost column is for 2.5ms spacing, the middle for 100ms and the last for 10s)

As it can be seen from the simulation results as they are shown in the figures 4 and 5, the resemblance between the data measured in the way described in section 3 and the original measurements shown in the figures 2 and 3, is remarkable. Apart from a slight deviation in the lower column of the measurement for videotelephony, the columns for different spacing have all nearly the same relative frequencies as they should have and they seem to agree quite well with the measurements in the figures 2 and 3. Possible reasons for the deviation in the videotelephony case have not been sought, but in any case it does not seem significant. It should be noted, that the spacing of 10s is considered the most likely to be used of the ones tried for applications not sending their information over too long distances.

5. CONCLUSION

First of all it can be concluded, that reduced cell counting as proposed above is effective. However it is needed only if we want to survey all the connections of a big enterprise in one single chip, subject to the restriction that it will be possible to implement a random number generator in hardware. Measurements in general will always be needed partly for charging purposes and partly because the requirements to network management have been increasing and will continue to do so for many years to come.

Acknowledgements

First I wish to thank my fellow researchers in the European research project COST 224 because the inspirational atmosphere they created made it possible to get the ideas needed to produce this work. Furthermore I am grateful because Willem Verbiest of Alcatel Bell provided the detailed data analysed in section 4.

REFERENCES

[CCI90]: CCITT SGXIII: Draft Recommendation I.361: ATM Layer Specification for B-ISDN, Study Group XVIII - Report R 23, Feb. 1990.

[ELL89]: Eckberg, A. E., Luan, D. T. and Lucantoni, D. M.: Bandwidth management: A Congestion control strategy for Broadband Packet Networks - Characterizing the Throughput Burstiness Filter. Comp. Netw. & ISDN Syst. v.20(1990) pp. 415-425

[Has72]: Haskell, B. G.: Buffer and Channel Sharing by Several Interframe Picturephone Coders. BSTJ, pp.261-289, Jan. 1972.

[Hua88]: Huang, S.: Source Modelling for Packet Video. ICC Session 38.7, Philadelphia 1988.

[Ive89]: Iversen, V. B.: Source Policing in ATM Networks. 8'th Nordic Teletraffic Seminar, Helsinki 1989.

[Kos90]: Kos Tutufour: On Admission Control and Policing in an ATM Based Network. ITC Specialist Seminar, Morristown 1990

[MAS88]: Maglaris, B., Anastassiou, D., Sen, P., Karlsson, G and Robbins, J. D.: Performance Models of Statistical Multiplexing in Packet Video Communications. IEEE Trans. Comm. Vol. 36 pp. 834-844, Jul 1988.

[MZT90]: Marsaglia, G., Zaman, A. and Tsang, W. W.: Toward a Universal Random Number Generator. Statistics and Probability Letters Vol. 8 pp. 35-39 Jan. 1990.

[NFO89]: Nomura, M., Tetsurou, F. and Otha, N.: Basic Characteristics of Variable Rate Video Coding in ATM Environment. IEEE JSAC Vol. 7 pp. 752-760, Jun. 1989.

[NRS91]: Norros, I., Roberts, J.W., Simonian, A. and Virtamo, J.: The Superposition of Variable Bit Rate Sources in an ATM Multiplexer. To appear in IEEE JSAC.

[RAC89]: RACE 1044: Preliminary evaluation of the seven EPF study cases. Delivery 44/EPF/TG0/DS/B/004/b2 from R1044 Evolutionary Prospects and Framework.

[RSK91]: Rasmussen, C., Sørensen, J., Kvols, K., and Sørensen S. B.: Source Independent Acceptance Control in ATM Networks. To appear in IEEE JSAC.

[RTh90]: Rathgeb, E. P. and Theimer, T. H.: The Policing Function in ATM Networks. ISS XIII, Stockholm 1990.

[SNR89]: Sen, P., Maglaris, B., Rikli, N. and Anastassiou, D.: Models for Packet Switching of Variable-Bit-Rate Video Sources. IEEE JSAC Vol. 7 pp. 865-869, Jun. 1989

[SrW86]: Sriram, K. and Whitt, W.: Characterizing Superposition Arrival Processes in Packet Multiplexers for Voice and Data. IEEE JSAC, Vol. SAC-4 pp. 833-846, Sep. 1986

[Tur86]: Turner, J. S.: New Directions in Communications (or which Way to the Information Age). IEEE Comm. Mag. Vol 24 pp 8-15, Oct 1986.

[Vei89]: Veirø, B.: Traffic Measurement on Variable Bit Rate (VBR) Sources with Application to Charging Principles. Comp. Netw. & ISDN Syst. v.20(1990) pp. 435-445

[VeP89]: Verbiest, W. and Pinnoo, L.: A Variable Bit Rate Video Codec for Asynchronous Transfer Mode Networks. IEEE JSAC Vol. SAC-7 pp. 761-770, Jun. 1989

[VPV88]: Verbiest, W., Pinnoo, L. and Voeten, B.: The Impact of the ATM Concept on Video Coding. IEEE JSAC Vol. SAC-6 pp. 1623-1632, Dec. 1988

Performance and Queueing Models

Performance Models for Hybrid Broadband Networks

Catherine Rosenberg[*] and André Le Bon[*]

[*]Département de Génie Electrique et de Génie Informatique, Ecole Polytechnique de Montréal, Montréal H3C 3A7, Canada

Abstract

In this paper we present a model for the performance analysis of hybrid broadband networks in which multi-rate circuit switched traffic as well as fast packet switched data traffic compete for the use of N slots or channels. The novelty is that we allow for several classes of circuit switched traffic based on the number of channels required as well as obtain approximations based on temporal decompositions due to the different time scales present.

1. INTRODUCTION

The emerging B-ISDN environment will offer unprecedented flexibility in networking. However, at the present time very few performance models for the analysis of hybrid architectures incorporating fast packet and multi-rate circuit switching methods are available. In this paper we will present exact as well as approximate model for the performance analysis of such systems.

There are two approaches towards the development of tractable models for such networks. They are both based on decomposition techniques. The first is the notion of spatial decomposition to isolate sub-networks or links which can be treated independently to reduce the state space and will not be dealt with here. The second method is via temporal decomposition based on the time-scales of the different underlying phenomena. This allows us to study the effect of a phenomenon acting at one time-scale on others acting at different time-scales. This is natural in the hybrid context where there are at least three orders of magnitude difference between the time-scales associated with fast packet switched traffic and circuit switched traffic. We use temporal decomposition to obtain an approximate model with the ability to compute means of measuring the limits of validity of the approximate results.

In the following we briefly describe the model and the methodology used.

2. MODEL DESCRIPTION AND METHODOLOGY

There are two distinct steps in our methodology. The first is the construction of a performance model describing the operation of the hybrid system. Due to the intractability of this model in realistic situations an approximate model is then obtained based on the temporal decomposition techniques developed in [1].

The model we consider is a hybrid system in which two modes coexist, namely an asynchronous data mode with variable-rate ATM connections (called the packet mode) and a synchronous multi-circuit mode (called the circuit mode). While the Grade of Service of a synchronous connection is determined at the call set-up, the successive ATM cells of a given call may encounter varying delays due to their bursty nature. Such models have been considered in the literature in the context of voice/data integration [2] and narrowband and wide-band traffic integration [3–5]. The circuit mode is used for delay sensitive traffic and is given preemptive priority over data traffic and is subject to blocking. The packet mode traffic is subject to queueing in a finite or infinite buffer. The key difference in our model over existing models is that the circuit mode traffic consists of N classes denoted $\{C_1, C_2, ..., C_N\}$ and these classes compete with the data traffic for N slots or channels. It is assumed that the class C_i requires i channels and the holding time is exponential with mean $1/\mu_i$. The i channels are seized and released simultaneously. A call of class C_i which cannot find i channels on arrival is assumed lost and a data packet which cannot find a channel or is preempted is queued in the buffer.

The above model is of great relevance in the context of networks with cross-connects where calls may require many slots simultaneously. The main aim of the study is to model the effect of the synchronous traffic on the performance of the asynchronous traffic.

The performance of the synchronous traffic (blocking probabilities per class) follows from the results of Kaufman [6]. We then obtain the infinitesimal generator matrix for the slot availability for the asynchronous traffic. This is used to obtain the performance model for the data traffic. It is seen that the this model is intractable for any realistic set of parameters and thus a simplified approximate model is constructed based on temporal decomposition [7] and [1].

The decomposition idea while analogous to the Courtois-type decomposition [8] differs in that it exploits the analyticity of the generating function to obtain a Mac-Laurin expansion around zero. The time-scales present due to the difference in holding times and arrival rates of the two types of traffic are shown to be such that the expansion is valid. In particular, we obtain a model which can approximate the true model with a good level of accuracy by means of a very simple and intuitive decomposition based on the fast-time scale process being in statistical equilibrium. The performance results obtained from the model will be compared to simulations of the system.

3. MODEL FORMULATION

We consider the following model for an hybrid system in which two modes coexist, namely an asynchronous data mode (packet mode) and synchronous multi-circuit mode (circuit mode in short). These two modes are competing for the use of a common resource which is composed of N slots or channels.

3.1. The circuit mode

Different services are included in this multi-circuit mode which we represent by different classes of "circuit" users. It is assumed that at any instant, no more than X channels are taken by all the classes of the circuit mode. Let X classes C_i of "circuit" users where a user is said to be of class C_i if it requires the simultaneous usage of i slots (not necessarily contiguous). The circuit mode is assumed to have preemptive priority over the packet mode as long as the constraint of no more than X channels used by the circuit mode is valid. If the request of a circuit customer of class C_i cannot be fully and immediately taken care of (i.e. if there are already more than X-i slots taken by the circuit mode), the customer is rejected. We assume that the "circuit" users of class C_i arrive according to a Poisson process of rate λ_i (where λ_i can be set equal to zero if we do not want to allow for a class of type C_i) and keep the i channels for a time distributed exponentially of parameter μ_i after which the i channels are released simultaneously. The analysis and the performance of the circuit mode does not depend at all upon the packet mode due to the preemptive priority of the circuit mode over the packet mode.

3.2. The packet mode

The data packets are assumed to arrive according to a Poisson process of rate λ_p and to have a length exponentially distributed with parameter μ_p. If there is no channel available when a packet arrives, it will enter a buffer which can be either finite or infinite. If a packet is preempted by a circuit mode user, it will join the head of the buffer and its service is resumed. Since the circuit mode cannot use more than X channels at a time, and since the circuit mode has preemption over the packet mode, the analysis of the packet mode corresponds to the analysis of an M/M/i/K where the number i of available servers can vary randomly between N-X and N depending on the multi-circuit mode, and K>i and can be infinite. This dependance of the packet mode on the circuit mode results in a Markov modulating process whose infinitesimal generator is given by the matrix $\mathbf{Q}=(q_{ij})$ where q_{ij} is the transition rate between the state where N-i slots are taken by the circuit mode and the state where N-j slots are taken (or between the state where i slots are available for the packet mode and the state where j are available (i, j=N-X,...,N)).

3.3. Discussion of the general assumptions

Let us discuss the main assumptions made at this point:

1. The synchronous or circuit mode has preemption over the asynchronous or packet mode. It will be more realistic to assume that the priority of the synchronous mode is without preemption but this complicates the analysis by introducing a dependency of the packet mode on the circuit mode.
2. The exponential distributions that we have chosen for the interarrival and holding times are not the most realistic assumptions but they allow us to validate relatively easily the temporal decomposition approach which is the main purpose of this paper.
3. The circuit mode is a loss mode, i.e. a customer of class C_i which cannot find immediately i channels available, is rejected. Any other assumption would necessitate a complex bandwith management scheme introducing more intractability.

4. ANALYSIS OF THE CIRCUIT MODE AND OBTENTION OF THE MATRIX Q

The assumption of the preemptive priority of the circuit traffic over the packet traffic leads to an independent analysis for the circuit mode. This has been studied by Kaufman [6] in which the stationary probability q(i) of having i slots occupied by the circuit mode and the blocking probabilities Pb_i for class C_i were obtained. Using these results we then obtain the infinitesimal generator matrix **Q** of the Markov chain which describes the impact of the circuit mode on the packet mode. Our main focus will be on the impact of the circuit mode on the performance of the packet mode.

4.1. General analysis of the circuit mode [6]

Let $\mathbf{n} = (n_1,...,n_X)$ where n_i represents the number of customers of class C_i being served, denote the state of the circuit mode. Then it can be shown that [6]:

$$P(\mathbf{n}) = \prod_{i=1}^{X} \frac{\rho_i^{n_i}}{n_i!} \cdot q(0) \qquad \text{for } \mathbf{n} \text{ such that } \sum_{i=1}^{X} i \cdot n_i \leq X \qquad (1)$$

and that:

$$j \cdot q(j) = \sum_{i=1}^{X} i \cdot \rho_i \cdot q(j-i) \qquad (2)$$

where $q(x) = 0$ for x<0, $\sum_{i=1}^{X} q(i) = 1$ and $\rho_i = \frac{\lambda_i}{\mu_i}$.

Hence we can calculate the blocking probability Pb_i for the class C_i (i = 1,...,X):

$$Pb_i = \sum_{j=0}^{i-1} q(X-j) \qquad (3)$$

Remark: The blocking probabilities Pb_i which are the performance measures for the circuit mode are easily obtained by computing the probabilities q(i) which are defined by (2) — a one-dimensional recursion.

4.2. Obtention of Q

The transition rate q_{ij} between the state where N-i slots are taken by the circuit mode and the state where N-j slots are taken (i.e. between the state where i slots are available for the packet mode and the state where j slots are available with i, j = N-X,...,N), can be easily computed since we have the complete knowledge of the distribution P(n) available by (1).

Part of this matrix is easily obtained since the q_{ij} (i > j) is the rate of arrival of a customer of class C_{i-j} which is assumed to be λ_{i-j}.

The q_{ij} with i < j, is the departure rate $d_{C_{j-i}}(N-i)$ of a customer of class C_{j-i} when there are N-i slots taken by the circuit mode. To obtain this, we need the probability $L_l^k(m)$ of having k customers of class C_l when m slots are taken by the circuit mode. Since:

$$d_{C_{j-i}}(N-i) = \sum_{v=1}^{\lfloor \frac{N-i}{j-i} \rfloor} L_{j-i}^v(N-i) \times v \times \mu_{j-i} = q_{ij} \quad for \quad i < j \qquad (4)$$

But by (1), the probability $L_l^k(m)$ can be easily computed for $k \times l \leq m$ since:

$$L_l^k(m) = \frac{\sum_{\Sigma n_q = m} P(n_1, n_2, ..., n_l = k, n_{l+1}, ..., n_X)}{\sum_{\Sigma n_q = m} P(n_1, n_2, ..., n_l, ..., n_X)} \qquad m \leq X \qquad (5)$$

Thus we can find q_{ij} and noting that $q_{ii} = -\sum_{j \neq i} q_{ij}$ and that the indices i, j of q_{ij} vary from N-X to N, **Q** is given by:

$$\begin{matrix} N-X \\ \\ \\ \vdots \\ \\ \\ N \end{matrix} \begin{bmatrix} S_{N-X} & d_{C_1}(X) & \cdots & d_{C_{X-1}}(X) & d_{C_X}(X) \\ \lambda_1 & S_{N-X+1} & \cdots & d_{C_{X-2}}(X-1) & d_{C_{X-1}}(X-1) \\ \lambda_2 & \lambda_1 & \cdots & d_{C_{X-3}}(X-2) & d_{C_{X-2}}(X-2) \\ \vdots & \vdots & \vdots \ddots & \vdots & \vdots \\ \lambda_{X-2} & \lambda_{X-3} & \cdots & 2\mu_1 & \mu_2 \\ \lambda_{X-1} & \lambda_{X-2} & \cdots & S_{N-1} & \mu_1 \\ \lambda_X & \lambda_{X-1} & \cdots & \lambda_1 & S_N \end{bmatrix} \qquad (6)$$

where to simplify the notation within the matrix we denote: $S_i = -\sum_{\substack{j=N-X \\ j \neq i}}^{N} q_{ij}$.

A simple check shows that **VQ=0**, where **V** is the stationary probability vector with i=N-X,...,N and $V(i) = q(N-i)$.

4.3. General analysis of the packet mode (infinite buffer case)

We are going to study in this section the infinite buffer case. The finite buffer case yields to the same kind of analysis. The state description for the packet mode is (k,i) where k represents the number of packets in the buffer or in service and i represents the number of slots (or servers) available for the packet mode. Let P(k,i,t) = prob[k packets in system and i slots available for the packet mode at time t]

Under the assumption that multiple events are not possible, we obtain:

$$P(k,i,t+\Delta t) = \\ \lambda_p \Delta t P(k-1,i,t) + \min(k+1,i)\mu_p \Delta t P(k+1,i,t) \\ + \sum_{\substack{j=N-X \\ j \neq i}}^{N} q_{ji} \Delta t P(k,j,t) \\ + P(k,i,t)\left(1 - \left(\lambda_p \Delta t + \mu_p \Delta t \min(k,i) + \sum_{\substack{j=N-X \\ j \neq i}}^{N} q_{ij}\Delta t\right)\right) \quad (7)$$

where q_{ij} is the transition rate to go from i slots available for the packet mode to j slots, given by the matrix **Q** in (6) with i, j = N-X,...,N.

At equilibrium, the stationary probabilities satisfy the following set of equations for k>0:

$$P(k,i)(\lambda_p + \min(k,i)\mu_p) = \\ \lambda_p P(k-1,i) + \min(k+1,i)\mu_p P(k+1,i) + \sum_{j=N-X}^{N} q_{ij}P(k,j) \quad (8)$$

Noting that $\sum_{k=0}^{\infty} P(k,i) = q(N-i) = V(i)$ we have a complete set of balance equations. This set of equations can be solved numerically but in practice it quickly becomes unwieldy. Hence some simplified approximate methods are needed. This is done in the next section taking into account the particular structure of the model.

5. DECOMPOSITION ANALYSIS

5.1. Introduction

In this section we present an approach for obtaining an approximate solution for the system of equations above (8). This is based on the temporal decomposition introduced in [1] where we take advantage of "slow time variations", i.e. differing time scales to obtain an approximation. In [1] the method was developed for M/G/1 queueing systems, and it was indicated on how to extend it to product-form type networks. This approach is similar to the quasi-decomposability approach of Courtois [8]. The different time scales can be viewed as "phase changes" which are slow with respect to the principal model parameters. The decomposition is based on the fact that the rate of departure from any circuit state i (meaning i slots taken by the circuit), called $d_c(i)$ is very small compared to the rate of departure from any packet state j, called $d_p(j)$; where $d_c(i) = -q_{ii}$ and $d_p(j) = \lambda_p + \mu_p$. Let us denote ϵ the relative scale of time-variation,

$$\epsilon = \frac{\max_i [-q_{ii}]}{\lambda_p + \mu_p} \qquad (9)$$

In the following, we suppose that ϵ is in a neighborhood of zero, i.e. that $\epsilon \leq 10^{-3}$. This assumption is reasonable in the context of a hybrid system where there are at least three orders of magnitude difference between the time scale involved in the synchronous mode and the asynchronous mode.

5.2. Methodology

Define:
$$A = \frac{Q}{\epsilon} \quad i.e. \quad a_{ij} = \frac{q_{ij}}{\epsilon} \qquad (10)$$

Then the system of equations (8) can be rewritten in the matrix form:

$$\mathcal{P}[\mathcal{B} + \epsilon \mathcal{A}] = 0 \qquad (11)$$

together with the appropriate boundary conditions.

Hence solving these equations in terms of ϵ gives:

$$p(k, i) = f_{k,i}(\epsilon) \qquad (12)$$

which is analytic in ϵ for small ϵ (as long as $f_{k,i}(0)$ exists). Expanding this analytic function in a Mac-Laurin expansion for ϵ small gives:

$$P(k, i) = \sum_{d=0}^{\infty} P_d(k, i) \epsilon^d \qquad (13)$$

where $P_d(k, i)$ is called the dth order of $P(k, i)$.

5.3. Approximate solution for the packet mode

Using (10) and (13) in equation (8), we find:

$$(\lambda_p + \min(k, i)\mu_p) \sum_{d=0}^{\infty} P_d(k, i)\epsilon^d =$$

$$\lambda_p \sum_{d=0}^{\infty} P_d(k-1, i)\epsilon^d + \min(k+1, i)\mu_p \sum_{d=0}^{\infty} P_d(k+1, i)\epsilon^d + \qquad (14)$$

$$\sum_{j=N-X}^{N} \epsilon a_{ij} \sum_{d=0}^{\infty} P_d(k, j)\epsilon^d$$

Since this should be true for any small ϵ, we have for d=0:
$$(\lambda_p + \min(k,i)\mu_p)P_0(k,i) = \\ \lambda_p P_0(k-1,i) + \min(k+1,i)\mu_p P_0(k+1,i) \quad (15)$$

and for d > 0:
$$(\lambda_p + \min(k,i)\mu_p)P_d(k,i) = \\ \lambda_p P_d(k-1,i) + \min(k+1,i)\mu_p P_d(k+1,i) + \sum_{j=N-X}^{N} a_{ij} P_{d-1}(k,j) \quad (16)$$

Noting that:
$$\sum_{k=0}^{\infty}\sum_{d=0}^{\infty} P_d(k,i)\epsilon^d = V(i) \quad \forall \epsilon \text{ small} \quad (17)$$

we obtain:
$$\sum_{k=0}^{\infty} P_0(k,i) = V(i) \quad (18)$$

and:
$$\sum_{k=0}^{\infty} P_d(k,i) = 0 \quad for \; d > 0 \quad (19)$$

(15) just represents the equations of an M/M/i queue, which indicates that the 0th order approximation's average queue length, N_0 behaves as the weighted sum of the individual queue lengths of the system in each of the governing states (weighted by V(i)):

$$N_0 = \sum_i V_i N_{M/M/i} \stackrel{\Delta}{=} \sum_i V_i N_0(i) \quad (20)$$

Remark: since we have only presented the results for the infinite buffer case here some restrictions apply for the existence of the approximate solution. The zeroth order approximation will exist iff all the decoupled M/M/i queues are stable (i.e. $f_{k,i}(0)$ exists) implying that:

$$\forall i = N-X, ..., N \text{ s.t. } V_i \neq 0, \quad \frac{\lambda_p}{i\mu_p} < 1 \quad (21)$$

whereas it can be easily shown that the model considered above is stable iff:

$$\frac{\lambda_p}{\mu_p} < \sum_{i=N-X}^{N} iV_i \stackrel{\Delta}{=} N_c \quad (22)$$

with N_c being the mean number of slots available for the packet mode. The condition (22) is a much less restrictive condition than (21). However this restriction does not apply to the finite buffer case.

To obtain higher order approximations some careful algebra is required which can be simplified by the use of generating functions. In particular, let the first order approximation for the average queue length be N_1:

$$N_1 = N_0 + \epsilon \sum_i \sum_k k P_1(k,i) = N_0 + \epsilon H_1 \qquad (23)$$

Let $\mathcal{P}_0(z,i)$ denote the generating function corresponding to $P_0(k,i)$ and let $\mathcal{P}_0''(1,i)$ denote the second derivative of this generating function for z=1, then after much algebra we obtain:

$$H_1 = \frac{1}{(\lambda_p - i\mu_p)} \left[\mu_p \sum_{k=0}^{i-1} k(k-i) P_1(k,i) - \sum_{j=N-X}^{N} a_{ij} \left(\frac{1}{2} \mathcal{P}_0''(1,j) + N_0(j) \right) \right] \qquad (24)$$

Let $\rho \triangleq \frac{\lambda_p}{\mu_p}$ and to simplify the equation let:

$$T(k,i) = \sum_{j=N-X}^{N} a_{ij} P_0(k,j) \qquad (25)$$

then for k<i:

$$P_1(k,i) = \frac{\rho^k}{k!} P_1(0,i) - \frac{1}{k! \mu_p} \left[\sum_{l=1}^{k} \left(T(k-l,i) \sum_{m=1}^{l} (k-m)! \rho^{m-1} \right) \right] \qquad (26)$$

with:

$$P_1(0,i) = \frac{\sum_{j=N-X}^{N} a_{ij} N_0(j) + \sum_{k=1}^{i-1} \frac{(k-i)}{k!} \left[\sum_{l=1}^{k} \left(T(k-l,i) \sum_{m=1}^{l} (k-m)! \rho^{m-1} \right) \right]}{\mu_p \sum_{k=0}^{i-1} (k-i) \frac{\rho^k}{k!}} \qquad (27)$$

6. PERFORMANCE RESULTS AND CONCLUSIONS

Figures 1–4 show N_s, the mean number of packets in the system obtained by simulation, N_0 and N_1 which are respectively the zeroth order and first order approximation of this mean number obtained by (20) and (23) and, N_m the mean number of packets obtained by considering the very natural approximate model consisting of an $M/M/N_c$ queue where N_c is the mean number of slots available

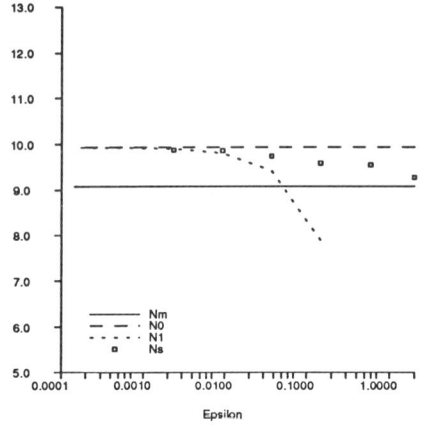

N=20, X=10, a=9.0, N_c=15.0, σ_c=9.1, σ_0=3.5

C_1: λ_1=1.000, μ_1=0.333
C_4: λ_4=0.033, μ_4=0.100
C_8: λ_8=0.005, μ_8=0.008

Figure 1

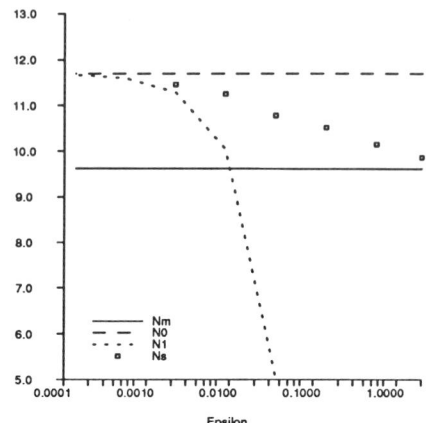

N=20, X=10, a=9.5, N_c=15.0, σ_c=9.1, σ_0=23.5

C_1: λ_1=1.000, μ_1=0.333
C_4: λ_4=0.033, μ_4=0.100
C_8: λ_8=0.005, μ_8=0.008

Figure 3

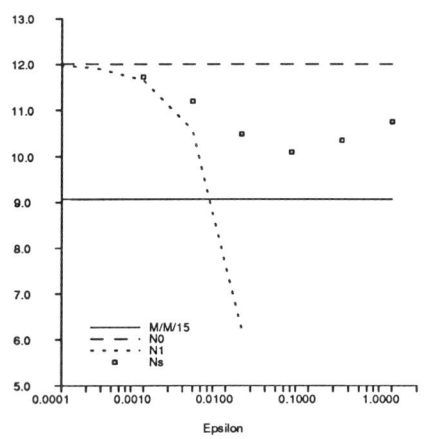

N=20, X=10, a=9.0, N_c=15.0, σ_c=25.0, σ_0=9.0

C_{10}: λ_1=1.000, μ_1=1.000

Figure 2

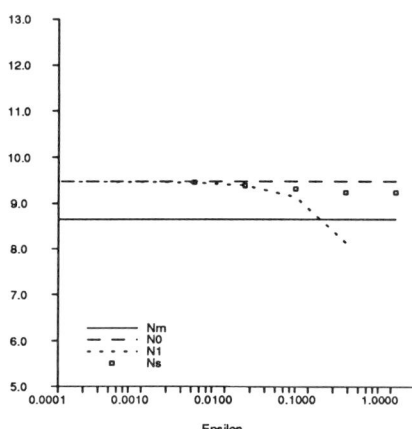

N=20, X=10, a=8.0, N_c=11.0, σ_c=9.0, σ_0=0.2

C_{10}: λ_1=1.000, μ_1=0.111

Figure 4

for the packet mode, all plotted versus ϵ. In these four cases, N the total number of slots was equal to 20 and X the maximum number of slots taken simultaneously by the circuit mode taken to be 10. The other parameters were chosen in such a way that, for Figures 1–3, N_c the mean number of slots available for the packet mode was 15 and was 11 for Figure 4. In the four figures ϵ is plotted on a logarithmic scale. In Figures 1 and 3, three types of circuit classes were assumed to coexist namely C_1, C_4, C_8 whereas in Figures 2 and 4 only one circuit class C_{10} was assumed to be present. These different choices of mixture of class C_i have an impact on the variance σ_c of the number of slots available for the packet mode. In Figures 1 and 3, $\sigma_c = 9.1$, in Figure 4 $\sigma_c = 9$ whereas in Figure 2 $\sigma_c = 25$. The parameter σ_c seems to be an important factor in the limits of validity of our approximations. Another parameter which seems to be an important factor is the variance of the mean number of packets per phase, i.e.:

$$\sigma_0 = \sum_i V_i \left(N_{M/M/i}\right)^2 - (N_0)^2 \qquad (28)$$

Finally, the parameter $a = \frac{\lambda_p}{\mu_p}$ (i.e. $\rho_p = \frac{\lambda_p}{\mu_p N_c}$) was chosen to be 9.0 (i.e. $\rho_p = 0.6$) for Figures 1 and 2, to be 9.5 (i.e. $\rho_p = 0.6333$) for Figure 3 and to be 8.0 (i.e. $\rho_p = \frac{8}{11} = 0.7272$) for Figure 4.

The first remark to be made is that the role of the first order approximation is not really to give a more accurate result than the zeroth order approximation but rather to indicate the limits of validity of this zeroth order approximation (i.e. when the first order starts to differ from the zeroth order, the validity of our approximations becomes questionable). Moreover if N$_0$ and N$_m$ are very close then any of these values will be a very good approximation for N$_s$ even for ϵ big (i.e. in the vicinity of 1). In this case there is no gain in computing N$_1$.

The limit of validity of our zeroth order approximation is very variable, it can go in the worst case to $\epsilon < 10^{-3}$ up to in the best case ϵ close to 1.

In short, for the zeroth order approximation to be good, a certain number of conditions should be met, namely:

- ρ_p not too big with the added condition that a not too close to N-X, this can be seen by comparing Figure 3 ($\rho_p = .6333$ and $a = 9.5 \simeq N - X = 10$) and Figure 4 ($\rho_p = .7272$ and $a = 8 < 10 = N - X$),
- σ_c should not be too big, this can be seen by comparing Figures 1 and 2 for which a and N$_c$ are the same and Figure 1 shows much better results than Figure 2 because $\sigma_c = 9.1$ (resp. 25) for Figure 1 (resp. Figure 2). In particular, one important observation is that the introduction of circuit classes C_i for large i, will tend to increase σ_c which has an adverse impact on the limit of validity of our approximations,
- σ_0 seems to be a good global measure of the validity of our approximation. This last remark has lead us to consider simple heuristic methods to evaluate the mean number of packets in queue (these methods will be presented in a forthcoming paper).

In conclusion, we have developed a simple approximation (the zeroth order one) for the packet mode together with a mean of measuring its validity. We have been able to point out which were the important factors for the validity of our approximations. Our objective was to find out a good approximation for the performances of the packet mode when $\epsilon \leq 10^{-3}$. This objective had to be revised since we have exhibited some configurations for which the initial objective was not met due to parameters which did not follow the criteria defined above.

1. Gelenbe, E. and Rosenberg, C.; Queues with Slowly Varying Arrival and Service Process, Management Science, Vol. 36, #8, August 1990.
2. Ghani, S. and Schwartz, M.; A Decomposition Approximation for the Analysis of Voice/Data Integration, Proceedings of the Indo-US Workshop on Systems and Signal processing, Bangalore, India Jan. 1988.
3. De Serres, Y. and Mason, L.; A Multiserver Queue with Narrowband and Wideband Restricted Access, IEEE Trans. on Communications, Vol. 36, No. 6, June 1988.
4. Liao, K. Q. and Mason, L.; An Approximate Performance Model for a Multi-slot Integrated Services System, IEEE Trans. Communications, Vol. 37, No. 3, March 1989.
5. Gimpelson, L. A.; Analysis of Mixtures of Wide and Narrow-band Traffic, IEEE Trans. on Comm. Technology, Vol 13, No. 3, Sept. 1965.
6. Kaufman, J.; Blocking in a Shared Resource Environment, IEEE Trans. on Comm., Vol 29., No. 10, October 1981.
7. Coderch, M., Willsky, A. S., Sastry, S. S. and Castanon, D. A.; Hierarchical Aggregation of Singularly Perturbed Finite State Markov Processes, Stochastics, Vol. 8, 1983.
8. Courtois, P. J.; Decomposibility: Queueing and Computer Applications, Academic Press, N.Y. 1977.

Buffer sizing in Bulk Service Systems

Gérard Hébuterne (*), Catherine Rosenberg (**)

(*) Centre National d'Etudes des Télécommunications, Lannion-A/SLC. Route de Trégastel, 22301 Lannion FRANCE.

(**) Ecole Polytechnique de Montréal, Case Postale 6079, succ "A", Montréal, Québec, H3C 3A7 CANADA.

Abstract. The paper addresses the problem of buffer sizing for queues with bulk service. The model arises in the context of the study of a hybrid frame where synchronous cells are first served and ATM cells are given access to the remaining slots. We derive loss probabilities by means of simple conservation arguments from the state probabilities at departure epochs obtained through classical Markov chain analysis.

1 - INTRODUCTION

In this paper we present two models which arise in the study of broadband networks although they are of larger interest. The context in which they arise is in the study of a hybrid frame, such as SONET, within which synchronous connections are inter-mixed with variable-rate ATM connections. While the Grade of Service of a synchronous connection is determined at call set-up, successive ATM cells of a given call may encounter varying delays and blocking situations at each frame building stage. This requires buffer dimensioning for which appropriate models have to be constructed.

At the cell level, the effect of the synchronous connections on the ATM cells (at for example the outgoing multiplexer of a switch) is modelled as follows: at periodically spaced epochs frames are generated. Some of the slots are dedicated to synchronous connections (they are called "synchronous slots"). The assignment holds at least for the call duration and can be considered at the cell level as semi-permanent. The remaining slots ("ATM slots") are shared among the ATM cells. Thus from an ATM flow point of view, the output multiplexer is a server with periodic vacations (being unavailable when synchronous cells are sent). The distribution of the vacation period depends on the frame management, the holding time of synchronous connections, among others, and may be fairly complex. Without synchronous connection present, a M/GI/1/N model would suffice to yield loss probabilities for ATM cells with Poisson arrivals, see eg. [1], [2].

Since the model with vacations are difficult to analyse, another approximate model has been proposed for the ATM system. Instead of considering the detailed scheme of the frame filling, i.e. the exact occurrence of ATM slots within it, we assume that each of the frames is constituted at the beginning of the frame; the number of idle slots (not used by synchronous connections) is denoted as N, i.e. up to N cells are read from the buffer in a single operation. This is equivalent to a finite capacity bulk service with cyclic visits.

While the above approximate model must be validated, in this paper we present the analysis of the finite bulk service system. A similar model but with infinite capacity has been studied in [3] in the context of real-time scheduling, with a possible application to TDMA channels.

In the context of such queueing models a common difficulty is that explicit computation of state probabilities can be done at certain epochs (departure) while Grade of Service parameter of interest might be linked with probabilities at other instants (arrival epochs). Hence there is a need to relate the two probabilities: we deal with this problem.

This is usually done by relating the departure epochs with the entry epochs and then with arrival epochs. The classical example is the M/G/1/K and the related Geo/G/1/K, see eg. [1] [4], where PASTA and conservation are used. The analysis presented here follows this general approach. The measure of interest is the loss probability and we obtain this via a conservation argument, which leads to substential simplification over the semi-regenerative approach [5].

The hybrid problem considered here leads to a cyclic server. Since the same kind of argument used here works as well for the traditional bulk server, we begin with the classical $M/GI^{[B]}/1/K$ system (Section 2). Section 3 is devoted to the cyclic server version.

2 - THE FINITE-QUEUE BULK-SERVICE SYSTEM

The first model considered is the usual bulk service queue. The infinite-size version of the problem is well known, see e.g. ref. [6]. Customers arrive according to a Poisson process with rate λ. The server serves customers in batches of size at most N. The service durations are i.i.d. random variables, independent of the bulk size. The service time distribution is arbitrary, let $E(S)$ denote its mean value. Let α_j be the number of customers arriving during the j's service time (whether the arrivals are accepted or not). Let $\rho = \lambda E(S)$.

Let X_j the number of customers left behind by the j's bulk departure, i.e. the queue size just before the j's departure. $\{X_j, j \geq 0\}$ is an irreducible finite-state Markov chain - this is obvious from relations (1) or (2) hereafter. Let P_n be the corresponding steady-state probability: $P_n = \mathrm{Lim}_{j \to \infty} P\{X_j = n\}$. Note that the P_n's are defined at <u>departure epochs</u>.

2.1 The recurrence equations

The recurrence equation governing the evolution of X_n depends on the way the limitation is performed.

1. the limit may be on the total number of customers in the system (waiting room and server); we denote K the maximum number. In this case, one sees easily that for $X_n \in [0, K-1]$:

$$X_{n+1} = \text{Min}\{\alpha_{n+1}, K-1\} \qquad \text{if } X_n = 0$$
$$= \text{Min}\{\alpha_{n+1} + X_n, K\} - \text{min}\{X_n, N\} \qquad \text{if } X_n \neq 0 \qquad (1)$$

2. the limitation is on the waiting room only - to avoid confusion, we denote its capacity as $\overline{K}$. In this case, for $X_n \in [0, \overline{K}]$:

$$X_{n+1} = \text{Min}\{\alpha_{n+1} + \text{Max}\{X_n - N, 0\}, \overline{K}\} \qquad (2)$$

In the usual M/G/1/K queue the two cases are equivalent provided one takes $\overline{K} = K-1$. Here this is no longer true, due to the possibility of variable server occupancies.

From the appropriate set of equations, the P_n's are numerically derived by a classical Markov chain analysis. The mean bulk size E(bulk) is obtained from the P_n:

- if $X_n = 0$, the next service begins with the first arrival, and the next bulk is of size 1;
- if $X_n \leq N$, the next bulk of size X_n empties the queue;
- if $X_n \geq N$, the next bulk is of size N.

So that:

$$E(\text{bulk}) = P_0 + \sum_{n=1}^{N} n.P_n + \sum_{n>N} N.P_n = N + P_0 - \sum_{n>N} (N-n)P_n \qquad (3)$$

Also, the mean time between departures $E(D)$ will be needed; if the n-th departure leaves a non-empty queue behind it, the $n+1$-th interval between departures is a service time. On the other hand if the queue is empty, an additional delay is incurred, corresponding to the first customer to enter the system and initiate the next service. Finally, the mean time between successive departures is:

$$E(D) = E(S)(1 - P_0) + [\frac{1}{\lambda} + E(S)]P_0 = \frac{P_0}{\lambda} + E(S) \qquad (4)$$

2.2 The Loss Probability

Let us observe the system for a time interval T, during which we register the arrivals, enterings and departures. During the interval T, we observe $b(T)$ end-of-service, corresponding to $d(T)$ departing customers.

The departure rate is simply $d(T)/T$. Now, as T increases,

$$\text{Lim}_{T \to \infty} \frac{d(T)}{T} = \text{Lim} \frac{d(T)}{b(T)} \times \text{Lim} \frac{b(T)}{T}$$

This relation holds since each of the right hand side terms has a finite limit. The first one is the "mean" bulk size, while the second one represents the inverse of the distance between bulk departures. They are given by (3) and (4), respectively. Finally,

$$\mathrm{Lim}_{T \to \infty} \frac{d(T)}{T} = \frac{E(\mathrm{bulk})}{E(S) + P_0/\lambda}$$

For the system to have a stationary distribution, the departure rate must equal the entrance rate. This last is simply $\lambda(1 - P_{loss})$. Finally

$$P_{loss} = 1 - \frac{E(\mathrm{bulk})}{\rho + P_0} = 1 - \frac{1}{\rho + P_0}\left\{N + P_0 - \sum_{n=0}^{N}(N-n)P_n\right\} \quad (5)$$

Note that the relation, which is a kind of conservation property, does not depend explicitly on the buffer size K, not even on the way the buffer size is limited. As a matter of fact, relation (5) holds whether the system obeys relations (1) or (2). Such conservation relations work because the initial modelling at departure epochs carries all the information about the whole system.

The relation (5) may be compared with relation (2.11) of reference [7] which analyses a system with bulk arrivals.

3 - THE CYCLIC QUEUE WITH BULK SERVICE

The second model is a slight modification of the bulk server. This modification is necessary to model more precisely the periodic nature of the ATM system based on a frame structure described in Section 1. In particular if there is no ATM cells in the buffer, the frame "starts" anyway implying a bulk service size zero. Hence in this second model the server inspects cyclically the queue, according to an arbitrary renewal process. The intervals of time elapsed between successive visits to the queue are i.i.d. random variables, independent of the queue occupancy; a cycle begins immediately after the end of the previous one even if the queue is empty. Let $E(S)$ be the mean inter-visit time, and α_j the number of customers arriving (accepted or not) during the j-th tour.

At each visit, the server processes up to N customers, which are served during the tour.

Let X_j be the number of customers in the queue just before the n-th visit. Here too, it is a finite-state Markov chain, and the stationary probability distribution is denoted as $P_n = \mathrm{Lim}_{j \to \infty} P\{X_j = n\}$.

The infinite version of this model has been examined in [2], and was shown to be partly equivalent with the previous $M/GI^{[B]}/1/K$ model.

3.1 The recurrence equations

As above, one has to distinguish carefully, according to the mechanism performing the limitation.

1. Limit on the total number of customers in the system (waiting room and server); we denote K the maximum number. Here, contrarily to the actual bulk service case, for $X_n \in [0,K]$:

$$X_{n+1} = \text{Min}\{\alpha_{n+1} + X_n, K\} - \text{min}\{X_n, N\} \tag{6}$$

No special equation is needed for $X_n = 0$; this is because an "empty" bulk is produced in this case.

2. Limitation on the waiting room only - to avoid confusion, we denote its capacity as $\overline{K}$. In this case, for $X_n \in [0, \overline{K}]$:

$$X_{n+1} = \text{Min}\{\alpha_{n+1} + \text{Max}\{X_n - N, 0\}, \overline{K}\} \tag{7}$$

In the case of a limited waiting room, the recurrence (7) is the same as (2). This is no longer the case if the limit is on the whole system.

Here again, once the P_n's are estimated, the mean bulk size is obtained from the P_n:

- if $X_n = 0$, the next tour is with an empty bulk;
- if $X_n \leq N$, the next bulk is of size X_n, while if $X_n \geq N$, it is of size N.

So that:

$$E(\text{bulk}) = \sum_{n=1}^{N} n.P_n + \sum_{n>N} N.P_n = N - \sum_{n=0}^{N} (N-n)P_n \tag{8}$$

3.2 Loss Probability

The argument is the same as in Section 2.2, with the simplification that the average cycle duration is $E(S)$, independently of the queue length. The entrance rate is $\lambda(1 - P_{loss})$, and must equal the departure rate $E(\text{bulk})/E(S)$. Thus,

$$P_{loss} = 1 - \frac{E(\text{bulk})}{\lambda E(S)} = 1 - \frac{1}{\lambda E(S)} \left\{ N - \sum_{n=0}^{N} (N-n)P_n \right\} \tag{9}$$

4 - CONCLUSION

Two variants of bulk service systems with finite capacity have been studied. The first one is the traditional $M/GI^{[B]}/1/K$ queue, the second one is a periodic server with bulk services.

Each of the finite capacity systems has been analysed using the embedded Markov chain at departure epochs. The rejection probability was obtained from the probabilities at departure epochs through a conservation argument and without any additional analysis.

Returning to the ATM problem described in the introduction, it is clearly the second model which is relevant. The inter-visit duration is deterministic and equal to the frame duration. N is the number of slots assigned to the ATM part of the frame. As was said above, it remains however to validate the model, using the results presented in this paper.

Note that for practical use, the set of equations giving the P_n's (recurrences (6) or (7), e.g.) must be handled with care: in ATM applications, the targeted loss is in the range of $10^{-6} - 10^{-10}$, requiring a high degree of precision since it is given by the difference of two terms in the range of the unity.

References :

[1] Louvion, J.R., Boyer, P., Gravey, A., : *A Discrete-time single server queue with Bernoulli arrivals and constant service time.* 12th ITC, Torino. North-Holland (1988), pp 1304-1312.

[2] Gravey, A., Hébuterne, G., Boyer, P., Louvion, J.P., : *Buffer Sizing in an ATM Switch for both ATM and STM Traffics.* International Journal of Digital & Analog Cabled Systems, Volume 2, pp 247-252, 1989.

[3] Hébuterne, G., : *A gate with periodic openings and bulk service.* 12th ITC, Torino. North-Holland (1988), pp 1501-1507.

[4] Reiser, M., Kobayashi, H., : *The effects of service time on system performance.* IFIP'74, North Holland (1974).

[5] Chaudhry, M.L., Templeton, J.G.C., : *The queueing system $M/GI^{[B]}/1$ and its ramifications.* European Journal of Operational Research, vol 6 pp 56-60 (1981).

[6] Chaudhry, M.L., Templeton, J.G.C., : *A First Course in Bulk Queues.* John Wiley, 1983.

[7] Miyazawa, M., : *Complementary generating functions for the $M^X/GI/1/K$ and $GI/M^Y/1/K$ queues and their application to the comparison of loss probabilities.* Journal of Applied Probabilities, vol 27 pp 684-692 (1990).

Performance analysis of the $H_2/G/1$ system

Ilya Ehriel

Leningrad Department of Research Institute
of Communication, Leningrad, USSR

Abstract
    New user-network interface which is recommended by CCITT has definite peculiarities. The main attention should be turn upon model selection for the signalling channel .The suggested model has a hyperexponential distribution of the incoming demand flow and general distribution of the service time . By the analysing of this queuing system an approach for the mean waiting time is offered.

    An increase of the subscriber's interest to the ISDN technique needs a creation of its operation model The model to be design must correctly describe ISDN features . It will be necessary to obtain time-probability characteristics for the such model .
    One of the main ISDN features is a user-network interface with separate channels for the transfer of user and signalling information (B and D channels accordingly) . A model design for the ISDN signalling channel needs analysis of applied protocols .
    Fig.1 presents the reference configuration for ISDN basic access (2B + D) , where point S is the interface for a number of terminals that are connected to the passive bus . It is possible to connect up to 8 TE to the single passive bus , which separate a resource of the

single physical D channel.D channel can be used both for signalling information and low-speed packet user information transfer , division of which is made with the usage of priorities . Using variable length frames an exchange of signalling messages is made in frame-oriented manner .

In order to support mentioned capabilities new (for the traditional telephony) protocol elements have been applied . Time-probability characteristics of these protocols have not been studied enough well. An access function to the signalling channel and signalling functions are divided on layers in accordance with 7-layer Open System Interaction model .

The D channel resource division between TE's on the layer 1 is made with CSMA/CD protocol (carrier sense multiply access with collision detection) . A multilink operation on the layer 2 for each TE is provided by LAP D protocol (link access protocol on the D channel) , which is X.25 CCITT protocol modification . Layer 3 messages are conformed in to layer 2 frames and transfered through physical D channel on the layer 1 .

Thus , establishment procedures of each connection represent the process of a number of layer 2 frames transfer through D channel .

Consider a frame generation process for the single TE . The possible realization of this process in simplified form is shown on Fig.2 , where

$1/\lambda_1$ mean time between message arrivals from the layer 3 onto the layer 2 ,

$1/\lambda_2$ mean time between frame arrivals from the layer 2 onto the layer 1 .

Even simplified analysis of signalling protocols do not allow to take Poisson process as an arrival process for the D channel. Erlang approximation is not also suitable, since such burst arrival process as a general rule has squared coefficient of variation more then 1. The service time in the D channel depends on frame length and can not be also described with Poisson process.

Having taken the D channel as a server with the layer 2 frame queue, which is generated (in common case) by a number of TE independently each from other, we could represent such model as a ⟨GI/G/1⟩ queuing system in Kendall classification.

Obvious results for the waiting time in the ⟨GI/G/1⟩ system is not managed to obtain [5]. Should make some parameters (for example arrival process parameters) precise, one can obtain an approach for the mean waiting time in the such system.

In [1] for the concretization of the arrival process a load function is proposed. This function is marked as $c_z^2(\rho)$ and characterizes the influence of the arrival process A(t) and load coefficient ρ

$$c_z^2(\rho) = (1-\rho)(c_s^2+1) + 2(1-\rho)w/m_b \quad , \tag{1}$$

where m_b - first moment of the servise time,

c_s^2 - squared coefficient of the service time variation,

w - mean waiting time.

The $c_z^2(\rho)$ function is also defined as

$$c_z^2(\rho) = c_s^2 + I_c[f(\rho)] \quad , \tag{2}$$

where $I_c[f(\rho)]$ - interval index for counts (IIC) $t \geq 0$,

$f(\rho) = \rho/(1-\rho)$ - empirical function reflecting argument transfer from the $t \geq 0$ region to the $0 \leq \rho \leq 1$ region.

In its turn

$$I_c[f(\rho)] = I_c[\rho t] = D[A(\rho t)]/E[A(\rho t)] \quad,$$

where E - mean and
       D - dispersion of the arrival process.

After that, the mean waiting time as a function from ρ is

$$\frac{w(\rho)}{m_b} = \frac{c_z^2(\rho) - (1-\rho)(c_s^2 + 1)}{2(1-\rho)} \quad. \qquad (3)$$

Shown on the Fig.2 process has been suggested to approximate with the hyperexponential distribution of the interarrival time [2]. Apply the result (2) to the $\langle H_2/G/1 \rangle$ system. As shown in [3] for the $A(t) = H_2$

$$I_{c\rho}(t) = c_a^2 - \frac{2\beta}{\gamma t}[1 - \exp(-\gamma t)] \quad, \qquad (4)$$

where c_a^2 - squared coefficient of interarrival time variation,

$$\gamma = (1-p)\lambda_1 + p\lambda_2$$

$$\beta = p(1-p)(\lambda_1-\lambda_2)^2/\gamma^2$$

Then changing t in (4) on $f(\rho)$ for the $\langle H_2/G/1 \rangle$ system

$$c_z^2(\rho) = c_s^2 + c_a^2 - \frac{2\beta}{\gamma f(\rho)} [1-\exp(-\gamma t)] = c_s^2 + c_a^2 - \Phi(\rho, p, \lambda_1, \lambda_2) \; .$$

Result (3) conforms as

$$\frac{w(\rho)}{m_b} = \frac{c_a^2 + \rho(c_s^2 + 1) - 1 - \Phi(\rho, p, \lambda_1, \lambda_2)}{2(1-\rho)} \; . \qquad (5)$$

The equality (5) defines the waiting time value in $\langle H_2/G/1 \rangle$ queuing system as a function of c_a^2, c_s^2, ρ, p, λ_1 and λ_2.

$\langle H_2/G/1 \rangle$ queuing system has been simulated ($G=H_2$) with GPSS/PC . Comparison of the simulated results and calculations with formula (5) is shown in Tabl.1. Results of calculation with Daley and Langenbach-Belz formulas for the simulated systems are also shown .As seen from the table, formula (5) gives better approach then Daley and Langegbach-Belz formulas for the c_a^2 and ρ regions which are interested us when we are dealt with ISDN basic acsess questions .

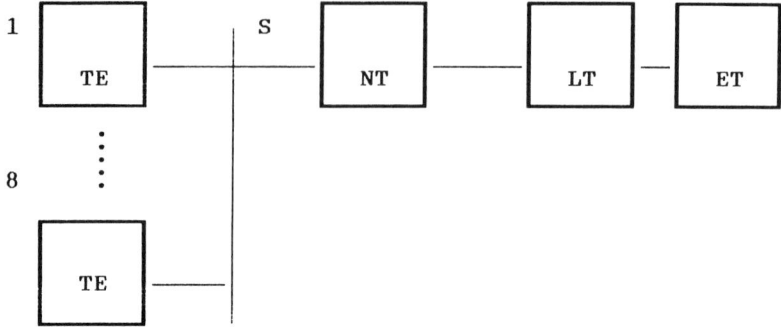

TE - Terminal Equipment
NT - Network Termination
LT - Line Termination
ET - Exchange Termination

Fig.1 User-network Interface Reference Configuration

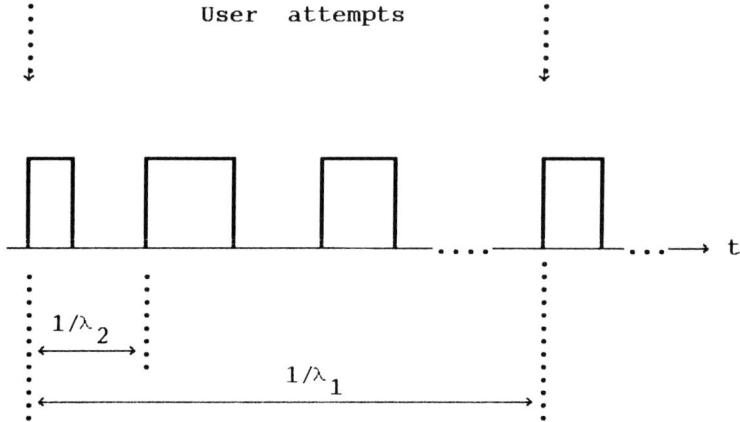

Fig.2 Frame Generation Process

Table 1

A				B	W			
					$c_s^2=1.5, 1/\mu_1=50, 1/\mu_2=150, p_1=0.5$			
c_a^2	ρ	$\frac{1}{\lambda_2}$	$\frac{1}{\lambda_1}$	P_1	sim	calculation		
						(5)	Lang. belz	Daley
1.17	0.68	100	200	0.5	281	260	283	430
1.21	0.33	200	400	0.5	69	59	64	229
1.82	0.46	75	250	0.5	150	129	93	240
3.47	0.68	125	400	0.3	670	552	388	956
17.5	0.24	20	4000	0.3	960	1120	176	2034

References.

1. K.W.Fendick,W.Whitt.Mesurements and Approximations to the Describe the Offered Traffic and Predict the averege Workload in Single-Server Queue .IEEE Proceedings v.77, N1,1989 .
2. H.-Dieter Suedhofen, Peter F. Pavlita . Modelling of compaund traffic streamsin computer communication networks . 11th ITC ,Keyoto , 1985 .
3. D.R. Cox .Reneval Theory .London ,1961.
4. CCITT , Blue Book .Recommendations I,Melbourne ,1988 .
5. L.Kleinrock . Queuing systems,London , 1965.

High Speed LAN and MAN

Telecommunication Services for Developing Economies
J. Filipiak (Editor)
© 1991 Elsevier Science Publishers B.V. All rights reserved

An Alternative Solution to the Electro-Optic and Service Bottleneck Problems in Multi Gbit/s LANs: the SUPERLAN Architecture

A. Popescu[a] and R. P. Singh[b]

[a]Department of Telecommunication and Computer Systems, The Royal Institute of Technology, P.O.Box 700 43, S - 100 44 Stockholm, Sweden

[b]Bell Communications Research, 331 Newman Springs Road, P.O.Box 7020, Red Bank, New Jersey 07701, USA

Abstract

This paper examines the two main bottleneck problems in the design of integrated multi Gbit/s local area networks (LAN), one from the network perspective - the electro-optic bottleneck problem - and other from the user application perspective - the service bottleneck problem. A novel architectural solution is proposed that makes use of the wavelength division multiplexing (WDM) principle. This architecture is based on a multi-class network model in which the total user traffic on the fiber is separated into two classes - isochronous and nonisochronous - each carried on a separate wavelength at multi Gbit/s rates. Each traffic class has a dedicated access control mechanism, and the control information is transported on a separate wavelength(s) with a lower rate(s). The principal features of this architecture are especially low processing rates that are commensurable with the node electronic speeds, simplicity of the media access control (MAC) protocols, and a minimal number of Gbit/s components at the physical layer (PHY).

1. INTRODUCTION

With the availability of the fiber optic technology, new network architectures are emerging to effectively utilize the abundant transmission capacity. The choice of WDM technique has further resulted in manifold increase in transmission capacity. However, the performance increase of the supporting nodal electronics needed for switching, buffering and control purposes has not matched this trend, so novel architectures must be devised in order to make use of the huge transmission capacity. This picture is further complicated by the fact that the future networks must provide support to a multitude of broadband traffic classes with conflicting service requirements. Performance and cost-effective LAN architectures need to employ efficient methods to share the system resources among the network stations in a manner that circumvents the mismatch between the transmission and processing speeds, and at the same time, provides adequate quality of service to all user applications. Generally, existing integrated LANs employ sophisticated mechanisms to control and coordinate access to the communication media, so that the service performance requirements of each user application are met. But because the needed processing speeds are limited, the network transmission resources are not efficiently utilized. This is the kind of approach in which the performance needs of the user applications are adapted to a given network environment. This means that the network architecture is designed first and then complex functions are incorporated in the MAC to satisfy the performance needs of the applications. Our approach to high speed LAN design is just the reverse, i.e., we adapt the underlying network transport mechanisms to the needs of the specific applications. Based on this consideration, we propose a new LAN model, in which we assign one wavelength to each application or traffic class, partitioned on the basis of some characteristics, such as loss or delay sensitivities, holding times, bit rates, etc. Moreover, this is a model in which each traffic class has a dedicated access mechanism, each used specifically to arbitrate access for resources on the separate wavelength carrying the traffic from the same application. Such an approach offers important advantages, such as low processing rates that are commensurable with the node electronic speeds, simplicity for the access protocols, and a minimal number of high speed components. In this context, we start with the simplest model. Based on the delay sensitivity, we divide the total user traffic on the network into two classes - isochronous (voice and video) and nonisochronous (data) - that are WDM integrated in a multi Gbit/s LAN at the low speed network interface.

In this paper we examine the two basic issues in the design of very high speed LANs - the electro-optic bottleneck and service bottleneck - and motivate our solutions. We propose and describe a novel LAN architecture, called SUPERLAN [1]. It is intended mainly to be used as an underlying transport environment in local areas to carry out further studies in the areas of high speed protocol design and analysis, as well as very high speed applications such as the development of interactive, distributed environments that provide 2D or 3D multi-media virtual worlds [2]. Further envisaged applications for SUPERLAN include the development of a future very high speed metropolitan area network (MAN) based on a combination of centralized and distributed control and, connected with this, the development of a future broadband switch fabric as well.

This paper is organized as follows. Section 2 presents some of the critical issues in the design of a multi Gbit/s network designated for broadband integrated services. In Section 3, we discuss the issue of electronic bottleneck and motivate our proposed solutions. In Section 4, we discuss the second bottleneck, i.e., the service bottleneck, and show how the SUPERLAN architecture obviates the need for complex mechanisms for resource allocation. Section 5 is concerned with a brief description of the SUPERLAN architecture, emphasizing its distinctive features and advantages. Reference [1] contains a more detailed design description of SUPERLAN. We conclude the paper in Section 6, where we outline our ongoing work.

2. BACKGROUND

A multi Gbit/s LAN must be designed to support a wide range of isochronous and nonisochronous applications with arbitrary bit rates, both narrowband and broadband. The increasing need for high bandwidth networking, under increasingly strict performance constraints, to provide integrated communication services ranging from the ubiquitous 64 kb/s voice up to multi Gbit/s video and graphics has posed fundamental challenges to LAN design. With the available fiber optic technology, the potentially usable bandwidth is enormous. The question, however, is what techniques should be employed for media access and how they should be implemented in order to divide the available raw bandwidth to support services with widely varying performance demands. Typically, sophisticated access procedures and data buffering must be executed very quickly to match the service requirements and transmission speeds on optical fiber. But the difficulty in designing and implementing electronic processors for handling media access protocols operating beyond Mbit/s rates and the lack of practical optical components with memory functions have resulted in a clear disparity between the rate at which data may be transmitted and the rate at which data may be processed and switched. Such implementation-related problems, created by the fundamental speed mismatch between the electronic and optical components of a very high speed LAN, have resulted in electro-optical interfaces being the real performance bottlenecks. So the design of new architectures and new access protocols targeted to remove the *electronic bottleneck* from network nodes represents the first issue to be resolved in designing multi Gbit/s networks.

The second bottleneck that prevents the high bandwidth of the optical fiber media from being available for hosts/users is the so-called *service bottleneck* between the MAC layer and the LLC layer (Fig. 1). The incoming traffic is highly heterogeneous in its characteristics and performance requirements, with different and contradictory bandwidth demands, holding times and call arrival rates. Moreover, each traffic will differ in its admissible access and transit delays, access throughput, etc. This traffic can be divided into two general classes, isochronous and nonisochronous, with different performance demands. As the term indicates, for the first class information is generated either in a steady time-synchronous mode in the case of synchronous traffic, or in a nearly synchronous mode in the case of isochronous traffic. This traffic, characterized by long holding times and modest setup times, may accept high access delay but requires low transit delay. Also, it can tolerate occasional information loss. In supporting synchronous or isochronous services, the network must be able to provide real or virtual connections with guaranteed performance/quality of service as negotiated at the call setup. The negotiated performance can be of various types, as for instance, an upper bound on call blocking probability, a lower bound on throughput, an upper bound on transit delay, or an upper bound on loss rate.

On the other hand, nonisochronous sources generate bursty information of random lengths at random times. The nonisochronous traffic may tolerate high transit delays, but generally no losses. This means that in this case the flow and congestion control mechanisms operate at the packet or burst level, and not at the call level as in the previous case. Moreover, since holding times of packets or bursts (the usual mode in which nonisochronous information is generated and transported) is small, packet or burst connections must be set up and torn down far more rapidly than in the case of isochronous traffic. Although packets or bursts can be buffered in the stations, awaiting access to the transmission medium, the access delay should be kept small to minimize the risk of buffer overflow.

In order to cope with a diverse mix of broadband services of hundreds Mbit/s rates and with conflicting performance requirements, as mentioned above, simply making use of the existing architectures and media interfaces do not provide any benefit. In fact, the integration of broadband services on the existing architectures adds more complexity to the already complex MACs. The added functionality further exacerbate the electro-optic bottleneck problem, resulting in even more inefficient resource utilization. Thus, there is a definite need to closely examine these two bottleneck issues and to explore new ways of dealing with them. These include a proper architecture, as well as a proper design of switching and transmission schemes, protocol mechanisms and network management policies.

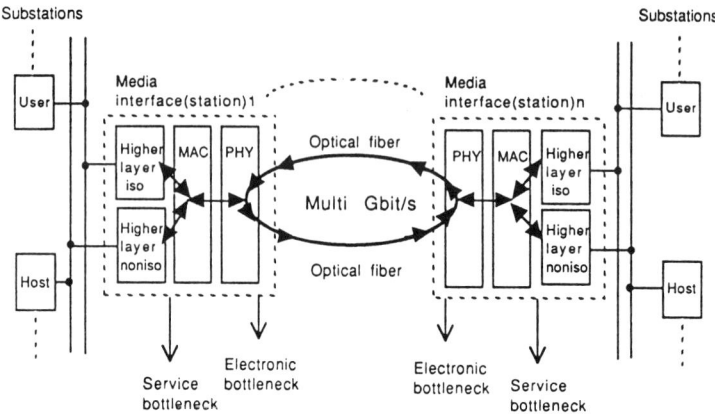

Fig.1 - Multi Gbit/s LAN configuration

3. ELECTRONIC BOTTLENECK

Although the optical fiber technology may support traffic with capacities up to Tbit/s, the electronic components at the network nodes, which typically operate at rates up to about 1 Gbit/s, drastically limit the total throughput. For instance, the new systems being designed and developed to take advantage of lightwave technology, such as the Fiber Distributed Data Interface (FDDI), the Distributed Queue Dual Bus (DQDB) and the Asynchronous Transfer Mode (ATM) in combination with the Synchronous Optical Network (SONET) can provide network throughputs only up to hundreds of Mbit/s. Such networks are inherently limited by the use of electronic components at stations and are of architectures that do not take advantage of the very high bandwidth of optical fiber.

There are many approaches to open up this electro-optic interface bottleneck and, among them, two seem most promising. The first one, the so-called "multihop" architecture, makes use of a new network architecture to achieve high capacity with existing devices. This architecture, used by several recently proposed networks, such as Manhattan Street Network (MSN) [3], ShuffleNet [4], Store-and-Forward With Integrated Frequency-Time (SWIFT) [5], Lightnet [6] and Wavelength-Division Optical Network (WON) [7], has a distributed topology, in which each station has access to a small number of fixed-wavelength transmitters and receivers. These wavelengths are assigned to stations in a manner that allows any pair of users to communicate with each other either directly (without hopping) or through one or more intermediate stations (with wavelength hopping). This approach presents, nevertheless, the inherent drawback of a so-called "deloading factor" [8]. The load imbalance, due to either traffic intensity variability or traffic pattern variability, has an effect to reduce throughput per station by a factor of 0.3 to 0.5 relative to the balanced-load situation, and this means that the load imbalance might have a serious negative impact as the network size increases. Also, the fact that the stations have a store-and-forward configuration means that such networks present rather high latencies, which should be avoided for isochronous traffic.

The second solution for this bottleneck problem is to use a pure WDM technique for transmission, combined with some form of WDM or TDM for switching. Examples of such alternative architectures are Lambdanet, the Hypass and Bhypass Switches [9], the Photonic Knockout Switch [10], and the Star-Track Switch [11]. Although such networks are capable of supporting multi Gbit/s traffic rates, they present fundamental difficulties when using distributed control. Typically, these networks have a star topology with a central passive node to and from which all communication takes place, and each user transmits its information on a unique wavelength. When setting up a connection, the transmitter and the receiver must first be tuned to the same wavelength. But in the absence of a central active controller, how do they find each other? A pretransmission coordination (i.e., a call setup procedure) between two users wishing to communicate is therefore required. Possible solutions to this are [10-13]: (i) scanning of the wavelength bands by users listening for callers; (ii) using a WDM switching function in a central active node; and (iii) connection setup through a separate, common channel. Despite differences in operation and performance, all these procedures are time-consuming and cause high access delays. Therefore, they are not suitable for bursty traffic with low access delay requirements. Moreover, the wavelength switching is a time-consuming process as well. The multiwavelength switch fabric can be performed in two ways, either by tunable receivers/optical filters (multiwavelength broadcast and select) or by tunable lasers (active wavelength routing). Both of them require wavelength-agile components which are not yet available in a rapidly-tunable form, and nor are likely to be in the near future [14-16].

Based on the above considerations, we advocate a new concept for opening up this bottleneck. We propose a WDM based network architecture, in which the wavelengths are no longer dedicated to different users, but to different applications, including one or more wavelengths dedicated for control/reservation purposes. The users have permanent access to all wavelengths. There can be as many control channels on different wavelengths as there are applications, each used specifically to arbitrate access for resources on the separate wavelength carrying the traffic from the particular application. Thus, two wavelengths are dedicated to each application or traffic class - a very high capacity channel to carry the user information and a low capacity channel to carry control information for media access. But since the needed bandwidth for control purposes would generally be quite small, several control channels can be combined into a TDM format on one or more (but less than the number of traffic classes or applications) channels on a fewer number of wavelengths. These concepts lead to a network architecture which can provide very high throughputs with simple and low delay access mechanisms. In addition, it is not necessary to switch from one wavelength to another as in the previous cases. Also, the a priori partitioning of the network resources and the separation of the access control protocols for different traffic classes mean that the impact of one traffic class on the performance of the other classes is greatly diminished or even eliminated when compared to single-class based network architectures mentioned above. Furthermore, because of a large amount of bandwidth available on the network, there is no need to employ complex resource-sharing mechanisms for resource allocation to different traffic classes, as the percentage improvement in network utilization will be small at the expense of protocol sophistication. In fact, each control channel can follow its own simple media access control mechanism, depending on the performance requirements of the particular traffic class it supports. In contrast to this, the WDM networks reported in the literature convey all traffic on the same wavelength. Because of this, the problem of contention resolution for the common network resources between different traffic streams, with contradictory performance requirements, becomes complex. Moreover, the contention resolution mechanisms must operate at the very high speed of the WDM network.

4. SERVICE BOTTLENECK

As mentioned in Section 2, the service bottleneck problem arises because different service applications or traffic classes have different, and often conflicting, performance requirements. Three main points are discussed below which reflect some of the fundamental technological premises that form the bases for the new approach to very high speed integrated LAN design discussed in this paper. On these bases, we examine three main components of a LAN, namely, transmission techniques, switching techniques and access mechanisms, and demonstrate how the SUPERLAN architecture provides a solution to the service bottleneck problem.

An integrated network, in which all traffic from diverse applications is carried in a single network, offers several advantages, which include flexibility, bit-rate transparency, universal access with bandwidth on demand, and a single network solution. In spite of these conceptual advantages, such an approach poses tremendous technical difficulties in the dimensioning of network elements and allocation of resources to competing user applications. It is becoming increasingly obvious that the issue of resource allocation and congestion control in a very high speed integrated network is generally difficult to handle. It includes a myriad of interacting preventive and reactive mechanisms that concern with guaranteeing adequate quality of service for each traffic class while preventing one user from degrading the performance of other users. The resulting equipments (hardware and real-time software) will turn out to be very complex, if all these mechanisms were to be ever realized. Moreover, electronic processing being the performance bottleneck, the goal of the network design should be to conserve processing by reducing the functional complexity of network elements to the necessary minimum, and by trading off communication capacity and buffers. Thus, there is a need to reconsider the approach to very high speed LAN design, which integrates diverse service applications on to a single network.

Secondly, the use of optical fiber as a transmission medium has provided much lower loss and error rates. Now most of the losses are due to the buffer overflow in congested processors. This means that it is possible, by a proper and careful design of the network, to completely ignore performance parameters like clipping (for real-time signals) and error rate (for nonisochronous data). Therefore, the main performance criteria that we consider in our design are: (i) throughput; (ii) access delay; and (iii) transit delay.

Finally, the third and the last point that must be taken into consideration is concerned with the increased nodal processing power, storage capacity and control data transfer rate capability required for very high speed networks. Taken together, these enhancements mean an increased bandwidth need that must be provided in each node for control purposes. To this end, it must be pointed out that the increased nodal bandwidth needs for control traffic should not lead to performance bottlenecks like unacceptable delay or loss performance for other traffic classes. Therefore, minimization of the impact of the control mechanisms on the network performance should be an important consideration in network design.

There are two types of contentions in an integrated network: (i) contention among classes of traffic for common resources; and (ii) contention among users/hosts inside the same class of traffic for common resources. In a LAN, the medium itself is a transport resource as well as a switch. There are two ways of providing transmission resource to different classes of traffic:

- all traffic classes compete for a common resource, in time or frequency/wavelength domain. We ignore other alternatives, like common resources in code domain, because of their sophistication. The main drawback of this contention policy stems from the need to employ complex and high speed resource-sharing mechanisms for resource allocation to different traffic streams, with different performance requirements. This heavily complicates the access protocol. The case of common resources in time domain, best illustrated by the evolving ATM networks, means that the stations have their transmissions scheduled to take place at different segments of time. A serious drawback of such an approach is the high impact of one traffic class on the performances of the other traffic classes. The case of common resources in wavelength domain, the so-called *wavelength dedicated to user* approach, is exemplified by networks like ShuffleNet [4], SWIFT [5], WON [7], Lambdanet [9], etc. Here the optical fiber bandwidth is split up into multiple disjoint frequency bands, and the stations are allocated one or more such channels for transmission purposes. They transmit all their traffic in a TDM form on these wavelength channels, so we still have contention among various traffic classes for common resources in time domain, with the disadvantage mentioned above.

- separate resources for different traffic classes. Given the large amount of resource available in a multi Gbit/s network, the advantages of the former policy, expressed in terms of service flexibility and network efficiency, do not motivate the increased complexity of the protocol processing, congestion control and packet switching. We consider, therefore, that an a priori partitioning of the network resources for different traffic classes should be used for very high speed networks. There are two choices to share the media resources among various traffic classes, in time domain or in wavelength domain. The first alternative, used by several networks like FDDI, MAGNET [17], etc., has the disadvantage of the dependency on very high speed components. These networks make use of an *in-band* model for integrating different traffic in the same TDM frame. Here the control and the signaling data are time-multiplexed with the information data in a very high speed TDM frame. These two sets of data have different requirements. The former has a higher priority and need more reliable transmission and greater processing capabilities rather than high speed or high throughput. For the latter, however, it is just the opposite, i.e., it requires high speed and high throughput. Such a TDM integrated approach needs very high speed components not only for general purposes, such as clock synchronization, but also for some other particular functions, such as frame/token synchronization, differentiation of control data and information data, etc. This problem can be obviated by using resource partitioning in the wavelength domain. This approach, the so-called *wavelength dedicated to application,* is proposed for SUPERLAN (described in the next section). Here the wavelengths are no longer dedicated to different users, but to different traffic classes, including one or more wavelengths dedicated for control/reservation purposes. The stations have permanent access to all wavelengths and the access for both user information data and control data does not occur at the very high speed electro-optical station interface with the optical fiber, but at the low speed electronic interface between the station/media interface and the users attached to it (Fig. 2). This provides simplicity and low processing rates for the MAC protocol, as well as a minimal number of high speed components.

The next issue that must be reconsidered in the network design is how to resolve contention among the users, i.e., what switching technique should be employed to switch information among users inside the same class of traffic. This can be provided in the time domain (like in FDDI, DQDB, MAGNET, etc.) or in the wavelength domain (like in Lambdanet, the Photonic Knockout Switch, etc.). In spite of the very high switching bandwidth (multi Gbit/s) provided by the second approach, this has, nevertheless, the disadvantage that it needs wavelength-agile components and, as a result, is not suitable for bursty data. We chose, therefore, a time switching technique for SUPERLAN. Also, because of the fact that the future multi Gbit/s LAN must provide switching support to a heterogeneous mixture of isochronous and nonisochronous sources, a proper switching approach must be chosen. This means that the switching architecture must provide transfer of data from one user to another user so that the desired grade of service, such as throughput, delays or blocking performance, are met for all traffic classes. There are two main switching techniques: (i) circuit-switching, with poor bandwidth utilization but with relative simplicity of the access mechanism; and (ii) packet-switching, with high service flexibility and network efficiency but with sophisticated control algorithms. Taking into consideration the fact that mixed networks (circuit-switching and packet-switching) are very likely to exist for a long time to come, as well as the fact that the internetworking between the existing networks and the future multi Gbit/s network must be as simple as possible, with no performance penalty and of low cost, an integrated use of the both switching approaches is suggested for the very high speed networks. SUPERLAN has, therefore, a mixed switching architecture, with time domain circuit-switching services provided in one wavelength band for the isochronous traffic, and with time domain packet-switching services provided in another wavelength band for the nonisochronous traffic.

Finally, the last element of the service bottleneck that must be reconsidered is the access control. At the present time there are a multitude of access control mechanisms in three different categories (fixed assignment, demand assignment or random access), and which are provided in various domains (time, frequency/wavelength, etc.). A common feature of most existing networks is the fact that they make use of a common media access procedure for all types of traffic, and this procedure acts either at the cell/packet level or at the cell and call/session levels. This approach is disadvantageous because of unnecessary protocol sophistication. For instance, it is not necessary to have an access mechanism acting at the cell level for isochronous traffic. Here an access mechanism acting at the call level is sufficient. Because of this sophistication, the access control mechanism becomes a real performance bottleneck for broadband services with hundreds Mbit/s rates. We consider, therefore, that the future very high speed network must have highly simple access protocols. Based on these considerations, we propose a network model in which each traffic class/application has its own application-oriented access protocol, with no interference from other applications. These access protocols can be separated in time or in wavelength domain. By this concept we get simpler and low speed access protocols with the penalty of hardware replication in each node. For SUPERLAN we propose, therefore, dedicated MACs for the isochronous traffic (acting at the call/session level) and for the nonisochronous traffic (acting at the burst level). In the current design, the access procedures for the isochronous and nonisochronous traffic are separated in the time domain on a single wavelength that is dedicated to carry only control data.

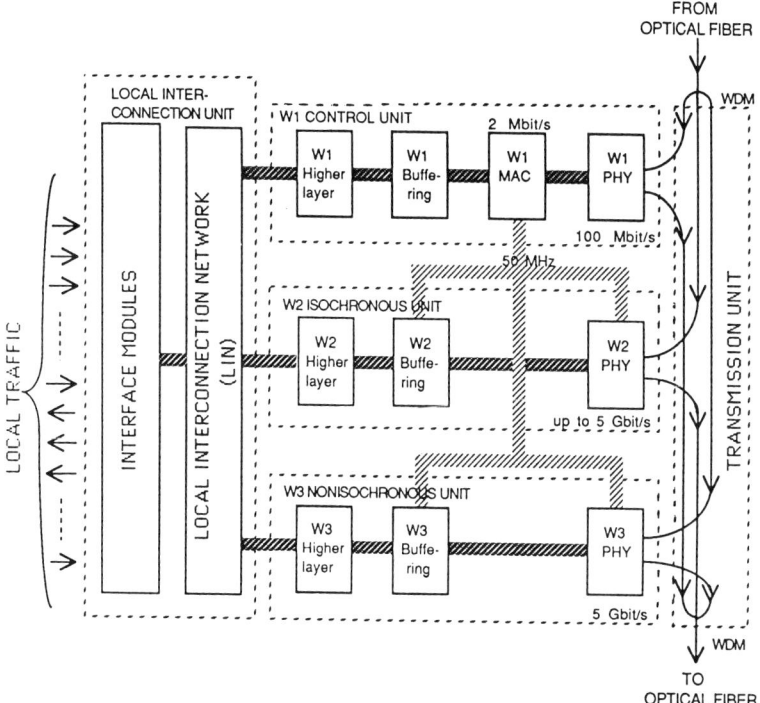

Fig.2 - Station block scheme

5. SUPERLAN: A MODEL FOR VERY HIGH SPEED LANS

SUPERLAN [1] is a very high speed local area network that makes use of a coarse WDM technique for allocating isochronous and nonisochronous traffic in an integrated network to two different wavelengths, i.e., the wavelength w2 conveys the isochronous traffic and the wavelength w3 conveys the nonisochronous traffic. In addition, there is a third wavelength (w1) that conveys, in a synchronous fashion, only the low speed control information that concerns with resource allocation and connection setup procedures. All the attached stations always have access to all wavelengths in the network and they provide separation of the local traffic into traffic classes (isochronous and nonisochronous) and control data before data are transmitted on the network (Fig. 3).

SUPERLAN is capable of supporting isochronous and nonisochronous traffic with arbitrary bit rates, both narrowband and broadband. It provides circuit-switching services for isochronous traffic, with perfect recovery at destination, i.e., the contention mechanism is based only on blocking. Moreover, this is a model in which each channel is able to adapt its speed and capacity to the specific bandwidth need of the particular application it supports. This is particularly useful for many types of isochronous traffic where the channel speed can be linked to the coding rate of the signal, reducing or eliminating the need for rate adaptation and multiplexing/demultiplexing functions inside the network. It also provides packet-switching services for nonisochronous traffic, that are based on delay-

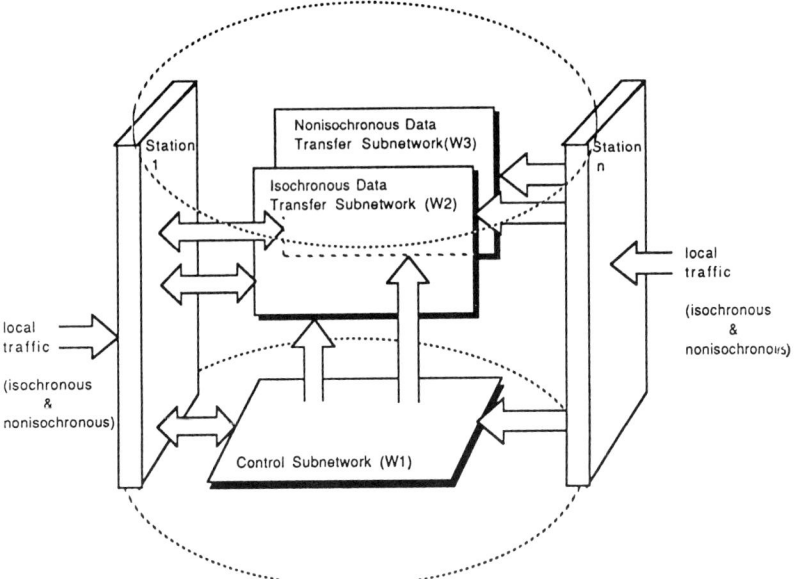

Fig.3 - Generic configuration of SUPERLAN

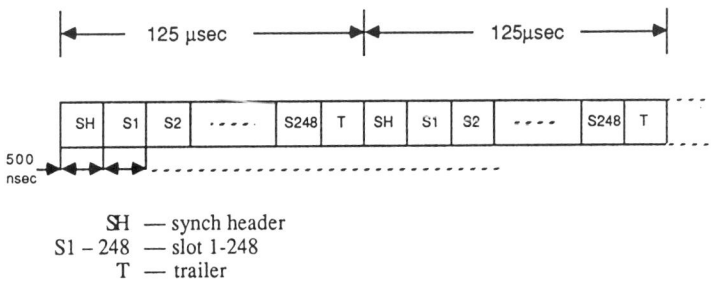

SH — synch header
S1 – 248 — slot 1-248
T — trailer

Fig.4 - W1 frame structure

throughput trade-off. The network can support switched services at rates up to 5 Gbit/s for each traffic class, for a total network throughput of over 10 Gbit/s. It is also able to provide communicative (for individual communication) and distributive (for mass communication) services that may include audio, video and data signals, and that need point to point and/or multipoint communications among a preliminary maximum number of 50 stations. Each of these stations provide a flexible interconnection to different networks and devices, such as classical data networks that support OSI type protocols, ISDN and B-ISDN/ATM networks, etc., with throughputs independent of data rates of SUPERLAN.

The network has a physical ring configuration with a master station (MS) and a number of ordinary stations (OS). It also has a logical hybrid topology, i.e., a logical bus topology for the w1 control subnetwork and logical ring topologies for the w2 isochronous and w3 nonisochronous subnetworks. These logical subnetworks run in a synchronous fashion that is based on a stretching/shrinking technique [1].

SUPERLAN is, therefore, comprised of three logically separate subnetworks, but provides users with the functionality of a single, multi Gbit/s network. It makes use of three parallel, wavelength-separated subnetworks that run in a synchronous fashion, as described below:

- the w1 control subnetwork, with a capacity of 100 Mbit/s, that supports resource management information flows as well as connection management and control information flows. It has a logical bus topology, with an insertion access method based on serial registers. The MS node acts as a transmitter and receiver, and the OS nodes act as repeaters and drop/insert intermediate stations. A synchronous TDM transport mode with constant-length frames has been chosen for transmission of control and signaling data. Fig. 4 shows the structure of w1 TDM frame. A train of 125 µsec. TDM frames with 100 Mbit/s rate flows continuously along the bus, originating and ending at the MS node. The w1 frame consists of 250 time slots of 500 nsec. temporal length. The first slot serves as a frame synchronization header (SH) that contains a fixed-bit pattern used for establishing frame synchronization in each node. The next 248 time slots contain network control information for actual communication among a maximum of 50 nodes, so that each station may have access to a control channel with a capacity of about 2 Mbit/s. The last time slot (T) is used for optical transmission line-error detection along the w1 bus as well as for auxiliary functions (reservation bits, bits for network management purposes, etc.). Every station has free access to a certain number of time slots in a frame, i.e., they can read and/or overwrite control data in different time slots every time the w1 TDM frame passes through the node. A duplicated fiber loop concept is adopted that, together with the station line switching capabilities, solves the problem of faulty portions of the SUPERLAN, and provides highly reliable working facilities for the w1 subnetwork.

- the w2 isochronous subnetwork, with a *maximum* capacity of 5 Gbit/s, that provides the transfer of isochronous user information. It has a logical ring topology with an insertion access method based on serial-parallel and parallel-serial registers. In this subnetwork all the stations have the same configuration and functions. The transmission form is still by constant-length (125 µsec.) TDM frames, but with *variable-capacity, fixed-time* slots. The slot length is of 20 nsec. duration, and each slot provides a variable capacity ranging from 0.5 Gbit/s to 5 Gbit/s rates. Thus, as shown in Fig. 5, the total number of bits in each time slot can vary. There is no explicit reading of addresses in the block of user information data; they are implicitly contained in the slot positions in each frame, which are decided through a call setup procedure. Fig. 6 shows an example of a w2 TDM frame. There are 6250 slots in a 125 µsec.

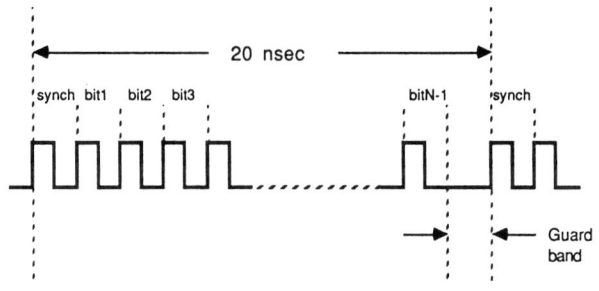

Where N = 9 bits, with a data rate of 0.5 Gbit/s
         or 17 bits, with a data rate of 0.9 Gbit/s
         or 25bits, with a data rate of 1.25 Gbit/s
         .
         .

         or 97 bits, with a data rate of 5 Gbit/s

Fig.5 - W2 or W3 (N=97) slot configuration

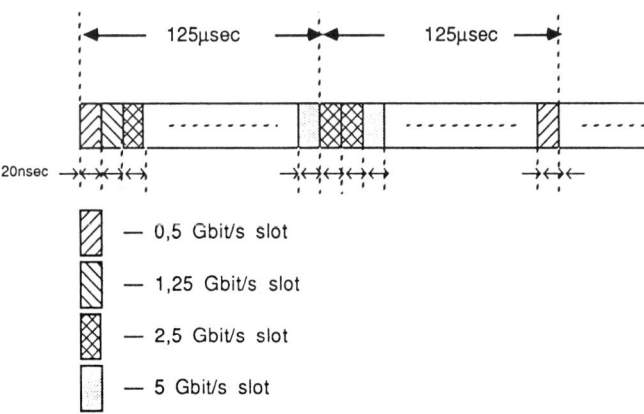

Fig.6 - W2/W3 frame structure example

frame and every station reads and/or overwrites information data in preallocated time slots each time the w2 TDM frame passes through the station.
- the w3 nonisochronous subnetwork, with a *fixed* capacity of 5 Gbit/s, that provides the transfer of nonisochronous user information. It is a subnetwork similar to the w2 subnetwork, in the sense that it has the same logical topology and the same TDM frame model, but with *fixed-capacity, fixed-time* slots. Once again, the slot length is of 20 nsec. duration. But now each slot carries a fixed number of bits corresponding to the 5 Gbit/s rate. There are, therefore, 6250 slots in a 125 μsec. frame.

These subnetworks are integrated at the transmission level, i.e., they share common transmission facilities among switching nodes, but maintain separate facilities for media access and buffering inside the stations (Fig. 2). User nodes/stations are connected, as a minimum, to the control subnetwork and, according to their service needs, they may be connected to the other subnetworks as well.

There are three phases of a connection-oriented communication procedure: connection establishment; data transfer; and disengagement/termination. These phases have different requirements for different types of traffic and they are supported by different subnetworks in SUPERLAN. The connection and termination phases are supported by the w1 control subnetwork, and the data transfer by the w2 isochronous and w3 nonisochronous subnetworks. Based on the characteristics of the traffic supported on w2 and w3 subnetworks, we propose the following two different access mechanisms provided in the w1 subnetwork [1]:

- a centralized access protocol for isochronous traffic on the w2 subnetwork. This traffic, characterized by long holding times and modest setup rates, may accept high access delay but requires low transit delay. Because of these properties, a centralized access protocol is proposed that is similar to the global scheduling multiple access (GSMA) scheme presented in [18].
- a distributed access protocol for nonisochronous bursty traffic on the w3 subnetwork. Since holding times of packets or bursts is small, packet or burst connections must be set up and torn down far more rapidly than before. This means that the access delay should be kept small. Therefore, it is essential that a distributed access mechanism be used, as in a ring topology with a centralized control scheme we have rather long access delays. This mechanism is based on a reservation scheme, under which the individual stations send reservation messages across the w1 subnetwork to inform the network about their needs for resources in the w3 subnetwork. The removal of these reservation messages is done at the source station.

The fairness issues are incorporated in control models which are chosen with regard to the main parameters of interest, i.e., blocking for isochronous traffic and access delay for the nonisochronous traffic. These models will be based on performance studies to be conducted later.

6. CONCLUDING REMARKS

A novel architecture for multi Gbit/s LANs that responds to the two main issues, i.e., the electro-optic and service bottlenecks, has been motivated and presented. The distinguishing features of this architecture include:
- utilization of a coarse WDM to open up the electro-optic bottleneck in the network nodes;

- separation of the low rate, highly reliable control data from the very high speed information data, and of the isochronous traffic from the nonisochronous traffic; and
- separated and simplified access protocols which can support packet- and circuit-switching.

The novelty of this model is given mainly by the idea of allocating different wavelengths to different applications, thereby making use of the abundant bandwidth available on the fiber. This design concept shows that the logic complexity of a Gbit/s LAN can be greatly minimized and reduced to low speed electronic elements. Equally important, this model is very suitable for applications that integrate narrowband and broadband services and that may include video, image, data, voice, etc., each of them with its own characteristics and requirements. Such a flexible arrangement permits growth in applications on the network without interfering with other existing applications, and avoids some of the performance pitfalls of other networks mentioned earlier in the paper. Other advantages are due to the following:

- it is possible to develop application-oriented access protocols, with no interference from other applications;
- the network capacity depends mainly on the access technique to the transmission media, i.e., the S/P and P/S registers, and not on protocol; and
- there are no wavelength-agile components in the network.

A factor which works against such an architecture is hardware replication in each node for each wavelength. But perhaps the SUPERLAN architecture may not represent a cost penalty over the single integrated LAN architectures when all the ancillary functions are considered.

The emerging multichannel dense WDM/coherent systems will favor this architecture. Then it is possible to develop a multi Gbit/s network with more subnetworks, each of them dedicated to a specific type of service (voice, video, graphics, data, facsimile, etc.) with simple, low-delay, application-oriented access protocols. Besides these, this network includes two more dedicated wavelengths, one for clock distribution and the other for network management purposes. This is also an area for further study.

ACKNOWLEDGEMENT

The work was performed while R. P. Singh was visiting the Department of Telecommunication and Computer Systems at the Royal Institute of Technology, Stockholm, Sweden. He gratefully acknowledges the financial support provided by the department and the Swedish Board of Technical Development under grant number STU-88-2323.

7. REFERENCES

[1] Popescu, A. and Singh, R.P., "SUPERLAN: A Model for Very High Speed Local Area Networks," in Proc. of the Second IFIP WG6.1/WG6.4 International Workshop on Protocols For High-Speed Networks, Palo Alto, California, Nov. 1990.

[2] Pehrson, B., "MultiG: Research on Distributed Multi Media Applications in a Multi Gbit/s Network," SICS Technical Report, the Swedish Institute of Computer Science, Stockholm, Sweden, 1990.

[3] Maxemchuk, N. F., "Regular Mesh Topologies in Local and Metropolitan Area Networks," *AT&T Technical Journal*, Vol. 64(7), Sept. 1985, pp 1,659-1,685.

[4] Acampora, A.S., "A Multichannel Multihop Local Lightwave Network," in Proc. of GLOBECOM '87, Tokyo, Japan, Nov. 1987, pp 37.5.1-37.5.9.

[5] Chlamtac, I. and Ganz, A., "Toward Alternative High Speed Networks: The SWIFT Architecture," in Proc. of IEEE INFOCOM '87, San Francisco, California, March 1987, pp 1,102-1,108.

[6] Chlamtac, I., Ganz, A. and Karmi, G., "Circuit Switching in Multi-Hop Lightwave Networks," in Proc. of the ACM SIGCOMM '88 Symposium, Stanford, California, Aug. 1988, pp 188-199.

[7] Bannister, J.A., *"The Wavelength-Division Optical Network: Architectures, Topologies, and Protocols,"* Ph. D. Dissertation, University of California, Los Angeles, 1990.

[8] Eisenberg, M. and Mehravari, N., "Performance of the Multichannel Multihop Lightwave Network Under Nonuniform Traffic," *IEEE Journal on Selected Areas in Communications*, Vol.6, No.7, Aug. 1988, pp 1,063-1,078.

[9] Goodman, M.S., "Multiwavelength Networks and New Approaches to Packet Switching," *IEEE Communications Magazine*, Vol.27, No.10, Oct. 1989, pp 27-35.

[10] Eng, K.Y., "A Photonic Knockout Switch for High-Speed Packet Networks," *IEEE Journal on Selected Areas in Communications*, Vol.6, No.7, Aug. 1988, pp 1,107-1,116.

[11] Lee, T.T., Goodman, M.S. and Arthurs, E., "A Broadband Optical Multicast Switch," in Proc. of ISS'90, Stockholm, Sweden, May 1990, Vol. III, pp 7-13.

[12] Henry, P.S., "High-Capacity Lightwave Local Area Networks," *IEEE Communications Magazine*, Vol.27, No.10, Oct. 1989, pp 20-26.

[13] Brackett, C.A., "Dense Wavelength Division Multiplexing Networks: Principles and Applications," *IEEE Journal on Selected Areas in Communications*, Vol.8, No.6, Aug. 1990, pp 948-964.

[14] Toba, H., Inoue, K. and Nosu, K., "A Conceptional Design on Optical Frequency-Division-Multiplexing Distribution Systems with Optical Tunable Filters," *IEEE Journal on Selected Areas in Communications*, Vol.SAC-4, No.9, Dec. 1986, pp 1,458-1,467.

[15] Lee, T.P. and Zah, C., "Wavelength-Tunable and Single-Frequency Semiconductor Lasers for Photonic Communications Networks," *IEEE Communications Magazine*, Vol.27, No.10, Oct. 1989, pp 42-52.

[16] Kobrinski, H. and Cheung, K., "Wavelength-Tunable Optical Filters: Applications and Technologies," *IEEE Communications Magazine*, Vol.27, No.10, Oct. 1989, pp 53-63.

[17] Lazar, A.A., Temple, A.T. and Gidron, R., "MAGNET II: A Metropolitan Area Network Based on Asynchronous Time Sharing," *IEEE Journal on Selected Areas in Communications*, Vol.8, No.8, Oct. 1990, pp 1,582-1,594.

[18] Mark, J.W., "Global Scheduling Approach to Conflict-Free Multiaccess via a Data Bus," *IEEE Transactions on Communications*, Vol.COM-26, No.9, Sept. 1978, pp 1,342-1,352.

A Study on Performance Improvement Algorithm in DQDB MAN

Tetsuya YOKOTANI, Hiroyuki SATO, Shigeo NAKATSUKA

Communication Systems Dev. Lab., Mitsubishi Electric Corp.
Ofuna, Kamakura, 247, Japan

Abstract

In this paper, performance improvement algorithm in DQDB MAN (Distributed Queue Dual Bus Metropolitan Area Networks) is proposed. DQDB which is standardized by IEEE 802 committee is one of protocols which provide medium access control in their main part. This protocol is adopted in metropolitan area networks. In DQDB MAN, a slot with a header similar to a header of ATM (Asynchronous Transfer Mode) cell is a unit for transmission and exchange. A network is configured in two unidirectional buses supporting transmission in opposite directions. An information is communicated by using slots. For this purpose, such slots are broadcasted to a downstream of bus including the destination node. Since a slot is not stored in each node, transit delay of a slot is very small. Here, it should be noted that transmission slots are never erased on halfway of bus and are terminated at end of the bus. In a word, the fulfilled slot cannot be reutilized in downstream nodes. This paper proposes algorithm for improving this point.

At first, a special node called erasure node is introduced in DQDB network for slot reutilization in downstream side and the operation of erasure node is described.

In the latter half of this paper, the performance of this algorithm is estimated using a queueing model, where M/G/1 queueing model is applied to the network model. Average and variance of queueing delay and of transmission delay are calculated by this method. The result shows the effectiveness of the proposed algorithm for improvement of the performance.

1. INTRODUCTION

Recently the study of local area network(LAN) and metropolitan area network (MAN) have been rapidly developed. In particular, they which harmonize with Broadband Integrated Services Digital Network (B-ISDN) using Asynchronous Transfer Mode (ATM) architecture being standardized by the CCITT have become of major interest. One candidacy of them is the Distributed Queue Dual Bus Metropolitan Area Network(DQDB MAN) which is standardized by IEEE 802 committee [1]. DQDB provides a protocol for medium access control in its main part and also includes parts corresponding to physical layer convergence and adaptation layers in ATM protocol. Although this protocol is adopted in MAN at first, it may be also adopted in access networks of B-ISDN, that is, distributed NT2 in the model defined by the CCITT [2], and backbone LANs. This paper discusses performance improvement of the above-mentioned DQDB protocol.

The structure of this paper is as follows. In chapter 2, the summary of this protocol, especially medium access control is described. In chapter 3, the problem requiring performance improvement of this protocol is presented and the concept of improvement is described. In chapter 4, the detailed operation of the proposed algorithm mentioned in chapter 3 is described. In chapter 5, the performance of this algorithm is estimated using a queueing model and the effectiveness of this algorithm in the performance improvement is described.

2. SUMMARY OF DQDB PROTOCOL

In DQDB, a slot with a header similar to header of ATM cell is a unit for transmission and exchange. A slot format is drawn in Fig.1. Functions of each field in slot header will be described later.

A network topology is configured in two unidirectional buses supporting communications in opposite directions as shown in Fig.2, BUS A and BUS B. Slot generator or terminator is located at the end of each bus. In slot generator, slots are generated without space.

DQDB protocol stack should be defined as Fig.3. As the detail of this protocol can be found in [1], only medium access control in DQDB protocol is summarized in this chapter. Medium access control is the most remarkable part in DQDB protocol. To share a medium, these functions provide the operation of transmission slot and receiving slot at each node attached to the transmission line. Medium access control operation consists of two access control modes, that is, Queued Arbitrated (QA) mode and Pre-Arbitrated (PA) mode which use QA and PA slots for access, respectively. QA and PA slots are identified by Slot type bit (ST) in slot header as in Fig.1 which is set or reset at slot generator. It is thought that QA slot is used for data communication service and PA slot is used for isochronous communication (voice and video) services. The operations of QA mode are as follows.

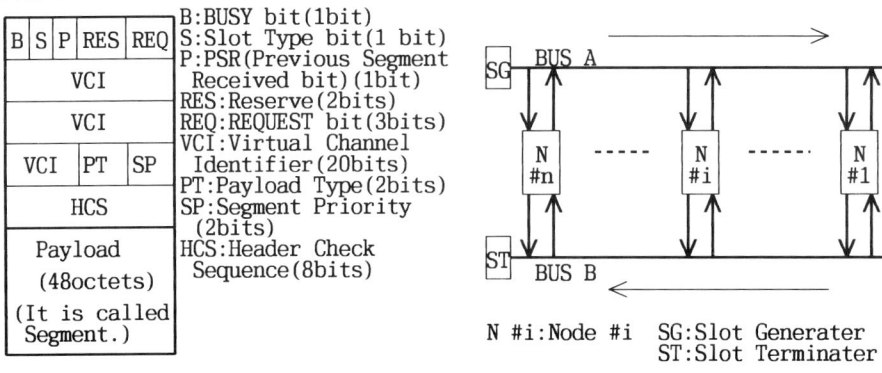

Fig.1 DQDB slot format

Fig.2 A network topology

N #i:Node #i   SG:Slot Generater
               ST:Slot Terminater

With regard to transmitting slots, the case that slots are transmitted to BUS A is explained and the case to BUS B is identical and independent of operations of BUS A.

This control operation is supported by two counters, Request counter (REQ.COUNT) and Count down counter(CD.COUNT), and two areas of slot header, Busy bit(BUSY) and Request bit (REQ).

In this control, two states are defined, that is, Idle state and Count down state. In Idle state, a node does not have a transmission information and the state is translated into Count down state when a transmission information is generated. In Count down state, it has a transmission information and the state is translated into Idle state when a transmission of information is finished. In the former, the node monitors the value of BUSY in receiving slot header from BUS A and REQ in receiving slot header from BUS B. The value of BUSY indicates whether the slot is used or not. REQ is used when a node in the downstream side has transmission information and reserves slots to transmit it to upstream side. A slot whose BUSY is set 1 is named BUSY slot and whose BUSY is set 0 is named FREE slot. Furthermore, a slot whose REQ is set 1 is named RESERVATION slot and whose REQ is set 0 is named NON RESERVATION slot. Therefore, if receiving slot from BUS B is RESERVATION slot, REQ.COUNT is increased by 1, and if such slot from BUS A is FREE slot, it is decreased by 1. In the latter, the transmission information is divided into segment size, that is 48 octets. The value of REQ.COUNT is loaded to CD.COUNT and REQ.COUNT is cleared. Then, if the node receives NON RESERVATION slot from BUS B, REQ will be set one and be passed to the next node. Also, if a FREE slot is received from BUS A, CD.COUNT will be decreased by 1. As the results, when CD.COUNT shows 0, the transmission of the information can be started to BUS A if the node detects FREE slots. The above-mentioned control operations are shown by state diagram in Fig.4. These are basic control operations for single priority case.

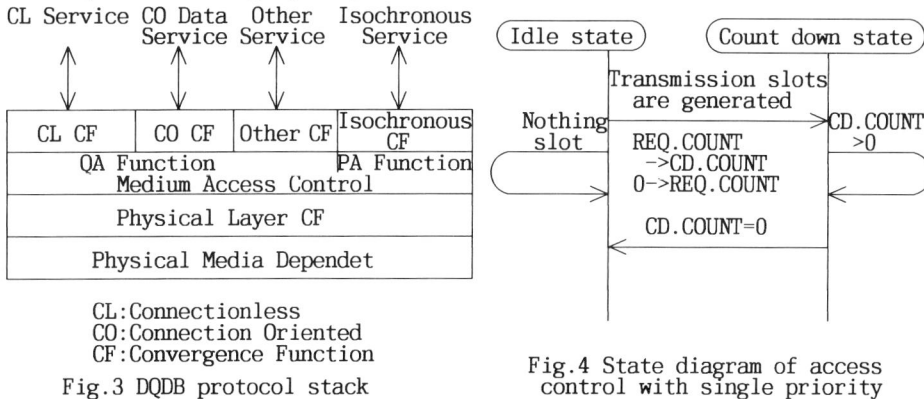

Fig.3 DQDB protocol stack

CL: Connectionless
CO: Connection Oriented
CF: Convergence Function

Fig.4 State diagram of access control with single priority

DQDB protocol supports the assignment of multiple priorities to access of QA slots. These priority levels must not exceed three levels. In a node, REQ.COUNT and CD.COUNT are provided for every priority level. The counters operate similarly to the basic case and independently at every priority level, excepting that account must be taken of REQs at higher levels and that CD.COUNT is increased if it received REQs of higher priority levels.

With regard to receiving slot, all slots are always copied into a node and they are passed to the next node in downstream side. In each node, VCI field in a slot header of copied slot is used for addressing information similar to ATM cells' VPI/VCI. This VCI field is checked in order to determine whether the node is destination or not. In each node except for the destination, this slot is disregarded. In the destination, segments which are information part of such slots are transferred to higher layer, that is, adaptation layer. Here, it should be noted that transmission slots are never erased on halfway of bus and terminated at slot terminator located at end of the bus.

By the way, the operation of PA is different from the operation of QA. PA looks like Time Division Multiplex (TDM) in a slot.

3. PROPOSAL OF ERASURE NODES

As described in chapter 2, DQDB protocol specifies simple medium access control. Furthermore, in this control, as all slots are stored and forwarded in each node, transit delay of the slot is very small. However, since transmission slots are never erased on halfway of bus, the fulfilled slot cannot be reutilized in downstream nodes. If improvement is made on this point, it is possible for the network to guarantee high performance. In this chapter, with regard to communication of QA slots, it is proposed that special nodes named erasure nodes be introduced and allocated on halfway of buses as in Fig.5 in order to improve performance by means of reutilizing fulfilled slots in downstream nodes (see [3] and [4]).

However, it should be noted that the succeeding analysis is based on a concept that reutilization of slots is carried out only for QA slots, because PA slots are scheduled by system management entity.

3.1 Additional operation of medium access control

Since any slots are released at erasure nodes located halfway of bus for reutilization, it is necessary to distinguish fulfilled slots, that is, worked slots which were used by communication between nodes located at upstream side, from other slots in the node. But, in DQDB protocol, it is impossible that such slots have mark which they were fulfilled in the node. Therefore, it is proposed that the previous segment received bit (PSR) of assigned slot header is used as follows instead of this marking. So, in a node, if it is found that receiving slot is the slot which is addressed to the node itself, the header of next slot is marked to indicate that the previous slot has been received by the destination node. PSR is set 1 for this purpose. Such operations are added to medium access control as mandatory operations.

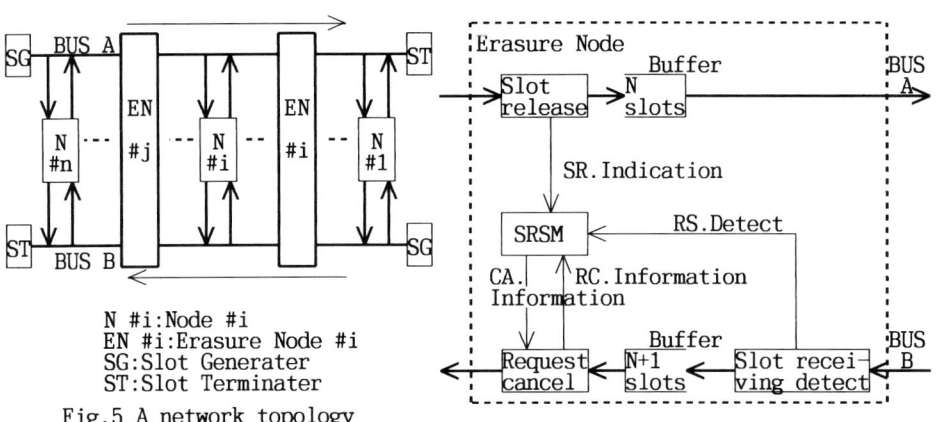

Fig.5 A network topology with erasure nodes

Fig.6 Constitution of erasure nodes

3.2 Outline of erasure nodes

Erasure nodes have two functions. They are the function which changes all BUSY slots into FREE slots and one which changes some RESERVATION slots into NON RESERVATION slots in the opposite bus in accordance with an algorithm described below. An erasure node has 4 functional blocks; Slot release, Request cancel, Slot receive detect and Slot Release State Machine (SRSM) as well as 4 primitives and buffers as in Fig.6. They will be explained as follows.

(1) Slot release functional block

It includes buffer having capacity of one slot stored because it decodes PSR of the current receiving slot and releases a stored previous slot. If the slot is released, this functional block sends SR.Indication to SRSM, resets BUSY of the previous slot and resets PSR of the current receiving slot. Therefore, a FREE slot is generated.

(2) Request cancel functional block

It cancels REQ of receiving slot if it receives RESERVATION slot and finds CA.Information from SRSM. And then, it sends RC.Information to SRSM.

(3) Slot receiving detect functional block
It sends RS.Detect to SRSM whenever it receives slot including PA slot. It has a function of timer.

(4) SRSM (Slot Release State Machine) functional block
This is the main part in erasure nodes.
It works to control REQ cancellation using information from the above-mentioned functional blocks. That is, it decides whether it sends CA.Information to Request cancel functional block by using SR.Indication, RS.Detect and RC.Information. The detailed operation of this state machine will be explained in the next chapter.

(5) Buffers
Some buffers are provided as in Fig.6 in order to increase the effectiveness of erasure nodes. These buffers have the capacity of N slots stored on BUS A and N+1 slots stored on BUS B in order to achieve symmetrical transit delay in an erasure node.

4. PERFORMANCE IMPROVEMENT ALGORITHM IN ERASURE NODES

This chapter describes detailed algorithm operations with an emphasis on SRSM.

4.1 Constitution of SRSM
Generally speaking, several slots may be collectively released. So, SRSM must record all time instants when each slot is released. Supporting this function, SRSM is implemented as Fig.7.

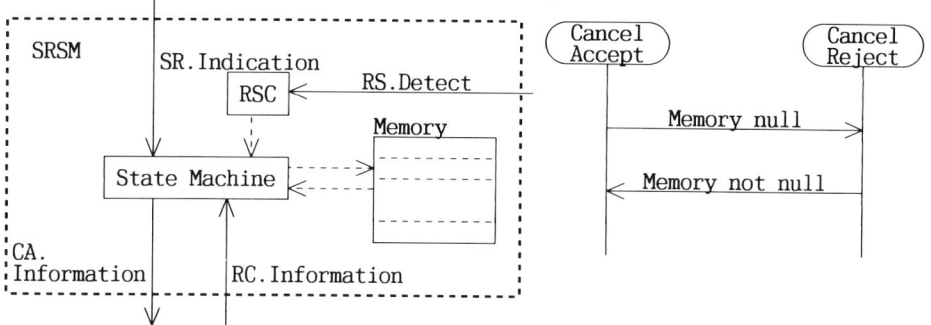

Fig.7 SRSM constitution   Fig.8 SRSM state diagram

SRSM consists of State Machine, Memory, and Receive Slot Counter (RSC).
RSC is increased by receiving RS.Detect and then this value is sent to State Machine. Memory records some RSC values which are selected by State Machine. State Machine operates by using primitives such as SR.Indication, CA.Information, RC.Information, RS.Detect and so on.

4.2 Operations of state machine
State Machine is shown in Fig.8 and Fig.9. It has two states, Cancel Accept and Cancel Reject. The transition between these states depends on state of Memory and initial state is Cancel Reject as in Fig.8. In Cancel Accept state, CA.Information is always issued from State Machine to Request Cancel functional block. In Cancel Reject state, CA.Information is not sent. In a word, CA.Information is set by State Machine in Cancel Accept and reset in Cancel Reject. Each state operation is explained by flow chart and the following context.

start (Cancel Reject State) initial state
if SR.Indication received from Slot release block
  then input Ct-RSC to Memory: set CA.Information to Request cancel block
    go to *Cancel Accept State*
  else go to *start (Cancel Reject State)*

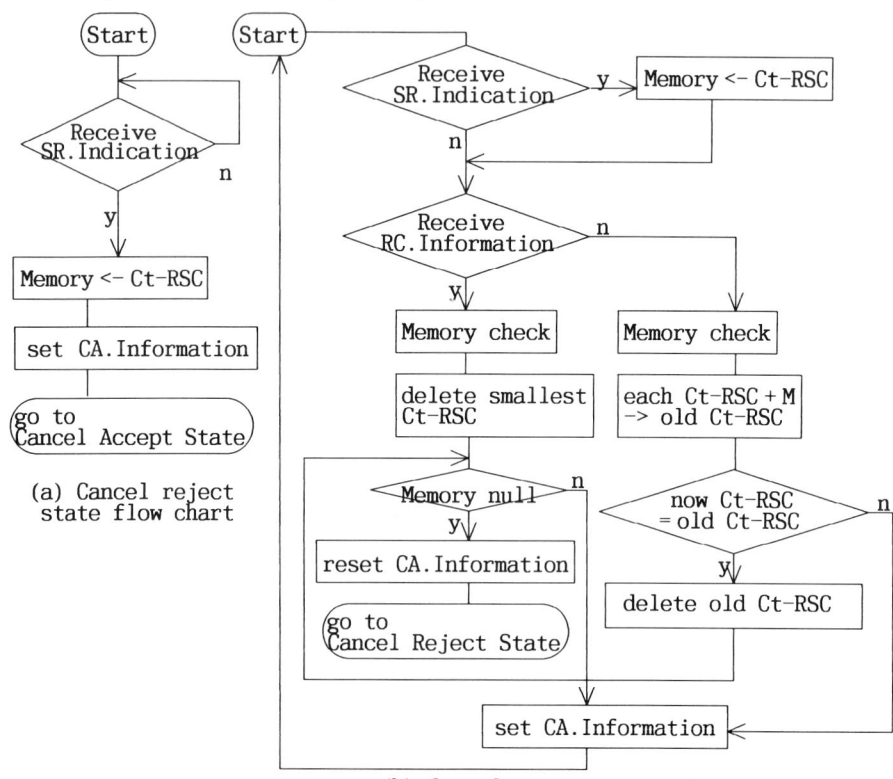

(a) Cancel reject state flow chart

(b) Cancel accept state flow chart

Fig.9 Flow chart of operations in each state

start (Cancel Accept State)
if SR.Indication received from Slot release block
  then input Ct-RSC to Memory
if RC.Information received from Request cancel block
  then Memory check: delete smallest Ct-RSC
    if Memory null
      then go to *Cancel Reject State*
      else set CA.Information to Request cancel block
  else Memory check
    for all Ct-RSCs in Memory *do*
      if now Ct-RSC = stored Ct-RSC + M
        then delete the stored Ct-RSC in Memory
          if Memory null
            then reset CA.Information

                    go to Cancel Reject State
                else set CA.Information to Request cancel block
            else set CA.Information to Request cancel block
go to start (Cancel Accept State)

where Ct−x is value of x.

In the above−mentioned description, it is necessary that system parameter M is fixed. M must satisfy the relation that it is more than stored delay in buffer with capability to store N+1 slots on BUS B within an erasure node and it is less than twice of the sum of stored delay and transmission delay between an erasure node and an adjacent downstream node on BUS A. In particular, upper bound of M, that is, twice of the sum of stored delay and transmission delay has very significant meaning. If this upper bound is not provided, the following problem occurs.

It is assumed that the upper bound is D. As in Fig.10, the slot is released at an erasure node and then becomes a FREE slot. This time is defined as T1. This slot is stored in buffer, and it is sent to the adjacent node on BUS A after N+1 slot time. Then the free slot arrives at the adjacent node at the time T2. A RESERVATION slot is sent from the adjacent node to the erasure node on BUS B at the time T3 before T2. The slot arrives at the erasure node on BUS B at the time T4. In this case, because the elapse time from T1 to T4 is less than M slots time, REQ of RESERVATION slot is reset and the slot is turned to NON RESERVATION slot. But, if the condition is not satisfied, that is, M is more than D, it is possible that T2 preceeds T3 and the FREE slot arriving at T2 is not availed at the adjacent node.

Thus, the upper bound is important for safe release of slots.

5. PERFORMANCE EVALUATION

In this chapter, the performance of the proposed erasure nodes algorithm is evaluated by a queueing model in the case where slots are transmitted on BUS A. But, it is assumed that there is a single priority level of QA.

5.1 System without erasure nodes

In the original system without erasure nodes, because it is not permitted that slots are reutilized, one FREE slot is shared among all nodes. It is assumed that parameters are defined as follows.

   I = (number of nodes)
   λ = (slot generating rate in each node)
   H = (transmission rate)
   p = (occupation ratio of QA slot)

Using the above−mentioned notation, the time required to send one slot, B, is shown as (1).

$$B = \frac{53 \times 8}{H} \quad (1)$$

where 53 denotes the number of octets in a slot (see Fig.1).

It may be assumed that QA slots and PA slots are mixed at random. Therefore, it can be considered that interarrival time of QA slots in each node obeys geometric distribution. So, the average transmission time of QA slot, the second moment of the transmission time and the third moment of it are respectively represented by MQ, VQ and KQ.

If the joint generation process of QA slots in all nodes is to obey a Poisson process and each node has a buffer with infinite capacity, the average queueing delay, Wq, and variance of the queueing delay, Vq, in each node are respectively derived by using

Pollaczek−Khinchin formula [5] as follows.

$$W_q = \frac{I\lambda VQ}{2(1-\rho)} \quad (2)$$

$$V_q = W_q^2 + \frac{I\lambda KQ}{3(1-\rho)} \quad (3)$$

$$\rho = I\lambda MQ \quad (4)$$

where (4) represents utilization rate of QA slot in this system.

Next, the propagation delay will be calculated as follows.
If the destination of each slot is uniformly scattered and the propagation delay between the adjacent nodes is represented by R, its average, Wp, and variance, Vp, are respectively represented as in (5) and (6). U means the utilization rate of isochronous communication using PA slot.

$$W_p = \sum_{k=1}^{I-1} \left[R\{p\rho + (1-p)U\}k \cdot \frac{2(I-k)}{I(I-1)} \right] \quad (5)$$

$$V_p = \sum_{k=1}^{I-1} \left[[(Rp\rho k)^2 + \{R(1-p)Uk\}^2] \cdot \frac{2(I-k)}{I(I-1)} \right] - W_p^2 \quad (6)$$

As the results, the average transit delay, Wt, and the variance of the transit delay, Vt are respectively derived by (7) and (8).

$$W_t = W_p + W_q \quad (7)$$
$$V_t = V_p + V_q \quad (8)$$

5.2 System with K erasure nodes

In the next discussion, the case of including K (=1,2,...) erasure nodes is considered. As in Fig.11, output and input ports in j−th erasure node are named Aj, Bj, Cj and Dj, and the number of nodes except for erasure nodes attached between j−th erasure node and j+1−th one is represented by Ij. Furthermore, the area in which Ij nodes are contained is named SAj.

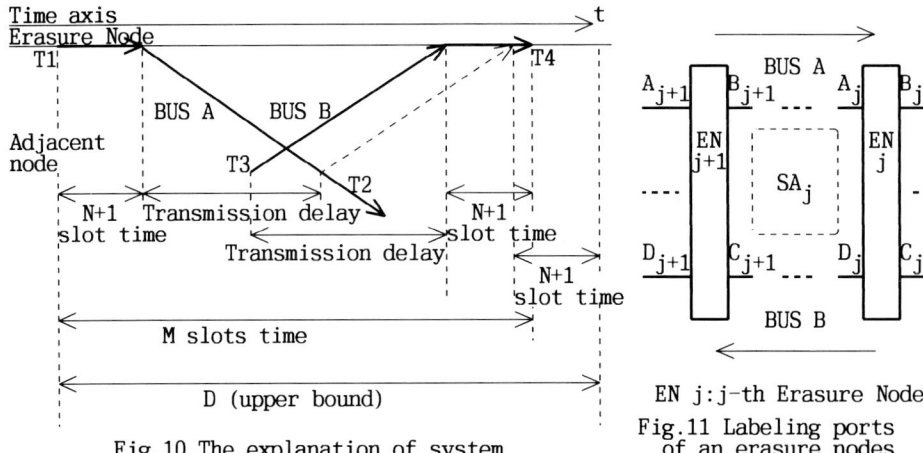

Fig.10 The explanation of system

EN j: j−th Erasure Node
Fig.11 Labeling ports of an erasure nodes

Now, if each node has a buffer with infinite capacity to store transmission QA slots, the probability that nodes within SAj have slots to be transmitted for carrying information is represented by P(1,j) as in (9).

$$P(1,j) = 1 - \left(1 - \frac{\lambda B}{P}\right)^{I_j} \quad (9)$$

Also, for this performance analysis, some probabilities are introduced as follows.

P(2,j)=(probability that RESERVATION slot received at port Cj on BUS B is converted to NON-RESERVATION slot; this is equal to probability which, in port Cj, RESERVATION slot is received on BUS B within M slots times from the time of a slot released on BUS A)
P(3,j)=(probability that, in port Cj, receiving slot is RESERVATION slot on BUS B)
P(4,j)=(probability that, in port Dj, transmission slot is RESERVATION slot on BUS B)
P(5,j)=(probability that, in port Aj, receiving slot is BUSY slot on BUS A)
P(6,j)=(probability that one slot arriving at j-th erasure node is released there on BUS A)

If it is assumed that RESERVATION slot is found at random on BUS B, P(2,j) obeys geometric distribution. So, making use of its memoryless property, the residual time at arbitraty time instant is identical distribution. Therefore, as RESERVATION slot is generated on SAj to transmit a slot, P(2,j) is represented as in (10).

$$P(2,j) = \sum_{k=1}^{M} P(3,j)(1-P(3,j))^{k-1} \quad (10)$$

M should be chosen as follows by using R and N.

$$M = 2(N+1) + \left[\frac{2R}{B}\right] \quad (11)$$

where [ ] is Gauss' symbol.

Also, P(3,j) and P(4,j) are respectively represented as in (12), and (13).
$$P(3,j) = P(1,j) + P(4,j-1) \quad P(4,0) = 0 \quad (12)$$
$$P(4,j) = P(3,j)(1-P(2,j)) \quad (13)$$

By the way, on BUS A, BUSY slot is found in port Aj when BUSY slot is transmitted by port B_{j+1}, or when FREE slot is transmitted by port B_{j+1} and this slot is utilized by SAj. Therefore, P(5,j) is represented as (14).

$$P(5,j) = 1 - P(6,j+1) + P(6,j+1)(1-P(4,j))P(1,j) \quad P(6,K+1) = 1 \quad (14)$$

If the destination node of each slot is uniformly scattered, the probability that one slot is BUSY slot and such slot is released by j-th erasure node, Q1(j), is defined by (15).

$$Q1(j) = \left[\left(I - \sum_{l=1}^{j} I_l\right)I_j + \sum_{m=1}^{I_j-1}(I_j - m)\right] \bigg/ \sum_{k=1}^{I-1}(I-k) \quad (15)$$

Using (14) and (15), P(6,j) is given as follows.
$$P(6,j) = P(5,j)Q1(j) \quad (16)$$

Based on (16), the rate of increase, FS, that some nodes are accessible to FREE slots, (that is, the probability that FREE slots are generated by erasure nodes whenever a BUSY slot is released) is derived as follows.

$$FS = \sum_{j=1}^{K} \frac{P(2,j) + P(6,j)}{P(6,j)} \quad (17)$$

Therefore, average queueing delay, WKq, and variance of queueing delay, VKq, are given as follows.

$$WK_q = \frac{I\lambda}{2(1-\gamma)} \frac{VQ}{(1+FS)^2} \quad (18)$$

$$VK_q = WK_q^2 + \frac{I\lambda}{3(1-\gamma)} \frac{KQ}{(1+FS)^3} \quad (19)$$

$$\gamma = I\lambda \cdot \frac{MQ}{1+FS} \quad (20)$$

where (20) represents utilization rate of QA slot in this system.

Next, when the transit delay of the case with erasure nodes is to be calculated, it is necessary to consider not only the propagation delay between nodes and the queueing delay in each node but also the delay to store in buffer of erasure nodes. The average, WKs, and the variance, VKs, of the latter delay are represented by (21) and (22).

$$WK_s = \left[\sum_{j=2}^{K+1} I_{j+1} \sum_{i=1}^{j} M\{p\gamma+(1-p)U\} i I_i\right] \bigg/ \sum_{k=1}^{I-1}(I-k) \quad (21)$$

$$VK_s = \left[\sum_{j=2}^{K+1} I_{j+1} \sum_{i=1}^{j} \{(Mpi\gamma)^2 (M(1-p)U)^2\} i^2 I_i\right] \bigg/ \sum_{k=1}^{J-1}(I-k) - WK_s^2 \quad (22)$$

As the propagation delay between nodes should be equal to the case of system without erasure nodes, the average, WKt, and the variance, VKt, of total transit delay are derived by the use of (5) and (6) as in (23) and (24).

$$WK_t = W_p + WK_s + WK_q \quad (23)$$

$$VK_t = V_p + VK_s + VK_q \quad (24)$$

5.3 Numerical results and consideration

Some numerical results using the above-mentioned technique are shown as follows.

In the first example, the result is shown as in Fig.12. It shows the relation between the number of generated slots in each node per second and the average or the variance of total transit delay given by (23) or (24), that is, the abscissa is the number of generated slots and the ordinate is the average or the variance of total delay. The values of parameters are set as follows.
 The number of erasure nodes, K; 1, 4, 9, 19
 The number of nodes except for erasure nodes, I; 20
 The number of buffer capacity in the case with erasure nodes on BUS A, N; 0

In the second example, the result is shown as in Fig.13. Ordinate and abscissa have the same meaning. The values of parameters are set as follows.
 The number of erasure nodes, K; 3
 The number of nodes except for erasure nodes, I; 20
 The number of buffer capacity in the case with erasure nodes on BUS A, N
 ; 0, 1, 5, 10

Each example shows the case without erasure nodes, and makes a comparison between systems without erasure nodes and with them.
By the way, common parameters in each example are set as follows.
 The transmission line; SDH STM-1 155.520 Mbps [6](including frame header)
 The rate of occupation of QA slot, p; 0.5

The utilization rate of isochronous communication using PA slot, U; 0.3
The propagation delay between nodes, R; 2.5 microseconds

Furthermore, it is assumed that the number of nodes between erasure nodes is identical, that is, Ii = Ij for all i and j in each example.

Considerations about each graph are as follows.

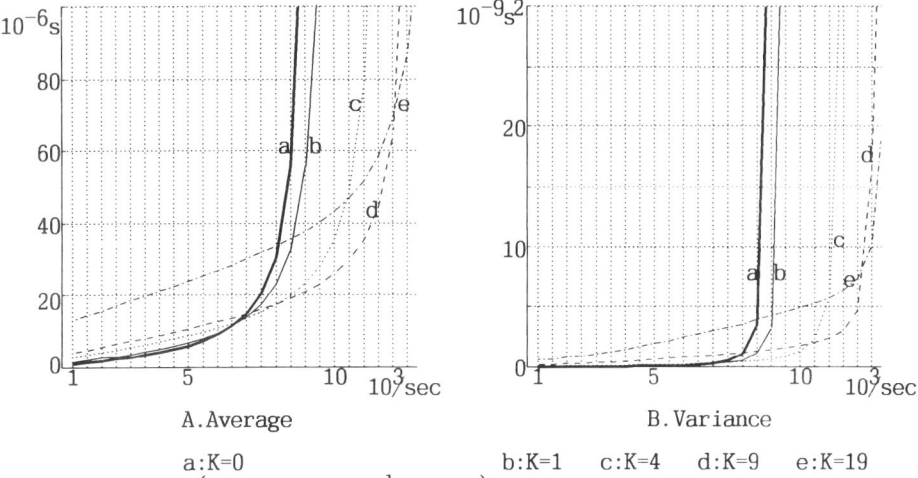

A. Average   B. Variance

a:K=0   b:K=1   c:K=4   d:K=9   e:K=19
(non erasure nodes case)

Fig.12 The case that the number of erasure nodes is variable

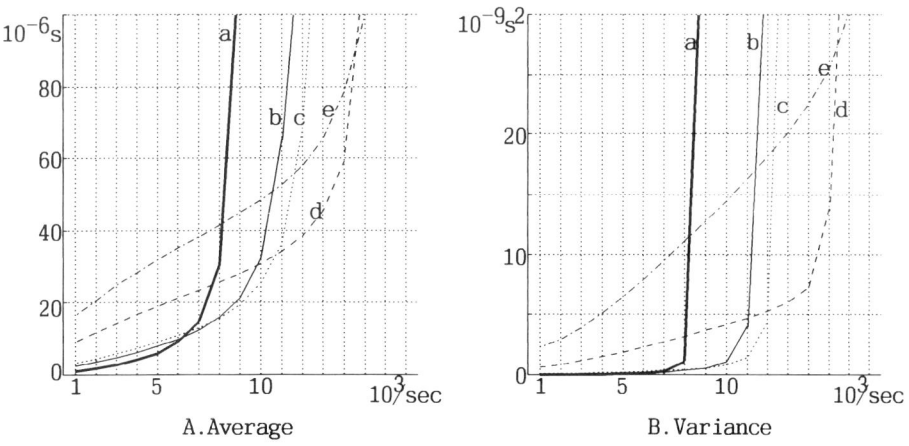

A. Average   B. Variance

a:non erasure nodes case   b:N=0   c:N=1   d:N=5   e:N=10

Fig.13 The case that the buffer size is variable

As seen from Fig.12, when the number of erasure nodes increases in light traffic, the overall transit delay becomes larger because the disadvantage that slots are stored by buffers in erasure nodes exceeds the advantage gained by release of slots. In heavy traffic, such relationship is reversed. When more than about 8,000 slots per second are transmitted in each node, erasure nodes become indispensable facilities from viewpoints of both average and variance of transit delay.

The same comment applies to Fig.13. But, when the buffer size is large, such phenomenon is not found except for the case of an extremely heavy traffic. Therefore, the effectiveness of erasure nodes is not significant when large buffer is attached. In short, small buffer is adequate to obtain their effectiveness.

6. CONCLUSION

This paper dealt with DQDB protocol which is standardized by IEEE 802 committee. At first, it described the system architecture adopting DQDB protocol and summary of this protocol. In medium access control of this protocol, it was pointed out that filfulled slots which had been used in communication was not removed on each bus. Next, This paper proposed an operation algorithm in order to improve this point. So, because slots can be reutilized in the same bus, it was proposed that erasure nodes were introduced and worked slots were released in such nodes. Also, the authors proposed the configuration of erasure nodes and the operation algorithm working in such nodes for slots releasing. Furthermore, this paper modeled DQDB protocol with and without such algorithm to M/G/1 queueing model and obtained the average and variance of transit delay. As the results, it was shown that the proposed algorithm was effective.

As new problems for future study, since the erasure node is only provided for QA slots in this paper, it is necessary to consider handling of PA slots.

ACKNOWLEDGMENT

Authors would like to thank Dr. M. Murotani, Mr. S. Aoyama and Mr. T. Shikama, who belong to Mitsubishi Electric Corporation, Communication System Development Laboratory, for valuable discussion and comments. Also, they wish to thank Dr. J. F. Mollenauer, who is IEEE 802 committee 802.6 working group chairman, for providing fruitful discussion.

REFERENCES

[1] "DQDB Subnetwork of a Metropolitan Area Network", IEEE Draft Standard P802. 6/D15, october, 1990
[2] "B-ISDN User-Network Interface", CCITT Recommendation I.413, May, 1990
[3] T. Yokotani, H. Sato, S. Nakatsuka, "Proposed Erasure Nodes Algorithm in DQDB", IEEE 802 committee contribution, 802.6 90/49, July, 1990
[4] T. Yokotani, H. Sato, S. Nakatsuka, "Modification of Erasure Nodes Constitution", IEEE 802 committee contribution, 802.6 90/83, November, 1990
[5] L. Kleinrock, "Queueing System, Volume 1: Theory", Jhon Wiley and Sons, 1975
[6] "Synchronous Multiplexing Structure", CCITT Recommendation G.709, May, 1990

THE ADAPTIVE LEAKY BUCKET: A NEW APPROACH FOR BANDWIDTH
ENFORCEMENT IN DQDB NETWORKS

A. Lombardo[*], S. Palazzo[*], D. Panno[*], R. Pignatelli[**],
L. Susanna[**]

(*) Ist. di Informatica e Telecomunicazioni, Catania (Italy)
(**) Scuola Sup. G. Reiss Romoli, L'Aquila (Italy)

ABSTRACT

In networks based on statistical multiplexing, it is important to monitor the actual bandwidth use of each virtual connection and to ensure that traffic parameters are within the limits agreed on during the call set-up. This prevents a non-controlled call from heavily delaying or even destroying the data units of other calls. In this paper the authors present and analyze a traffic source bandwidth enforcement mechanism conceived for DQDB networks. This mechanism, which is referred to as the Adaptive Leaky Bucket, exploits the access mechanisms provided for in the DQDB standard, without introducing any extra network signalling load and, therefore, without affecting network performance.

1. INTRODUCTION

Current technology is so mature that in the near future Metropolitan Area Networks (MANs) will be implemented and provide for a first step towards Broadband Integrated Service Digital Networks (B-ISDNs). In particular, standardization work on MANs has now nearly been achieved by the project IEEE 802.6, which has defined the DQDB (Distributed Queue Dual Bus) standard [1]. This standard explicitly provides for packet-based information transport, according to the emerging trends for the asynchronous transfer mode (ATM) in B-ISDNs.

However, in this kind of network exploitation of statistical multiplexing emphasizes congestion control problems. In fact, proper congestion control is needed to allocate the shared bandwidth fairly in such a way that performance requirements of all network users are satisfied. A suitable set of congestion controls may include features for routing, admission control, buffer and queue management, reactive control and bandwidth enforcement. In particular, bandwidth enforcement (or policing) mechanisms are applied at the access points of each user connection to guarantee that traffic flows admitted into the networks conform to the parameters declared in the connection phase.

Among the various mechanisms for bandwidth enforcement, the best-known is the Leaky Bucket (L.B.) [2]. This scheme adopts a coercitive measure: it immediately discards the data units considered to be in excess. To this aim (see Fig.1) a buffer is filled with the data units (cells) of the controlled connection and emptied at a constant rate. When the buffer is full data units in excess are lost. Let us say that the buffer is emptied at the rate of B_e bit/s and has a finite length of Q bits. The values of the control parameters B_e and Q are fixed according to the characteristics of the controlled traffic flow, with the aim of maintaining high levels of network efficiency while ensuring the quality of the service offered to the user.

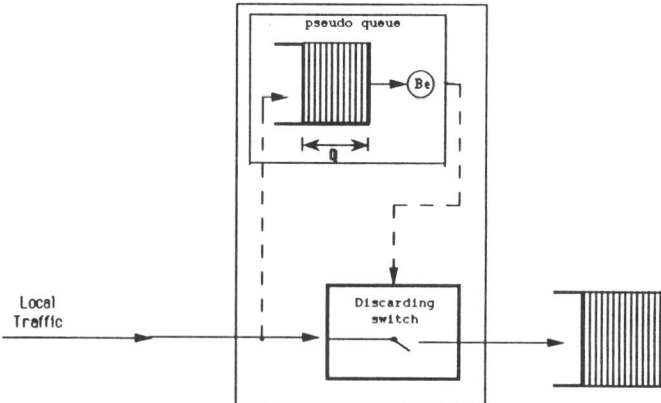

Fig.1 Functional scheme of the Leaky Bucket mechanism

However, previous studies [3][4][5] have shown that it is not possible to implement a Leaky Bucket mechanism that effectively prevents the danger of network congestion and at the same time is not too restrictive for the user. In particular, as the L.B. acts taking into account only the traffic measured locally, it may cause an unacceptable cell dropping rate, even in conditions of light network load.

Better results can be obtained by combining the policing mechanism applied on each source with congestion control exercised over the multiplexed network traffic.

To this purpose, the Virtual Leaky Bucket (V.L.B.) has been proposed [4][6][7], which is a modified version of the L.B. mechanism. The V.L.B. mechanism uses the same pseudo-queue scheme as the L.B. to recognise cells in excess, but instead of discarding them it marks their header and transmits them over the network. Only if congestion occurs along the virtual circuit route the marked cells are discarded by the transit nodes.

Unfortunately, it only makes sense to implement the V.L.B. mechanism in switching node based long-haul networks. In networks with distributed switching like standardized MANs and LANs, this policing mechanism cannot be used as once the data units in excess, even if marked, have entered the network they irrevocably occupy the bandwidth.

In this paper, we introduce a novel efficient policing mechanism, hereafter referred to as the Adaptive Leaky Bucket (A.L.B.), which is specifically conceived to be implemented in DQDB networks. In Section 2 the A.L.B. mechanism is outlined. Sect. 3 presents the traffic monitoring algorithm which is implemented in each user station to estimate the current network load. Finally, Sect. 4 shows the simulation results and discusses the effectiveness of the mechanism.

2. THE ADAPTIVE LEAKY BUCKET MECHANISM

The A.L.B. enforcement mechanism is illustrated in Fig. 2.

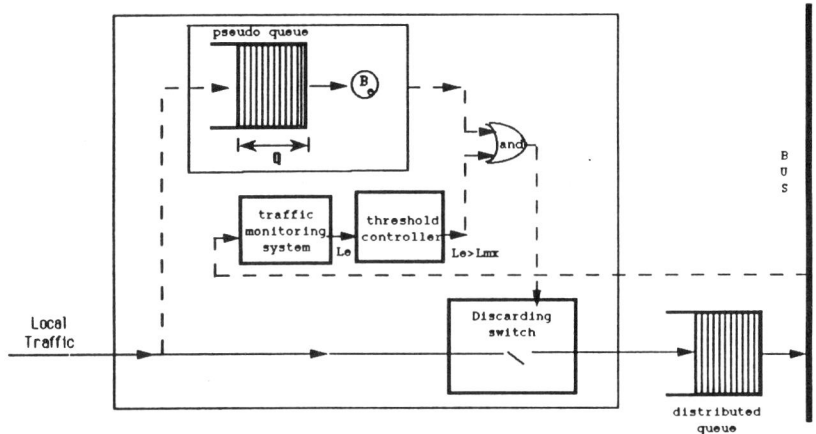

Fig. 2 Functional scheme of the Adaptive Leaky Bucket mechanism

Like the L.B., the A.L.B. uses a pseudo-queue scheme to recognize cells in excess, but these are discarded only if the measured global bus load, L_e, exceeds the maximum admitted load, L_{Mx}, that the channel, operating in statistical multiplexing, is able to support while preserving a certain degree of network performance. Of course, the maximum admitted load, L_{Mx}, depends on the statistic characteristics of the traffic to be multiplexed as well as the network performance required. This is shown in Fig. 3, which has been obtained for bursty traffic and different values of peak bandwidth, burstiness and average burst length.

In order to implement the above mechanism, each station must be able to monitor and estimate the global bus load, L_e. To this aim we have defined a suitable traffic monitoring algorithm that can be effectively applied to DQDB networks. It is introduced in the next section.

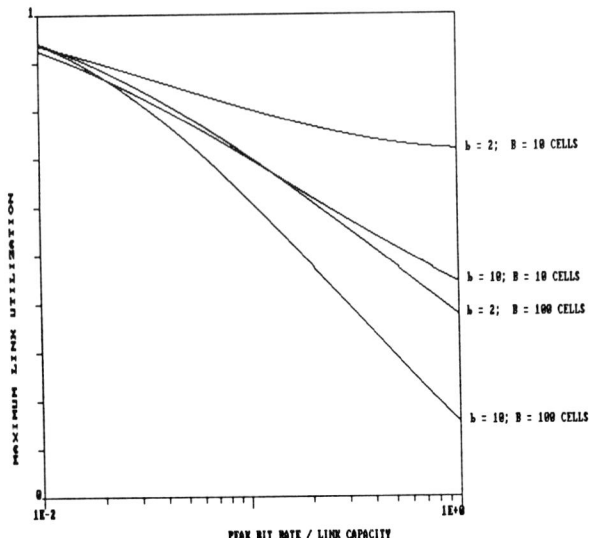

Fig. 3 Maximum link utilization calculated to meet the requirement : Pr (delay > 40 cells) = 10^{-2}

3. THE TRAFFIC MONITORING ALGORITHM

The traffic monitoring algorithm is implemented in any station ("access unit" or AU in Fig. 4) of the DQDB network.

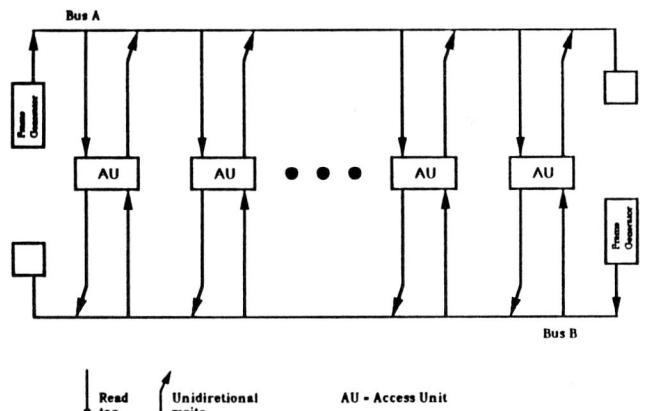

Fig. 4 Architecture of the DQDB MAN

Although it is distributed, it does not require any ad hoc signalling. To this purpose, in fact, information conveyed by the queued-arbitrated access protocol defined for the DQDB standard is exploited. Referring to the basic operations of

the MAC protocol [1], the BUSY bits monitored on the forward bus give a measure of the bus load due to the upstream stations, while the REQ bits monitored on the reverse bus give a measure of the bus load queued for transmission by the downstream stations. In each station, the traffic monitoring system is equipped, as in [8], with two Load Counters, LC_A and LC_B, one for each of the two buses A and B. Referring for example to bus A, the counter LC_A is increased by the station every time one of the following events occurs:
- a BUSY slot is monitored on bus A, that is, an upstream station is loading the bus;
- the REQ_Counter or the Countdown_Counter is decreased, that is, a downstream station is going to load the bus;
- the station passes from the Countdown to the Idle state, that is, the station itself is loading the bus.

At time intervals of T secs, the global load on bus A is measured as:

$$L_e = n * LC_A / T \qquad [bit/s]$$

where n indicates the slot length in bits; then the LC_A counter is reset.

The same calculation is made through the counter LC_B to estimate independently the load on bus B, and both the measurements are repeated ciclically.

It is worth noting that the interval T must be selected carefully to prevent traffic peaks on the bus from remaining undetected (T too long) or temporary traffic bursts due to natural statistical bus load variations from being mistaken for network congestion occurences (T too short). At a first glance, the above choice seems to raise the same problems already mentioned about the determination of the L.B. parameters. However, as the traffic multiplexed in the network varies much more slowly than that of a single bursty source, we can conclude than the choice in the case we are now dealing with is much less critical than that of the L.B. parameter. Therefore, in the following, we simply assume the interval T as equal to the propagation delay of a signal flowing along the whole bus.

4. SIMULATION RESULTS

The A.L.B. mechanism has been assessed through simulation. The DQDB network is assumed as loaded in such a way that the probability that a station transmits a segment on each bus is proportional to the number of the downstream stations. The traffic offered by each station is bursty and is characterized by the following parameters: B_p (peak bit rate), B_m (mean peak rate), T_p (mean peak duration).

A traffic source with burstiness $b = B_p/B_m$ has been modeled by a two-state cell generator. In the ON state the cell generator transmits cells of n bytes at a rate of B_p/n and in the OFF state it is silent. The distribution of the length of peak and silent periods is assumed to be exponential, so that transitions between states occur with a frequency of $1/T_p$ (from ON to OFF) and $1/[T_p(b-1)]$ (from OFF to ON).

The parameter values selected for the simulation runs are shown in Tab. 1a.

In each DQDB station the scheme of Fig. 2 has been implemented in such a way that the enforcement mechanism controls the input of user segments into the local queue of the station (subsequent insertion into the distributed queue being ruled by the DQDB MAC protocol).

In order to control the sudden variations of the source peak bit rate effectively, the parameter B_e has been fixed close to the nominal peak bit rate of the source, and precisely $B_e = 0.9 \cdot B_p$. Finally, assuming an admitted cell loss rate P_1 of 10^{-5}, we have sized the pseudo-queue according to the expression derived in [5]:

$$Q = \frac{1}{\lambda_1 - \lambda_2} \cdot \ln \frac{\lambda_1 - \lambda_2 \cdot (1 - P_1)}{P_1 \cdot \lambda_1}$$

where λ_1, λ_2 are given by:
$\lambda_1 = 1 / [T_p \cdot (B_p - B_e)]$
$\lambda_2 = 1 / [T_p \cdot (b - 1) \cdot B_e]$

The L.B. parameter values thus obtained are shown i Tab.1b.

NODE	B_m [Mb/s]	b	B_p [Mb/s]	T_p [ms]	$T_p(b-1)$ [ms]	B_e [Mb/s]	Q Cells
1	18	3	54	5	10	48.6	772
2	16	3	48	5	10	43.2	686
3	14	3	42	5	10	37.8	601
4	12	3	36	5	10	32.4	515
5	10	3	30	5	10	27.0	429
6	8	3	24	5	10	21.6	343
7	6	3	18	5	10	16.2	257
8	4	3	12	5	10	10.8	172
9	2	3	3	5	10	5.4	86
10	0	-	0	-	-	0	0

a)   b)

Table 1: Simulation parameters

In the simulation, the possibility to disable the threshold controller has been provided in order to compare the behaviour of the A.L.B. with that of the traditional L.B. under the same conditions. Moreover, various simulation runs have been carried out for different values of the threshold L_{Mx}.

As far as the workload variations are concerned, the anomalous behaviour of each station has been considered as due only to increased values of the peak bit rate, the other parameters being assumed as constant.

Simulation results are shown in Fig. 5, which represents the cell loss rate as a function of the ratio σ between the actual peak bit rate of the source and the nominal peak bit rate declared. In particular, for the sake of simplicity, in Fig. 5 we have reported only two significant curves which refer to the traditional L.B. and the A.L.B. with $L_{Mx}=0.95$ respectively.

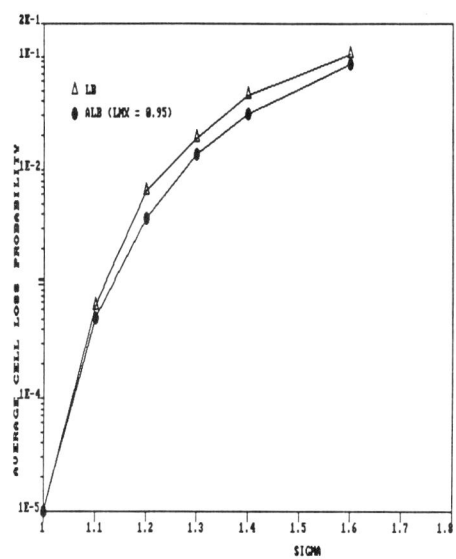

Fig. 5   Performance comparison between the A.L.B. and L.B. mechanisms

As expected, simulation has confirmed that the cell loss rate measured on the source traffic is lower in the A.L.B. scheme than in the simple L.B. one, and, among the A.L.B. variants, it decreases as the threshold L_{Mx} is increased. Such an improvement in the performance, and, consequently, in the global throughput, does not affect network efficiency at all, on account of the congestion control intrinsically achieved by the A.L.B. mechanism. Simulation has in fact shown that, even for high values of σ, the average bus access delay exhibits no notable variations in the range from the traditional L.B. to the A.L.B. with increasing threshold values up to $L_{Mx}=0.95$.

5. CONCLUSIONS

In this paper we have introduced a novel approach for bandwidth enforcement in DQDB networks, which aims at subordinating the effects of the policing mechanism to the actual network status in terms of congestion. The mechanism we propose, referred to as the Adaptive Leaky Bucket, assumes that in each DQDB station a suitable traffic monitoring algorithm allows the station to estimate locally the global average network load. Only in case of network congestion, that is, when the estimated load is higher than a fixed threshold, user cells considered in excess by the Leaky Bucket mechanism are dropped; otherwise they are processed by the MAC protocol and admitted into the network. As a consequence, the user cell loss probability decreases and, at the same time, the danger of congestion occurrences is averted. The effectiveness of the method introduced has been assessed by simulation.

REFERENCES

[1] IEEE: Proposed Standard - Distributed Queue Dual Bus (DQDB) Metropolitan Area Network (MAN), October 1989.

[2] J. Turner: "The Challenge of Multipoint Communication", Proc. 5th ITC Seminar on Traffic Engineering for ISDN Design and Planning, Lake Como, May 1987.

[3] W. E. Leland: "Window-Based Congestion Management in Broadband ATM networks: The Performance of Three Access-Control Policies", Proc. GLOBECOM '89, Dallas, November 1989.

[4] G. Gallassi, G. Rigolio, L. Fratta: "ATM: Bandwidth assignment and bandwidth enforcement policies", Proc. GLOBECOM '89, Dallas, November 1989.

[5] M. Butto', E. Cavallero, A. Tonietti: "Effectiveness of the Leaky-Bucket policing mechanism in ATM networks", to appear in IEEE J-SAC, Vol. 9, n. 3, April 1991.

[6] M. Hirano, N. Watanabe: "Characteristic of a Cell Multiplexer for Bursty ATM Traffic", Proc. GLOBECOM '89, Dallas, November 1989.

[7] A. Eckberg, D. Luan, D. Lucantoni: "Meeting the Challenge: Congestion and Flow Control Strategies for Broadband Information Trasport", Proc. GLOBECOM '89, Dallas, November 1989.

[8] A. Lombardo, S. Palazzo: "An Architecture for a Pure ATM Metropolitan Area Network", Proc. Globecom '89, Dallas, November 1989.

The Cambridge Backbone Network
An Overview and Preliminary Performance

K. Zieliński
Institute of Computer Science
University of Mining & Metallurgy
David J. Greaves
Olivetti Research Ltd.
University of Cambridge, Computer Laboratory

1 Introduction

The CBN (Cambridge Backbone Network) is a collaborative project between Olivetti Research Limited and the University of Cambridge Computer Laboratory. The project has designed and build an experimental communication network for integrated computer data, voice and real-time video applications.

The CBN architecture can be characterized in general terms as follows:

- Backbone network 500 to 1500 Mbit/sec slotted ring topology,

- ATM and multicast modes of operation,

- 200 km of fibre – metropolitan dimensions,

- 30 km link length easily supported,

- Expected configuration typically 30 stations offering 30 Mbit/sec each,

- Direct compatibility with Cambridge Fast Ring [4].

The CBN offers an ATM LAN/Man architecture based around a source release slotted ring [1,4]. The network operates on monomode optical fibre and is designed for an eventual line rate of 1000 Mbit/sec. The bandwidth of the fibre optic channel can be partitioned into a number (currently four) TDM channels. This enables stations of varying cost and bandwidth to be attached to one network, parameterised by the number of channels a station can use concurrently.

This paper gives an overview of the CBN station hardware and software design and reports preliminary performance measurements of the first implementation in terms of throughput and response time. The main purpose of these studies was to identify the potential bottlenecks in the current CBN hardware and software architecture to give some ideas for future development.

2 Backbone Network Station Hardware Design

The hardware configuration of a simple half-duplex station [1] is shown in Figure 1. The Backbone Network uses 4B5B/NRZI modulation. Each station has a local, crystal-locked transmit clock and retimes the received data using an elastic buffer. The received data from the optical fibre is passed to an ECL access chip which performs two types of function. The first concerns the transformation of the serial data to eight bits wide and provision of clock and byte synchronisation. This requires the following functions:

- Decide whether the current input voltage is a one or zero using D-type flip-flop,
- Convert from NRZI to NRZ,
- Decode 5 bit blocks into 4 bit data word or special code 'syn' used for synchronization,
- Gain bit synchronization using a slip method,
- Signal a code error on codebook violation,
- Gain frame synchronization on frame header, and
- Encode output stream into 4B5B and thence NRZI.

The second type of function provided by the access chip is at the frame level. The format of a Backbone Network frame is shown in Figure 2 .

Each frame starts with a header block. The frame then has the full/empty (F/E), monitor passed (M) and type (T) flag bits of each slot. The frame trailer contains response (R) and qualifier (Q) for each slot and a 12 bit CRC that covers all fields from F/E to Q inclusive.

The incorporated functions include:

- Check CRC of received frame,
- Examine and update F/E bits to decide which slot is to be read or written,
- Generate data strobes for reading or writing the desired channel,
- Check and update response and qualifier bits of outgoing frame, and
- Write new valid CRC to outgoing frame.

The 8 bit data streams from the ECL, access chip are connected to the RAM packet buffer through the semi-custom demultiplexer devices. Their purpose is to widen the data from 8 to 32 bits.

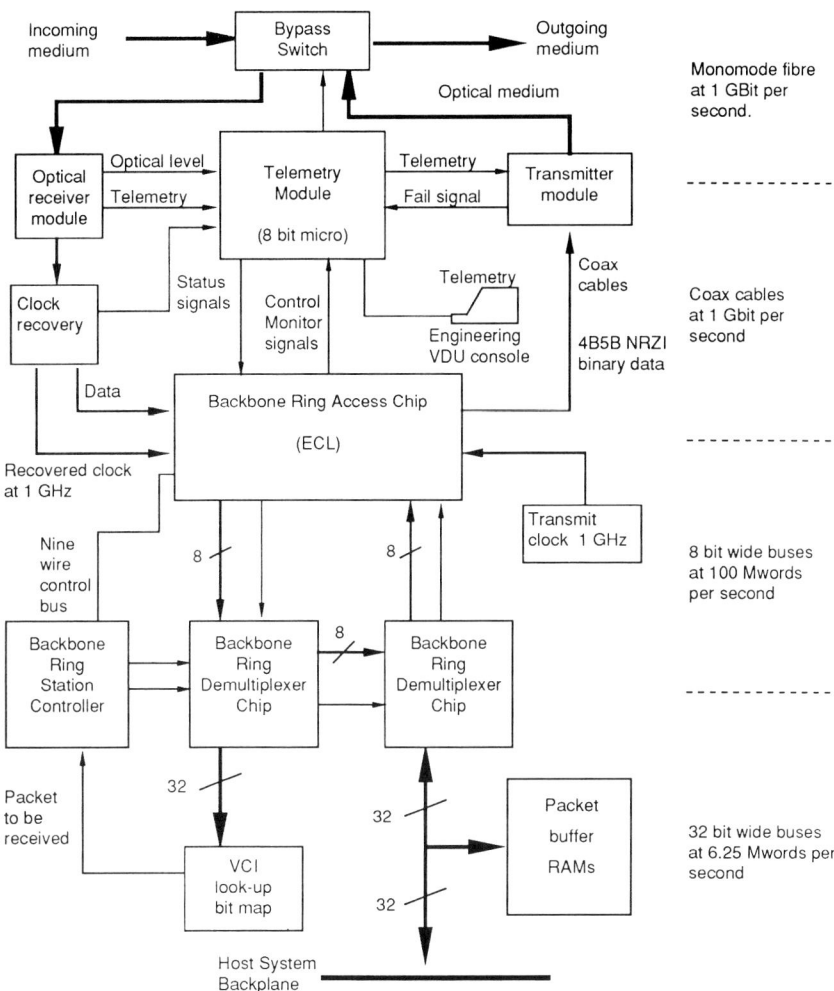

Figure 1: Block diagram of a simple CBN station.

Header	Full Monitor Type	Data area for four mini-packets. Each contains nine 32 bit words and including the routing tages.	Response Qualifier	CRC
(4)	(4 + 4 + 4)	($4 \times 9 \times 32 = 1152$bits)	(4 + 4)	(12)

Figure 2: CBN frame. It contains four CFR size slots.

3 V1S Interface

The first implementation of the CBN station uses the V1S CBN station interface. This interface provides access to the network over the VME bus being used with the Motorola MVME147 68030 (20 MHz) card. This interface was designed for minimum complexity, and their raw performance is simply determined by the bandwidth of their buffer RAMs.

All cells waiting to go onto the network or which have just been received share the same RAM array, which consists of four single-ported bytewide CMOS static RAMs, operating at a fixed rate 3.2 MHz. Therefore their terminal bandwidth is 100 Mb/sec, which is also the peak rate that data will be received off the CBN ring, owing to its multi-channel architecture. Each V1S stations receives from only one channel, allocated on a per station basis, and always transmits a cell on the appropriate channel for the destination station.

From the functional point of view, the RAM array is organized in one RX FIFO and four transmit FIFOs one for the every channel. Each FIFO is implemented as a circular buffer 256 cells long.

The RAMs are shared by both the receive and transmit sides of the station, and also no word of data can be sent or received without first passing, in FIFO fashion, through them. The essential point is that there is contention for the RAM, and the maximum data rate through the stations, when simultaneously transmitting and receiving, is 25 Mb/sec.

The V1S interface performs no protocol processing; cell headers are copied over the VME interface, when generated by or for checking by the host processor board. With V1S there is no DMA, each 32 bit word must be copied by the processor.

The V1S interfaces help the CPU in one way: they allow received cells to accumulate in the interface, in the receive FIFO, until a cell is encountered which generates an interrupt condition. The interface looks in the cell data fields which are specific to the MSDL segmentation and reassembly protocol (details in the next section). The interrupt condition is programmable on a per VCI (Virtual Circuit Identifier) basis, and can set to interrupt on:

1. end fragment cell from an MSDL PDU only,

2. beginning MSDL fragment cell only,

3. the both mentioned above,

4. a per cell basis (every cell on that VCI).

No interrupts are required for transmitting; the transmit side is assumed to be always ready, which it is, owing the light loads on our current network.

4 Protocol Architecture

The CBN is currently running under the MSNA (Multi-Service Network Architecture) protocol stack supported by the Wanda micro-kernel. This was originally developed

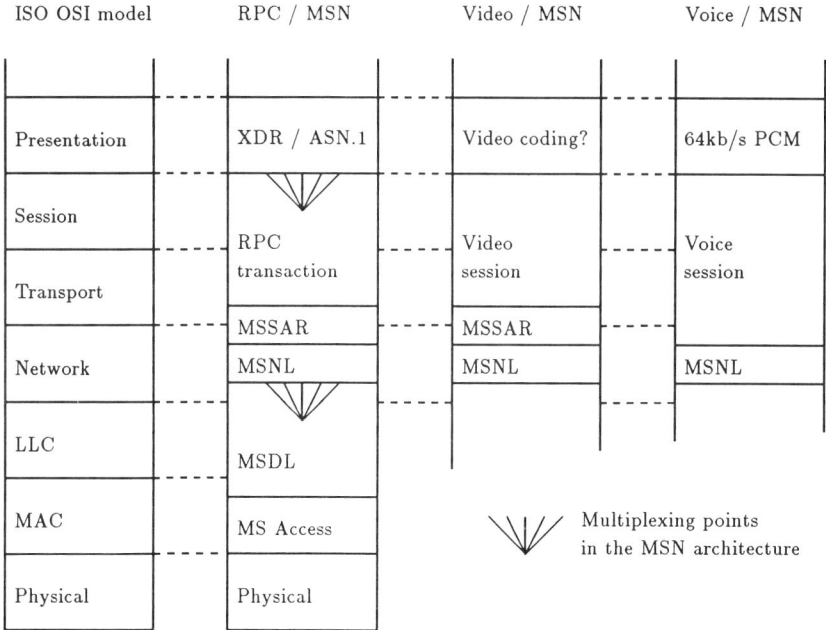

Figure 3: Relationship of MSNA elements to OSI.

in the Cambridge Computer Laboratory and is currently ported onto many popular workstations and VME boards.

The layered reference model [3] for the MSNA architecture is presented in Figure 3 Certain aspects of this model are similar to those of the ISO OSI reference model. The use of different layers to provide services is still present.

The network layer protocol MSDL (Multi-Service Data Link) performs fragmentation and re-assembly on a per-VCI basis. MSNL (Multi-Service Network Level) has no impact on the performance results reported here, since in the MSNA, network level interconnections are not multiplexed over MSDL virtual circuits. MSNL simply performs out-of-band connection set up. The MSDL layer is based on lightweight virtual circuits, referred to as associations, where there is no hop-by-hop error recovery, and any node involved in the circuit is free to unilaterally terminate it at any time.

The MSDL protocols can accommodate any cell size, but when used over fixed size cell network such as the CBN, MSDL must use the physical network size. The CBN was designed with the cell size the same as was used on the CFR [4], since both preceded the CCITT adoption of 5 + 48 bytes [5]. The CBN (and CFR) use a 4 + 32 format, with the header including 16 bit VCI and 16 bits of SAR information in the header (Figure 4 and 5).

This paper refers to the CBN V1S interface evaluation, so the main issue is of the efficiency of the MSDL implementation on the current hardware. So that the MSNA

Type	VCI	Data Field	Response
1	16	256 + 16 = 272	1

Figure 4: Backbone Ring Cell Format. Field sizes are in bits. The response and type bits are supported in the hardware, but not used by MSDL.

VCI	Sequence	RID	Part	Start
16	8	6	1	1

Figure 5: The format of the first 32 bits of a CBN cell when MSDL is being used for block fragmentation. Field sizes are in bits.

software suite could easily be ported onto multiple different network types and configurations, the subset of functions which have to be implemented in the network driver was limited only to the fragmentation operation and the hardware dependent send and receive operations. The more sophisticated re-assembly operation, which is not as hardware dependent, was performed in the separately compiled MSDL layer. Hence on the transmit side, the driver software has direct access to the association's I/O buffers, but the same is not true for the receive side. This approach is fully justified by the demand of easy portability, but in the context of very high speed networking, needs more careful consideration.

The implementation of MSDL for the Backbone Ring uses the first two data bytes of a cell for fragmentation and re-assembly information. As shown in Figure 4, only the first 32 bit word of the cell need be manipulated by MSDL protocol. The RID is a re-assembly indentifier which is common for all cells from one block. The Start flag is set for the first cell from a block and for this first cell the Sequence number contains the number of cells that compose the block. The Sequence number counts down for each cell so that a value of one indicates the last cell of a fragment. If the Part full flag is set, then the cell body does not all contain valid data and the number of valid bytes is stored in the last byte of the cell.

5 Preliminary Performance of V1S Station

The performance results presented in this section were measured over a CBN of approximately 10 km long working at a frequency 512 Mhz. The rotational latency was 22.7 microsecond the physical ring contained 7 CBN frames, giving a total of 28 slots. The ring had no other traffic during the experiments. The experiments concerned:

1. Measurement of the maximum transmit operation speed,

2. Analysis of the data transfer speed achieved between two CBN stations,

3. Measurement of the two-way delay transfer time (ping time).

According to the former description, the MSDL block of data is directly copied by the driver software from process I/O buffer space, hence the final transmit operation is

no.cell	Block length [Byte]	Rate [Mbps]
1	32	1.2
2	64	2.28
5	160	4.76
10	320	7.31
20	640	9.88
30	960	11.27
40	1280	13.33
44	1408	14.54
63	2016	13.33
125	3200	14.54
250	6400	15.23

Table 1: Transmit data rate versus number of cell per block.

No cell	Block length [Byte]	stage 0 rate [Mbps]	stage 1 rate [Mbps]	stage 2 rate [Mbps]	stage 3 rate [Mbps]
1	32	0.87	0.90	0.91	0.91
2	64	1.51	1.62	1.68	1.74
3	160	2.78	3.04	3.44	3.6
10	320	3.8	4.3	5.15	5.8
20	640	4.65	5.4	6.88	8.15
30	960	5.03	5.99	7.77	9.48
40	1280	5.25	6.28	8.23	8.88
44	1408	5.31	6.36	8.40	10.4
63	2016	5.38	6.74	8.75	11.4
125	3200	5.8	7.1	9.48	12.5
250	6400	6.0	7.34	9.89	13.33

Table 2: Transfer data rate versus number of cells per block for different stages of the receive side modification.

determined by: buffer handling overhead, fragmentation operation overhead, the speed of copying the data to the FIFO in the interface and the underlying performance of the hardware communication subsystem itself. The obtained results have been shown in Table 1.

The increase of transmit speed speed with data block length is explained by the amortization of buffer handling cost and the reduction in frequency of interaction with the driver. For the long data blocks the obtained in-block transmit data rate was 16 Mbps. This corresponds to over 18 Mbps actual transmit rate (headers + data).

The first measurements of the data transfer speed between two CBN stations were rather disappointing. A careful review of the receive side software made clear that a few improvements were possible:

1. Removing the additional copy operation from the CBN receive FIFO to the buffer inside the CBN driver by direct copying to the association buffer.

No. cell	Block length [Byte]	First delay [μs]	Last delay [μs]	Each delay [μs]
1	32	562	562	562
2	64	587	625	587
5	160	694	818	694
10	320	875	1150	881
20	640	1225	1813	1256
30	960	1581	2469	1619
40	1280	1937	3125	2019
44	1408	2075	3394	2150

Table 3: Two-way time delay versus length of data block for the three different interrupt condition settings.

2. Merging of the association handling and re-assembly operation within driver receive interrupt handler, that eliminates an additional up-call.

3. More efficient coding of the combined association handling, re-assembly, and copy algorithms.

The obtained transfer rate after each step of improvement is given in Table 2. Stage 0 corresponds to the original version of software. For the longest data blocks, the obtained transfer data rate was over 14.2 Mbps, which corresponds to over 15.6 Mbps transfer rate and was limited by the speed of receive process.

The two-way delay transfer time (ping time) is the time for a block data to be sent to a process in the remote station and back to the sending process. This parameter characterises the latency of the communication subsystem and speed of the Wanda kernel. The obtained results for the improved version of software and the different setting of the receive VCI condition are shown in Table 3. The table shows that the response time is almost independent of whether the interrupt condition is set for start of block or on every cell. In contrast, setting the interrupt condition for the end of data block breaks the receive-transmit process pipelining and leads to a substantial increase of the latency. Evidently there is a trade off between context switching overhead and increase of latency caused by the breaking of the transmit/receive pipeline and the consequent loss of cut-through.

6 Future Work and Conclusion

On the measurement side, we need to perform similar experiments on a more heavily loaded network. The V1S interfaces include false traffic generators, so that the the load can be artificially increased. Another interesting case is the duplex communication situation, where the interference between simultaneous receive and transmit processing needs examining. Finally, the effect of the ATM substrata on higher level protocol performance needs measuring.

As far as hardware development goes, we must bear in mind that V1S interfaces were designed as a cheap interface with about 30 Mbps simultaneous receive and transmit throughput. There has been an impact on host throughput as a result of the current CBNs being operated at only about half their design speed, that is 512 MHz instead of 1000 MHz. This has increased the contention resolution time for FIFO buffer access and the currently obtained 12 μs copy time might have been expected to be about 8 μs in a full speed system.

A simplistic, single cell DMA engine is being designed. This should be able to copy a cell in about 4 μs, and if this copying time becomes parallel with the per-cell protocol processing activity on the microprocessor, an overall speed up of about 3 times may be envisaged.

In general, it seems that the only satisfactory solution needs some form of dual-processor architecture, where the 'host' processor is relieved of as much context switching overhead as possible. The results in this paper enable us to predict that a current technology, general purpose processor might be able to keep up with per-cell processing for rates up to 50 Mbps (75 Mbps with 48 byte cells).

Acknowledgements

Many thanks for invaluable help are due to Joe Dixon, Mark Hayter and the rest of the Wanda group at the Computer Laboratory. Thanks are due to the rest of the systems group, and especially to D (Mac) McAuley for the MSNA picture. Dimitris Lioupis, David Milway and Andy Hopper greatly contributed to the Backbone Network project. The work in this paper was funded by Olivetti Research Ltd. and Esprit Project OSI95.

References

[1] 'The Cambridge Backbone Network.' DJ Greaves D Lioupis and A Hopper. IEEE Infocom 90, San Francisco June 1990. Also in Proceedings European Fibre Optic Conference (EFOC/LAN 88) Amsterdam, June 1988.

[2] 'Backbone Ring V1s Station Interface Specification.' DJ Greaves. Olivetti Research Limited 1990.

[3] 'Protocol Design for High Speed Networks.' University of Cambridge TR 186. DR McAuley, December 1989.

[4] 'The Cambridge Fast Ring Networking System.' A Hopper and RM Needham. IEEE Transaction on Computers, Vol.37 no 10, October 1988.

[5] CCITT I series recommendations, especially I.321, Geneva May 1990.

[6] 'An analysis of TCP processing overhead.' DD Clark, V Jacobson, J Romkey and Salwen, IEEE Communication Magazine, June 1989.

Telecommunication Services for Developing Economies
J. Filipiak (Editor)
© 1991 Elsevier Science Publishers B.V. All rights reserved

Simulation Study of Some Aspects of Fairness in DQDB

G. Dallos

Technical University of Budapest, Institute of Communication Electronics, Stoczek u 2, Budapest, H1111 HUNGARY

Abstract
The paper deals with the fairness problem of DQDB. But instead of aiming bandwidth balancing between simultaneously transmitted messages, it looks for some other method which can guarantee the same average message delay for any station along the bus. The results are obtained from simulation using a traffic model appearing sensible.

1. INTRODUCTION

The demand on fairness in DQDB [1] resulted in a solution, namely: Bandwidth Balancing. It is easy to show that the original aim turned into a little funny solution or at least the result can generate strange situations.
Look at the next simple example! Let two long messages overlap each other, i.e.,

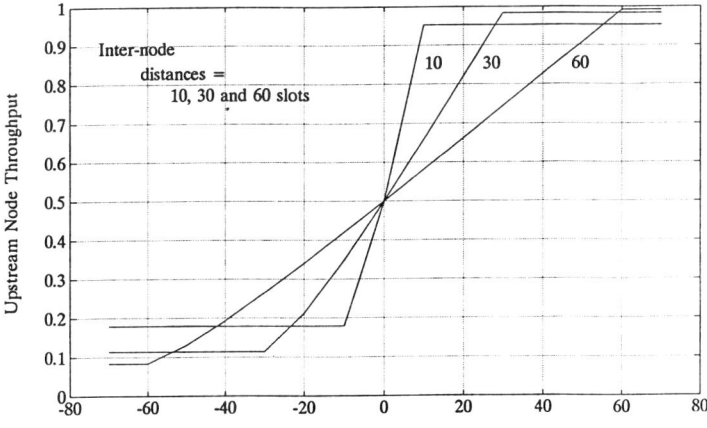

Figure 1. The "soft" FIFO service of a "pure" DQDB

during the transmission of the first message a second one arrives at an other node. The service of the two messages in a simple DQDB, that works without Bandwidth Balancing, depends on the time difference of the arrivals of the messages and the positions of nodes on the bus. In Figure 1 [2] we illustrate how is the total throughput of a DQDB bus divided between two nodes by showing the upstream node throughput vs. start time difference between message arrivals.

We can say that the service type of two simultaneous long messages is the so called FIFO but it must be taken into account that the interpretation of the arrival times is necessarily different from the real arrivals in a concentrated queue. Really the arrivals regarding each other are always interpreted "softly" and never exclusively.

However, what happens in the case of bandwidth balancing? Shortly after the second arrival, the two messages are served equally, independently of the arrivals' order. Why? Because of the misinterpretation of fairness!

We wanted to avoid hogging of the bus by anybody and the result is an unfair service method which, and it is more important, cause general "unhappiness", in some cases.

How will the two overlapping messages be served in a DQDB with Bandwidth Balancing? As we said, shortly after the second arrival the service of the first message will be decreased to half speed and the second message service will be started with the same half-speed, too. What is the result? The complete service of the first message will be delayed but without any advantage of the second one. The first message got some disadvantage, considering a FIFO type service, while the second message service will be finished exactly at the same time as it was served in FIFO order. Thus we can say that regarding all participants the method results in a general "unhappiness".

The subsequent part of the paper contains six more sections. Section 2 deals with a measure of a parameter of quality of service. It gives an idea to investigate the order of service of DQDB comparing to FIFO. Section 3 tries to make clear what we mean on hogging. Section 4 specifies the assumptions of the study. Section 6 gives the summary of simulation results and finally Section 7 sums up our conclusions.

2. A MEASURE OF A PARAMETER OF QOS

It is probably true that in a queuing system customers regard a FIFO service to be fair, but they are happy if they can have any advantage on others. The advantage can be got by priority schemes, and of course anybody who want to got an advantage must pay for it, but it can be got by a type of politeness. A customer can give an advantage over itself in the case when an other customer needs a short service period relatively to its own needs. In this case the first customer, who gives the advantage, receives a slight disadvantage but the other customer, who arrives later and gets the advantage, can receive a really notable advantage and therefore it will be really happy. We can formulate the total "happiness" i.e. the closeness to FIFO service in the system as follows:

$$H = \left[\lim_{n \to \infty} 1/n \sum_n \frac{\text{actual system time}}{\text{FIFO system time}} \right]^{-1} - 1 .$$

Sum up the actual length of time passed during each customers stayed in the system but it must be compared to those which would be needed in an exact FIFO queue. If any part of the sum is divided by the number of customers involved in an actual "change", then 1/"sum"-1 equals to the "happiness". The negative value means "sadness", i.e., a worse service compared to FIFO.

To illustrate this idea let us show two typical cases. The first case shows a "meeting" of two long messages in Fig. 2. In the first two line the messages' arrival time are shown, together with their lengths. The next line shows a clear FIFO service, illustrating the messages' actual system times. In the next line the service of DQDB with Bandwidth Balancing is sketched. It shows that the service of message B starts at its arrival time but, together with message A just at half speed. More precisely, Bandwidth Balancing has a short transient time during the "balancing" period and a true bandwidth sharing starts just after this time period. We neglected in these figures that transient time periods.

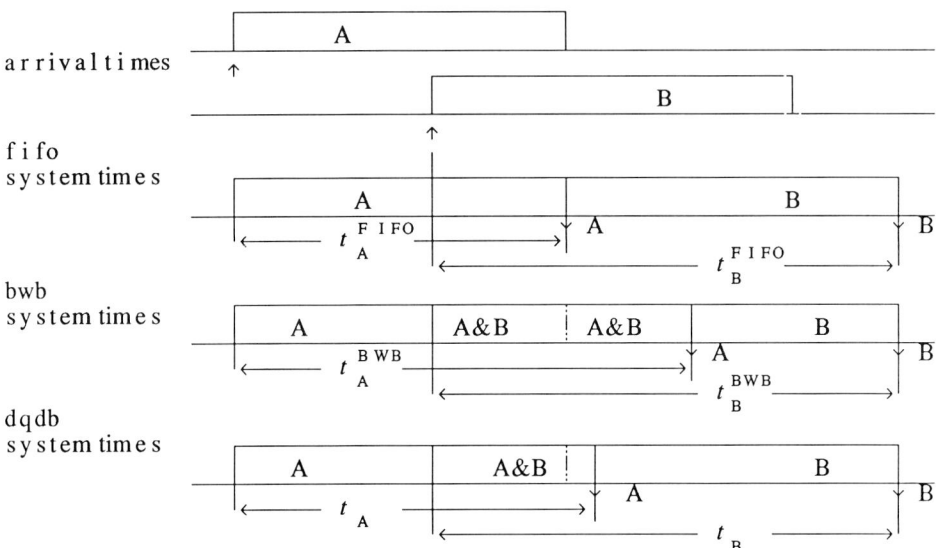

Figure 2. Services of two simultaneous long messages

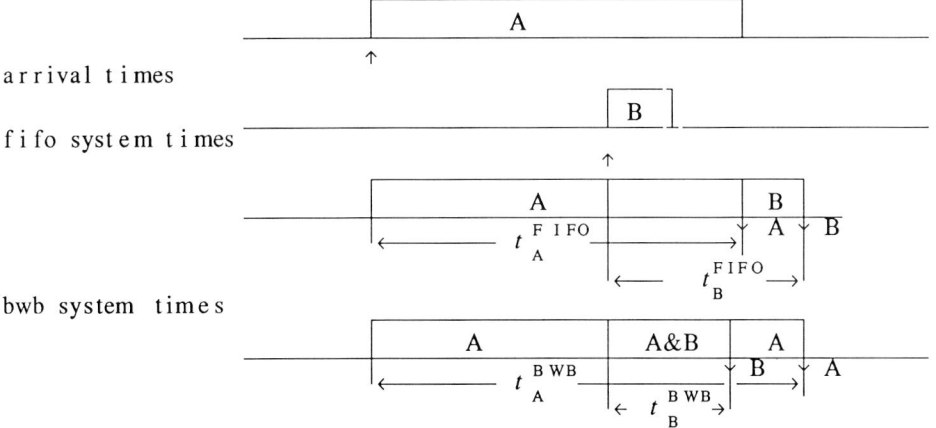

Figure 3. Service of a long and a short messages simultaneously

The last line of Figure 2. illustrates the service of two long messages by DQDB without Bandwidth Balancing. Since the second message arrived much later than the first one, it gets a very small fraktion of bandwidth and therefore it does not hinder the transmisson of the first message significantly. It corresponds to the "soft" FIFO service of DQDB illustreted in Fig. 1.

The second case shows a "meeting" of a long message and a short one. The service of the short message starts promtly after its arrival and it delays the service of long message moderately. However, the short message's gain is considerable. It is easy to see that without Bandwidth Balancing, the pure DQDB service results in a very similar service in this case. A very short, eventually a single segment message will probably be served during the service of the long message and the respective system times of the messages will be very close to those of Bandwidth Balancing.

Thus we can conclude that Bandwidth Balancing can cause disadvantages in some cases. Our first example shows a typical case when Bandwidth Balancing is non-preferable.

3. THE HOGGING

We must make clear distinction between two different cases:
Case 1: any node starts to transmit a long message,
Case 2: during a transmission of a given node, one or more other messages are queued up in the same message transmission queue.

In the first case there is no hogging at all. If any other node wants to transmit a message during the service of the long one and it must wait to start transmission, it

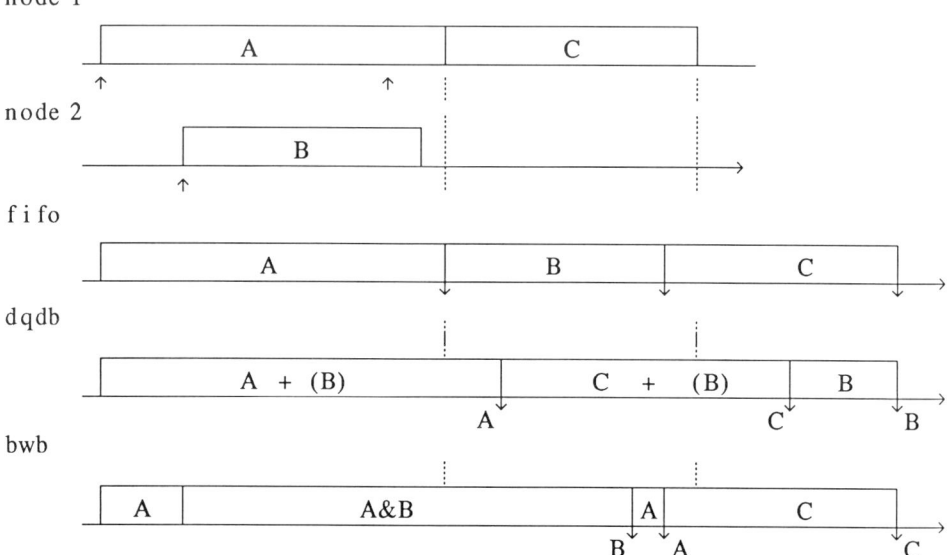

Figure 4. Illustration of hogging

simply means that there is a FIFO service. On the contrary, in the second case, if any other node's transmission will be unfairly delayed by messages queued up at a transmitting node, it must be regarded as hogging.

To illustrate this idea we sketched a case in Fig. 4 when messages arrive at two nodes in a fairly typical manner. Let Node 1 be the upstream node and let the distance of nodes be equal to the arrival time difference between message A and B. Considering the possible services, we get an interesting result shown at the lower part of the figure.

Thus we can say, if we want to fight again unfairness caused by hogging, then we must find algorithms to avoid the sticking of messages together, keeping back messages, arrived earlier at other nodes, for getting service in time.

4. ASSUMPTIONS

All of our view is based upon two assumptions:
- The messages to be transmitted are such natural units of communication between parties that their fragments are useless.
- Any node is able to make use of a bus transmission capability at its maximum speed.

The first assumption means that the parties taking place in a communication can act iff they received a whole message. For example, let the message be a picture to be transmitted and the parties want to discuss some details of the picture in their conversation. It is clear that usually fragments of the picture is useless for them. Thus if in two simultaneous conversations the parties want to send pictures (i.e. big files) to each other, it is clearly disadvantageous if the first picture transmission is slowed down by the second picture's transmission started afterwards.

The second assumption is important from practical point of view. It is far from trivial that any node can be able to transmit (and receive) with the high speed of a DQDB bus i.e. eventually at 150 Mbps. To mention an example for proving it, Bellcore introduced some access classes for customers of its SMDS network, i.e. a customer connected to SMDS at DS3 rate can restrict himself as well as the network to transmit and receive only at lower speeds on the average.

Thus the second assumption says that nodes are capable for bursty transmission, even more during the transmission of segments of long messages, a higher layer of the node is able to prepare one or more messages for transmission which can results in a hogging.

5. TRAFFIC MODEL

In our simulation study two features of the traffic are taken into account.

First of all, we assume that users of multimedia communication networks generate messages which can be collected into two different groups. Messages belonging to the first group can be called as short ones. These messages can be extremely short control messages, short letters or side informations of a conversation. The other group contains long messages, much longer than those belonging to the first group. Examples of long messages can be long data files, long articles or chapters of a book transmitted for editing, or high resolution graphics as medical diagnostic pictures transmitted for consultation.

In Figure 5. we illustrated our traffic model by a distribution density function of length of messages. The next parameters can be set for simulation: the minimum and maximum values of the density function plus the ratio of the number of long messages to the number of short ones.

On the other hand, we assume that all the traffic is "internal" concerning a DQDB subnetwork and each node sends and receives the same amount of messages on the average. Since the buses of DQDB are unidirectional, a given connection between two nodes can take place only on the appropriate bus. Thus the traffic, generated by nodes located downstream on one of the buses, decreases linearly with the nodes' positions and the

Figure 5. Illustration of the distribution of length of messages

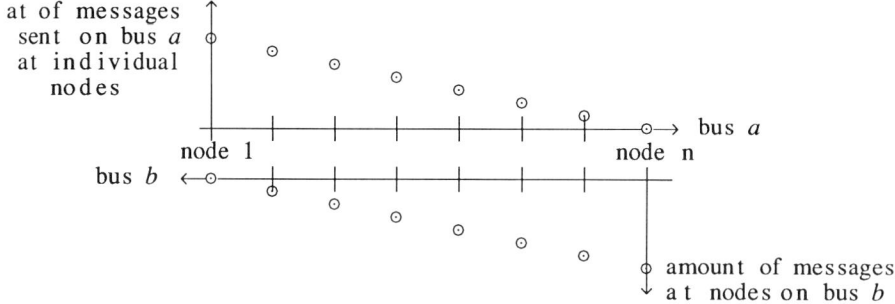

Figure 6. Illustration of the distribution of message generated at nodes

last node does not send any message on this bus but sends all messages on the other one. This is illustrated on Fig. 6.

It is worth to note that the delay figures depend severely on the distribution of messages.

6. SIMULATION RESULTS

The following figures show our simulated results. We choosed from the exhausting mass of results some typical figures. For the sake of simplicity, the number of nodes along the buses are only seven. The length of the buses equals 120 slots. The minimum length of short messages equals one segment, and maximum of short messages is 200 segments. The length of long messages was between 9500 and 10500 segments. The average of the latter ones correspond to half a megabyte long files. The short messages represented .95 % of total traffic and only 5 % of the traffic were long messages. The total traffic was 50 % of the bus capacity on each of the buses.

Fig. 7 and Fig. 8 show the delay "curves" for DQDB. We used a solid line but it has no sense between nodes.

Figure 7. shows the average delay at individual nodes along Bus A. Node 7 does not send any message on this bus but sends all of its messages on Bus B. It is clear from the figure that the average delay of messages at individual nodes inrease downstream.

Figure 8. shows the average delays of messages at nodes on both buses. The result of the increasing delay is a delay "hill" at the middle positioned nodes. But we must note that the delay differences are not severe.

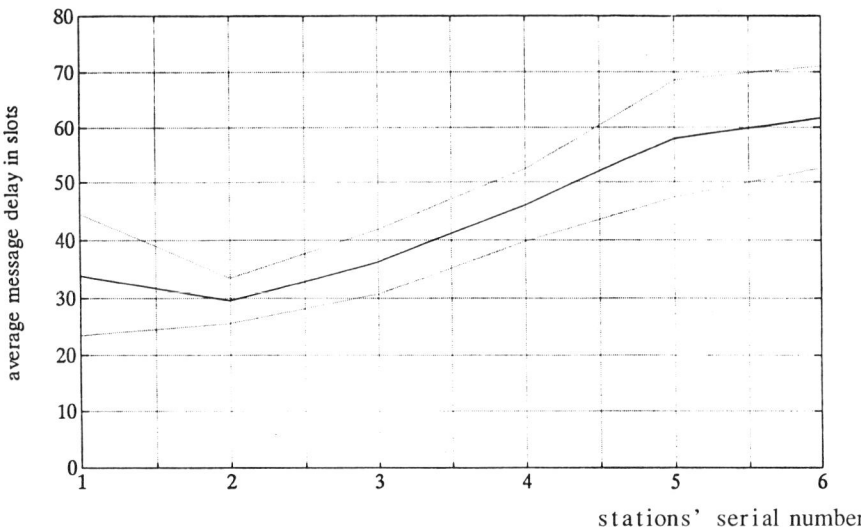

Figure 7. Average message delay of DQDB at nodes on bus a

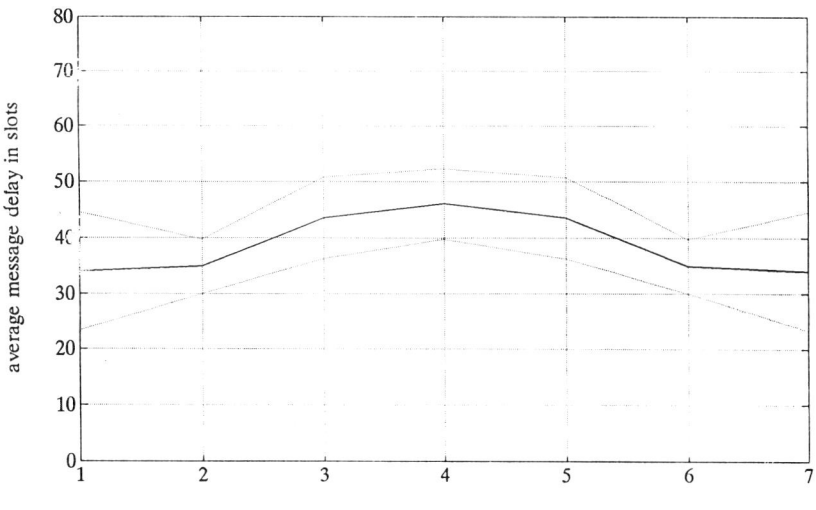

Figure 8. Average message delay of DQDB at nodes on both buses

Fig. 9 and Fig. 10 show the delay "curves" for Bandwidth Balancing. The result shows that BWB with a parameter $\beta = 8$ resulting in a delay curve changed dramatically. Fig. 10 shows a valley of delay in the middle of bus but it is pretty deep.

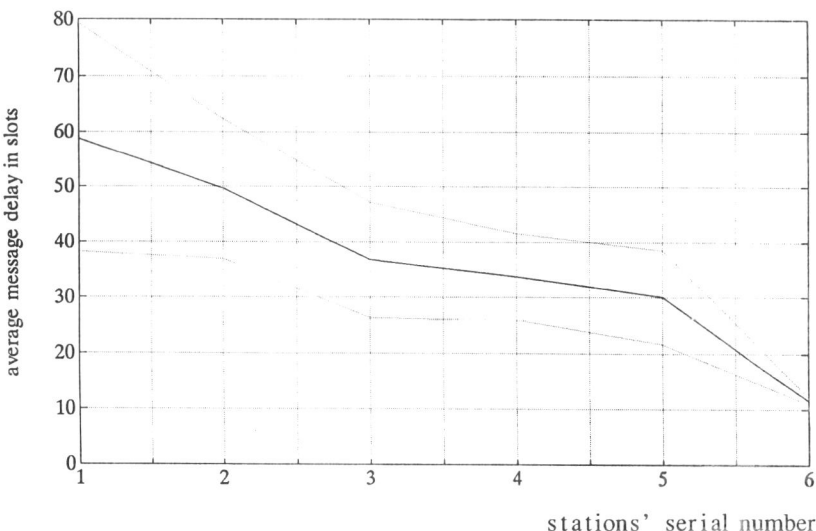

Figure 9. Average message delay of BWB at nodes on bus a

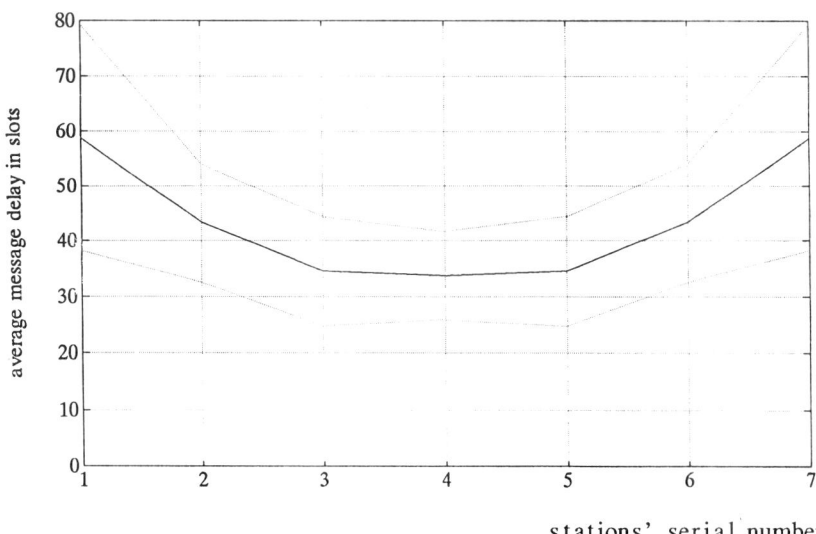

Figure 10. Average message delay of BWB at nodes on both buses

Finally Fig. 11 and Fig. 12 show the result of a modified version of Bandwidth Balancing. In this case, the nodes leave free slots only between messages. This method aims the avoidance of hogging. The wasted capacity can be much less than that of BWB.

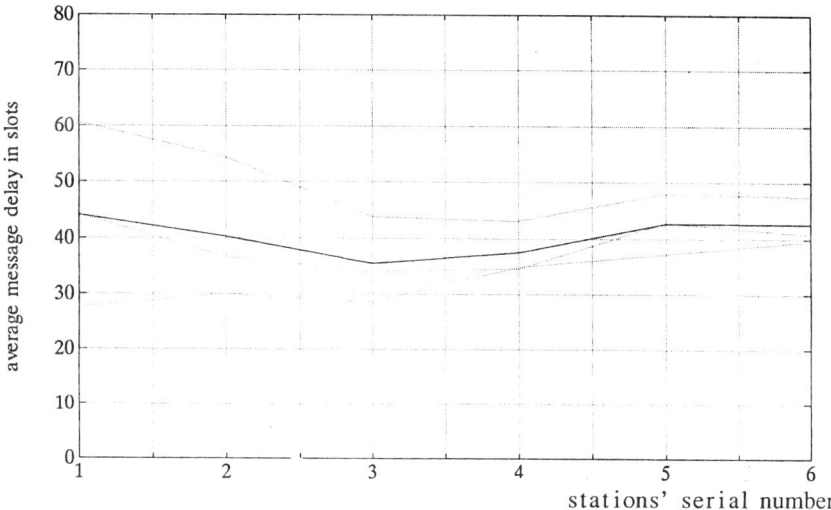

Figure 11. Average message delay of modified BWB at nodes on bus *a*

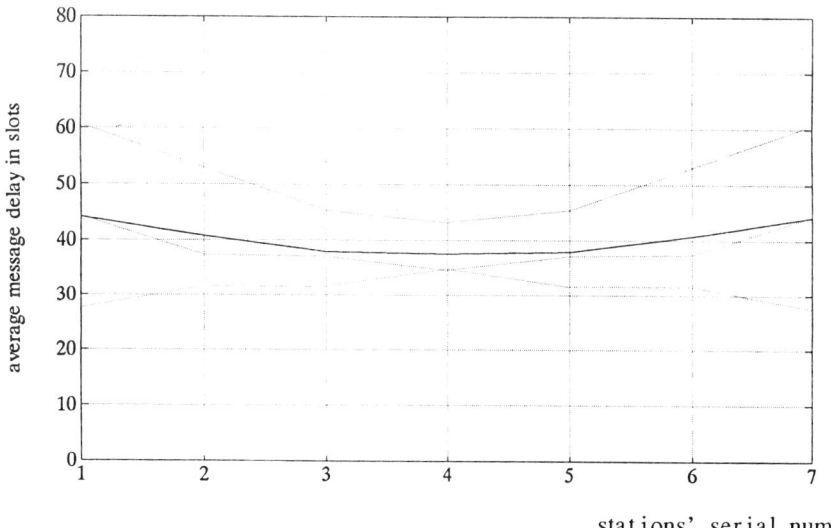

Figure 12. Average message delay of modified BWB at nodes on both buses

Whereas, we must note that this method does not aim to share the capacity of bus between simultaneous messages. It rather prefers the "soft" FIFO service provided by DQDB.

Fig. 12 shows that the suggested modification results in a rather fair distribution of message delay along the buses.

7. CONCLUSIONS

We can conclude that the delay inequity of DQDB is tolerable.

The BWB cause unnecessary delay increase at nodes positioned near the head of buses. It is understandable if take into account the distribution of traffic generated by nodes along the bus.

It is possible to use the so called Idle Capacity scheme of Bandwidth Balancing efficiently for avoiding the hogging effect but it can be questionable to do anything.

8. ACKNOWLEDGEMENT

The author wish to thank Mr. Laszlo Gyalog for his great help in getting the simulation results.

9. REFERENCES

1 IEEE 802.6 Working Group, "Distributed Queue Dual Bus (DQDB) Subnetwork of a Metropolitan Area Network (MAN)", Draft D15 of Approved Standard,
2 E.L. Hahne, A.K. Choudhury and N.F. Maxemchuk, "Improving the Fairness of Distributed Queue Dual Bus Networks," IEEE INFOCOM '90, San Francisco, 1990, pp. 175-184.

Switching Techniques

ON RELIABILITY AND THROUGHPUT OF INTERNAL INTERCONNECTION NETWORKS OF ATM - SWITCHING NODES

H. Dahms

Institute for Communications Engineering, Fachhochschule Lübeck, Germany

Abstract

This paper deals with the analysis of four different structures of interconnection networks, which are based on a Baseline-network. Investigations have been made concerning their throughput and reliability to answer the question whether these networks are appropriate for ATM-switching nodes, too. The characteristic values are calculated by means of closed form solutions and simulations. Results are presented in examples and diagrams.

1. INTRODUCTION

Interconnection networks for ATM-switching nodes should show the following characteristic features:
- high throughput, i.e.
  1. no central control unit for routing and switching functions,
  2. the number of switches, waiting time in front-end-processors and network-delay be as small as possible,
- good capability of extension
- less expenditure of hardware
- high fault-tolerance

In the sequel, an interconnection network be called k-fault-tolerant, if there is at least one link between any pair of input-/output-ports for the case that j≤k switches have broken down. The fault-tolerance is mainly based upon the existence of conjugated switches.

Two switches of stage k of a Baseline-network be called conjugated, if they are linked to the identical switch of the stage k+1. Their binary addresses be $(x_{n-2}...x_0)$ and $(y_{n-2}...y_0)$. If the equations $x_i = y_i$, $1 \le i \le (n-2)$ and $x_0 <> y_0$ are valid, these switches be conjugated.

If one switch fails, the preceding switches try to transmit the packets via the corresponding conjugated switch to the addressed destination port. Therefore switches of the stages 0 to (n-2) must be able to choose between at least two output-ports.

The system configuration presented in section 2 deals with the cell layer of an ATM-node. Models for the characteristic values are derived in section 3. Finally, in section 4, Baseline and extended Baseline-networks are compared to each other. A conclusion follows in section 5.

2. SYSTEM CONFIGURATIONS

2.1 Baseline-network

Baseline-networks are a subset of the binary Delta- and Banyan-networks with the following characteristic features:

- Banyan attribute
- self-switching (self-routing)
- 2x2-switching elements
- n stages between $N=2^n$ input- and output-ports
- each stage consists of N/2 switching elements
- the total number of 2x2-switches is given by $N/2*n$

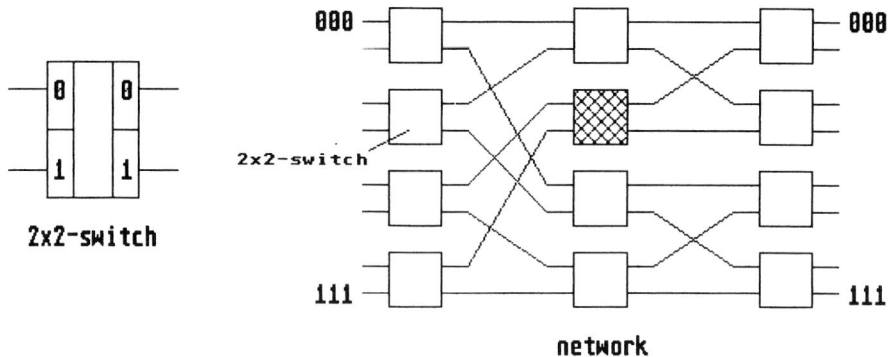

Figure 1. Baseline-network with n=3 stages

In virtue of the Banyan attribute, however, Baseline-networks are not fault-tolerant. The lower part of input-circuits of Figure 1, e.g., can not be connected with the upper part of output-circuits, if the cross-hatched switching element will break down [1]. Therefore some modified networks be investigated, too.

2.2 Extended Baseline-networks

Generally you can get a better fault-tolerance and a higher throughput of a network by extra stages, lines, switches, networks or extra passes and an increased degree of the switches or finally by a combination of the above mentioned steps [2]. In this paper modified Baseline-networks with extra lines and an increased degree of the switches are investigated which have the following characteristic features:

- no central control unit
- small number of stages which have to be passed
- good capability of being extended
- fault-tolerance
- the same destination addresses as the original Baseline-network

2.2.1 Augmented Baseline-Network, ABN-0

The network ABN-0 can be developed from the Baseline-network by means of the following rules [3]:

- each input-port of the switches of the first stage is connected in series with a 2x1-multiplexer with the same address
- an auxiliary output-port a-out of switch $(x_{n-2}...x_0)$ is connected with a-in of switch $(y_{n-2}...y_0)$, if the following equations are valid:
  $x_i = y_i$, $i <> 1$ and $x_i <> y_i$, $i = 1$
- all other ports are connected with each other in the same way as it was shown for Baseline-networks (section 2.1)
- each output-port of the last stage is followed by an 1x2-multiplexer

2.2.2 Augmented Baseline-Network, ABN-1

Compounding those two switches of the network ABN-0 which are connected by an auxiliary link to one 4x4-switch yields the second modified Baseline-network, ABN-1 (Fig. 2). This arrangement leads to a reduced and constant number of switches between any pair of input-/output-ports [1].

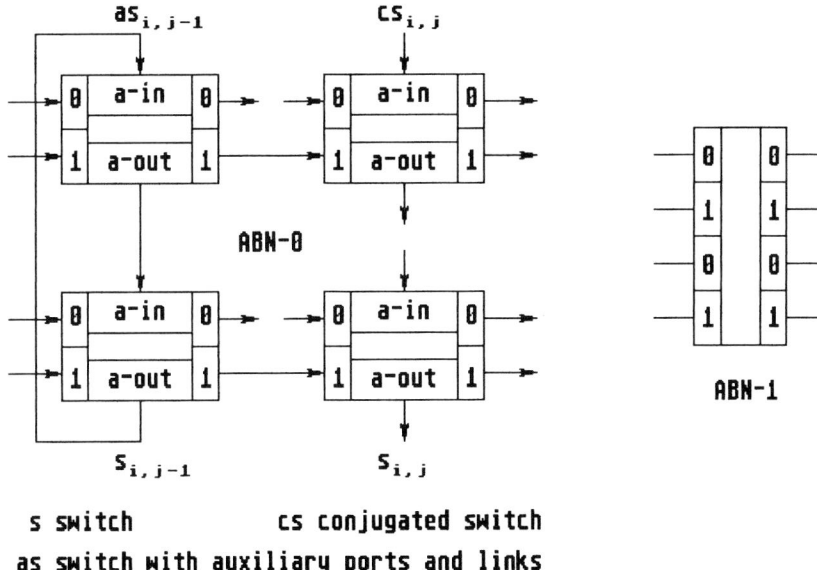

s switch   cs conjugated switch
as switch with auxiliary ports and links

Figure 2. Switches and links of the modified Baseline-networks ABN-0 and ABN-1

2.2.3 Merged Baseline-Network, MBN

The rules for developing an MBN [4] out of a Baseline-network are:

- Replace each 2x2-switch by a 4x4-switch.
  Additional input-/output-ports are called C-ports, original ones S-ports.
- S-ports are linked like the ports of a Baseline-network. C-output-ports are connected to the C-input of the conjugated switches (Fig. 3).
- The last stage is connected in series with N 2x2- and 2x1-multiplexer.

The network MBN is fault tolerant but it requires twice as much links between its stages as the original Baseline-network [1].

2.3 Routing algorithm

As mentioned above all these networks are self-routing or self-switching, respectively. The original Baseline-network has only one path through the total network between each combination of input-/output-ports. As each switch of the modified networks has extra lines or ports we have to regard their special routing algorithms in more detail.

2.3.1 ABN-0

In order to increase the performance, additional links are used. Each switch tries to transmit the cells to that output-port which corresponds to the destination address. If this is impossible, each switch will try to use its auxiliary output-port. If there is no breakdown of the input multiplexer and the corresponding switch each source is sending its cell to the multiplexer with the equivalent address, otherwise to the corresponding second multiplexer.

Each switch of stage k, ($0 \leq k \leq n-2$), is receiving cells from the input-ports 0/1 and a-in, whereby the ports 0/1 have a higher priority than the port a-in. The switches of the last stage choose the link to the output-multiplexer corresponding to the n-th bit of the destination address. The output-multiplexer have to interpret the total destination address and to switch a link to the equivalent destination point. If cells can't be transmitted by one of these possibilities, they will get lost.

2.3.2 ABN-1

The routing algorithm is quite similar to that of the network ABN-0 with the exeption that the 4x4-switches of stage k ($0 \leq k \leq n-2$) consist of two logical 2x2-switches and may receive cells at their 4 input-ports (Fig. 2). Each logical switch tries to transmit cells to that output-port which corresponds to the k-th bit of the destination address. If it is impossible, it tries to use the corresponding output-port of its second logical switch. If it is going wrong the cell will get lost.

2.3.3 MBN

Each front-end-processor is connected to a switching element of stage 0 by an S-link and a C-link (Fig. 3). At first an input-port tries to transmit cells via an S-link. If the corresponding switch works wrong, it may try to switch a C-link. Each switch prefers to transport cells which have been received at a S{C}-input-port to a corresponding S{C}-output-port. In the case of blocking or breakdown an alternative connection to the C{S}-output-port will be erected. If these attempts fail, the cell will get lost.

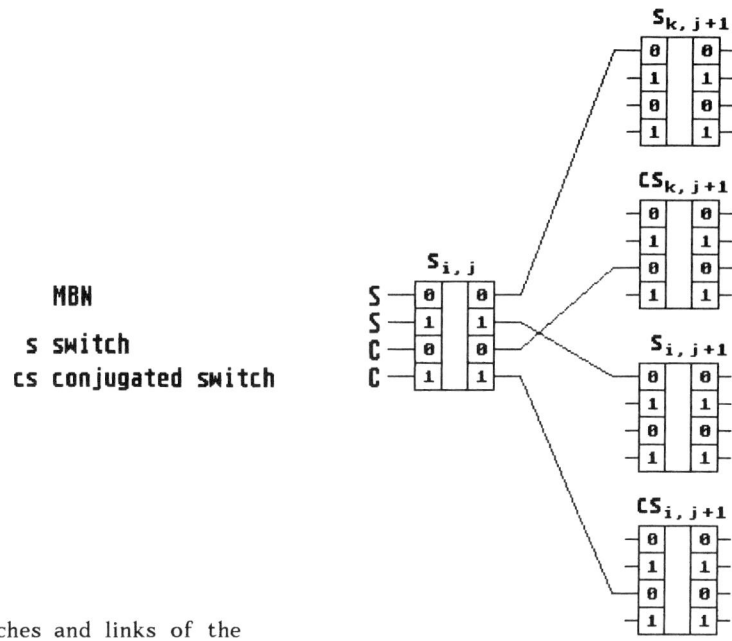

MBN
s switch
cs conjugated switch

Figure 3. Switches and links of the modified Baseline-network MBN

3. CHARACTERISTIC VALUES

The influence of topological modifications on the throughput, the reliability and the terminal availability of the interconnection network has been analysed. The reliability be the probability that a module doesn't break down within a time interval [0,t]. The terminal availability be the probability for a faultless path between any pair of input-/output-ports.

3.1 Throughput

The analysis of the throughput is based on a model with the following basic assumptions [5]:

- networks are investigated in a faultless state
- cells arrive with a probability p at the front-end-processors
- input- and output-addresses are distributed uniformly
- cells meeting a failing switch are getting lost
- the probability pa that a cell will be accepted at its destination-point be an attribute for the throughput

3.1.1 Baseline-network

The following notations are used in the sequel, additionally:

- $p(k)$ be the probability that a cell will arrive at the input-port of a switch of stage k
- $b(k)$ be the probability for a busy output-port of a switch of stage k.

It can be shown [1] that b(k) is given by

$$b(k) = p_0/2 + p_1/2 - 1/4 * p_0 p_1 \tag{1}$$

where p_0 and p_1 be the probabilities that a cell will arrive at the input-port 0 or 1, respectively, of a switch of stage k. As the destination addresses are uniformly distributed, it holds $p_0 = p_1 = p(k)$ and with eq. (1)

$$b(k) = p(k) - 1/4 * p^2(k) \tag{2}$$

Furthermore is $p(k+1) = b(k)$ so that we obtain the following recursion:

$$p(k+1) = p(k) - 1/4 * p^2(k) \tag{3}$$

As cells which have been accepted by the last switches of the Baseline-network will arrive at their destination, too, the probability pa be defined as

$$pa = p(n)/p \tag{4}$$

3.1.2 ABN-0

The analysis of the throughput is based on the model of section 3.1.1 but it has to be extended by the fact that the ordinary input-ports 0/1 have a higher priority than the auxiliary input-port a-in. Therefore the operation of a switching element is observed at two succeeding cycles φ_0 and φ_1 at which cells are processed coming from the input-ports 0/1 or from the auxiliary-port, respectively.

In addition to the notations of section 3.1.1 pa be the probability that a switch of stage k will send a cell at the end of cycle φ_0 via a-out. This probability is equivalent to the case that a switch of stage k will receive a cell at the beginning of cycle φ_1 at a-in. Furthermore, b(k) is the probability that an output-port of a switch of stage k, different from a-out, is busy at the end of φ_0. As only cells from port 0/1 are transmitted during the first period, we can use eq. (2) for the calculation of b(k).

Let us consider the probability p(k+1), $1 \leq k \leq (n-2)$, and an output-port of stage k in more detail. p(k+1) is the probability that a cell will arrive at an input-port of a switch of stage k+1. This probability is therefore given by the probability that an output-port of stage k is busy during the first period and the compound probability that an output-port of stage k is not busy and a cell will arrive at a-in, which has to be switched to an output-port during the second period. These facts yield equation (5)

$$p(k+1) = b(k) + pc \tag{5}$$

with $pc = \{1-b(k)\} * p_{aux}(k) * 1/2$.

A cell at one of the input-ports 0/1 is only switched with the aid of a-out if there is a collision with a cell from the other input-port during the first cycle. Therefore $p_{aux}(j)$ is given by

$$p_{aux}(j) = 1/2 * p(j) * 1/2 * p(j) * 2 = 1/2 * p^2(j) \tag{6}$$

Equations (2), (5) and (6) yield:

$$p(k+1) = p(k) - p^3(k)/4 + p^4(k)/16 \tag{7}$$

As the last stage of the ABN-0 is identical with that one of the Baseline-network, it holds with eq. (3):

$$p(n) = p(n-1) - 1/4 * p^2(n-1) \tag{8}$$

p(n) is again the probability that a cell has been accepted at the last stage. In contrast to the Baseline-network p(n) is not identical with the probability that a terminal has accepted this cell, too, because collisions may occur at the output multiplexer. Output-multiplexer of the last stage of the network and input-multiplexer of the terminals be combined to a 2x2-switch, which can be analysed like a stage of a Baseline-network.

3.1.3 ABN-1

The analysis is quite similar to that of the ABN-0, if p_{aux} will be considered as the probability that one of the two logical switches of the 4x4-switch will use one of the output-ports of the second logical switching element.

3.1.4 MBN

To pay regard to the service-discipline of the input-ports (S/C-inputs are allocated by priority to S/C-output-ports), the mode of operation has to be analysed at two successive cycles again. The following actions will be carried out during these two steps:

During the first one, cells are received at the S/C-input—ports and switched to the corresponding S/C-output-ports. If this is not possible the switching elements try to pass the cells to the C/S-output-ports during the second period. If this fails, too, the cells will get lost. At the end of this second cycle all accepted cells are transmitted to the next stage. The analysis of these switches is following the reasoning of section 3.1.2 bearing in mind that an MBN-switch consists of two pairs of input- and output-ports.

In doing so, we receive the following equations concerning the probability that a cell coming from stage (k-1) has arrived at an S-input-port or a C-input-port of stage k:

$$ps(k+1) = bs(k) + \{1-bs(k)\} * pcs(k) * 1/2 \tag{9}$$

$$pc(k+1) = bc(k) + \{1-bc(k)\} * psc(k) * 1/2 \tag{10}$$

pcs(k) and psc(k) be the probabilities for a collision and can be calculated by equation (6)

$$pcs(k) = 1/2 * pc^2(k) \quad \text{and} \quad psc(k) = 1/2 * ps^2(k) \tag{11}$$

bs(j) and bc(j) be the probabilities for the case that an S/C-output-port of stage j is busy at the end of the first cycle:

$$bs(j) = ps(j) - 1/4 * ps^2(j) \quad \text{and} \quad bc(j) = pc(j) - 1/4 * pc^2(j) \tag{12}$$

The input-ports of each of the n 2x2-multiplexer after the last stage of the network are connected with the S-output- and the corresponding C-output-port of a switch. Therefore the probability that an output-port of this first multiplexer is busy is given by

$$bM1 = 1/2*ps(n)\{1-1/2*pc(n)\} + 1/2*pc(n)\{1-1/2*ps(n)\} + 1/4*ps(n)*pc(n) \tag{13}$$

The 2x1-multiplexer of the destination points are linked with the output-port of a 2x2-multiplexer. An output-port of these second multiplexer is busy with the probability

$$bM2 = bM1*(1-bM1) + bM1*(1-bM1) + bM1*bM1 \tag{14}$$

3.2 Reliability and MTTF (Mean Time To Failure)

The reliability of the interconnection network essentially depends on the reliability $rs(t)$ of a switching element which is for example [6] given by

$$rs(t) = e^{-\mu t} \tag{15}$$

with μ the mean rate of a breakdown.

If M is the smallest number of switching units which can generate a breakdown and if $ru(t)$ is their reliability, the reliability-function of the interconnection network is given by

$$r_{in}(t) = ru(t)^M \tag{16}$$

The MTTF can be calculated by means of equation (17).

$$MTTF = \int_0^\infty r_{in}(t)\,dt \tag{17}$$

Table 1
Reliabilty functions and MTTF (Eqs. 15 to 17)

interconnection network	M	$ru(t)$	MTTF
Baseline	$n*N/2$	$e^{-\mu t}$	$1/(\mu M)$
ABN-0, MBN	$n*N/4$	$1-(1-e^{-\mu t})^2$	$\sum_{i=0}^{M} \binom{M}{i} (-1)^i * 2^{M-i}/\{(M+i)\mu\}$
ABN-1	$(n+1)*N/8$	"	"

3.3 Terminal availability

The terminal availability a_t is the probability with which a faultless path is placed at the cell's disposal. The availability of each switch be denoted by p_{AS} and is defined as

$$p_{AS} = MTTF/MTBF \tag{18}$$

with MTBF, the Mean Time Between Failure [6].

In Baseline-networks each cell is transmitted via n stages and one switch per stage. To get a faultless connection, none of the n switches may break down. The terminal availability is therefore given by

$$a_t = p_{AS}^n \tag{19}$$

For ABN-0-networks you can assume an essentially higher terminal availability compared to that one of the Baseline-network, because there are two different paths between any pair of input-/output-ports. As it is shown in [1], the terminal availability can be calculated by means of equation (20).

$$a_t = \{p_{AS} + (1-p_{AS})*p_{AS}\} * \{p_{AS}+(1-p_{AS})*p_{AS}^2\}^{n-1} \tag{20}$$

The difference of these networks compared to ABN-0 is that cells will pass each switch only once. With regard to this fact, it holds [1]:

$$a_t = \{p_{AS} + (1-p_{AS}) * p_{AS}\}^n \tag{21}$$

3.4 Waiting time and transmission delay

These two characteristic values have been investigated by means of simulations. The model of each switching element is shown in Figure 4. Each input-port 0/1 of a switch can choose to send a received cell either directly to the corresponding output-port 0/1 or to the input-queue 0/1. If an output-port of stage k begins to transmit a cell, the preceding switch of stage k-1 will get an acknowledge-signal that this switch of stage k is ready to accept the following cell. Furthermore, the well known principles of cut through switching have been applied to the simulated cell layer. Because of lack of space the simulated operation mode of the switches can't be described in more detail.

Each input-port 0/1 of a switch can choose to send a received cell either directly to the corresponding output-port 0/1 or to the input-queue 0/1. If an output-port of stage k begins to transmit a cell, the preceding switch of stage k-1 will get an acknowledge-signal that this switch of stage k is ready to accept the following cell. Furthermore, the well known principles of cut through switching have been applied to the simulated cell layer. Because of lack of space the simulated operation mode of the switches can't be described in more detail.

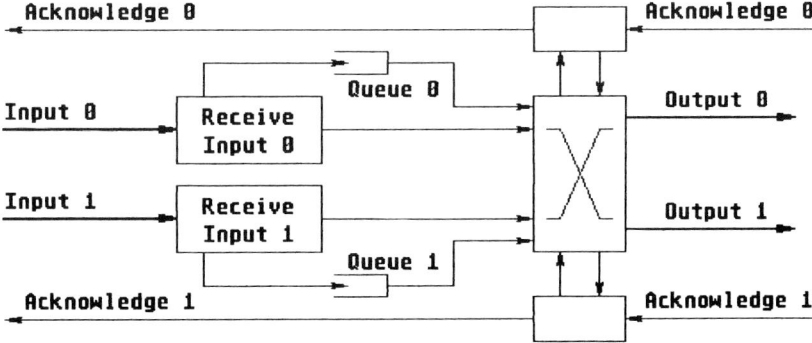

Figure 4. Model of a switching element

4. COMPARISON BETWEEN NETWORK STRUCTURES

Figure 5 is showing the probability of acceptance pa (eq.4) for p=1 as a function of the seize of the interconnection networks. The throughput of the networks ABN-0 and ABN-1 are identical and a little bit greater than that one of the corresponding Baseline-network. The throughput of the MBN-system is essentially greater, particularly, if no multiplexer are used. The throughput increases with decreasing load (p<1).

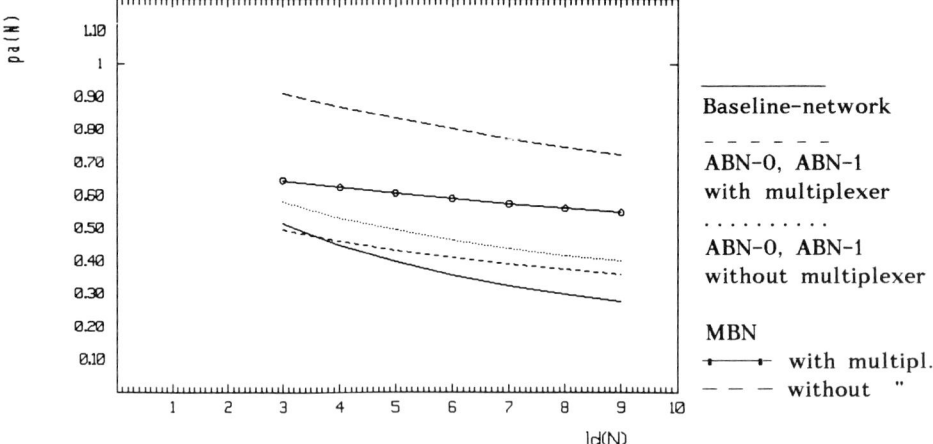

Figure 5. Probability of acceptance pa(N)

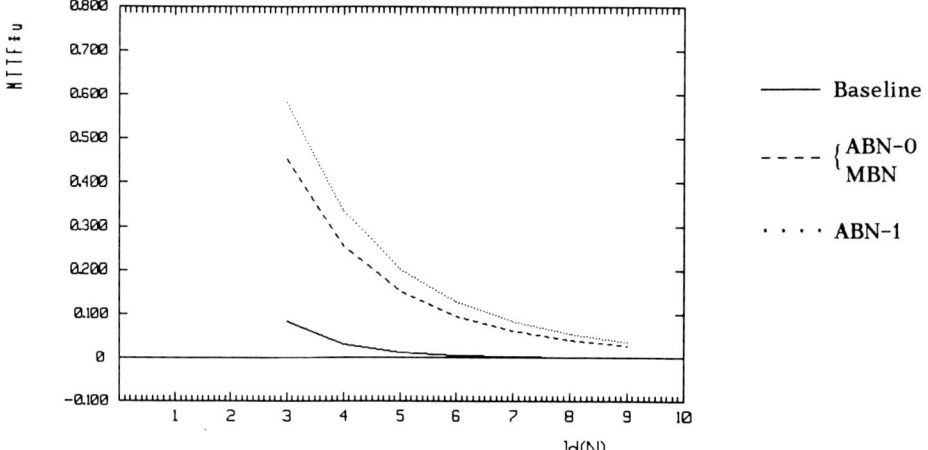

Figure 6. Mean Time To Failure (MTTF)

The reliability (MTTF) was raised by at least a factor five (ANB-0) or seven (ABN-1, MBN) compared to that one of the Baseline-network (Fig. 6).

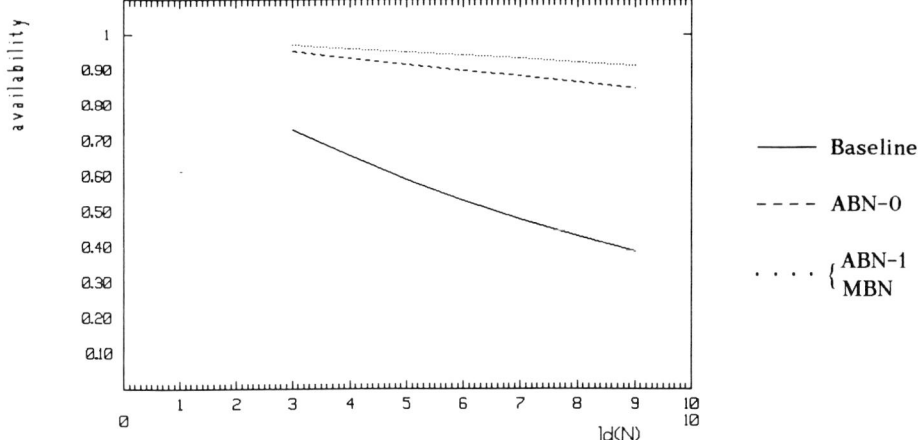

Figure 7. Terminal availability

The terminal availability (Fig. 7) is greater than 80-90 per cent for all extended networks and seizes, whereas that one of the corresponding Baseline-networks is always evidently less than 80 per cent. It was calculated for a reliability of a single switch of 90 per cent.

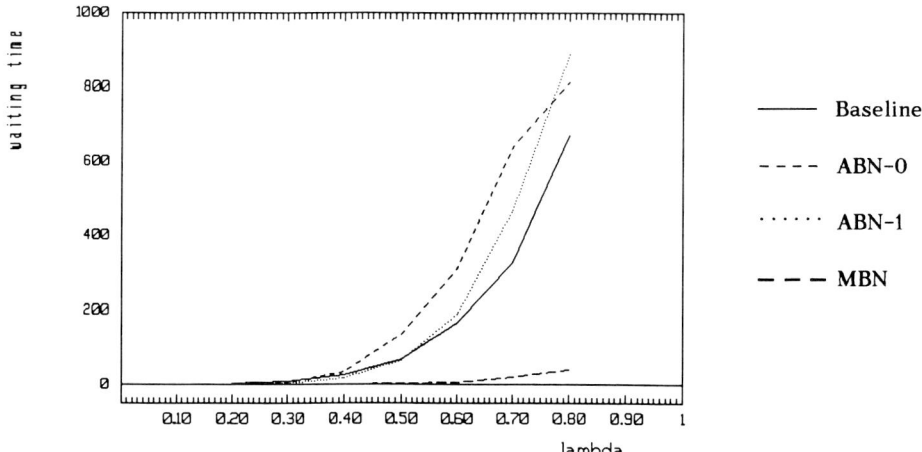

Figure 8. Waiting time as multiples of the internal clock

The mean values of the waiting times and the network delay are represented as multiples of the internal clock in Fig. 8 and 9. The MBN-network has the same number of switches as a Baseline-network but twice as much links. That is why it has the least network delay and waiting time.

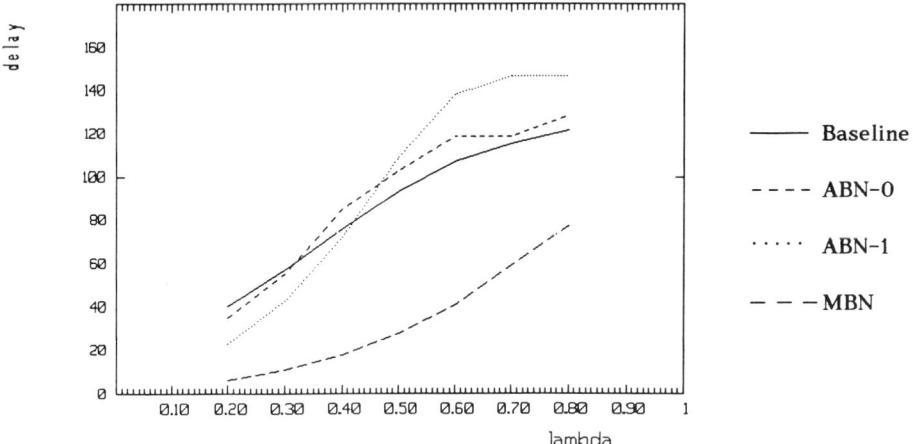

Figure 9. Network delay as multiples of the internal clock

The probability p(x) that a cell has to pass x switches of the ABN-0-network has been calculated, additionally. The results are shown in Fig. 10 as a function of the seize of this network.

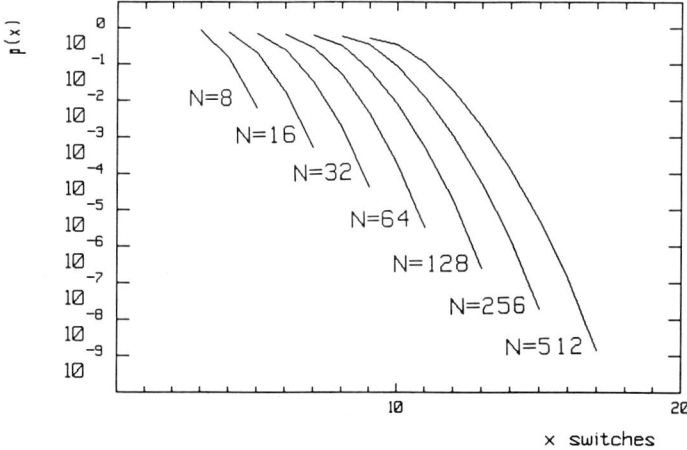

Figure 10. Probability p(x) of passing x switches, ABN-0

5. CONCLUSION

In this paper four different structures of interconnection networks have been analysed by means of calculations and simulations. The first extension (ABN-0) of the Baseline-network has an essentially higher reliability but the worst throughput of all networks, presented. The second one (ABN-1), based upon ABN-0, has the greatest reliability and the least expenditure of hardware. The throughput of the ABN-1 is comparable to that of the Baseline-network. The MBN-network does lead to the same reliability as ABN-0 but to the highest hardware-costs, too. Simulations have shown that it has the best throughput of these four interconnection networks.

REFERENCES

1 C. Roggenland, Netzwerke für die Paketvermittlung, Diplomarbeit, University of Paderborn
2 Fehlertoleranz in Systemen, Inf. Spektrum, Springer, 1968
3 S.M. Reddy and V.P. Kumar, Augmented Shuffle-Exchange Multistage Interconnection Networks, Computer, June 1987, pp. 30-40
4 S.M. Reddy and V.P. Kumar, On Fault-Tolerant Multistage Interconnection Networks, Proc. Int. Conf. Parallel Processing, Aug. 1984, pp. 155-164
5 J.H. Patel, Performance of Processor-Memory Interconnections for Multiprocessors, IEEE Trans. on Comp., vol. C-30, no. 10, 1981, pp. 771-780
6 Qualitätsbegriffe für elektronische Bauelemente, Siemens AG

COMBINATORIAL FEATURES OF THE CLOS TYPE SWITCHING NETWORK

W.Kromołowski and M.Szymanowski

Technical University of Wrocław, 27 Wybrzeże Wyspiańskiego, 50-370 Wrocław, Poland

Abstract
In the report influence of the inputs set {x} on outputs set {y} on the inner blocking phenomena and trial of the graphs theory application for description of these conversion have been shown.

1. INTRODUCTION

Each digital switching network has its space equivalent [1]. This equivalent can be a basis for concluding of properties of this switching network. The equivalents of all pracically used digital networks are Clos switching networks. That is the reason for which Clos networks analysis is so important.
The properties of the switching networks can be considered in two aspects: probabilistic and combinatorial. Probabilistic properties of the networks determine their ability of traffic handling and can be described by various traffic charakteristics, particularly by characteristic $B = f(A)$ of the traffic losses vs mean traffic intesity. On the other hand, combinatorial properties of the switching networks are determined by realization of defined functions of calls.
In the next parts of the report some combinatorial properties of the switching networks will be considered.

2. CONVERSION OF THE SET OF INPUTS ON THE SET OF OUTPUTS IN THE CLOS SWITCHING NETWORK

Let's examine Clos switching network $\nu(m,n,r)$ [2] like that on the Figure 2.1.
State of this network, that is quantity and conventional number of the switching paths, established as the physical realization of the calls service process, can be described as one of many possible conversions of inputs set {x} on outputs

set {y}. If the momentary paths, which are established in the preparatory phase of the calls service process to generators and receivers of various signals are not taken into account, each of these conversions uniquely describe state of switching network.

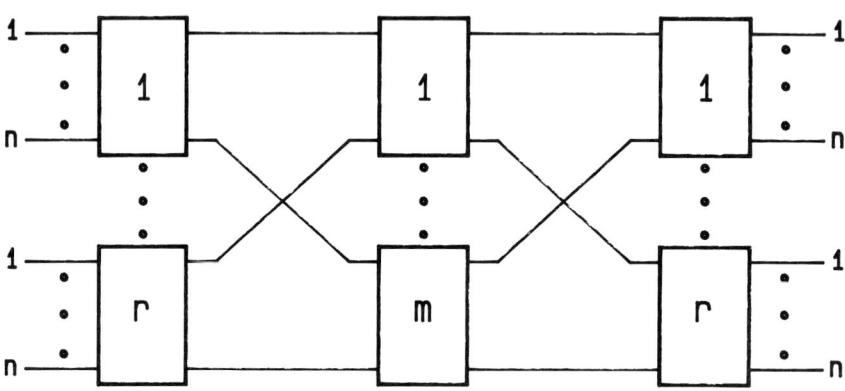

Figure.2.1. Clos switching network $\nu(m,n,r)$

For the analysis of the switching paths setting possibility of the succesively flowing calls, it will be sufficient to give numbers of the input and ouput switches, if only configuration of the switching network secures existing of the switching path between them. Certainly, the fact of setting of particular switching path affects on possibility of inner blocking phenomenon occurence. The minimization of this possibility is subject of the various switching paths selection algorithms analysis.

As recapitulation, you can state that the switching paths existing in the switching network can be described as the set $\{O\}$ of all conversions $X_i - Y_j$ representing, with the accuracy as to single switch, connections of the input switches with the output switches. Notice, that set $\{O\}$ can be shared on separable sets $\{O_{ij}\}$. Each of these sets describes connections between i-th input switch and j-th output switch and these sets have to satisfy the condition:

$$\{O_{ij}\}_1 \vee \{O_{ij}\}_2 \vee \ldots = \{O\} \qquad (C1)$$
$$\{O_{ij}\}_1 \wedge \{O_{ij}\}_2 \wedge \ldots = \emptyset$$

Now, the following theorem can be formulated:
Theorem 2.1:
If set {O} of all defined with accuracy as to switch conversions of set {x} of inputs on set {y} of outputs can be shared on no more than r subsets $\{O_{ij}\}$ satisfying condition (C1), Clos switching network $\nu(n,n,r)$ is the network without blocking.

The proof of this theorem is very simple and consists in arguing analogous to Clos proof, proving necessary number of the second section switches for making switching network without blocking.

It is clear the size of each subset $\{O_{ij}\}$ can be equal at most n because each input switch has n inputs. In consequence all connections represented by subset $\{O_{ij}\}$ can be established using at most n switches of the second section of network, as it shows Figure 2.2.

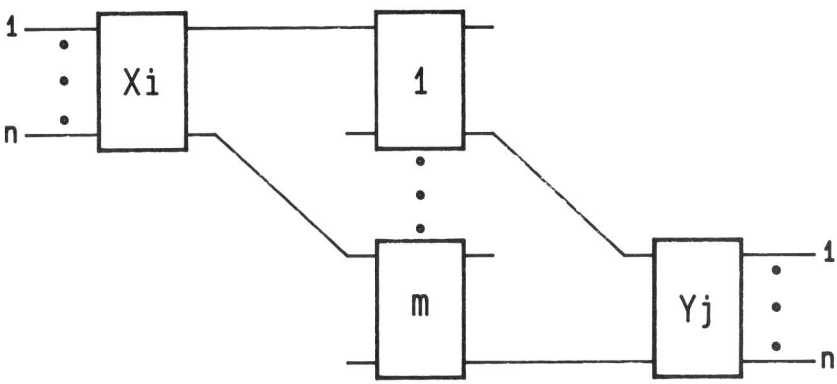

Fig. 2.2. Illustration of the Theorem 2.1 proof

The above remark is of course true for each couple of switches $X_i - Y_j$ i,j = 1,2,...,r, because n switching paths exist between them. Then, if number of subsets $\{O_{ij}\}$ of set $\{O\}$ is no greater r, it has to exist at least one free switching path between arbitrary couple of switches $X_i - Y_j$ in which not all inputs and outputs are busy.

The above property can serve for working out an usable algorithm of the switching paths establishing in the Clos switching networks. Proposed algorithm can be formulated as follows:

Searching of the switching path in the Clos switching network should be always started at the trial of establishing it to the output switch having the same number as input switch.

The proposed algorithm is analogous to well known Beneš "stuff of traffic" algorithm. However, it has considerably disadvantage consisting in necessity of continuous measuring, registering and modifying of the actual load of each second section switch of the network. The algorithm shown above hasn't this disadvantage and thanks to it the control device can be less complex and searching of the switching path needs less time consumption.

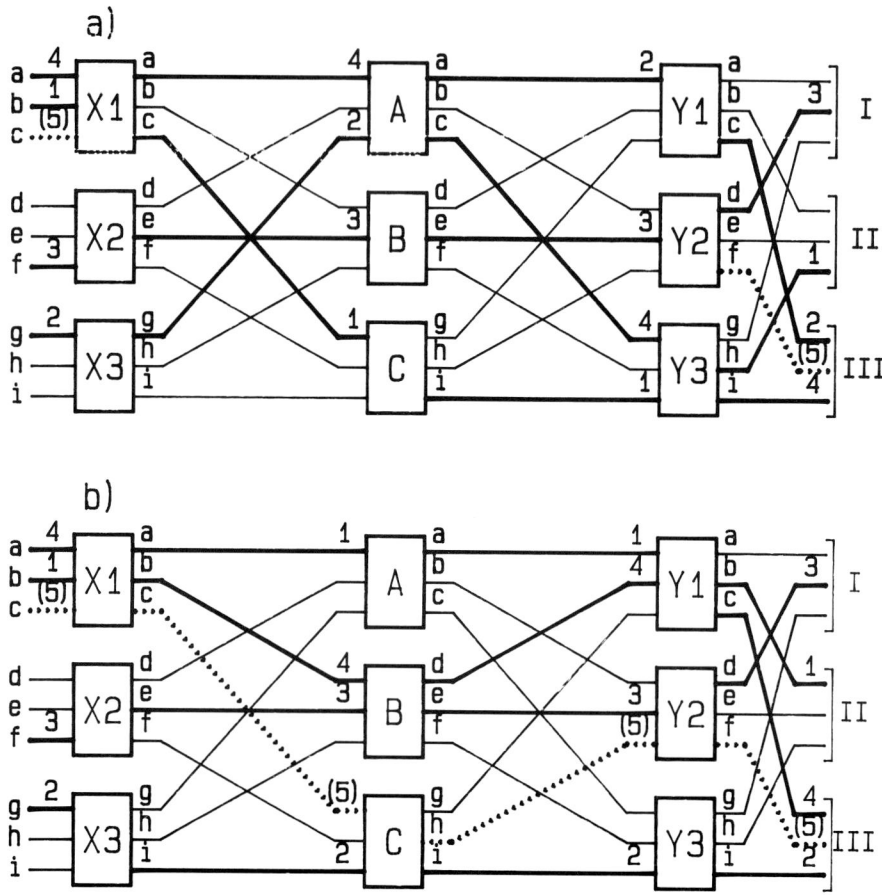

Figure 2.3. Illustration of proposed algorithm in the Clos switching network $\nu(3,3,3)$: a) switching paths 1,2,3 and 4 established according to random selection of the switching paths algorithm, b) switching paths 1,2,3 and 4 established according to proposed algorithm of path selection.

Let's check the run of the above mentioned algorithm on the simple example and consider the three-section Clos switching network shown on Figure 2.3. Then let's assume that the four paths are established in this network, according to, for example, random selection of the switching paths algorithm.
The above paths are on Figure 2.3 marked as 1,2,3 and 4:
- path 1 between input line b of X1 switch and arbitrary chosen line of direction II,
- path 2 between input line a of X3 switch and arbitrary chosen line of direction III,

- path 3 between input line c of X2 switch and arbitrary chosen line of direction I,
- path 4 between input line a of X1 switch and arbirtary chosen line of direction III,

In the situation shown on Figure 2.3 a), switching path (5) leading from line c of the switch X1 to free line of the direction III can't be established because of the inner blocking. This blocking can be avoided if shown above paths will be successively establishing according to proposed algorithm. Then the desired path (5) can be set, as it is shown in Figure 2.3 b).

It seems that above explained algorithm can be applied in concentration and sharing networks, that is in these switching networks in which group or free selection of the switching paths can be used. Good example of its possible application can be digital switching networks of S-T-S or T-S-T type.

For search of usability of the foregoing algorithm in switching networks, in Telecommunication and Acoustics Institute of Technical University of Wrocław, corresponding simulation tests are actually beeing prepared. These tests will be helpful for evaluation of quantitative influence of algorithm on blocking probability depending on various switching networks structures and various loads.

In the Clos networks $\nu(n,n,r)$ can be observed one more property, which can be expressed as Theorem 2.2.

Theorem 2.2:
If number of the switching paths between input switch X and output switch Y in the Clos switching network $\nu(n,n,r)$ is less n, blocking can't occure between these switches.

The proof of this theorem directly results from Figure.2.4.

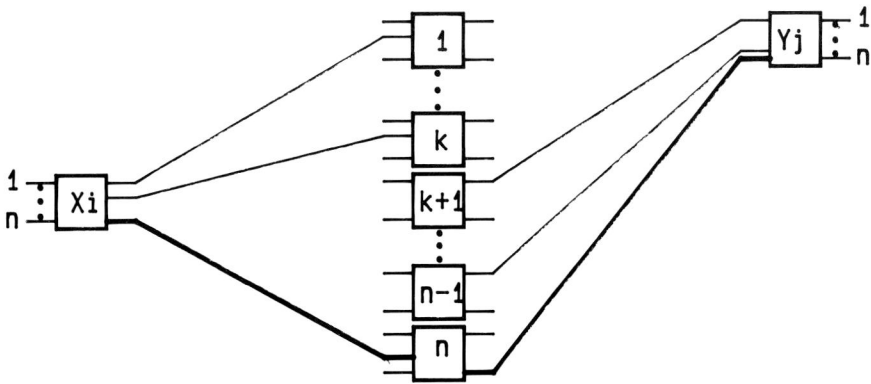

Figure. 2.4. Illustration of the Theorem 2.2. proof

Let's assume k paths actually begin in X_i switch and these paths lead to various Y switches. The refered paths engage of course k switches of the second section of the network. On the other hand, if n-k-1 paths finish in Y_j switch, they engage n-k-1 other switches of the second section. Then, if n paths exist between switches X_i and Y_j, one free path exists between these switches.

It can be derived an important conclusion from Theorem 2.2.
Conclusion:
If to each of switches of arbitrary couple of switches of the first and third section of the Clos switching network $\nu(n,n,r)$ will be connected at most n/2 lines, then there is no blocking between them in arbitrary state of the network.

Practical significance of this conclusion can be probably noted in ISDN networks, in which, in connection with switching of various telecommunication signals, it can be neccessary to do particular directions non-blocked.

3. APPLICATION OF THE GRAPHS THEORY FOR THE DESCRIPTION OF THE CONVERSION OF SET {X} ON SET {Y}

The theory of graphs seems to be a helpful tool for analysis of various phenomena occuring in the switching networks.

In the graph representing switching network the vertices of graph mostly represent links and branches - switching points. An example of the Clos switching network $\nu(3,3,3)$ graph is shown in Figure 3.1. In this figure the state of network, in which all switching paths are free (there is no switching paths established in the switching network) is shown.

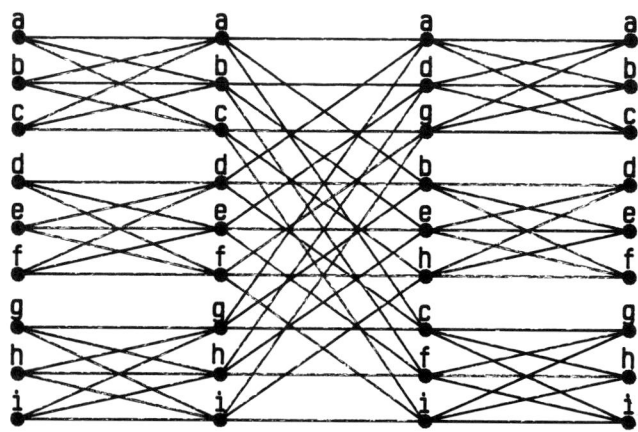

Figure 3.1. Graph G(0) of the Clos switching network $\nu(3,3,3)$

Let's mark as G(k) graph of the switching network, in which k paths are established.

Graph G(k) is performed by removing from graph G(0) all

vertices representing links beeing the parts of the particular switching paths and all branches, adjoined to these vertices. Obtained in that way graph G(k) is of course one of many possible subgraphs of the graph G(0) and can be named as graph of free switching paths in the switching network.

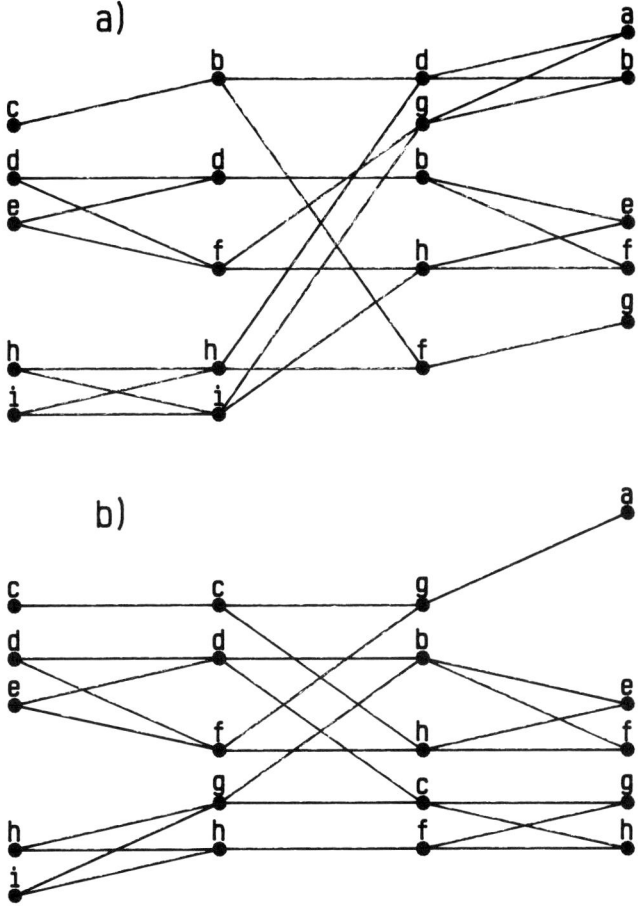

Figure.3.2.Graphs G(4) of the swithing network $\nu(3,3,3)$ a) for the situation shown on the Figure.2.3 a), b) for the situation shown on the Figure.2.3 b)

Figures 3.2 a) and 3.2 b) show graphs G(4) of the free switching paths of the network $\nu(3,3,3)$ for states shown respectively in the Figures 2.3 a) and 2.3 b).
One of the most important parameters characterizing graphs is its cyclomatic number $\lambda(G)$ [3], which can be calculated from

the following formula:
$$\lambda(G) = n(G) - \omega(G) + s(G) \qquad (F.1)$$
where: $n(G)$ - number of branches of the graph G,
$\omega(G)$ - number of vertices of the graph G,
$s(G)$ - number of components of the graph G.

Cyclomatic number of graph G representing switching network has following property:

Theorem 3.1:
Cyclomatic number of the graph $G(k)$ of the free switching paths in the Clos switching network $\nu(m,n,r)$ decreases with increasing network load measured as the number of busy paths.

The proof of the Theorem 3.1 results simply from formula (F.1). To do this let's notice at first that the number of components of the graph G(k) fulfills the relation $1 \le s(G) \le r$. In the cases of $k \le r$ and each free path is established between switches of different couples $X_i - Y_j$, $s(G)=k \le r$.

Then, for Clos switching network $\nu(m,n,r)$, $\omega(G) = 4k$, $n(G) = 3k$ and $\lambda(G) = 0$. In all other states of network, number of components of the graph G(k) is 1. Taking into account the way of performing of the graph G(k), number of removed branches is always greater than number of removed vertices. In consequence, the relation $G(k+1) < G(k)$ is true.

This theorem, which has been just proved as true, can be a basis for formulating the following hypothesis:

Hypothesis:
Cyclomatic number $\lambda(G)$ of the graph $G(k)$ representing all free switching paths can be a measure of "distance" of the Clos switching network $\nu(m,n,r)$ from the state of inner blocking.

4. CONCLUSIONS

The providing of the above hypothesis as true will allow for working out the algorithm, checking the compliance of the Clos switching network beeing in the particular state on the appearance of the inner blocking. Application of that algorithm together with proposed algorithm of the swithing path choice should allow on raising the standard of efficiency of the switching network control.

5. REFERENCES

1 M.Szymanowski, Pola komutacyjne EACT, Politechnika Wrocławska Wrocław 1979
2 V.E. Benes, Matematical Theory of Connecting Network and Telephone Traffic
3 B.Korzan, Elementy Teorii grafów i sieci, metody i zastosowania WNT Warszawa 1978

COMPARISON OF SOME COMBINATORIAL PROPERTIES OF DIRECT AND INDIRECT BINARY N-CUBE INTERCONNECTION NETWORKS

G.G. Veselovsky[a] and M.V. Kupryanova[a]

[a]Institute of Control Sciences, Profsoyuznaya 65, 117806 Moscow, USSR

Abstract

The combinatorial features of the two main types of n-cube networks are compared. Investigations are carried out for such well-known permutations as perfect shuffle, cyclic shift with amplitude a, bit reversal and flip permutation. The analysis is done with the help of the number theory methods. The conditions under which conflicts in the networks occur are formulated.

1. INTRODUCTION

Multiprocessor systems of the SIMD (single instruction stream-multiple data stream) type occupy a significant place among the architectures of parallel computing systems. In such systems N processing elements (PE) fulfil one and the same instructions but with various operands. In the systems of the SIMD type connection between PEs is established by means of an interconnection network, which can realize certain permutations on the contents of registers. At the same time memory is assumed to be distributed in the system, i.e. each PE has the main memory of a sufficient capacity. In accordance with the Lenfant classification the permutations, oftenly used in parallel processing, are grouped in families [1]. These families are based on such well-known permutations as perfect shuffle, cyclic shift with amplitude a, bit reversal and flip permutation. At present, most of processor interconnections in parallel systems belong to one of the two types: multistage networks of the shuffle exchange type and binary hypercube "direct" and "indirect" respectively [2]. The direct binary n-cube or hypercube network consists of $N=2^n$ nodes interconnected by links in accordance with the following rule:

two nodes whose binary addresses differ in exactly one bit possition are connected by a link, i.e. Hamming distance between them is egual to 1 [2]. An example of a binary 4-cube with 16 nodes is shown in Fig. 1.(a)
A node in the structure consists of a PE and means to communicate with its neighbors. The cube routing algorithm consists in the following. Let a message be transferred from one node with a binary number $(s_{n-1}s_{n-2}...s_0)$ to a node with the number $(d_{n-1}d_{n-2}...d_0)$. The number of a source-node S is to be referred as a source tag while the number of a destination node D is to be called a destination tag. In routing comparison starts with the least significant digits s_0 and d_0 of the source tags and destination tags, respectively. With $s_0 \neq d_0$ a message is sent from the node $(s_{n-1}s_{n-2}...s_0)$ to the node $(s_{n-1}s_{n-2}...d_0)$. With $s_0 = d_0$ comparing is continued up to the first unidentical pair of tags bits and only then a message is sent to the corresponding neighbor node. As a result of n bit-by-bit comparison a sequential substitution of source tag bits for destination tag bits takes place with the message transfer from node to node till a message reaches a node with the number assigned by a destination tag. Simultaneously any number of a bit in which a source tag differs from a destination tag determines a dimension according to which a message leaves a given node. It is assumed, that positions of the tag bits are numbered from left to right.

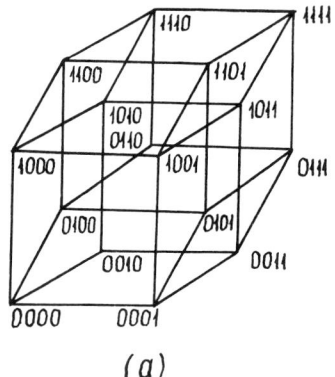

(a)

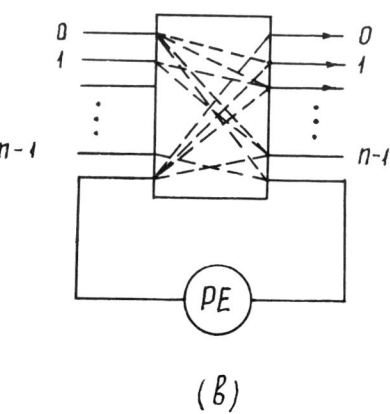

(b)

Figure 1. (a) Structure the direct binary 4-cube network
(b) Architecture of a node

Fig. 1.(b) presents the architecture of a node in the direct n-cube. Since a message is supplied to a given node with respect to a dimension p, and tag bits, which correspond to dimensions 0,...,p, were already corrected, then in this case this message may leave the node only when dimension is q > p. As to a multistage or indirect n-cube communication network, it is realized on the basis of switches of a dimension 2×2 with the stages n = $\log_2 N$. Each switch can be in one of the two states: the upper input is connected to the upper output, lower input is connected to the lower output (straight position); the upper input is connected to the lower output, the lower input is connected to the upper output (cross position). In many cases it is useful to provide the possibility of connecting one input to both outputs (broadcast position). Interconnections of indirect n-cube networks are realized in such a way that when setting the switches in the stage k to the cross position and all the remaining switches to the straight position the connectoins set up between inputs and outputs, correspond to the dimension k of the direct cube network.The indirect binary n-cube network 16×16 is shown in Fig.2. For routing of connections in the indirect n-cube an algorithm which similar to the above one for the direct n-cube can be used. Substitution of the next bit of the source tag by a bit of the destination tag corresponds to a progress to the next network stage but in such a way, that in the case of this bit inverting it is required to switch over this switch input to the upper output if the input is lower and vice versa to the lower output if the input is upper one.

In Fig.2 the dotted line shows the path from input number 0100 to the output number 0111 routed by making use of the above aigorithm. A unit of time of a message delay is assumed a message delay time on one stage in the multistage network or message delay time in transferring between two neighboring nodes of a static hypercube which is the same. This unit of time will be in the seguel called a time step.

This work present comparison of effectivenes of various permutations elaboration using two types of the n-cube networks. Investigations are carried out with application of the number theory methods. Conditions involving conflicts in the network are formulated. Conflict is understood as a situation which is inadmissible by definition and indicates that two different messages claim the same link in the network.

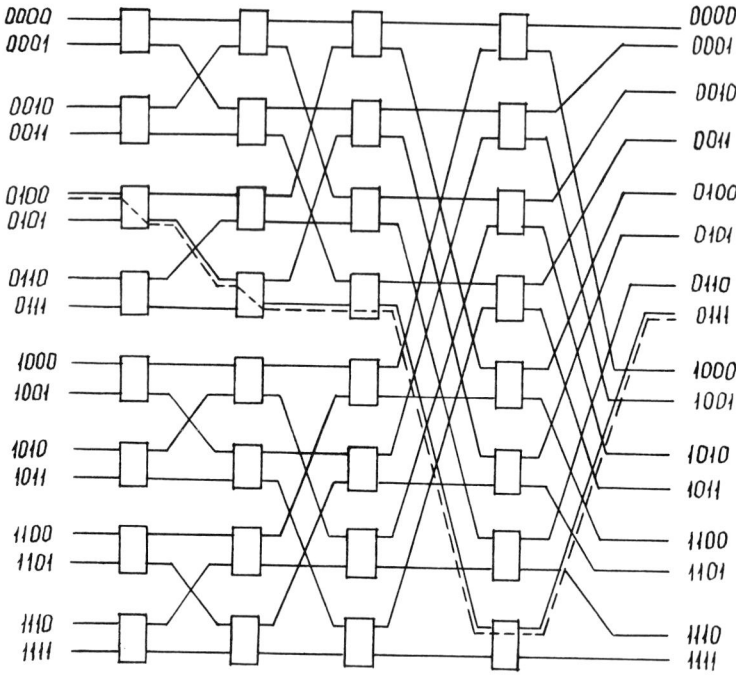

Figure 2. The indirect binary 4-cube network.

2. SOME COMBINATORIAL PROPERTIES OF INDIRECT N-CUBE INTERCONNECTION NETWORK

Determine the permutation P_N as a set of the integer pairs $P_N = \{ (S_i, D_i) / 0 \leq i < N \}$, which represent a mapping of inputs to outputs $S_0 \to D_0$, $S_1 \to D_1, \ldots, S_{N-1} \to D_{N-1}$. Then for an indirect n-cube the state of a tag after a k-th stage with the above routing algorithm taken into account has the following form: $s_{n-1} \ldots s_k d_{k-1} \ldots d_1 d_0$, where $n = \log N$. Note that a conflict occurs when two different messages claim the same communication link between the stages. It signifies that for some input-output pairs (S_i, D_i), $(S_j, D_j) \in P_N$ there exist the following equalities:

$$s_{i,n-1}\cdots s_{i,k} = s_{j,n-1}\cdots s_{j,k}$$

$$d_{i,k-1}\cdots d_{i,1}d_0 = d_{j,k-1}\cdots d_{j,1}d_0$$

These equalities signify identity of the least significant k+1 bits of destination tags and most significant n-(k+1) bits of the source tags. In the sequel some definitions and methods of the number theory are being used [3].

Definition: let $a = a^*m + \alpha$ and $b = b^*m + \beta$ where $\alpha, \beta < m$ and $a, b, \alpha, \beta, a^*, b^*$ and m are all nonnegative integers. We say $a \equiv_m b$ if and only if $\alpha = \beta$. This is the common definition of "a is congruent to b modulo m".

Definition: assume m is a factor of N. With a and b as above, we say $a \equiv_N^m b$ if and only if $a^*m \equiv_N b^*m$, i.e., $m\lfloor a/m \rfloor \bmod N = m\lfloor b/m \rfloor \bmod N$.

For example, let 11010 (binary), b = 01010, and c = 10010. Then $a \equiv_4 b \equiv_4 c$, $a \equiv_{16}^2 b$ (since 101=101) and $a \not\equiv_{16}^2 c$ (since 101≠001)

Then for some $m = 2^{k+1}$, $(0 \leqslant k \leqslant n-1)$ the conflict conditions can be written in the following way: $D_i \equiv_m D_j$, $S_i \equiv_N^m S_j$.

Theorem 1: For an indirect n-cube with this routing algorithm conflict takes place iff there exist such k ($0 \leqslant k \leqslant n-1$), that for some $m = 2^{k+1}$ the following conditions are fulfilled:

$$S_i \neq_N S_j, \quad D_i \equiv_m D_j, \quad S_i \equiv_N^m S_j \qquad (1)$$

Since the indirect n-cube properties are well investigated [4],[5], there is no necessity to describe them in detail. It is known, for example, that permutations frequently occurred in practice as a perfect shuffle and a bit reversal in realization cause conflicts in the network while such permutations as a cyclic shift with the amplitude a and a flip permutation are realized without conflicts.

3. ANALYSYS OF SOME PERMUTATIONAL PROPERTIES OF THE DIRECT BINARY N-CUBE INTERCONNECTION STRUCTURE

3.1. The analysis of the network in the syncronous mode of the routing algorithm execution

Consider a possibility to realize the main types of permutation for a direct n-cube by using the number theory methods mentioned previously. In our case the architecture of the node provides, when it is necessary, the simultaneous information transfer between two neighbouring nodes in the opposite directions. Analyzing the structure of a direct n-cube node (fig.1.(b)) a conclusion can be made on a possibility of the existence of a conflict. Actually it means, that two different messages reached the same node simultaneously and leave it using the same edge in the next time step. To describe this situation using the notions of congruence which were previously indicated, determine a synchronous mode for routing fulfilment that consists in the following: in each time step a current pair of source and destination tag bits is compared starting with the least significant ones for all pairs source-destination, but in this time step only those messages for which these bits values are not identical are transferred to the corresponding nodes. Thus, transfer of all these messages from the assigned list or, in other words, assigned permutation realization is completed simultaneously with the n time steps, i.e. after analysis of the most significant tag bits of all source-destination pairs. It should be noted that the synchronous as well as asynchronous modes considered in the next paragraph, belong to the routing algorithm, described in the introduction.

Let it be pairs (S_i, D_i), $(S_j, D_j) \in P_N$. The state of tags after the k-th step with the routing mode taken into account, has the following form:

$$s_{i,n-1} \cdots s_{i,k} d_{i,k-1} \cdots d_{i,1} d_{i,0};$$
$$s_{j,n-1} \cdots s_{j,k} d_{j,k-1} \cdots d_{j,1} d_{j,0};$$

Note that conditions of simultaneous reception of two different messages in one and the same node coincide with the condition of theorem 1, and namely $S_i \equiv^m_N S_j$, $D_i \equiv_m D_j$. Write down the states of tags after the (k+1)-st time step:

$$s_{i,n-1} \cdots s_{i,k+1} d_{i,k} d_{i,k-1} \cdots d_{i,1} d_{i,0};$$
$$s_{j,n-1} \cdots s_{j,k+1} d_{j,k} d_{j,k-1} \cdots d_{j,1} d_{j,0};$$

Then conditions of transfers along the same edge have the following form: $S_i \neq {}_N S_j$, $S_i \equiv {}_N^{2m} S_j$, $D_i \equiv {}_{2m} D_j$, where $m = 2^{k+1}$ ($0 \leqslant k \leqslant n-1$). Hence, taking into account conditions (1) of theorem 1, we come to final conclusions.

Theorem 2: For a direct n-cube in the synchronous routing mode a conflict takes place iff there exists such k ($0 \leqslant k \leqslant n-1$), that for $m = 2^{k+1}$ the following conditions are fulfilled:

$$S_i \neq {}_N S_j; \quad D_i \equiv {}_{2m} D_j; \quad S_i \equiv {}_N^m S_j; \qquad (2)$$

Theorem 3: Permutations of the $\pi^{(n)}$ type, a cyclic shift with an amplitude α, on a static direct n-cube are realized in the synchronous routing mode without conflicts.

Proof: By contradiction assume, that a conflict exists, then conditions (2) of theorem 2 should be fulfilled. Since $D_i = S_i + \alpha$; $D_j = S_j + \alpha$, then $S_i + \alpha \equiv {}_{2m} S_j + \alpha \rightarrow S_i \equiv {}_{2m} S_j$, but $S_i \equiv {}_N^m S_j$, then $S_i \equiv {}_N S_j$ contradicts the initial condition.

Theorem 4: Permutation of the type $\tau^{(n)}$ ($S \rightarrow S \oplus \lambda$), called flip permutation, on a direct n-cube in a synchronous routing mode is realized without conflicts.

The proof is evident and similar to that of theorem 3 and therefore its presentation is not necessary. Note one important property for this permutation which is useful in its further applications. If $\lambda = 2^i$ ($0 \leqslant i \leqslant n-1$), then the permutation $\tau^{(n)}$ on a direct n-cube is realized in one time step.

Theorem 5: Permutation of the $\sigma^{(n)}$ type, $(s_{n-1} s_{n-2} \ldots s_0) \rightarrow (s_{n-2} \ldots s_0 s_{n-1})$, called a perfect shuffle on a direct n-cube in the synchronous routing mode is realized without conflicts.

Proof: By contradiction assume, that conflict exists and it is equal to conditions (2) of theorem 2. Thus $D_i \equiv {}_{2m} D_j$, then $S_i \equiv {}_m S_j$, since D_i, D_j is obtained from S_i, S_j by one bit cyclic shift to the left. Then from $S_i \equiv {}_m S_j$ and $S_i \equiv {}_N^m S_j$ it follows that $S_i \equiv {}_N S_j$, which contradicts the initial condition.

Theorem 6: Permutation of the $\rho^{(n)}$ type, bit reversal $(s_{n-1} s_{n-2} \ldots s_0) \rightarrow (s_0 \ldots s_{n-2} s_{n-1})$, on a direct n-cube in a synchronous routing mode causes conflicts in the network.

Proof: Let $S_i \neq {}_N S_j$, then choose such $к+2 \geq \lceil n/2 \rceil$, that for $m = 2^{k+1}$ the condition $S_i \equiv {}_N^{2^{n-(k+2)}} S_j$ is fulfilled, i.e. k+2 of the most significant bits of the source tags S_i and S_j are identical, then $D_i \equiv {}_{2m} D_j$, i.e. к+2 of the least significant bits of the destination tags are also identical. However from $S_i \equiv {}_N^{2^{n-(k+2)}} S_j$ follows $S_i \equiv {}_N^m S_j$, i.e. all conditions of theorem 2 are fulfilled.

3.2. Analysis of the network in the asynchronous routing algorithm mode

Now consider the asynchronous mode to control information exchange in the direct n-cube, which consists in the following. Realization of the assigned permutation for all source - destination pairs starts simultaneously. However unlike the above considered synchronous mode the bits of source and destination tags are being analyzed independently in each pair (S_i, D_i) till the first unidentity with the consequent transfer of a message corresponding to this pair to one of the neighboring nodes according to the route formed by this pair. Then in the next time step all pair analysis is carried out independently up to the next unindentity, etc. In this way the transfer of various messages in this mode can proceed in various number of time steps, in a general case less than n.

Thus let there be the pairs (S_i, D_i), $(S_j, D_j) \in P_N$. Introduce the notations:

$l_1(k)$ - a set of dimensions where bits of the S_i source tag are not identical to the bits of the D_i destination tag;

$l_2(k)$ — a set of dimensions where the bits of the S_j source tag are not identical to the bits of the D_j destination tag ($0 \leqslant k \leqslant n-1$, $0 \leqslant l_1(k), l_2(k) \leqslant n-1$).

Assume that there exist $l_1(k)$ and $l_2(k)$ such that $l_2(k) \geqslant l_1(k)$. Then for two messages to reach the same node after the k-th time step it is necessary to fulfil the following conditions:

$$D_i \equiv_{2^{l_1(k)+1}} D_j; \quad S_i \equiv \frac{2^{l_2(k)+1}}{N} S_j; \quad S_i \equiv_{2^{l_2(k)+1}} 2^{l_1(k)+1} D_j;$$

For two messages to leave the node along one and the same edge it is necessary to have identity of tags intermediate states in the (k+1)-st time step. Since the condition $S_i \equiv \frac{2^{l_2(k)+1}}{N} S_j$ should be fulfilled in the k-th time step it causes the necessity of identity of the values $l_2(k+1)$ and $l_1(k+1)$ in the (k+1)-st time step since only in this case switching-over takes place simultaneously in one and and the same dimension. Therefore conditions of transfer along one and the same edge in the (k+1)-st time step have the following form:

$$l_2(k+1) = l_1(k+1); \quad D_i \equiv_{2^{l_2(k+1)+1}} D_j; \quad S_i \equiv \frac{2^{l_2(k+1)+1}}{N} S_j;$$

To simplify further conclusions and make writing convenient we shall redenote $2^{l_1(k)+1}$, $2^{l_2(k)+1}$, $2^{l_1(k+1)+1}$, $2^{l_2(k+1)+1}$ by $L_1(k)$, $L_2(k)$, $L_1(k+1)$, $L_2(k+1)$.

Theorem 7: For the direct n-cube in the asynchronous routing mode a conflict will take place if and only if there exists such k and $l_1(k)$, $l_2(k)$, $l_1(k+1)$, $l_2(k+1)$ ($0 \leqslant k \leqslant n-1$), ($1 \leqslant l_1(k), l_2(k) \leqslant n$), with $l_1(k+1) = l_2(k+1)$, that the following conditions are fulfilled:

$$S_i \neq {}_N S_j; \quad S_i \equiv_{L_2(k)}^{L_1(k)} D_j; \quad S_i \equiv^{L_2(k)}_N S_j; \quad D_i \equiv_{L_2(k+1)} D_j; \qquad (3)$$

Theorem 8: Permutation of the $\sigma^{(n)}$ type, a perfect shuffle, for the direct n-cube, in the asynchronous routing mode is realized without conflicts.

Proof follows from theorem 7.

Theorem 9: Permutation of the $\rho^{(n)}$ type, bit reversal, for the direct n-cube in the asynchronous routing mode is realized without conflicts.

Proof: Let there be such $l_1(k)$, $l_2(k)$, $l_1(k+1) = l_2(k+1)$, that all conflict conditions (3) are fulfilled. In addition note, that from the idea of bit reversal it follows that $l_1(k) = l_2(k)$. Since $S_i \equiv_{L_1(k)} D_i$, then $D_i \equiv_{L_1(k)} D_j$, but $S_j \equiv_{L_1(k)} D_j$, then $S_i \equiv_{L_1(k)} S_j$. Thus, if $S_i \equiv_N^{L_1(k)} S_j$ then theorem 7 is valid only if conditions $S_i \equiv_N S_j$, $D_i \equiv_N D_j$ are fulfilled.

Theorem 10: If on a direct n-cube in the synchronous routing mode a realized permutation causes conflicts then this permutation will also cause conflicts on the indirect n-cube.

The proof of this evident fact is the corollary of theorem 1, 2. If $D_i \equiv_{2m} D_j$ is fullfiled, then the more so $D_i \equiv_m D_j$.

Theorem 11: If permutation on the indirect n-cube is realized without conflicts, then this permutation is also conflict-free on a direct n-cube in the synchronous and asynchronous routing modes.

Proof: Let $l_2(k+1) = l_2(k) + \gamma$ $(0 \leq \gamma < n-1)$, $l_2(k) \geq k$, i.e. $l_2(k) = k + \mu$ $(0 \leq \mu < n-1)$, then $L_2(k+1) = m2^{\gamma+\mu}$. If $D_i \neq_m D_j$, then the more so $D_i \neq_{L_2(k+1)} D_j$.

Theorem 12: If permutation on the direct n-cube in the asynchronous routing mode causes conflicts in the network, then this permutation will also cause conflicts on the direct n-cube in the synchronous routing mode as well as on the indirect n-cube.

Proof: Let $l_2(k+1) = l_2(k) + \gamma$ $(0 \leq \gamma < n-1)$, $l_2(k) \geq k$, i.e. $l_2(k) = k + \mu$ $(0 \leq \mu < n-1)$, then $L_2(k+1) = m2^{\gamma+\mu}$,

$L_2(k) = m2^\mu$. If $D_i \equiv {}_{L_2(k+1)}D_j$ then $D_i \equiv {}_{L_2(k)}D_j$ ($\gamma = 0$) for the indirect n-cube; and $D_i \equiv {}_{2L_2(k)}D_j$ ($\gamma = 1$) for the direct n-cube in the synchronous routing mode, i.e. all conditions of theorems 1,2 are fulfilled.

Besides from theorem 11 it follows that a cyclic shift with the amplitude a and a flip permutation in the assigned routing mode are realized without conflicts.

4. CONCLUSION

The carried-out investigations and theorems 1, 2, 7 used as the basis made it possible to compare elaboration effectiveness of the most frequently used permutations with application of two types of the n-cube networks. It is known that for an indirect n-cube such permutations as a cyclic shift with the amplitude a, flip permutation are realized without conflict in n number of time steps where n is hypercube dimension and such permutations as a perfect shuffle and bit reversal cause conflicts in the network and they are realized in two passes through the network, i.e. in 2n time steps. Perfect shuffle is realized without conflict on the direct n-cube both in the synchronous and asynchronous routing modes. Bit reversal when elaborated on a direct n-cube in the synchronous routing mode causes conflicts and therefore it requires 2n time steps to be realized. Application of the asynchronous routing mode enables the elimination of conflicts when realizing the bit reversal and realization of this permutation not more than in n time steps. Thus, a conclusion can be made that the use of a direct n-cube in parallel computing systems of the SIMD type with a distributed memory in realization of such important permutations as a perfect shuffle and a bit reversal provides two times advantage of their fulfilment in comparison with the use of an indirect n-cube. Making use of the obtained results it can be shown that elaboration of the fast Fourier transform (FFT) in the SIMD type computing system using a direct n-cube as the means of interprocessor communication provides very essential advantage of time reduction in realization of information transfers in comparison with the use of an indirect n-cube for the same purposes. It is connected with the pequliarities of realization of the main permutations required in the FFT and, namely, flip permutation with the $\lambda = 2^i$ and bit reversal in that network and in the other one. Information transfers, as it was defined in the above accepted terminology for a direct n-cube, require 2n time steps for their fulfilment while for an indirect n-cube - n^2+2n time steps, respectively. (It is assumed that the number of points for the FFT is equal to the number of processing elements (PE) $N=2^n$). With N=256 for the FFT application of a direct n-cube provides 5 times advantage in fulfilment of information transfers in comparison with an indirect n-cube. At the same time realization of a direct n-cube for the values N in the ranges, presenting practical interest, requires hardware whose cost is only two times greater than that of an indirect n-cube.

5. REFERENCES

1 J. Lenfant, IEEE Trans. Comput., vol.c-34, No. 6 (1985) 506.
2 K. Padmanabhan, Commun. of the ACM, vol. 33, No. 1 (1990) 43.
3 D. Lawrie, IEEE Trans. Comput., vol. c-24, No. 12 (1975) 1145.
4 J. Siegel, IEEE Trans. Comput., vol. c-26, No. 2 (1977) 153.
5 Z. Cvetanovic, IBM J.Res. Develop., vol.31, No. 7 (1987) 452.

MULTISTAGE INTERCONNECTION NETWORK AS A FAST PACKET SWITCH

Z. HULICKI

Telecommunications Department, *University of Mining & Metallurgy*, al. Mickiewicza 30, PL 30-059 Kraków, Poland

Abstract
A fast packet switch design supported by multistage interconnection networks is examined. Properties of networks based on different interconnection functions are studied in terms of the switch performance measures and capabilities to satisfy the requirements of multi-service traffic. A problem of traffic concentration is also taken into account. It has stated that tradeoff between the network cost and its performance exists and topology as well as different control mechanisms and decisions rules also result in them.

1. INTRODUCTION

All over the world, it can be observed an enormous work to modernize and upgrade telecommunication systems to meet the demands of communication networks in the 1990's. These new networks, which embrace the latest technologies for communications services, demand sophisticated switching, transmission and management systems. These systems enable to handle rapidly increasing telecommunications traffic with the flexibility to respond to major innovations in every relevant area of telecommunications.

For many years to support a single class of traffic only a single switching mechanism was required. Resently, with the rapid drop in the cost of computer based equipment has come the requirement to support many more communications services, including voice, electronic mail, videotex and other forms of video, image, facsimile, text and many forms of data transmission. The existing practice of providing a separate network, with its own switching mechanism, for every class of traffic cannot be extended to support the demand forecast for many new communications services [1], [10]. The most flexible solution is that of an integrated communications network with a single switching mechanism capable of handling all classes of traffic. Fast packet switching has been suggested as a possible switching mechanism to support integrated services.

If fast packet switching is to support the growth in demand for communications services for the foreseeable future, the design and implementation of a fast packet switch must be investigated. Modularity and growth are important properties for such switch since it may vary in size.

A design which leads to a very simple hardware implementation is investigated in this contribution. A simple implementation offers flexibility, a wide range of potential applications and operation at both conventional speeds and also at possibly very high speeds.

The design features a number of parameters that have an effect upon the performance of the switch. An exploration of this effect should allow the selection of the preferred design parameters.

2. THE SWITCH ENVIRONMENT PROBLEMS

In order to achieve communication between source and destination across a network, a switching function is necessary. Each class of source traffic, however, exerts various requirements on the performance of the communications network. Multi-service traffic comprises voice, electronic mail, videotex, Hi-Fi audio, broadcast quality video, compressed video, text, image, facsimile, telemetry and many forms of data transmission [7]. Some services may tolerate high and variable delay across the network. Other services, however, such as voice and video, place stringed requirements upon the upper bound and variance of delay across the network.

In various systems a high capacity communication medium is typically required if an interfacility communication is not to be a system bottleneck. If the communicaions channel is fast enough to avoid being a bottleneck then most forms of data traffic become bursty due to user behaviour. Most forms of non-data traffic are also bursty [9]. Therefore, in order to carry out different traffic types, such as voice, data, video, interactive text or image, and efficiently use the communication resources new switching techniques have been proposed [1], [2].

The integration of multi-service traffic is currently proposed in a hybrid manner. When a single switching mechanism handles all classes of traffic is integration at the switching level achieved. Then a network may be considered fully integrated. The greatest advantage of full integration is the flexibility to adapt quickly to the changing traffic requirements of new communications services. Other advantages include transmission efficiency, independence of the switching mechanism from the characteristics of the source traffic, and the need to support and maintain only a single integrated network. Fast packet switching offers one possible solution for a fully integrated network. It enables us to seek structures permitting the efficient sharing of network resources by many types of traffic. Packet traffic, however, permits evolution to the support of multi-service traffic upon a single integrated switching mechanism. Thus the use of high-throughput / low-delay packet switching systems is concerned to support the interconnection of high speed links.

Various solutions can be obtained when simple protocol processing and high-speed links are used. However, as the number of the information processing facilities grows, interconnection using a traditional method such as the crossbar switch becomes quite expensive in terms of hardware cost [3], [7], [9], [11]. Moreover, a multi-service environment of network switch implies a complex structure of traffic flow and of user requirements, concerning an assurance of high operation reliability [9]. Therefore, a central issue in the design of a fast packet switch is the interconnection medium which provides communication paths between network facilities.

In order to interconnect large numbers of high speed links and support a number of various services as well as different classes of traffic, resulting from emerging wideband services, such as image and video, multistage interconnection networks are being developed for use as a fast packet switch [7], [8], [9]. These networks offer a bandwidth for both conventional computer communications applications and graphics, image or video applications.

3. TOPOLOGY AND CAPABILITIES OF A SWITCH

Because the bandwidth of a single path interconnection medium imposes a limit upon the switch capacity that may be achieved, some form of multi-path interconnection medium have been proposed to support communication between a large number of peripheral ports concurrently. Design decisions that are essential in choosing a cost-effective communication network should be examined now. One shall also survey the various topologies and communications protocols, and discuss issues related to concurrent information transfer.

In order to satisfy the need of interconnection a wide variety of the designs of a fast packet switch have been developed [1], [2], [7], [9], [10], [11]. Many of designs are based upon the use of a multi-stage interconnection network which appears to be the most attractive and forms a central feature of the design examined in this contribution.

In selecting the architecture of an interconnection network, four design decisions can be identified [3], [4]. They concern operation mode, control strategy, switching method, and network topology. Two types of communication can be applied: synchronous and asynchronous. Synchronous communication is needed for processing in which communication paths are established synchronously. Asynchronous communication is needed for multiprocessing in which connection requests are issued dynamically. A system may also be designed to facilitate both types of communication.

In determining a suitable architectural structure a key factor is network topology. The network topologies tend to be regular and can be grouped into two categories: static and dynamic. In a static topology, links between two data processing installations are passive and dedicated buses can not be reconfigured for direct connections to other terminals. On the other hand, links in the dynamic category can be reconfigured by setting the network's active switching elements. These are three topological classes in the dynamic category: single-stage, multi-stage, and crossbar (see Fig.1.).

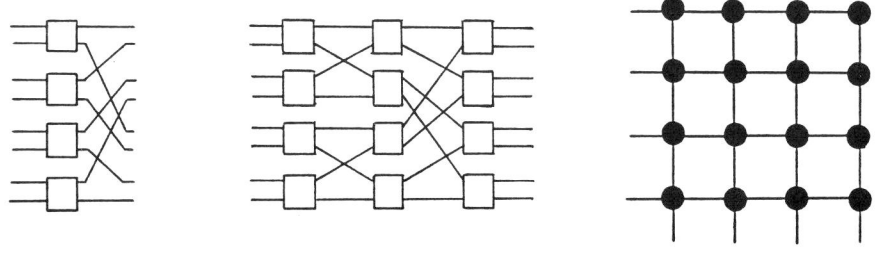

single-stage   multi-stage   crossbar

Fig.1. A dynamic category of network topology.

A single-stage network is composed of a stage of switching elements cascaded to a link connection pattern. Packets may have to recirculate through the single stage several times before reaching their final destination.

In the absence of a switch with reasonable cost and performance which has prevented the growth of the mentioned above communications systems, several cheaper multistage interconnection networks have been proposed (see [3], [7], [8]) to circumvent the high cost of switch. A multi-stage network consists of more than one stage of switching elements and is usually capable of connecting an arbitrary input terminal to an arbitrary output terminal. The two-sided multi-stage networks, which usually have an input side and an output side can be divided into three classes: blocking, rearrangeable and nonblocking. Unfortunately, the rearrangeable nonblocking networks as well as nonblocking networks are very expensive in hardware costs [9]. Therefore, from the practical point of view only blocking networks can be taken into account. Unfortunately, in such the networks simultaneous connections of more than one terminal pair may result in conflicts in the use of network communication links. Examples of this type of network include data manipulator (see Fig.2) and generalized cube [3], [8], baseline [3], and

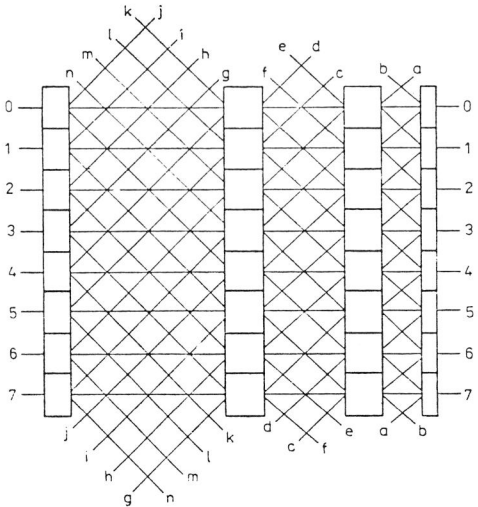

Fig.2. The *data manipulator* network (for N = 8)

SW-banyan (see Fig.3), and delta [1], [3], [7], [8], [11].

The interconnection networks are usually designed in a way that they can be constructed modularly in terms of a single type of building block called switching module and interconnecting links. The switching methodology and the control strategy are implemented in switching elements according to required communication protocols. The control-setting function can be managed by a centralized controller or by the individual switching element, enabling both circuit and packet switching. These networks use fewer gates than the crossbar network. Typically they are constructed of $N=2^n$ input and output ports, where n is an integer, and consist of n stages (or columns) of switching modules. The networks differ in the way the stages are connected.

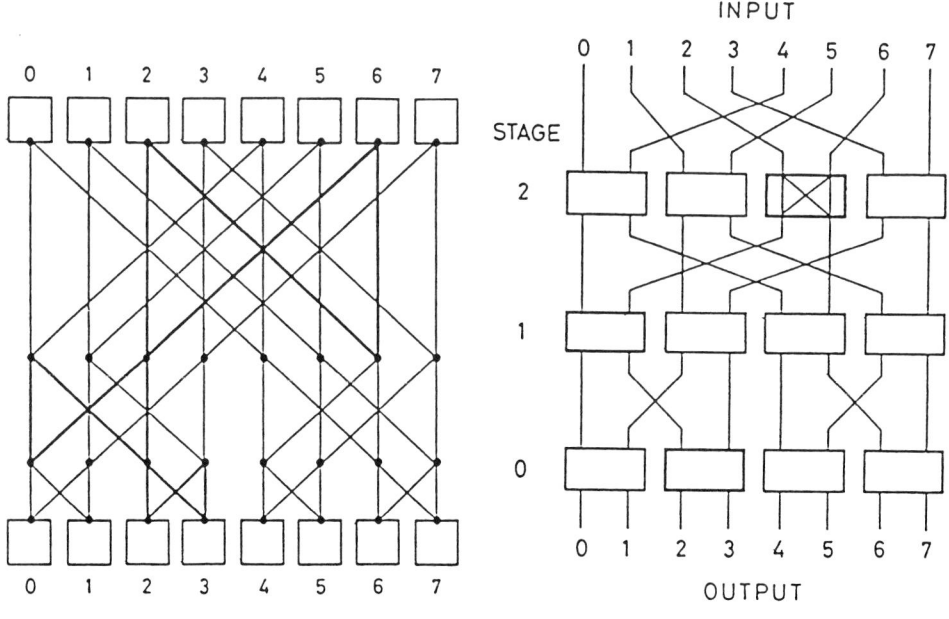

SW-banyan   SW-banyan as *generalized cube* network

Fig.3. The *SW-banyan* network (N=8, S=F=2, L=3)

In almost all of the networks two inputs/two outputs switching modules have been discussed as basic elements. However, the Gamma network [8] considered use three input/three output ports switching modules instead of the usual 2 x 2 crossbar switches. Unfortunately these cheaper networks suffer some drawbacks. First, inputs require more time to pass the network [4], [5]. Besides, there usually exists only one path between any particular information source - information destination pair. Therefore, if any one basic switching module fails, the whole network can fail. Moreover, the networks introduce complexities into connection. On the other hand, the reliable operations of interconnection networks are important to the overall system performance. Thus, a crucial practical aspect of an interconnection network used to meet system communication needs is fault tolerance. It is important to design a network that combines full connection capability with graceful degradation - in spite of the existence of faults. A number of different networks which offer redundant paths between the input and output terminals have been proposed to provide fast, efficient and reliable communications at a reasonable cost [3], [8]. Switching elements can be buffered or unbuffered , but the use of self-routing interconnecting networks is favored as particularly small delays can be achieved [4], [5], [7], [9], [10]. However, no single network is generally preferred because the cost-effectiveness of a particular design varies with such factors as its tasks, the desired speed of interfacility data transfers, the actual hardware implementation of the network, the number of facilities in the system, and the cost constraints on construction.

4. TRAFFIC CONTESTION AND THE SWITCH PERFORMANCE ISSUES

The communication protocols ensure reliable data routing from source to destination (i.e. routing technique depending on the network topology and the operation mode used) and provide the handshaking process among switching elements. In fast packet switching, the functions of flow control and error control are implemented on an end-to-end basis. Thus services that require error detection and correction may implement a retransmission strategy on an end-to-end basis whereas services, such as voice, that may tolerate a certain degree of error may take advantage of the low delay [1], [6], [7], [9]. As the protocol requirements of each switch are reduced, packets may be processed entirely in hardware. Thus switches of much greater capacity may be constructed and the switch may become more transparent to the data it carries than for conventional packet switching. Fast packet switching is in general connection-oriented. Thus once a virtual circuit is established across the network very short packet headers may be used to distinguish between each of the virtual circuits multiplexed over a single link. Also the routing of each packet may be performed in hardware by table look-up. As the packet overhead has been significantly reduced, the delay across the switch may be reduced to levels comparable with that of circuit switching. Fast packet switching, however, is able to spread the effects of delay and loss over all calls or over a selected class or classes of calls. Fast packet switching is also able to vary the allocation of bandwidth to individual sources instantaneously and can thus allow much greater flexibility. Moreover, it may also give a better performance for data traffic as end-to-end retransmissions are carried using the entire bandwidth of the switch fabric.

During information transfer through a fast packet switch a traffic stream can be combined with other traffic streams. Thus, in a general case network design issues should refer to different portions of information (packets, bursts, calls (e.g. file transfer)) which will travel through a switch.

Since the bus cannot provide sufficient throughput and the crossbar switch is too expensive, it is particularly interesting to know what kind of throughput various interconnection networks can provide. An open question is how to select an interconnection network that will best be suited to the needs of a fast packet switch.

As far as multistage interconnection networks are concerned, they can be built up using a variety of structures. Two significant examples of topological structures based on switching matrices are given in Fig.4. These matrices are open to either circuit- or message- oriented solution. Banyan networks are often referred to as a self-routing networks because the path selection can be simply performed on the basis of the binary address of the required outlet.. The drawbacks of such networks lie in the measures required to reduce blocking. A class of regular banyan networks includes the generalized-cube network structure, the omega network, the indirect binary n-cube, the baseline networks and the flip network. The PM2I network structure is used as the basis for interconnection networks, including the data manipulator network, the ADM network, the IADM network and the gamma network [8]. These interconnection networks have only one path between a source inlet S and destination outlet D when S=D, and have more than two paths between some source / destination pairs. The multiple-path property is exploited to allow dynamic rerouting of connections to avoid busy or faulty nodes. Their routing algorithm is more complex when compared to the multistage cube-type interconnection networks. Therefore, an open question

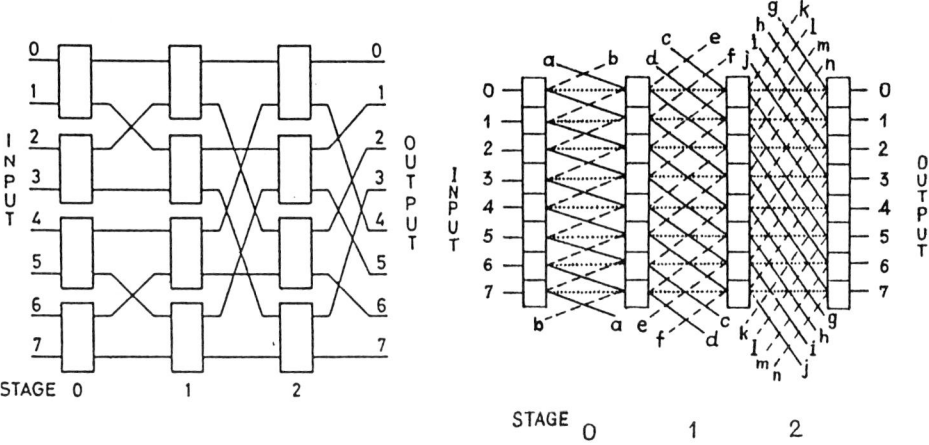

indirect binary n-cube network (N=8)   IADM network (N=8)

Fig.4.

is the improvement in performance that a multiple-path structure might offer,for various distributed control algorithms and decision rules, when compared to a cube-type network. The other question at this time is an amount of the hardware required to implement the given class of interconnection network.

5. A SWITCH MODEL AND SIMULATION RESULTS

In order to characterize the interrelationships of many of interconnection networks a unified theory has been developed and employed in [4] and [5] to carry out the performance analysis of the *generalized cube* networks as well as the *data manipulator networks* [5]. The main results from [4] and [5]are used to examine the performance of the abovementioned interconnection networks and compare them to the simulation results.

In order to quantify the various properties of a particular multistage interconnection network, some performance measures should be employed. The performance measures of interconnection networks include three group of characteristics [4], but for practical applications the traffic characteristics are of major importance [5]. This group comprise the following measures: capacity of the interconnection network, average packet delay, interconnection network throughput, interconnection network efficiency, and average length of link input queue. All this measures have been defined in detail in [4]. For the purpose of this contribution some of them will only be used. This two useful measures are the mean packet delay and the switch efficiency. To compare the average delay through different network structures, the slotted traffic source model is used with a uniform random distribution of packet destinations. It is assumed that the buffering is external to the switch fabric, i.e. a non-buffered switch fabrics formed from non-buffered switching elements are under considerations. Therefore, the queueing effects are left out of account in this model. A conflict is said to occure if more than one packet arrives to the same output link at a given time.

Two classes of packet switched multistage interconnection networks without and with redundant interconnection paths are under consideration. A class of multistage interconnection networks known as the regular *SW-banyan* networks have only one path for any inlet - outlet pair and the path selection can be simply performed on the basis of the binary address of the required outlet. On the other hand, a class of multistage interconnection networks called *data manipulators* have multiple path between a given network input and output. This creates additional interconnection capabilities, additional implementation costs, and additional control complexity [5]. A problem of central importance in the design of a fast packet switch is that, what are the relative speeds that can be achieved when implementing an interconnection network using different control schemes

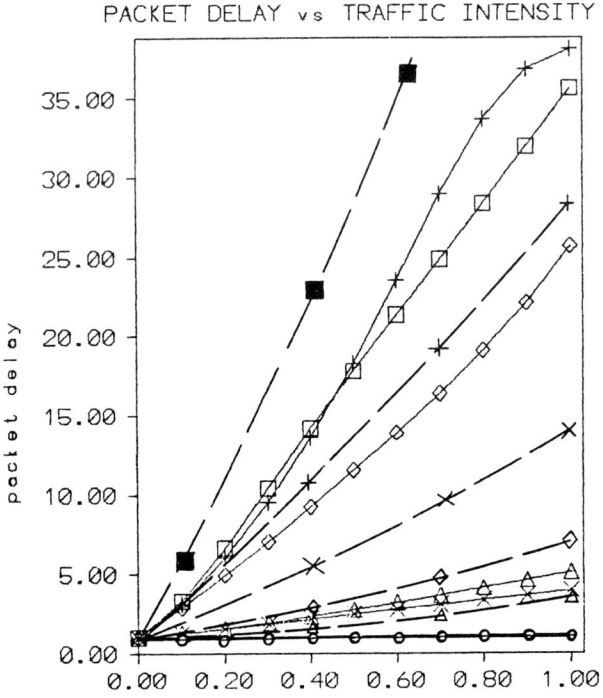

Fig.5.

and what is the trade-off between fault-tolerance and the network performance measures. One can expect that various traffic capabilities of the switching modules as well as properties of the used control mechanisms and decision rules can result crucially in the trade-off between the network costs and its performance.

Selected results from a study of the interconnection network performance are shown in Fig.5 and Fig.6. In the both switch structures the normalized average packet delay increases nonlinearly with the growth of traffic intensity (see Fig.5). The reason for this phenomenon is an augmentation of the retransmission rate according to increasing network workload. A dependence between the switch efficiency and the traffic intensity is depicted in Fig.6 . An exponential decrease of the switch efficiency can be

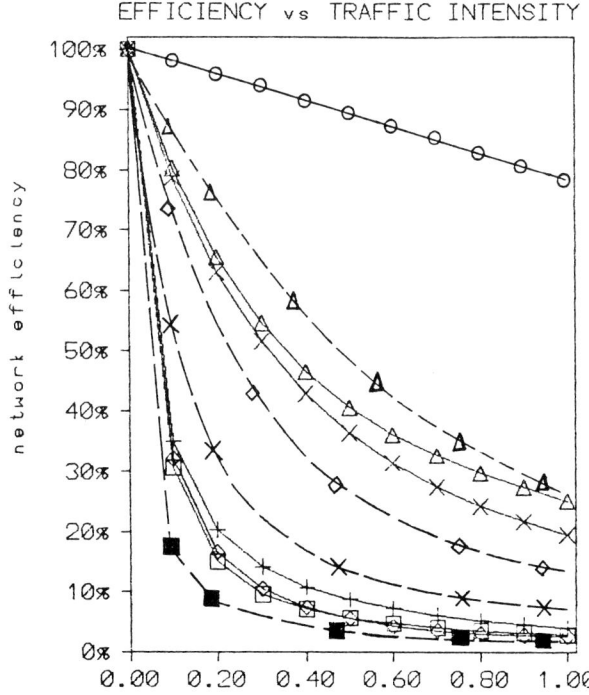

Fig.6.

observe together with the increase of traffic intensity but a decline of the efficiency is more sharp for the lower values of network load. This phenomenon is induced by the same reason as the one given for the packet delay.

At the same time one can point out a competition between switch structures based on different interconnection networks. It is obvious that a packet is to be suffered delay of smaller size in the banyan type network than to pass through the data manipulator type network. Therefore, the network throughput as well as the switch efficiency of the banyan structure are better than those of the data manipulator architecture. This follows from the differences occuring in the control mechanisms as well as in the construction of the switching modules employed in the both switch structures. Thus, the greater the number of conflict transitions across the module, the greater the packet delay in a network and the less the network throughput and the lower the switch efficiency.

Decision rules also crucially influence the network performance measures. In the both types of multistage interconnection network a system control unit can operate in accordance with one of two various decision rules [4], [5]: a random decision rule and a majority decision rule (denoted *rr* and *mr* in the attached figures respectively). It is evident (see Fig.5 and Fig.6), that the lower packet delay and the greater network throughput and network efficiency are obtained at the prize of a more sophisticated control concept employed in the network, to be used as the fast packet switch. Thus, more or less elaborated network control logic can result in the switch costs, i.e. the implementation costs will also vary.

In order to gain an insight into the effect of the various switch parameters on performance of the switch fabric a simulation study has been carried out [6]. The simulation results of various aspects of a simple model of multi-service traffic applied to the switch enable to select the appropriate switch design parameters. Performance measures of the analytical model are compared to the simulation results and discussed.

A simulation model deals with performance of the switch based on different types of interconnection network in terms of the troughput and average packet delay. It is a discrete time, event driven simulator [6]. All sources are set to the same value of applied load so that a new packet of a fixed length is always available at every input port on completion of the preceding packet transmission. All packets are given a random destination distribution and all output ports perform the function of a perfect sink. Each packet set-up attempt is modelled as an instantaneous event on a stage by stage basis. In a case of centralized control and the multiple paths, all routes to the destination are examined until a free path is traced. If no free path is available the set-up attempt is considered blocked and a new set-up attempt is planed out in the next cycle. The release of the path on completion of packet transmission is instantaneous, though there may be a stage by stage release mechanism.

Selected results from the simulation study [6] are shown in figures: Fig.7, Fig.8, Fig.9 and Fig.10. One can observe qualitatively the same or similar nature of the performance measures of the *extra stage cube (ESC)* network as those of the *augmented data manipulator (ADM)* network (see Fig.7 and Fig.8). It is obvious that faultiness of a switching module has crucial impact on both the mean packet delay and the network efficiency, i.e. on the switch performance. Besides, if the sources are set to different values of applied workload, then alterations of the switch performance measures can be also stated. A question is, however, what is the switch performance when the packets are given a nonuniform destination distribution, i.e. when a concentration of packets at the given output port occurs. It can be noticed (see Fig.9 and Fig.10) that the switch performance deteriorates together with the increase of the probability of packet concentration at the given output port of the switch.

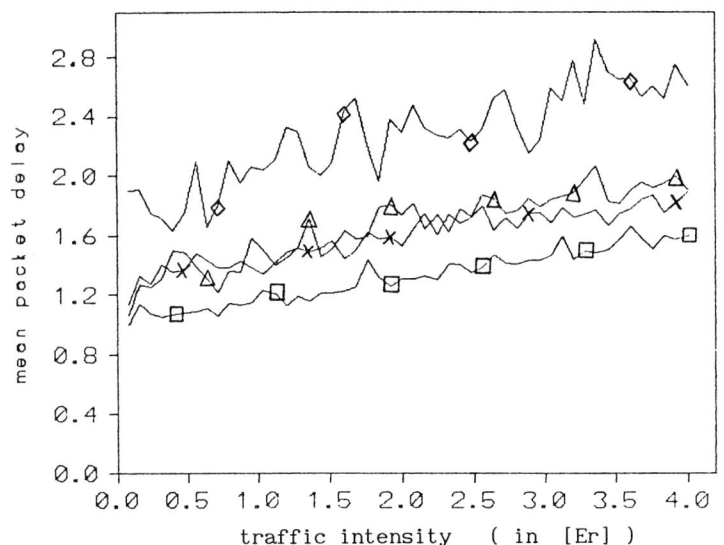

□ ESC, N=4, p_f=0.02   △ p_f=0.1   x p=$_f$0.1, var.pr.distrib.   ◊ ADM, N=4, p_f=0.1

Fig.7.

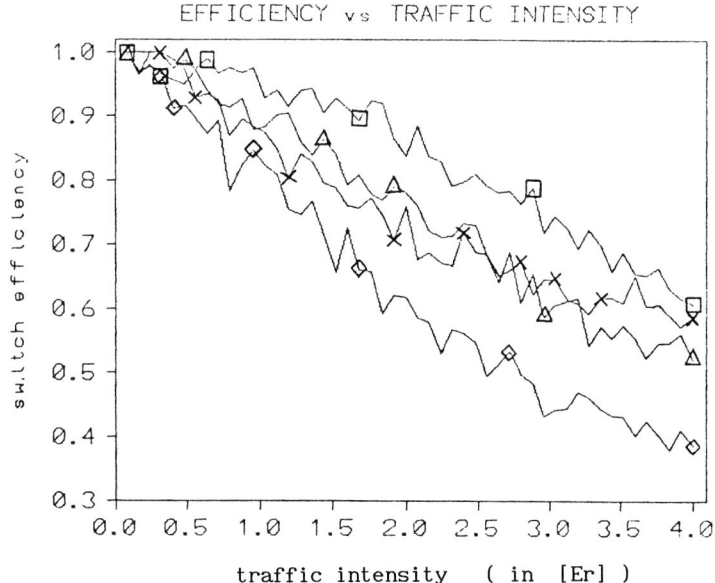

□ ESC, N=4, p_f=0.02   △ p_f=0.1   x p=$_f$0.1 , var.pr.distrib.   ◊ ADM, N=4, p_f=0.1

Fig.8.

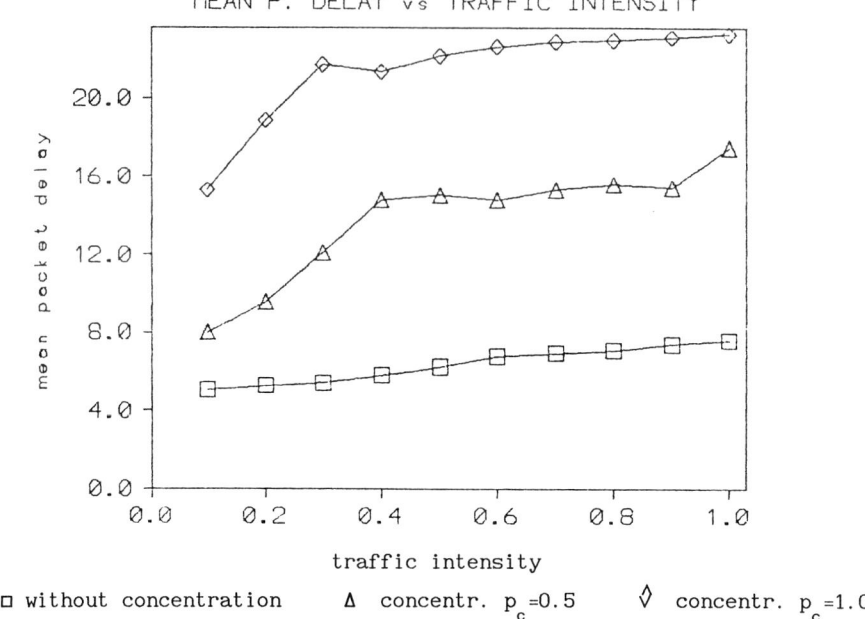

Fig.9. The mean packet delay in the *indirect binary n-cube* network (N=32)

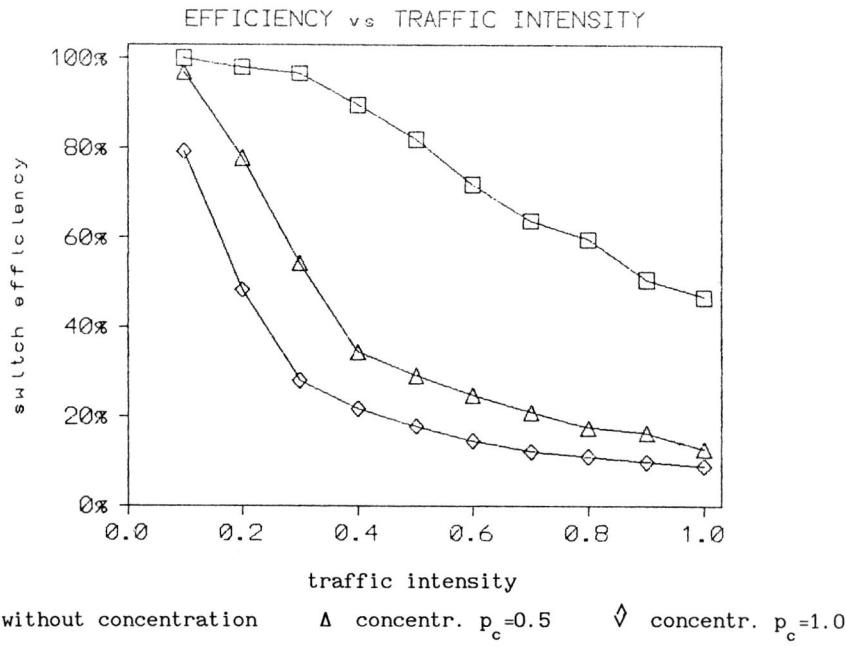

Fig.10. The efficiency of the *indirect binary n-cube* network (N=32).

A comparison of the performance measures of the analytical model to the simulation results validates the carried out analysis and confirms the expectations concerning the performance of a fast packet switch.

6. SUMMARY

The relationships between two classes of multistage interconnection networks have been explored in terms of switching module capabilities, control mechanisms, decision rules, topology and above all traffic performance. The effect of the various parameters on performance of the fast packet switch has been discussed and the switch performance measures have been compared to the simulation results.

It has stated that exist the classic tradeoff between cost and performance when choosing between different multistage interconnection networks for the fast packet switch design, i.e. an improvement of network performance can be obtained at the cost of a complication and enlargement of network control. Moreover, fault tolerance for a network may be obtained at the price of a reduction of network performance.

7. REFERENCES

[1] Ahmadi H., Denzel W.E.: "A survey of modern high-performance switching techniques", IEEE J. Selec. Areas Commun., vol. SAC-7, No. 7, Sept. 1989, pp. 1091 - 1103.
[2] Dieudonne M., Quinquis M.: Switching techniques for asynchronous time division multiplexing (or fast packet switching). Proc. ISS 87, paper B5.1, pp. 1-6.
[3] Fang T-Y.: A survey of interconnection networks. Computer, vol. 14, No 12, December 1981, pp. 12-27.
[4] Hulicki Z.: Performance comparison of multistage cube - type interconnection networks. Proc. Int'l Zurich Sem. on Digit. Commun., March 1988, pp. 213-219.
[5] Hulicki Z.: Performance of packet switching in data manipulator network. Proc. Int'l Zurich Sem. on Digit. Commun., March 1990, pp. 387-409.
[6] Hulicki Z.: Performance evaluation of a fast packet switch design by simulation techniques. To be published in the IEEE J. Selec. Areas Commun.
[7] Newman P.: A fast packet switch for the integrated services backbone network. IEEE J. Selec. Areas Commun., vol. SAC-6, Nr 9, December 1988, pp.
[8] Siegel H.J.: Interconnection networks for large-scale parallel processing. Lexington Books, Lexington 1983.
[9] Sincoskie W.D.: "Frontiers in switching technology. Part two: Broadband packet switching", Bellcore Exchange, vol.3, No 6, November / December 1987, pp. 22 - 27.
[10] Turner J.S.: Design of an integrated services packet network. Dept. Comput. Scien., Washington Univ., Tech. Rep. WUCS-85-3, March 1985.
[11] Turner J.S.: Design of a broadcast packet switching network. Washington Univ., Comput. Scien. Dept., Tech. Rep. WUCS-85-4, April 1985.

Packet and Fast Packet Switching Networks

Packet network structures based on multilink interfaces

F.E. Martín, E. Granel

TELEFONICA DE ESPAÑA, Madrid, Spain

Abstract
   This paper presents a qualitative performance analysis of multilink interfaces in data packet networks by comparing with other systems based on several single links. Both alternatives are useful for increasing throughput capacity without changing the bitrate.

1. INTRODUCTION

   Multilink procedure is an optional facility of the packet networks defined in the X.25 CCITT Recommendation where it is considered as a sublayer between the Data Link Layer and the Packet Layer.
   According to CCITT the multilink procedure performs the functions of accepting packets from the Packet Layer, distributing them across different Single Link Procedures (SLP) to be transmitted, and resequencing the packets received from these SLP's for delivery to the corresponding Packet Layer.
   In this way it is possible the data interchanging over several single links in parallel between a DCE and a DTE, or between a DCE and another DCE with one or more logical channels. Two or more packets of the same virtual circuit can be transmitted simultaneously by different links of the same multilink interface.
   Multilink is used in many networks in order to build reliable connections between nodes and also to increase their throughput capacity when no high speed links are available .These reasons also lead the customers to suscribe multilink access interfaces when it is allowed by the packet network.
   The multilink interface performance is analysed in this paper comparing it with other single interfaces which are considered as equivalent from the throughput capacity point of view. The comparison is carried out in terms of efficiency, delay, buffers occupation, availability and blocking probability.

2. ANALYSIS, SCENARIES AND MODELS

2.1. Scenaries

Aiming to achieve conclusions on the performance of multilink and single link interfaces, four systems are compared. (fig. 1):
- System 1 (S1): Multilink interface consisting of n links with bitrate V.
- System 2 (S2): Multilink interface with 2 n links of bitrate V/2.
- System 3 (S3): n single link interfaces of bitrate V.
- System 4 (S4): Single link interface of bitrate nV.

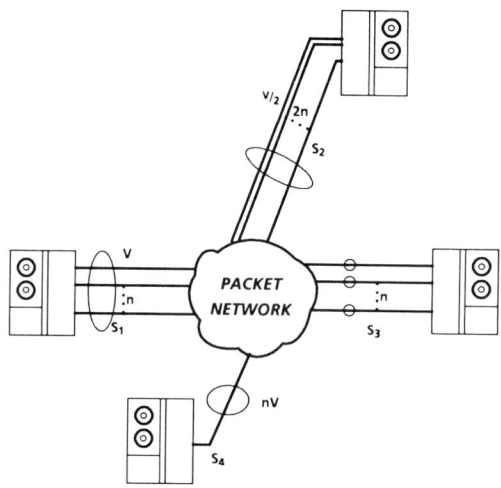

Figure 1. Compared systems (S_1, S_2, S_3, S_4,)

Actually, four different ways for getting the same throughput are compared:
- Multilink with a few links of high capacity (system 1).
- Multilink with more links of low capacity (system 2).
- Several single link interfaces of low capacity (system 3)
- One single link interface of high capacity (system 4)

In order to obtain conclusions in quantitative terms we will consider the values n = 2 and v = 9600 b/s in this study.

2.2. Analitical models

Figure 2 shows the queue models representing the analysed systems.

We could assume as a first approach the arrival process to be poissonian and the service time to be distributed according to an Exponential function. That is, we could consider the M/M/n model. However, markovian conditions are not fulfilled by packet networks if we take into account the following facts:

a) Users generate whole messages that are splitted into packets. Even if the message generation could be distributed by a poissonian law (variation coefficient of the interarrival time is equal to 1), the packet generation

will not. Packets will arrive in burst and the variation coefficient of the packet interarrival time is greater than 1. This effect is more evident in the user interface.
b) Flow control in the network may introduce distortions in the arrival process.
c) Service time in our case is the time required for transmitting a packet through a link, which is proportional to the packet size. When a message is splitted into packets, as many as possible of these packets will have the maximum allowed size. Therefore service time will present less variation than the Exponential distribution.

Markovian assumption for the service process is pessimistic and may be accepted as a conservative case. However, the same assumption for the arrival process is too optimistic. We propose to use in its place a 2-moments approach considering an Hyperexponential distribution of order 2. Therefore the model adopted here is the H2/M/n.

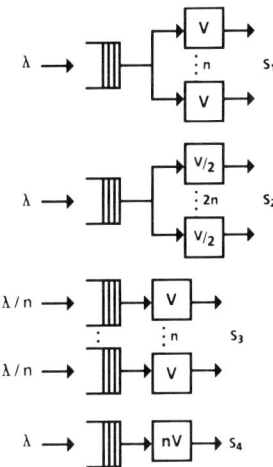

Figure 2. Equivalent models

2.3. Resolution of H2/M/n equations

The arrival process is characterised by two parameters:
δ : arrival rate
c : variation coefficient of the interarrival time.
The probability density function of the H2 distribution is:

$$f(t) = p.\delta_1.e^{-\delta_1 t} + (1 - p).\delta_2.e^{-\delta_2 t} \qquad [2.1]$$

where,

$$p = [1 + \sqrt{1 - 2/(c^2 + 1)}]/2 \qquad [2.2]$$

$$\delta_1 = 2.\delta.p$$

$$\delta_2 = 2.\delta.(1-p)$$

The service process is characterised by two parameters : number of servers (n) and service rate (μ).
Let us call :

$a_1 = \delta 1 / \mu$ and $a_2 = \delta 2 / \mu$ [2.3]

The model H2/M/n is equivalent to a 2-states system. In the state s (s = 1,2) the arrival process is exponential with rate δs. After each arrival the system may switch to states 1 or 2 with probabilities p and 1-p respectively. Defining the variable x (number of customer in the system), we have a two-dimensional system with states (s,x). Figure 3 shows the state-diagram transition of such a system, where we introduce the notation :

$m_i = \min(i,n)$ [2.4]

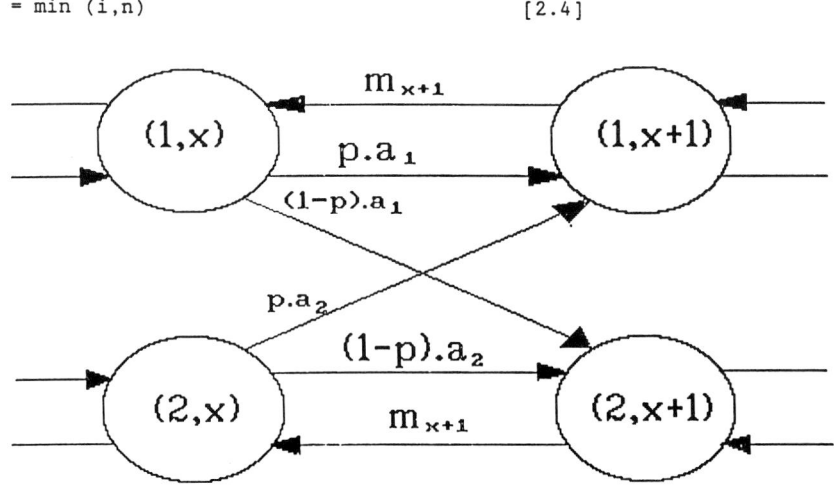

Figure 3. State-Transition Diagram

Let us call :

$f_x = \text{Prob}(1,x)$, $g_x = \text{Prob}(2,x)$, $p_x = f_x + g_x$ [2.5]

An elaborated analysis of fig.3 may lead to the following expressions:

$f_{n+1} = B^1 . f_n$ [2.6]

where :

$B = (1/2 + A/n) [1 - \sqrt{1 - 8.(c^2.A/n + (A/n)^2)/(c^2+1).(1+2A/n)^2}]$
$f_{n-1} = [a_1.(n-1)+a_2.n+(1-B).n.(n-1)].(c^2+1).f_n/[2A^2+2.(n-1).A.c^2]$
[2.7]

$0 < x < n - 1$

$$f_x = [a_1.x + a_2(x+1) + x.(x+1)].(c^2+1).f_{x+1}/[2A^2 + 2.x.A.c^2] -$$
$$[x.(x+2).(c^2+1)].f_{x+2}/[2A^2 + 2.x.A c^2] \qquad [2.8]$$

$$f_0 = f_1/a_1 \qquad [2.9]$$

$$\sum_{x=0}^{\infty} f_x = 1/2 \qquad [2.10]$$

A similar set of expressions may be derived for the g's interchanging a_1 with a_2.

As a consecuence of [2.6] we have:

$$\sum_{x=n}^{\infty} f_i = f_n/(1-B) \qquad [2.11]$$

$$\sum_{x=n}^{\infty} g_i = g_n/(1-B)$$

The system may then be solved very quickly:

STEP 1: Express f_x ($0 \leq x < n$) as a function of f_x by means [2.7]-[2.9].
STEP 2 : Compute f_n satisfying [2.10], taking into account [2.11].
STEP 3 : Repeat the process for the g_x's.
STEP 4 : $p_x = f_x + g_x$.

Finally, we can compute mean packet transfer delay in this interface:

$$D = 1/\mu + (1/\delta).\sum_{i=0}^{\infty} i.p_{n+i} = (1/\mu).[1 + (B/A).p_n/(1-B)^2] \qquad [2.12]$$

3. SYSTEM EFFICIENCY

In order to make an efficiency analysis of the four considered systems, we will study the situation in which the same packet transfer delay (T) is obtained in each of them. Then, the carried traffic in these conditions is considered as the reference efficiency parameter.

In fig.4 a graphic including a throughput comparison is shown. This graphic has been obtained by considering two different delay situations: T = 300 ms and T = 500 ms.

Although the graphic has been calculated under the hypothesis that the variation coefficient of the arrival process is c = 1, the conclusions are general and can be applied to any situations.

Notice, looking at fig.4 that :
- The single system number 4 (only one line) is the most efficient.

- When the same lines are used the multilink interface (system 1) works better than the interfaces based on single independent lines (system 3).
- The multilink efficiency is better in system 1 (high capacity lines) than in system 2 (greatest number of lines).

On base on these conclusions the followig considerations must be taken into account in order to design a packet network :
- A network designed for a given traffic and a target delay needs less transmission resources using single interfaces than using interfaces with several lines of smaller capacity.
- When two or more lines are required for reliability or throughput considerations the multilink procedure provides a better resource saving than the systems based on several single interfaces.

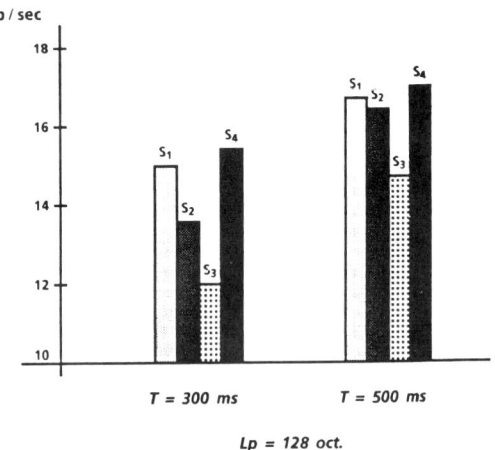

Figure 4. Systems Efficiency

Nevertheless in order to complete the efficiency analysis of the multilink interfaces, we must consider other factors related to the parameters and variables of the protocol working on the network.

If this protocol is nearly x.25 or x.75, two basic aspects have to be considered :
- Packet Layer Flow Control is based on window credits asigned to every logical channel of the multilink interface. When the window size is lower than the interface number of lines it is not possible to achieve the best multilink efficiency.

The graphic simulation of a three lines (L_1, L_2, L_3) multilink interface is shown on fig.5. As can be observed, the flow controlled situations lead to silence periods in which at least one of the lines are idle. Obviously, the higher the number of virtual circuits on the interface is, the less is the described effect.

- When a complete user message consisting of several packets is transmitted by the multilink interface a time overlapping effect exists, thus different packets of any communication can be simultaneously transmitted over the interface lines in the same interval of time.

The packet transmission in system 1 has been represented in fig.6. Two situations have been considered : message size Lm = 2Lp, and Lm = Lp ; where Lp represents the maximum packet size in the network.

In the first case the transmission efficiency is better than in the second one. Obviously this effect depends directly on the packet size.

At this point we must take into account that the number of packets of a given user message is in opposite proportion to their size, and that the flow controlled situation probability increases when a great number of packets are interchanged over a virtual communication. So using short packets could have a double effect : improving the transmission efficiency but increasing the flow controlled situations probability.

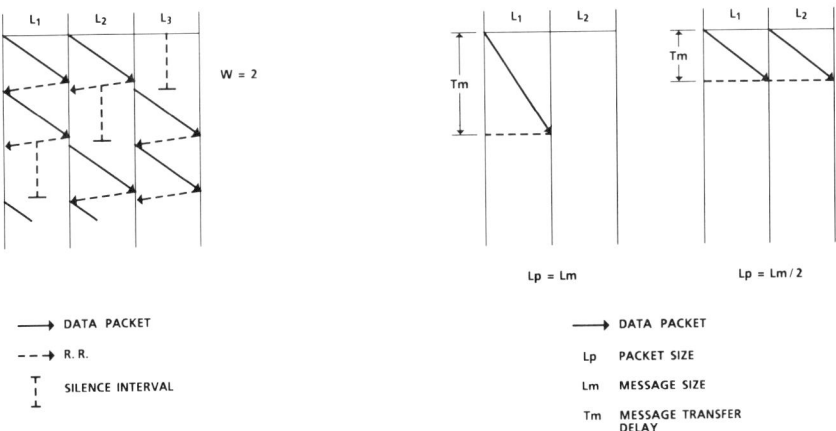

Figure 5. Layer 3 Window Influence   Figure 6. Packet Size Influence

4. THROUGHPUT, TRANSFER DELAY AND BUFFER OCCUPATION

First of all, it must be said that the maximum throughput capacity is similar in the four considered systems for an ideal situation. However, the multilink procedure adds several functions to the normal process in the packet switchers : multilink variables control, resequencing, etc. This process time increase leads to a smaller throughput capacity in the network nodes.

Related to the transfer delay, two parameters may be considered : "Data Packet Transfer Delay" (DPTD) and "Message Transfer Delay" (MTD).

Data Packet Transfer Delay as a function of the offered traffic in every system is represented in fig.7 for two arrival process (c = 1 and c = 1.5).

Notice in this figure that:
- Systems based on greatest line capacities provide the best packet delay, though in high traffic situations delay tends to be the same in all the systems.
- Systems based on greatest line capacities provide the best transmission delay, while the queue delay is smaller in systems with larger number of lines. Both effects tend to be compensated when the line utilization is nearly 100%.
- Comparing the multilink interface (system 1) with its equivalent based on several single interfaces (system 3), notice that the queue delay in the first case is smaller than in the second one, what causes a better global delay.

Related to de Message Transfer Delay, let us consider a user transaction consisting of N packets of size L_p. The parameter we are looking for can be obtained from the next expression :

$$MTD = [(N-1) L_p] / \Gamma + T_p$$

where Γ represents the interface throughput capacity and T_p the last packet transfer delay (see fig.8).

Because N, L_p and Γ are identical in the four systems, the delay differences are given by the transfer delay of a packet of size L_p.

When large messages are considered ($N.L_p / \Gamma \gg T_p$), MTD tends to be similar in all the compared systems. As an exception to this rule, in a system type 3 formed by two single interfaces, a given virtual circuit is not simultaneously supported by both of them but only by one.

In this situation, the throughput of a virtual communication is only the half of the obtained in the rest of the systems. So system 3 provides the worst performance from the point of view of the Message Transfer Delay.

The qualitative results obtained earlier from the delay analysis may be applied to the buffer occupation. Nevertheless the maximum buffer occupation in each of the compared systems depends on the flow control procedure, and on the number of established virtual circuits.

In case of flow control window based, the worst situation memory occupation is given by $S = N_v.W$, where N_v is the number of virtual channels working on the system, and W the flow control window size.

The S value decides the maximum number of virtual connections allowed in a packet switcher.

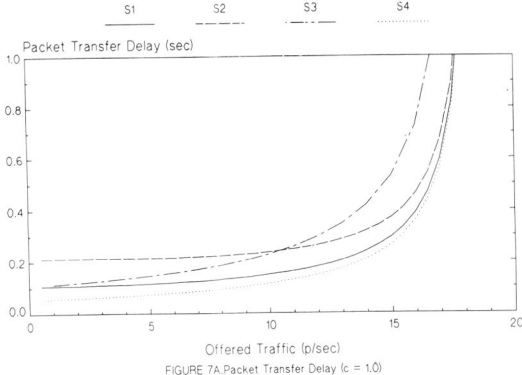

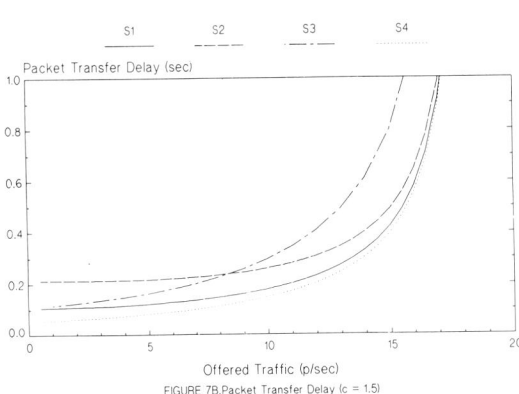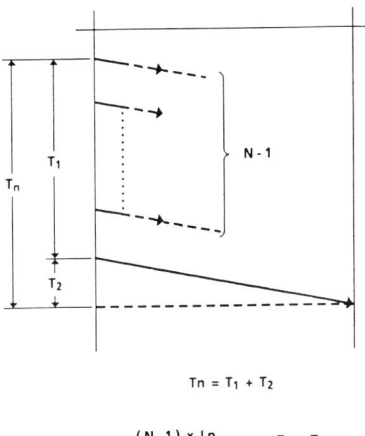

Figure 7. Packet Transfer Delay   Figure 8. Message Transfer Delay

5 .INTERFACE AVAILABILITY

For a user establishing a virtual communication over a given interface, the system 3 and 4 availabilities are A_L, where A_L coincides with the single line availability. The system 1 and 2 availability is :

- System 1 $A_1 = 1 - (1 - A_L)^2$

- System 2 $A_2 = 1 - (1 - A_L)^4$

If we consider for instance $A_L = 0'89$:

- System 1 $A_1 = 0.987$
- System 2 $A_2 = 0.999$
- System 3 $A_3 = 0.890$
- System 4 $A_4 = 0.890$

It is obvious that the system availability is greater by using multilink procedure.

A network with single connections must be based on reliable architectures with alternative paths.

Two network architectures are represented in fig.9. The former is based on multilink interfaces, while the second one exhibits single lines.

If the availability is evaluated in both cases by means the probability that node 1 to be disconnected from node 2, we would obtain the following results :

- Architecture 1 $P_1 = 1 - A_n^2 \cdot [1 - (1 - A_L)^2]^3$
- Architecture 2 $P_2 = (1 - A_n \cdot A_L^2)^2$

where A_n and A_L are the network node an line availability respectively.

As an example let us consider $A_n = 0.99$, $A_L = 0.89$. Then $P_1 = 0.0550$ and $P_2 = 0.0465$.

Notice that the second architecture is even cheaper (less number of lines) than the first one, but requires some additional network functions in order to maintain a good performance : alternative routing, automatic reconnection in case of failure, etc.

As a conclusion we can say that it is possible to design reliable and cheap architectures without using multilink interfaces.

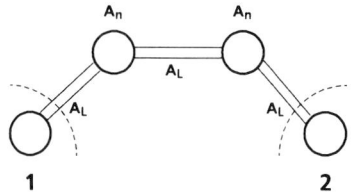

ARCHITECTURE 1

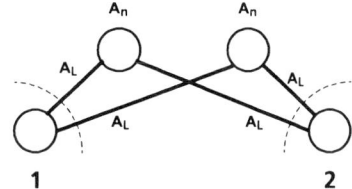

ARCHITECTURE 2

Figure 9. Network Architectures

6. BLOCKING PROBABILITY

The blocking probability value mainly depends on the calls offered traffic and on the number of logical channels equipped in the system.

For a given traffic intensity u_c and n_c logical channels, the blocking probability value can be computed as follows :
- In systems with only one interface the blocking probability P_{B1} is calculated by using the Erlang-B model, considering n_c servers and u_c erlangs as traffic intensity.
- When n_i interfaces are considered (system 3, $n_i = 2$), the blocking probability is calculated by means the Erlang-B model, considering n_c/n_i servers and u_c/n_i erlangs.

Generally $P_{B2} > P_{B1}$, so in our case considering multilink interfaces would be better from the point of view of the blocking probability. Nevertheless as a difference to the physical channels in circuit networks, a logical channel is a cheap concept and can be generously supplied in order to decrease the blocking probability.

Some values of P_{B1} and P_{B2} are shown in Table 1 for a given n_c and several u_c values.

n_c	u_c (Erl.)	P_{B1}	P_{B2}
100	10	2.025 E-41	1.967 E-32
100	50	1.630 E-10	3.602 E-6
100	75	9.227 E-4	8.739 E-3
100	100	7.570 E-2	1.047 E-1

Table 1. Blocking Probability.

7. CONCLUSIONS

The main conclusions derived from this study are the following :

- In order to achieve a given throughput it is generally more efficient using interfaces with few high capacity lines than using many lines with a lower capacity.

- When several lines are required the multilink interface is the best solution. Howewer, the extra process time needed in this situation decreases the throughput capacity of the packet switchers.

- The multilink interfaces exhibit a good performance from the availability point of view, but it is also possible to achieve reliable network architectures by using only single interfaces.

- In terms of blocking probability, the multilink systems work better than those which are based on several single interfaces. In this systems more logical channels have to be supplied in order to achieve the same blocking probability value.

8 .REFERENCES

1 X-25, CCITT Recommendation.
2 Kleinrock, "Queueing Systems",Volume I and II, J.Wiley & Sons, New York,1975.
3 Paul Kühn, "Tables on Delay Systems", Institute of Switching and Data Technics, University of Stuttgart, 1976.
4 J.F.Hayes, "Modeling and Analisys of Computer Communications Networks", Plenum Press, New York, 1984.
5 C.Macchi,J.F.Guilbert, "Teleinformatique. Transport et traitement de l'information dans les resaux et systemes teleinformatiques", Bordas-C.N.E.T - E.N.S.T.,Paris, 1983.
6 M.Schwartz, "Telecommunication Networks : Protocols, Modeling and Analysis", Addison-Wesley, 1987.
7 A.Alabau, J.Riera, "Teleinformática y redes de computadores", Marcombo, Barcelona, 1984.
8 F.E.Martín, L.Lavandera, "Las comunicaciones de datos. Génesis, realizaciones y tendencias", AHCIET - ICI,Madrid, 1989.

Study of S-ALOHA packet radio networks with a split-channel configuration

Jozef Woźniak

Institute of Telecommunications, Technical University of Gdansk,
80-952 Gdansk, POLAND

Abstract
In this paper we study an S-ALOHA packet radio network that employs a split - channel configuration. In particular we investigate two station operational schemes with almost instantaneous acknowledgments. To ensure a proper station operation we assume that the stations are equipped with buffers assigned to local and transit packets and their copies. Applying an approximate Markov chain approach we analyse the behaviour of station buffers and the station throughput - delay performances.

1. INTRODUCTION

The ALOHA type protocols are still considered to be very attractive channel access schemes in many applications and environments. These protocols, due to their simplicity, can be easily implemented in satellite VSAT systems [1], [2] and ground MAN/WAN multihop packet radio networks [3], [4]. S-ALOHA algorithms are also thoroughly tested for possible mobile applications [5], [6].
   In this paper we investigate the performance of slotted ALOHA multihop packet radio networks (PRNETs). We are particularly interested in the 2-nd layer station protocols that include channel access, acknowledgement scheme and buffers management.
   The method in which the stations get the right of access to the common channel is one of the most important issues of the broadcast networks. When the channel access demand is unpredictable, distributed and bursty random access schemes are successfully applied. In these schemes, users are allowed to transmit their packets at random. This may cause collisions among packets. Because of collisions, stations must be acknowledged whether their transmission attempts were successful or not in order to undertake appropriate actions.
   In the paper we consider two operational S-ALOHA channel access schemes with the split-channel configuration. In this configuration the channel bandwidth is divided in time into two separate channels for data and acknowledgement traffic [7].
   According to the models developed in the paper we assume that acknowledgements (ACKs) are sent back to source stations either in the second part of the time slot following the successful data (information) packet (PI) reception or in the second part of the next time slot. These models will be referred to as models A and B, respectively. In both models we get almost instantaneous ACKs.

In contrast to some previous works we assume that all stations can not only generate and receive packets but also relay packets in transit [8], [9]. To manage these functions the stations are equipped with buffers assigned to local and transit packets and their copies.

The main purpose of this paper is to provide the analysis of proposed station operational schemes. We describe two station models. Given station operational schemes we compute a number of important performance measures. The primary performance measure is the station throughput. Other measures include the average packet delay and the average number of packets in buffers. We make use of the discrete-time Markov chain approach.

2. STATION MODELS

Let us consider a PRNET consisting of a certain number of stations (nodes), distributed over a large geographical area, which are connected via a broadcast radio medium. We assume that all stations are capable of estabilishing end - to - end routes with the tandem stations storing and forwarding the packets to their destinations. The PRNET topology (connectivity) is strongly affected by the radio range. In the paper we restrict ourselves to the analysis of a local, single - hop subset of m homogeneous stations located around a chosen station. For the sake of simplicity we assume that each of these neighbouring stations transmits packets (PI) with probability P and refrains from transmission with probability 1-P independently from the actual situation at the station in question. Each of the stations from the local cluster distributes PI to each of its m neighbouring stations with probability $p_m = P/m$. Thus the total traffic directed to a chosen station is $P = mp_m$. A single station acts as a tranceiver (transmitter/receiver) which can either transmit or receive at one time, but not do both simultaneously. Packets interferences are the only cause of transmission error. In the paper (both) primary and secondary interferences are considered jointly. The capture effect and stability conditions are not discussed.

The PRNET behaviour is analysed using approximation models that make heavy use of some other simplifying assumptions, such as the independence of station queues and memoryless properties of buffer input processes.

We assume that all stations are equipped with two basic buffers as shown in Fig. 1.

Buffer B is assigned to local and transit packets. Data packets PI may be inserted into this buffer in time slots not occupied by PI transmissions from the station. Priority in entering buffer B is given to transit packets arriving from other stations. A packet from buffer B can be transmitted only when a single packet buffer of copies (BC) is empty. In the following we discuss two versions of the station operational schemes referred to as models A and B. These models illustrated in Fig. 2 state:

Model A
- PI transmissions (or retransmissions) start at the beginning of time slots of length T_A. Simultaneously with a PI transmission, from buffer B the packet is inserted into buffer BC (single packet buffer of copies).

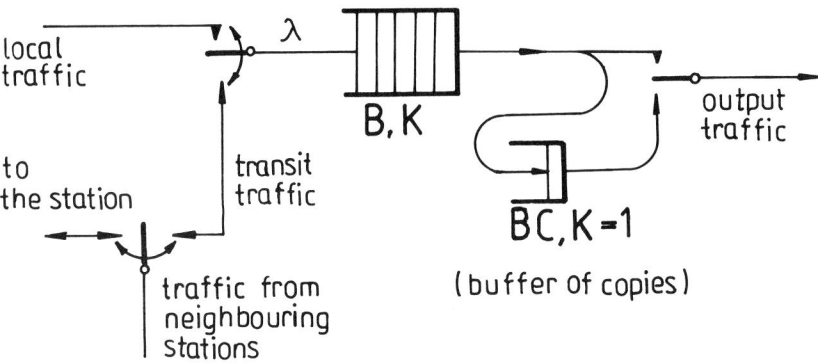

Fig. 1. Buffers configuration at a station.

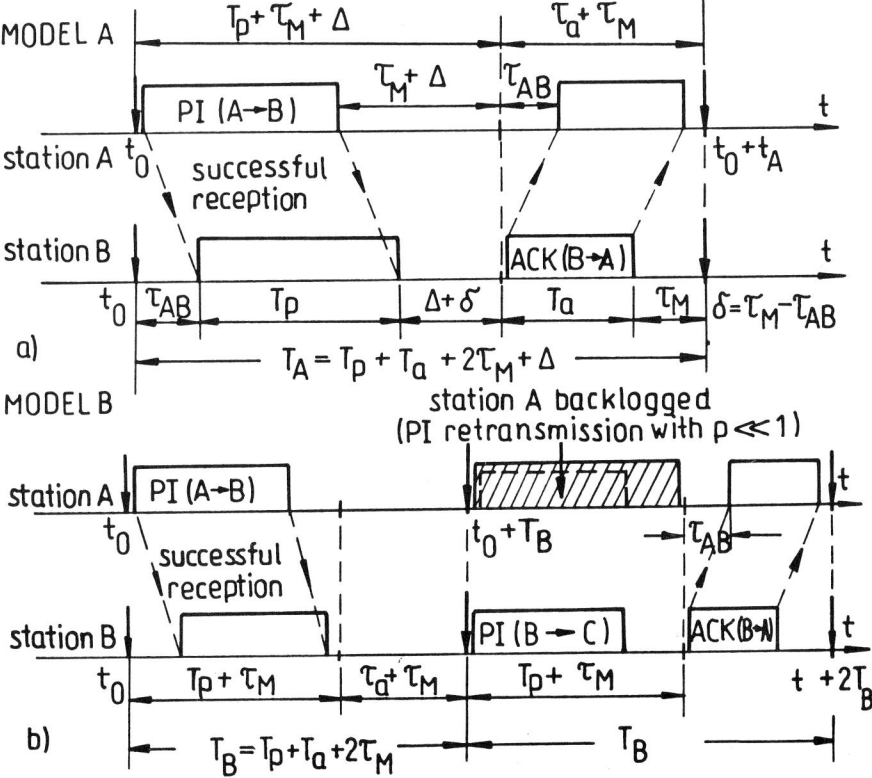

Fig. 2. Examples of channel operation for : a) model A, b) model B.
$\tau_M = \max \tau_{AB}$, $T_A - T_B = \Delta$.

- In the case of sucessful PI transmission a destination station acknowledges this reception in the second part of the same slot. Each station reserves, within a slot, a special time interval Δ necessary for undertaking a proper decision (station reaction time).
- A lack of ACK in the second part of a time slot signals the source station its unsuccessful transmission. The station then becomes backlogged.
- To avoid continuous PI interferences, retransmissions from a BC are realized with the probability $p \ll 1$ in subsequent slots.

Model B
- The transmission and retransmission procedure is almost the same as in model A. The only difference relates to the fact that the decision about a PI acceptance is undertaken with one slot delay (as presented in Fig.2b). In this model we eliminate the reaction time Δ (packet processing time) from a slot duration.
- To limit the waiting time, that packets spend in BC, each packet is retransmitted with the probability $p \ll 1$, immediately after its first transmission. If the first attempt is successful one the second will be ignored.

Taking into account the maximum propagation delay $\tau_M = \max \tau_{AB}$ and PI as well as ACK transmission times (T_p and T_a, respectively) we get a slot duration in model A as

$$T_A = T_p + T_a + 2\tau_M + \Delta \qquad (1)$$

while in model B

$$T_B = T_p + T_a + 2\tau_M \qquad (2)$$

Thus

$$T_A - T_B = \Delta \qquad (3)$$

The transmission process from buffers B and BC in both presented models can be illustrated in a general way using Petri nets. An example of such an illustration is shown in Fig. 3. From Fig. 3 it follows that the new transmission from B can be arranged only after an ACK reception.

When no transmission from B or BC is realized a station automatically switches to the reception state.

A received packet can be adopted by a station when the station
(1) is the destination node for PI,
(2) is a transit node and there is enough room in buffer B to accomodate PI.

We shall assume that the conditional probability that a received PI is a transit one equals

$$\Pr(\text{transit PI}) = 1 - \frac{1}{\bar{l}} \qquad (4)$$

where

$\bar{l}$ - is the average transmission length in the network, expressed in the number of hops.

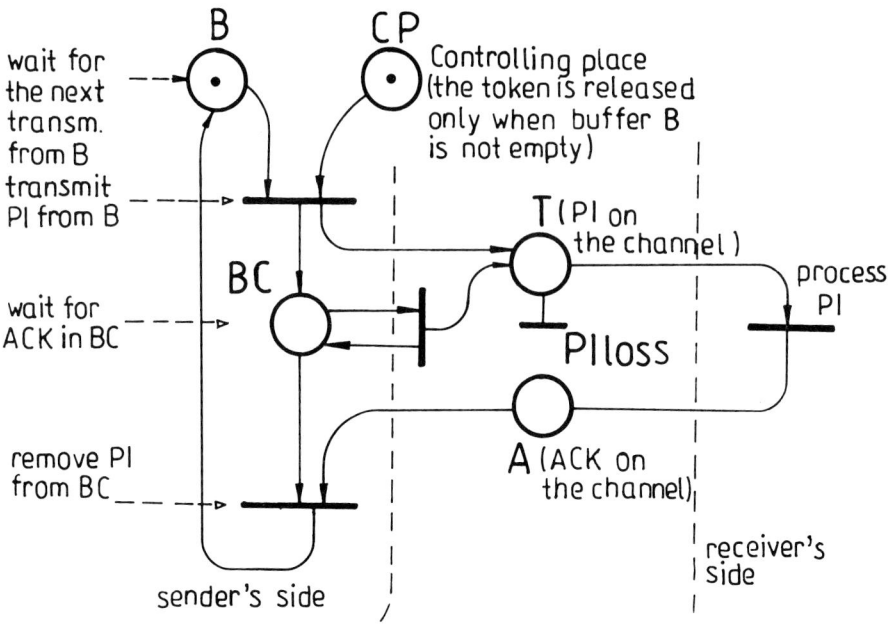

Fig. 3. Petri net representing the transmission process from a station. Packets reception is not depicted in the diagram.

For $\bar{I} = 1$ we have no transit traffic, while for $\bar{I} \gg 1$ this traffic dominates.

In the following discussion we analyse the behaviour of station buffers and the station throughput - delay performances. We compare the effectiveness of both models for different values of the time slot components and traffic intensities.

3. ANALYSIS OF STATION MODELS

In both described models we can consider buffers BC as servers in appropriate queueing systems. Thus, the time which packets spend in buffer BC is equivalent to the packet service time. We shall analyse the behaviour of station buffers B and BC using the Markov chain approach assuming memoryless properties of buffer input processes. To stress the difference between models A and B, buffers B and BC will be analysed separately, as if they were independent units.

Buffer B is assumed to have the capability to store up to K packets, while the BC is able to accommodate a single PI copy only.

To investigate the behaviour of B we shall analyse the state-transition diagram presented in Fig. 4. States of the buffer will be described by the number of packets waiting for transmission.

Depending on the model we consider, the exact formulae of the transition probabilities (input and output rates) λ and μ will vary slightly.

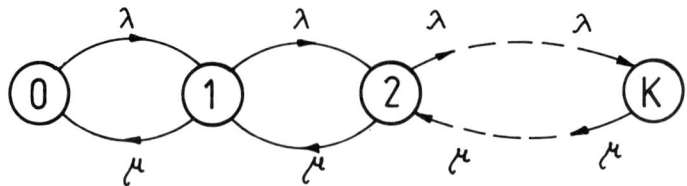

Fig. 4. State - transition diagram for buffer B.

However in both models μ corresponds to the probability $P_0^{(BC)}$ that buffer BC of copies is empty, i.e.,

$$\mu = P_0^{(BC)} \qquad (5)$$

On the other hand λ is the probability that either a transit packet (λ_t) or a local one (λ_l) is inserted into buffer B in a slot not occupied by a PI transmission from the station in question

$$\lambda = [\lambda_t + (1 - \lambda_t) \lambda_l] \, \Pr(\text{no PI transmission}) \qquad (6)$$

where

$$\lambda_t = \binom{m}{1} P_m (1-P)^{m-1} (1 - \frac{1}{l}) \qquad (7)$$

Using the above notations we get the average output rate g_{BK} from buffer B, and the average number N_{BK} of packets stored in B as

$$g_{BK} = \mu [1 - P_0^{(B)}] = \lambda \frac{1 - \varsigma^K}{1 - \varsigma^{K+1}} \qquad (8)$$

$$N_{BK} = \frac{1}{1 - \varsigma^{K+1}} \left(\frac{1 - \varsigma^K}{1 - \varsigma} - K \varsigma^{K+1} \right) \qquad (9)$$

where

$$\varsigma = \frac{\lambda}{\mu} \qquad (10)$$

The average time that packets spend in B is obtained by Little's result and is given by

$$D_{BK} = \frac{N_{BK}}{g_{BK}} \qquad (11)$$

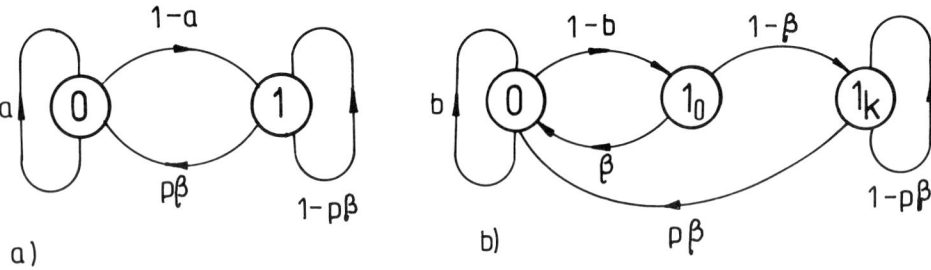

Fig. 5. Markov chain model for buffer BC, a) model A, b) model B.

To find the proper expressions of λ and μ we must analyse the behaviour of BC. Markov - chain models for buffers BC in models A and B are presented in Fig. 5.

Model A
Analysing the diagram shown in Fig. 5a under the steady - state conditions we get the stationary state probabilities $P_i^{(BC)}$, i = 0,1 as

$$P_0^{(BC)} = \frac{p\beta}{p\beta + 1-a} \qquad (12)$$

$$P_1^{(BC)} = \frac{1-a}{p\beta + 1-a} \qquad (13)$$

where

β - is the conditional probability of successful PI transmission

$$\beta = (1 - P)^m \qquad (14)$$

a - denotes the joint conditional probability of either no transmission from B or successful PI transmission (in the first attempt) under the condition that buffer BC is in state k = 0 at the beginning of a time slot

$$a = P_0^{(B)} + (1 - P_0^{(B)})\beta \qquad (15)$$

Thus the average output rate (traffic) from BC is

$$g_{BC} = p\, P_1^{(BC)} \qquad (16)$$

while the average time a PI copy spends in BC takes the form

$$D_{BC} = \frac{P_1^{(BC)}}{p\beta} \qquad (17)$$

Model B

In the case of model B we get the stationary state probabilities $P_1^{(BC)}$ and $P_{1k}^{(BC)}$ as

$$P_1^{(BC)} = \frac{(1-b)\beta p + (1-b)(1-\beta)}{(2-b)p\beta + (1-\beta)(1-b)} \qquad (18)$$

and

$$P_{1k}^{(BC)} = \frac{(1-b)(1-\beta)}{(2-b)p\beta + (1-\beta)(1-b)} \qquad (19)$$

where

$$b = P_0^{(B)} \qquad (20)$$

By a simple argument we can obtain probabilities $P_0^{(BC)}$ and $P_{10}^{(BC)}$ taking into account that $P_1^{(BC)} + P_0^{(BC)} = 1$ and $P_1^{(BC)} = P_{10}^{(BC)} + P_{1k}^{(BC)}$. The output rate from buffer BC, i.e., g_{BC} is as follows

$$g_{BC} = pP_{1k}^{(BC)} \qquad (21)$$

while the average delay D_{BC} is,

$$D_{BC} = \frac{P_1^{(BC)}}{p\beta} + 1 \qquad (22)$$

To complete our analysis we should present the formulae on the probability of no PI transmission. These probabilities take the form:

$$Pr^{(A)}(\text{no PI transmission}) = P_A = P_0^{(BC)} P_0^{(B)} + P_1^{(BC)}(1-p) \qquad (23)$$

and

$$Pr^{(B)}(\text{no PI transmission}) = P_B = P_0^{(BC)} P_0^{(B)} + P_{10}^{(BC)} + P_{1k}^{(BC)}(1-p) \qquad (24)$$

4. RESULTS

Analysing the behaviour of station buffers B and BC it becomes evident that these buffers affect each other very strongly. This also refers to the whole network of station queues. In fact to find the exact final formulae on buffer parameters it would be necessary to solve very complex equations. However, the main goal of this paper is to present and compare the effectiveness of two station operational schemes. Thus in the following we introduce some simplifying assumptions to help alleviate the complexity of the analysis. First of all we accccept the existence of stable conditions in the local subnet, apart from the fact that the output rates from m neighbouring stations and the station in question can be different. Next, for the sake of tractability we assume infinite capacity K of buffer B.

The stable station operation requires that the average output rate g_{BK}

from buffer B must be equal to the total station throughput s, i.e.,

$$s = g_{BK}\beta + g_{BC}\beta = g_{BK} \qquad (25)$$

Taking into account the above assumptions, we consider below the throughput and packet delay variations versus the network and local subnet parameters (m, P, $\bar{l}$) as well as internal station parameters (p, λ_l).

The comparision of station performance parameters for both considered models is presented in Figures 6 - 8. In all Figures it has been assumed that the probability p of packets retransmission from buffer BC is p=0.05. For this value we get the average length of retransmission interval equal to $\bar{R}$=20, i.e., large enough to accept memoryless properties of the input stream. Additionally we accept that $m = 10$ and the average transmission length in hops is $\bar{l} = 2$.

In Figure 6 we present the total station throughput s variations versus the total traffic offered by m neighbouring stations (of the station in question). From the Figure one can see that model A ensures higher station throughput. However we must remember that in this model a slot duration is longer than in model B by Δ. In both models higher values of s are observed for higher values of the intensity λ_l of local packets generation. This corresponds entirely to our expectations.

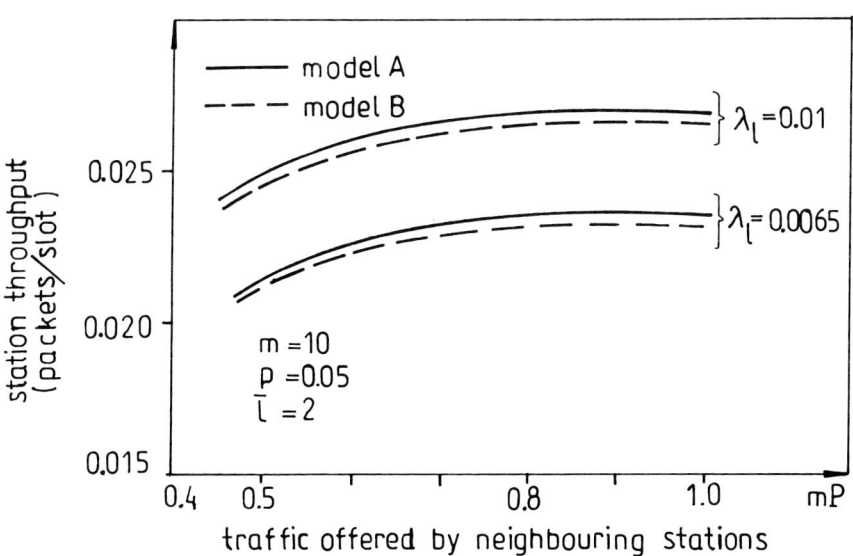

Fig. 6. The station throughput vs. mP.

Next Figure (Fig.7) shows variations of the effective station throughput versus mP. Here we accept that the reference slot duration is the same as in model B (i.e., T_B). From this Figure it follows that the effectiveness of model A decreases when the packet processing time (station reaction time) Δ increases. Nevertheless, for $\Delta \ll T_B$ model A presents its advantage over model B.

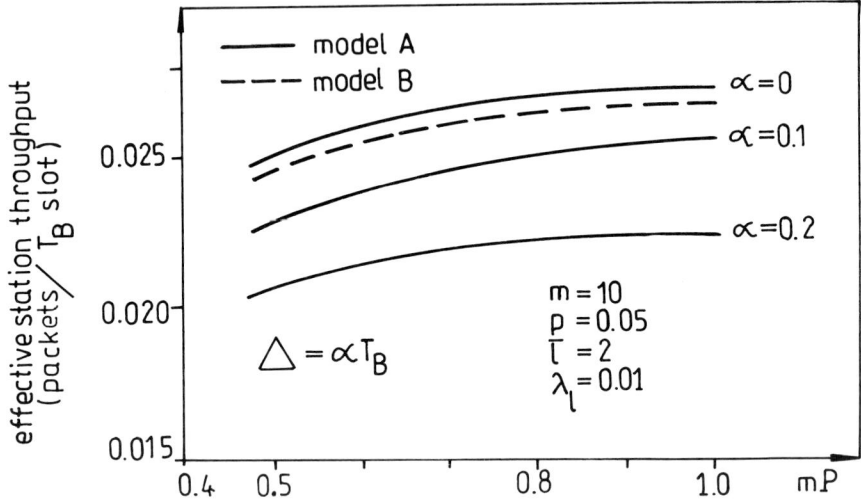

Fig. 7. Effective station throughput vs. mP.

The benefits related to model A operation are also evident when we consider packet delay - station throughput characteristics. From Fig. 8 one can see that model A guarantees less packet delay as compared to model B. (Even when we consider the effective slot duration). At the same time we observe that for constant values of s an increase of the local component s (achieved by the increase of λ_L) causes a significant decrease of the average delay packets spend at station buffers. This is due to the fact that in such a case a decrease of the transit component related to lower values of mP takes place.

5. CONCLUSIONS

In this paper we have presented two S-ALOHA type operational schemes useful for multihop packet radio networks. Both these schemes employ a split - channel configuration. We analysed the behaviour of station buffers applying an approximate Markov chain approach. We restricted our analysis to the local, single - hop subnet of m homogeneous stations located around a chosen one. For the sake of simplicity we assume that neighbouring stations operate independently of the actual situation at the station in question. From the analysis it follows that model A, which corresponds to the acknowledgement scheme with instantaneous ACKs, shows certain advantages over model B with one slot delayed ACKs.

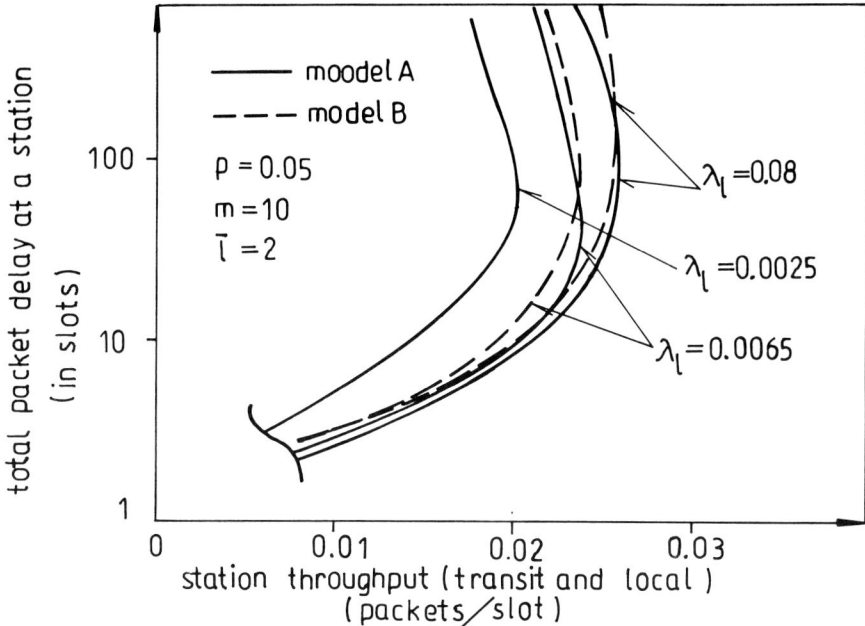

Fig. 8. Packet delay - station throughput tradeoffs.

6. REFERENCES

[1] Chakraborty D.: VSAT communication Networks. -An overview. Communications Magazine. Vol.26, No.5, 1988, pp. 10 - 23.
[2] Crespo E., Nelson A.: Digital satellite services and applications via the EUTELSAT satellite multiservices system. Proc. ICDSC-7, 1986, pp. 759-764.
[3] Abramson N.: Development of the ALOHANET. IEEE Trans. Inf. Theory. Vol. 31, 1985, pp. 119-123.
[4] Shacham N., Westcott J.: Future directions in packet radio architectures and protocols. Proceedings of the IEEE. Vol.75, 1987, pp. 83-99.
[5] Gotthardt C., Perz H.-J.:S-ALOHA multihop networks with adjustable transmission power. Proc. Intern. Workshop on Mobile Commun., Aachen 1990, pp. 460-472.
[6] Namislo C.: Analysis of mobile radio slotted ALOHA networks. IEEE Journal on Selected Areas in Commun. Vol. 2, 1984, pp. 583-588.
[7] Tobagi F., Kleinrock L.: The effect of acknowledgement traffic on the capacity of packet - switched radio channels. IEEE Trans. Commun. Vol. 26, 1978, pp. 815-826.
[8] Takagi H., Kleinrock L.: Throughput - delay characteristics of some slotted ALOHA multi - hop packet radio networks. IEEE Trans. Commun. Vol. 33, 1985, pp. 1200-1207.
[9] Wozniak J.: An approximate analysis of multihop S-ALOHA networks with or without immediate acknowledgments. Computer networking. Proc. COMNET'90. North-Holland 1990, pp.93-102.

Telecommunication Services for Developing Economies
J. Filipiak (Editor)
© 1991 Elsevier Science Publishers B.V. All rights reserved

Adaptive Isarithmic Flow Control in Fast Packet Switching Networks - Heavy Traffic Case

Marc Cotton P.Eng. and Lorne G. Mason

INRS-Télécommunications, 3 Place du Commerce, Verdun, Québec, Canada, H3E 1H2

Abstract

We consider the centralized adaptive isarithmic flow control scheme proposed by Mason and Gu [1,2] and studied further by Coderre [3]. These previous studies were done in the context of "traditional" packet switched networks (low speeds). This flow control technique is studied in the context of fast packet switching networks. The high transmission rates involved in such networks make it necessary to specifically consider propagation delays. This is done through the addition of $M/D/\infty$ queues on all trunks. The previous studies presented heuristic adaptive algorithms for the controller that attempted to maximize a performance criteria: the product of powers. We present an adaptive algorithm for the controller which is a very good approximation to the *optimal operation* in a fast packet switching network under *heavy* traffic. The claim is supported by analytic results (expected operation of system based on product form) and Monte Carlo simulation results. Our frame of analysis is product form networks [4].

1. INTRODUCTION

Packet switching networks are usually dimensioned to achieve message transport delay objectives under normal traffic conditions. From time to time circumstances arise which place more load on the network than it was designed to handle. These overloads may be general in nature, or they may be focused, where certain nodes or links become congested and the resulting delays become excessive. Under such conditions it is desirable to control the rate at which traffic enters the network in order to avoid performance degradation and unfair allocation of capacity. To do so, the network uses *network access level* flow control. One technique for such control is the *isarithmic* scheme first proposed by Davies [5].

The isarithmic flow control technique limits the total number of packets in the network. This is accomplished by allowing packets to enter the network only after they acquire a *permit*. With a limited number of permits in circulation, say W, the total number of packets in the network will never exceed W.

Mason and Gu [1,2] proposed adaptive versions of this scheme, one of which is studied further in this paper. A central controller is used to distribute permits to the nodes. When a permit is awarded to a node by the controller it is stored in a *source queue* associated with each node. When an exogenous packet arrives at a node it acquires the permit at the head of the source queue. When the packet arrives at destination, the permit is returned to the controller and redistributed. If the source queue is empty, the packet is lost. The exogenous arrival process at a node "serves" the source queue. If a permit arrives to an empty source queue its service time will be the residual time until the next exogenous arrival. If a permit arrives to a non empty source queue its service time will be the exogenous interarrival time. Assuming that the exogenous arrivals follow a Poisson process, the residual and interarrival times both follow an identical exponential distribution. Hence the source queues are $\cdot/M/1$. Since there is only one source queue per node the probability of a packet being lost (blocking probability) is the same for all traffic originating at the same node.

The current paper addresses performance oriented adaptive isarithmic flow control in fast packet switching networks using the central controller approach in networks under heavy traffic (the meaning of heavy traffic is discussed in section 6).

2. PRODUCT OF POWERS

Power, defined as the ratio of global throughput to mean message transfer delay, was originally proposed by Giessler et al. [6] as a means of unifying the antithetic objectives of high throughput and low transfer delay. Kleinrock [7] pointed out that the maximum power point corresponds to the point at the knee of the delay throughput curve where a line segment passing through the origin is tangent to the delay vs throughput curve. This is an intuitively reasonable compromise between the conflicting objectives to high throughput and low delay.

While the power criterion has a number of appealing attributes, there are certain undesirable features. As pointed out in [8], the maximum power operating point can be unfair in that certain users may have zero throughput. To circumvent this, Bharath-Kumar proposed the product of powers measure as an alternative to power for performance oriented flow control. Mazumdar et al. [9] have shown that the product of powers criterion corresponds to the Nash arbitration strategy among the users of a network where each user is attempting to maximize his own power. This result justifies the use of product of powers in that it is an operating point which is Pareto optimal and satisfies Nash's axioms of fairness. The product of powers is given by:

$$P(\cdot) = \prod_{i,d|\Lambda_{i,d}\neq 0} \frac{\lambda_{i,d}}{T_{i,d}}, \tag{1}$$

where $\Lambda_{i,d}$: exogenous offered traffic from node i to node d; $\lambda_{i,d}$: throughput of $i-d$ traffic; $T_{i,d}$: mean end-to-end delay for $i-d$ traffic that is not blocked.

3. CONTROLLER DESCRIPTION

The controller has two components: a *learning automaton* (hereafter referred to as automaton) and a *processor*. The automaton is based on the structure proposed by Mason [10]. The automaton has N actions: the ith action, a_i, is to route a permit to node i. Action a_k at time t ($a(t) = a_k$) is selected with probability $\pi_k(t)$ ($\sum_k \pi_k(t) = 1$). The processor receives feedback from the network for an action and computes a normalized response $b(t) \in [0, 1]$. The automaton's action probabilities are updated with:

$$\pi_i(t+1) = \pi_i(t) + G(\delta_{ik} - \pi_i(t))b(t) \qquad i = 0, \ldots, N-1, \; a(t) = a_k, \tag{2}$$

where G: gain of the automaton; δ_{ik}: Kronecker delta; N: number of nodes.

The feedback consists of delays and throughputs from which a function $H_k(\cdot)$ is computed for every node k. This function appears in the Kuhn-Tucker conditions of an optimization problem (discussed in section 6). $H_k(\cdot)$ is a positive function and is normalized to $[0, 1]$ with:

$$b(t) = 1 - \frac{H_k(\cdot)}{\max_j H_j(\cdot)}, \tag{3}$$

given that $a(t) = a_k$.

4. EQUILIBRIUM DISTRIBUTION OF NETWORK

In order to write down the equilibrium distribution of the network, it is assumed that the automaton's action probabilities, $\pi_i(t)$, do not change with time. Let π_i denote these probabilities.

In analyzing fast packet switching networks, propagation delay must be considered. This delay is constant on a trunk and is modelled by an $M/D/\infty$ queue. The $M/D/\infty$ queue is a special case of the $M/G/\infty$ queue for which the product form solution [4] is valid.

A network under isarithmic flow control has a constant number of permits circulating in it and is therefore closed with respect to permit flow. A methodology similar to the one used in [1,3] is used to write down the equilibrium distribution.

Each trunk is modelled by an $M/M/1$ queue (trunk processor) and an $M/D/\infty$ queue (propagation delay). To account for non-zero permit distribution times the controller is modelled by an $M/D/\infty$ queue with delay p_c. A permit/packet combination is assumed to have a size exponentially distributed with mean $\alpha = 1024 \times 8$ bits. A trunk between nodes j and l has a capacity of $C_{j,l}$ bits per second and a propagation delay of $p_{j,l}$ sec. The source queue at node i is served by the aggregate exogenous Poisson arrivals at node i. Hence, its service rate is $\sum_d \Lambda_{i,d} = \Lambda_i$. Routing in the network is random with $r_{j,l}^d$ being the probability that a permit/packet at node j destined for node d goes to node l.

A permit visits the controller once per "cycle" (a cycle is a trajectory from the controller, to a source queue, through the network, and back to the controller). We compute the relative number of visits of a permit to the different queues based on this visit. The relative number of visits to source queue i by a permit is π_i. The probability that a permit carries a $j - d$ packet in a cycle is:

$$s_j^d = \pi_j \frac{\Lambda_{j,d}}{\Lambda_j}. \tag{4}$$

The relative number of visits of a permit carrying a packet destined for node d at node j, v_j^d, is given by:

$$v_j^d = s_j^d + \sum_i v_i^d \, r_{i,j}^d. \tag{5}$$

With v_j^d, the relative number of visits of a permit to trunk $j-l$ is computed with:

$$f_{jl} = \sum_d v_j^d\, r_{j,l}^d. \tag{6}$$

The stationary distribution of the network is given by:

$$Pr\{\mathbf{x}\} = \frac{1}{G(W)} \cdot \prod_{i \in S} \left(\frac{\pi_i}{\Lambda_i}\right)^{x_i} \cdot \prod_{jl \in T} \left(\frac{f_{jl}\,\alpha}{C_{j,l}}\right)^{x_{jl}} \cdot \prod_{jl \in P} \frac{(f_{jl}\,p_{j,l})^{x_{jl}}}{x_{jl}!} \cdot \frac{p_d^{x_c}}{x_c!}, \tag{7}$$

where c: controller queue; S: set of all source queues; T: set of all trunk processor queues; P: set of all propagation delay queues; $G(W)$: normalization constant with W permits in the network; $\mathbf{x}$: vector that contains the population of each queue (the components of $\mathbf{x}$ must sum to W).

LBANC [11] and dynamic scaling [12] are used to compute $G(W)$. The controller's throughput (permits per unit time) is [13]:

$$\lambda_c = \frac{G(W-1)}{G(W)} \tag{8}$$

and the throughput for every $i-d$ pair is simply $\lambda_{i,d} = \lambda_c s_i^d$. The mean total delay at every queue (t_{jl}) is given by Reiser's mean value analysis method [14]. The average delay for an $i-d$ packet is:

$$T_{i,d} = \sum_{jl \in T \cup P} t_{jl}\, y_{i,j}^d\, r_{j,l}^d, \tag{9}$$

where $y_{i,j}^d$: average number of visits at node j in an $i-d$ trajectory through the network.

5. OPTIMIZATION PROBLEM

The formulæ derived in the foregoing section show that throughputs and delays are completely determined by the probabilities π_i and the permit population W. The optimum operating point is computed with:

$$\max_{\{\pi_i, W\}} P(\boldsymbol{\pi}, W) = \prod_{i,d|\Lambda_{i,d}\neq 0} \frac{\lambda_{i,d}}{T_{i,d}},$$

subject to:

$$\sum_{i,d|\Lambda_{i,d}\neq 0} \lambda_{i,d}\, y_{i,j}^d\, r_{j,l}^d \leq \frac{C_{j,l}}{\alpha} \quad \forall \text{ trunks } (j,l), \tag{10}$$

$$\sum_i \pi_i = 1, \quad W > 0, \quad 0 \leq \pi_i.$$

This is a mixed integer-real non-linear optimization problem with linear constraints. Solving (10) directly is made difficult by the integer variable W. Since W and π_i completely determine the throughputs $\lambda_{i,d}$ and vice-versa, we may circumvent the difficulty posed by the integer variable by performing the optimization with the throughputs as variables, then computing the corresponding W and π_i which realize the optimal throughput vector (this is a variation of Coderre's [3] technique). Since there is only one source queue per node we must have:

$$\frac{\lambda_{i,j}}{\Lambda_{i,j}} = \frac{\lambda_{i,d}}{\Lambda_{i,d}}. \tag{11}$$

The throughput of a single node pair originating at node i (say $i-j$) completely specifies all throughputs originating at node i through (11) as long as $\Lambda_{i,j} \neq 0$. Let λ'_i denote the throughput of a node pair originating at i, with an offered traffic greater than 0. The optimization problem (10) is reformulated:

$$\max_{\{\lambda'_i\}} P(\boldsymbol{\lambda}') = \prod_{i,d | \Lambda_{i,d} \neq 0} \lambda'_i \frac{\Lambda_{i,d}}{\Lambda'_i} \frac{1}{T_{i,d}},$$

subject to: $\tag{12}$

$$\sum_{i,d|\Lambda_{i,d}\neq 0} \lambda'_i \frac{\Lambda_{i,d}}{\Lambda'_i} y^d_{i,j} r^d_{j,l} \leq \frac{C_{j,l}}{\alpha} \quad \forall \text{ trunks } (j,l),$$

$$0 \leq \lambda'_i \leq \Lambda'_i.$$

With λ'^*_i, * denotes an optimal quantity, $\lambda^*_{i,d}$ is computed with (11) and the optimal distribution probabilities with:

$$\pi^*_i = \sum_j \frac{\lambda^*_{i,j}}{\sum_{k,l} \lambda^*_{k,l}}. \tag{13}$$

To compute the optimal permit population LBANC is iterated with increasing w until:

$$\frac{G(w-2)}{G(w-1)} < \sum_{i,j} \lambda^*_{i,j} \leq \frac{G(w-1)}{G(w)} \tag{14}$$

and $W^* = w$ for which (14) is satisfied.

During the optimization process w must be computed for non-optimal sets of throughputs to compute delays. As discussed in [3] using w integer may result in errors. To relax w to be real, and compute delays in a closed network with a non integer number of permits, a linear extrapolation for the average queue lengths is used as described by Coderre [3].

6. CHARACTERIZATION OF THE OPTIMAL SOLUTION

We now proceed to write the Lagrange function for (12). A multiplier $\theta_{j,l}$ is associated with each capacity constraint and the bound constraints are modified (with the corresponding multipliers in parenthesis):

$$\lambda'_i \geq 0 \quad (\psi_i), \quad \Lambda'_i - \lambda'_i \geq 0 \quad (\beta_i). \tag{15}$$

The Lagrange function is:

$$\mathcal{L} = -P(\boldsymbol{\lambda}') + \sum_{j,l} \theta_{j,l}\left[\left(\sum_{i,d|\Lambda_{i,d}\neq 0} \lambda_i' \frac{\Lambda_{i,d}}{\Lambda_i'} y_{i,j}^d r_{j,l}^d\right) - \frac{C_{j,l}}{\alpha}\right] - \sum_i \beta_i(\Lambda_i' - \lambda_i') - \sum_i \psi_i \lambda_i'. \quad (16)$$

If $\lambda_x'^* = 0$ for any x, then $P(\boldsymbol{\lambda}') = 0$. Any feasible vector $\boldsymbol{\lambda}'$ with $\lambda_x' > 0$ $\forall x$ will yield $P(\boldsymbol{\lambda}') > 0$. Thus $\lambda_x'^* > 0$ and, by the complementary slackness conditions, $\psi_x = 0$. We are studying networks under heavy demand which means that $\lambda_x'^* < \Lambda_x'$ and it follows that $\beta_x = 0$. It is easily shown [15] that in a closed network with more than one queue and finite population, the throughput of a single server queue **cannot** be equal to the service rate. Henceforth capacity constraints cannot be saturated and $\theta_{j,l} = 0$. The first order Kuhn-Tucker conditions are then:

$$\frac{\partial \mathcal{L}}{\partial \lambda_x'} = -\frac{\partial}{\partial \lambda_x'} P(\boldsymbol{\lambda}') = 0 \qquad x = 0, \ldots, N-1. \quad (17)$$

Let

$$h(\boldsymbol{\lambda}') = \prod_{i,d|\Lambda_{i,d}\neq 0} \frac{1}{T_{i,d}}.$$

Equation (17) then reduces to [15]:

$$N - 1 = -\frac{\lambda_x'}{h(\boldsymbol{\lambda}')} \frac{\partial}{\partial \lambda_x'} h(\boldsymbol{\lambda}') \qquad x = 0, \ldots, N-1. \quad (18)$$

The L.H.S. of (18) does not depend on x. Hence the R.H.S. must be same for all nodes at the optimal operating point. Let $H_x(\boldsymbol{\lambda}')$ be equal to the R.H.S. of (18). It follows that for a point to be a solution of (12) it is necessary to have

$$H_x(\boldsymbol{\lambda}') = H_y(\boldsymbol{\lambda}') \qquad \forall\, x, y. \quad (19)$$

The set of equations given by (19) form the necessary conditions for a point to be a solution of (12).

Expanding $H_x(\boldsymbol{\lambda}')$ in a closed network context is very complex. We can however approximate the necessary conditions by computing the derivative in an open network context. $H_x(\boldsymbol{\lambda}')$ then becomes [15]:

$$H_x(\boldsymbol{\lambda}') = \sum_{k|\Lambda_{x,k}\neq 0} \frac{\lambda_x'}{T_{x,k}} \left(\sum_{j,l\in A(x,k)} y_{x,j}^k r_{j,l}^k Y_{x,(j,l)} tp_{jl}^2\right) \quad (20)$$

$$+ \sum_{a,b|(a,b)\in C(x)} \frac{\lambda_x'}{T_{a,b}} \left(\sum_{j,l\in A(a,b)} y_{a,j}^b r_{j,l}^b Y_{x,(j,l)} tp_{jl}^2\right),$$

where $Y_{x,(j,l)} = \sum_d \frac{\Lambda_{x,d}}{\Lambda_x'} y_{x,j}^d r_{j,l}^d$; $A(i,d)$: set of trunks used by $i-d$ traffic; $C(x)$: set of all node pairs that use a trunk also used by traffic originating at x; tp_{jl}: total delay at the jl trunk processor.

Equations (19) and (20) form the necessary conditions for a point to be a solution of (12) in an open network context.

7. EXPECTED BEHAVIOUR ANALYSIS

If we define $g_i(t) \triangleq E[\pi_i(t)]$ and $F_i(t) \triangleq E[b(t)|a(t) = a_i]$ the approximate expected behaviour of the automaton [10] is:

$$g_i(t+1) = g_i(t)\Big[1 + G\Big(F_i(t) - \sum_j g_j(t)F_j(t)\Big)\Big], \qquad i = 0, \ldots, N-1. \tag{21}$$

If $F_i(t) > E[F.(t)] = \sum_j g_j(t)F_j(t)$ then $g_i(t+1) > g_i(t)$ and vice-versa. In addition, if $F_i(t) = E[F.(t)]$ then $g_i(t+1) = g_i(t)$. For a system where an increase in $g_i(\cdot)$ yields a decrease in $F_i(\cdot)$ and vice-versa, the automaton tries to select the probabilities $g_i(\cdot)$ to equate $F_i(t)$ for all i. By selecting this function properly the automaton will solve the open network Kuhn-Tucker conditions. Close examination of $H_x(\boldsymbol{\lambda}')$ as given by (20) reveals that if $g_x(\cdot)$ increases then $H_x(\boldsymbol{\lambda}')$ increases and vice-versa. By selecting:

$$F_x(\cdot) = 1 - \frac{H_x(\boldsymbol{\lambda}')}{\max_y H_y(\boldsymbol{\lambda}')} \tag{22}$$

the automaton selects probabilities to solve the Kuhn-Tucker conditions.

8. EXPERIMENTAL DATA

Tests are done with two networks, N1 and N2, and two traffic demands, T1 and T2. A combination identified TtNn refers to network n under traffic demand t.

The two networks have the same topology depicted in figure 1. N1 has $C_{j,l} = 45$ Mbps and all trunks cover a distance of 1000 Km. N2 has trunk pairs 1 to 3 and 12 to 16 at 1700 Mbps and 25 Km (all other trunk pairs do not change).

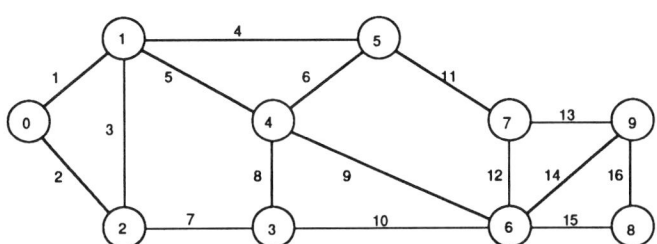

Fig. 1 10 Node Topology

For both traffic matrices we have $\Lambda_{i,i} = 0 \ \forall \ i$. T1 has $\Lambda_{i,d} = 15000 \ i \neq d$, and T2 was selected at random with $750 \leq \Lambda_{i,d} \leq 1500 \ i \neq d$.

9. OPTIMIZATION RESULTS

Table 1 contains the optimization results obtained by solving (12). Because of the magnitude of the product of powers a logarithm in base $b = 1.25$ is used.

Table 1
Optimization Results

	T1N1	T1N2	T2N1	T2N2
W^*	612	441	623	462
$\log_b(P(\cdot)^*)$	4573.50	4975.76	4561.11	4962.30

10. EXPECTED BEHAVIOUR OF AUTOMATON

The goal of computing the expected behaviour of the automaton is to determine the fixed point $(\mathbf{g}(t)|g_i(t+1) = g_i(t))$ that satisfies (21) in a network with optimal population (as given by table 1). This fixed point will give an idea of the performance that can be achieved by the automaton in selecting the distribution probabilities given that the optimal population of permits is in circulation. We have studied a permit population controller in series with the automaton which attempts to find W^*. This is not discussed in this paper.

Since we are not as yet concerned with the transient of the automaton, a relatively large automaton gain G is selected to speed up the convergence. For each of the four cases the algorithm is repeated twice, each time with a different initial distribution probability vector:

$$\mathbf{g}_1 = [\frac{1}{N}, \ldots, \frac{1}{N}] \text{ and } \mathbf{g}_2 = [0.3, 0.2, 0.0625, \ldots, 0.0625]. \tag{23}$$

Successive values of $\mathbf{g}$ are computed with (21) and (22) until the average g_i change is less then 10^{-8}. The throughputs and delays of the final iteration are compared with the optimal solution. We report the error in the product of powers ϵ_{PP}, and the absolute average error and standard deviation of the individual (origin - destination) throughputs ϵ_λ, delays ϵ_T, and power terms ϵ_P. A summary of the results for the four combinations demand/network and the number of iterations required are found in table 2. The errors for both runs with a given network and demand are the same, within the precision used in the table.

Table 2 shows that the automaton is basically obtaining the optimal operating point. In three out of the four cases the automaton yielded a higher product of powers than the optimization process! To insure that the optimization results are valid, the throughputs obtained by the automaton are used as starting point in the optimization program. The program terminates with this initial point as the optimal point, meaning that the optimization program did not get this point because of the precision of the computations. Figure 2 shows the product of powers at each iteration of the automaton for the T1N1 runs.

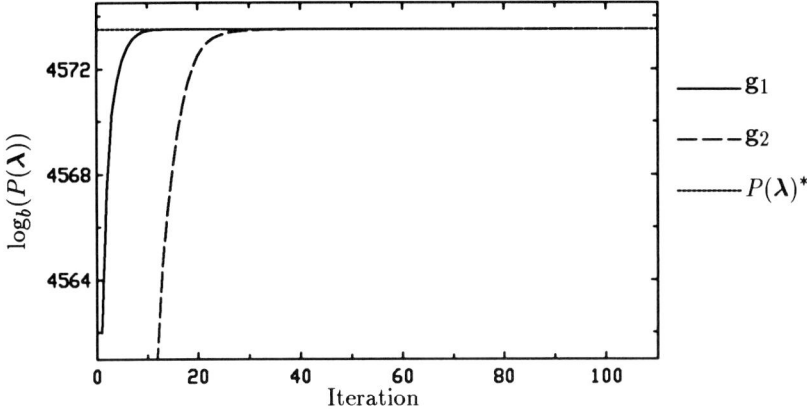

Fig. 2 Product of Powers vs iteration - T1N1

Table 2
Errors in the expected behaviour results

	$\log_b(P(\lambda))$	ϵ_{PP} (%)	$\epsilon_{\lambda_{i,d}}$ (%)		$\epsilon_{T_{i,d}}$ (%)		$\epsilon_{P_{i,d}}$ (%)		Iterations	
			ϵ_λ	σ_λ	ϵ_T	σ_T	ϵ_P	σ_P	g1	g2
T1N1	4573.50	-0.05	0.18	0.13	0.08	0.09	0.14	0.10	81	109
T1N2	4975.78	0.48	0.29	0.23	0.10	0.12	0.23	0.20	85	120
T2N1	4561.12	0.24	0.40	0.25	0.15	0.18	0.34	0.25	87	110
T2N2	4962.31	0.22	0.34	0.20	0.11	0.15	0.31	0.21	94	122

The expected transient behaviour can be computed with a small gain G (same as used in simulations). The time between two updates is given by $\frac{1}{\lambda_c}$. The expected transient behaviour of the automaton is assessed by plotting the error between the distribution probabilities ($g_i(t)$) and the optimal ones (π_i^*) versus time. Figure 3 shows the expected transient behaviour of the automaton for T1N1 with g_2 as starting point. For clarity purposes, only three of the ten error trajectories are plotted: $\epsilon_{g_0}(t)$, $\epsilon_{g_1}(t)$, and $\epsilon_{g_2}(t)$.

11. SIMULATIONS

The results presented in the previous section have shown that the automaton is capable of achieving near optimal performance in controlling the entry of traffic into a fast packet switching network under heavy traffic demand. During the simulations the network is made *quasi-stationnary* by selecting a very small automaton gain so that network dynamics are quick compared to changes in π.

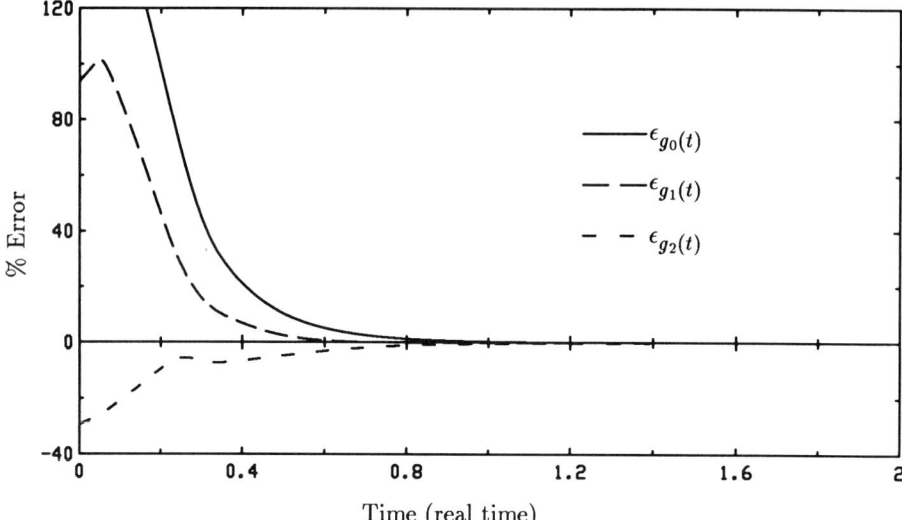

Fig. 3 Expected errors in distribution probabilities - T1N1

When a permit arrives at the controller the origin i, destination d, and time the permit left the source queue t_s are known (this information is written in the permit by the source queue). Knowing the distance between the destination node and the controller (and hence the delay - $p_{d,c}$) the end-to-end delay of the packet is $T_{i,d}$ = current_time $- t_s - p_{d,c}$. The $i - d$ throughput is computed by taking the inverse of the time since the last $i - d$ permit was received. And, if $i - d$ is also a trunk, the total trunk processing delay of trunk $i - d$ is $tp_{id} = (T_{i,d} - p_{i,d})$. Computations for every $i - d$ pair can be done in parallel. Using the foregoing results, $H_i(\cdot)$ is computed with (20) and b(t) with (3). The automaton then updates its action probabilities with (2).

The error between $\pi_i(t)$ and π_i^* is plotted in Figure 4 for the simulation of T1N1. Also of interest is the average performance seen by a user. It is evaluated by computing the average value of the distribution probabilities, $\overline{\pi_i(\cdot)}$. These averages are used to compute stationary throughputs and delays which are compared with the optimal results.

The simulations use $\boldsymbol{\pi}(0) = \mathbf{g}_2$. The number of permits in the network is W^* (as per table 1) and $G = .0001$. Table 3 contains the different errors between the stationary values found with the average probabilities $\overline{\pi_i(\cdot)}$ and the optimal results given by the optimization process. A new quantity in this table is $\epsilon_{\overline{\pi}}$, which is the average absolute error between $\overline{\pi_i(\cdot)}$ and π_i^*.

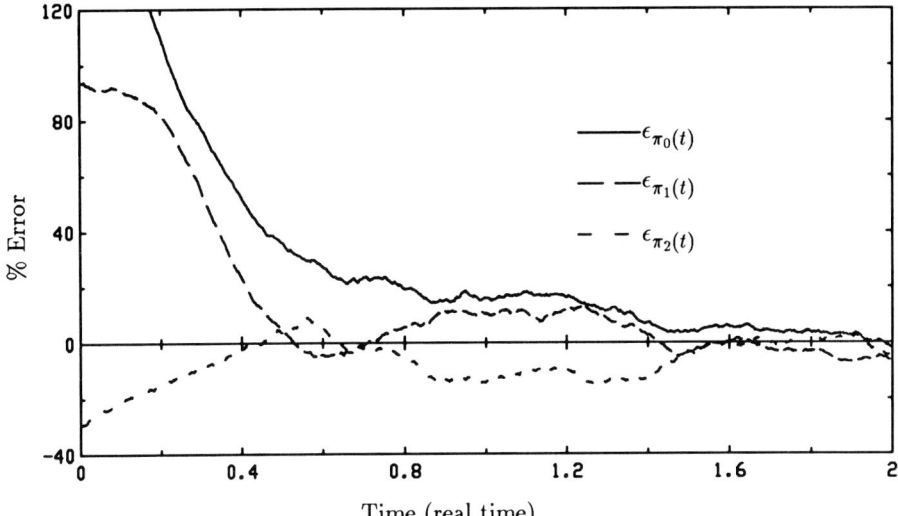

Fig. 4 Expected errors in distribution probabilities - T1N1

Table 3
Simulation Results

	$\log_b(P(\boldsymbol{\lambda}))$	ϵ_{PP} (%)	$\epsilon_{\lambda_{i,d}}$ (%)		$\epsilon_{T_{i,d}}$ (%)		$\epsilon_{P_{i,d}}$ (%)		$\epsilon_{\bar{\pi}}$ (%)
			ϵ_λ	σ_λ	ϵ_T	σ_T	ϵ_P	σ_P	
T1N1	4573.12	-8.13	2.31	1.54	0.59	0.81	1.95	1.33	2.35
T1N2	4975.44	-6.89	2.90	1.63	0.84	1.07	2.45	1.51	2.92
T2N1	4560.61	-10.56	2.50	1.18	0.89	1.44	2.16	1.18	2.54
T2N2	4961.63	-13.98	2.79	1.77	1.15	1.21	2.33	1.65	2.82

12. CONCLUSION

An adaptive controller capable of distributing permits to achieve near optimal performance in a fast packet switching network under heavy demand was presented.

The analytic model presented was used to formulate an optimization problem. With the first-order Kuhn-Tucker conditions the optimal solution was characterized: the optimal vector of throughputs is such that

$$H_x(\boldsymbol{\lambda}') = H_y(\boldsymbol{\lambda}') \quad \forall\, x, y.$$

Using open network formulæ for delays allowed us to obtain a closed form expression for $H_x(\boldsymbol{\lambda}')$. The automaton was designed to solve the first order Kuhn-Tucker conditions.

Monte Carlo simulations where then used to verify the transient and average performance of the automaton. These have shown that the automaton can adapt very rapidly and is near optimal. In a network under heavy demand, where traffic variations are in the order of minutes, the automaton could adjust its distribution probabilities to follow traffic variations.

The main contributions of this paper are the inclusion of propagation delay in existing analytic models to model and analyze fast packet switching networks under adaptive isarithmic flow control, the derivation of a feedback function used by the automaton in order to achieve near optimal performance, and the discovery that open network Kuhn-Tucker conditions are a very good approximation to the closed network Kuhn-Tucker conditions.

13. REFERENCES

[1] L.G. Mason, and XueDuo Gu, "Adaptive Isarithmic Flow Control," Rapport technique de l'INRS Télécommunications no. 85-18, Mai 1985.

[2] L.G. Mason, and XueDuo Gu, "Learning Automata Models for Adaptive Flow Control in Packet Switching Networks," in *Adaptive and Learning Systems - Theory and Applications*, K.S. Narendra New-York:Plenum Press, 1986, pp. 213-227.

[3] J.R.J. Coderre, "Modèle optimal d'un contrôle global de flux appliqué aux réseaux à commutation de paquets," Rapport technique de l'INRS-Télécommunications no. 89-15, 1989.

[4] F. Baskett, K.M. Chandy, R.R. Muntz, and F. Palacios-Gomez, "Open, closed and mixed networks of queues with different classes of customers," *Journal of the ACM*, vol. 2, pp. 248-260, 1975.

[5] D.W. Davies, "The Control of Congestion in Packet Switching Networks," *IEEE Trans. Commun.*, pp. 546-550, 1972.

[6] A. Giessler, J. Hanle, A. Konig, and E. Pade, "Free buffer allocation-An investigation by simulation," *Computer Networks*, pp. 553-574, 1978.

[7] L. Kleinrock, "Power and deterministic rules of thumb for probabilistic problems in computer communications," in *Proc. Int. Conf. Commun.*, June 1979, pp. 43.1.1-43.1.10.

[8] K. Bharath-Kumar, and J.M. Jaffe, "A New Approach to Performance Oriented Flow Control," *IEEE Trans. Commun.*, pp. 427-435, Avril 1981.

[9] R.Mazumdar, L.G. Mason, and C. Douligeris, "Fairness in Network Optimal Flow Control," in *ITS - Rio de Janeiro*, Sept 1990, pp. 25.1.1 - 25.1.7.

[10] L.G. Mason, "An Optimal Learning Algorithm for S-Model Environments," *IEEE Trans. on Automatic Control*, pp. 493-496, 1973.

[11] K.M. Chandy, and C.H. Sauer, "Computational Algorithms for Product Form Queueing Networks," *Communications of the ACM*, vol. 23, pp. 573-583, 1980.

[12] S.S. Lam, "Dynamic Scaling and Growth Behavior of Queueing Network Normalisation Constants," *Journal of the ACM*, vol. 29, pp. 492-513, 1982.

[13] E. Gelenbe, and I. Mitrani, *Analysis And Synthesis of Computer Systems*. New-York:Academic Press, 1980.

[14] M. Reiser, "A Queueing Network Analysis of Computer Communication Networks with Window Flow Control," *IEEE Trans. Commun.*, pp. 1199-1209, 1979.

[15] M. Cotton P.Eng., M.Sc. Thesis, To be published in 1991.

PERFORMANCE ANALYSIS OF MULTI-LAYER TOKEN RING LOCAL AREA NETWORKS

M. Kwiatkowski

Main Computer Center of the Mining Industry (COIG), 40-065 Katowice, Mikolowska 100, Poland

Abstract
We consider queueing network performance model of a token ring LAN (local area network) in which multi-layer OSI (open systems interconnection) architecture is used. For such queueing network, several performance submodels were developed for each layer working in isolation. Statistical independence and Poisson character of exogenous streams of messages arriving in layer are usually assumed in these submodels. It is shown that these assumptions can be violated if layers cooperate, thus invalidating predictions of simple decomposition method proposed in some papers. We also present some modifications improving accuracy of calculating mean window queue times for the transport layer which uses sliding window flow control.

1. INTRODUCTION

After several years of efforts the applicability of the OSI (open systems interconnection) seven layer reference model to LAN (local area network) environment was established. Within this model, four lower layers are of special interest because they are concerned with reliable and cost-effective data transfer. In this paper, we use OSI model to token ring LAN (see Figure 1).

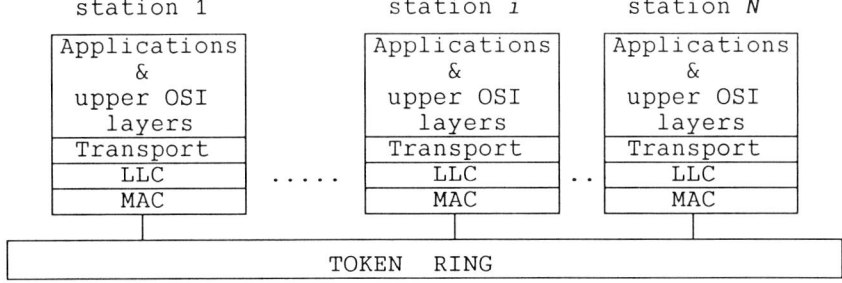

Figure 1. OSI layers in token ring network.

We consider standard IEEE 802.5 [1] protocol for MAC (medium access control) sublayer (in the paper jointly considered with physical layer), unacknowledged connectionless protocol for LLC (logical link control) sublayer and connection oriented protocol for transport layer. Flow control using sliding window protocol is only performed within transport layer. Network layer is by-passed as the routing function is not necessary in this type of LAN.

Many investigations were made in order to evaluate performance measures for individual data transfer layers. Only a few works were published for analytical performance prediction of several layers cooperating with each other. They use queueing network models. However, for analytical tractability, a typical approximate approach is based on decomposition of this network model along the layers thus creating layer submodels. Each performance submodel is then solved in isolation. In an upper layer, delay introduced by lower layers are represented by flow-equivalent servers. The major assumption is usually made that messages arriving to these submodels form independent Poisson processes.

Mitchell and Lide [2] proposed hierarchical modeling methodology for quantifying delays in multi-layer LAN systems. Unfortunately the accuracy of the proposed method was not demonstrated. Recently Murata and Takagi [3] developed a new iterative method to compute performance measures for multi-layer token ring LAN. In fact our layered model is similar to that analyzed in [3]. However, they made an assumption for transport layer that arriving messages facing a full window are discarded (*loss system*). In our approach, as it is shown below, we assume that such messages wait in a special window queue (*wait system*).

The aim of the paper is to examine accuracy of calculating mean waiting times at network queues using decomposition technique proposed in [2]. This paper is organized as follows. Section 2 presents general assumptions relating to class of token ring LANs under the study in this paper. Section 3 describes solution procedures for deriving mean message waiting times at queues of each layer. Numerical and simulation results as well as short discussion of the results are presented in section 4. The description of problems which remain to be solved are given in section 5.

2. GENERAL ASSUMPTIONS

There are N stations interconnected by a token ring LAN. Figure 2 presents queueing model of four lower layers of a station. Queueing systems ST_i, LT_i, ML_i and LT_i represent protocol processing times in transport and LLC layers. These times are independently distributed with known first two moments. Pairs of session entities exchange messages using services and functions of lower layers. Messages are not fragmented/reassembled within lower layers. In the transport layer, these messages are sent as *data messages* by means of unidirectional *virtual data channels*. We introduce the notation:

- C : Set of virtual data channels in transport layer.
- C_i : Set of virtual data channels in transport layer which function in the i-th station.
- C_{ij} : Set of virtual data channels which connect session entities in stations i,j; data messages are sent from i to j.

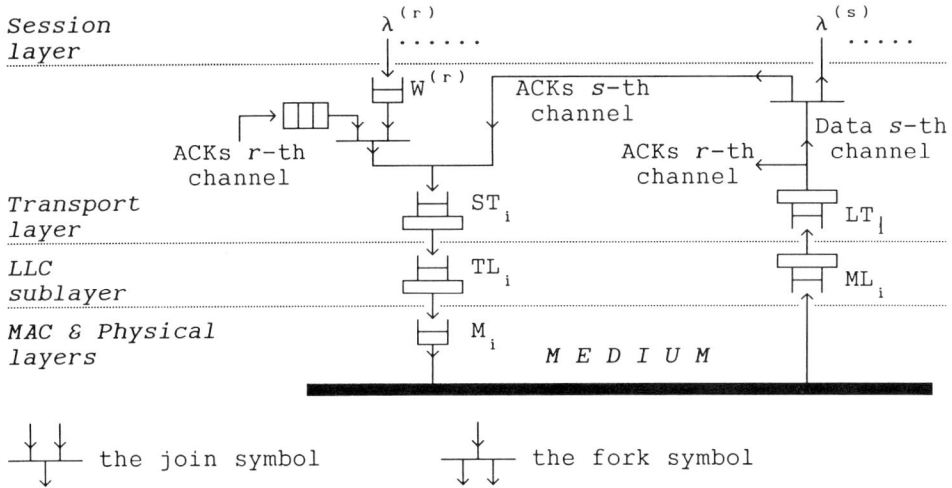

the join symbol   the fork symbol

Figure 2. Queueing model of a station.

For a channel c, $c \in C$, the length of messages is identically distributed with first two moments $El^{(c)}$ and $El^{(c)(2)}$ respectively. Session entities send messages to the transport layer in streams which are mapped one-to-one onto transport virtual data channels. These streams form independent Poisson processes with mean rates $\lambda^{(c)}$. Each channel use sliding window flow control protocol with window size of $R^{(c)}$. Meassages facing full window wait in a window queue $W^{(c)}$. Each data message is individually acknowledged by *ACK message* of constant length El_{ACK}. The ACK messages are sent through *ACK virtual channels*.

For the MAC sublayer, the mean and variance of overhead delay (physical ring propagation delay between stations plus data holding time at each station) are Er_{Mi} and Vr_{Mi}, respectively. Data transmission rate of the ring is expressed as μ_M. We also assume high data transmission reliability.

The network is considered to be in the stationary state. We assume infinite capacity of queues.

3. MODEL DECOMPOSITION

The queueing network model is of open type. In general, the analysis of such model is complex. In the absence of exact analytical methods, we decompose the model into smaller interconnected submodels which represent behaviour of layers/sublayers. Each submodel is analyzed in isolation. We make approximation that messages arrive at each queue of submodel according to independent Poisson processes.

3.1. Modeling of a MAC sublayer

We model this sublayer as a single-server multiqueue system with nonexhaustive cyclic service fed by independent Poisson arrivals. The arrival rates λ_{Mi} are given by

$$\lambda_{Mi} = \sum_{c \in C_i} \lambda^{(c)} + \sum_{j=1}^{N} \sum_{c \in C_{ji}} \lambda^{(c)}, \qquad i = 1, \ldots, N \qquad (1)$$

The mean time Es_{Mi} spent by i-th station transmitting a message is given by

$$Es_{Mi} = \frac{1}{\mu_M \lambda_{Mi}} \left(\sum_{c \in C_i} \lambda^{(c)} El^{(c)} + \sum_{j=1}^{N} \sum_{c \in C_{ji}} \lambda^{(c)} El_{ACK} \right)$$

$$i = 1, \ldots, N \qquad (2)$$

and the second moment $Es_{Mi}^{(2)}$ of the transmission time is

$$Es_{Mi}^{(2)} = \frac{1}{\mu_M^2 \lambda_{Mi}} \left(\sum_{c \in C_i} \lambda^{(c)} El^{(c)(2)} + \sum_{j=1}^{N} \sum_{c \in C_{ji}} \lambda^{(c)} El_{ACK}^2 \right)$$

$$i = 1, \ldots, N \qquad (3)$$

The utilization of the ring ρ_{Mi} by station i is defined as

$$\rho_{Mi} = \lambda_{Mi} \, Es_{Mi}, \qquad i = 1, \ldots, N \qquad (4)$$

and utilization of the entire ring ρ_M is then

$$\rho_M = \sum_{i=1}^{N} \rho_{Mi} \qquad (5)$$

The following conditions are necessary and sufficient for the stability of MAC layer [4]:

$$\rho_M < 1 \quad \text{and} \quad \max[\lambda_{Mi}] \cdot ER_M < 1 - \rho_M \qquad (6)$$

where ER_M is defined as

$$ER_M = \sum_{i=1}^{N} Er_{Mi} \qquad (7)$$

For obtaining the analytical expressions for mean waiting times Ew_{Mi} at MAC queues we make use of results found by Boxma and Meister [5], i.e.

$$Ew_{Mi} = \frac{1 - \rho_M + \rho_{Mi}}{1 - \rho_M - \lambda_{Mi} ER_M} * \frac{1 - \rho_M}{(1 - \rho_M)\rho_M + \sum_{j=1}^{N} \rho_{Mj}^2}$$

$$* \left[\frac{\rho_M}{2(1 - \rho_M)} \sum_{j=1}^{N} \lambda_{Mj} Es_{Mj}^2 + \frac{\rho_M}{2\, ER_M} \sum_{j=1}^{N} Vr_{Mj} \right.$$

$$\left. + \frac{ER_M}{2(1 - \rho_M)} \sum_{j=1}^{N} \rho_{Mj}(1 + \rho_{Mj}) \right], \qquad i = 1, \ldots, N \qquad (8)$$

3.2. Modeling of a LLC sublayer

Under the assumptions made in section 2, mean arrival rates at ML_i and TL_i queues are both equal to λ_{Mi}. With an assumption of Poisson arrivals, each LLC queueing system is modelled as an M/G/1 system. First two moments of the service time for these systems are given so the mean waiting times can be computed using the well known Polloczek-Khintchine formula [6].

3.3. Modeling of a transport layer

Using the same arguments as for the LLC sublayer, we model ST_i and LT_i queueing systems as an M/G/1 systems with mean arrival rates equal to λ_{Mi}. We again apply Polloczek-Khintchine formula to calculate mean waiting times for these systems.

In order to determine mean waiting times at window queues, we use a modified algorithm proposed by Varghese *et al* [7]. Here, we present this algorithm in short for the sake of brevity.

Virtual data channel and corresponding virtual ACK channel form *virtual chain* which is modelled as a tandem of servers. A critical assumption of the approach is that of the exponential distribution of service times of all servers along the chain.

In order to eliminate the background traffic, the method of *adjusted rates* [8] can be applied. To do so, the service rate μ_i of the i-th server is adjusted by background traffic λ_i, i.e., a new service rate μ^* is given by

$$\mu_i^* = \mu_i - \lambda_i \qquad (9)$$

Each virtual chain is modelled as an $M/\mu(n)/1$ queueing system with Poisson arrivals and state dependent service rates, where n denotes number of messages inside the chain. The mean waiting time Ew_w at the the window queue is approximately the same as the waiting time in this queueing system. In order to calculate Ew_w, service rates $\mu(n)$ should be found.

If the number of servers in virtual chain is equal to K and if the number of messages n inside the chain does not reach window size R (i.e. $n \leq R$), the rate $\mu(n)$ can be obtained as the throughput of a closed tandem queueing network with R exponential servers and population of n messages. Mean-value analysis [9] can be used to compute the throughputs.

For $n > R$, service rates $\mu(n)$ can effectively be obtained by modeling the virtual chain as an $M/C_K/1$ system (the C_K denotes Coxian-K distribution) and next by rewriting the Coxian system as an equivalent birth-death system.

The mean waiting time Ew_w can easily be computed for an $M/\mu(n)/1$ system if arrival rate λ at the window queue and service rates $\mu(n)$, $n = 1,\ldots,\infty$ are given [6].

In the situation we have each virtual chain can be modelled as shown in Figure 3.

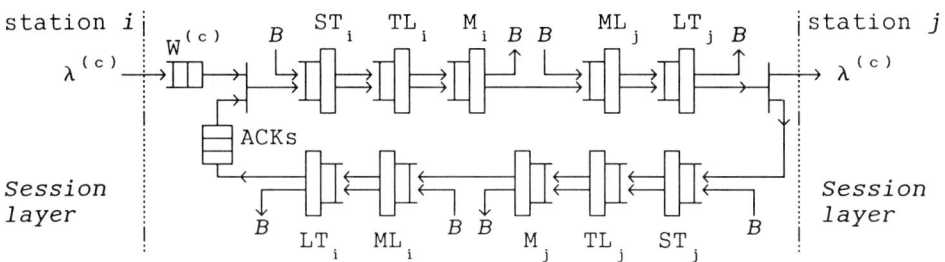

B - background traffic

Figure 3. Transport layer flow control submodel.

In this figure, M_i queuing systems represent the efect of MAC sublayer function. The mean service time of this server corresponds to mean message transmission time at station i. We model M_i system as an M/G/1 system with different classes of messages of different mean lengths. Mean waiting time Ew_{Mi} is given by (8). Apparent coefficient of variation of service time Cs_{Mi} may be found using the Polloczek-Khintchine formula with equations (1), (4) and (8), i.e. solving

$$Ew_{Mi} = \frac{\rho_{Mi}^2 (1 + Cs_{Mi}^2)}{2 \lambda_{Mi} (1 - \rho_{Mi})}, \qquad i = 1, \ldots, N \qquad (10)$$

for Cs_{Mi}.

In Figure 3, queueing systems may belong to several chains. The service time distributions of these systems may be nonexponential. In order to separate one chain from the others, we use more general method of eliminating background traffic than that proposed by Varghese *et al* (see above). This new approximate method is developed for M/G/1 system with different classes of customers and presented in Appendix for the sake of clearness. In case of exponential server and one class of customers this new method simplifies to the method of adjusted rates presented above.

After the seperation of chains, we obtain networks having only single-class queues with general service time distributions. In order to calculate window queueing delays, we model each virtual chain as an $M/\mu(n)/1$ queueing system. In order to calculate throughputs of the closed networks, we use Marie's device-complement procedure [10] instead of mean-value analysis procedure proposed by Varghese *et al*. Marie's method is one of the most accurate approximations for single-class queueing networks with nonexponential servers [11].

4. NUMERICAL RESULTS

In this section, we show the results of employing the method presented in section 3. We use the following parameter values for the model:

- $N = 3$.
- There are 4 virtual data channels, i.e., $C = \{1,2,3,4\}$, $C_{1,2}= \{1\}$, $C_{1,3}= \{2\}$, $C_{2,1}= \{3\}$, $C_{3,1}= \{4\}$.
- $El^{(c)} = 5000$ *bits*, $c \in C$; all data messages are exponentially distributed.
- $El_{ACK} = 48$ *bits*.
- $Es_{STi} = Es_{LTi} = 2$ *ms*, $Es_{MLi} = Es_{TLi} = 1$ *ms*; $i = 1,\ldots,N$. all protocol service times are expoentially distributed.
- $\mu_M = 1$ *Mbit/sek*.
- $Er_{Mi} = 3$ μs; $i = 1,\ldots,N$; overhead delays are constant.
- $R^{(c)} = R$; $c \in C$, i.e., window size is equal for all virtual data channels.

We consider two cases of asymmetric load of the network. In the first case, arrival rates are: $\lambda^{(1)} = \lambda^{(2)} = 10$ *msg/sec* and $\lambda^{(3)} = \lambda^{(4)} = 5$ *msg/sec*. In the second case, these rates are: $\lambda^{(1)} = \lambda^{(2)} = 20$ *msg/sec* and $\lambda^{(3)} = \lambda^{(4)} = 10$ *msg/sec*. The utilization of entire ring for these cases are $\rho_M = 0.15$ and $\rho_M = 0.3$, respectively.

Mean waiting times at queues of the network versus the window size *R* are presented in Figures 4 and 5. Simulation results for stations 2 and 3 are averaged over corresponding groups of queues. Regenerative method of simulation [12] with sequential stopping rules [13] were used in order to estimate the confidance intervals. The intervals are at the 90 % of confidance. Only sample means, denoted by dashing lines, are plotted in figures.

For small window sizes, Figures 4 and 5(a-b, f) demonstrate significant overestimation of analytical results. The bigger window size the better accuracy of the method. Overestimation of waiting times Ew_{Mi} may be explained as follows. The algorithm proposed by Boxma and Meister assumes that messages arrive at all queues according to independant Poisson processes. In case of small window sizes these processes are correlated strongly. For bigger window sizes, the blocking effect of flow control decreases, thus the correlation is smaller.

Figure 5(c-d) reveals that accuracy of analytical estimations of mean waiting times Ew_{LTi} and Ew_{MLi} are better for smaller window sizes. This is probably caused by cancellation of errors introduced both by flow control and blocking effect of a ring. For bigger window sizes, the latter becomes dominant and the results are much worse (underestimated).

Figure 5(e-f) shows that modified algorithm (denoted as I) for estimating mean waiting time at window queues performs better than original one proposed by Varghese *et al* (denoted as II).

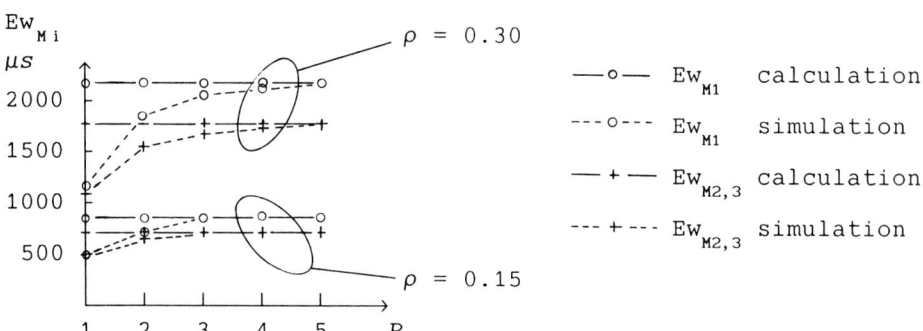

Figure 4. Mean waiting times for a MAC layer versus window size.

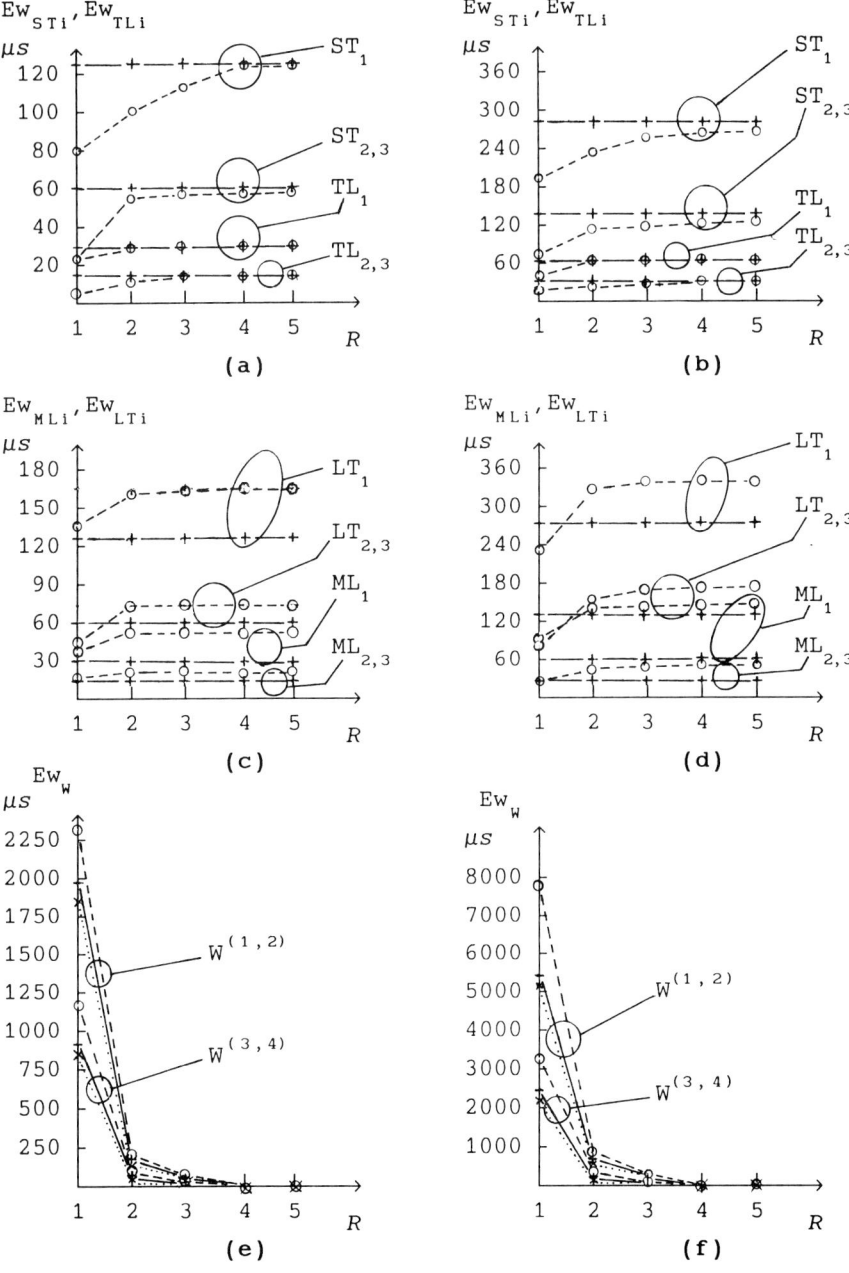

Figure 5. Mean waiting times for LLC and transport layers versus window size. (a,c,e) - $\rho_M = 0.15$. (b,d,f) - $\rho_M = 0.30$.
– – – Simulation. ——— Calculation (I). ·········· Calculation (II).

5. CONCLUSIONS

The results presented in previous section lead to the following conclusions:
- The accuracy of the decomposition method [2] used for the token ring network performance analysis is rather poor.
- The modifications proposed in this paper to the method of Varghese *et al.* improve the accuracy of calculating the mean waiting times at window queues.
- Futher studies still should be made to develop more accurate general approach to multi-layer token ring model.

6. REFERENCES

1. Standard IEEE/ANSI 802.5, Token Ring Access Method and Physical Layer Specifications (1985).
2. L.C. Mitchell, D.A. Lide, End-to-End Performance Modeling of Local Area Networks, IEEE Journal on Selected Areas in Communication, Vol. SAC-4, No. 6 (1986).
3. M. Murata, H. Takagi, Two Layer Modeling for Local Area Networks, IEEE Trans. on Comm., Vol.36, No.9 (1988).
4. P.J. Kuehn, Multi-queue systems with non-exhaustive cyclic service, Bell Syst. Tech. Journal, No. 58 (1979).
5. O. Boxma, B.M. Meister, Waiting-Time Approximations for Cyclic-Service Systems with Switchover Times, Performance Evaluation 7 (1987).
6. L. Kleinrock, Queueing systems, Vol. 1 : Theory, Wiley Interscience, 1975.
7. G. Varghese, W. Chou, A.A. Nilsson, Queueing Delays on Virtual Circuits Using a Sliding Window Flow Control Scheme, ACM Performance Evaluation Review (1983).
8. M. Reiser, Performance Evaluation of Data Communication Systems, Proceedings of the IEEE, Vol. 70, No. 2 (1982).
9. M. Reiser, A Queueing Network Analysis of Computer Communication Networks with Window Flow Control, IEEE Transactions on Communications, Vol. COM-27, No. 8 (1979).
10. R. Marie, An Approximate Analytical Method for General Queueing Networks, IEEE Transaction on Software Engineering, No. 5 (1979).
11. A.B. Bondi, W. Whitt, The Influence of Service-Time Variability in Closed Network of Queues, Performance Evaluation Review, No. 6 (1986).
12. S. Lavenberg, D. Slutz, Introduction to Regenerative Simulation, IBM J. Res. Develop., No. 19 (1975).
13. S. Lavenberg, C. Sauer, Sequential Stopping Rules for the Regenerative Method of Simulation, IBM J. Res. Develop., Nov. 1977.
14. P. Kuehn, Approximate Analysis of General Queueing Network by Decomposition, IEEE Transactions on Communications, Vol. COM-27, No. 1 (1979).

APPENDIX - LOAD CONCEALMENT METHOD FOR M/G/1 SYSTEMS WITH DIFFERENT CLASSES OF CUSTOMERS

We consider queueing system, given in Figure A1(a), of type M/G/1 with L classes of customers.

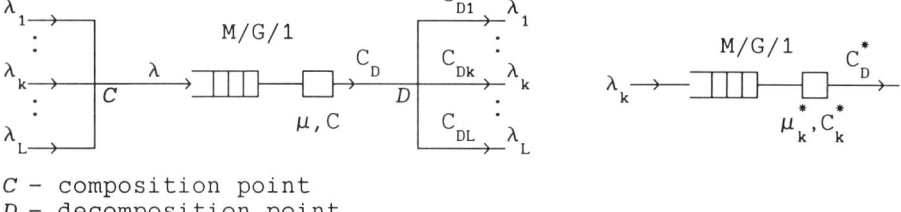

C - composition point
D - decomposition point

(a)   (b)

Figure A1. (a) The original M/G/1 system with different classes of customers. (b) Equivalent system with one class of customers.

The customers arrive according with independent Poisson processes with mean flow rates λ_k. Therefore, the arrival rate at the server is

$$\lambda = \sum_{k=1}^{L} \lambda_k \qquad (A1)$$

Each class is characterized by its mean service rate μ_k. The service time coefficient of variation C is the same for all classes. The mean service time $1/\mu$ can be calculated from the following equation

$$\frac{1}{\mu} = \frac{\sum_{k=1}^{L} \frac{\lambda_k}{\mu_k}}{\lambda} \qquad (A2)$$

Let us consider customers of the k-th class. We are interested in the parameters μ_k^* and C_k^* of a substitute equivalent one-class server. These parameters should be so calculated as to take into account a background traffic (see Figure A1(b)).

The squared coefficient of variation C_D^2 of the interdeparture time for an M/G/1 system is [14]

$$C_D^2 = 1 + \left(\frac{\lambda}{\mu}\right)^2 (C^2 - 1) \tag{A5}$$

At the decomposition point D, the k-th stream has squared coefficient of variation C_{Dk}^2 given by [14]

$$C_{Dk}^2 = \frac{\lambda_k}{\lambda} C_D^2 + 1 - \frac{\lambda_k}{\lambda}, \qquad k = 1, \ldots, L \tag{A6}$$

Substituting (A5) into (A6) we obtain

$$C_{Dk}^2 = \frac{\lambda_k \lambda}{\mu^2} (C^2 - 1) + 1, \qquad k = 1, \ldots, L \tag{A7}$$

For the substitute server, coefficient of variation C_D^* of the interdeparture time should be equal to C_{Dk}. If we take this into account and, additionally, assume that the mean time spent by a customer in substitute system (i.e. waiting time plus service time) should be the same as in original system, we obtain the following parameters of the substitute server

$$C_k^{*2} = \frac{a^2 \lambda}{\lambda_k} (C^2 - 1) + 1, \qquad k = 1, \ldots, L \tag{A8}$$

$$\mu_k^* = \mu\, a, \qquad k = 1, \ldots, L \tag{A9}$$

where

$$a = \frac{2\lambda \mu_k (\mu - \lambda) + 2 \mu_k \lambda (\mu - \lambda) + \lambda_k \lambda \mu_k (1 + C^2)}{2 \mu^2 (\mu - \lambda) + \lambda \mu \mu_k (1 + C^2) - \lambda \mu_k (\mu - \lambda)(C^2 - 1)} \tag{A10}$$

For $C^2 < 1$, right side of (A10) can be negative. In such case, we set C_k to zero.

If the original server is exponential with one class of customers, equations (A8) and (A9) reduce to

$$C_k^{*2} = 1 \tag{A11}$$

$$\mu_k^* = \mu - (\lambda - \lambda_k) \tag{A12}$$

(compare with equation (9), section 3.3.).

Management and Planning of ISDN and Circuit Switched Networks

Telecommunication Services for Developing Economies
J. Filipiak (Editor)
© 1991 Elsevier Science Publishers B.V. All rights reserved

Routing Optimization and Dimensioning of Networks with Revenues: Numerical Results

A. Girard[*]   B. Liau[†]   N. Boumzebra[‡]

Abstract

The mathematical model for routing optimization and dimensioning of telephone networks operating with revenue maximization and alternate routing has been described in [3]. From this description a numerical solution procedure was proposed that is in many ways similar to the classical ECCS dimensioning method commonly used for hierarchical networks and to the flow deviation technique used for routing optimization in packet networks. We describe here the first numerical results obtained from various numerical implementations of these algorithms.

We show that the routing optimization algorithm is stable but that the classical flow deviation technique is too slow and we propose a heuristic that does almost as well. We also show that the dimensioning stage, although quite simple in theory, turns out to be more difficult than expected and is very sensitive to the particular order chosen for the operations. Finally, we present some preliminary comparisons of small networks dimensioned under variations of the basic routing methods and with the residual capacity adaptive routing.

The network model examined here is an extension of Kelly's [9] revenue maximization. The basic assumption is that a call generates a given revenue that depends only on the particular path on which the call is connected. The problem considered there was how to split each traffic stream onto a set of available paths in order to maximize the total revenue generated by the network.

First we briefly review the two routing techniques considered here. Then we recall the mathematical model used for routing optimization, describe two problems related to its numerical solution and present computational results related to the behavior of the algorithms as well as to the merit of the routing methods themselves.

We then present a similar discussion for the combined routing optimization and dimensioning problems for the two routing methods.

1 Routing Techniques

Kelly's model dealt with the case where a call could be offered to a single, multi-hop path and would be lost if the path was not available. We have extended Kelly's model in [4] in three ways: 1) The routing is more general since it allows alternate routing, but only two-hop paths are now permitted, 2) Dimensioning was explicitly included into the model and a unified

---

[*]INRS-Télécommunications, 3 Place du Commerce, Verdun Québec, Canada H3E 1H6
[†]Centre National d'Études des Télécommunications, 38–40 rue du Général Leclerc, 92131 Issy-les-Moulineaux, France
[‡]INRS-Télécommunications, 3 Place du Commerce, Verdun Québec, Canada H3E 1H6

description was given of the joint routing optimization and dimensioning and 3) a new method was proposed for the resolution of the Erlang fixed point within the optimization model itself, giving a new interpretation of the induced costs defined in [9]. We now describe explicitly the routings considered here.

1.1 Alternate Routing and Load Sharing

The routing used in the networks considered here is a mixture of load sharing and alternate routing. In pure load sharing — also called here *basic load sharing* — calls that belong to a particular origin-destination pair (o, d) are allowed to use one of a given set $\{\mathcal{P}^{o,d}\}$ of $k^{o,d}$ paths each with fixed probability $\alpha_j^{o,d}$, here called the *load sharing coefficients*. When a call arrives, a particular path is selected randomly according to the $\alpha_j^{o,d}$s and the call is offered to this path; if the path is busy at this time — that is, if all the circuits on one or more of the trunk groups that compose this path are busy — then the call is lost.

Alternate routing, on the other hand, operates with an *ordered* set of paths. An arriving call is offered to each path in the prescribed order until a free path is found; at this point, the call is connected.

The routing used here is a mixture of load sharing and alternate routing: a call that arrives to the network is systematically offered to the direct link (o, d). If the link is busy, then the call is offered to one of the available alternate routes with a fixed probability represented by the load sharing coefficients. We assume throughout that the alternate paths contain only two links and that all potential alternate paths are present. These assumptions are not strong ones since two-link routing, being more robust than multilink routing in the presence of traffic surges, is the norm in nonhierarchical networks. The assumption of a fully connected graph and the assumption that the call should always be offered to the direct link first is somewhat stronger. In practice, it is not difficult to modify the theory to account for a different first choice on a preferred two-link path. We have made this assumption only for the sake of simplicity and in does not affect the validity of the theoretical results and, we think, of the numerical results that will be described later.

1.2 Residual Capacity Adaptive Routing

Load sharing with alternate routing is also a good model for the residual capacity adaptive routing technique, as implemented in the Dynamic Control Routing method [1]. The calculation of the traffic patterns produced by this method has been described in [11] and turns out to be quite complex. A simpler algorithm has been proposed in [5] based on the realization that if the system were viewed in the long term, each traffic stream would seem to be separated with some proportion $\alpha_j^{o,d}$ which should be roughly proportional to the expected residual capacity of the paths. It was found in [5] that the accuracy of this method, although not as good as that of [11], was adequate for network dimensioning purposes while being at the same time easier to evaluate.

2 Routing Optimization

Routing optimization applies only to the revenue maximization routing since for adaptive routing, the load sharing coefficients are prescribed by the mode of operation of the routing method. This optimization problem is based on the transformation of the original routing problem into a larger model in which the link blocking probabilities are added as free variables and the Erlang fixed point equations as constraints. It is stated as

$$\min_{\alpha,B} - \sum_{i,j} \sum_k w_k^{i,j} \overline{A}_k^{i,j} \qquad (1)$$

$$\sum_{k \geq 1} \alpha_k^{i,j} = 1, \quad (v^{i,j}) \qquad (2)$$

$$\alpha_k^{i,j} \geq 0, \quad (u_k^{i,j}) \qquad (3)$$

$$E(a_s, N_s) = B_s, \quad (y_s). \qquad (4)$$

where we have defined

w_k^{imj} revenue generated per call carried on the path from i to j through k. If $k = 0$, this is the direct link.

$\overline{A}_k^{i,j}$ the amount of traffic carried on path (i,k,j).

a_s the total traffic offered to link k.

B_s the blocking probability on link s.

N_s the number of circuits on link s.

$E(a, N)$ the Erlang-B function.

u, v, y the Kuhn-Tucker multipliers of the constraints.

$\eta_s \quad E(a_s, N_s - 1) - E(a_s, N_s)$

The solution algorithm is based on the solution of the Kuhn-Tucker equations. The first set of equations is obtained by taking the gradient of the Lagrange function in the direction of α. We get

$$v^{i,j} = A^{i,j} B_{i,j} Q_k^{i,j} \left[w_k^{i,j} - (y_{i,k}\eta_{i,k} + y_{k,j}\eta_{k,j}) \right]. \qquad (5)$$

Note that the left-hand term is independent of the path index k. This means that the right-hand side must be *equal* for all overflow paths that carry some (i,j) traffic. Since the term $A^{i,j} B_{i,j}$ is identical for all these paths, we must conclude that $Q_k^{i,j} \left[w_k^{i,j} - (y_{i,k}\eta_{i,k} + y_{k,j}\eta_{k,j}) \right]$ is the same for all these overflow paths. Since we know that $w_k^{i,j}$ is the revenue generated by a call carried on path (i,k,j), we can naturally interpret the term $[y_{i,k}\eta_{i,k} + y_{k,j}\eta_{k,j}]$ as the loss of revenue produced on the rest of the network by this call and the difference in the two as the net revenue per call on the path. Hence we can state that *the optimal routing is the one that equalizes the weighted net revenue on all alternate paths for each origin-destination pair.*

The second equation is obtained by taking the gradient of the Lagrange function in the direction of B. We get

$$y_{l,n} = A^{l,n} \left[w_0^{l,n} + \sum_k \alpha_k^{l,n} Q_k^{l,n} \left(y_{l,k}\eta_{l,k} + y_{k,n}\eta_{k,n} - w_k^{l,n} \right) \right]$$
$$+ \sum_j a_{l,n}^{l,j} \left(w_n^{l,j} - y_{n,j}\eta_{n,j} \right) + \sum_i a_{l,n}^{i,n} \left(w_l^{i,n} - y_{i,l}\eta_{i,l} \right) \qquad (6)$$

We have introduced a new quantity $a_{l,n}^{i,j}$ which is the traffic offered to trunk group (l, n) that belongs to flow (i, j). In our case, we have

$$a_{l,n}^{l,j} = A^{l,j} B_{l,j} \alpha_n^{l,j} (1 - B_{n,j}) \qquad (7)$$

$$a_{l,n}^{i,n} = A^{i,n} B_{i,n} \alpha_l^{i,n} (1 - B_{i,l}). \qquad (8)$$

We see that the optimal routing is obtained by equalizing the net revenue on all paths and the optimal multipliers by solving a linear system of equations.

2.1 Computation Order

Although Eqs. (5-6) must in principle be solved simultaneously, an obvious heuristic method would be to solve them alternatively within an iterative procedure. Also, the feasibility conditions of the Erlang fixed point must also be solved at the final solution. We can imagine two different ways to solve Eqs. (5-6).

Algo 1 Solve Eq. (5) to optimality for a fixed value of **y**. After optimality has been reached, solve Eq. (6) to obtain a new value of **y**. Iterate until convergence

Algo 2 The procedure for solving Eq. (5) is iterative. Hence we might argue that it is not really very useful to solve Eq. (5) to optimality since the value of **y** will be recomputed any way. We could then imagine an algorithm where only a few iterations of the solution algorithm for Eq. (5) are performed before a new **y** is recomputed.

Which of these two methods is the more accurate and the faster can only be examined through numerical calculation which will be discussed later.

2.2 Flow Deviation

The calculation of the optimal load sharing coefficients is to equalize the marginal net revenue on all paths that carry some flow. This is very similar to the optimal routing problem found in vehicle routing and in packet networks. A well-known method for these problems is the Frank-Wolfe algorithm adapted to the network model under the name "flow deviation" [2]. Unfortunately, the asymptotic convergence of the Frank-Wolfe method is quite poor and we must examine the advantages of using a faster heuristic in place of the classical Frank-Wolfe algorithm.

In the standard flow deviation method, a direction of descent **d** in the space of the multicommodity flow $\mathbf{f}^k$ is computed by finding the shortest path for each commodity k. In the present case, the link length is the net revenue for the current **y**. A step of size δ is then taken in this direction solving $\min g(\mathbf{f} + \delta \mathbf{d})$. In this method, the *same* step is taken for all commodities k, a fact that contributes to the poor convergence of the method.

A plausible heuristic is to take a different step for each commodity. In the particular technique we have investigated here, the shortest and longest paths are evaluated for each commodity. Then, flow is moved from the longest to the shortest path differently for each commodity in *fixed* step sizes; in other words, we do not try to reach an exact minimum in the selected direction but think that an approximate position should be sufficient.

2.3 Algorithm Performance

We now discuss the performance of the routing optimization algorithm with regard to the two options indicated above: the resolution of the flow deviation subproblem and the multiplier update policy.

Network	Exact		Algo 1		Algo 2	
	Obj.	Cpu Secs.	Obj.	Cpu Secs.	Obj.	Cpu Secs.
4-Node	601	17	601	1	601	1
6-Node	51558	311	51558	24	515549	6
9-Node	133247	4996	133247	106	133232	44
13-Node		*	347959	302	347845	112

Table 1: Comparison of Load Sharing Optimization: 13-Node Network, Value of Objective Function. * Did not run to completion due to excessive Cpu time.

2.3.1 Flow Deviation Computation

The convergence rate of the flow deviation method can be quite slow depending on the position of the optimal solution in the feasible set. If the point is interior then the convergence is linear with a relatively poor rate. If, however, the solution lies at an extreme point of the domain, then the flow deviation will converge in a few iterations. This is because the method searches for a minimum on the segment that links the current solution to an extreme point of the domain. For this reason, the convergence rate of the method is not uniform across all problems and we cannot give a guaranteed performance. We have tried both cases and found that when the solution is at an interior point hundreds of iterations are required to achieve a minimum. For this reason we recommend using the heuristic.

2.3.2 Multiplier Update Policy

The effect of the multiplier update policy is summarized in Table 1. We can see that updating the multipliers at each iteration is more accurate than doing the update only after convergence but is much more time consuming. Hence we recommend updating the multipliers only after the flow deviation step has converged.

Note, however, that this recommendation depends on the relative speed of the two steps. We think that the heuristic flow deviation is quite fast but we have used a straightforward Gaussian elimination technique for solving the linear system. It turns out, however, that this system is quite sparse. Thus we can expect that using a solution algorithm specifically designed for sparse system might change the relative times of the two steps and that in this case our conclusion might be different. This is currently being investigated.

Consider finally the results shown on Figs. 1 and 2. We have plotted on the first one the value of the objective function as a function of the number of iteration and on the second the difference between two consecutive values also as a function of the number of iterations. Note that there are some large steps in the second graph corresponding to abrupt changes in the value of the objective from one iteration to the next. This happens when the multipliers are recalculated each time the flow deviation algorithm has converged. The important point is that the objective function changes very little at these points. This is an indication that the optimal objective is not highly sensitive to the multiplier value and that a large-scale heuristic method could probably get some near-optimal results with perhaps one or two iterations of the multipliers. As we will see later, this has some implication on the adaptive implementation of the method.

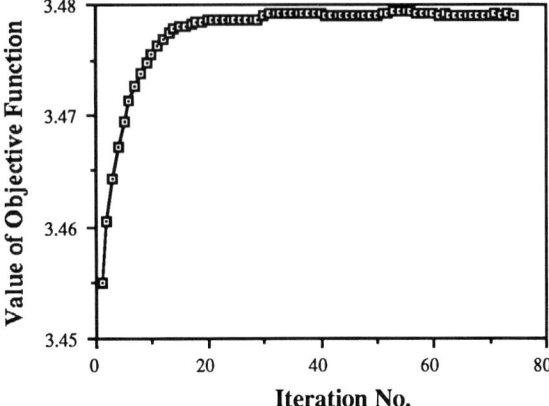

Figure 1: Objective Function vs No. of Iterations: Algorithm 2.

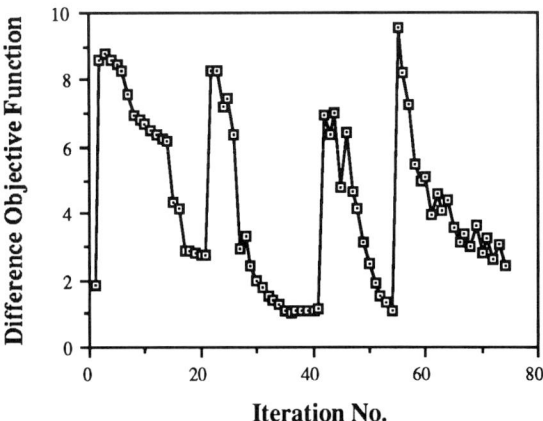

Figure 2: Difference in Objective Function vs No. of Iterations: Algorithm 2.

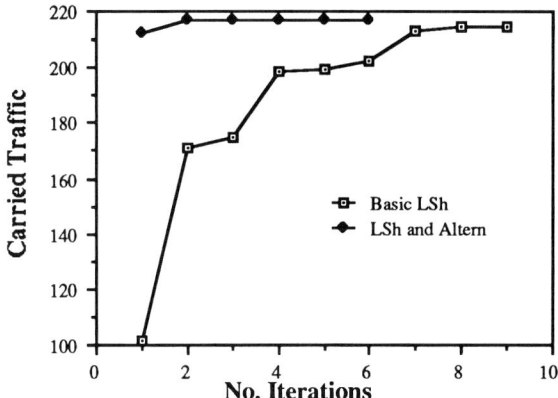

Figure 3: Value of Alternate Routing with Load Sharing: Toronto Network

2.4 Comparison of Routing Methods

We now present some typical results concerning the convergence of these algorithms and at the same time some comparison indicating the value of alternate routing with revenue optimization. First we show on Fig. 3 the total carried traffic is plotted as a function of the number of iterations of the routing optimization algorithm. The network is the subnetwork of the Toronto network that was used in the field trial of the residual capacity adaptive routing method [1]. The figure shows quite clearly that the convergence rate is quite good.

The same type of result is presented in Fig. 4 where the comparison is made on a much larger 272-group network — a part of the Marseille metropolitan network — that was originally dimensioned for a hierarchical routing with a multihour engineering method. The three routing methods shown on the graph are

Basic: Basic load sharing on all two-hop paths and the direct high usage group. The initial routing is to route all calls on the direct high-usage group.

Paths Hier: Basic load sharing on all paths permitted by the hierarchical routing rule: direct, two two-hop paths and one three-hop path through the two tandem nodes. The starting solution is the same as in the previous case.

Alternate: Load sharing on the overflow from the direct high-usage group. Only two-hop paths are considered here. The initial solution is to route all calls equally on all possible alternate paths.

Here again we see a good convergence of the routing optimization algorithm. More interestingly, however, is the fact that even the simple load sharing method gives a significant improvement in carried traffic over the hierarchical routing.

They are typical of what was obtained in many more computational experiments and should give the reader a feeling for the computational behavior of these algorithms. Needless to say, our conclusions are supported by more than what is shown here due to the space limitation.

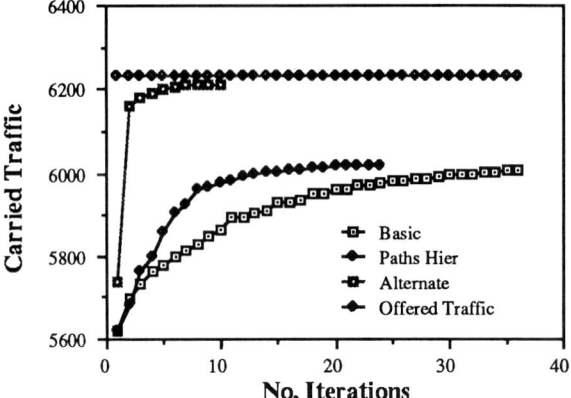

Figure 4: Value of Alternate Routing with Load Sharing: Marseille Network

	Offered	Res. Cap	Load Sh.	Load Sh. + Altern
Toronto	296	211.18	216.79	216.99
Marseille	6237	6032.7	6018.3	6213.3

Table 2: Performance Comparison: Adaptive vs Load Sharing

2.5 Comparison of Adaptive Routing and Load Sharing

We can also use these algorithms to compare the performance of the residual capacity adaptive routing and the load sharing routing methods. We have done this for the Toronto and the Marseille networks (both with one-way trunk groups) and the results are presented in Table 2. Note that the Toronto network has been dimensioned for adaptive routing while the Marseille network is dimensioned for hierarchical routing. In the first case, the network is designed for nonhierarchical routing which explains why the different routings have roughly the same performance. This is not the case for the Marseille network where the routing has to adapt to the configuration of the network. Although the load sharing routing seems to be outperforming the adaptive routing, we should not forget that the load sharing is not adaptive. It outperforms the adaptive method because it has specifically been reoptimized for the current network and traffic patterns. In a real situation, this would not be possible unless an adaptive version of the method is implemented, a subject that will be briefly discussed later on. Nevertheless, this superior behavior of the revenue-based method is worth some further investigation.

3 Dimensioning

We have investigated the dimensioning problem both for the load sharing method and for the residual capacity adaptive routing. In the first case, we have selected a dimensioning model where the revenues generated by the network are balanced with the cost of dimensioning. The model is stated as

$$\min_{\alpha, B, N} \sum_s C_s(N_s) - \sum_{i,j,k} \overline{A}_k^{i,j} w_k^{i,j} \qquad (9)$$

$$\sum_k \alpha_k^{i,j} = 1, \quad (v^{i,j}) \tag{10}$$

$$\alpha_k^{i,j} \geq 0, \quad (u_k^{i,j}) \tag{11}$$

$$N_{i,j} \geq 0, \quad (z_{i,j}) \tag{12}$$

$$E(a_s, N_s) = B_s, \quad (y_s). \tag{13}$$

$$\tag{14}$$

where we define the additional quantities

$C_s(N_s)$ the cost of installing N_s circuits on link s

$L^{i,j}$ the probability that a call originating at i for destination j cannot be connected. This is the end-to-end loss probability.

This is not the only cost model possible and other variations would still be amenable to the same solution algorithm. Note in particular that we do not have explicit grade of service constraints. The solution is governed purely by economic factors, mostly the ratio of the dimensioning costs to the revenues generated by the network. The Kuhn-Tucker equations for the dimensioning problem are Eqs. (5–6) and an additional equation for the trunk group dimensioning

$$C'_s = z_s - y_s \frac{\partial E}{\partial N_s}. \tag{15}$$

In the case of the adaptive routing, there is no routing optimization phase and the objective is only the dimensioning cost. We give here the multihour version of the problem where the dimensioning is done taking into account a number of T traffic matrices occurring at different times of day. We have

$$\min_{\alpha, B, N} \sum_s C_s(N_s) \tag{16}$$

$$L^{i,j}(t) \leq \overline{L}^{i,j}(t), \quad (x^{i,j}(t)) \tag{17}$$

$$\sum_k \alpha_k^{i,j}(t) = 1, \quad (v^{i,j}(t)) \tag{18}$$

$$\alpha_k^{i,j}(t) \geq 0, \quad (u_k^{i,j}(t)) \tag{19}$$

$$N_{i,j} \geq 0, \quad (z_{i,j}) \tag{20}$$

$$E(a_s(t), N_s) = B_s(t), \quad (y_s). \tag{21}$$

$$\tag{22}$$

Here we have added grade of service constraints in the form of end-to-end loss probabilities for each time period. This complicates somewhat the solution method because in this case we do not have an explicit set of equations governing the x multipliers for these constraints. They have to be calculated separately by some other method. In our case, we have simply dualized these constraints and included them in the objective function. We have then used an outer optimization loop of the subgradient type [8] to obtain an approximate solution for these multipliers.

The routing optimization equations separate by period and are of the same form as Eqs. (5–6). The dimensioning equations for group s takes the form

$$C'_s = \sum_{t=1}^T y_s(t) \frac{\partial E(a_s(t), N_s)}{\partial N_s} \tag{23}$$

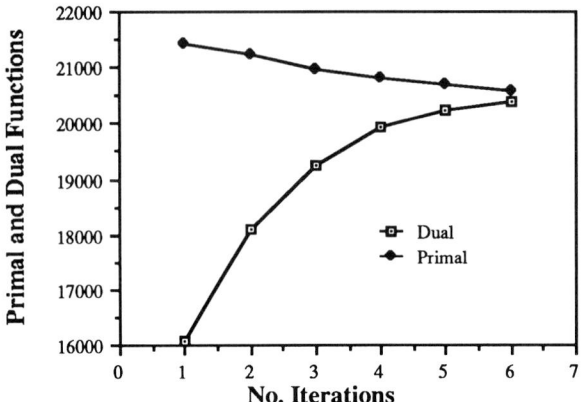

Figure 5: Dimensioning of the Marseille Network

It should be noted that these reduce to the correct single-hour dimensioning equations when $T = 1$.

3.1 Computation Order

The question of the order of calculation is compounded by the presence of the dimensioning equations which have to be solved simultaneously with the other Kuhn-Tucker equations. The same remark holds concerning their position in the overall iterative procedure.

In the case of the revenue maximization method, however, we have found that the particular choice of a position for the dimensioning step within the loop is crucial for the convergence of the method. The case where the value of y is computed along with the routing variables is more stable than other permutations of the algorithm steps.

3.2 Algorithm Performance

We have found that the subgradient method has good convergence provided we do not try to obtain a very high accuracy. The value of the dual function and of a primal feasible function are shown as a function of the iteration number on Fig. 5 for the single-hour dimensioning of the Marseille network. We see that the duality gap at the end is quite small and this behavior is typical of what we have found in other cases.

3.3 Value of Multihour Dimensioning

We have also examined the effect of using a multihour dimensioning method. We have chosen a subnetwork of the Marseille network for which we had two traffic matrices corresponding to a daytime and nighttime load. We have dimensioned the network with the adaptive routing method with one-way and two-way trunk groups. The results are presented in Table 3. The single-hour solutions are computed for the corresponding traffic demand and are not necessarily feasible for the other period. The multihour solution, on the other hand, is feasible for *both* periods simultaneously. We see that we can achieve a multihour feasible solution for a small increase in cost of approximately 5%.

Network			Dual	Primal	CPu (secs)
S	D	O	20183	20536	229
S	D	T	18682	18986	131
S	N	O	18337	18569	285
S	N	T	16877	17124	166
M		O	20945	21233	569
M		T	19166	19222	289

Table 3: Marseille Network Dimensioning S: Single-Hour Dimensioning algorithm M: Multi-hour Algorithm D: Daytime traffic matrix N: Nighttime traffic matrix O: One-way trunk groups T: Two-way trunk groups

An interesting question is whether the multihour method produces networks that are cheaper than the networks produced by other heuristic multihour dimensioning methods. We know that for hierarchical networks there is a definite advantage — between 10 to 15% in cost — in using a multihour method over a heuristic [7]. Because adaptive routing is more flexible than the standard hierarchical routing we may expect that this advantage is going to be smaller. In other words, the routing may make up for the demand variation even though the network has not been dimensioned exactly for the multiple demand set. The results presented in [6] indicated that in this particular case there is an advantage of about 5% in cost in using the multihour version of the algorithm instead of a heuristic based on a single demand matrix. It would be an interesting question to check whether this is also the case for the Marseille network used here. Work is currently in progress to answer this question.

4 Conclusion

The conclusions of this work can be summarized as follows. The two-step iterative procedure for optimizing routing via revenue maximization is a viable method and has good convergence properties. The tests that were performed indicate that a carefully optimized program — which our programs are certainly not at this time — could conceivably solve problems with more that 1000 trunk groups. The corresponding dimensioning problem is somewhat more difficult to compute although the computation time is not significantly larger; what seems difficult with the dimensioning problem is to obtain a stable algorithm, something that we have been able to achieve. An interesting question is how these networks compare with routing methods based on markov models such as the ones described in [10].

Further work is continuing to verify the use of these techniques in an adaptive environment where the values for the routing parameters are not computed from traffic forecast but are updated from measurements periodically made in the network. The results presented here are encouraging in the sense that the evaluation of the multipliers, which is the more time-consuming part of the algorithm, does not seem to be critical to the value of the final solution. Hence we could envisage a two-stage adaptive method where the routing variables are updated more frequently based on fast traffic measurements while the multiplier update would be made much more infrequently to take into account slow variations in the traffic patterns — typically the ones that are related to multihour dimensioning methods or to infrequent events such as failures. These will be the object of a future communication.

References

[1] W.H. Cameron, P. Galloy, and W.J. Graham. Report on the Toronto advanced routing concept trial. In *1st Network Planning Symposium*, pages 228–236, 1980.

[2] L. Fratta, M. Gerla, and L. Kleinrock. The flow deviation method: An approach to store-and-forward communication network design. *Networks*, 3:97–133, 1973.

[3] A. Girard. Optimisation des réseaux à acheminement à revenu maximum. Technical Report 89–24, INRS-Télécommunications, 3 Place du Commerce Verdun Qué. Canada H3E 1H6, 1989.

[4] A. Girard. *Routing and Dimensioning in Circuit-switched Networks*. Addison-Wesley, 1990.

[5] A. Girard and M.A. Bell. Blocking evaluation for networks with residual capacity adaptive routing. *IEEE Transactions on Communications*, COM–37:1372–1380, 1989.

[6] A. Girard, B.Liau, and N.Melki. Dimensioning telephone networks by revenue maximization. In *Proc. Sem. on Design and Control of a Worldwide Intelligent Network*, June 1990. CNET–PAA/ATR, 38-40 rue du Général Leclerc, Issy-les-Moulineaux, 92131 France.

[7] A. Girard, P.D. Lansard, and B. Liau. Generalization of the ECCS method to the multihour case. *IEEE Transactions on Communications*, To appear 1991.

[8] M. Held, P. Wolfe, and H.P. Crowder. Validation of subgradient algorithm. *Mathematical Programming*, 6:62–88, 1974.

[9] F.P. Kelly. Routing in circuit-switched networks: Optimization, shadow prices and decentralization. *Adv. Appl. Prob.*, 20:112–144, 1988.

[10] T.J. Ott and K.R. Krishnan. State dependent routing of telephone traffic and the use of separable routing schemes. In *11th International Teletraffic Congress*, pages 5.1A(5.1–5.6), 1985.

[11] J. Régnier, P. Blondeau, and W.H. Cameron. Grade of service of a dynamic call-routing system. In *10th International Teletraffic Congress*, 1983.

TRAFFIC MODELING IN NETWORKS WITH INCOMPLETE DATA

James A. Schmitt

AT&T Bell Laboratories, 3B-508, Crawfords Corner Road, Holmdel, New Jersey 07733, U.S.A.

Abstract
This paper describes an algorithm that infers a set of direct traffic demands between switches that can be routed on a given alternate routing network with specified trunk group sizes. The theoretical basis of the algorithm and the bounds on the nonunique solutions (traffic demands) are developed.

1. INTRODUCTION

The determination of the trunk group sizes (number of circuits) from the loads (forecasted traffic demand) on a given network is a standard activity for many telecommunication engineers[1] [2]. The solution of this trunk engineering problem is a set of trunk group sizes that allow the specified loads to be carried over the given network in a fashion consistent with a grade of service and routing rules. The trunk engineering problem is typically solved in network planning where the loads for each planning period are part of the input. In each planning period, the architecture of the network can change, switches can be deleted, relocated or installed, and the alternate routing patterns can be rearranged to improve network efficiency and reliability. The result is a network that differs from the network in the base year. In particular, the size of the trunk groups in the new network can differ substantially from those in the original network.

In some telecommunication networks, the number of circuits between pairs of switches in the base network is known but the first routed trunk group loads between them are not. The first route trunk group load (all the traffic streams that use the trunk group in their first choice route) represents a critical input for various network planning functions, as we have seen. This lack of first route trunk group loads occurs in some domestic networks because the process to generate first route trunk group loads from traffic data can be both costly and time consuming, and in international networks because two administrations negotiate the size of the international trunk groups between them. For example, in developing countries where telecommunication planning budgets are small, the cost of determining direct demand profiles can be avoided by simply not generating them. In this case, the administration adds circuits as needed to maintain a grade-of-service rather than according to a traffic demand forecast. Using the algorithm described in this paper, the administration can now infer the first route trunk group loads that can be supported on a given network and use them as the base loads in planning. A similar situation can exist in international networks but for entirely different reasons. Unlike in a domestic network where one telecommunication company controls the entire trunk group, two administrations share the control of international trunk groups between the administrations' gateways. The number of circuits in these trunk groups are negotiated by the administrations for each year in the multi-year planning horizon. The forecasts by each administration can differ substantially from each other and from the final negotiated size. Consequently, the first route trunk group loads between the gateways that can be supported on the decided trunk groups is unknown.

Thus, occasionally the inverse of the trunk engineering problem needs to be solved, that is, to determine the first offered trunk group loads between switch pairs given the set of trunk group sizes for a given network architecture, engineering criteria, and routing rules. The load values are inferred from the trunk group sizes. In this paper, we address this problem, and we call it the *inverse trunk sizing* problem. Once this is solved, network planning can begin[1], and the questions on the network efficiency and robustness can be addressed.

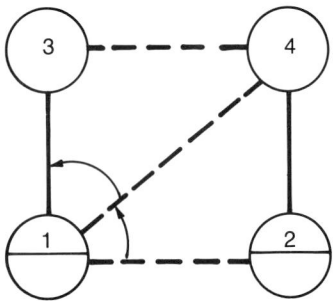

FIGURE 1   ROUTING OF OVERFLOWS IN A SECTION OF A HIERARCHICAL NETWORK

Before we discuss the inverse trunk sizing problem, certain terminology[3] used in hierarchical networks should be explained. Consider a section of a hierarchical network in Figure 1: switch 1 homes on switch 3 forming a part of one hierarchical chain and switch 2 homes on switch 4 forming a part of another hierarchical chain. Switches homing on another switch are connected by a *final* trunk group (solid lines). Final trunk groups are the final routes traffic can take. Finals also interconnect the highest switches in the different chains. Finals are engineered at a given grade of service or blocking, typically 1%. In a network with alternate routing, the switch first offers the call to a high usage (HU) group, if it exists. The HU group is a direct route between the switches and are denoted by dashed lines. If all the trunks in the HU trunk group are busy, the call is routed through another switch using a final trunk group. For example, if a call between switches 1 and 2 is blocked on the direct path (HU 1-2), then it is routed on the alternate path consisting of HU 1-4 and final 2-4. The HUs are sized or engineered based on economics using

ECCS engineering[2]. The grade of service or blocking on a HU group varies with the economics of the route: cheaper HU groups have lower blocking than expensive ones.

The terminology for the traffic demand is also important. The first route trunk group loads, denoted by L, are all the traffic streams that use the trunk group in their first choice route. For example, the first route trunk group loads between switches 1-2 and 1-3 are only the point-to-point demands between switches 1-2 and 1-3, respectively. However, the first route trunk group load between switches 3 and 4 is the sum of the point-to-point demand between switches 3 and 4 plus the point-to-point demand between switches 2 and 3 (because no HU exists between 2 and 3). This is also true on link 2-4 because the point-to-point demand between switches 2 and 3 is *first* routed on links 2-4 and 3-4. First route trunk group traffic is random. The actual demand on a trunk group (first offered trunk group load plus any overflow load[3] from another trunk group) is called the offered load, which is denoted by a. At places, the values of L and a on a group are identical, for example, on group 1-2 because no overflow is routed to this group. At other places, the values of L and a for a group are different, for example, on groups 1-4 and 1-3 because they receive overflow from links 1-2 and 1-4, respectively.

The solution of the inverse trunk sizing problem is defined as a set of first route trunk group loads such that when trunk groups are sized according to the hierarchical engineering rules, their sizes match the prescribed set of trunk group values. We restrict our attention to hierarchical networks because most of the international networks and typically domestic networks without extensive traffic data are of this type. We note that in a partially connected network with n switches, the number of two-way point-to-point loads (unknowns) would be $n(n-1)/2$, and the number of trunk groups (equations) would be less than $n(n-1)/2$. By solving for the first route trunk group loads, we insure that the number of equations and unknowns are the same. The first route trunk group load values are inferred from the prescribed trunk group sizes. This inference is complicated by the nonuniqueness and the effect of alternate routing in the network.

The nonuniqueness of the load values can be generated by any of the following factors: the need for integer trunk group sizes, the methods used in sizing high usage (HU) groups, and, most dramatically, trunk group modularity. For example, consider a hypothetical final group with no overflow load routed to it, of size twenty-four circuits, and engineered to 1% blocking. Let the modular rule be the following: one final DS1 group (twenty-four circuits) is installed for any offered load that engineers to less than twenty-five circuits. Using the Erlang B formula, $b = B(T,a)$

---

1. For networks which are not fully interconnected, an additional step that reduces the first routed trunk group loads to point-to-point loads, may be necessary.
2. ECCS engineering is a way of sizing a HU group whereby the carried load on the last added trunk in the group is economically justified based on the cost ratio of the direct and alternate routes.
3. Overflow load is the load not carried on the trunk group that is then routed to another trunk group. Overflow load is nonrandom.

where *b* is the blocking, *a* is the offered load in erlangs and *T* is the trunk group size and the modularity rule; we determine that any load value up to 16.1 erlangs results in a trunk group of 24 circuits if engineered to 1% blocking.

Alternate routing also adds complexity. If only direct final groups are allowed between switch pairs, then no alternating routing and, consequently, no overflow would exist. Thus, the offered load would always equal the first route trunk group load and only the nonuniqueness would arise. However, in the presence of alternate routing (overflows of possibly many nonrandom traffic onto random load parcel) in complex hierarchical structures involving many switches, the problem of inferring the first offered trunk group loads from the offered load quickly becomes difficult.

The paper develops the theoretical basis of the algorithm and precise bounds for the inferred loads. These bounds are independent of the particular method that is used in the solution algorithm. The bounds demonstrate, when modularization is not used, the variation among the possible load values can be remarkably small. The particular solution algorithm, which is described in the paper, is iterative between a traditional trunk sizing module and a new *load estimation module*. Computational procedures for the trunk engineering problem appear in Eisenberg[4] and Elsner[5]. The algorithm in the load estimation module, which determines the maximum first route trunk group loads that can be routed on the network, is explained. Because the solution set is not unique, the values of first route trunk group loads depend on the particular algorithm used. In an example, the computed results are shown to be within the theoretical bounds of the inferred load values and, therefore, represent accurate approximations for the first route trunk group loads.

The paper is organized as follows. Section 2 contains the theoretical basis of the algorithm and bounds of the inferred loads. These bounds are independent of the particular engineering modules that are used in the solution algorithm discussed in Section 3. This solution technique is iterative between an existing trunk engineering module and a simple load estimation module. Section 4 discusses an example and compares the computed results with the theoretical bounds developed in Section 2. The conclusions form Section 5.

2. THEORY AND THEORETICAL BOUNDS

The theory and theoretical bounds for the solution to the inverse trunk sizing problem are developed in this section. The formulas that provide the theoretical basis for hierarchical trunk sizing are reviewed and then are used inversely to generate bounds on the computed first route trunk group loads. These bounds are independent of the algorithm described in Section 3.

2.1 Review of Methods in the Trunk Sizing Module

In a trunk engineering module, high usage groups are sized to economically divide the offered load between lower cost direct route trunks, and the more efficient but higher cost alternate routes via final groups. The cost ratio of the direct route to the alternate route is related to the distance between the switches, and is combined into a constant referred to as the economic CCS (ECCS). The quantity ECCS represents the economically justified load on the last circuit of a trunk group assuming random traffic (Erlang B model applies). The optimal group size is the one that has the last trunk carrying the ECCS load. Mathematically, this trunk group size T (a positive integer) satisfies the two inequalities

$$a^*B(T,a) - a^*B(T+1,a) \leq ECCS \tag{1}$$

$$a^*B(T-1,a) - a^*B(T,a) > ECCS, \tag{2}$$

where the offered load, a and the constant ECCS are known, and B is the Erlang B formula. In a trunk sizing algorithm, the offered load a is known because the overflow loads, which are accumulated during the switch pair sequencing scheme (discussed below), are added to the first offered trunk group load. From the locations of the switches defining the pair, the distance between them can be computed, and then the ECCS value can be determined. Final groups are sized more simply than HU groups. If the traffic is random, the Erlang B formula can be used to calculate the trunks necessary to carry the offered load a at the grade of service b via

$$T = B^{-1}(a,b). \tag{3}$$

The formulas (1-3) for sizing the network trunk groups can be extended to nonrandom traffic via the Equivalent Random Method[3]. Although the sizing of the trunk groups in the trunk sizing module are performed via different formulas, which are computationally more efficient, the theoretical basis for determining trunk group sizes are the formulas (1)-(3).

The last important issue in the review of hierarchical networks is the sequencing scheme used in the trunk sizing module. This scheme allows the overflow traffic generated between switches lower in the hierarchy to be accumulated and added to the first offered trunk group load of the given switch pair to form the offered load before the trunk group is engineered. The existence of such a scheme is critical to the trunk sizing module as well as to the load estimation module. Consider the section of the hierarchical network in Figure 1. The first offered trunk group loads are denoted by L_{12}, L_{13}, L_{14}, L_{23}, L_{24}, and L_{34}. Assume that the switch pair (1,2) is the lowest in this two level hierarchy so no other traffic besides the first offered trunk group load is offered to the 1-2 HU group, $a_{12} = L_{12}$. The load actually carried on the 1-2 HU trunk group is less than the offered load, and the difference is the overflow load, denoted by O_{12}. The overflow load on any trunk group between switch pair i and j is equal to the product of the blocking on that trunk group, denoted by b_{ij}, and the offered load, that is, $O_{ij} = b_{ij} * a_{ij}$. The path of the overflow traffic from the 1-2 HU group is on the path 1-4 and 2-4, so that O_{12} is added to O^A_{14} and O^A_{24}. The variable O^A_{ik} denotes the accumulated overflow onto the i-k link. The initial value of O^A_{ik} is zero. The offered load for the switch pair 1-4, a_{14}, is L_{14} plus O^A_{14}. Following the same procedure the offered load on the final trunk group 1-3 is given by $a_{13} = L_{13} + O^A_{13}$, where $O^A_{13} = O_{14} = b_{14} * a_{14}$. No HU group exists between switches 2 and 3 because the demand between them is not sufficient to economically justify a trunk group. In this case, the L_{23} is zero. The offered load between switches 2-4 is given by the sum $a_{24} = L_{24} + O^A_{24}$, where $O^A_{24} = O_{12}$.

Between switches 3 and 4, the offered load is the sum $a_{34} = L_{34} + O_{34}^A$, where $O_{34}^A = O_{14}$. In summary, the sequencing scheme allows the overflow loads to accumulate properly by insuring that all the links that could contribute overflow to a given link are engineered before the given link.

This review illustrates only the most elementary concepts in trunk group sizing. Complicating issues that must be addressed in the trunk sizing module are the nonrandomness of the overflow traffic, the day-day variation in the sizing of final trunk groups, more detailed estimates of the overflow actually offered on the link of the alternative path, one-level-inhibit rule, etc. However, these added difficulties are not essential for the understanding of the solution of the inverse trunk sizing problem.

2.2 Theory for the Load Estimation

Consider a HU group between two switches with the number of *prescribed* trunks denoted by T^P. We wish to determine the offered load needed for the trunk group to be sized to T^P trunks. Since the locations of the switches are known, the ECCS value can be determined exactly as it was in the sizing module. Because the trunk size is known, Equations (1)-(2) can be solved for a. The values of the offered load that satisfy Equation (1) are bounded by $0 \le a \le a_{UPPER}$. The values of the offered load that satisfy Equation (2) forms the second interval which is is bounded by $a_{LOWER} < a$, where a_{LOWER} is a positive number. The properties of the Erlang B formula insures that $a_{LOWER} \le a_{UPPER}$. For any offered load contained in the interval,

$$a_{LOWER} < a \le a_{UPPER}, \tag{5}$$

the trunking algorithm would size to the value T^P for the given switch pair. If the offered load was random, for example, the load between switches 1 and 2 in Figure 1, then the bound given by expression (5) is the theoretical bound for the first offered trunk group load (L) for that switch pair. Thus, the solution of the inverse sizing problem on a HU group cannot be unique because it could be any value satisfying (5). Bounds of the overflow from this trunk group can also be computed:

$$O_{LOWER} < O \le O_{UPPER} \tag{6}$$

where
$$O_{LOWER} = a_{LOWER} * B(a_{LOWER}, T^P), \tag{7}$$

$$O_{UPPER} = a_{UPPER} * B(a_{UPPER}, T^P). \tag{8}$$

To infer a bound of the first route trunk group load, we must know all the overflows to the trunk group. This is guaranteed by the switch pair sequencing scheme. The bounds for all the overflow parcels can be computed via formulas (6-8). Because the first offered trunk group load can be expressed as a sum,

$$L = a + [-O^A] \tag{9}$$

a bound of the first offered trunk group load can be determined from the bounds of a, and $-O^A$. The lower and upper bounds, L_{LOWER} and L_{UPPER}, are the sums of the lower and upper bounds of each term on the right-hand side of Equation (9), respectively. Clearly then, the interval in which the offered load lies is a subset of the bounding interval for the first offered trunk group loads.

For finals, the offered load is a single value given by the formula

$$a = B^{-1}(T^P, b), \qquad (10)$$

where b is the grade of service and T^P is the prescribed trunk group size if the load a is random and that the day-day variation is neglected. If only first offered trunk group load is offered to the final, the value of the first offered trunk group load is given by Equation (10), namely $L = a$. However, the value given by Equation (10) is not the only load value that satisfies $B^{-1}(a, b) = T^P$ (as shown in the Introduction). As in the case of HU groups, the bound on the first route trunk group load for final groups can be constructed from Equation (9).

When the offered load is random, for example, when a contains no overflow load, formulas (5-10) can be used directly. However, when the offered load is not random (it contains overflow load), the calculation of the bounds become more complex but can be derived.

3. AN ITERATIVE ALGORITHM FOR INFERRING THE LOADS

The determination of a set of first offered trunk group loads from a specified set of trunk group sizes for a given hierarchical network is an iterative process between a trunk sizing module and a load estimation module. The overview of the iterative algorithm is given in Figure 2. The input includes the hierarchical network configuration, the set of prescribed trunk group sizes between the various switches, and a set of initial load values. The switches are denoted by subscripts, and the prescribed trunk group size between the switch pair i and j by T^P_{ij}. Initial value of the load between the switch pair i and j is denoted by L^o_{ij}. The value, L^o_{ij}, may be arbitrary, or, if possible, some realistic value. In Section 4, the almost negligible dependence of the consistent load values on the initial load values is shown for an example network.

The first step in the algorithm is to calculate via a given trunk group sizing module the set of trunk group sizes, T^o, that is consistent with the initial set of load values, L^o. A simple iteration loop is then entered. The kth iteration values of L^k and T^k, and the prescribed values T^P for each switch pair in the entire network are used in the load estimation module to compute the next iteration of the first offered trunk group load values, L^{k+1}. Subsequently, the values L^{k+1} are used by the trunk sizing module to determine the new iteration of trunk group sizes T^{k+1}. The convergence test for ending the iteration loop is done in a convergence iteration module. Because the purpose of the algorithm is to compute loads that engineer to the T^P values, the stopping or convergence criterion is expressed in terms of the difference between the T^P quantities, and the trunk group size T^{k+1}. The strictest convergence criterion, and the one used in the examples, is that the computed trunk group size T^{k+1}_{ij} must be equal to the prescribed trunk group size T^P_{ij} for every switch pair (i,j). If the convergence test is satisfied, the last computed first offered trunk group loads become the consistent load values, denoted by L^P, and the iteration procedure stops. If the convergence test is not satisfied, another iteration is started with the most current calculated values, that is, the (k+1) iterate of the loads and trunk group sizes.

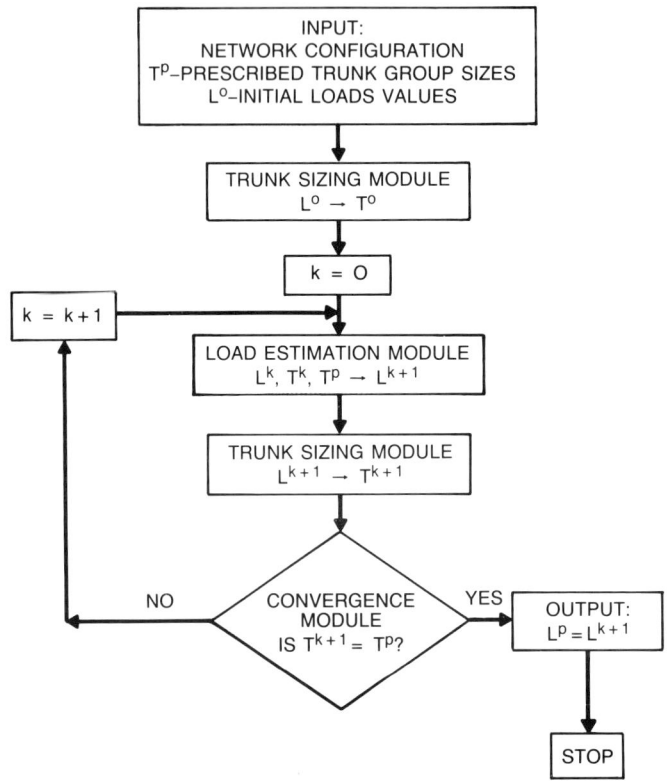

FIGURE 2 ITERATIVE ALGORITHM FOR DETERMINING A SET OF FIRST OFFERED TRUNK GROUP LOAD VALUES FROM A SET OF TRUNK GROUP SIZES IN A HIERARCHICAL NETWORK

Because the trunk group sizing module exists, only an explanation of the load estimation module is necessary to complete the description of the algorithm. This four step method is depicted in Figure 3. Using the switch pair sequencing algorithm that is used in the trunk sizing module, each link in the network starting at the bottom of the hierarchical network is considered. Let the current switch pair be the i-j pair. The first step is to approximate the offered load a_{ij}^k by adding any overflow loads to the first offered trunk group loads. The overflow load needed for the determination of the offered load for links higher in the network is approximated in step 4. Once the offered load, a_{ij}^k, is computed, it and the trunk group size, T_{ij}^k, which was computed previously by the trunk sizing module, are used to calculate an *effective blocking value*, b_{ij}^k, via the Erlang B formula. This blocking value is not the actual blocking value that a trunk sizing module would compute because the value of the offered load is only approximated and is assumed random (the use of the Erlang B formula) which is most often wrong. As will be demonstrated with the examples, the precise calculation of the overflow values, and the modeling of the nonrandom behavior of the offered load are not important. Furthermore, these approximations greatly simplify the load estimation module. What is important is that the alternate paths that are used by the sizing algorithm are exactly those used in the loading estimation module. In doing so, the overflow load will be placed on the same links in both algorithms. The second step uses the effective blocking value b_{ij}^k obtained in step 1 along with the prescribed trunk group size T_{ij}^p to compute a new iterate value of the offered load a_{ij}^{k+1} via the inverse of the Erlang B formula $a_{ij}^{k+1} = B^{-1}\left[T_{ij}^p, b_{ij}^k\right]$.

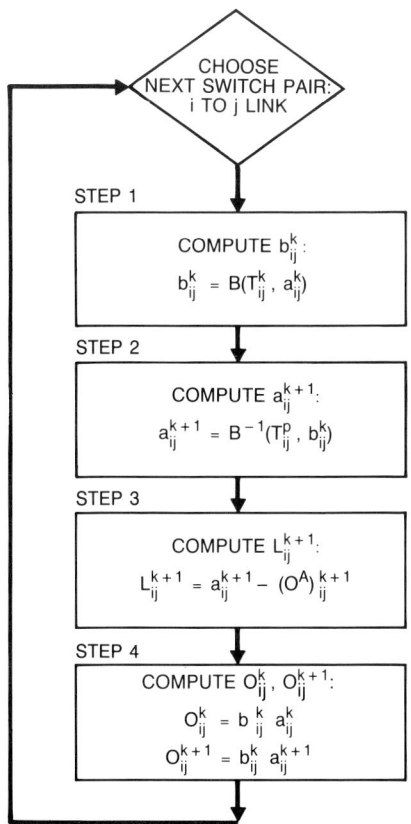

FIGURE 3   BASIC ALGORITHM OF THE LOAD ESTIMATION MODULE WHEN $T_{ij}^p \neq 0$

The third step is to deduce the new iterate L_{ij}^{k+1} from the a_{ij}^{k+1}. Instead of combining the overflow load with the first offered trunk group load to compute the offered load as in step 1, the new load, L^{k+1}, is calculated by disaggregating the new overflow load from the new offered load. This procedure is possible because of the switch pair sequencing scheme. At the lowest links in the hierarchy $L_{ij}^{k+1} = a_{ij}^{k+1}$. Thereafter, the overflow load alternately routed to the link i-j at the $(k+1)$iteration level must be subtracted from the a_{ij}^{k+1} value. The overflow load needed for the determination of the first offered trunk group load for the $(k+1)$ iterate for links higher in the network is approximated in step 4. The fourth and final step is to approximate the overflow from the link i-j at both iteration levels, k and $(k+1)$, and to assign these values to the links that make up the alternate path for link i-j. The estimation of the values of overflow loads are simple products $O_{ij}^k = b_{ij}^k a_{ij}^k$ and $O_{ij}^{k+1} = b_{ij}^k a_{ij}^{k+1}$. Using the same alternate path as in the trunking algorithm insures that the overflow loads will accumulate on the same links in both algorithms.

4. EXAMPLES

The hypothetical network consists of six switches arranged in a three level hierarchy given in Figure 4. The homing of the switches are indicated by solid lines (final trunk groups) connecting the lower ranking switches to the higher ranking switches. To insure a path exists between any switch pair in the network, a final trunk group exists between switches 4 and 5. The network is fully interconnected with HU's and finals. The prescribed trunk group sizes are: $T_{12} = 96$, $T_{13} = 26$, $T_{14} = 30$, $T_{15} = 86$, $T_{16} = 60$, $T_{17} = 55$, $T_{18} = 72$, $T_{23} = 49$, $T_{24} = 69$, $T_{25} = 74$, $T_{26} = 80$, $T_{27} = 48$, $T_{28} = 21$, $T_{34} = 49$, $T_{35} = 30$, $T_{36} = 71$, $T_{37} = 74$, $T_{38} = 18$, $T_{45} = 170$, $T_{46} = 66$, $T_{47} = 24$, $T_{48} = 15$, $T_{56} = 90$, $T_{57} = 175$, $T_{58} = 44$, $T_{67} = 77$, $T_{68} = 110$, and $T_{78} = 39$. The problem is to infer 28 first offered trunk group load values from these 28 prescribed trunk group sizes, that is, determine a set of first offered trunk group loads which when engineered result in the prescribed trunk group size on each of the 28 links. A series of runs with five different initial conditions are executed for this example. The initial sets of load values differ greatly. For example, for one run, every initial value of the first offered trunk group loads is 10 erlangs, and in another they are 250 erlangs. However, the variability between the computed loads is quite small. The maximum variation of the load values, L^P, consistent with the T^P values, for all the five cases between any switch pair is less than 1.5 erlangs, i.e.,

$$\max_{\substack{k=l+1,\ldots,5 \\ l=1,2,3,4}} | \left(L_{ij}^P\right)^k - \left(L_{ij}^P\right)^l | < 1.5,$$

where (i,j) is any switch pair in the network. In all the cases, the algorithm converged in five or less iterations taking less than 0.27 seconds of CPU time on an IBM 3033.

The theoretical bounds of the loads between switches connected by HU trunk groups are compared with the variation in the five sets of calculated load values for the 8 node network. For a given switch pair, all the first offered trunk group loads computed for the five cases lie between the theoretical bounds. The results for

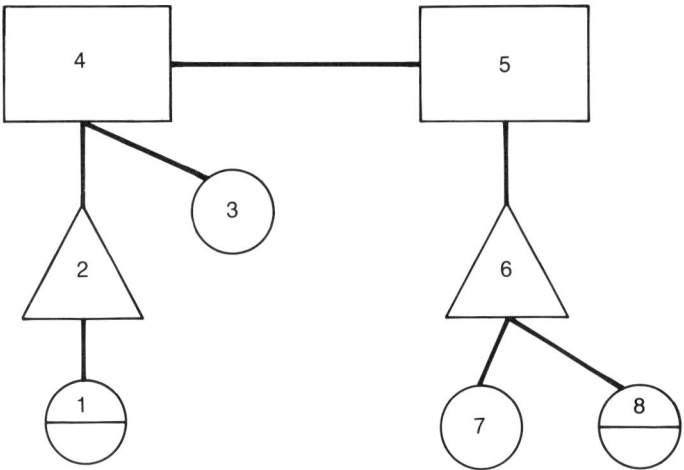

FIGURE 4 HYPOTHETICAL NETWORK OF EIGHT SWITCHES

TABLE 1 LOAD BOUNDS FOR SELECTED HU LINKS WITH VARIOUS PARAMETERS. VARIABLE Z^p DENOTES AVERAGE PEAKEDNESS OF OFFERED LOAD CORRESPONDING TO L^p

SWITCH PAIR	T^p	ECCS	Z^p	BOUNDS OF CALCULATED FIRST OFFERED TRUNK GROUP LOADS		THEORETICAL BOUNDS OF FIRST OFFERED TRUNK GROUP LOADS	
				LOWER	UPPER	LOWER	UPPER
1-3	26	16	1.0	22.3	23.0	22.1	23.0
3-5	30	18	1.72	21.9	23.0	20.5	23.4
3-6	71	19	1.28	60.4	61.3	60.2	64.2
3-7	74	19	1.0	70.1	71.0	70.0	71.0
1-6	90	19	1.47	48.6	49.0	47.8	49.9
5-7	175	17	1.0	166.5	166.9	166.5	167.5

some representative switch pairs are given in Table 1. The table shows typical results for a variety of prescribed trunk group sizes, of trunk groups with and without overflow (peakedness value greater than one and one, respectively), and of consistent load values. When the peakedness is one (links 1-3, 3-7 and 5-7 in Table 1), the offered load bounds are the first offered trunk group load bounds. For the entries in Table 1 except for the 3-5 link, formulas (1)-(2) and (5)-(9) are used to compute the theoretical bounds because either the offered loads themselves were random (links 1-3, 3-7, and 5-7), or the overflow originated from random offered load (links 1-6 and 3-6). The computation of the overflow from the non-random offered load of the 3-6 trunk group depends on the peakedness of the 3-6 offered load. In this case, the overflow is computed using extensions of formulas (5-9) that account for peaked traffic. The overflow is alternately routed on the 3-5 and 5-6 links.

5. CONCLUSIONS

The solution of the generic problem of determining the first offered trunk group loads between switch pairs that a given hierarchical network with alternate routing can support for specified trunk group sizes can now be solved. There are no restrictions as to the complexity of the network, where high usage (HU) trunks need to be placed, or on the size of any feasible trunk group. The algorithm can be extended to handle minimum group size requirements, modular trunk group engineering, and aggregated nodes representing more than one physical switch. Furthermore, the method can determine a distribution of the loads across multiple load-set-periods. In this case, the algorithm uses cluster busy hour engineering for HU links, and group busy hour engineering for finals.

6. REFERENCES

1. A.J. David and N. Farber, "The Switch Planning System for the Dynamic Nonhierarchical Routing Network", Proceedings of the 10th ITC, Session 3.2, Paper 8, Montreal, Canada, 1983.

2. S.G. Low and J.A. Schmitt, "The Integration of Domestic and International Networks for Switch Planning", Proceedings of the 12th ITC, Session 2.1, Paper 5, Torino, Italy, 1988.

3. Engineering and Operations in the Bell System, 2nd Ed., R. F. Rey (Ed), AT&T Bell Laboratories, Murray Hill, NJ, 1983, Chp. 5.

4. M. Eisenberg, "Engineering Traffic Networks for More Than One Busy Hour", Bell System Technical Journal, Vol. 56, pp. 1-20, 1977.

5. W. Elsner, "A Descent Algorithm for the Multihour Sizing of Traffic Networks", Bell System Technical Journal, Vol. 56, pp. 1405-1430, 1977.

A general purpose model for circuit-switched networks

Marc Lebourges

France Telecom, CNET/PAA/ATR, 38-40 rue du General Leclerc
92131 Issy-les-Moulineaux, FRANCE

Abstract
The general purpose circuit-switched networks model integrates known models for particular systems or phenomena, e.g. multi-services traffics, repeated attempts, exchange blockings, adaptive routing, in a single network model. The aim is not to have a precise modelling of each system, but to get correct qualitative behaviour and quantitative averages at the network level.

I. Introduction

The general purpose analytical model presented here has been developped for the Sagesse project [Leb.89], the aim of which is to simulate the interaction between a circuit-switched network evolution and its servicing and provisioning processes. We first precise the types of networks concerned, the requirements and the input data of the model. Then we give a description of the algorithm, stressing its particularities as compared with existing models. Finally we conclude on the numerical experimentations we have done.

II. Scope

We deal with circuit-switched networks of any type of routing strategy. Telecommunication services may need N x 64Kbits end-to-end connexion and be switched at M x 64Kbits.

II.1. Requirements

In a software, like Sagesse, where the routing architecture is parametrable, the model must be correct for any kind of network: hierarchical or not, using fixed, alternate, load-sharing, DCR or DAR-type adaptive routing, possibly at the same time on different parts or different streams of the network. This excludes one-moment models, well suited for meshed non-hierarchical networks, but inadequate for hierarchical networks. Reversely, the model must take account of the impact of non-hierarchical architecture on two moments models.

The modelling of adaptive routing should reflect its impact on network servicing and provisioning processes: thus modelling adaptive routing as fixed load sharing with coefficients iteratively calculated in the evaluation procedure simulates the interest of adaptive routing in case of incorrect forecasts or bad capacity allocation.

The model distinguishes offered Erlangs and offered calls for two reasons: first, exchange planning processes are simulated in Sagesse and exchange blockings, which depend on the call attempts, need to be evaluated. Second, the evaluation results feed a function which simulates traffic measurements, and some of these measurements concern call attempts. Repeated attempts, which have a bigger effect on call demand than on traffic demand, must be modelled.

II.2. Input data

The network is defined by the following input data. All networks which may be described this way are modelled:

There are E exchanges, e=1 to E, with a call attempts capacity U_e and an Erlang capacity X_e, N national numbers, n = 1 to N, and T telecommunication services, t = 1 to T. Each national number is related to an exchange for a telecommunication service : e = e(n,t).

A telecommunication service t has a service bit-rate S_t and a switching bit-rate W_t, where bit-rate 1 corresponds to 64kbits/sec. If $S_t = W_t$ it means that all liaisons necessary for a call of this service are switched together along its switching path. The service has a U.C. requirement per call and per exchange, H_t, where $H_t = 1$ corresponds to a standard telephone call.

The network is offered S Poisson traffic streams of mean Erlang intensities v_s and fresh call attempts rate α_s, s=1 to S. Blocked calls reattempts with probability τ_s. A traffic stream has an origin national number, o_s, a destination national number, d_s, and a service t_s. The Erlang intensity v_s corresponds to the average number of calls of the stream, the service bit-rate not being taken into account. The offered traffic of a stream, in circuits, is then $v_s . S_{ts}$.

When a call of service t and destination number n is at exchange e, and if the call has not reached its destination (e is different from $e(d_s,t_s)$), it is routed on a traffic routing number r(e,n,t). There are R routing numbers, r = 1 to R and K trunk-groups, k = 1 to K.

To each routing number corresponds the exchange e_r to which it belongs.and a routing number type, h_r, which characterize whether the call is simply routed on a single trunk-group ($h_r = S$), or if it is multiply routed on a set routing numbers ($h_r \neq S$). In the latter case, h_r may designate fixed load sharing routing (h_r = F), DCR-type adaptive routing based on residual capacity (h_r = C), or DAR-type adaptive routing based on blocking (h_r = A).

If h_r = S, k_r is the trunk-group on which the call is routed with a state-protection parameter s_r and a traffic filtering parameter f_r. To a routing number r is also attached the routing number d_r on which r overflows (if it does, otherwise d_r is null), the routing number g_r to be taken in sequence (if it can be specified, otherwise g_r is null) and the maximum number m_r of overflows authorized at the current exchange.

If $h_r \neq S$, $L_r = \{l_{1,r}, l_{2,r}, ..., l_{mr,r}\}$ is the list of m_r simple-type routings to which the routing r points to. If the load sharing coefficients are fixed, the algorithm also needs the list of the load sharing coefficients $\{c_{1,r}, c_{2,r}, ..., c_{mr,r}\}$.

Each trunk-group k has its extremity exchanges, $e_{1,k}$ and $e_{2,k}$ and a capacity C_k.

III. Characteristics of the model

After a general description, we detail the specific models included in algorithm.

III.1. General description

It is basically a two-moments version of Manfield and Downs one-moment model [Man.79], with several adaptations:
- The account taken of repeated attempts within the network model.
- The modelling of N x 64 trafic as peaky traffic.
- The modelling of call attempts loads.
- The integration of exchange call and traffic blockings in the network model.
- The possibility to study networks which combine different routing strategies: fixed, alternate, DCR-type adaptive routing, DAR-type adaptive routing, fixed load-sharing.
- The concept of routing blockings, instead of trunk-group blockings. Fresh and overflow traffics received by each traffic routing are distinghuished, as recommended by Pioro.
- The second moment a routing overflow is characterized by a parameter equal to its peakedness minus one over the mean. This allows first a direct calculation of the peakedness of the overflow of each stream using the considered traffic routing, taking baulking into account, and second to sum up differently the variances of correlated (in this model, overflows of a single traffic routing number) and of independant overflows (other cases), when they merge to be offered to a trunk-group. In this context the formula $(z_i - 1) = coef_i \cdot (z - 1)$, which give the peakedness of a parcel $coef_i$ of a peaked traffic, is extensively used.

The initialization is on blockings values per traffic routings and loadsharing coefficient for adaptive routings. The values iteratively calculated by the algorithm are shown on Figure 1.

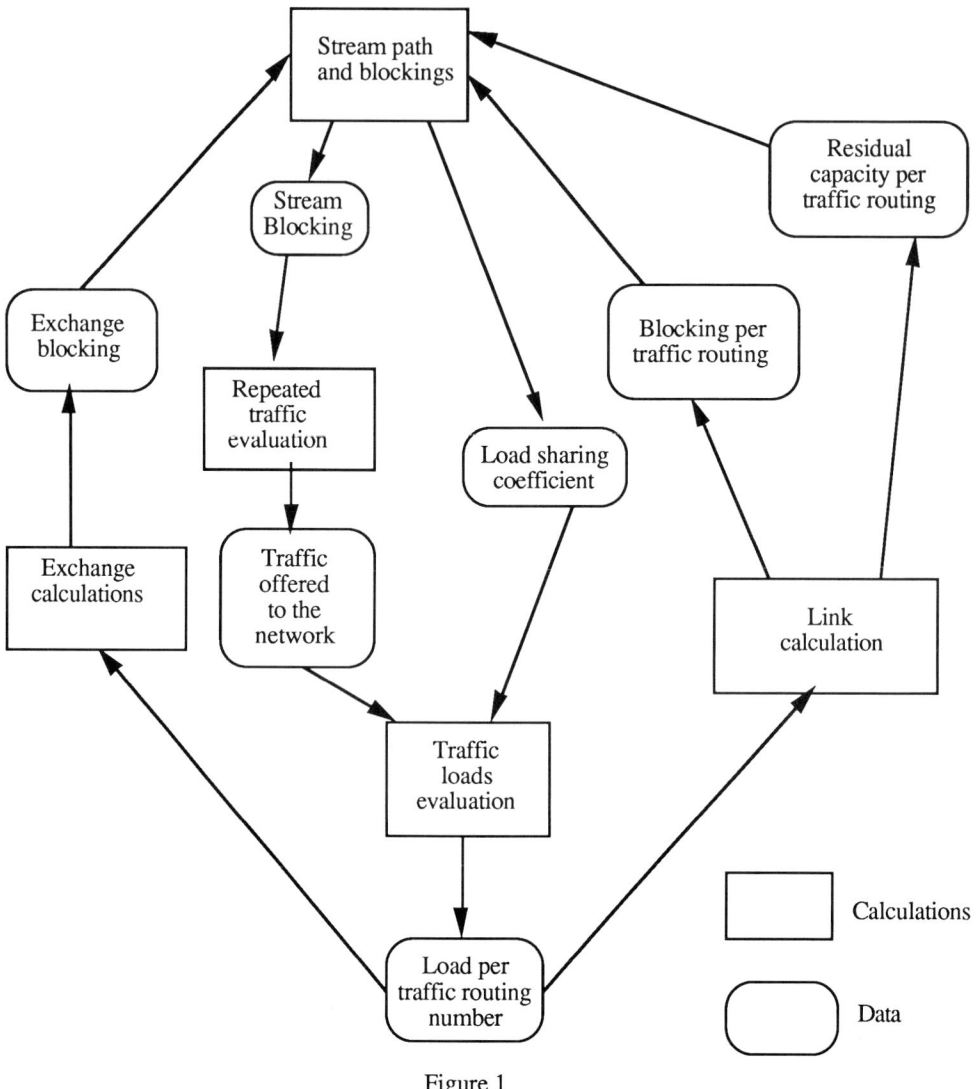

Figure 1

III. 2. Specific models

The specific models are grouped in three: trunk-group, exchange model and stream.

a) Trunk-group model

Each trunk-group k receive traffic from a list of R_k, $r_k = 1$ to R_k, traffic routing numbers. Each traffic routing adresses the trunk group with a specific trunk-reservation parameter s_{rk}. For each traffic routing and each telecommunication service, we distinguish traffic offered to k, fresh or in serie in the one hand, and traffic offered to k as overflow and modelled by mean and varaince in the other hand. This distinction allows to model the most significant differences in blockings, while not having to large computer data structure. It was shown by Pioro that this distinction was necessary and sufficent to get a correct overall behaviour for non-hierarchical network. If actual stream blockings are considered necessary, it may be done by creating the necessary traffic routing numbers in the network input data.

In the present paragraph, we will drop the index k of the current trunk-group, in order to have lighter notations. Let:

$\rho_{1,r,t}$ Mean intensity of the fresh or serie traffic offered to r,

$\varpi_{1,r,t}$ Variance of the fresh or serie traffic offered to r. After numerical tests, the smoothing of carried traffic has not been included in the model, but $\varpi 1r,t$ may be different of $\rho 1r,t$ because the service bit-rate may be different from the switching bit-rate for the considered service and the peakedness of fresh or serie traffic can thus differ from 1.

$\rho_{2,r,t}$ Mean of the overflow traffic offered to r,

$\varpi_{2,r,t}$ Variance of the overflow traffic offered to r,

for r such that r : list of trunk-group k.

These trafics take account of the ratio between the service and the switching bit-rate, but not of the switching bit-rate itself. The offered traffic per routing and service, in terms of 64kbit/s circuits, is given by:

$$\rho\chi_{1,r,t} = \rho_{1,r,t} \cdot W_t , \varpi\chi_{1,r,t} = \varpi_{1,r,t} \cdot W_t , \rho\chi_{2,r,t} = \rho_{2,r,t} \cdot W_t , \varpi\chi_{2,r,t} = \varpi_{2,r,t} \cdot W_t$$

as the switching rate is a multiplicative factor of the peakedness. If we group the traffic routing per increasing trunk-reservation parameters s_p, $p = 0$ to P, where $(P+1)$ is the number of priority level of the trunk-group, the agregated traffic offered to each level of priority if then, with an independance assumption:

$$\rho_p = \Sigma_t \Sigma_{sr<=sp} (\rho\chi_{1,r,t} + \rho\chi_{2,r,t}) \text{ and } \varpi_p = \Sigma_t \Sigma_{sr<=sp} (\varpi\chi_{1,r,t} + \varpi\chi_{2,r,t})$$

We know make the following approximation: when the occupancy of the studied trunk-group S_0 is equal or below $C_k - s_p - 1$, the system behaves as a truncated system S_p of $C_k - s_p$ circuits. It is an approximation because it assumes equal the statistical states of the two systems, conditional to the entrance in the $C_k - s_p - 1$ state, which is not rigorous. With this approximation the mean duration, and thus the outgoing rate, of the macro-state "less than ($C_k -$

$s_p - 1$) occupied circuits", is the same for both system. This approximation works recursively for the different trunk-reservation level, for states from $(C_k - s_P)$ to C_k.

Here, the non-Poisson nature of the traffics is modelled by Wilkinson's ERT. So to each agregated traffic corresponds a fictitious overflowing trunk-group of mean ρ^*_p and capacity C^*_p, numerically estimared with Rapp's approximation. We do not consider the states when some trunks are free on the fictious overflowing while the studied system is blocked for the considered traffic. Doing this, we consider as blocking states states where the fictitious traffic can be carried, and thus we overestimate the blocking. But we also overestimate the mean duration of the macro-state during which the fictitious traffic may carried and thus we underestimate the blocking. These two approximations of inverse impact may then lead to a reasonable accuracy.

From the preceeding points, we can write equilibrium equations between state probabilities of the studied and the P truncated systems. These equations are formally similar to the ones of a standard Erlang loss trunk-group and can be solved recursively, as described in [Leb.88]. Let B^t_p and Bi^t_p be respectively the global blocking of traffic (ρ_p, ϖ_p) and the individual blocking of traffic $(\rho_{p-1}, \varpi_{p-1})$ on the truncated system S_p. For $p = P$, B^t_P is given by an ERT calculation. For $p = P$ to 1, Bi^t_p is derived from B^t_p with Akimaru's approximation for individual blockings [Aki.83]. The B^t_p is derived from Bi^t_{p+1} by the recursion:

$x(0) = \rho^*_p / (Bi^t_{p+1} \cdot \rho_p)$, inverse of the blocking of traffic ρ^*_p on the $(C^*_p + S_{p+1})$ system,

$x(l) = 1 + x(l-1) \cdot (1 + C^*_p + C_k - s_{p+1})$ and

$B^t_p = \rho^*_p / (x(s_{p+1} - s_p) \cdot \rho_p)$, blocking of ρ_p on the S_p system.

To each traffic $(\rho \chi_{i,r,t}, \varpi \chi_{i,r,t})$, where $i = 1$ or 2, corresponds ERT parameters $(\rho^*_{i,r,t}, C^*_{i,r,t})$, derived with Rapp's approximation. If traffic $(\rho \chi_{i,r,t}, \varpi \chi_{i,r,t})$ has a reservation level p, its blocking is given by:

$$B_{i,r,t} = \rho^*_{i,r,t} \cdot (\sum_0^{s_p} \pi_i) / \rho_{i,r,t} \quad \text{where} \quad \pi_l = \frac{1 + l + (C_k - s_p) + C^*_{pl+1} + C^*_{i,r,t}}{\rho^*_{pl+1}} \cdot \pi_{l+1}$$

where pl is the level of priority corresponding to an occupancy of $(C_k - s_p) + 1$. Probability π_{sp}, needed to calculate the other π_l is given by an analoguous recursion:

$x(0) = \rho^*_{i,r,t} / (\rho \chi_{i,r,t} \cdot B^t_{i,r,t})$, where $B^t_{i,r,t}$ is the blocking on the truncated system S_p, derived from B^t_p with Akimaru's approximation. (This is a correction of a printing error in [Leb. 88]).

$x(l) = 1 + x(l-1) (1 + (C_k - s_p) + C^*_{pl+1} + C^*_{i,r,t}) / \rho^*_{pl+1}$

$\pi_{sp} = x(s_p)$

The accuracy of this calculation has been shown in [Leb. 88].

As is well known, the second moment of the overflow of a trunk group can be expressed in terms of the average number of occupied circuits on the overflow trunk-group, conditional to the fact that the first choice is blocked. The averages, conditionned on the different states of the first choice group are related by linear equations (e.g. see [Mats.] or [Wilk.56]), similar to the equations for the blocking probability. With the approximations presented above, we can write this system of equations, first on the truncated systems, then on the studied system. The different variances can then be found, not by a closed formula but by the recursive and linear resolution of the system, as in [Mats.], but in a much simpler way, first because the overflowing trunk-group state is here scalar, not vectorial, second because the variances corresponding to the starting truncated system S_P are directly given by the classical Nyquist formula.

Individual overflow peakednesses for fresh and overflow type traffics are derived from overflow peakednesses for each level of priority, considering that overflows are a randomly splitted and using the $(z_{i,r,t} - 1) = \alpha_{i,r,t} \cdot (z_p - 1)$ formula, where $\alpha_{i,r,t}$ is the probability that an overflowing call of priority p overflows from routing r, is of service t, an was offered to r as a fresh-type call (if $i = 1$) or as an overflow type call (if $i = 2$). Peakedness due to the switching bit-rate is then deduced for each service, and "peakedness ratios" are calculated and memorized to be used in the stream model:

$$zr_{i,r,t} = ((z_{i,r,t}/W_t) - 1)/(\rho_{1,r,t} \cdot B_{i,r,t})$$

Time-average residual capacities per priority level cr_p, to be used to model DCR-type adaptive routing, are easily calculated recursively on the truncated systems S_P to S_0. From these results, time-average residual capacities, conditional to the non-saturation of truncated systems for the traffic of corresponding priority are straifghtforward. Time-average residual capacities per priority level on the actual system are then derived from this and the time-blocking of each level of priority, which is estimated with Akimaru's approximation from the call blocking already calculated. The time-average residual capacity per routing and per service is equal to the time-average residual capacity of the priority of the routing.

b) Exchange model

The exchange model is very simplified in order not to make assumptions on the caracteristics of the exchange, and just reflect the overall interrelation between exchanges and network grade of services in a stationnary situation.

1. Offered calls and traffic per service

The traffic offered to an exchange for each service is the sum of two computations:
- the sum of fresh-type and overflow-type carried calls and traffics of all traffic routings which points to trunk-groups the other extremity of which is the considered exchange. Blockings on these traffic routings have to be taken into account in this summation, in the average and the peakedness calculations.
- the sum of offered traffics to the exchange by streams originating at the exchange.

Taking the service switching bit-rate into account (the ratio between service and switching bit-rate already being taken into account in stream and traffic routings offered traffics), we get in formal terms for the offered traffic in 64kbits connexions:

$$\rho\chi_{e,t} = W_t \cdot [\sum_{\{r\}_e} (\rho_{1,r,t} \cdot (1 - B_{1,r,t}) + \rho_{2,r,t} \cdot (1 - B_{2,r,t})) \; + $$
$$\sum_{\{s\}_{e,t}} m_s \cdot (1 - L_s) / (1 - B_{e,t}) \;], \text{ for the mean and}$$

$$\varpi\chi_{e,t} = W_t [\sum_{\{r\}_e} (\rho_{1,r,t} \cdot (1 + (\varpi_{1,r,t}/\rho_{1,r,t} - 1) \cdot (1 - B_{1,r,t})) + \rho_{2,r,t} \cdot (1 + (\varpi_{2,r,t}/\rho_{2,r,t} - 1) \cdot (1 - B_{2,r,t})) \;) \; +$$
$$\sum_{\{s\}_{e,t}} m_s \cdot (1 + (z_s - 1) \cdot (1 - L_s) / (1 - B_{e,t}))\;], \text{ for the variance}$$

where L_s is stream s network end-to-end blocking, $B_{e,t}$ is exchange e blocking for service t, m_s is the stream offered traffic (account taken of repeated attempts as is described in the stream model), $\{s\}_{e,t}$ is the set of streams originating at e and of telecommunication service t, that is so that $e = e(o_s,t_s)$, and $\{r\}_e$ is the set of routings pointing to trunk-groups the other extremity of which is e. The total circuit traffic is then:

$$\rho\chi_e = \sum_t \rho\chi_{e,t} \quad \text{and} \quad \varpi\chi_e = \sum_t \varpi\chi_{e,t}$$

The call rate offered to an exchange is the result of the same type of summation, with the difference that baulkings must not be taken into account and that the services switching bit-rate are replaced by the services U.C. switching requirements. Also the peakedness of call arrivals is not considered: so only the mean call rate is needed:

$$\rho\alpha_e = \sum_t H_t \cdot [\sum_{\{r\}_e} (\rho\alpha_{1,r,t} \cdot (1 - B_{1,r,t}) + \rho\alpha_{2,r,t} \cdot (1 - B_{2,r,t})) \; + \sum_{\{s\}_{e,t}} \alpha_s \;]$$

where $\rho\alpha_{i,r,t}$, $i = 1$ to 2, is offered call rate, fresh-type and overflow type, of service t on routing r and α_s is offered call rate of stream s, account taken of repeated attempts.

2. U.C. constraints

Blockings due to U.C. saturation happen before connexion blockings. We have chosen to keep the model very simple and assume a single call blocking, due to U.C. saturation, for all traffics using the exchange. The blocking $B\alpha$ is estimated with Erlang B formula:

$B\alpha_e = \text{Erl}(\rho\alpha_e, U_e)$

3. Connexion constraints

The connexion traffic is first slimmed by U.C. blocking $B\alpha_e$. We then consider the connexion unit as a full-availibility trunk-group. The average exchange traffic blocking is derived by Hayward's formula from the connexion capacity X_e and the two moments of the offered traffic: $(1 - B\alpha_e)\rho\chi_e$ for the mean and $(1 + ((\varpi\chi_e/\rho\chi_e) - 1) \cdot (1 - B\alpha_e)) \cdot \rho\chi_e$ for the variance.

Individual service blockings $B\chi_{e,t}$ are derived by Akimaru's approximation from individual services offered traffics:
$(1 - B\alpha_e)\rho\chi_{e,t}$ for the mean and $(1 + ((\varpi\chi_{e,t}/\rho\chi_{e,t}) - 1) \cdot (1 - B\alpha_e)) \cdot \rho\chi_{e,t}$ for the variance.

A better blocking estimation could be derived from exact formulae for N x 64 traffics [Rob.81]. But it would be inconsistent with this simplified exchange modelling.

4. Global exchange blocking per service

The global exchange blocking per service is then: $B_{e,t} = 1 - (1 - B\alpha_e) \cdot (1 - B\chi_{e,t})$

c) Stream model

As shows Figure 1, the stream model has three submodels: one for to calculate stream blockings, one to model repeated traffic and one to report stream traffics on traffic routings.

1. Stream blockings

The end-to-end path of each stream is first derived from the initial routing description. It is described as an ordered sequence of routing numbers, non redondant, that is with any routing number being present only once. Each routing number has associated parameters, such as its overflow and sequence routing numbers, its indicator on whether the stream is offered fresh or overflow, and if relevant its overflowing routing number. When a stream takes a routing both as a fresh and as an overflow, for different paths, it is considered as fresh-type. When a stream takes, for different paths the same routing as overflow with different overflowing routing, the stream is considered as overflow on the routing but without overflowing routing number.

The Gaudreau's classical recursive formulae [Gau.80] are applied for stream end-to-end blockings, with some adaptations:

- exchange blockings for service of the stream are included in the formulae.
- the indicator on the fresh or overflow type of the stream is used to select the relevant routing blocking.
- for a multiple routing, blocking is the average, weighted by the load sharing coefficients, of the blockings of the simple routings to which it points to. In case of adaptive routing, the load sharing coefficients are recalculated each time the stream blocking is considered, based on the two routings in sequence average residual capacity for DCR-type, and based on the two routings in sequence blocking for DAR-type:

the two routings in sequence residual capacities are estimated from single routings average residual capacitie assuming independent Poisson distribution for routings residual capacities probability law. This assumption is not supposed to be a realistic model, but is only used for this calculation.

the two routings in sequence blockings are straightforwaedly derived from routings mean blockings, assuming routing independance.

As a by-product of the recursive calculation which gives the end-to-end stream blocking L_s, we get for all the routings of the list, the blocking of the stream from each routing until the destination, overflow path included. These blockings until the destination are used in the traffic load evaluation to take account of baulking effects.

2. Stream offered traffic to the network

The difference which may exist between the stream service bit-rate S_t and the stream switching bit-rate W_{ts} is modelled applying a peakedness factor $z_s = S_{ts}/W_{ts}$ to the traffic freshly offered by the stream.

Repeated attempts are classicaly modelled. The global blocking G_s, probability that a call intent is lost after all repetitions, is given by:

$G_s = L_s (1-\tau_s) / (1 - L_s \tau_s)$

The call rate offered to the network derives from the fresh offered call by:

$a_s = \alpha_s /(1 - \tau_s . L_s)$

The fictitious offered traffic m_s, which integrates the effects of repeated attempts is derived from the global G_s and the actual L_s end-to-end blockings. It is the traffic which must be offered to the network, without repetitions, so that the stream carried traffic is consistent with the global blocking, which include repetitions:

$m_s = v_s . (1 - G_s) / (1 - L_s)$

3. Stream traffic loads

Knowing the mean and variance of the stream offered traffic, the individual blocking and the blocking until the destination of all routing numbers of its routing sequence, the traffic offered by the stream on each routing number is calculated. Offered traffics are influenced by baulking effects but not call rates. Traffics offered by each stream to its routings are agregated to compute each routing mean offered traffic.

When the stream average overflow from a routing of the stream routing sequence is multiplied by the routing peakedness ratio $zr_{i,r,t}$, $i = 1$ or 2 depending on whether the stream is fresh or overflow on the routing, it gives the peakedness minus one of the stream overflow of the considered traffic routing.

The stream traffic parcels received by routing numbers without overflowing routing are considered as independant. Their variances are thus additionned to get the routing variance.

The stream traffic parcel received by routing numbers with an overflow number are considered as correlated to parcels overflowing from the same routing number. The summation is on the "peakedness minus one" and on the average. When all streams are studied, we can calculate, from the agregatted average and "peakedness minus one", the variance of the overflow from the overflowing routing, and to agregated it to the variance of the overflow routing.

IV. Numerical results

The algorithm has been tested on standard cases (hierarchical and fixed routings, with no repeated attempts) on a 12 nodes test network which has been elaborated during the international Eva project [Gue.88]. Its accuracy has thus been compared with the diverse one and two-moments models tested during the Eva project, in particular Manfield and Downs one moment method, under different traffic conditions. In all cases, it proved to be the most accurate for individual stream blocking: better than standard two-moments models in low-blocking situations and better than one-moment models in high-blocking situations. It also gives a reliable estimation of overall blocking and showed no systematic biais in the tested cases. Tests of the algorithm on other types of networks (with adaptive routings, repeated attempts, multi-services traffics) have only be qualitative: whether the algorithm converge correctly and gives meaningful results. The results were not compared to reference numerical results.

V. Conclusion

As described in the present paper, the general purpose model meets functionnaly its requirements. All the phenomena of interest for the analysis of an advanced circuit-switched network are modelled with what seems a priori a reasonable compromise between accuracy and complexity and have been integrated in a single framework.

The accuracy and computation effort necessary for this modelling still have to be investigated in most of the cases covered by the model. The model framework being quite modular, specific improvements may be introduced without changing the architecture of the algorithm.

VI. References

[Leb.89] M. Lebourges, D. Petit: " Sagesse: a Simulator of the Network Servicing and Provisioning Processes." 4th. International Network Planning Symposium, 3.2, Palma de Majorca. 1989

[Man.79] D.R. Manfield, T. Downs: "On the one-moment analysis of telephone traffic networks", IEEE Trans. on Com., 27, 1979

[Leb.88] M. Lebourges: " Individual Blockings on a Telephone Trunk-Group with Multiple Inputs and Trunk-Reservation Parameters." 12th International Teletraffic Congress, 2.1B.6.1, Torino 1988

[Aki.83] H. Akimaru, H. Takahashi, "An Approximate Formula for Individual Call Losses in Overflow Systems", in IEEE Trans. on Comm. Vol. Com. 31, N°6, June 1983.

[Mats.83] J. Matsumoto, Y. Watanabe, "Analysis of Individual Traffic Characteristics for Queuing Systems with Multiple Poisson and Overflow Inputs", in 10th International Teletraffic Congress, Montréal 1983.

[Wilk.56] R.I. Wilkinson, "Theories for Toll Traffic Engineering in the U.S.A.", Annex of J. Riordan, in Bell System Technical Journal, Vol. 35, N°2, 1956, pp. 421-514

[Rob.81] J.W. Roberts, "A service system with heterogeneous user requirements" in Performance of Data Communications Systems and their Applications, 1981, pp. 423-431.

[Gau.80] M.D. Gaudreau: "Recursive formulas for the calculation of point-to-point congestion", IEEE Trans. on Comm. Vol. Com-28 N°3 - pp. 313, 316 - Mars 1980

[Gue.88] J.P. Guérineau, M. Dao, C. Becque: "Otarie: an interactive tool for performance evaluation study in telephone networks" 12th International Teletraffic Congress, 3.2.B.5, Torino 1988

A dynamic non-hierarchical routing with incomplete data

Konvit, M. and Borik, M.

Department of Information Systems, University of Transport and Communications Zilina, Czecho-Slovakia

Abstract
A decision making proces of the dynamic adaptive routing (DAR) is based on the know network data (ND). The analysis of the behaviour of the DAR in abnormal situations when non or only an incomplete set of the ND is available is given in the article. It is shown that the combined use of the instantaneous number of busy trunks and the short-time average of the offered traffic load improves operation of DAR in case of failure of the ND actualisation. The problem is illustrated on the DAR algorithm described in [6]. As it follows from simulation experiment missing ND can be substituted by their estimates without substantial decrease of quality of routing process during relatively long time period.

1. Introduction

A great deal of work has been dedicated to the different aspects of the DAR techniques in the last decade [1, 2, 3, 4, 5]. At the very beginning the DAR problem has been stated as follows:
Subject to the GOS constrains, to find the DAR mechanism that ensures maximization of the traffic carried by the network.
With a new technological development (ISDN, B-ISDN) the multiservice networks with a statistical multiplexing brought new views on the problem of the DAR. As a consequence, the DAR problem is nowadays recognized in the following from:
Problem P1: Subject to the GOS constrains to find the revenue effective allocation of the network resources.
In order to share the network resources in the optimal way the actual ND have to be available for the DAR mechanism. The problem of DAR with the incomplete ND can be stated in the following form:
Problem P2:
Subject to the GOS constrains to find the allocation of the network resources which will maximize the revenue when complete ND are available and with incomplete ND will be better or at least the same as the one received by fixed alternative routing.
The problem of substitution of incomplete ND is discussed in section 2 for single service circuit switching networks. The simulation of the DAR with incomplete ND is described in section 3. The comments on the simulation results and conclusions are given in section 4.

2. Network data substitution

Let us suppose a distributive DAR. The number of links in any route is restricted to two. No restrictions are placed on the number of routes for any source-destination pair of nodes.

The traffic which is subject of routing consists of the fresh traffic offered to the network from its neighbourhood and the traffic already existing in the network. It implies the ND should be chosen in such a way that they will describe the instantaneous composition of the traffic. On the other hand the idea that the set up of individual call has only negligeble influence on the network state is commonly accepted. Then there is no need to change the management decisions after each accepted call demand. As a consequence, the separation of the traffic management and traffic control functions is possible solution. It is reported in [3] that the instant numbers of busy trunks arc the best ND from the real-time traffic control point of view. The offered traffic averages are the ND traditionally used for the traffic management. So, an instant number of busy trunks n together with 15 minut averages of the offered traffic Y are sufficient and necessary ND.

The decision about the call acceptance can be based either on the detailed knowledge of the idle/busy state of all network resources or on a suitable chosen control criterion. In our case it is the probability that during the existence of the alternate route (AR) call there will be more than one direct call blocked due to the establishment of this particular AR connection. As it is shown in [5, 6] such a criterion can be evaluated from the above mentioned ND.

Let us now consider the fragment of the network shown in Fig. 1.

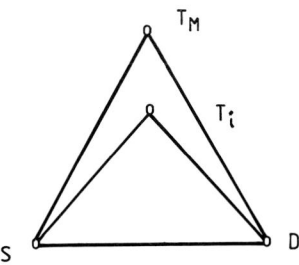

Figure 1.

If the demand on astablishing of the AR connection form the source node S to the destination node D occurs the traffic offered to the transit nodes T_i, $i = 1, \ldots, M$ and the numbers of busy trunks in the links $T_i D$ have to be know in the source node S. Due to failures some of these quantities can be unavailable.

Let t_e is the time of the last actualisation of the ND, T be the ND actualisation interval and $\Delta t = t - t_e$ be the duration of the failure of the ND actualisation. Then we can define the critical time interval Δt_k such that if $\Delta t < \Delta t_k$ the missing values is can be substituted by their estimates. Otherwise, if

Δt > Δt different approach has to be taken. It is clear that due to different character of the time series for n and Y, respectively the lenght of the Δt will be different for missing n and Y.
 There are two possible approaches to the problem of incomplete ND:
1. Solution based on the theory of Markov chain and the use of mathematical models.
2. Solution based on the simulation study.
 We will proceed in terms of the second solution. The following rules have been defined for a completition of the incomplete ND:
- if $\Delta t < \Delta t_k$ the linear Kalman filter based on know two state model [8] is used for the missing value (s) estimation. In order to overcome the problem of unknown covariance matrices Q and R two complementary equations for evaluating Q, R were added to the set of ordinary Kalman filter equations similarly as it was suggested in [7]. This way the filter produces the k steps ahead forecasts evaluated from the last know value.
- if $\Delta t > \Delta t_k$ the mean value of Y_o is used instead of true offered traffic. The number of busy trunks n is evaluated in the following way:
Let v be the random variable expressing the losses in the link in question caused by the existence of the AR connection. The set of the possible values of v is v(1), ..., v(N). Variable v takes value v(i) with probability p(i). The probability is evaluated according to the known Erlang formula. The mean value of v is

$$\bar{v} = \sum_{i=1}^{N} v(i) \cdot p(i) \tag{1}$$

This mean value is used for an estimation of missing value(s) of n. Let $w_i = \bar{v} - v_i$. Then the value of n is found from the equation

$$w_i = \min_{i} \{w_i\} \tag{2}$$

If the failure of actualisation lasts for a longer time i. e. $\Delta t \gg \Delta t_k$ the missing ND have to be obtained on the different principle, for example on the base of the traffic load values from the previous day.

4. Simulation experiment

 The revenue effective control criterion has the from [6]

$$R = \min_{p=i \ldots P} \{R_p\} \tag{3}$$

where

$$R_p = \sum_{i=1}^{m_p} h \cdot c_i^p \cdot X_i^p (Y_0, n) \qquad (4)$$

In the expression (4) P is the number of possible ARs; m_p is the number of links in the AR p; n is the number of busy trunks in the link i, (i = 1, 2) of the AR p; c_i^p is the tariff per time unit in the link i of the AR p; h is the mean duration of the overflow call. The call is allowed to overflow only on condition that

$$R_p < c \cdot h \qquad (5)$$

is found.

For the simulation of DAR we have used the model described in [6]. The aim of the experiment was to compare the behavior of the different types of routing in situation when there are incomplete ND. The simulations were carried over an interval where statistically significant influence of the routing method on the network revenue losses was expected. Over this interval 12 independent simulation experiments with step 0,5 erl were carried out. In order to avoid a dependance of the results on the chosen tariffs the tariff matrix was changed at random.

4.1. The results of simulation

The results of simulation for direct routing, fixed adaptive routing, revenue-effective adaptive routing and the routing with incomplete ND are shown in Fig. 2 and Fig. 3.

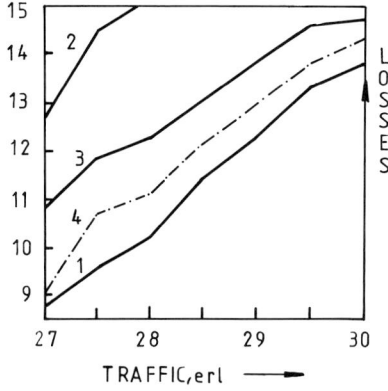

Figure 2. A comparison of different routing methods

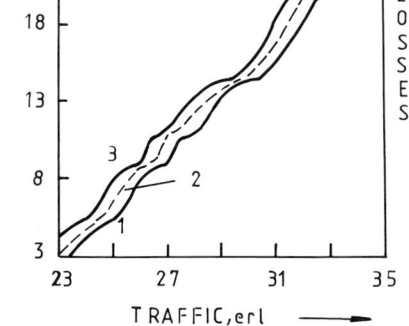

Figure 3. The DAR with incomplete network data

Figure 2 shows the traffic loss curves versus the traffic offered to the network. The curve No. 1 represents performance of the intercity network controlled by the revenue-effective adaptive routing algorithm. Curve No 2 represents performance of the same intercity network controlled by the direct routing. The curve No. 3 belongs to the fixed alternative routing. Finally, the curve No. 4 describes the revenue-effective adaptive routing with incomplete ND in case when the number of busy trunks is unknown. Figure 3 shows the behavior of the DAR for different number of unknown values of the offered traffic load.

5. Conclusions

The DAR ensures effective use of all network resources especially in cases when proper steps against inefective use of the AR are taken. In this paper we have described the behaviour of the revenue-effective adaptive routing algorithm in situations when the failures of the ND actualisation have occured. The simulation results proved the algorithm is well behaving with insufficient ND for relatively long time (of order of tens minutes). It has been proved the critical time interval Δt has not the constant lenght but depends on the shape and the amount of traffic.

6. References

1 European Regional Project for Development of International telecommunication. "Workshop type seminar on inteligent routing strategies". Zruc/Czecho-Slovakia, April 1986
2 Ash G. R., Kafker A. H., Krishnan K. R.: "Servicing and Real-Time Control of Networks with Dynamic Routing ", Bell Syst. Tech. Journal, vol. 60 n 8 1981
3 Chemouil P., Filipiak J.: "Kalman Filtering of Traffic Fluctuations for Real-Time Network Management", 4th IFAC/IFORS Symp on Large Scale Systems, Zürich 1986
4 Akinpelu J.M.: "The Overload Performance of Engineered Networks with Non-hierachical and Hierachical Routing", Prac. of Int. Teletraf. Conf., ITC 10 Montreal 1985
5 Blunar K., Konvit M.: An adaptive Routing Algorithm for Telephone Networks, ITC 12, Torino 1988
6. Borik M. j.: Traffic routing algorithm for intercity progressive rate telephone network, in print
7 Knottnerus P.: Forecasting: Kalman filtering and prediction intervals, ITC 12, Torino 1988
8 Chui C.K., Chen G.: Kalman filtering with real-time applications, Springer Verlag, Berlin 1987

End-to-End Blocking in an Integrated Services Network with Link Capacity Allocation Control

Hideaki YOSHINO[†] and Yoko HOSHIAI

NTT Communication Switching Laboratories,
3-9-11, Midori-cho, Musashino-shi, Tokyo 180, Japan

Abstract

This paper examines the end-to-end blocking performance in an integrated services nonhierarchical network with alternate routing. First, we show imbalance and bistable behavior of individual end-to-end blocking probabilities without control for a multi-class network model. To prevent the imbalance and suppress the instability, link capacity allocation control is introduced. The effects of this control on a nonhierarchical alternate routing network are clarified by approximate analysis for various traffic mixes and allocation control parameters.

1. INTRODUCTION

B-ISDN will allow the integration of various bandwidth services such as voice, data, and video. Since a service requiring a large bandwidth is more frequently blocked than one requiring a small bandwidth, various capacity allocation controls such as reservation control and class limitation control, have been studied for single-link models. The intent with this study is to extend the control to network models, and to clarify the end-to-end blocking performance of multi-class alternate routing networks.

For single-class nonhierarchical networks with alternate routing, Nakagome and Mori[1] carried out approximation analyses and revealed the existence of network instabilities. That is, for certain loads the network operates in two states : a low network blocking state and a congested state. This bistable behavior causes a high blocking rate during overload. Later, Krupp[2] showed that the bistable behavior can be stabilized by reserving a small number of trunks for first-routed calls. The effect of the trunk reservation for more general nonsymmetric networks can be found in [3] and [4].

In this paper, first, we show that the imbalance of individual end-to-end blocking probabilities and network instabilities exist in multi-class nonhierarchical networks with alternate routing, where the class of calls is differentiated by the required bandwidth and the offered load. To prevent the imbalance and suppress the instability, a link capacity allocation control, which is a combination of reservation control for wideband calls

---

[†] Currently with Institute of Communications Switching and Data Technics, University of Stuttgart, FRG

and for first-routed calls, is introduced. End-to-end blocking probabilities and bistable behavior in a nonhierarchical alternate routing network are clarified by approximate analysis for various traffic mixes and allocation control parameters. In addition, the setting of allocation control parameters to suppress and to ensure maximum useful network throughput under various traffic conditions is discussed.

2. THE MODEL

2.1 The network model

Fig.1 shows an example of the network model used in this study. The model is assumed to be a fully connected, symmetric, and nonhierarchical network with N nodes. The link capacity of every one-way link is assumed to be the same. And the offered load to each node-pair is assumed to be the same. The routing procedure is fixed alternate routing. That is, every fresh call attempts the direct link first. And if the link is busy, the call will attempt the alternate path up to m times. The calls overflowing the mth alternate path are lost and cleared. The handover number of the alternate paths is assumed to be 2 (all the alternate paths with length 2 are allowable). We use the following notations :

K : number of traffic classes
L : link capacity
$w^{(k)}$: bandwidth to serve a k-th class call
$a_f^{(k)}$: offered load of the k-th class first-routed calls for each link.

2.2 Link capacity allocation control

Under the link capacity allocation control, allowable capacity in each link is predefined for first- and alternate-routed call for each class (see Fig. 2). The link capacity allocation control is described as following :

Link Capacity Allocation Control : The k-th class first-routed call arriving upon link l is accepted iff [total capacity in use in link l] $\leq c_f^{(k)} - w^{(k)}$, similarly, the k-th class alternate-routed call arriving upon link l is accepted iff [total capacity in use in link l] $\leq c_f^{(k)} - w^{(k)}$, where

$c_f^{(k)}$: allocated capacity for k-class first-routed calls
$c_o^{(k)}$: allocated capacity for k-class alternate-routed calls

This control scheme can be considered as a generalization of **trunk reservation** control for both wideband calls and first-routed calls.

With this scheme, the following effects can be achieved.
(1) Balancing of blocking probabilities for different traffic classes by reducing the allowable capacity for narrowband calls. (The effect of trunk reservation for wideband calls).
(2) Eliminating instabilities due to alternate routing by reducing the allowable capacity for alternate-routed calls. (The effect of trunk reservation for first-routed calls).

3. APPROXIMATION ANALYSIS

To simplify the analysis, we assume that

(A1) The arrival processes of both first- and alternate-routed calls are assumed to be Poisson and call holding time is to be exponentialy distributed,

(A2) The control times to connect and to disconnect calls are assumed to be very short compared to the call holding times and are negligible.

The end-to-end blocking probability, $B_E^{(k)}$ for class-k calls can be found by solving two nonlinear equations expressed with the following variables simultaneously.

(1) $a_o^{(k)}$: offered load of class-k alternate-routed calls to each link,

(2) $B_f^{(k)}(B_o^{(k)})$: link blocking probability for class-k first- (alternate-) routed calls.

In the following we derive the equations on the above variables.

The offered load of alternate-routed calls is considered to be the sum of 1st through mth alternately routed traffic as shown in Fig.3. The offered load of the ith alternately routed traffic can be defined as the amount of traffic overflowing the $(i-1)$st alternate path and offered to the ith alternate path. Note that an overflowing call is offered to the link only if the other link of a two-link alternate path is free, i.e. with probability $(1 - B_o^{(k)})$ for class-k calls. The offered load of class-k alternate-routed calls is thus given by

$$\begin{aligned}
a_o^{(k)} &= 2a_f^{(k)} B_f^{(k)}(1 - B_o^{(k)}) \\
&+ 2a_f^{(k)} B_f^{(k)}\{1 - (1 - B_o^{(k)})^2\}(1 - B_o^{(k)}) \\
&+ 2a_f^{(k)} B_f^{(k)}\{1 - (1 - B_o^{(k)})^2\}^2(1 - B_o^{(k)}) \\
&+ \cdots \\
&+ 2a_f^{(k)} B_f^{(k)}\{1 - (1 - B_o^{(k)})^2\}^{m-1}(1 - B_o^{(k)}) \\
&= \frac{2B_f^{(k)}}{1 - B_o^{(k)}}[1 - \{1 - (1 - B_o^{(k)})^2\}^m] a_f^{(k)}.
\end{aligned} \quad (1)$$

Since we assumed that the overflow traffic is also Poisson, the link blocking probability can be derived using the heterogeneous state-dependent input loss model with trunk reservation. In the model, the exact analysis requires solving the system of balance equations of a huge state space. Therefore we use the approximation based on the following equation [5].

$$nQ(n) = \sum_{k=1}^{2K} a^{(k)} S_k(n - w^{(k)}) Q(n - w^{(k)}), \quad n = 1, 2, \cdots, L. \quad (2)$$

where $Q(n)$ is the probability that n servers are busy (the link capacity, L is assumed to be integer), and

$$S_k(n) = \begin{cases} w^{(k)}, & n < c^{(k)} - w^{(k)} \\ 0, & otherwise, \end{cases} \quad (3)$$

$$a^{(k)} = \begin{cases} a_f^{(k)}, & k = 1, 2, \cdots, K \\ a_o^{(k)}, & k = K+1, K+2, \cdots, 2K, \end{cases} \quad (4)$$

$$c^{(k)} = \begin{cases} c_f^{(k)}, & k = 1, 2, \cdots, K \\ c_o^{(k)}, & k = K+1, K+2, \cdots, 2K, \end{cases} \quad (5)$$

$$w^{(K+k)} = w^{(k)}, \quad k = 1, 2, \cdots, K. \quad (6)$$

This approximation is very good not only for the case of equal holding times [5], but also for the case of different holding times [6].

The distribution $Q(n)$ yields the link blocking probability for each class first- and alternate-routed calls as follows.

$$B_f^{(k)} = \sum_{n=c_f^{(k)}-w^{(k)}+1}^{L} Q(n), \quad k = 1, 2, \cdot, K$$

$$B_o^{(k)} = \sum_{n=c_o^{(k)}-w^{(k)}+1}^{L} Q(n), \quad k = 1, 2, \cdot, K \quad (7)$$

After solving equations (1) and (7) simultaneously, the end-to-end blocking probability for class-k calls is given by

$$B_E^{(k)} = B_f^{(k)}\{1 - (1 - B_o^{(k)})^2\}^m. \quad (8)$$

4. VALIDATION

In this section, we evaluate the approximation by means of comparisons with simulation results under the following conditions,

$L = 6.144$ [Mb/s] : Link Capacity
$m = 5$: number of alternate routes
$K = 2$: number of classes (wideband and narrowband calls)
$w^{(1)} = 512$ [Kb/s] : Bandwidth of wideband calls
$w^{(2)} = 32$ [Kb/s] : Bandwidth of narrowband calls
$A = a_f^{(1)}w^{(1)} + a_f^{(2)}w^{(2)}$ [erl · Mb/s] : Total offered load
$a_f^{(1)}w^{(1)} : a_f^{(2)}w^{(2)} = 1 : 1$
Allocated link capacity : $c_f^{(1)} = L$, $c_o^{(1)} = L - R_d$,
$c_f^{(2)} = L - (w^{(1)} - w^{(2)})$, $c_o^{(2)} = L - R_d - (w^{(1)} - w^{(2)})$,

where R_d represents the reservation capacity for first-routed calls (see Sect.5).

Comparisons with simulation results for the end-to-end blocking probabilities are displayed in Fig.4. In this experiment, the reservation capacity for first-routed calls was varied from 0.0 to L (6.144 [Mb/s]) for $A = 4.2, 4.4$, and 5.0.

The results indicate that the approximation predicts the end-to-end blocking probabilities in multi-class alternate routing networks with reasonable accuracy. And we can see that the approximation captures the effect of control parameters on the network performance. The accuracy of the approximation relys on the following assumptions,

(1) Poissonian assumption of the arrival processes of alternate-routed calls,

(2) Independency assumption of call blocking between links in the network,

(3) Adoption of the Roberts' approximation.

5. NUMERICAL RESULTS

In this section, we describe the numerical results and clarify the end-to-end blocking probabilities in a multi-class nonhierarchical alternate routing network for various traffic mixes and control parameters.

The following conditions and notations are used in this section.

$k = 2$, $w^{(2)} = 32[Kb/s]$,
$m = 5$ (7 node symmetric network),
$r_a = a_f^{(1)} w^{(1)} / a_f^{(2)} w^{(2)}$: offered load ratio,
$r_w = w^{(1)}/w^{(2)}$: bandwidth ratio,
$A_f = (a_f^{(1)} w^{(1)} + a_f^{(2)} w^{(2)})/L$: normalized total offered load.

5.1 Results without controls

To begin with, we describe the end-to-end blocking probabilities of a multi-class alternate routing network without controls. The results are obtained by the approximation described in Sect. 4 setting the control parameters for each class call as follows,

$$c_f^{(1)} = c_f^{(2)} = c_o^{(1)} = c_o^{(2)} = L$$

Fig.5 shows the end-to-end blocking probabilities for each class versus the normalized total offered load, A_f, with $L = 156$ [Mb/s], $r_a = 1$, and $r_w = 64$.

It is clearly shown that the wideband calls experience the higher call blocking. Furthermore, the bistable blocking behavior is observed as in the case of single class alternate routing network [1],[2]. The bistable region of both classes occures at the same value of offered load. This type of network instabilities causes a high blocking especially for wideband calls under heavy traffic conditions.

5.2 Equalized blocking

We now apply the link capacity allocation control to balance the end-to-end blocking probabilities for different call classes. A control equalizing the blocking probabilities for each class is considered here. This control can be achieved by setting the control parameters as follows,

$$c_f^{(1)} = c_o^{(1)} = L,$$
$$c_f^{(2)} = c_o^{(2)} = L - (w^{(1)} - w^{(2)}).$$

Fig.6 plots the equalized end-to-end blocking probabilities under the same traffic conditions of Fig.5.

It can be seen that the equalization of blocking probabilities for different call classes can be achieved by this control. However, the control does not improve the blocking performance for wideband calls. What is worse is that the range of the bistable region becomes wider under the control.

5.3 Stabilization of end-to-end blocking

For single-class alternate routing networks, Krupp[2] and Akinpelu[3] showed that the instability behavior disappeared and the blocking performance was improved by using trunk reservation control for first-routed calls. We now examine the effect of trunk reservation for first-routed calls on the performance of multi-class routing networks. A control equalizing the blocking for each class and reserving link capacity for first-routed calls are considered here. Specifically, we set

$$c_f^{(1)} = L$$
$$c_o^{(1)} = L - R_d$$
$$c_f^{(2)} = L - (w^{(1)} - w^{(2)})$$
$$c_o^{(2)} = L - R_d - (w^{(1)} - w^{(2)}),$$

where R_d represents a reservation capacity for first-routed calls.

Fig.7 shows the end-to-end blocking probabilities with this control with L=156 [Mb/s], r_a=1, and r_w=64. In this figure, five curves are plotted with $R_d = nw^{(1)}$ (n=0, 0.5, 1, 2, and 3). It is clearly shown that the instability behavior seen without controls disappears with the use of the reservation control for first-routed calls. The upper bound of bistable region is not sensitive to R_d. In the case that offered load is lower than the upper bound, the blocking probability increases as R_d increases. On the other hand, if the traffic is higher than the bound, the blocking probability decreases as R_d increases. Therefore, the reservation control for first-routed calls turn out to yield a smooth and gradual increase in the end-to-end blocking with the traffic load.

Fig.8 shows the end-to-end blocking probabilities plotted as a function of the parameter r_a with L=156 [Mb/s], r_w=64, R_d=6.144 [Mb/s], and A_f=0.9, 0.92, 0.95, and 1.0. We can see that the end-to-end blocking probability is insensitive to the offered load ratio (traffic mix) if B_E is between 10^{-2} to 10^{-1} which can be thought as a typical GOS.

Fig.9 shows the end-to-end blocking probability plotted as a function of the parameter r_w with L=156 [Mb/s], r_a=1.0, R_d=10 [Mb/s], and A_f=0.95, 0.98, and 1.0. Again, the network performance is insensitive to the bandwidth ratio if B_E is between 10^{-2} to 10^{-1}.

5.4 Setting of control parameters

We now discuss the setting problem of link capacity allocation parameters which maximizes network's total carried traffic. Unfortunately, it is quite difficult to solve the optimization problem directly because the end-to-end blocking probabilities are non-linear function of offered load, and total offered load (first- and alternate-routed traffic) depends on the allocated capacity.

Here we explore the setting problem of reservation capacities for first-routed calls under the condition that end-to-end GOS constraint is the same for all call classes. Fig.10 shows the end-to-end blocking probabilities versus the reservation capacity for first-routed calls, R_d, with L=156 [Mb/s], r_a=1.0, r_w=64, and A_f=0.85, 0.90, and 0.95. Notice that the values of $R_d = 0$ and $R_d = L(156[Mb/s])$ correspond to the end-to-end blocking probabilities of alternate and nonalternate routing, respectively. It can be seen that there exists an optimal reservation parameter between $R_d = 0$ and $R_d = L$ for each offered load. The difference between the optimal value and the value of $R_d = L$ can be considered as *routing gain*. Therefore, R_d must be small enough so that the routing gain can be achieved and R_d must be large enough so that bistable behavior will not appear. Numerical experiment with various traffic parameters ($r_a = 0.1 \sim 10$, $r_w = 1 \sim 100$) shows that if R_d is set to be 2 or 3 times of $w^{(1)}$, no bistable behavior was found and the routing gain was achieved efficiently.

6. SUMMARY

We have examined the end-to-end performance of a multi-class alternate routing network. Results obtained indicate that, without control, imbalance and instabilities of individual blocking probabilities occur for certain offered load. We have applied a control method named link capacity allocation control to the network model to prevent the imbalance and suppress the instabilities. The effectiveness of this control has been demonstrated by approximate analysis of some specific network examples. We have also discussed the setting of control parameters to ensure maximum useful network throughput under various traffic conditions. We recommend that such a control be considered in the introduction of integrated services networks with alternate routing.

ACKNOWLEDGMENT

The authors wish to thank Prof. P. Kühn for assistance in preparing this paper.

REFERENCES

[1] Y. Nakagome and H. Mori, "Flexible routing in the global communication network," Proc. ITC7, 426.1-8 (1976).

[2] R.S. Krupp, "Stabilization of alternate routing network," Proc. ICC'82, 31.2.1-5 (1982).

[3] J.M. Akinpelu, "The overload performance of engineered networks with nonhierarchical and hierarchical routing," Proc. ITC10, 3.2.4 (1982).
[4] T.G. Yum and M. Schwartz, "Comparison of routing procedures for circuit-switched traffic in nonhierarchical networks," IEEE Trans. Commun., **COM-35**, 5, pp.535-544 (1987).
[5] J.W. Roberts, "Teletraffic models for the telecom 1 integrated services network," Proc. ITC10, 1.1.2 (1982).
[6] M.Pióro, J. Lubacz, and U. Körner, "Traffic engineering problems in multiservice circuit switched networks," 11.3, ITC Specialist Seminar, Adelaide (1989)

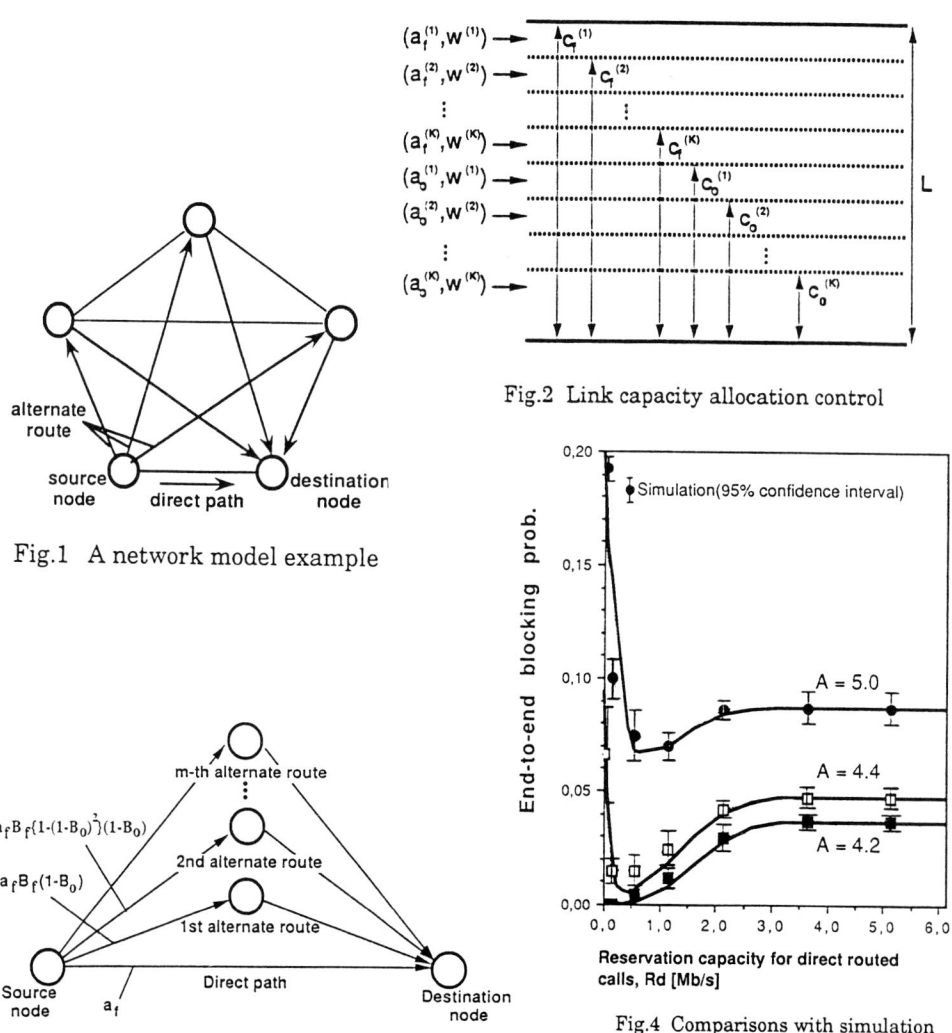

Fig.1 A network model example

Fig.2 Link capacity allocation control

Fig.3 Offered load to each link

Fig.4 Comparisons with simulation resultsfor the equalized end-to-end blocking probabilities

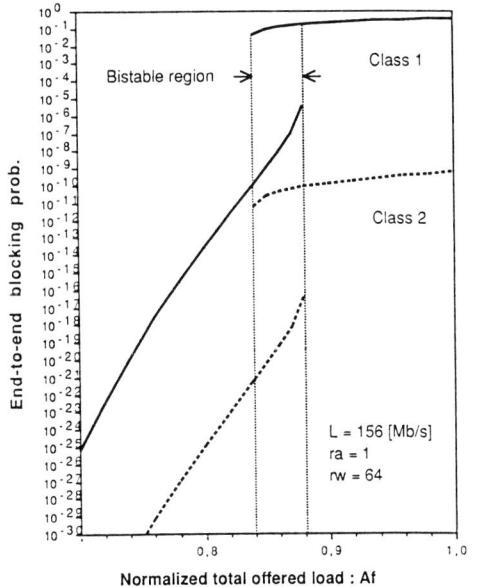

Fig.5 End-to-end blocking probabilities without controls

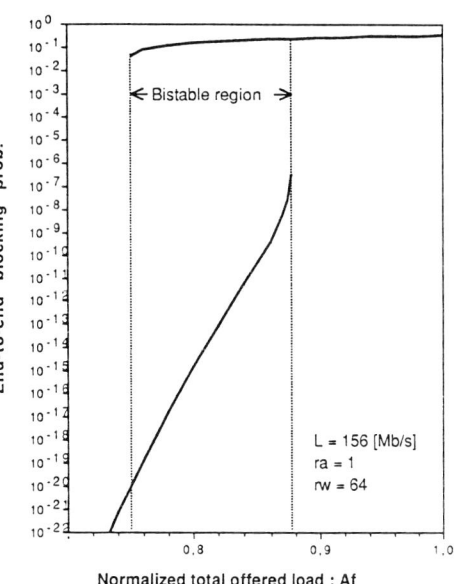

Fig.6 Equalized blocking probabilities

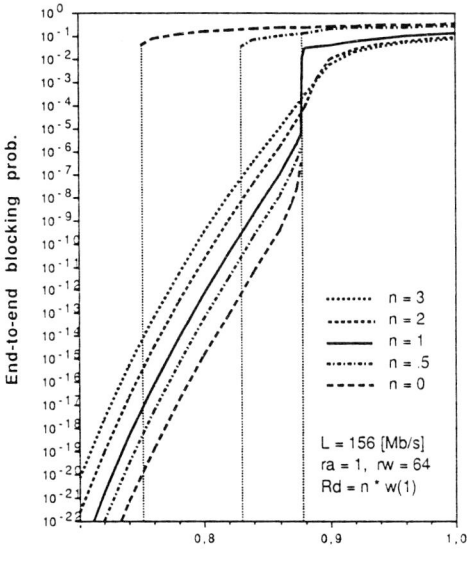

Fig. 7 End-to-end blocking probabilities of reservation control for first-routed calls

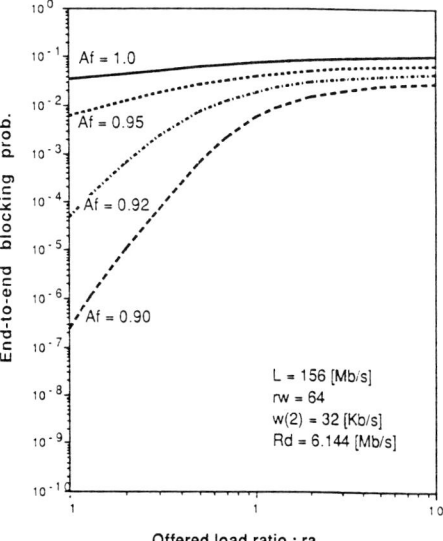

Fig.8 End-to-end blocking probabilities versus offered load ratio

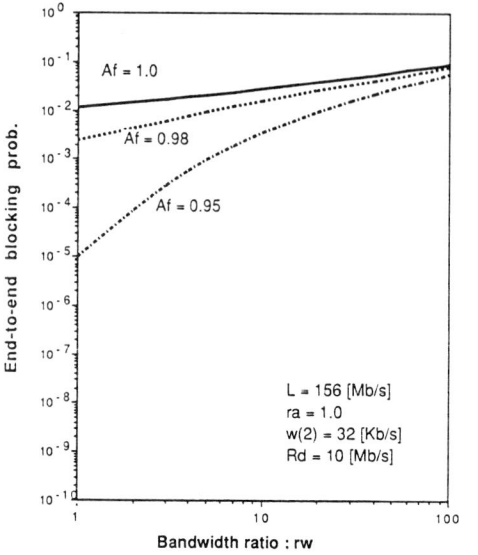

Fig.9 End-to-end blocking probabilities versus bandwidth ratio

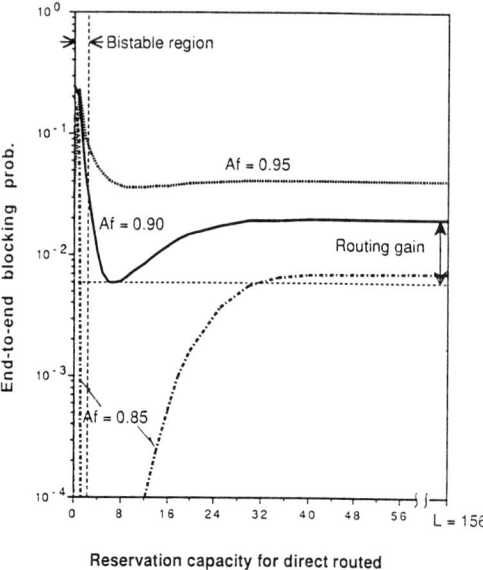

Fig.10 End-to-end blocking probabilities versus reservation capacity for first-routed calls

Dynamic adaptive routing algorithm for intercity progressive rate telephone network

M.J. Borik and M. Konvit

Department of Information Systems, University of Transport and Communication, Zilina, Czech and Slovak Federative Republik

Abstract
A new routing algorithm for the intercity progresive rate network is derived in this article. The aim is to minimize the losses of revenue. The mean value of losses caused due to rejection of some overflow is compared with the mean value of losses caused by the existence of this overflow call.

1. INTRODUCTION

Traffic routing techniques in circuit-switched networks (CSTN) can be generally divided into folowing cathegories:
- fixed alternative routing,
- dynamic adaptive routing.

According to [1],[2],[3] we can find the adaptive routing be the most proper control technique for relatively heavy traffic load offered to the network. The adaptive control of network flows maximalizes the carried traffic by using of all the network capacities in more efficient way. Numerous algorithms for the adaptive routing have been published till today [1],[2],...,[15]. The works of Ash [4],[5] are the basic works in this field where the concept of non-hierarchical routing and the network status map was introduced into network control. Narendra [6] aplied learning automata to telephone traffic routing and control. Akinpelu and Haensche [7],[8] have described an overload performance of network with adaptive routing. Pioro [9] has worked out method of multihour optimization. Chemouil and Filipiak [10] sugested to use the future network state forecast to improve real time adaptive routing efficiency. Various cathegories of network control methods are shown also in [11],[12]. The works [2],[13],[3] compare the performance of several network routing and control methods.
All control algorithms mentioned above use node or network state information in present time to route the traffic but do not take into account the influence of decision to the possible future evolution of the situation in the network. A small attention is devoted to network revenue in these works,[14] is an exception.

Therefore the aim of this article is to derive dynamic adaptive non-hierarchical routing algorithm that could provide maximal network revenue. The algorithm need know the values of the offered traffic and numbers of busy trunks in all network nodes. These state information were chosen as the network state representants because they allow quickly and exactly react to the network traffic changes [15].

Mathematical model is created and routing algorithm is derived in Chapter 2. A simulation model of the network is described in Chapter 3. The results are discused in Chapter 4.

2. MATHEMATICAL MODEL OF NONHIERARCHICAL NETWORK

2.1. General

Let us suppose that following assumptions are valid for the non-hierarchical communication network :
- the network is in a statistical equilibrium,
- the offered calls form a Poissonian input stream,
- holding times are independent and negative exponentially distributed with unit mean,
- dedicated data links or CCITT No 7 subnetwork is available,
- a computer is used in each SPC node to work out an optimal routing.

The goal si to maximalize network revenues. The main task is to establish an alternate route call so that minimize the losses of revenue caused due to potencial rejections of direct calls.

2.2. Formulation of the problem.

The offered call is accepted if there is a free trunk in the direct group. Else the demand is rejected or an overflow is established. The overflow route consist of M links. Let us consider the overflow link j be a trunk group of N_j trunks ($j=1,..,M$). Let c_j be the tariff per time unit for calls in the link j of the overflow route and h be the mean duration of overflow call. Let the random variable zo_j denotes the total number of unsuccessfull calls in the overflow link j. Let the random variable z_j denotes the number of unsuccessfull calls in this link caused due to the use of this link for the overflow call. Let the random variable u_j denotes losses of revenue in this link caused due to the use of this link for the overflow call. Let the random variable u denotes total losses of revenue in all links of the overflow route caused due to the use of those links for the overflow call. Let $P(z_j=k)$ be the probability fuction of the random variable z_j for $j=1,..,M$

$P(z_j=0) = P(zo_j=0)$

$P(z_j=1) = P(zo_j \geq 0)$ \hfill (1)

The random variable z_j has binomical distribution with possible set of values $\{1,0\}$. Let us try to express the mean value $E(u)$ of the random variable u :

$$E(u) = E(\sum_{j=1}^{M} u_j) = \ldots = \sum_{j=1}^{m} h \cdot c_j \cdot \sum_{k=0}^{1} k \cdot P(z_j=k) = \sum_{j=1}^{M} h \cdot c_j \cdot P(z_j=1) \quad (2)$$

To evaluate $E(u)$ we need calculate $P(z_j=1)$ e.g. probability of exactly one lossed call due to the use of the link for overflow call in each of M links of overflow route.

2.3. State model.

Let N denotes the number of trunks in the group which is offered Poissonian stream of demands with intensity y. Let us assume just one trunk in the group be busy due to the use of this link for overflow route.

Definitions:
- the system is in the state n when n trunks are busy in the group for $n = 0,1,\ldots,N-1$.
- the system is in the state N when N-1 trunks are busy in the group and a new demand is comming. This one is rejected due to the existence the overflow call in the grup.
- random variable T_y denotes the time to the arrivall of the next demand offered to the link and F_{Ty} is its distribution function.
- random variable H denotes the duratin of the call with unit mean and F_H is its distribution function..
- random variable T_n ($n = 0,1,\ldots,N$) denotes the time to the next call departure in the link, when the system is in the state n, and F_{Tn} is its distribution function.
- random variable S_n ($n = 0,1,\ldots,N$) denotes the time of the transition from the state n to the state N. F_{Sn} denotes its distribution function.
- random variable $T_{i,j}$ ($i=0,1,\ldots,N-1; j=0,1,\ldots,N;$) denotes the time interval when the system is in the state i on condition that in the end of this period the system reaches the state j, and F_{Tij} is its distribution function.

It can be shown that T_y and T_n are independent negative exponentially distributed with the mean y and n respectively:

$$F_{Ty}(x) = 1 - e^{-yx} \quad (3)$$

$$F_{Tn}(x) = 1 - e^{-nx} \quad (4)$$

There are only two possible state changes. The system in the state n can reach the state n-1 (some departure of call) e.g. $T_n < T_y$ or the state n+1 (new demand is comming) e.g. $T_y \leq T_n$. No other state can be reached from the state n by one step. Let us express the probabilities of the above changes by using the theory of measure and integral [16].

$$P(T_y < T_n) = \int_0^\infty F_{Ty}(x) dF_{Tn}(x) = \int_0^\infty (1-e^{-yx}) d(1-e^{-nx}) = \ldots = \frac{y}{y+n} \quad (5)$$

$$P(T_n<T_y) = \int_0^\infty F_{Tn}(x)dF_{Ty}(x) = \int_0^\infty (1-e^{-nx})d(1-e^{-yx}) = \ldots = \frac{n}{y+n} \quad (6)$$

Let us describe the system using the results (5) and (6) and an assumption that random variables S_j and $T_{i,j}$ are independent.

$S_0 = T_{01} + S_1$

$S_n = T_{n,n-1} + S_{n-1}$ with probability $\dfrac{n}{y+n}$

$S_n = T_{n,n+1} + S_{n+1}$ with probability $\dfrac{y}{y+n}$ (7)

$n = 1,\ldots,N-1$

Let us compute distribution of random variable $T_{i,j}$

$F_{T0,1}(x) = P(T_{0,1} \leq x) = P(T_y \leq x) = 1 - e^{-yx}$

$F_{Tn,n-1}(x) = P(T_n \leq x\ /\ T_n < T_y) = \dfrac{y+n}{n} \cdot P(T_n \leq x, T_n < T_y) =$
$= 1-e^{-x \cdot (y+n)}$ (8)

$F_{Tn,n+1}(x) = P(T_y \leq x\ /\ T_y < T_n) = \dfrac{y+n}{n} \cdot P(T_y \leq x, T_y < T_n) =$
$= 1-e^{-x \cdot (y+n)}$ (9)

According to the results (8) and (9) we can express distribution of the random variable $T_{i,j}$

$F_{Ti,j}(x) = 1-e^{-x \cdot (y+i)}$ (10)

Let us compute the distribution of the random variable S_n ($n = 0,1,\ldots,N$) using results (7) and (10).

$P(S_n \leq x) = \dfrac{n}{y+n} \cdot P(T_{n,n-1}+S_{n-1} \leq x) + \dfrac{y}{y+n} \cdot P(T_{n,n+1}+S_{n+1} \leq x)$ (11)

for $n=1,\ldots,N-1$

$P(S_0 \leq x) = P(T_{0,1} + S_1 < x)$

Let us apply Laplace-Stielties transformation to (11).

$$\int_0^\infty e^{-sx}dP(S_n \leq x) =$$

$$= \int_0^\infty e^{-sx} d[\frac{n}{y+n} \cdot P(T_{n,n-1}+S_{n-1} \leq x) + \frac{y}{y+n} \cdot P(T_{n,n+1}+S_{n+1} \leq x)] =$$

$$= \frac{n}{y+n} \cdot \int_0^\infty e^{-sx} dP(T_{n,n-1}+S_{n-1} \leq x) + \frac{n}{y+n} \cdot \int_0^\infty e^{-sx} dP(T_{n,n+1}+S_{n+1} \leq x)$$

It can be shown that the folowing equations are valid:

$$\int_0^\infty e^{-sx} dF_{sn}(x) = \frac{n+y}{s+n+y} \cdot \int_0^\infty e^{-sx} dF_{sn-1}(x) + \frac{n+y}{s+n+y} \cdot \int_0^\infty e^{-sx} dF_{sn+1}(x)$$

$$\int_0^\infty e^{-sx} dF_{s0}(x) = \frac{y}{s+y} \cdot \int_0^\infty dF_{s1}(x)$$

$$\int_0^\infty e^{-sx} dF_{sN}(x) = 1 \qquad (12)$$

Let us define

$$X_n(s) = \int_0^\infty e^{-sx} dF_{sn}(x)$$

Then the following system of the equations is valid:

$$X_0(s) = \frac{y}{s+y} \cdot X_1(s)$$

$$X_n(s) = \frac{n}{s+n+y} \cdot X_{n-1}(s) + \frac{y}{s+n+y} \cdot X_{n+1}(s)$$

$$X_{N-1}(s) = \frac{N-1}{s+N-1+y} \cdot X_{N-2}(s) + \frac{y}{s+N-1+y} \qquad (13)$$

2.4. The losses caused due to an overflow call in an overflow link.

Now we need to express the probability that the time to a first refused call (denoted S_n) is less then the overflow call duration (H).

$$P(S_n<H)=1-P(S_n>H)=1-\int_0^\infty F_H(x)dF_{sn}(x)=1-\int_0^\infty (1-e^{-x})dF_{sn}(x)=X_n(1) \quad (14)$$

Let us define

$$X_n = X_n(1) \quad (15)$$

In fact, this is the probability of at least one lost call in the overflow link during the overflow call. According (1) it is the probability of exactly one lost call in the overflow link during an overflow call caused by establishing of the overflow call when n links are busy. Probabilities X_n (n=0,1,..,N-1) can be computed from the system of linear equations, which we can express from (13) for s=1:

$$X_0 = \frac{y}{1+y} \cdot X_1$$

$$X_n = \frac{n}{1+n+y} \cdot X_{n-1} + \frac{y}{1+n+y} \cdot X_{n+1}$$

$$X_{N-1} = \frac{N-1}{N+y} \cdot X_{N-2} + \frac{y}{N+y} \quad (16)$$

2.5. Mean losses of revenues.

Let use define X_{nj} (j=1,..,M) be the variable X_n and n_j be the number of busy trunks in link j of an overflow route. According (14) and (15) the following equation is valid:

$$P(z_j=1) = X_{nj}$$

So X_{nj} is the probability needed for computing (2). We can write

$$E(u) = \sum_{j=1}^{m} h \cdot c_j \cdot X_{nj} \quad (17)$$

2.6. The traffic routing algorithm.

The algorithm is based on evaluation of the characteristic values R_p of all trunk groups of all possible overflow routes. Let us define

$$R_p = \sum_{i=1}^{m_p} h \cdot c_i^p \cdot Xn_i^p; \quad p = 1,2,..,P \quad (18)$$

where P is number of possible alternative routes,
       m_p is number of links in an alternative route p,
       n_i^p is number of busy trunks in the link i of an alternative route p,
       Xn_i^p is the probability of exactly one lost call in the link i of an alternative route p during an overflow call caused due to establishing the overflow call when n_i^p links are busy. It is computed from (16).
       c_i^p is the tariff per time unit for calls in the link i of an alternative route p,
       h is the mean duration of an overflow call,
Let us define criterion

$$R = \min_{p=1,\ldots,P} \{ R_p \} \qquad (19)$$

The rules for taking decision on the use of particular alternative route are:
1. When M=0 or $R \geq c.h$ no alternative route is chosen and call is rejected.
2. When $M \geq 0$ and $R < c.h$ such an index p is found that $R_p = R$ is valid. The alternative route p is chosen for call establishment, where c is the tariff per time unit for direct calls.

3. SIMULATION MODEL OF NON-HIERARCHICAL NETWORK

A simulation model of fully connected symetrical network has been written in *Simula* language on the IBM AT type personal computer. The aim of an experiment has been to compare the different types of routing controll mechanizm.

3.1. The model description

We have used the symetrical 5-node network in our study. A trunk matrix, alternate path matrices, tariff matrix, offered load matrices, a matrix of random generator seed values for the call arrival process generation and other random generator seed values have been defined.
  The simulations were carried out with the above inputs to determine an interval of interest on the offered load axis first. An interval between 21 erl and 36 erl proved to be the very interval of the offered load where a statisticaly significant influence of the routing method on the network revenue losses can be expected. In this case this interval we carried 12 independent simulation experiments with step 0.5 erl. We had to change the tariff matrix to avoid a dependance of our results on chosen tariffs. It was changed at random. Obtained mean values of network losses of revenues in percentage for certain average offered traffic loads and certain random generator seed value for adaptive routing method and for direct routing method resulted from simulation model are compared below.

3.2. Experiment results discusion.

Figure 1 shows the traffic loss curves versus the mean network offered traffic. Curve No 1 represents performance of the intercity network controlled by algorithm derived above in chapter 1. Curve No 2 represents the performance of the same intercity network controlled by the direct routing. Hence we made statistical calculations to prove significance of the difference between these two routing methods. Figure 2 depicts the lowest significance level where the hypothesis on the equality of the mean values can be rejected. As it follows from this figure we can (at least on the significance level α =0.02) consider this adaptive routing to be statistically significanly better than direct routing over the interval from 23 erl to 31 erl. We can even consider this adaptive routing to be better on the significance level α = 0.25 over all intervals of the offered load.

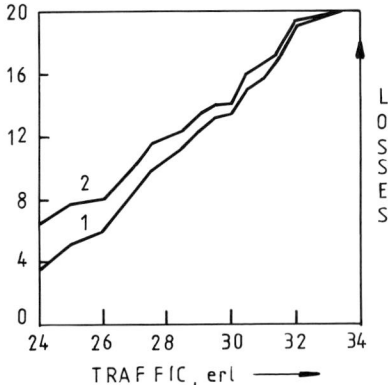
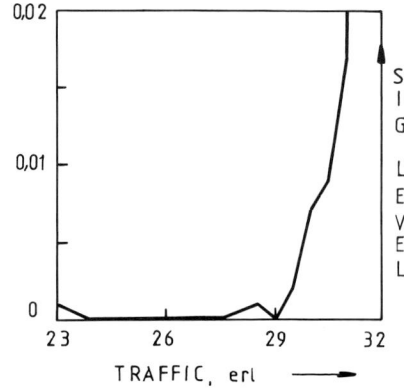

Figure 1. The losses revenue versus offered traffic (adaptive and direct routing)

Figure 2. Significance level versus offered traffic (adaptive and direct routing)

4. CONCLUSIONS

It is commonly known that alternative routing is more effcient than a direct routing only for some (lower) traffic load. In this paper we have described an alternative routing algorithm suitable for intercity networks or,when using constant tariffs for metropolitan networks, too.The results based on simulation studyies here and in [3] show that it proves the same advatages as dynamic alternate non-hierarchical routing with adaptive blocking described in [3]. It means that the dynamic alternate routing with adaptive external blocking of overflow calls derived in this article allows to increase the efficiency of intercity progresive rate network very remarkabely.

5. REFERENCES

1. European Regional Project for Development of International telecommunication."Workshop type seminar on inteligent routing strategies", Zruc/Czechoslovakia, April 1986
2. Garcia J.M, Le Gall F., Castel C., Chemouil P., Gauthier P., Lechermeier G. : "Comparative Evaluation on Centralized/ Decentralized Traffic Routing Policies in Telephone Networks", Proc. of Int. Teletraf. Conf., ITC'11, Kyoto,1985
3. Konvit M. , Borik M.J. : "A Comparison of Routing Strategies for Telephone Networks", ITU seminar on network planning, Greece 1990
4. Ash G.R., Kafker A.H., Krishnan K.R. :"servicing and Real-Time Control of Networks with Dynamic Routing",Bell Syst. Tech. Journal, vol. 60 n°8 1981
5. Ash G.R. :"Use of a Trunk Status Map for Real-Time DNHR",Proc. Int. Teletraf. Conf., ITC'11, Kyoto, 1985
6. Narendra K.S., Wright E.A., Mason L.G. :"Aplications of Learning Automata to Telephone Traffic Routing and Control", IEEE Trans. SMC, vol.7 n°11 1977
7. Akinpelu J.M. :"The Overload Performance of Engeneered Networks with Non-hierarchical and Hierarchical Routing",Proc. of Int. Teletraf. Conf., ITC'10 Montreal 1985
8. Haenschke D.G., Kettler D.A., Oberer E. :"Network Management and Congestion in the U.S. Telecommunications Network", IEEE Trans. Com. vol. 29 n°4 1981
9. Pioro M., Wallström B. :"Multihour Otimization of Non-hierarchical Circuit Switched Communication Networks with Sequential routing", Proc. of Int. Teletraf. Conf., ITC'11, Kyoto, 1985
10. Chemouil P., Filipiak J. :"KAlman Filtering of Traffic Fluctuations for Real-Time Network Management",4th IFAC/IFORS Symp on Large Scale Systems, Zürich 1986
11. Pioro M., Wallström B.,Raneby KH L. :"Routing Principles in Non-hierarchical Networks - an Introductory Study", Technical report, Lund Institute of Technology, Dept. of Com. Syst., 1984
12. Grandjean Ch.:"Call Routing Strategies in Telecommunication Networks", Proc. of Int. Teletraf. Conf.,ITC'5 New York 1967
13. Cameron W.H., Galloy P., Graham W.J. :"Report on the Toronto Advanced routing Concept Trial", Proc. NETWORKS'80 Conf., Paris 1980
14. Szybicki E. : "Adaptive, Tarif Dependent Traffic Routing and Network Management in Multi-Service Telecommunicatins Networks", Proc. of Int. Teletraf. Conf., ITC'11, Kyoto,1985
15. Filipiak J., Chemouil P.: " Modelling and Prediction of Traffic Fluctuation in telephhone Networks", IEEE Trans. Com. vol. 35 n°9 1987
16. Neubrun T., Riečan B. :" Measure and Integral", VEDA, Bratislava, 1981

Performance of hybrid switching networks with priority: movable boundary case

M.S. Moustafa[a] and M.I. Marie[b]

[a]Visiting Assistant Professor, The American University in Cairo, Math. Unit, Science Department

[b]Assistant Professor, Al-Azhar University, Faculty of Engineering, Computer and System Department

Abstract
This paper deals with performance of hybrid (voice and data) switching networks. A certain portion of the frame is allocated to voice traffic, while the remaining frame capacity is assigned to the data traffic. To achieve a better transmission utilization than the fixed boundary case, data are allowed to use any residual voice capacity available; i.e., movable boundary case. The voice traffic is treated as a loss system, while the data traffic are considered as multiple classes of different priorities. Numerical results and comparisons with nonpriority (FIFO) case and fixed boundary with nonpreemptive priority case [1] are given.

1. INTRODUCTION

This paper studies the performance of a multiplex structure in which two types of traffic (voice and data) will be treated. This structure utilizes a master frame format of a time division statistical multiplex facility. This integrated system provides a higher transmission utilization efficiency than the separated systems. The overall concept and the historical background that gave rise to this approach have been given by Coviello and Vena [3] the frame structure subdivided into regions one for each traffic type is assumed to exist within the mode for illustrative purposes (Fig. 1). Voice calls will be circuit switched by assigning synchronous slots while data traffic are statistically multiplexed by asynchronous packet distribution. The frame length is divided by a boundary, this boundary may be fixed or movable. We are interested in the movable case under priority for data traffic. In this paper we consider the data traffic have multiple classes of different priorities. The expected waiting time for each class is our target. The voice traffic is treated as a loss system and is loaded into the frame on FIFO served basis at the beginning of the frame period. The structure in Fig. (1) illustrates how voice and data can be combined into one master structure. This represents 10 ms frame taken from a T_1 carrier which gives a frame length of 15440 bits. The frame period of 10 ms was chosen as an acceptable delay. We consider the data traffic have three classes of nonpreemptive priority, see [4].

In section 2 we develop a mathematical model that can be used to evaluate the performance of this structure. In section 3 we give the result which illustrates the performance of data under movable boundary case with three classes of priority. In section 4 the conclusions and some extensions will be given.

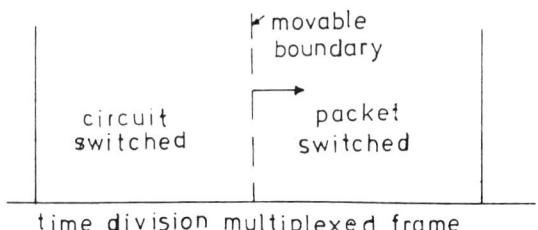

Fig. (1) Hybrid Switched Frame

2. MATHEMATICAL MODEL

The multiplex structure described in section 1, allows the voice and data customers to access to the system only at a fixed interval of time (a gate in queueing theory). We assume that there are N channels available for the voice traffic and C channels for the data. Every b time units, these channels are filled with waiting customers of both types.

If the number of free voice channels is less than the number of voice customers a head of an arriving voice customer, is lost and leaves the system without receiving any service. We assume that the arrival process is Poisson with parameter ν and voice holding time is negative exponential with parameter μ. The probability that a call is lost can be approximated by Erlang's loss formula, provided that the voice traffic offer (νb) is sufficient small (see [5], [6]).

For the data traffic, we assume that the classes arrival in independent Poisson processes with parameters λ_j, $j = 1,2,3$. The service time is constant and equal to b seconds for all classes. The mean delay, in steady state, for each class of data depends on the voice traffic since the unused portion of the voice channel is made available to data traffic. According to these assumptions, the mean delay, in steady state, for each class of data is given by the following theorem:

Theorem
Consider a system with J nonpreemptive data classes. For each class j, assume the arrival process is Poisson with parameter λ_j and
$\lambda = \sum_{j=1}^{J} \lambda_j$. The service time is constant and equals to b seconds for all classes. There are $S = C+N - \sum_{k=1}^{N} K p_K$ available servers, where p_K is the probability of K busy voice channels just after the gate opens. If $\lambda b < S$ then the steady state mean delay of a class j customers, $j = 1,2,\ldots,J$ is given by

$$E[W_j] = \frac{1}{2S\lambda(1-\sigma_j)(1-\sigma_{j-1})}[2S(1-\rho)\sum_{r=1}^{M-1}\frac{1}{1-Z_r} + \sum_{k=2}^{N} k(k-1)p_k + M + (\lambda b - M)(M-S)] \quad (2.1)$$

where $\rho = \lambda b/S$, $M = N+C$, $\sigma_j = \sum_{i=1}^{j} \lambda_i b/S$, $S = M - \sum_{k=1}^{N} k p_k$,
Z_r ($r = 1, 2, \ldots, M-1$) are the M-1 unique roots of $Z^M = \exp(-\Theta(1-Z))$,
$\Theta = \nu/\mu + \lambda b$ inside the unit disk.

Proof

The delay of a j-customer is $\sum_{i=1}^{j}$ (service times of i-customers in queue at the j-customer's arrival epoch) + $\sum_{i=1}^{j-1}$ (service times of i-customers who arrive while j-customer is in queue) + expected remaining service time of customers in service at customer's arrival epoch, if any.

Let the random variable Q_i be the number of i-customers in queue in the steady state. Since Poisson arrivals, see time average, the expected number of i-customers in the queue at the arrival epoch of a j-customer is $E[Q_1]$. The time between departure of each of these customers is b/S, so the expected value of first sum is

$$\sum_{i=1}^{j} E(Q_i) b/S = \sum_{i=1}^{j} \lambda_i b E[W_i]/S$$

Similarly, the expected value of the second sum is

$$\sum_{i=1}^{j-1} \lambda_i E(Q_i) b/S = E[W_j]\sigma_{j-1}$$

Let E[R] denote the expected remaining service time of customers in service at customer's arrival epoch, we have

$$E[W_j] = \sum_{i=1}^{j-1} \lambda_i b E[W_i]/S + E[W_j]\sigma_{j-1} + E(R)$$

Solving for $E[W_j]$, we get

$$E[W_j] = \frac{E(R)}{(1-\sigma_j)(1-\sigma_{j-1})} \quad (2.2)$$

To obtain E(R), we consider the expected waiting time, E[W] for movable boundary non-priority case given by [2], namely

$$E[W] = \frac{1}{\lambda}[\sum_{r=1}^{N+C-1}\frac{1}{1-Z_r} + \frac{\sum_{k=2}^{N}k(k-1)p_k - (N+C)(N+C-1) - (\lambda b)^2 + 2\lambda b(N+C)}{2(N+C-\lambda b - \sum_{k=1}^{N}k p_k)}] - b \quad (2.3)$$

where Z_r ($r = 1, 2, \ldots, N+C-1$) are the roots of $Z^{N+C} = \exp(-\Theta(1-Z))$ inside the unit disk.

The last equation can be written in the following form:

$$E[W] = \frac{1}{2S\lambda(1-\rho)}[2S(1-\rho)\sum_{r=1}^{M-1}\frac{1}{1-Z_r} + \sum_{k=2}^{N}k(k-1)P_k + M + (\lambda b-M)(M-S)] \quad (2.4)$$

Comparing equations (2.2) and (2.4), we obtain equation (2.1).

3. COMPUTATIONAL RESULTS

The numerical results, shown below, are used to demonstrate a typical application of the model presented in this paper, i.e., the hybrid switching with movable boundary.

We consider a system of $N = 10$ channels and $C = 5$ channels such that $S=N+C-\Sigma k\ p_k = 10$ channels with $b = 10^{-3}$ seconds and $\lambda = 5$ customers per second. The data traffic have three classes (interactive, control, and bulk), see [4]. First, we assume that the arrival rate for these classes are 0.1, 0.2 and 0.3 respectively. The results are shown in Fig. (2). Secondly, we assume that the arrival rates are equal, and the results are shown in Fig. (3).

The results show that the expected waiting times for high priority classes (the first and the second) are smaller, while the expected waiting for the third class (low priority) is greater than the expected waiting time under the FIFO discipline.

A comparison of the fixed and movable boundary cases for the first class which has the highest priority is given in Fig. (4).

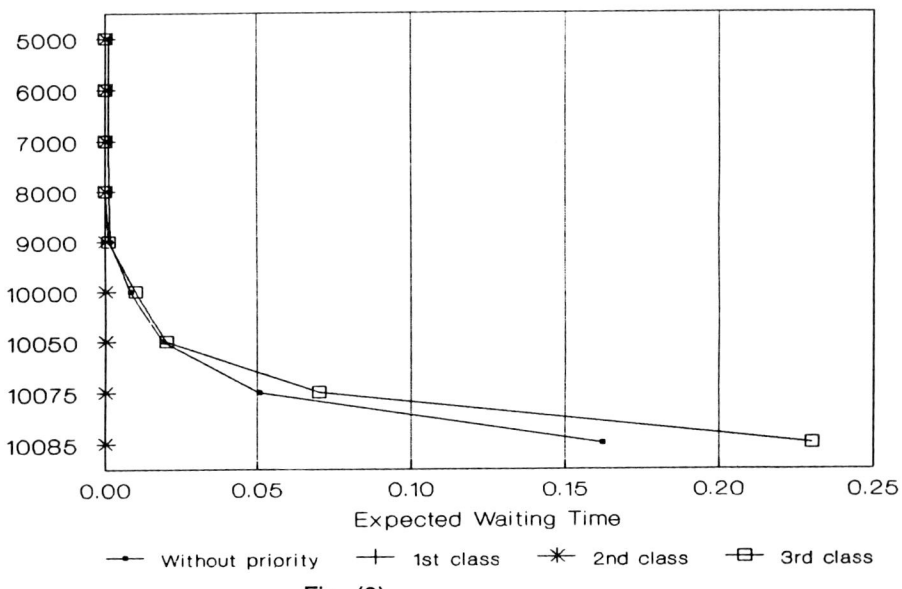

Fig. (2)

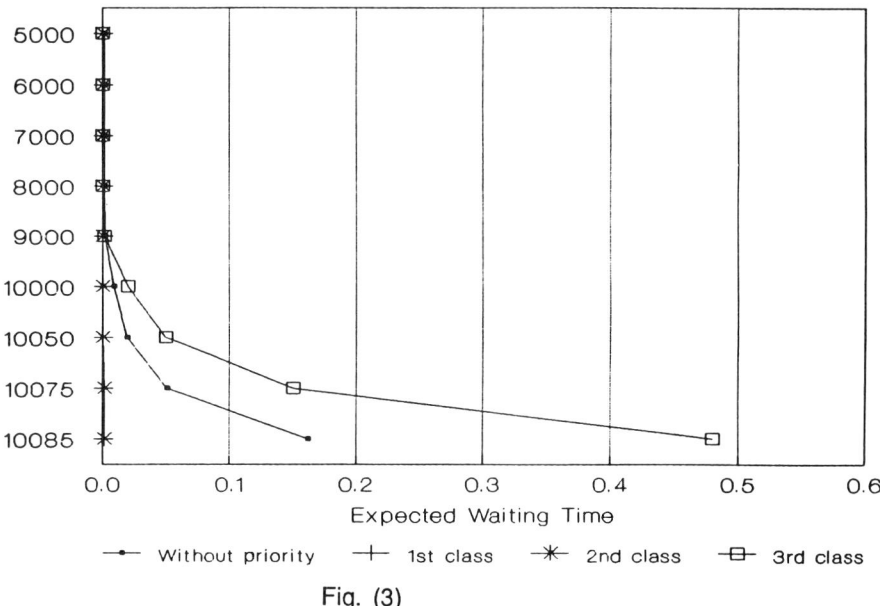

Fig. (3)

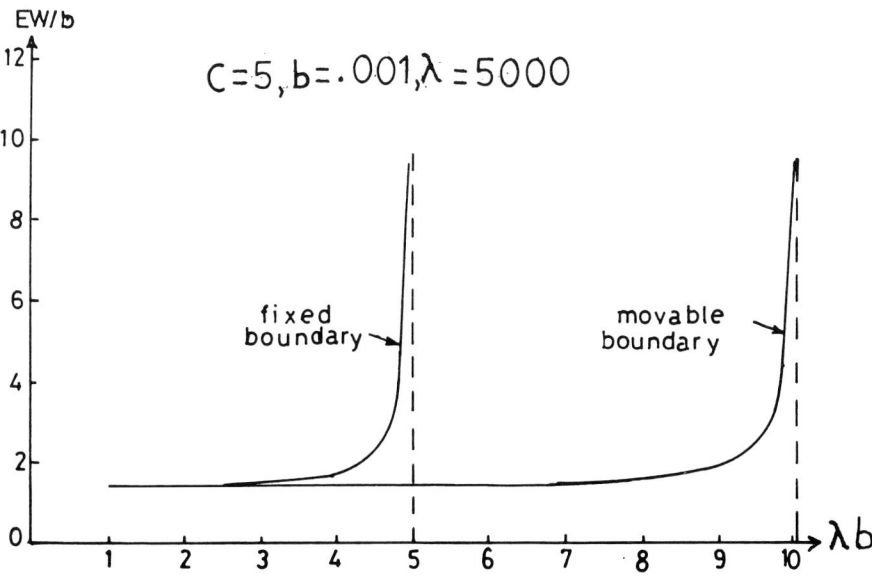

Fig. (4) Movable versus fixed boundary

4. CONCLUSION

We have presented a mathematical model for evaluating the performance of an integrated circuit and packet switching multiplexer with movable boundary. The data traffic have been treated as a nonpreemptive priority system. The data have three classes of different priorities. The results show that the benefits, of smaller expected waiting times that occur to the first two classes with high priority are gained at the expense of increased delays for low priority class (class three). It is also shown that a better transmission utilization is achieved as a result of using a movable boundary rather than a fixed one. As an extension to this work, we can obtain the priority analysis for movable boundary case under batch input data.

5. REFERENCES

1. Marie, M.I., and M.S. Moustafa (1991), "Performance of Hybrid Switching Networks with Priority: Fixed Boundary Case", submitted for publication.
2. Fisher, M.J., and T.C. Harris (1976), "A Model for Evaluating the Performance of an Integrated Circuit and Packet Switched Multiplex Structure", IEEE Trans. Comm. Com-24, 195-202.
3. Coviello, G., and P. Vena (1975). "Integration of Circuit/Packet Switching by (Slotted Envelope Network) Concept. "Telecommunication Conference, New Orleans LA., p. 12-17.
4. Israel G., N.H. Wen, and J.O. Benediot (1981). "Analysis and Design of Hybrid Switching Networks", IEEE Trans. Comm. Vol. Com-29. 1290-1300.
5. Gross, D., and C.M. Harris (1985). Fundamentals of Queueing Theory, New York, John Wiley.

ON SOME TELETRAFFIC PROBLEMS OF RSU CONTROL SYSTEM INVESTIGATION IN TELECOMMUNICATIONS

B.Goldshtein, S.Brusilovsky, R.Rerle

Central Research Institute of Telecommunications, Leningrad Branch (LONIIS), Warschawskaya 11, Leningrad, 196128, USSR

Abstract
  This report introduces the waiting-time approximations for Queueing Model with two feedbacks and also gives the simulation results. This Queueing Model represents the interactive operation between Remote Subscriber Unit Control Element (RSU CE) and the control elements in the exchange. The result of the analysis of this Oueueing Model is the approximation formula for average time spent by a customer in the system. The obtain results are compared with simulation which was executed by simulation language GPSS.

  The evolution of microelectronic technology allows to design the Digital Switching System (DSS) with microprocessor control for Telephone Networks. One of the distinctive features of this control systems is the great number of there architectures and distribution of call processing functions between the control elements in the control system of DSS. The reduction of cost, size and power consumption of microprocessor control elements has permitted to design the cost-effective RSU. Application of these RSU is cutting down the charge for line equipment in Telephone Networks.
  Economic efficiency of RSU introduction in existing Telephone Networks is defined by rational architecture of RSU and there practical application. The architecture of RSU depends on the type of terminal and call processing functions associated with it. The connection RSU with Digital Switching System DX/200 is illustrated on Fig.1. RSU CE with DX/200 control elements (SSU,OMC,CM etc.) realize the call processing functions. Let's consider the abstract diagram of the realization of call processing functions executed by the system RSU-DX/200 shown on Fig.2. Call processing in the system RSU-DX/200 is a complex set of functions executed by RSU and DX/200 software.

The RSU software is organized in a number of hierarhical processes. Fig.2 shows the names of these processes. Not considering in detail the interaction between software blocks we'll describe the simple example of RSU's subscriber call processing procedure.

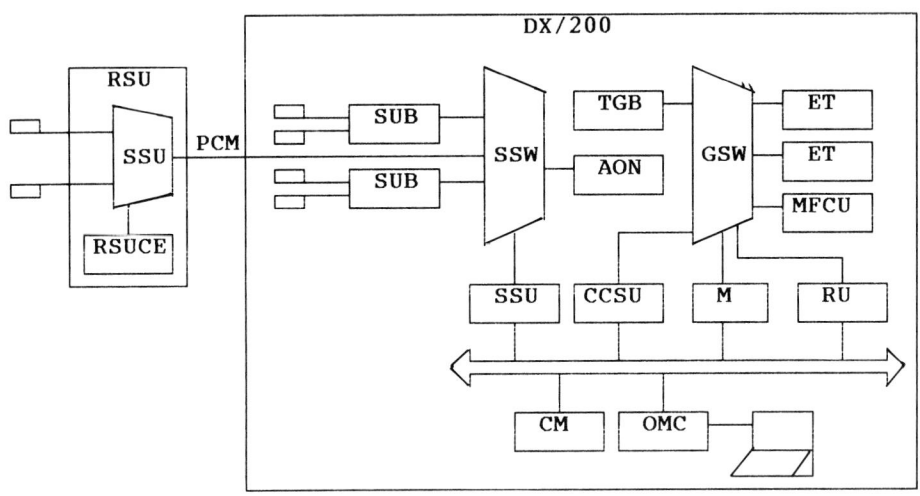

Figure 1.Distributed Control System RSU-DX/200.

The software provides the 16-ms filtering of a customer loop.Note, that this filtering is provided by a software on each changing of a customer loop state in order to decrease a number of errors which appears due to interferences in the subscriber's line. After having been filtered the signal is analyzed with the purpose of detecting "offhook" state. The message about analysis result is sent over the calling party's line to DX/200 it is connected. The station replies with a steady "DIAL TONE" indicating that the equipment is ready for dialing. The calling customer then starts dialling. The first 1-3 digits are required to select an appropriate outgoing trunk.

The result of first three digits analysis is one or the other possible responses: "BUSY" (i.e. network resourses not available) or "O'k TO SEND".

Each response from the called interface causes the corresponding task to be generated as indicated on Fig.2. After an "O'k TO SEND" response has been received from the exchange DX/200 the resolution for setup speech path is applyed in RSU. Then the other digits of a calling party customer are decoded. In the event that three dialing digits are not completed provision must be made to abort the call and release any processes assigned ("RELEASE CONNECT").

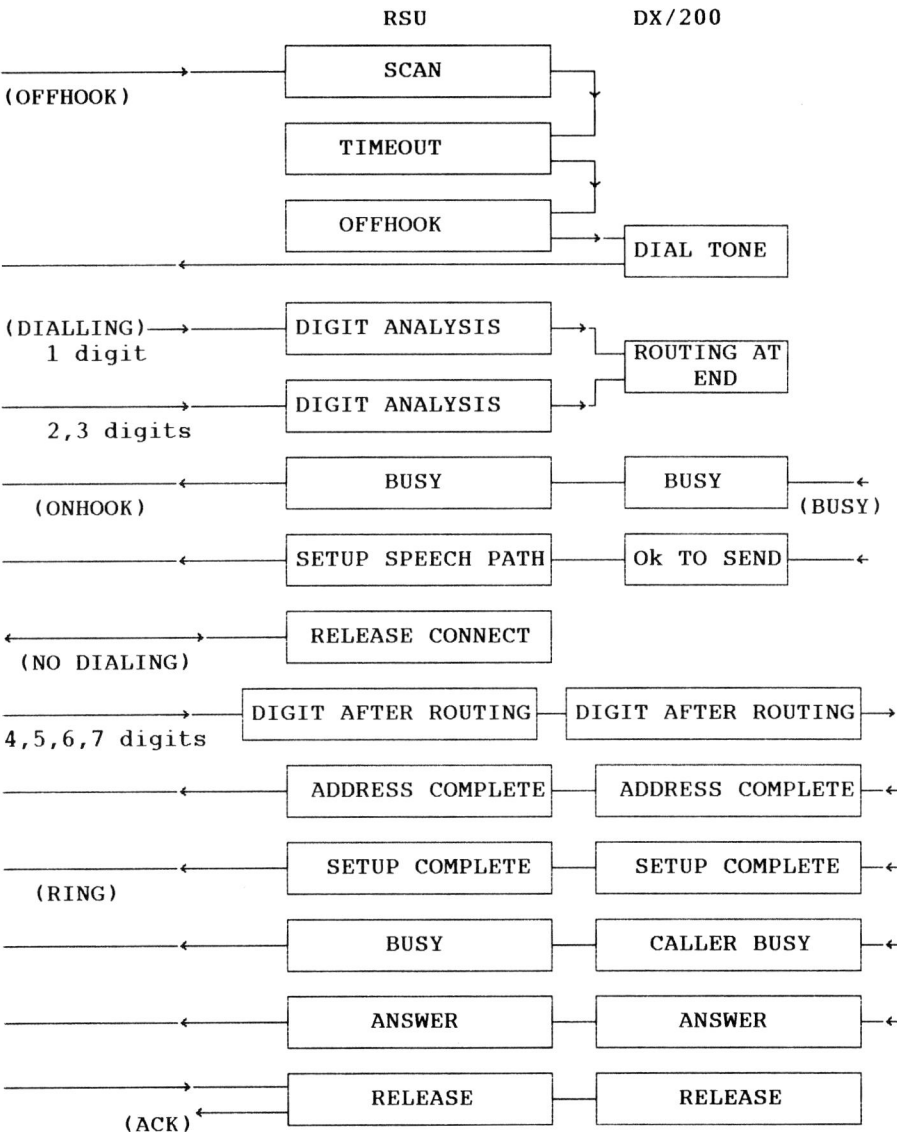

Figure 2. Call processing by the system RSU-DX/200.

On completion of dialing the following sequence of messages are sent to RSU:
- "ADDRESS COMPLETE;
- "SETUP COMPLETE;
- "RING".

In case if the called customer line is busy the message "CALLER BUSY" is sent to calling party (RSU). In another case a speech path is set up. On completion of the talking either of customers goes on hook and a hang up signal is returned to the other party and disconnection takes place ("RELEASE", "CLEAR").

As shown on Fig.2 the call processing is a streamline set of call service stages (CSS). Each stage include the execution of definite sequence of programs.

It is necessary to know the delays in the different CSS to determine the time parameters of control system. The time parameters give some qualitative values of call processing. Rec.CCITT Q.500 normalize the quality of call processing about average time separately for each CSS.

Note, that in order to determine the performance of RSU CE service quality it is necessary to determine the average waiting time of call service due to analysis of probability-time characteristics. Research of this characteristics is based on Queueing Theory.

The specific properties of operation of distributed microprocessor control system are such that each microprocessor repeatably executes the definite sequence of functions, generates the intermediate results and sends them to other microprocessors. Thus the service of each call is a multiple and many stage execution of the processes set realized by different microprocessors.

The precise analysis of such distributed control system with interconnected service is possible only in particular case, e.g. in Jackson Network. On the other hand the problem is facilitated that the investigation object is usually one control element from the distributed control system (in this case it is RSU CE).

The interaction technique RSU CE with exchange's control elements is based on the control function distribution between them. In conformity with the interaction technique some service functions are executed by RSU and some are transmitted to the exchange's control elements.

The information processing realized by different control elements and correspondingly by the different program modules which are interchanged by the information with each other. An activating of the requests between RSU CE and the control elements in the exchange is went on for the completion of arriving call service.

In connection with this it is suggested the research of distributed control system RSU-DX/200 as a Queueing System with two feedbacks (Fig.3). The approach used is to study each microprocessor separately.

RSU is studied as a Single Server Queue (SSQ) with internal (determinate) and external (exponential) feedbacks. The internal feedback with SSQ3 simulates the timeout which is used for limitation of subscribers and devices response-times and also for controlling the time processes. External feedback based on SSQ2 simulates call processing in DX/200.

The interactive operation between RSU CE and exchange's control elements could be described by the following simplified model (Fig.3).

Incoming calls arrive according Poisson process with intensity λ . In this report the calls mean the requests between the

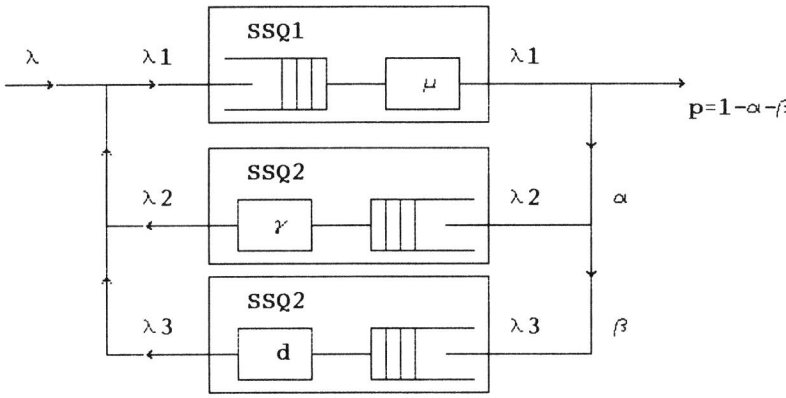

Figure 3. Queueing System RSU-DX/200

program modules in distributed control system RSU-DX/200 during different stages of service. The customer joins the FIFO queue with an infinit waiting room. The service time in the RSU CE is exponential with mean μ^{-1}. After service the customer either leave the system with probability $p=1-\alpha-\beta$ or is sent through the delayed feedback loop (d) with probability β and is placed in the tail RSU CE queue or is sent with probability α to the SSQ with an infinit waiting room and exponential service time with mean γ^{-1} which represents the exchange's control elements. Having been served by exchange's control elements the customer joins the RSU CE queue again. Thus to pass the system above the customer has to circulate a random number of times through two feedback loops. Naturally, this representation of distributed control system is based on some assumptions.

The stream of incoming calls served by RSU CE is successfully approximated with Poisson distribution. The substitution of this fact is the theorem of Grigelionis claiming that the superposition of great number of independent streams with low intensity is approached under the some regular conditions to the Poisson stream.

As note before the call processing times in distributed control system are taken as exponentially distributed. Generally speaking the program execution times of call processing are fixed values for any control elements.

Executing program times in all program modules in assembly can be considered as random values with a peakedness factor smaller than but not far from 1. This fact was proved by Takach for the case when a number of sequentialy executed call processing programs are geometrically distributed.

The control function distribution model RSU-DX/200 is the multiphase Queueing System. Jackson's theorem follows that for such system a number of customers in all queues in current instant is independent and the expression for the customers number distribution in each of queues is the same for M/M/1 Queueing System. Note, that the feedback loops streams aren't Poisson becouse the customers leaving SSQ1 arrive again in SSQ1 with probabilities α and β .Therefore the total stream into SSQ1 also isn't Poisson. However according to Jackson theorem the distribution of customer number in the state M/M/1 system queue is the state distribution for the SSQ1 queue even though the total arriving process into the SSQ queue isn't Poisson. Thus as follows from Burke's theorem incoming Poisson stream through the server with the exponential service time distribution generates outgoing Poisson stream. In other words if the arriving stream is Poisson with parameter λ the depature stream is also Poisson with the same parameter.Therefore the viewing model is analized as a simple sequence of SSQ.

Denote $\lambda 1$ the intensity of customer stream arriving to the SSQ1 as a result of random number of the customer passings through feedback loops. Note that this stream is the sum of three streams arriving to SSQ1: incoming stream with intensity λ and two feedback streams, i.e.

$\lambda 1 = \lambda + \alpha \lambda 1 + \beta \lambda 1$; $\lambda 1 = \lambda / p$,where

$p = 1 - \alpha - \beta$ - the departure probability;
α - the probability of service in SSQ2;
β - the probability of service in SSQ3.

Denote $\lambda 2$ the intensity of the customer stream arriving to SSQ2, i.e. $\lambda 2 = \lambda 1 \alpha = \lambda \alpha / p$.

Denote $\lambda 3$ the intensity of the customer stream arriving to SSQ3, i.e. $\lambda 3 = \lambda 1 \beta = \lambda \beta / p$.

It follows from [6] that the number of the customer service cycle in the M/M/1 system without delay in feedback loop has geometric distribution.

Denote $\bar{t}1, \bar{t}2, \bar{t}3$ - the average times of single service in SSQ1, SSQ2, SSQ3 respectively.

Based upon foregoing it is possible to write the expression for the average time spent by a customer in the Queueing System:

$$\bar{T}=\bar{t}1+\alpha(\bar{t}1+\bar{t}2)+\alpha^2(\bar{t}1+\bar{t}2)+...+\alpha^n(\bar{t}1+\bar{t}2)+...$$
$$+\bar{t}1+\beta(\bar{t}1+\bar{t}3)+\beta^2(\bar{t}1+\bar{t}3)+...+\beta^n(\bar{t}1+\bar{t}3)+...=$$
$$=(\bar{t}1+\bar{t}2)/(1-\alpha)+(\bar{t}1+\bar{t}3)/(1-\beta)-\bar{t}2-\bar{t}3 \qquad (1)$$

Using now Polyachek-Hinchin formula one can calculate $\bar{t}1$, $\bar{t}2$ and $\bar{t}3$. Thus assuming SSQ as a M/M/1 (i.e. the standard deviation of service time $C_b^2=1$) we get:

$$\bar{t}1=1/\mu + \lambda/\mu(p\mu-\lambda) \qquad (2)$$

Analogous to SSQ1 we get for SSQ2:

$$\bar{t}2=1/\gamma + \lambda\alpha/\gamma(p\gamma-\lambda\alpha) \qquad (3)$$

At last for SSQ3 which is a M/D/1 system ($C_b^2=0$) we get:

$$\bar{t}3=d + \lambda\beta d^2/2(p-\lambda\beta d) \qquad (4)$$

The values $\mu, \gamma, p, \lambda, \alpha, \beta$, d in the expressions (2), (3) and (4) is defined above.

Using (2), (3) and (4) in (1) we have:

$$\bar{T}=(1/\mu+\lambda/\mu(p\mu-\lambda))(1+p)/(1-\alpha)(1-\beta)+(1/\gamma+\lambda\alpha/\gamma(p\gamma-\lambda\alpha))\alpha/(1-\alpha)+$$
$$+(d+\lambda\beta d^2/2(p-\lambda\beta d))\beta/(1-\beta) \qquad (5)$$

For existing the stationary operating mode of queueing system it is necessary to observe the following condition:

$$\lambda \leq p(\mu+\gamma/\alpha+1/\beta d)/3 \qquad (6)$$

Formula (5) permits to estimate the average service time in RSU CE and in the exchange's control elements with determined $\lambda, \mu, \gamma,$ and p. The analysis of this formula enables to distribute in optimal way the call processing functions between RSU CE and exchange's control elements.

The sequencce of the call processing operations executed by the distributed control system software one can approximately assume independent from the redistribution of call processing functions between RSU CE and DX/200 control elements. That's why the future analysis expediently to focus in the range of the summing constant performance of RSU CE and DX/200 control elements, i.e. $\mu + \gamma$ = const. Fig. 4 shows the function $T=f(\mu+\gamma=\text{const},p=\text{var})$.

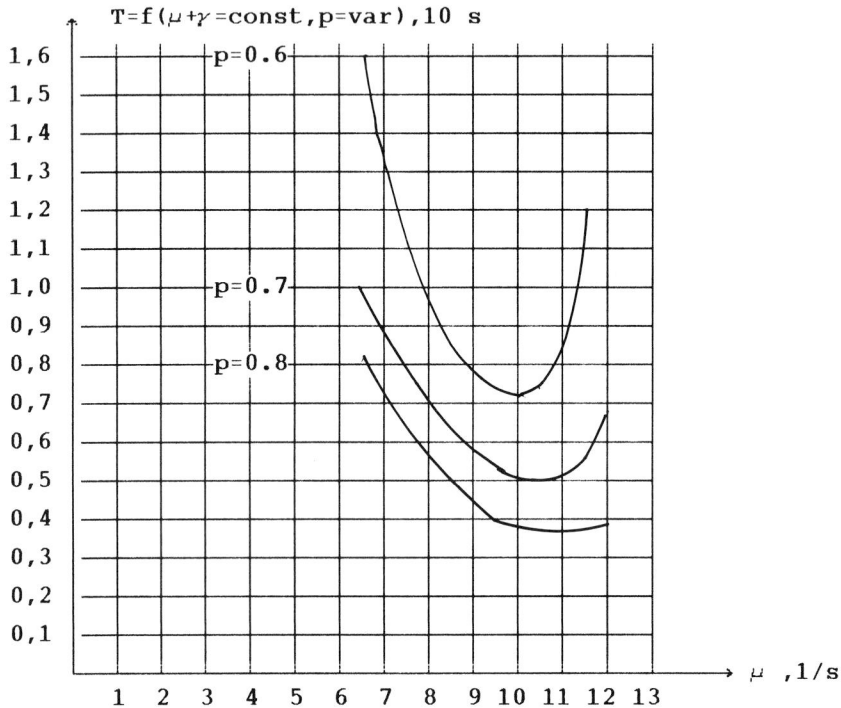

Figure 4. Function $T=f(\mu+\gamma=\text{const},p=\text{var})$.

The obtained results are compared with simulation which was executed by simulation language GPSS. The fit is quite good. The suggested approach and obtained numerical results can be useful in the practical engineering of DSS. In particular, in order to distribute call processing functions between RSU CE and exchange's control elements taking the acceptable service time into consideration.

REFERENCES

1. Ершова Э.Б., Ершов В.А. Цифровые системы распределения информации. - М.:Радио и связь, 1983.
2. CCITT.Blue book.Rec.Q.500.-Vol.VI.5.-Geneva,1989.
3. Берлин А.Н. Алгоритмическое обеспечение АТС.-М.: Радио и связь,1986.
4. Яшков С.Ф. Анализ очередей в ЭВМ.- М.: Радио и связь, 1989.
5. Кузьмин И.В. и др. Синтез вычислительных алгоритмов управления и контроля. - Киев: Техника,1975.
6. Takach L. A Single-Server Queue with Feedback.- Bell Syst.Tech. J., vol.42, 1963.-pp 505-519.
7. Walrand J.Probabilistic Look at Networks of Quasi-Reversible Queues.-IEEE Trans.Inf.Theory, IT-29,825-831,1983.
8. Burke P.J. The Output of a Queueing System. Operations Research.-1956.-N 4.-pp 699-704.

DIGITAL CROSS CONNECTS APPLICATION FOR THE FUTURE SUBSCRIBER NETWORK

N. Sokolov

Central Research Institute of Telecommunications, Leningrad Branch (LONIIS), Warschawskaya 11, Leningrad, 196128, USSR

Abstract

Digital Cross Connects may become the basis element of the future Subscribe Networks. Taking into account the Digital Cross Connect advantages, the cost reduction of a Subscriber Network may be achieved.

1. Introduction

Subscriber Network hasn't undergone any essential modification during telephone communication development from manual systems to modern digital exchanges. However, the next step in electrical communication progress, concerning creation of Integrated Services Digital Network (ISDN) will result in essential modification of the Subscriber Network structure basic principles. Requirements to the future Subscriber Network are very diverse.
It may be illustrated by the following examples:
- future Subscriber Networks have to be cheap, since their present cost is very high;

- future Subscriber Networks capabilities have to meet the wide range of the digital transmission bit rates requirements (up to some hundreds Mb/s);
- Subscriber Networks structure has to provide highly reliable connections between terminals and an exchange.

Moreover, the control system supposed to be implemented at Subscriber Network must be able to detect failures and overload conditions.

In most cases future Subscriber Network comes from the existing subscriber loops. For this reason the existing Subscriber Network duct structure should be used as the basis for future development.

2. Digital Cross Connect Functions

Digital Cross Connect (DCC) may become the basic element for the future Subscriber Network structure mentioned above. DCCs may be installed in places where existing cable cabinets are located. At the future Subscriber Networks, DCC will carry out the following functions:
- multiplexing of subscriber interfaces to form standard digital streams;
- Subscriber Network management under network enlargement, traffic increasing and cable lines failure conditions;
- interfacing of different transmission media (for example fiber optic cable and coaxial one).

Some other Subscriber Network problems could be solved using DCC.

3. Subscriber Network Structure

The hypothetical model of the proposed Subscriber Network Structure is shown in figure 1.

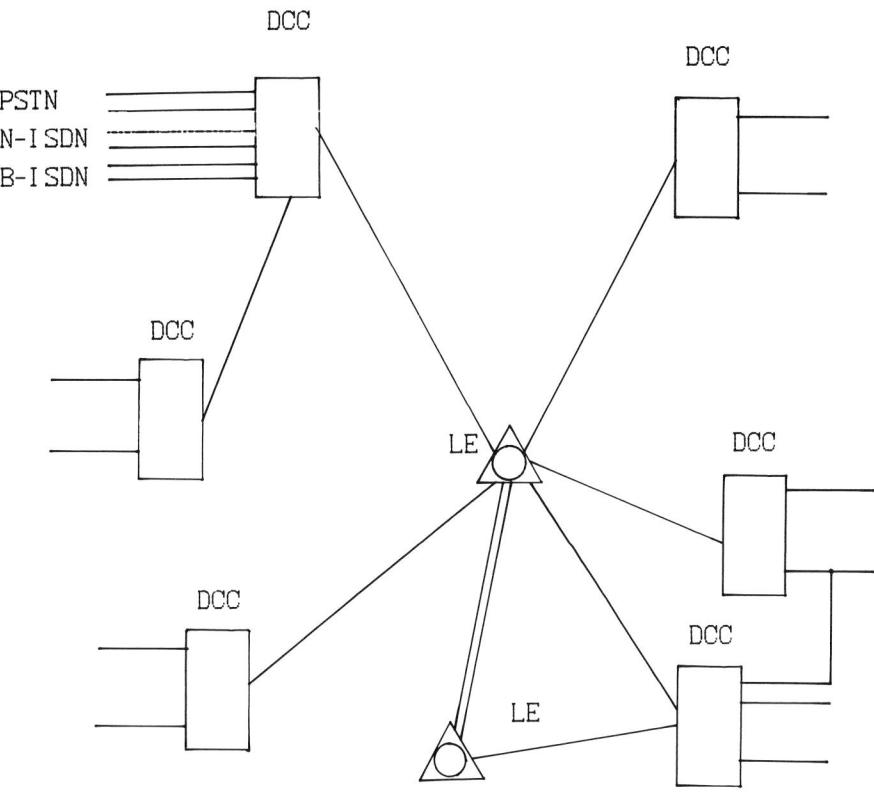

Figure 1. Model of the Subscriber Network structure

Obviously, all kinds of the subscriber interfaces must be taken into consideration. Three cases (PSTN, N-ISDN, B-ISDN) are shown in figure 1.

Subscriber terminal is connected to the nearest DCC by individual line. In special cases, subscriber terminal may be connected to the two DCCs for the increase of the reliability.

DCC may be connected to the local exchange (LE) directly or across other DCC. Moreover, DCC may be connected to the two local exchanges. This opportunity will increase the network reliability. In addition, this solution permits to use the specific traffic control procedures.

Other structures may be studied too.

4. General Advantages

To implement the proposed Subscriber Network Structure, the DCCs should be installed. The expenses for the DCC introduction will repay themselves due to the reduction of the cable lines total length. The relative saving in cable products will amount 42% for the Network Subscriber model with six DCCs.

The second advantage is the smoothing of a traffic in some cases. For example, some different closely located DCCs may be connected to the different local exchanges. It is useful if the DCCs combine subscriber lines having the same busy hour time position. This strategy gives the reduction of traffic fluctuations and local exchanges performance improving.

The third advantage consists in a possibility of the reliability increase mentioned above.

5. Conclusions

Subscriber Network structure with DCCs, presented in this paper, shows the good outlook. It seems, that this approach must be taken into consideration under future Subscriber Network investigations.

Design and Control
of Telephone Systems

ON PRIORITY ASSIGNMENT PROBLEMS IN SPC SYSTEMS

D. Bursztynowski, W. Burakowski & W. Syski

Institute of Telecommunications, Warsaw University of Technology,
Nowowiejska 15/19, Warszawa, Poland

ABSTRACT

The paper addresses some performance-oriented design probems in a class of SPC systems with central control modeled by an M/G/1 system with priorities and feedback. In this paper, we focus on performance optimization by task prioritizing. In the course, two different types of objectives are identified: capacity-oriented and average response time-oriented ones. We discuss briefly the problem of minimization of the average holding cost in the system, and then present an approach to the call-handling capacity optimization problem. A sample of numerical results is included to show the influence of various priority assignments on system performance characteristics. Also, the problem of dependence of the optimal character of the solutions on call mix changes is discussed.

1. INTRODUCTION

Increasing complexity of switching components has brought about the need for performance-oriented design methods. These are commonly understood as rules and/or tools enabling the designer to take right decisions at a minimum cost and time expenditure. However, the detailed objectives and the methods used in different stages of the project vary according to the data available as the input for the analysis and decision-making process. Thus we feel that developing a set of tools/rules, each one referring to a particular question, rather than pursuing a fully self-contained method seems to be a realistic way to approach the problem. From among the factors affecting the overall switching system control performance, task prioritizing is of our interest in the present study.

The problem of scheduling the service to customers by means of priorities under various cost criteria is widely discussed in the literature. This fact suggests potential usefulness of the existing solutions in optimization of switching systems. Our observation, however, is that the features of switching systems, in particular specific requirements on them, considerably limit the applicability of many of the known results in practice.

In the case of switching systems controls, the call handling capacity is usually the key objective in performance studies. Since capacity is an implicit function of task priorities, a realistic approach is then to choose a substitute for capacity which would be both reliable and handy. We say that a substitute is reliable if its optimization leads also to the optimization of the capacity; it is handy if efficient methods exist to optimize it. The choice of the substitute is thus an important issue requiring that the factor(s) actually limiting the capacity be recognized. In this context, response times and buffer occupancies (preferably their distributions) are usually of greatest interest. The ultimate success depends on how these quantities are used in the optimization process.

To the authors' knowledge, [1], [2], and [3] were the first contributions to address the problem of developing task prioritization rules in systems we focus on. While in [1] and [2] the traffic model of the system was only loosely defined, a model known in the literature as an M/G/1 queue with feedback and priorities (see e.g. [7] for a detiled model description) was studied in [3]. Results presented in [4] are an important extension to those of [3] due to having been derived in a systematic way via formal considerations for a wide range of priority disciplines (non-preemptive, preemptive, and mixed preemptive/non-preemptive).

The papers cited above have two elements in common: 1) the average response time (ART) in the system is the objective (substitute), and 2) the results take the form of simple rules, also referred to as real-time design guidelines. The simplicity of the results renders them useful in various applications provided that the objective chosen (ART) is the right one. However, we have found them hard to apply due to various constraints while optimizing the traffic-handling related part of the program of a telephone exchange. In the case of the particular system, it has been also recognized that there exist other performance metrics (more reliable, handy) leading to better results (higher capacities) than the ART. These items stimulated our research.

An approach to cope with the average response time minimization problems based on priority assignment optimization and using a discrete-optimization simulated annealing algorithm (SAA) was suggested by the authors for the basic (non-preemptive priorities) version of the model in [5]. In this contribution, we partly follow the methodology introduced in [5], extend it to the case of average holding cost in the system, and apply it to directly optimize the system call-handling capacity.

Section 2 gives a short description of the considered traffic model. A sample of analytical results on minimization of the average holding cost in the system is presented in section 3. In section 3, we also clarify the notion of constraints used thereafter. In section 4, we justify the use of SAA, and describe briefly the adopted version of the algorithm. Then we present an approach to the central processor call-handling capacity optimization problem in a medium-size telephone switch with the aid of SAA (section 5). In section 6, we give a sample of numerical results to compare system performance characteristics for priority assignments obtained to optimize various objectives; we also comment on the issue of the dependence of the optimal character of the solutions on the call mix changes. Finally, conclusions are drawn and suggestions for future work are given.

2. TRAFFIC MODEL OF THE SYSTEM CONTROL

The traffic model of the swiching processor we consider in the paper is a version of the general M/G/1 system with priorities and feedback. The model has proved to be very useful in performance analysis of various switching systems (e.g. System 12 [6], [7], ITX switching system [8], [9]). In the following, we give a short description of the model. More details including some extensions to the model considered here can be found in the references.

The system is fed by a number of Poisson sources, each one with its own arrival rate $\lambda(i)$, $i=1,...,N$. An arrival from source of type i may correspond to an "external event", e.g. a specific subscriber action (off-hook, digit reception, etc.), or to a call to a system-oriented function (administration, fault processing, diagnostic, etc.), and demands an ordered set of $T(i)$ tasks (forming a so-called task sequence) to be executed. Each task (i,j) (i.e. j-th task in i-th sequence) is assigned an arbitrary priority $p(i,j)$ drawn from the set $\{1,...,P\}$ (1 is the highest priority and P is the lowest one), and has service time drawn from general distribution $H(i,j)$ with mean $h(i,j)$. Upon activation, a task joins an appropriate priority queue, waits its turn, and obtains service (which can not be preempted in our case). The first task of sequence i (i.e. task (i,1)) is activated at the moment of arrival to the system.

Immediately after task (i,j) service completion, its successor (i.e. task (i,j+1)) is activated. If a task is the last one in the sequence it leaves the system at service completion. Within each priority queue, the FIFO mechanism is employed.

The following notation will be used:
- n(i,j)   average number of tasks (i,j) queued in the system at a random moment;
- s(i,j)   average number of tasks (i,j) served at a random moment;
- w(i,j)   average waiting time of task (i,j); we thus have n(i,j)=λ(i)w(i,j);
- PA   priority assignment;
- {PA}   the set of all feasible priority assignments.

3. AVERAGE HOLDING COST MINIMIZATION PROBLEM

Let c(i,j) denote the holding cost of task (i,j). Then the average holding cost in the system, C, can be expressed as

$$C = \sum_{(i,j)} c(i,j)[n(i,j)+s(i,j)] \tag{1}$$

We also define the remaining service time of sequence i with task (i,j) in queue, $h_r(i,j)$, as follows:

$$h_r(i,j) = \sum_{k=j}^{T(i)} h(i,k) \tag{2}$$

Since the system we consider is work-conserving (see e.g. [10]), the following equality holds:

$$\sum_{(i,j)} n(i,j)h_r(i,j) + \sum_{(i,j)} s(i,j)[h_{res}(i,j) + h_r(i,j+1)] = \text{const}(PA)$$

where $h_{res}(i,j)$ stands for the residual service time of task (i,j). The values s(i,j) are constant (do not depend on priority assignment), so the above expression reduces to

$$\sum_{(i,j)} n(i,j)h_r(i,j) = \text{const}'(PA) = U \tag{3}$$

Thus the generic priority assignment problem of minimizing the average holding cost in the system can be stated as follows:

find

$$PA_0 : C(PA_0) = \min_{PA \in \{PA\}} \{C(PA)\} \tag{4.a}$$

with

$$\sum_{(i,j)} n(i,j)h_r(i,j) = U \tag{4.b}$$

3.1 The problem without constraints

The problem expressed in (5) is referred to as priority-unconstrained when we impose no restrictions on priorities of tasks in the system, i.e. assuming that each task can be assigned priority regardless of priorities of remaining tasks in the system. The {PA} is then a P^T space (with T being the total number of tasks over all sequences in the system), and the optimal priority assignment rule is as follows:

$$\text{if } c(i,j)/h_r(i,j) > c(k,l)/h_r(k,l) \quad (5)$$

then assign
$$p(i,j) < p(k,l)$$

Rule (5) can be derived from the result given in [11], where a dynamic programming approach was used to solve the problem in a system with multiple probabilistic feedback (branching) and batch arrivals.

Average holding cost function C may be given various interpretations according to the values of the holding costs c(i,j). Here we comment on the most common ones:

1. For $c(i,j)=1$, C denotes the average queue length. The minimization of C leads also to minimization of the average waiting time in the system (through the Little's law), and consequently, to the minimum of the average response time in the system (the time spent in service is not affected by priorities).
2. For $c(i,j)=1/\lambda(i)$, C denotes the sum of sequence waiting times over all sequences. It can be used as a relatively good substitute for capacity in certain cases but this has to be validated for each particular application.
3. For c(i,j)=signal memory requirement, C is the average signal buffer occupancy and can be used in buffer size optimization.

An important feature of the unconstrained problems with the interpretations 1 and 3 just described is that the optimal assignments are independent of the arrival rates.

3.2 The problem with constraints

We have assumed so far that each task may be given an arbitrary priority. This is however rarely the case in practice. The limit on the number of priority levels in the system is the simplest but a common example of constraint. We will treat it a bit later.

An important type of constraint is the requirement that some different tasks be assigned the same priority. This may occur for instance when a (physically the same) part of code is run during the execution of (logically) different tasks. In this case we have been able to derive a very robust, although heuristic, algorithm to obtain suboptimal assignments. A short description of this follows. For a given subset S consisting of tasks which are to be assigned the same priority, let us define the weighted holding cost of subset S as

$$CW(S) = \frac{\sum_{(i,j) \in S} \lambda(i) c(i,j)}{\sum_{(i,j) \in S} \lambda(i)} \quad (6.a)$$

and the weighted average service time of subset S as

$$HW(S) = \frac{\sum_{(i,j) \in S} \lambda(i) h_r(i,j)}{\sum_{(i,j) \in S} \lambda(i)} \quad (6.b)$$

Then the proposed priority assignment rule reads:

$$\begin{array}{l} \text{if} \\ \quad (k,l) \in S_i, \ (m,n) \in S_j, \ S_i \neq S_j \\ \text{and} \\ \quad CW(S_i)/HW(S_i) > CW(S_j)/HW(S_j) \\ \text{then assign} \\ \quad p(k,l) < p(m,n) \end{array} \quad (7)$$

Note that each task in the model belongs to one and only one subsetset. In particular, if we do not constrain the priority of a task then it constitutes a one-element subset.

Assignments generated using (7) were compared with the optimal solutions obtained by the exhaustive search method for a number of relatively simple examples and an excellent performance of (7) was stated in all cases (e.g. for the ART criterion, the relative difference of the objective never exceeded 0.5% and always tended to zero with increasing occupancy).

The great number of priority levels resulting from the use of (7) is a practical disadvantage of the method. We are thus interested in the possibility to deal with a mixture of the two types of constraints mentioned so far. Let $\{S_i\}$ denote the set of all the subsets which describe the constraints imposed on the task priorities in the system. Let $|\{S_j\}|$ stand for the number of subsets, and let P^* denote the required number of priority levels. Let us also define $R(S_i) = CW(S_i)/HW(S_i)$. We propose a two-stage heuristic algorithm to generate a suboptimal priority assignment. In the first stage we reduce the number of subsets to the required value P^*. Then, in the second stage, we assign priorities to the tasks according to the rule (7). The algorithm is as follows.

Algorithm 3.2.

>   begin
>   /* STAGE 1 */
>   step 1.   divide the set of tasks according to the initial partition $\{S_k\}$;
>
>   step 2.   if $|\{S_k\}| \leq P^*$ then go to step 6
>   else continue with step 3;
>
>   step 3.   find S_i, S_j such that   /* main step */
> $R(S_i+S_j) = \min \{ R(S_m+S_n) \}$;
> $(S_m, S_n \in \{S_k\}, S_m \neq S_n)$
>
>   step 4.   /* join S_i and S_j */
>   4.1. $S_i = S_i \cup S_j$;   /* join subsets */
>   4.2. $\{S_k\} = \{S_k\} - S_j$;   /* adjust $\{S_k\}$ */
>
>   step 5.   go to step 2;
>
>   /* STAGE 2 */
>   step 6.   apply rule (7) to assign priorities to the subsets;
>   end.

Algorithm 3.2 is very fast and it performs well provided that the degree of reduction of the number of priorities does not exceed 4÷5. It is thus useful in the prelimiary studies on the system.

We stress that the optimal assignments depend in general on the arrival intensities in the case of constraints. This property may be a potential source of difficulties in application of the results in practice since a change in the sequence arrival intensity pattern (caused e.g. by a change in the call mix) may violate optimality of the priority assignment employed. Thus additional studies on the priority assignment performance assuming various call mix schemes may be necessary in such cases. Note also that if we dispose of the constraints, the rule (7) and algorithm 3.2 both converge to the optimal rule (5).

In general, there are a large number of potential sources of constraints and we can not anticipate all of them in advance. Moreover, we probably would not able to quickly find a solution to every new problem. Thus another approach, capable of covering a wide spectrum of problems, is needed.

4. PRIORITY ASSIGNMENT OPTIMIZATION ALGORITHM

The approach to the priority assignment problem presented in this paper bases on a deeper exploitation of the existing analytical methods for performance evaluation of the considered traffic model. The problem resolves itself to developing a general-purpose tool to embed these methods in the optimization process. For more information on the analytical methods, the interested reader is referred to the references cited earlier.

Because of the large cardinality of the set of feasible solutions {PA}, an exhaustive search methodology is not suitable for our problem and another approach must be chosen.

In [5], an approach to the priority assignment problem with the aid of simulated annealing algorithm (SAA) and the mean queue length as the objective (equivalent to ART) was proposed. With SAA, the general priority optimization problem can be stated similarly as we did in (4); the only difference refers to the scope of applications since as we have decided to exploit the analytical methods in our approach, we now relax restrictions on the kind of the objective function (it may be of any type provided that it can be efficiently evaluated) and we do not make any explicit use of constraint (4.b).

The simulated annealing approach is based on ideas from statistical mechanics and is motivated by an analogy to the behaviour of physical systems in the presence of a heat bath [12]. However, it requires strong assumptions to provide convergence to a global solution of (4) (satisfaction of these may lead to a large number of iterations comparable to the exhaustive search method). Therefore a new heuristic version of SAA has been proposed in [13] to obtain the optimization result in resonable time. The new method does not guarantee that the global optimal solution of (4) will be found; the algorithm has no longer strict theoretical justification. Yet, as the numerical examples show, the results of that version of SAA are usually quite acceptable.

5. CENTRAL PROCESSOR CALL-HANDLING CAPACITY OPTIMIZATION

5.1 General

In this section, a case study on an SPC system is presented along with an extension of the method [5] to directly optimize the central processor call-handling capacity.

The definition of the processor call-handling capacity adopted in this paper is based on the concept of the dominant service constraint (see [14]), i.e. the service constraint exceeded first at increasing call intensity. The call-handling capacity is then defined as the greatest call rate at which dominant service constraint is still satisfied. We use the expected delays of call setup and call release phases. Some of the actual values of corresponding service constraints for these phases are drawn from the CCITT recommendations and some are system-specific.

We start with the common-sense assumption that the call-handling capacity reaches its upper limit if there is no outstanding "dominant service constraint", i.e. if at certain call intensity (the capacity) all service constraints are barely satisfied. The optimal (suboptimal) priority assignment then gives a result which is the closest (a close) one to the ideal result described above. Therefore the objective function to be used in (4) is as follows:

$$\lambda(PA) = \max_{x}\{ x : D_i(x,PA) \leq SC_i,\ i=1,...,N_c \} \tag{8}$$

where N_c denotes the number of constraints, subscript i refers to i-th delay, and $D_i(x,PA)$ and SC_i stand for the expected value of the i-th delay given call intensity x under priority assignment PA, and the service constraint for the i-th delay, respectively.

5.2 Numerical example

The numerical example concerns the call processing level of a medium-size telephone exchange [8]. The generic queueing model comprises over 100 sequences (much over 200 tasks), although this varies in a certain range according to the actual configuration.

Apart from the priority assignment PA1 obtained with the objective (8), we also study the following assignments:

- PA2 obtained to minimize the ART in the system (problem (4) with $c(i,j)=1$);
- PA3 obtained to minimize the sum of response times over all call-setup and call-release phases (problem (4) with $c(i,j)=1/\lambda(i)$);
- PA4 in which all tasks have the same priority (that is, priorityless case) and serving as a basis for comparison.

We consider an end-office configuration of the sytem with one type of local subscribers (connected via remote subscriber units) and two types of signalling for incoming and outgoing calls, namely SxS and R2 signalling.

As we mentioned in section 4, the optimal character of priority assignment can be affected by variations in the load profile. It is of interest to what extent this property is important in practice. An exhaustive study on the considered configuration was carried out during which realistic changes in the call mix were assumed. We stated independence of capacity-oriented solutions from variations in the call mix (the same assignment PA1 was obtained in all cases) and insignificant influence of call mix changes on solutions under remaining objectives. This can be explained by the fact that the scenarios for different calls at high level of call control are in great part similar to each other, and this level consumes a significant part of the processor time. Priority assignments PA2÷PA4 were obtained assuming equal intensities for all types of calls. In Table 1, the results for call-handling capacity of the system for these assignments are presented.

Table 1.

Assignment	Capacity call/sec
PA1	6.72
PA2	6.45
PA3	6.67
PA4	6.22

We note a gain of more than 8% in call-handling capacity for assignment PA1 compared with PA4. Corresponding gains in the case of assignments PA2 and PA3 amount to 3.7% and 7.2%, respectively.

The formulation of the capacity-optimization problem assumes rather high utilization of the processor, that is, the traffic conditions typical for heavy load/overload. In fact, the capacity of 6.72 calls/sec corresponds to processor utilization approximately 0.97 (a fraction 0.2 of the processor time is reserved for maintenance, so the effective utilization due to traffic handling amounts to 0.77). It can be thus supposed that the results given in Table 1 should remain valid (at least in terms of tendencies) in sustained overload with the overload control mechanism tunned to regulate the processor utilization on the level 0.97. To validate this conjecture, in Figs 1÷5, two types of curves (theoretical and obtained via simulation) are drawn for some of the characteristics considered. For the clarity-of-drawing reasons, only simulation results are presented in Figs 2÷5. Dashed horizontal lines refer to service constraints.

In Fig.1, the ART characteristics in the system for assignments PA1-PA4 are drawn against the offered load. Note the strong dependence of the ART characteristics on priority assignment and good accordance of theoretical results with the simulation (for assignments PA1-PA3; for clarity reasons, only theoretical curve is drawn for assignment PA4). Referring to Table 1 we conclude that capacity-oriented objectives (capacity and sum of average response times over set-up and release pheses) are quite different in their nature compared to the ART-oriented ones.

Good confirmation of the latter statement can be found in Figs 2-5 where curves for the average value of some selected delays are drawn against the offered load. In Fig.1, average dial-tone delays are drawn. In Fig.3 and Fig.4, curves for the average post-dialing delay for local calls (1) and for outgoing calls (2, R2 signalling assumed) are presented.

We find that only curves for assignment PA1 are all below the corresponding service constraints. Assignment PA3 is relatively good, too. Curves for assignment PA2 are very unstable in the sense that some of them are much below their limit (see Figs 2, 3) while the others unacceptably exceed the limit (Figs 4, 5). A lot also remains to be improved in the priorityless case (see Fig.5).

We conclude that the ART criterion generally is not the proper choice in capacity optimization - it is unreliable in the sense from chapter 1. This is mainly due to the fact that it does not guarantee proper balance between particular delays in exceeding of the corresponding service constraints.

Capacity as the explicit objective is the best choice, although much longer execution times are required. We use the bisection method to evaluate the capacity for a given assignment which, assuming the relative error of 0.1%, results in the execution time approximately ten times longer than in the case of the ART. The sum of response times is more handy as a substitute but poorer reliability is observed compared to capacity evaluated directly.

6. CONCLUSIONS

In the paper, we have identified two different types of objectives which can be of interest in priority assignment optimization considerations in SPC systems. They were described in sections 3 and 5, respectively. To prevent or at least to limit compromising other quantities, one should be aware of the possible tradeoffs imposed by a particular objective to be used. The relation between capacity and the ART in the system described in section 5 is a good example of the above. The priority optimization process is usually subject to various constraints. We have presented a method to cope with the problem in a class of systems modeled by the M/G/1 queue with priorities and feedback. The use of a version of the simulated annealing algorithm is the key to developing an efficient search procedure. However, analytical method to evaluate the objective must exist to take advantage of the algorithm. The latter seems to be the most serious limitation in applying our approach to other models.

Capacity evaluated via 95 percentiles of delays in the system has been chosen as the next step in the development of the approach. Application of the method in more diverse environments (e.g. No 7 signalling) and formulation of general priority assignment rules concerning the 95 percentiles of delays in the system would also be of great interest.

REFERENCES

[1] Richards P. ISDN Traffic Performance Issues in an Evolving Network Environment. **Proc. 5th ITC Specialist Seminar**, Lake Como, Italy, 1987.
[2] Krym M., Richards P. Designing for Performance in Telecommunications Systems. **Proc. 12th ITC**, Torino, 1988.
[3] Petit G., Andries R. Performance Optimization in an M/G/1 System Based on Priority Assignment. **Proc. 12th ITC**, Torino, 1988.
[4] Villen-Altamirano M. Response Time Optimization by Task Priority Assignment. **Electrical Communication**, Vol. 62, No 3/4, 1988.
[5] Bursztynowski D., et. al. A Discrete-Optimization Approach to Priority Assignment in a Switching System. **Proc. 2nd ITC Sem. Teletraff. Th. Comp. Model.**, Moscow, 1989.
[6] Fontana B., et al. Models and Tools for Evaluating the Traffic Handling Performance of System 12 ISDN Exchanges. **Electrical Communication**, Vol. 61, No 1, 1987.
[7] Villen-Altamirano M., Fontana B. Models to Evaluate Response Times in Single-Processor Systems and their Application to a Multiprocessor System. **Proc. 12th ITC**, Torino, 1988.
[8] Dąbrowski M., et al. Hardware and Software Design of a Medium Size Telephone Exchange. **Proc. ISS'84**.
[9] Burakowski W., Villen-Altamirano M. Queue Size Distributions in M/D/1 Queue with Non-Preemptive priorities and Deterministic Feedback. **Proc. 1st ITC Sem. Teletraff. Th. Comp. Model.**, Sophia, 1988.
[10] Kleinrock L. Queueing Systems II. John Wiley & Sons, 1976.
[11] Meilijson I., Weiss G. Multiple Feedback at a Single-Server Station. **Stoch. Proc. Appl.** 5 (1977), pp. 195-205.
[12] Kirkpatrick S., et al. Optimization by Simulated Annealing. **Science** 220 (1983), pp. 671-680.
[13] Syski W. Simulated Annealing Discrete Optimization Package -User Manual. **Technical Report**, Project RPI 02.15, Warsaw University of Technology, Warsaw, 1989.
[14] Farber N. A Model for Estimating the Real-Time Capacity of Certain Classes of Central Processors. **Proc. 6th ITC**, Munich, 1970.

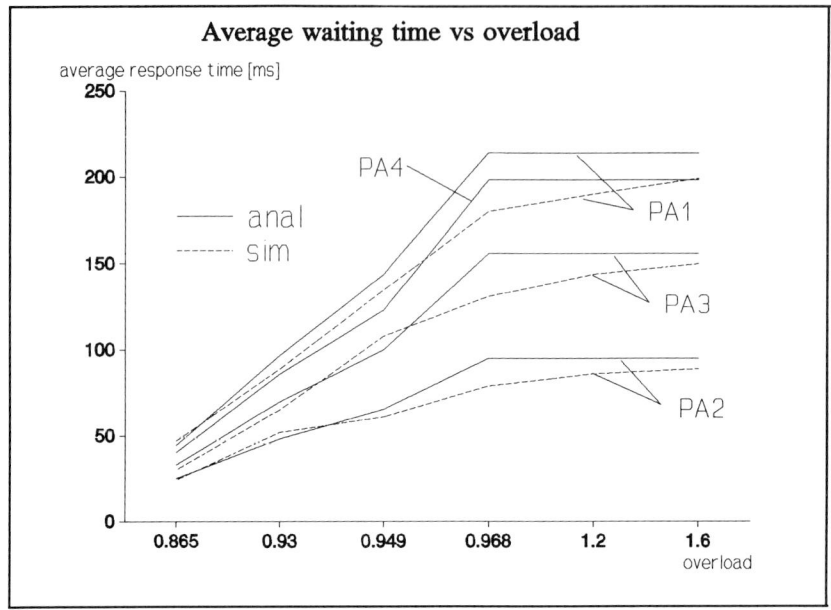

Fig.1.

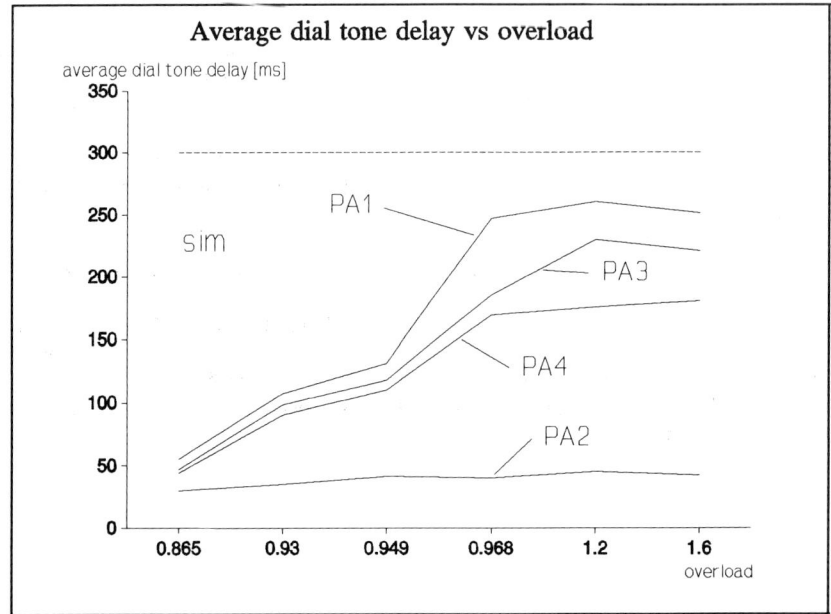

Fig.2.

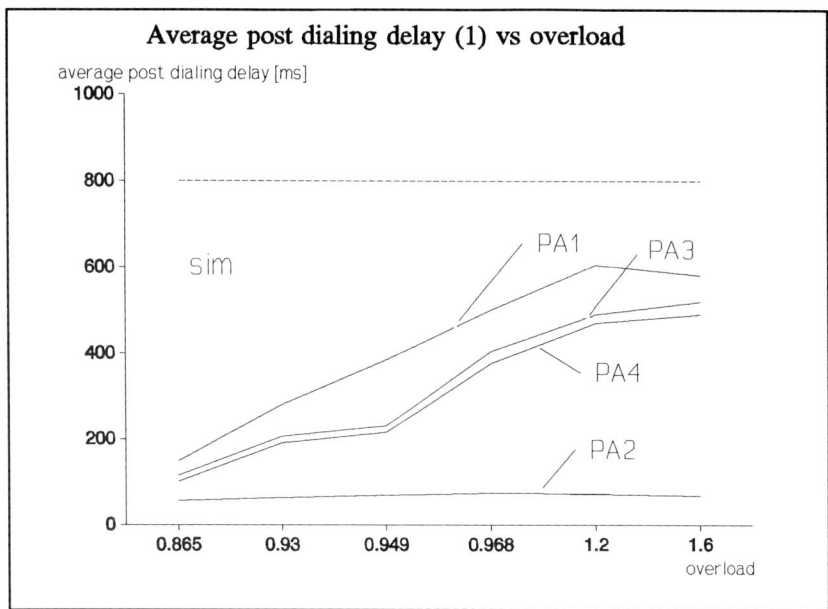

Fig.3.

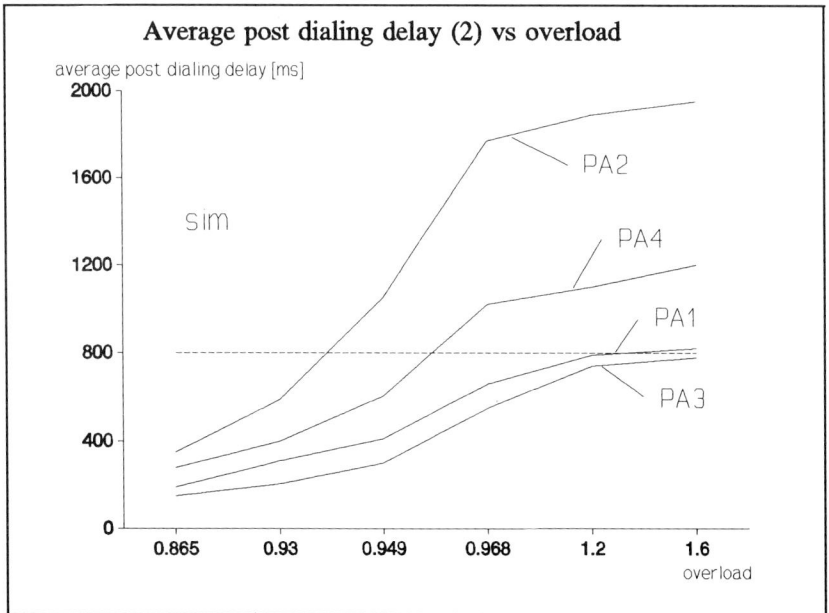

Fig.4.

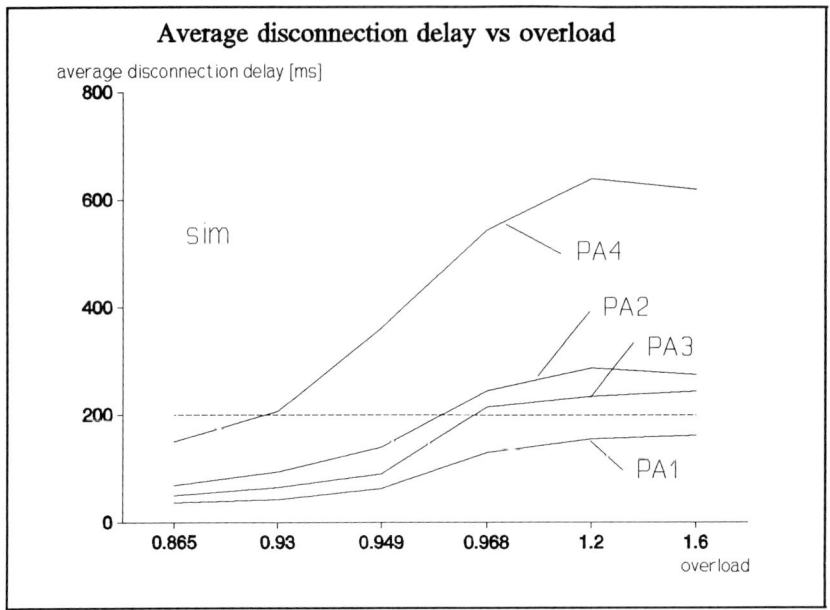

Fig.5.

Overload Control of SPC-switches Using Optimal Alarming

Tobias Rydén[†] and Georg Lindgren[‡]

[†] Department of Communication Systems, Lund Institute of Technology
Box 118, S-221 00 Lund, Sweden

[‡] Department of Mathematical Statistics, Lund Institute of Technology
Box 118, S-221 00 Lund, Sweden

Abstract

A decade ago a theory of optimal alarms for unwanted events in stochastic processes was developed. This paper describes how this theory can be used to derive load regulators for SPC-switches. The results show that regulators based on this theory give better system performance (in a certain sense) compared to simple regulators developed without knowledge of the system structure, but the differences are small for the studied arrival processes. The new regulators are calculated to optimize the stationary performance of the switch, but transient properties are also studied.

1 Introduction

The introduction of fast signalling systems and new services has increased the importance of overload control in telecommunication systems. One of the main problems within the subject is transient performance. These problems arise when the offered load to a switch increases from a low level to a high level very fast, e.g. due to a trunk failure. Earlier studies which have taken transient properties into account, like [3], often have a starting point like "how shall we construct an overload control algorithm which gives good transient performance?". Note that the word "good" is used here, not "optimal", since it is hard to define what is meant by optimal transient performance. The question asked in this paper is "how shall we construct an algorithm which gives optimal *stationary* performance?". It turns out that in this case it is easy to define an optimality criterion that makes sense. When this stationary optimal algorithm has been calculated, its transient properties are investigated.

References to earlier works within the subject overload control can be found in [3], while the paper by Karlstedt [2] gives an overview of the main problems.

1.1 The queueing network model

To analyze and control the behaviour of an SPC-switch, we must first build a mathematical model of the system, based on a number of assumptions and simplifications. We first assume that the switch is controlled by a single *central processor* (CP). There may be additional *regional processors*, but the CP is always involved in the work that needs "intelligence", such as setting up calls. We also assume that the CP is the bottleneck when the traffic carried by the switch increases. Jobs needing service from the CP are placed in a queue in front of the CP and served in order of arrival (FIFO). Next we note that a subscriber making a call needs service from the CP during three well defined phases of the call.

Pre-dialling The CP gives dial-tone to the subscriber immediately after off-hook.

Call set-up When the dialling is completed, the CP sets up the call.

Post-dialling After on-hook the CP clears the connection through the switch that was used during the call.

In existing SPC-switches, each of the three phases consists of a large number of small jobs executed by the CP, but we here make the assumption that each phase consists of *one* larger job. Hence every call needs three services from the CP, so the queueing network shown in fig. 1 constitutes a model of the switch. The server and the queue are the CP and the CP-queue, respectively. An incoming job needs service twice, the second time after being delayed for a random time T, which corresponds to the time needed to complete the dialling. The service needed after on-hook has been omitted because it adds nothing of interest to the model; a discussion of this can be found in [3]. A basic fact from control theory is that systems with delays are harder to control than systems without delays. This is also true for this queueing network model, as some load control algorithms cause load oscillations when the offered load changes rapidly [5]. The gate in front of the switch is used to control the carried load, but we defer a detailed discussion of how this is done until the exact stochastic properties of the model have been given.

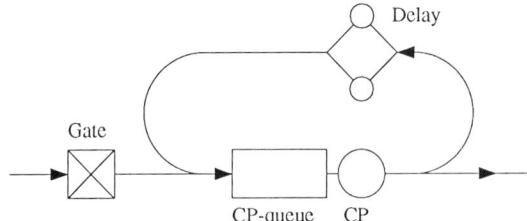

Figure 1: *The queueing network model of the SPC-switch.*

1.2 The arrival process and service times

The service times are assumed to be exponentially distributed with means $1/\mu_1$ and $1/\mu_2$ for the first and second service respectively. As we will see, none of these assumptions are crucial for the calculations done later. Throughout this paper the values $1/\mu_1 = 3$ ms and $1/\mu_2 = 6$ ms will be used. Further, we let $1/\mu = 1/\mu_1 + 1/\mu_2$ denote the mean total service time for a call. One of the main difficulties when trying to control the server load is that the mean of the second service time is so much larger than the mean of the first one. The delay distribution is assumed to be deterministic with length 5 s, which is not very realistic since the time needed for dialling certainly varies from one call to another and from one subscriber to another. However, a deterministic delay is from a transient point of view a worst case since it may cause poor transient performance [5]. As we will see later, simulations of the system with Erlang-r, or gamma, distributed delay show that the effects of this assumption are not drastic.

The arrival process is assumed to be Poissonian with an intensity that at any time may be either "low" or "high". These two intensities will be denoted by Λ_{lo} and Λ_{hi}. Thus the arrival process is a modulated Poisson process, but the process is assumed to be "slow" in the sense that the probabilities of changes from low to high intensity and vice versa can be neglected in the calculations done later on. Thus the process is completely characterized by Λ_{lo}, Λ_{hi} and the probabilities p_{lo} and p_{hi}, where p_{lo} = P(the intensity of the arrival process is low), since neither the transient nor the stationary performance depend on the exact intensity switch mechanism. Four different arrival processes will be studied, and the characteristics of these models are summarized in table 1. This table also shows the offered loads $\hat{\rho}_{\text{lo}} = \Lambda_{\text{lo}}/\mu$ and $\hat{\rho}_{\text{hi}} = \Lambda_{\text{hi}}/\mu$ corresponding to the different intensities, and the probability p_{acc} which will be explained in the next section.

Model	Λ_{lo}	Λ_{hi}	p_{lo}	p_{hi}	$\hat{\rho}_{\text{lo}}$	$\hat{\rho}_{\text{hi}}$	p_{acc}
1	50	150	0.9	0.1	0.45	1.35	0.6296
2	50	110	0.9	0.1	0.45	0.99	0.8586
3	70	105	0.9	0.1	0.63	0.95	0.8995
4	70	105	0.5	0.5	0.63	0.95	0.8995

Table 1: *The different arrival processes used in the paper and the corresponding offered loads and acceptance probabilities.*

1.3 Fundamentals of the load regulators

The gate in front of the server queue is used to control the carried load in the following way. The gate can be in one of two possible states; it can be turned *off*, which means that all new calls are allowed to enter the switch, or it can be turned *on*, which means that each new call is accepted with a fixed probability p_{acc}, independently of other calls.

All regulators treated in this paper are discrete-time regulators, i.e. the state of the gate can only be changed at equidistant time points t_k, called *control times*. The time between two such times is called the *control interval* and is denoted by τ_c. Throughout this paper we will use $\tau_c = 0.5$ s. This choice was made mainly by heuristic reasons; if the arrival intensity changes from low to high and τ_c is too long, a very serious overload scenario may take place. A discussion of the choice of τ_c can be found in [3]. At each control time t_k the regulator must decide in which state the gate shall be until t_{k+1}. This decision must be based only on the gate state between t_{k-1} and t_k, denoted by G_k, and the number of *accepted* arrivals between t_{k-1} and t_k, denoted by A_k. The reason why the arrivals are counted after the gate instead of before will become clear later. It might seem foolish, and certainly is, not to use more information, but this would increase the complexity of the calculations a lot.

Finally we have to decide upon a value for p_{acc}. Before doing so we will take a closer look on the concept of "overload". So far we have been rather vague on this. Now to be precise we say that if the server load during an interval of fixed length τ_ℓ exceeds a specified level ρ_{max}, overload has occurred in that interval. Hence the overload probability is simply

$$p_{\text{rol}} = \text{P(the load during an interval of length } \tau_\ell \text{ exceeds } \rho_{\text{max}}). \tag{1}$$

The above definition does not make sense unless we assume that the queueing network is in statistical equilibrium, i.e. the queueing process is (strictly) stationary, and consequently we do not specify values of p_{rol} unless this is the case. It should be noted though that stationarity only holds "modulo τ_c", i.e. the distribution of the number of jobs in the queue, e.g., is not the same at control times as at times somewhere between two control times.

As numerical values of the variables defining overload we will use $\tau_\ell = 2\tau_c = 1$ s and $\rho_{\max} = 0.95$. This choice of τ_ℓ will turn out to be very convenient later in the calculations. In addition to p_{rol} we will also need another overload probability, denoted by p_{ol} and defined as

$$p_{\text{ol}} = \text{P(the load during an interval of the the form } [t_{k-2}, t_k] \text{ exceeds } \rho_{\max}). \tag{2}$$

As above, we use it only when stationarity holds. Apparently p_{ol} is the probability that overload occurs in an interval of length τ_ℓ starting and ending at control times, whereas p_{rol} is the probability that overload occurs in an interval of length τ_ℓ starting at a random instant. The reason for using two probabilities is that p_{ol} is the one which will be minimized later, but p_{rol} may be at least as interesting. However, it will turn out that they do not differ much.

When the offered load is 0.85 and the gate is turned off, p_{ol} and p_{rol} are both approximately 0.13, which is reasonably low. Hence p_{acc} should be chosen so that the intensity $\Lambda_{\text{hi}} p_{\text{acc}}$ corresponds to the load 0.85. The resulting values of p_{acc} for the different models are shown in table 1. The probabilities p_{ol} and p_{rol} are unfortunately very difficult to calculate exactly or numerically, and the value 0.13 is a simulation result.

1.4 A simple load regulator

From a heuristic point of view, it is dangerous to have the gate turned off when the arrival intensity is high. Therefore, it is natural to consider a regulator which at every control time tests the hypothesis $H: \lambda_k = \Lambda_{\text{hi}}$ versus the alternative $K: \lambda_k = \Lambda_{\text{lo}}$, where λ_k is the arrival intensity in the interval $[t_{k-1}, t_k]$. If H is rejected, the gate is turned off in the interval $[t_k, t_{k+1}]$. As mentioned above, the only information that may be used in the test is A_k, the number of accepted arrivals in $[t_{k-1}, t_k]$, and G_k, the gate state during the same interval. The test used was the Neyman-Pearson most powerful test [1] which rejects H if $A_k \leq n$, where n is a suitable limit, which may depend on the gate state G_k.

Note that we must take G_k into account, since the expectation of A_k under H is $\Lambda_{\text{hi}} \tau_c$ when the gate was turned off, and $\Lambda_{\text{hi}} p_{\text{acc}} \tau_c$ when the gate was turned on. Thus we get two limits, denoted by n_{off} and n_{on}, and the regulator turns the gate off during the interval $[t_k, t_{k+1}]$ if $A_k \leq n_q$, where $q = G_k$. The limits are given by the *level* of the test, denoted by α and defined by $\alpha = \text{P}(H \text{ is rejected}|H \text{ is true})$. In a similar fashion the *power* of the test versus K is defined by $1 - \beta = \text{P}(H \text{ is rejected}|K \text{ is true})$. Table 2 shows the values of n_{off} and n_{on} obtained for the different arrival process models. Note that α has not been chosen the same in all cases, since a too low level makes the power versus K too low. Summing up we have

Model	α	n_{off}	n_{on}	$1-\beta$ G_k=off	$1-\beta$ G_k=on
1	0.001	49	26	≈ 1	0.994
2	0.01	38	31	0.994	0.980
3	0.05	40	35	0.825	0.768
4	0.05	40	35	0.825	0.768

Table 2: *The limits for the simple load regulators, the levels of the tests and the powers versus K.*

Load control algorithm 1: THE SIMPLE REGULATOR. At t_k, turn the gate off until t_{k+1} if $A_k \leq n_q$, where $q = G_k$ and the limits n_{off} and n_{on} are given by table 2. Otherwise turn the gate on.

2 Optimal alarming in the queueing network model

We will now develop a regulator that at every control time makes the same type of choice as the simple regulator developed in section 1.4, i.e. decides whether the gate shall be turned off during the next control interval or not. The information used in the decision is also the same, A_k and G_k. This new regulator, however, tries to avoid overload in an *optimal way*.

2.1 Basic assumptions, simplifications and calculations

Before we derive the new regulator, we note that if the gate is turned off when the arrival intensity is high, the processor load will increase a little at once, but even more after a period T, when the jobs need service again. This is because $1/\mu_2 > 1/\mu_1$. When trying to decide whether the gate should be turned off or not, it is therefore natural to take a look at what might happen after a time T.

The arrivals in the interval $[t_{k-1}, t_k]$ will return from the delay in $[t_{k+9}, t_{k+10}]$ since T is constant and equal to $10\tau_c$. This is an approximation, as we have neglected the queueing delay before the first service. However, when the load is 0.85, the "target load", the mean of this delay is about 17 ms, which is 29 times smaller than τ_c. Letting B_k denote the number of jobs returning from the delay in $[t_{k-1}, t_k]$ we can write the approximation as $B_k = A_{k-10}$. We also make the approximation that the delayed arrivals form a Poisson process. This assumption is also far from true, but makes it easier to do the simulations described in the next section. Since the total number of arrivals to the queue in the interval $[t_{k+9}, t_{k+11}]$ (of length τ_ℓ) equals $A_{k+10} + B_{k+10} + A_{k+11} + B_{k+11}$ we can control the load in that interval by turning the gate off or on during $[t_k, t_{k+1}]$, thus affecting $A_{k+1} = B_{k+11}$. Now, to avoid overload in the interval $[t_{k+9}, t_{k+11}]$ in a kind of optimal way we turn the gate on if

$$\phi(n,q) = \frac{P(A_k = n, G_k = q | C'_{k+11})}{P(A_k = n, G_k = q | C_{k+11})} \leq \phi_{\max} \qquad (3)$$

where C_{k+11} is the event that overload occurs in $[t_{k+9}, t_{k+11}]$ and $\phi_{\max}$ is a suitable limit. The reason for choosing this rule is that if the switch was not regulated, this ratio is the *optimal alarm* for the event C_{t+11} [4], and if we get an alarm we turn the gate on. Optimal is here used in the sense that the test above gives the largest probability of overload when alarming out of all tests with the same overload detection ability. The theory of optimal alarm systems does unfortunately not carry over to the case of controlled processes, and the approach taken here can thus be seen as "experimental".

It is obvious that both the overload probability $P(C_{k+11} | A_k, G_k)$ and the posterior probability that the arrival intensity is high, are increasing in A_k, implying that $\phi(n, q)$ is non-increasing in n for each q. In other words, the gate will be turned off during the next interval if A_k is less than or equal to a limit n_q, where $q = G_k$. Apparently this new regulator works exactly as the simple regulator, but the limits n_{off} and n_{on} are calculated in a different way. From a heuristic point of view the function of the regulator can be explained as follows; since $\tau_\ell = 2\tau_c$ we may "afford" to turn the gate off during the second half of the load measurement interval if there were few arrivals during the first half. Throughout this section we assume that stationarity holds in the sense of section 1.3.

The main difficulty when using this kind of regulators is to calculate the likelihood ratio, and we devote the rest of the section to this problem. First we apply Bayes rule to the denominator in Eq. (3) and get

$$P(A_k = n, G_k = q | C_{k+11}) = \frac{P(C_{k+11} | A_k = n, G_k = q) P(A_k = n, G_k = q)}{\sum_{m,r} P(C_{k+11} | A_k = m, G_k = r) P(A_k = m, G_k = r)}. \qquad (4)$$

Obviously it suffices to calculate two types of probabilities, $P(C_{k+11} | A_k = n, G_k = q)$ and $P(A_k = n, G_k = q)$.

Calculation of $P(C_{k+11}|A_k = n, G_k = q)$

Conditioning on what is of interest in $[t_{k+9}, t_{k+11}]$ we get

$$P(C_{k+11}|A_k = n, G_k = q) = \sum_{r,s,\Lambda} P(C_{k+11}|A_k = n, G_k = q, \lambda_k = \Lambda, G_{k+10} = r, G_{k+11} = s) \qquad (5)$$
$$\cdot P(\lambda_k = \Lambda, G_{k+10} = r, G_{k+11} = s|A_k = n, G_k = q).$$

The probability $P(C_{k+11}|A_k = n, G_k = q, \lambda_k = \Lambda, G_{k+10} = r, G_{k+11} = s)$ is obtained through simulation, a procedure described in detail in the next section. We proceed as

$$P(\lambda_k = \Lambda, G_{k+10} = r, G_{k+11} = s|A_k = n, G_k = q) = \qquad (6)$$
$$P(G_{k+10} = r, G_{k+11} = s|\lambda_k = \Lambda, A_k = n, G_k = q)P(\lambda_k = \Lambda|A_k = n, G_k = q).$$

Since the arrival process is "slow" the occurrence of the event $\lambda_k = \Lambda$ implies $\lambda_i = \Lambda$ for all i. To obtain $P(G_{k+10} = r, G_{k+11} = s|\lambda_k = \Lambda, A_k = n, G_k = q)$, note that the gate states $\{G_{k-1}, G_k\}$ form a discrete-time Markov chain according to fig. 2. In fact we have two chains, one for each

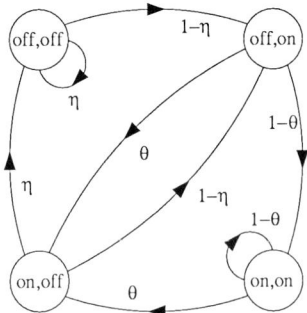

Figure 2: *The Markov chain that describes the gate states $\{G_{k-1}, G_k\}$. The transition probabilities of the chain are $\eta = \eta(\Lambda) = Po(\Lambda\tau_c; n_{\text{off}})$ and $\theta = \theta(\Lambda) = Po(\Lambda p_{\text{acc}}\tau_c; n_{\text{on}})$, where $Po(u; n)$ denotes the probability that a Poisson random variable with mean u is less than or equal to n, and Λ is the arrival intensity.*

arrival intensity. Also note that in order to establish the transition probabilities we must know n_{off} and n_{on}. But the purpose of the calculations in this section is to obtain the optimal choice of these quantities! Thus we must iterate the algorithm; we start with some values of n_{off} and n_{on} which we use to calculate new values of these limits, and so on. For the arrival process models in this paper, this procedure converges when the limits of the simple regulators are used as starting values.

We now have, suppressing the dependence on the arrival intensity,

$$P(G_{k+10} = r, G_{k+11} = s|\lambda_k = \Lambda, A_k = n, G_k = q) = \begin{cases} p^{(10)}_{\text{off,off;rs}} & \text{if } n \leq n_q \\ p^{(10)}_{\text{off,on;rs}} & \text{if } n > n_q \end{cases} \qquad (7)$$

where $p^{(10)}$ are the ten-step transition probabilities of the Markov chain. To calculate $P(\lambda_k = \Lambda|A_k = n, G_k = q)$ we rewrite this probability as

$$P(\lambda_k = \Lambda|A_k = n, G_k = q) = \frac{P(A_k = n|\lambda_k = \Lambda, G_k = q)P(\lambda_k = \Lambda, G_k = q)}{\sum_{\Gamma} P(A_k = n|\lambda_k = \Gamma, G_k = q)P(\lambda_k = \Gamma, G_k = q)}. \qquad (8)$$

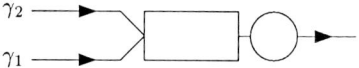

Figure 3: *The simplified queueing model used to simulate the overload probabilities* $P(C_{k+11}|A_k = n, G_k = q, \lambda_k = \Lambda, G_{k+10} = r, G_{k+11} = s)$.

We first note that $P(A_k = n|\lambda_k = \Lambda, G_k = q)$ is known since A_k, given λ_k and G_k, has a Poisson distribution. Further, $P(\lambda_k = \Lambda, G_k = q) = P(G_k = q|\lambda_k = \Lambda)P(\lambda_k = \Lambda)$, where the probability $P(\lambda_k = \Lambda)$ is known since it is part of the arrival process model, and, suppressing the dependence on Λ, $P(G_k = q|\lambda_k = \Lambda) = p_{\text{off},q} + p_{\text{on},q}$, where $p_{\text{off,off}}$ etc. are the equilibrium state probabilities of the Markov chain.

Calculation of $P(A_k = n, G_k = q)$

The second kind of probabilities in Eq. (4) are now easily obtained. We have

$$P(A_k = n, G_k = q) = \sum_\Lambda P(A_k = n|G_k = q, \lambda_k = \Lambda)P(G_k = q|\lambda_k = \Lambda)P(\lambda_k = \Lambda) \qquad (9)$$

and all these probabilities are known. Since $P(C'_{k+11}|A_k = n, G_k = q) = 1 - P(C_{k+11}|A_k = n, G_k = q)$, we now have an expression for the likelihood ratio $\phi(n, q)$.

2.2 Simulation of overload probabilities

In the previous section we were able to analytically derive all probabilities but one, namely $P(C_{k+11}|A_k = n, G_k = q, \lambda_k = \Lambda, G_{k+10} = r, G_{k+11} = s)$, which must be simulated. To do this a simplified model of the queueing network was used, according to fig. 3. In this model there are two arrival streams with intensities γ_1 and γ_2, corresponding to new and delayed jobs respectively. It is clear that the stream of new jobs is a Poisson process, and as mentioned in the previous section we assume that this also holds for the stream of delayed jobs. Since we want to find out the value of the probability above this single server queue was simulated during a time of length $\tau_\ell = 2\tau_c$.

As mentioned above, $\lambda_k = \Lambda$ implies $\lambda_i = \Lambda$ for all i, and consequently we have $\gamma_1 = \Lambda$ during the first half of the simulation interval if G_{k+10}=off and $\gamma_1 = \Lambda p_{\text{acc}}$ otherwise, and similarly for the second half of the interval if we replace G_{k+10} by G_{k+11}. Since $B_{k+10} = A_k$ we have n delayed arrivals during the first half of the interval, and by well-known properties of the Poisson process, the instants of these arrivals are independent and uniformly distributed. This property is shared by the double stochastic Poisson processes, but not by any other stationary point process, hence assuming the process of delayed arrivals be non-Poissonian would make the simulations considerably harder to perform. During the second half of the simulation interval the delayed arrivals form a Poisson process with intensity Λ. Thus we assume that G_{k+1}=off, regardless of the actual values of A_k and G_k. This choice was made since at t_k we want to test "overload will occur in $[t_{k+9}, t_{k+11}]$ if G_{k+1}=off" versus "overload will not occur if G_{k+1}=off".

Now it remains to specify the distribution of the number of jobs in the queue at the start of the simulation. If $\Lambda = \Lambda_{\text{lo}}$, this distribution was chosen as the equilibrium distribution for $\gamma_1 = \gamma_2 = \Lambda_{\text{lo}}$ since, intuitively, the regulator should seldom turn the gate on when the arrival intensity is low. When $\Lambda = \Lambda_{\text{hi}}$ the distribution was chosen as the equilibrium distribution for $\gamma_1 = \gamma_2 = \Lambda_{\text{hi}} p_{\text{acc}}$, since the regulator should seldom turn the gate off when the arrival intensity is high. Though it is possible to obtain these equilibrium distributions [6], these calculations were not performed. Instead the queue was run for a while to reach equilibrium before the "real" simulation was started.

2.3 Carried load

The carried load, given the arrival intensity, is

$$\rho_x = \hat{\rho}_x \{ P(G_k = \text{off} | \lambda_k = \Lambda_x) + P(G_k = \text{on} | \lambda_k = \Lambda_x) p_{acc} \} \qquad (10)$$

and since it depends on n_{off} and n_{on} only, this expression also holds when a simple regulator is used.

The optimal test was used to find limits n_{off} and n_{on} for the different arrival processes and the result is shown in table 3. This table also shows the values of ϕ_{max} that were used to calculate the limits of

Model	Simple regulator		Complex regulator		ϕ_{max}
	n_{off}	n_{on}	n_{off}	n_{on}	
1	49	26	50	26	10
2	38	31	47	20	30
3	40	35	45	27	160
4	40	35	44	30	70

Table 3: *The limits for the different simple and complex regulators and the corresponding values of ϕ_{max}.*

the complex regulators. These values were not chosen according to any rule, but to get the stationary loads for the simple and complex regulators approximately equal. Summing up we have

Load control algorithm 2: THE COMPLEX REGULATOR. At t_k, turn the gate off until t_{k+1} if $A_k \leq n_q$, where $q = G_k$ and the limits n_{off} and n_{on} are given by table 3. Otherwise turn the gate on.

3 Comparison of different regulators

In this section we compare the simple and complex regulators for the different arrival processes. Since the only performance measure of interest for which we have an analytic expression is carried load, most of the results were obtained through simulation.

3.1 Stationary properties

Table 4 summarizes the stationary properties of the different regulators. The regulators for model 1 were never compared since they are almost identical. The general result seen from the table is that the complex regulators give better performance than the simple ones, but the differences are very small. The largest differences are found for models 3 and 4. For these models the powers of the tests performed by the simple regulators are small compared to models 1 and 2. It seems natural that this property should be in favour of the complex regulator.

Model	$\rho_{\text{lo}}^{(\text{sim})}$	$\rho_{\text{hi}}^{(\text{sim})}$	$\rho_{\text{tot}}^{(\text{sim})}$	$\rho_{\text{lo}}^{(\text{com})}$	$\rho_{\text{hi}}^{(\text{com})}$	$\rho_{\text{tot}}^{(\text{com})}$	$\rho_{\text{tot}}^{(\text{com})} - \rho_{\text{tot}}^{(\text{sim})}$
2	0.44963	0.85113	0.48978	0.45000	0.85002	0.49000	0.022%
3	0.61824	0.85377	0.64179	0.62060	0.85014	0.64355	0.176%
4	0.61824	0.85377	0.73601	0.62260	0.85057	0.73658	0.057%
Model	$p_{\text{ol,lo}}^{(\text{sim})}$	$p_{\text{ol,hi}}^{(\text{sim})}$	$p_{\text{ol,tot}}^{(\text{sim})}$	$p_{\text{ol,lo}}^{(\text{com})}$	$p_{\text{ol,hi}}^{(\text{com})}$	$p_{\text{ol,tot}}^{(\text{com})}$	$p_{\text{ol,tot}}^{(\text{com})} - p_{\text{ol,tot}}^{(\text{sim})}$
2	0.00000	0.13510	0.01351	0.00000	0.13050	0.01305	-0.046%
3	0.00000	0.13946	0.01395	0.00004	0.13091	0.01313	-0.082%
4	0.00000	0.13946	0.06973	0.00004	0.13189	0.06597	-0.376%
Model	$p_{\text{rol,lo}}^{(\text{sim})}$	$p_{\text{rol,hi}}^{(\text{sim})}$	$p_{\text{rol,tot}}^{(\text{sim})}$	$p_{\text{rol,lo}}^{(\text{com})}$	$p_{\text{rol,hi}}^{(\text{com})}$	$p_{\text{rol,tot}}^{(\text{com})}$	$p_{\text{rol,tot}}^{(\text{com})} - p_{\text{rol,tot}}^{(\text{sim})}$
2	0.00000	0.13256	0.01326	0.00000	0.13212	0.01321	-0.005%
3	0.00004	0.13911	0.01395	0.00008	0.12791	0.01286	-0.109%
4	0.00004	0.13911	0.06958	0.00000	0.13235	0.06618	-0.340%

Table 4: *The stationary performance measures carried load (ρ) and the overload probabilities p_{ol} and p_{rol} for the different regulators. Figures with superscript* (sim) *refer to simple regulators while figures with superscript* (com) *refer to complex regulators. Figures with subscript* tot *are calculated from the corresponding* lo *and* hi *entries with respect to p_{lo} and p_{hi}, i.e. $\rho_{\text{tot}}^{(\text{sim})} = \rho_{\text{lo}}^{(\text{sim})} p_{\text{lo}} + \rho_{\text{hi}}^{(\text{sim})} p_{\text{hi}}$ etc. The differences $p_{\text{ol,tot}}^{(\text{com})} - p_{\text{ol,tot}}^{(\text{sim})}$ and $p_{\text{rol,tot}}^{(\text{com})} - p_{\text{rol,tot}}^{(\text{sim})}$ are negative with statistical significance (two-sided, $p \leq 5\%$) for models 3 and 4.*

3.2 Transient properties

To evaluate the transient properties of the different regulators, simulations with a step change in the arrival intensity were run. The results for model 2 are shown in fig. 4. In both simulations the arrival intensity increases from 50 s^{-1} to 110 s^{-1} at $t=10$ s and the diagrams show the average server load of 1000 independent runs. Apparently the differences in performance are negligible and this is the

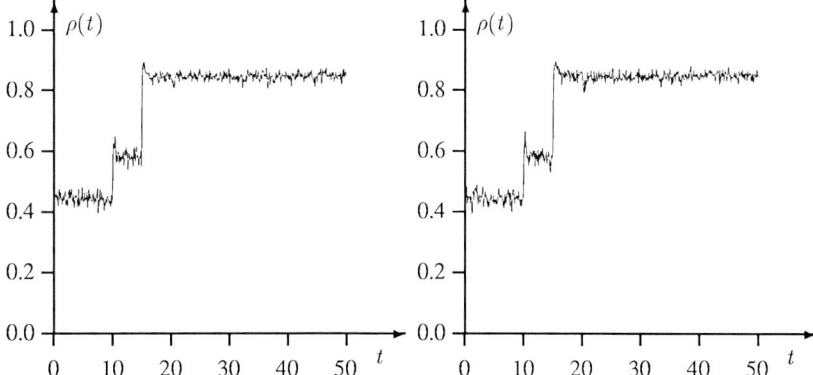

Figure 4: *The transient performance of model 2 when using the simple regulator (left) and the complex regulator (right). At $t=10$ s the arrival intensity increases from 50 s^{-1} to 110 s^{-1}.*

case for all arrival process models. Note that the load does not reach the level 0.85 until $t=15$ s, the time when the intensity from the delay increases.

3.3 Non-deterministic delay

As mentioned in the beginning of the paper, the deterministic delay T does not reflect the function of a real SPC-switch very well. A more realistic delay distribution is for example the Erlang-r distribution or, equivalently, the gamma distribution. It is unfortunately more or less impossible to

Model	$\rho_{lo}^{(sim)}$	$\rho_{hi}^{(sim)}$	$\rho_{tot}^{(sim)}$	$\rho_{lo}^{(com)}$	$\rho_{hi}^{(com)}$	$\rho_{tot}^{(com)}$	$\rho_{tot}^{(com)} - \rho_{tot}^{(sim)}$
3	0.61824	0.85377	0.64179	0.62060	0.85014	0.64355	0.176%
4	0.61824	0.85377	0.73601	0.62260	0.85057	0.73658	0.057%
Model	$p_{ol,lo}^{(sim)}$	$p_{ol,hi}^{(sim)}$	$p_{ol,tot}^{(sim)}$	$p_{ol,lo}^{(com)}$	$p_{ol,hi}^{(com)}$	$p_{ol,tot}^{(com)}$	$p_{ol,tot}^{(com)} - p_{ol,tot}^{(sim)}$
3	0.00004	0.13501	0.01354	0.00000	0.12679	0.01268	-0.086%
4	0.00004	0.13501	0.06753	0.00000	0.12676	0.06338	-0.415%
Model	$p_{rol,lo}^{(sim)}$	$p_{rol,hi}^{(sim)}$	$p_{rol,tot}^{(sim)}$	$p_{rol,lo}^{(com)}$	$p_{rol,hi}^{(com)}$	$p_{rol,tot}^{(com)}$	$p_{rol,tot}^{(com)} - p_{rol,tot}^{(sim)}$
3	0.00004	0.13469	0.01351	0.00001	0.12809	0.01282	-0.069%
4	0.00004	0.13469	0.06737	0.00001	0.12828	0.06415	-0.322%

Table 5: *The stationary performance measures carried load (ρ) and the overload probabilities p_{ol} and p_{rol} for the different regulators of model 3 and 4 with E_{40}-distributed delay. Figures with superscript (sim) refer to simple regulators while figures with superscript (com) refer to complex regulators. Figures with subscript tot are calculated from the corresponding lo and hi entries with respect to p_{lo} and p_{hi}, i.e. $\rho_{tot}^{(sim)} = \rho_{lo}^{(sim)} p_{lo} + \rho_{hi}^{(sim)} p_{hi}$ etc. The differences $p_{ol,tot}^{(com)} - p_{ol,tot}^{(sim)}$ and $p_{rol,tot}^{(com)} - p_{rol,tot}^{(sim)}$ are all negative with statistical significance (two-sided, $p \leq 5\%$).*

calculate the optimal test for a system with this delay, since the new arrivals during a control interval return from the delay during more than one control interval. Instead simulations of the system with E_{40}-distributed delay were run to study the stationary performance of the different regulators obtained for deterministic delay, and the results of these simulations are shown in table 5. As can be seen from the table, the figures for the complex regulators are still better.

3.4 A more complex arrival process

To investigate the performance of the regulators for more complex arrival processes than above, the system was simulated with Markov modulated Poisson arrival processes. These processes had three intensity levels and were modulated in discrete time in the sense that the intensity changed only at control times, according to a discrete time Markov chain. The regulators used was the regulators of models 3 and 4 above, and the results are shown in table 6. As can be seen, the differences in performance between the simple and complex regulators are now even smaller. The complex regulators give slightly larger loads and the simulations indicate smaller overload probabilities, but these latter improvements are not statistically significant.

Regulator	3, Simple	3, Complex	4, Simple	4, Complex
Intensities	$(70, 87.5, 105)$ s^{-1}		$(70, 87.5, 105)$ s^{-1}	
Transition matrix	$\begin{pmatrix} 0.99 & 0.01 & 0.00 \\ 0.20 & 0.20 & 0.60 \\ 0.02 & 0.03 & 0.95 \end{pmatrix}$		$\begin{pmatrix} 0.99 & 0.01 & 0.00 \\ 0.10 & 0.20 & 0.70 \\ 0.01 & 0.01 & 0.98 \end{pmatrix}$	
Stationary state probabilities	$(0.772, 0.018, 0.210)$		$(0.556, 0.012, 0.432)$	
ρ	0.66992	0.67016	0.72150	0.72231
$\rho^{(com)} - \rho^{(sim)}$	0.024%		0.081%	
p_{ol}	0.02317	0.02205	0.05336	0.05269
$p_{ol}^{(com)} - p_{ol}^{(sim)}$	-0.112%		-0.067%	
p_{rol}	0.02248	0.02141	0.05403	0.05272
$p_{rol}^{(com)} - p_{rol}^{(sim)}$	-0.107%		-0.131%	

Table 6: *The stationary performance measures carried load (ρ) and the overload probabilities p_{ol} and p_{rol} when a Markov modulated Poisson process was offered the system. Figures with superscript* (sim) *refer to simple regulators while figures with superscript* (com) *refer to complex regulators. None of the differences* $p_{ol}^{(com)} - p_{ol}^{(sim)}$ *and* $p_{rol}^{(com)} - p_{rol}^{(sim)}$ *is negative with statistical significance.*

4 Conclusions

We have seen that it is possible to improve the performance of the switch by using regulators based on the theory of optimal alarms. However, the differences are small and these regulators have obvious drawbacks; to construct them one must know the internal structure of the switch and the structure of the arrival process, but having the simulation results for the modified models in mind, these drawbacks may not be as serious as at first sight. Of course, one may also take the opposite point of view and say that the simple regulators are indeed very good. It should be noted though that the arrival processes in this paper are badly suited for prediction; since they are "slow" and since the Poisson process has independent increments one can just predict the global behaviour of the process (the intensity), but not the local behaviour (the exact arrival instants during the next control interval).

Acknowledgement

We would like to thank professors Ulf Körner and Bengt Wallström for many discussions on overload control.

References

[1] Bickel, P. J., Doksum, K. *Mathematical Statistics*, Holden-Day, 1977.

[2] Karlstedt, T. *Overload Control of SPC Systems — A Survey*, The 7:th Nordic Teletraffic Seminar, Lund, 1987.

[3] Körner, U., Wallström, B., Nyberg, C. *Overload Control of SPC-systems*, Technical Report 101, Department of Communication Systems, Lund University, 1989.

[4] de Maré, J. *Optimal Prediction of Catastrophes with Applications to Gaussian Processes*, Annals of Probability, vol 8, 1980, pp 841–850.

[5] Nyberg, C., Körner, U., Wallström, B. *Two Short Papers on Overload Control of Switching Nodes*, Technical Report 102, Department of Communication Systems, Lund University, 1990.

[6] Wallström, B. *On the M/G/1 Queue with Several Classes of Customers Having Different Service Time Distributions*, Technical Report, Department of Communication Systems, Lund University, 1980.

DESIGN OF TWO-LEVEL PSTN

M.Pióro, A.Tomaszewski, J.Lubacz, M.Jarociński

Institute of Telecommunications, Warsaw University of Technology
Nowowiejska 15/19, 00-665 Warszawa, Poland

Abstract

The paper concerns engineering of two-level telephone networks. A methodology of network design is proposed, and a network design support system TOOLNET employing it is described. The design of the Warsaw metropolitan network is used as an illustration of the application of the system.

1. INTRODUCTION

Network digitalization, together with network-wide introduction of common channel signalling (SS7), provide favourable conditions for structural and functional changes of PSTN. The theoretical and practical advantages of a non-hierarchical network structure and traffic routing are well recognized. One of the aspects of structural changes is the potential decrease of the number of levels in network hierarchy (network plains).

In this paper we consider a structure in which network nodes (exchanges) belong to one of two levels: a transit level and a local level. The transit level is assumed to be of a non-hierarchical, meshed structure; nodes of the local level are connected to one or more nodes of the transit level with low loss links. Connections between local level nodes are established through direct links, if such exist, or routed through the transit level. Nodes of the transit level may be assumed to generate their own traffic.

Clearly, such a network structure requires that nodes of the transit level be digital and employ SS7, so it could be considered for PSTN in which the digitalization process is well advanced. Somewhat surprisingly, the two-level network structure is also of considerable interest for environments in which digitalization is just at its starting point. An example are network development projects considered in Poland; construction of a two-level network is currently initiated in Warsaw, and a similar structure is also considered for the Polish intercity network.

The application of a two-level network structure in an analog environment was in the first place motivated by technical circumstances; traffic carrying effectiveness was an important, but secondary aspect. The idea is to create, from scratch, a fully digital transit network level; all existing analog exchanges become local level nodes. The number of nodes in the transit level may be relatively small; e.g. for the 50-node Warsaw area only

8 transit exchanges are foreseen. In spite of this, after the transit level is created, each newly implemented digital local exchange will be introduced into an environment which is virtually digital (as the majority of traffic will be carried and switched by digital equipment). In this way the inherent problems of an overlay strategy of network digitalization is considerably relaxed. The concept is especially attractive if a rapid growth of telephone density is planned in an obsolete analog environment.

The paper is organized as follows. In Sec.2 we present in short an approach to engineering two-level networks. Sec.3 describes the network design support environment TOOLNET which was developed at the Institute of Telecommunication, Warsaw University of Technology; TOOLNET was developed under contract with from the Polish Ministry of Telecommunications. Sec.4 presents sample results concerning the application of the proposed methodology and TOOLNET to the design of the two-level metropolitan network of Warsaw.

2. TWO-LEVEL NETWORK DESIGN ISSUES

The design process assumes that network nodes location and the traffic generated between each origin-destination (OD) node pair are given.

The design process of a two-level network comprises the following tasks:

T1: a) distributing the network nodes between the transit and the local level,
 b) connecting local level nodes to transit level nodes,

T2: a) planning trunk groups of the local level, together with
 b) planning traffic routing schemes for all OD pairs starting or terminating at the local level,

T3: a) planning trunk groups of the transit level, together with
 b) planning alternative routing patterns.

The tasks are strongly interdependent. In practical design problems several constraints are imposed on the initial and eventual network structure. For example the number of nodes in the transit level may be limited, the flexibility of structuring the local network level may be limited by network operator's requirements concerning utilisation of existing transmission and switching resources, etc.

Tasks T1 and T2 both involve the usual tradeoff concerning hierarchical network planning. On one hand, if a call is not set up on direct trunk group it uses more switching and transmission resources. On the other hand, because of PCM systems modularity, providing OD pairs with direct trunk groups may result in poor trunk group's capacity utilization. Thus concentrating traffic is necessary. In the two-level network structure this implies concentrating traffic on trunks connecting local level nodes with the transit level. The overall effectiveness of traffic carrying is additionally increased if alternative routing is applied in the transit level.

Task T3 involves the problems of selecting the most economical routing strategy in the transit level and designing the transit network for the chosen strategy. The problem of making the right choice of a routing strategy for a newly formed transit level is addressed in a companion paper [1]. It is demonstrated that a simple fixed alternative routing with a limited access to overflow paths, though with an off-line reconfiguration facility, provides an economical yet still simple solution.

We have implemented a set of algorithms which provide solutions to problems underlying tasks T1-T3. A detailed description of the algorithms is given in [2]. Here we shall only briefly summarise the main ideas.

Task T1 is approached through a monotonic optimisation method. The starting point is a network structure with all nodes belonging to the transit level. A node is "pushed down" to the local level if this provides reduction of the total network cost. The selection of nodes to be pushed down is performed in an iterative process. In each iteration cycle all nodes are tried and the node that yields greatest cost reduction is located in the local level. The "pushing down" process involves choosing the best (cheapest) connection of the node being pushed down with a transit level node ("parent" node). The evaluation of the network cost is done with the assumption that the whole traffic between local level nodes is routed through the transit level.

Task T2 is performed with traditional hierarchical networks dimensioning methods (cf.[3]). Direct traffic routing in the transit level is assumed; the marginal costs of the transit level trunk groups are modified so as to take into account the efficiency of alternative routing. The dimension of both local and transit level trunk groups are multiples of a basic module.

Task T3 is performed in two steps. First, a continuous trunk group dimensioning is accomplished. The optimisation is approached with Rosen gradient method. Blocking probabilities for fresh and overflow traffic are treated as the optimisation variables. Trunk group dimensions, sequences of alternative paths and dynamic trunk reservation parameters are computed as to minimise transit level cost while providing the required quality of service.

This algorithm yields a good initial solution for modular dimensioning. Modular dimensioning employs a simulated annealing approach. A set of network modification operations for tuning the current solution is defined; for each OD pair the modification concerns the routing pattern and the size of the direct trunk group. The sequence of network modifications is generated by choosing randomly network OD pairs and performing the modification which yields minimal expected cost of the modified network. Heuristic approximations are used to assess this cost.

A separate algorithm, based on a modification of the basic modular dimensioning algorithm, is used to minimise the dispersion of end-to-end losses in transit level OD pairs.

Tasks T1-T3 are basically performed sequentially. A practical design process requires however that the tasks are iterated. This follows not only from the complex interdependence of elementary design tasks, but also from the way network components cost is assessed. In intermediate phases of a design the cost is evaluated upon currently assumed configuration of trunk groups. The configuration is however an object of optimisation and thus is cost sensitive. To resolve this mutual interdependence we have chosen to use a transmission network planning algorithm to evaluate average incremental costs of trunk groups. The algorithm determines the set of node disjoint paths for each required trunk group and allocates the trunk group's capacity among them. The number of paths in the set depends on the chosen degree of spatial diversity. The method employs a shortest path labelling algorithm.

3. TOOLNET

The two-level network design methodology described in the previous section is supported by system TOOLNET created at the Warsaw University of Technology under contract from Polish Ministry of Telecommunications.

TOOLNET has a modular multi-layer functional structure. Two main layers are distinguished, both built around Project Database (Fig.3.1): the User Interface layer and the Design Tools layer.

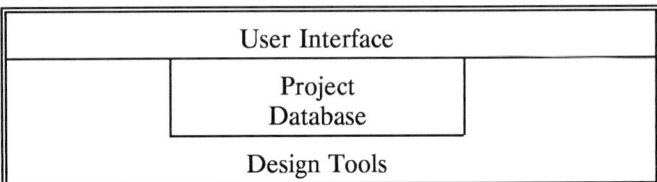

Fig.3.1. TOOLNET system structure

The **User Interface** is the only visible part of the system; its main task is to make the system user-friendly and to integrate the set of system functions (rather then programs). It contains a menu-driven Manager program which activates particular system functions/tools and a number of input/output programs which serve to examine and modify data at any stage of a network design process.

In general, the network data are presented in a graphic form. The type of network elements to be displayed can be selected, e.g. "local-level exchanges and trunk groups connecting them to transit-level exchanges", or "transmission nodes only". The user can then invoke some additional information about a chosen object, e.g. "the total node traffic capacity", or "alternative route sequence and associated loss probability". Some data can be edited manually on the graphic screen, e.g. the graphic representation of a node can be moved to a new position. All data is stored in text files, so they can also be edited with the help of a built-in text editor.

The **Design Tools** layer contains two packets of network planning programs: the LogDes and TraDes. Each packet is managed by its own supervisor program; this helps in determining calculation options and invoking particular tasks.

The LogDes pack serves for designing the logical (trunk) network. It is split into five task/program groups:
- optimization of the distribution of network nodes between the two network levels,
- optimization of the local level trunk groups, together with traffic routing schemes of local level OD pairs,
- dimensioning of the transit level trunk groups, together with optimization of alternative paths sequences,
- re-optimization of alternative path sequences for a transit level configuration,
- analysis of traffic characteristics of a given network configuration.

The TraDes pack serves for designing the physical (transmission) network. It is split into three task/program groups:
- optimization of the transmission graph,
- optimization of trunk groups layout within the transmission network,
- calculation of total and residual cost of particular elements of the network.

Project Database is created separately for each network project. The database assists the whole life-cycle of a network development process; hence the database structure is tree-like to reflect the multi-branch (multi-decision) nature of the design process. Project data is dispersed among the database tree levels; the following tree levels are distinguished:

- Project level, which is the root of a database specific to a network project being developed.
- Network level, which contains data specific to a network version (e.g. switching node list, transmission node list, traffic matrix, transmission graph outline, etc.).
- Variant level, which comprises description of a network splitting into the transit and local levels. For each network variant the trunk network is dimensioned for direct traffic routing in the transit level, and trunk groups are routed within the transmission graph.
- Generation level, which mainly concerns different mutations of the transit level optimized for chosen alternative routing schemes; also the final trunk groups routing within the transmission graph is stored here.

The system has been programmed partially in FORTRAN (mainly the optimization part) and partially in Pascal, and implemented on PC/386 class computers. The mix of programming languages is achieved thanks to the open, modular and loosely-coupled system internal structure adopted. For a typical PC/386 configuration the system is capable of both storing data and performing all calculations needed for a network consisting of up to about 100 nodes; calculation time is kept in the range of single hours (the most time consuming task is the optimization of the distribution of nodes between the two network levels).

4. APPLICATION EXAMPLE

We consider here as an example the design of the Warsaw metropolitan network. The network will soon undergo a revolutionary structural and technical transformation. In its present state the network consists mostly of analog exchanges and transmission systems, located in a 50 km diameter zone. Over 40% of the 46 existing exchanges is of step-by-step type, most of the remaining are of crossbar/register type. The total generated traffic is about 31,000 Erl.; traffic generated in individual nodes ranges from 150 to 2,700 Erl., origin-destination traffic ranges from 1 to 150 Erl.

The network has a complex functional structure in which the step-by-step exchanges - the oldest equipment - play a critical role in the distribution of traffic. The existing structure and equipment yield severe difficulties in applying a classical, overlay-based approach to digitalization. On the other hand intensive digitalization is necessary if an essential increase of telephone density is to be achieved in a short time (the PTT plans to double the density in just a few years).

In this situation an alternative approach to the overlay-based digitalization/development was adopted. Namely it was decided to adopt the two-level network structure. All existing exchanges will form the local level, while a purely digital transit level will be built from scratch (System 12 toll/local exchanges, mesh network structure, fiber-optic transmission systems). All existing exchanges will be connected to nodes from the transit level. Existing analog transmission facilities will be used for direct interconnecting local level exchanges, if economically justified.

For the first, initial phase of the network evolution it was decided to form the transit level of 8 nodes (to be implemented in 1992). Next, existing old exchanges will be gradually replaced with digital ones and the transmission plant of the local level will be digitalized. A question arises whether the newly introduced exchanges should be appended to the local or to the transit level of the network. To solve this network optimization problem we have used TOOLNET and the associated network design methodology.

The design process, based on real network components data, has shown that the initially assumed 8-node configuration of the transit network is not optimal. An optimisation process leads to a conclusion that a 24-node configuration is more cost effective. This indicates that the newly introduced digital exchanges (to substitute old analog systems of the local level) could be appended to the transit level. The final decision should however take into account not only traffic carrying effectiveness issues but also an affordable network development strategy and OAM aspects.

The dependence of the total network cost on the type of traffic routing employed in the transit level was analyzed; direct and fixed alternative routing were considered.

Fig. 4.1 illustrates relative performance of the 8-node and the optimal 24-node configuration, with and without alternative routing, with respect to: total number of PCM modules, total number of km-modules, transmission cost, total cost (transmission plus switching).

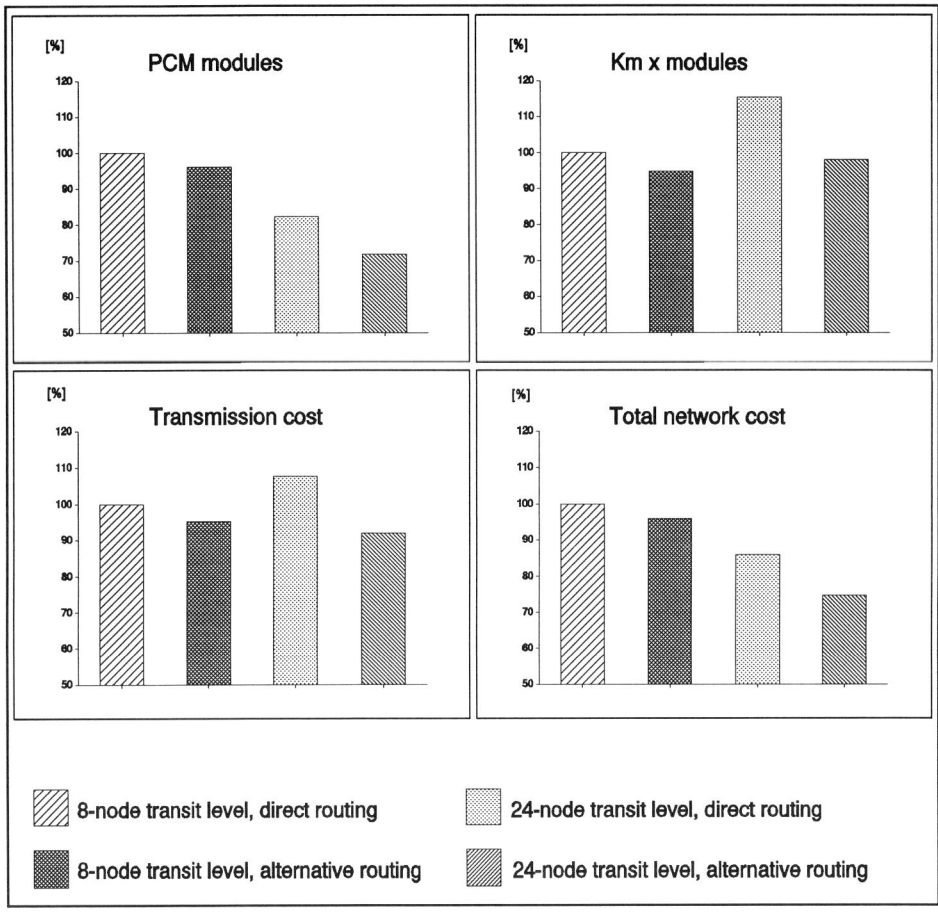

Fig.4.1 Comparison of network designs, related to 8-node transit level with direct routing (100%)

5. REFERENCES

[1] M. Pióro, M. de Miguel, I. Pita: Telecom Networks Evolution Towards Secure Dynamic Structures; *8th ITC Specialists' Seminar*, Cracow, 1991, paper 1.3

[2] M. Pióro, A. Tomaszewski: Modular Engineering of Telephone Networks with Dynamic Routing, paper accepted for *ITC13*, Copenhagen, 1991

[3] S.W. Levine, M.A. Wernander: Modular Engineering of Trunk Groups for Traffic Requirements, *ITC5 Proceedings,* New York, 1967

AGENT SCHEDULING FOR ACD SWITCHES

P. Nowikow and K. Wajda

Department of Telecommunications, The University of Mining and Metallurgy, al. Mickiewicza 30, 30-059 Kraków, POLAND

ABSTRACT

Paper describes system for staff scheduling in Automatic Call Distribution PBX exchanges based on gathered statistical data. General traffic engineering aspects for ACD exchanges employing Erlang C formula was presented. Algorithm for choosing number of active operators-agents implementing an idea of "daily call profile" was also included. Real trials - carried for ROLM 9000 ACD exchange - proved that system is technologically feasible and economically attractive.

1. INTRODUCTION

The PBX ACD-type telephone exchange automatically routes incoming calls in sequence to available agents. If all agents are busy, the exchange greets callers with a recorded message while putting them in line (creating a queue) for the next agent.
Typical applications of this ACD-type exchanges are:
- answering problems with credit card services,
- air tickets booking, etc.

Modern ACD exchanges have capability of collecting and printing limited statistics on the day events which are organized as summary and detailed reports. Not all gathered data are useful for scheduling purposes. This excess of statistical data is a basis for experienced supervisor of agent group.

Selection of number of agents is a fundamental problem in ACD exchange management. It means settlement of number of active agents in succeeding short time periods (e.g. 15, 30 minutes). From theoretical point of view, choosing of agent's group size is a problem close to basic traffic engineering task: dimensioning of telephone link. General differences consist in:
- possibility of quick change of agent's group size in ACD exchange, enabling adaptation to predicted traffic conditions,
- possibility of incoming calls queueing.

2. SCHEDULING SYSTEM IN ACD EXCHANGE

Scheduling system accomplishes multistep processing of statistical data obtained from ACD exchange and assigns optimal agent's work time - based on traffic conditions and existing personal limitations.

There are subsequent elements of the scheduling system:
1. Collecting of data representing traffic offered to ACD group.
2. Computing of linear long-term trend coefficients.
3. Prediction of offered traffic based on gathered statistical data with optional implementation of linear long-term trend.
4. Computing of minimal recommended number of agents according to required service quality.
5. Scheduling of agent's work time.

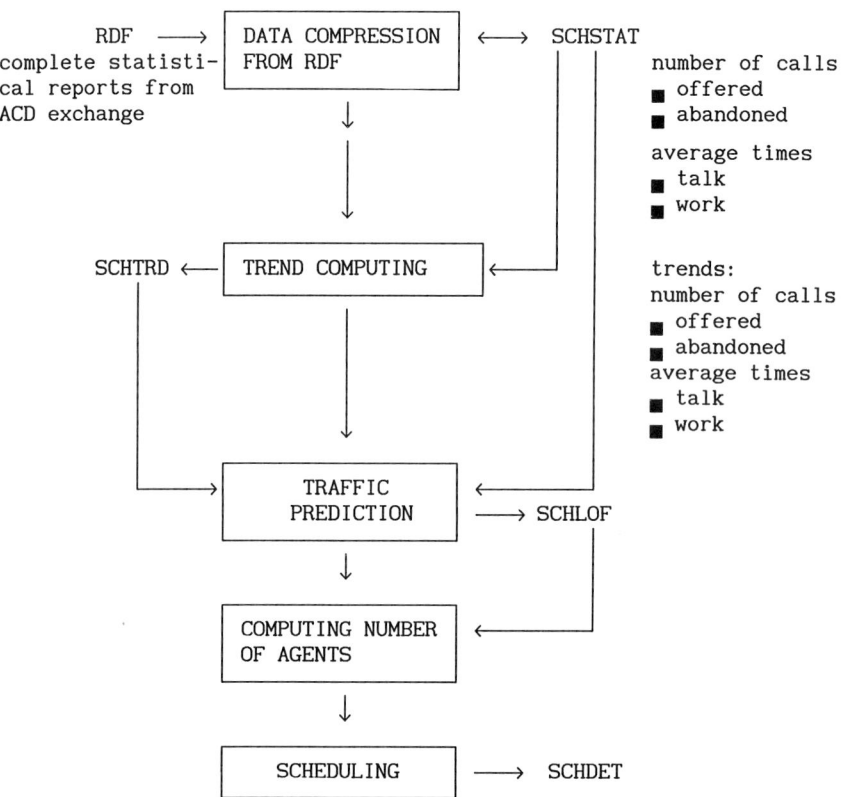

RDF - Raw Data File containing reports from ACD exchange, written in a cyclic way,

SCHSTAT - processed RDF, containing selected data,
SCHTRD - file containing linear trend coefficients,
SCHLOF - file containing predicted traffic daily profiles,
SCHDET - file with detailed scheduling.

Fig.1. Functional diagram of scheduling system with regard to used information and created file's contents.

3. ACD AGENT GROUP AS SERVICE SYSTEM

ACD exchange ROLM 9000, produced by IBM, has modular structure enabling arrangement of proper configuration for large scale of traffic intensities.

Fundamental organization unit is a group of agents (ACD group) identified by unique name.

Telephone traffic is typically directed to appropriate group by dedicated trunks. If service is possible - call is accepted and directed to agent, if not - directed to queue. Calls may be also interflowed between offices via "trunk" circuits or transferred to an ACD group from any phone connected to the PBX (also from any agent position in another ACD group).

4. "DAILY CALL PROFILE"

Processing of large amount of statistical data from telephone network carried by different countries suggests that there exists characteristic "daily call profile" [11]. Detailed shape of offered traffic profile reveals changes due to day of week and season of year. Comparison of "daily profiles" is possible after their normalization. In "normalized profile" each point of graph is expressed as part of highest point in a day. Detailed investigations suggest that one can recognize two different kinds of daily profiles: for working day (from Monday till Friday) and for weekend day (Saturday and Sunday). We assume "daily call profile" as traffic unit of higher level.

5. DATA SELECTION FROM STATISTICAL REPORTS

Activity of ACD exchange is registered as special reports created according to specified program. We propose two-step attempt to offered traffic prediction:
1. Processing of statistical data from the past and calculation of offered traffic profile.
2. Implementation of given traffic profile taking into account linear long-term trend.

To describe behaviour of the M/M/m system with queue it is sufficient to know offered traffic and average call holding time. These parameters are not included directly in reports. Assignments that we make below are only proposals and can be modified according to supervisor's suggestions.

Statistical reports are gathered according to following time units: superior unit is a *shift* (several hours period) in which summary reports are collected, unit of lower layer is *time of day* (commonly time period 30 or 15 minutes).

Present version of data registration for offered traffic estimation purposes is as follows:

---
number of calls offered to ACD group during *time of day* =
number of calls primarily offered to ACD group +
number of abandoned calls

Assuming above version we leave out of look occurrence of overflow between ACD groups.

> average call holding time =
> average talk time + average work time (after call completion)

After data selection from statistical reports we have "number of offered calls profile" at our disposal. Next using definition of offered traffic A:

$$A = \frac{\text{number of offered calls}}{\text{observation period}} \times \text{average holding time} \quad (1)$$

we get finally "predicted daily profile of predicted traffic".

6. COMPUTING MINIMAL RECOMMENDED NUMBER OF AGENTS

Service quality for system with queue one can describe independently using following parameters:

1. Average time spent in queue - *ASA (Average Speed of Answer)*,

Using Erlang theory following expression is valid:

$$ASA(m,A) = E_{2,m}(A) \frac{\tau}{m-A}, \quad m > A \quad (2)$$

where:
$E_{2,m}$ - probability that any call is directed to queue,
m - number of agents,
A - offered traffic,
τ - average holding time.

Provided ASA=a parameter is given the task of choosing minimal number of agents can be then formulated as:
choose minimal number of agents m fulfilling conditions:

$$ASA(m,A) \leq a, \quad m > A \quad (3)$$

then:

$$m \geq A + E_{2,m}(A)\frac{\tau}{a} \quad (4)$$

2. Probability g that any call will be served during settled time period t (equivalent formulation: 100g% calls served during t second):
then

$$m \geq A - \frac{\tau}{t} \ln \frac{1-g}{E_{2,m}(A)}, \quad m > A \quad (5)$$

7. SCHEDULING OF AGENT'S WORK TIME

Before scheduling "daily profile of minimum recommended agent number" is computed. On this basis multistep scheduling algorithm matches work time of agent to traffic needs of ACD exchange. Scheduling is closely connected with ACD agent's group organization so below we mention only general assumptions. Detailed solutions undergo considerations and system supervisor's suggestions.

1. Agents are divided into groups with different priorities connected with their experience and value for supervisor:
   full-time - agents with highest priority, most experienced, always taken into considerations with their personal suggestions and limitations,
   split-time - agents with medium priority, working in separable time periods, also always used,
   part-time - agents with lowest priority, taken into account only in a case of necessity.

2. In each agent's schedule one can distinguish following kinds of activity:
   work,
   lunch,
   break1, break2,
   CKA - Closed Key Activity ⎫ *planned work but other then*
   PSA - Prescheduled Activity ⎭ *incoming call service*

3. Multistep algorithm for agent insertion works in following way: before scheduling minimum number of agents is known for each short period (15, 30 minutes). One after the other, agents are taken into account beginning from full-time and system settles their work time (using existing limitations) in such way to obtain minimum difference between recommended and present set number of agents (this is integrated directly, using square root and after rising to the second power).

8. SPECIFICITY OF PROPOSED SOLUTION

Proposed system for ACD staff scheduling has following distinctive features:
1. Implementation of equations originating from Erlang theory.
2. Putting great stress on collection and comprehensive verification of real data from ACD exchange.
3. Reduction of data amount for ACD agent group.
4. Computing of linear trend for number of calls primarily offered and abandoned and for call holding time with possibility of selective trend implementation.
5. Assuming that "daily call profile" is characteristic representation of offered traffic when time scale is of hours order.

ACKNOWLEDGEMENTS

We would like to thank engineers from firms MTC-Systems (Toronto) and Account-A-Call (Los Angeles) for constructive comments and programming support.

REFERENCES

[1] T. Abe and H. Saito: Bayesian forecasting with multiple state space model, International Teletraffic Congress ITC'85, Kyoto 1985.
[2] P. Chemouil and B. Garnier: An Adaptive Short-Term Traffic Forecasting Procedure Using Kalman Filtering, International Teletraffic Congress ITC'85, Kyoto, Japan.
[3] A.A. Frederics and G.A. Reisner: Approximations to Stochastic Service Systems, with an Application to a Retrial Model, Bell System Technical Journal, Vol.58, 1979, pp. 557-576.
[4] A. Jajszczyk: Introduction into switching theory, Technical University of Poznań Press, Poznań 1984 (in Polish).
[5] L. Kleinrock: Queueing Systems Vol.2: Computer Applications, John Wiley & Sons, New York.
[6] A.M. Lee and P.A. Longton: Queueing Processes Associated with Airline Passenger Check-in, Opnl. Res. Quart, Vol. 10, 1959, pp.56-71.
[7] G.F. Newell: An Approximate Stochastic Behaviour of n-Server Systems with Large n, Springer-Verlag, New York, 1973.
[8] P. Nowikow and K. Wajda: System for staff scheduling in ACD-type telephone exchanges, Technical University of Poznań, Report TR-53/6/1990/P, December 1990, (in Polish).
[9] R.K. Otnes and L. Enochson: Digital time series analysis, John Wiley & Sons, New York, 1972.
[10] ROLM 9000, Reference Manual, IBM.
[11] J. Rubas J.: A Course in Teletraffic Engineering, Telecom Australia, 1978.
[12] D.Y. Sze: A Queueing Model for Telephone Operator Staffing, Operations Research, Vol.32, No.2 March-April 1984.
[13] C.R. Szelag: A Short-Term Forecasting Algorithm for TrunkDemand Servicing, The Bell System Technical Journal, Vol. 61, No.1, January 1982.
[14] Y. Takahashi: An Approximation Formula for the Mean Waiting Time of an M/G/c Queue, J. Opns. Res. Soc. Jap., Vol. 20, pp.150-163.

Analysis and Design
of Teletraffic Systems

STUDY OF MESSAGE DELAYS IN THE PRESENCE OF LONG MESSAGES AND CORRELATED ARRIVALS IN SIGNALLING SYSTEM NO. 7 NETWORKS

M. Ghassemi and R. A. Skoog

AT&T Bell Laboratories, New Jersey, USA

Abstract

If some of the proposed uses of SS No. 7 networks materialize, SS No.7 will work in an environment where the signalling traffic is different from call set up traffic. Call set up traffic has comparatively small Message Signal Unit (MSU) lengths. These varied applications that are planned or being contemplated result in non-traditional signalling traffic characteristics. These applications use connection-oriented SCCP (Signalling Connection Control Part). Two of these applications are User to User Information (UUI) Transport and possible use of the signalling network to transport Telecommunication Management Network data. Generally, it is expected that the MSUs pertaining to these applications will be substantially longer than traditional signalling traffic and their arrival process is non-Poisson. In this paper, we have studied the problem of transport of long MSUs in the signalling network and calculated the delay characteristics and its impact on signalling link engineering.

The results show that indeed long MSU traffic has radical deleterious effect on delay characteristics. We show that, for a relatively small amount of long MSU traffic, and even with Poisson arrivals, engineering the network to meet the delay characteristics pertaining to a traditional traffic mix has a significant economic cost. Therefore, either long MSU traffic must be controlled to very low levels or significant additional network capacity must be provided. Otherwise, the delay performance of all signalling traffic will deteriorate.

We explore different solutions to this problem for their merits and shortcomings and suggest some solutions which we believe are appropriate for SS No. 7 networks.

1. INTRODUCTION

A well known distinguishing characteristic of SS No. 7 networks is that they are real time sensitive, i.e., the delay characteristics of the network is of utmost importance. This is so because of the desire for short post dialing delays. With the introduction of new and more complex services, complex transactions need to be completed within a reasonable post dialing delay. Therefore, such services are only realizable if different component delays in the network are not unreasonably large. Unfortunately, complex services usually introduce long Message Signal Units (MSUs) in the signalling network which contribute to excessive delay. Therefore, some mechanism to minimize these delays is essential as more complex services become commonplace in the signalling networks.

If some of the proposed uses of SS No. 7 networks materialize, SS No.7 will work in an environment where the traffic has different attributes compared to call setup traffic. Call setup traffic has comparatively small MSU lengths, while these applications generate long MSUs and correlated arrivals. These varied applications that are planned or being contemplated result in non-traditional signalling traffic characteristics. These applications use connection-oriented SCCP. Two of these applications are User to User Information (UUI) Transport and possible use of the signalling network to transport Telecommunication Management Network (TMN) data. Also, proposals have been introduced in Study Group XI to accommodate segmentation in SCCP-connectionless services. Some proposals suggest large amounts of data (possibly tens of kilobytes) be segmented and transported using the SCCP Connectionless service. The effect of such protocols on the signalling traffic mix is similar to that of Connection-Oriented Services, i.e., higher likelyhood of highly correlated arrivals of long MSUs. Generally, it is expected that the MSUs pertaining to these applications will be substantially longer than traditional signalling traffic. In this paper, we call such MSUs *Long MSUs*. Under conditions where long MSU traffic is not a negligible fraction of total signalling traffic, a question to consider is whether present delay requirements could be economically met. In this paper, we have studied this problem for different assumed traffic mixes and calculated delay characteristics and its impact on signalling link engineering.

The results show that indeed long MSU traffic has radical deleterious effect on delay characteristics. We show that, with even a relatively small amount of long MSU traffic, engineering the network to meet the delay characteristics pertaining to a traditional traffic mix has a significant economic cost.

Also, a related problem encountered in SS No. 7 environments which have to transport non-call-related signalling traffic (such as TMN traffic) is the correlated arrival of MSUs. This problem will be encountered when segmentation (connection-oriented or connectionless) is used. Similar to the long MSUs, the effect of correlated arrivals is adverse effect on the delay characteristics.

We explore different solutions to the long MSU problem and the related problem of correlated arrivals (non-Poisson) of MSUs for their merits and shortcomings and suggest solutions which we believe are appropriate for SS No. 7 networks.

In Section 2, we examine the delay characteristics of a CCS No.7 signalling link and show how the performance will change with different levels of long MSU traffic present. In Section 3, we show how the engineered utilization of the signalling link must be reduced to meet the traditional delay requirements. In Sections 4 and 5, we examine different possible solutions and identify the most technically and economically viable solutions. In Section 6 we have summarized our conclusions.

2. STEADY STATE SIGNALING LINK DELAY PERFORMANCE

We will consider a single signalling link in the SS No.7 network and examine its delay performance as the level of offered long MSU traffic is increased. The simplest situation to consider is a link operating under a steady state traffic load. For this situation, we will characterize how the signalling link queueing delays are affected by the introduction of long MSU traffic. More complex situations involve transient queuing behavior. For example, an important transient occurs after a signalling link failure. Transients are not studied in this paper.

The two major characteristics of long MSU traffic affecting the signalling link queueing behavior are the message lengths and correlated arrival process. In this paper, we consider the effect of MSU length analytically and the effect correlated arrivals phenomenologically. We assume that the signalling traffic in a typical signalling link has three distinct components.

1. The traditionally short MSU length ISDN User Part (ISUP) call set up traffic;

2. The medium MSU length traffic, such as the TCAP traffic for Free-Phone (800) services and the like;

3. The long MSU length traffic, such as UUI or Network Management Information.

We use the M/G/1[1] non-preemptive priority queuing discipline to analyze the signalling link delay characteristics. We assume that message service time is proportional to the number of octets in the message. Therefore, the service time distribution can be obtained using the frequency of messages of different lengths. We have assumed 20% of calls need database access, thereby giving rise to medium length MSU traffic. We denote the ratio of the number of long MSUs to the total number of MSUs by P. Therefore, the service time distribution and its moments are functions of P. We assume that the above three types of traffic are independent Poisson streams. Also, we have assumed the ISUP IAM message is 67 octets long, the average of all other ISUP messages is 19 octets and medium length TCAP messages are 100 octets. We have performed the analysis for two values of long MSU message length, 150 octets for cases that the information transported fits in a single MSU, and 278 octets for the cases that user information is segmented by SCCP and full length MSUs are used.

To examine the impact of long MSU traffic on the message delay characteristics, the link utilization, ρ, will be held constant, and we will look at the behavior of the ratios $R_Q(P)$ and $R_\sigma(P)$ defined by

$$R_Q(P) = Q(P)/Q(0) \tag{1a}$$
$$R_\sigma(P) = \sigma_Q(P)/\sigma_Q(0) \tag{1b}$$

where Q(P) is the mean queuing delay at long MSU ratio P and $\sigma_Q(P)$ is the standard deviation of queuing delay at long MSU ratio P. $R_Q(P)$ and $R_\sigma(P)$ reflect the increases in expected queuing delay and queuing delay variability of signalling messages due to the introduction of long MSU traffic.

We have used an analytic model to approximate Q and σ_Q. Using an M/G/1 nonpreemptive priority queuing model to approximate the signalling link behavior,[1] one obtains

$$Q(P) = \frac{T_f}{2} + \frac{\lambda E[S^2]}{2(1-\rho)} \tag{2a}$$

$$\sigma_Q^2(P) = \frac{T_f^2}{12} + \frac{\lambda E[S^3]}{3(1-\rho)} + \frac{\lambda^2 E^2[S^2]}{4(1-\rho)^2} \tag{2b}$$

where

S = message service time (a random variable)
λ = message arrival rate
ρ = link utilization = λE[S]
C = link speed (octets/sec)
T_f = service time of a fill-in signal unit

and the moments of S depend on P. Equation (2) agrees with the relationships provided in Q.706[2].

To show illustrative results, we will use a link speed of C = 8 octets/msec and nominal link utilizations of ρ=0.2 and ρ=0.4. A fill-in signal unit is 6 octets long, so the service time of a fill-in signal unit, T_f, is 0.75 msec. Figure 1 shows the behavior of $R_Q(P)$ and $R_\sigma(P)$ obtained from (2). As we see from the graphs the average queuing delay and, to a much larger extent, the standard deviation of the queuing delay significantly increase with the percentage of long MSU traffic. The increase is especially pronounced for a long MSU length of 278 octets. It seems that even moderate levels of long MSU traffic produces significant delay performance degradation. For example, at a 10% level of long MSU traffic (of length 278 octets), queuing delay standard deviation is more than three fold larger.

---

1. The fill-in signal units are modeled as lower priority traffic with an arrival rate $\lambda_f = (1-\rho)/T_f$; so the link is fully utilized by messages and fill-in signal units, with message signal units having nonpreemptive priority.

3. STUDY OF ADJUSTMENT OF SIGNALING LINK UTILIZATION TO ACCOMMODATE FOR LONG MSU TRAFFIC

The problem of accommodating long MSU traffic can be stated as follows. Given that there is a certain amount of long MSU traffic in the network, we want to reengineer the network (i.e., add more signalling links) such that the delay characteristics do not exceed those of an existing network with no long MSU traffic.

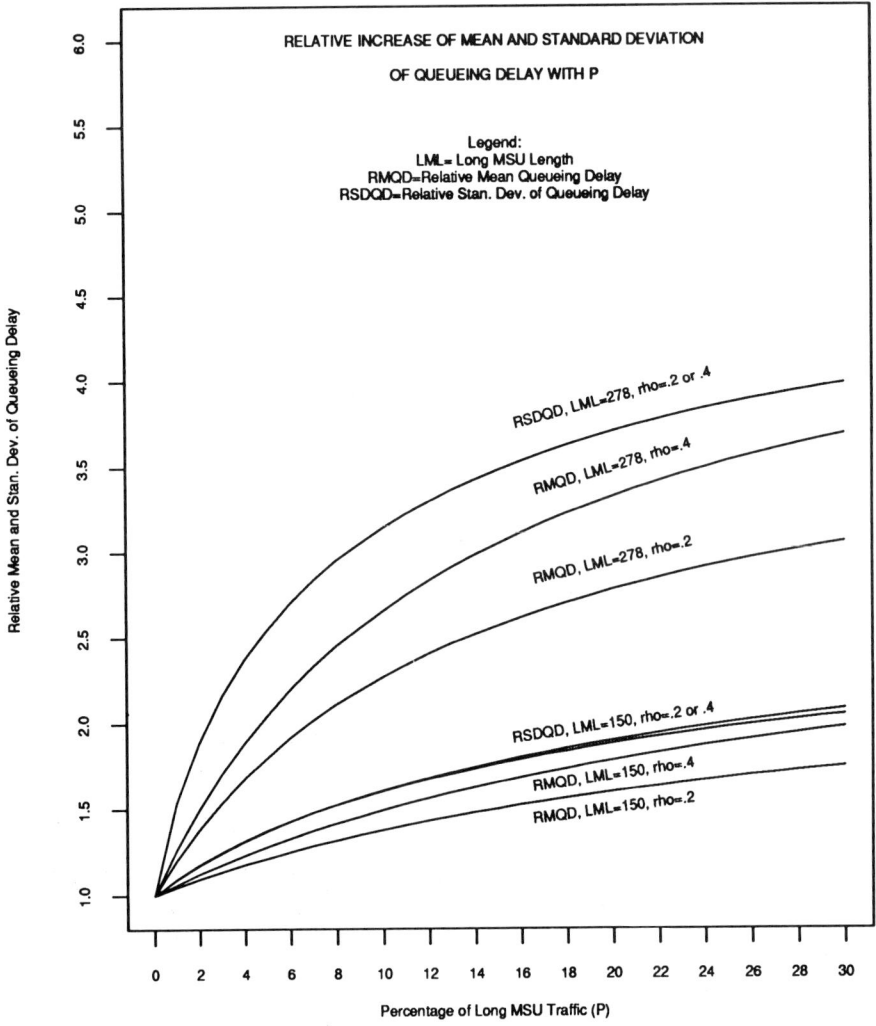

Figure 1. Relative Increase of the Queueing Delay Mean and Standard Deviation as a Function of Percentage of Long MSU Traffic

The problem could be approached in two ways as follows. We could keep the mean queueing delay constant, vary P and calculate what ρ must be to have that mean queueing delay. This method does not give desirable results, because the standard deviation of the queuing delay increases with P beyond acceptable values. This is shown in Figures 2 and 3. In the second approach the signalling link utilization is calculated as a function of P to keep standard deviation of the queuing delay constant at its value for $P = 0$.

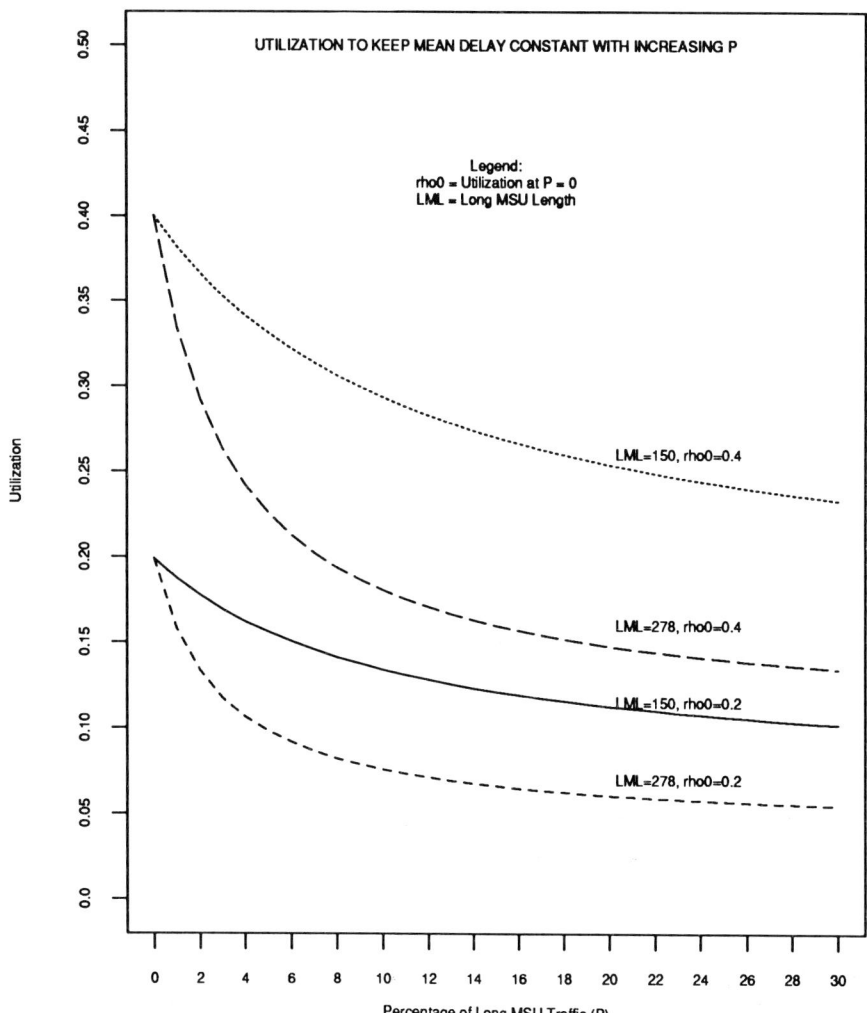

Figure 2. Utilization of the Signalling as a Function of Percentage of Long MSU Traffic to Keep the Mean Queueing Delay Constant

In this case, the mean queueing delay remains below its value for $P = 0$ for different values of P. These are shown in Figures 4 and 5.

Figure 4 shows how the introduction of long MSU traffic affects the economics of the network, if we want to have acceptable delay characteristics. For example, for 5% long

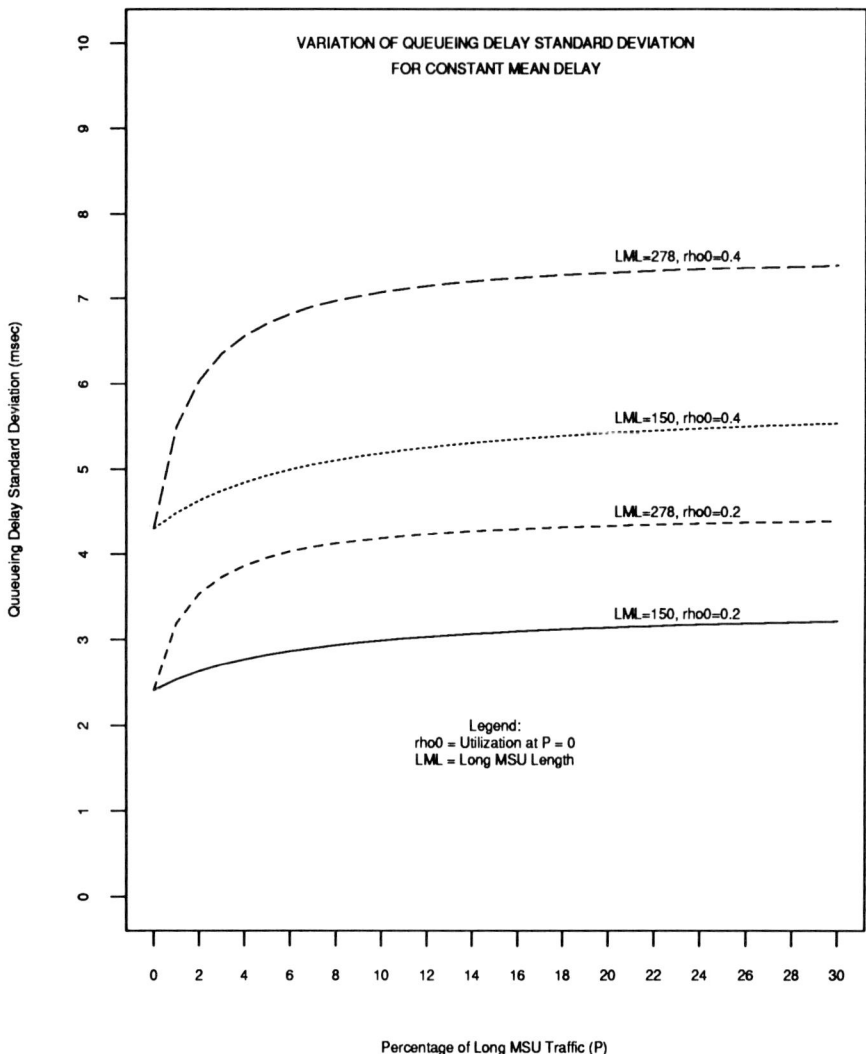

Figure 3. Variation of the Queueing Delay Standard Deviation as a Function of Percentage of Long MSU Traffic with Mean Queueing Delay Kept Constant

MSU traffic of length 278 octets, we have to reduce the signalling link utilization from 0.2 to less than 0.05 to keep the delay characteristics constant. This means that for each original signalling link we need to add at least 3 more signalling links.

We should note that the above analysis does not take into account other deleterious factors such as traffic peakedness or non stationary behaviour. These factors combined with long MSU traffic could deteriorate the network behaviour even more.

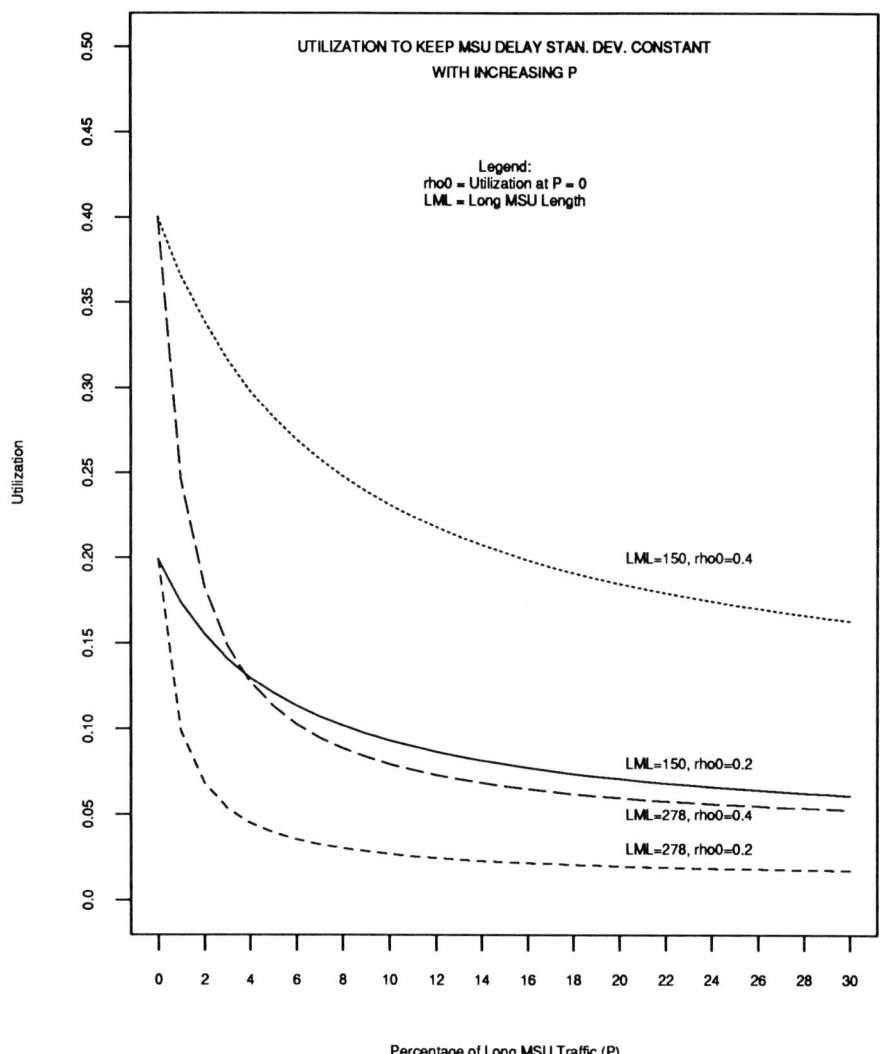

Figure 4. Utilization of the Signalling as a Function of Percentage of Long MSU Traffic to Keep the Standard Deviation of the Queueing Delay Constant

4. SOLUTIONS TO THE DELAY PROBLEM CAUSED BY LONG/CORRELATED MSUS

We can distinguish two types of sources for long MSU traffic as follows.

1. There is a non-negligible amount of long MSU traffic generated but the arrival distribution is close to Poisson. This approximates the case of small amounts of call-related User to User Information (UUI) traffic.

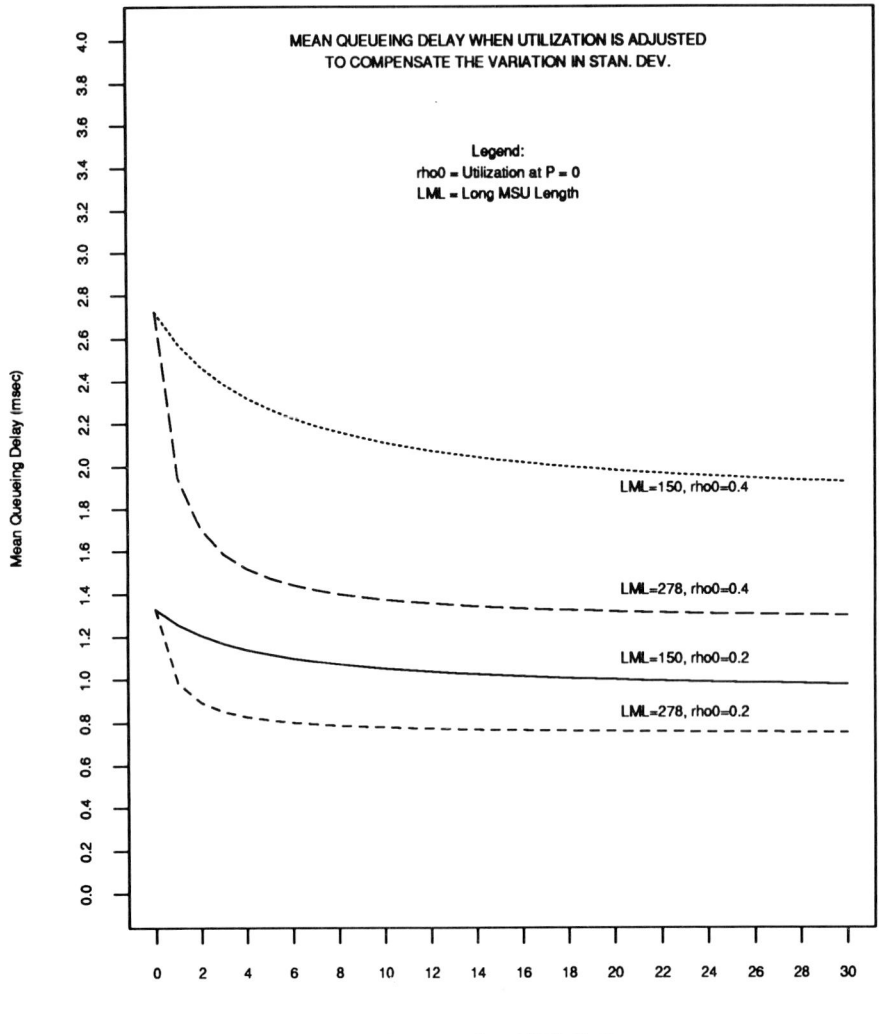

Figure 5. Variation of Mean Queueing Delay as a Function of Percentage of Long MSU Traffic with Standard Deviation of Queueing Delay Kept Constant

We call this type of traffic *Uncorrelated Long MSU Traffic*.

2. There is a non-negligible amount of correlated arrivals in the network. In this case, the adverse effect on the delay characteristics will be observed irrespective to the length of the MSUs. This case applies to the circumstances where segmentation (connection-oriented or connectionless) is employed. We call this type of traffic *Correlated MSU Traffic*.

The latter type of traffic has the most adverse effect on the signalling network delay characteristics because of the following factors:

- Usually it has the Longest MSUs: segmentation cuts the information into the largest possible lengths;
- The message arrival times are much more correlated, i.e., the arrival process is far from Poisson, i.e., highly bunched;
- The amount of data transferred per connection could be excessively large (e.g., TMN applications).

Because of the differences between the above two types of traffic the solutions for the two cases could be different. The most important solution to the excessive delay due to long MSUs must involve controls on Correlated MSU Traffic. Also, solutions should be found to alleviate the problem of long message delays due to Uncorrelated Long MSU Traffic.

In the following we explore different possibilities in solving the above problem for their merits and shortcomings for each of above two cases. The possible solutions are as follows:

1. Adding more signalling links to the network;
2. Increasing the bit rate of the signalling links;
3. Segmentation into smaller length MSUs;
4. Time of the day restriction of Connection Oriented and Segmented Connectionless traffic;
5. Flow control of long and/or Correlated MSU traffic;
   - Control by the user of network services;
   - Control in the network layer function;

We discuss each of the above options in the following:

4.1 Increasing the number of signalling links

Increasing the number of signalling links causes a reduction of the utilization of each signalling link and improves the delay performance. As we discussed before, adding signalling links will be an expensive solution to the long MSU delay problem. It was shown that under no failure condition, to compensate for delay characteristics deterioration due to only five per cent long MSU traffic, with uncorrelated arrival distribution (Poisson) of total traffic, one needs to quadruple the number of signalling links. Given that if there is no other control on long MSU traffic, its percentage could rise well beyond five per cent, and for the Correlated MSU Traffic delay performance will be even worse, this solution does not seem viable either for Uncorrelated Long MSU Traffic or for Correlated MSU Traffic.

Another limitation in this case is due to the limited number of Signalling Link Selection (SLS) codes available to be assigned to the signalling links according to the MTP (Message Transfer Part) level 3 protocol. The SLS field is only four bits long, i.e., the maximum number of links accommodated within a combined link set cannot be larger than 16. Once this limit is reached, other solutions must be explored.

4.2 Increasing the bit rate of the signalling links

Increasing the bit rate of the signalling links while keeping the traffic offering constant will reduce the utilization of the signalling link. Also, because the message emission time is reduced for higher speed links the service time distribution is improved accordingly and delay performance is improved. The effectiveness of this solution depends on the level of increase in the signalling link bit rate. This solution could be applied for both Uncorrelated and Correlated Long MSU Traffic, although it still may not be enough for the latter case.

Although provision of increased bandwidth by increasing the bit rate is not as expensive as provisioning the same bandwidth by adding multiple 64 kb/s signalling links, it may still be economically unattractive. Also, increasing the bit rate on signalling links implies changes in the MTP level 2 protocol implementations. For example, the values of error thresholds or timers will need to be changed. This may not be acceptable to the administrations who have present level 2 implementations.

This could be a viable solution for specific national networks and should be studied further.

4.3 Segmentation into smaller length MSUs

Segmenting the connection-oriented (or segmented connectionless) messages into smaller segments (carried in separate MSUs) without controlling the rate of the segments entering the network is not a viable solution. This is because although the length of the individual MSUs becomes smaller, the total amount of message data to be transported has not been changed and larger network layer overhead is imposed because of more segmentation. The result is a worsening of the problem because more bits must be sent and the segments of a message are still bunched together.

4.4 Time of the day restriction of Connection Oriented and Segmented Connectionless traffic

This solution is discussed in Section 5.

4.5 Throttling the Long/Correlated MSU Traffic

The control of the Long/Correlated MSU traffic offered to the signalling network through SCCP can be accomplished in two different ways as follows:

- Control by the user of network services:
  In this case, the user of network services will control the long MSU traffic offered to the network layer voluntarily according to certain pre-established guidelines. The advantage of this method is that it does not imply any network layer protocol (SCCP) changes. The disadvantage of this method is that its enforcement is not guaranteed. If certain users do not obey the established guidelines, they deteriorate the delay performance of the network for all traffic.

- Traffic Control in the Network Layer:
  In this case, the control function is implemented as part of the network layer functions; therefore, it can be better enforced. The disadvantage of this method is that it implies network layer protocol (SCCP or MTP) enhancements, which may not be desirable at the present specification stage of these protocols.

5. ANALYSIS OF TRAFFIC CONTROLS

Considering the above alternatives, the Traffic Control in the Network Layer is the most viable solution to the delay performance problem caused by Long/Correlated MSU Traffic. In the following, we provide three versions of network layer traffic control as straw proposals. We propose that Study Group II and IX jointly work on this problem and come up with a solution which is viable and acceptable to both Study Groups. More detailed analytical and simulation studies of these methods will be done in future. All proposals are chosen as enhancements to SCCP Routing Control procedures rather than MTP, so that MTP, which is a more stable set of protocols, is not impacted.

- *Solution 1: Time of the day restriction of Long/Correlated MSU Traffic:*

  In this procedure, a time of day clock is used to accept the Long/Correlated MSU traffic in low traffic hours of the day and reject them in peak traffic hours when the control is turned on. The Correlated MSUs are identified in SCCP by their usage of Segmentation/Reassembly functions. The criterion for rejection of Long/Correlated MSUs will be as follows: If within certain time T0 after reception of an MSU pertaining to certain connection, a number of MSUs more than N are received pertaining to that connection all MSUs N+1 ... will be discarded until T0 expires. In other words, a high rate of long MSU arrivals from a particular user causes discard of its MSUs belonging the that specific signalling connection or signalling relation.

It is the responsibility of the user to control the traffic it is offering to the network in such a way that its traffic does not become subject to the above network layer control.

This procedure can be realized in the following two different ways:

1. The control is exercised at the interface between SCCP and its users, i.e., the control is applied by SCCP to the messages offered to the network by the network layer users;

2. The control is applied at the interface between SCCP and MTP to the Long/Correlated MSU Traffic which is delivered by MTP to SCCP, i.e., to the MSUs whose sources are elsewhere in the network.

   The disadvantage of latter method is that under certain circumstances when the system starts discarding messages, connections are reset and this may be too disruptive. Also, the control is applied *after* the MSUs have used the network. Its advantage is that the enforcement of the control is guaranteed because it is done in a signalling point other than the one generating the long MSU traffic. This procedure could be enhanced by the introduction of a new SCCP management message to inform the user generating the Long/Correlated MSU traffic of the activation of such control. This message is needed to inform the other end of the connection to throttle its offered traffic. Also, it should be made sure that when connections are reset, the reattempts to re-establish connections and sending the same traffic (or retransmission of discarded traffic when higher classes of SCCP connection-oriented service is used) does not result in oscillation and worsening the situation. This is for further study.

- *Solution 2: Long/Correlated MSU Control Procedure: Acceptance of long MSUs, on every connection or from certain destination, only after certain throttle period:*

In this case, the SCCP will apply the long/correlated MSU control before such traffic is delivered to the MTP processes. Certain amount of buffer space is allocated to connection-oriented traffic (or connectionless signalling traffic using segmentation). This buffer will be called the "Long/Correlated MSU Control Buffer". The Long/Correlated MSU Control Procedure (i.e., the present procedure) will control this buffer. As long/correlated information is received from the SCCP user it is segmented and stored in Long/Correlated MSU Control Buffer. Then MSUs are released every $T_{throttle}$ seconds from the Long/Correlated MSU Control Buffer and delivered to MTP. If the contents of the Long/Correlated MSU Control Buffer exceeds a certain threshold, SCCP generates an Indication Primitive to its users to stop generation of traffic or risk message discard. When the Long/Correlated MSU Control Buffer is full, the received correlated traffic from the SCCP users will be discarded.

- *Solution 3: Long/Correlated MSU Control Procedure: Throttled serving of individual transmit buffers pertaining to individual connections:*

Like the previous solution, this solution incorporates the control in the same node which is the source of the controlled traffic. In this case, a buffer is assigned to each specific SCCP connection (or signalling connection in the case of connectionless segmented traffic). The traffic received from the SCCP user for each connection is stored in the corresponding buffer. A process is added to SCCP which limits serving of these buffers to a fraction of total time. For example, every second only 1/10 of a second is allocated to serving these buffers. If the buffer occupancy for a connection is more than a certain limit, an Indication Primitive is issued to the user to stop or reduce the rate of message generation. If the rate of generation of connection-oriented messages is not controlled by the user, SCCP will discard the messages when the corresponding buffer is full.

The advantage of solutions co-located with the traffic generating node is twofold. Firstly, the control is exercised before the Long/Correlated MSU Traffic enters the network, and secondly, resetting connections caused by message discard and ensuing difficulties is prevented. Its shortcoming is that the generation and control are co-located and enforcement of controls is not guaranteed, unless different SCCP implementations used for the international SS No. 7 network are carefully tested for their compliance with the chosen procedure.

Combinations of Solutions 1 and Solutions 2 or 3 are also possible. In this case, the time of the day clock will activate/deactivate the corresponding procedures.

Also, one could restrict the message length in conjunction with these controls.

6. CONCLUSION

Introduction of long MSU traffic, even in limited amounts, deteriorates the network delay performance radically. Reduction of signalling link utilization to compensate for this deterioration has a significant economic cost. Therefore, either the amount of long MSU traffic must be restricted to very low levels by implementing one of the SCCP options discussed, or significant additional network capacity must be provided, or a combination of both, in order to avoid deterioration of the delay performance of all signalling traffic.

Various options have been examined for reducing the impact of carrying long MSUs on SS No. 7 networks. Three options have been identified as viable options needing further study; these options are using traffic controls for long MSUs, increasing the signalling link bit rate and reducing maximum message length for uncorrelated long MSU traffic. Three possible methods for traffic control for connection-oriented long MSU traffic are provided. These methods are compared for their feasibility and effectiveness. It is suggested that one of these three methods (or some variation of them) for SCCP control of such traffic be standardized. Also, we suggest higher bit rates for signalling data links be studied.

7. Acknowledgements

We are thankful to K. C. Hsu, A. Modarressi, and D. Schmidt for their valuable help.

REFERENCES

1. S. S. Lavenberg, "Computer Performance Modeling Handbook, Academic Press, New York, 1983, p.75.

2. CCITT Blue Book, Volume VI, Fascicle VI.7, Q.706, Table 1/Q.706.

ACCELERATED SIMULATION OF RARE EVENTS USING RESTART METHOD WITH HYSTERESIS

Manuel VILLEN-ALTAMIRANO
Telefónica I+D
Madrid, Spain

José VILLEN-ALTAMIRANO
Universidad Politécnica de Madrid
Madrid, Spain

Abstract

RESTART (REpetitive Simulations Trials After Reaching Threshold) is a method for accelarating simulations to estimate the probability of rare events. It allows dramatic reductions of simulation time for an equal confidence of results. This paper considers the impact of possible restoration costs in its implementation and extends the method by the introduction of hysteresis. It shows how hysteresis can reduce such impact and maintain the advantages of RESTART even in cases with significant restoration costs.

1 INTRODUCTION

Normal simulations are impracticable for estimating the probability of occurrence of rare events. The need of coping with this type of events often arises with new technologies; e.g., in asyncronous transmission mode (ATM), cell loss probability should be less than 10^{-9}. Computer time in the order of months would be required to estimate such low probabilities with an acceptable confidence.

Some techniques have been developed to accelerate those rare event simulations. A review of them was made in 1988 [1]. More recent techniques are reverse-time models [2] and RESTART method [3].

The RESTART (REpetitive Simulations Trials After Reaching Threshold) method, which has a straightforward application to most simulation models, basically consists of the following: Given the rare event A, an event C, such that $C \supset A$ and $1 >> P\{C\} >> P\{A\}$, is defined, and many repetitive simulation trials are made of those time intervals in which C occurs. As shown in [3], this simple procedure leads to a dramatic reduction of the simulation time for an equal confidence of the results or, vice versa, to a dramatic increase of confidence for the same simulation time.

To repeat the simulation of a certain interval it is necessary to restore the system state at the starting point of the interval for each retrial. In simulations of complex systems, this restoration may have a significant cost, in terms of computer time, which might reduce the efficiency of RESTART. The objective of this paper is to describe how to cope with this restoration cost. Proposals of two types are made:

- Ones for reducing the impact of the restoration cost on the gain obtained with RESTART by means of the use of hysteresis and by an appropriate adjustment of the RESTART parameters.
- The other ones for reducing the restoration cost itself.

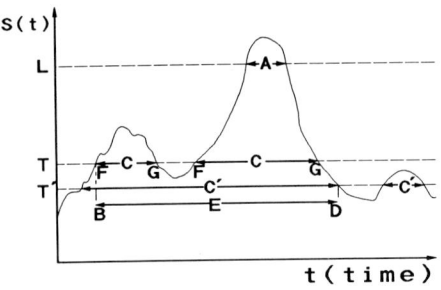
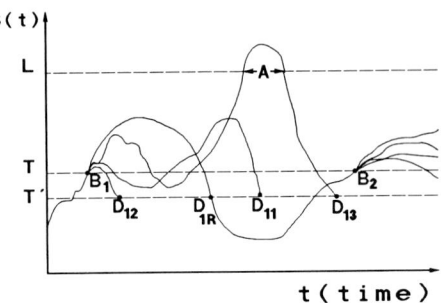

Figure 1: Example of system evolution along time

Figure 2: Illustration of the application of RESTART

This paper is organized as follows: the RESTART method with hysteresis is described in section 2 and analyzed in section 3. Based on this analysis, section 4 provides rules to optimize its parameters considering the restoration cost, and section 5 reports the gain obtained. Finally, section 6 gives recommendations to reduce the restoration cost, and section 7 shows an application example.

2 DESCRIPTION OF RESTART WITH HYSTERESIS

Let us see in Figure 1 an example of system evolution along time. The occurrence of event A, the probability of which must be estimated, has been associated with a system state parameter S taking values equal to or higher than L. Two events C' and C, being $C' \supset C \supset A$, have been defined and associated with values of S equal to or higher than intermediate thresholds T' and T, respectively.

For the sake of clarity, as in [3], this description will assume that events A, C and C' can only occur at certain instants (e.g., at call arrivals) which will be called instants of interest; however, this method can also be applied to estimate continuous time probabilities.

Calling I_i to the ith instant of interest occurred in the simulation, events E, B, D, F and G are defined as follows:

- $I_i \in E \iff \exists j \leq i/(I_j \in C) \cap (\forall k/j \leq k \leq i \implies I_k \in C')$
- $I_i \in B \iff (I_i \in E) \cap (I_{i-1} \notin E)$
- $I_i \in D \iff (I_i \notin E) \cap (I_{i-1} \in E)$
- $I_i \in F \iff (I_i \in C) \cap (I_{i-1} \notin C)$
- $I_i \in G \iff (I_i \notin C) \cap (I_{i-1} \in C)$

Figure 1 can represent a normal simulation, while figure 2 represents a simulation made with RESTART. It is a normal simulation with these modifications:

- When an event B occurs, the system state is saved.
- When an event D occurs, the system state of last event B is restored, and the interval $[B, D)$ is simulated again.
- The above process is repeated R times, as illustrated in Fig. 2. The starting event of every trial is always the same, B_1, while the ending event is different, $D_{11}, D_{12},..., D_{1R}$, depending on system evolution in the trial.

- When the event D_{1R} occurs, simulation continues in the normal way until event B_2 occurs; then the same process is applied.
- The statistics should be accordingly corrected, either in real time or once the simulation has finished.

3 STATISTICAL ANALYSIS

In this section, the variance of the estimator of $P\{A\}$ is calculated with RESTART and without it. The analysis is made for steady state simulations, but its conclusions can be extended to transitory state simulations.

3.1 NOTATION

The following notation is used:
- $P = P\{A\}$
- $P_1 = P\{C\}$
- $P_2 = P\{A/C\}$
- $P'_1 = P\{B\}$
- P'_2 = expected value of the number of events A in an interval $[B, D)$;
- $Q = P\{C'\}/P\{C\}$
- l = expected value of the number of instants of interests in an interval $[B, D)$;
- m = expected value of the number of intervals $[F, G)$ in an interval $[B, D)$;
- μ = expected value of the number of instants of interests in an interval $[F, G)$;
- $a = m \cdot \mu$
- N = number of instants of interest simulated without counting those of the retrials;
- N_1 = number of events B which occur in the simulation without counting the retrials;
- N_2 = number of events A which occur in the simulation counting the retrials.

The following relations are straighforward:

$$P = P_1 \cdot P_2 = P'_1 \cdot P'_2 \quad ; \quad P_1 = a \cdot P'_1 \quad ; \quad P_2 = P'_2/a \tag{1}$$

3.2 ANALYSIS OF SIMULATIONS WITHOUT RESTART

In a simulation without RESTART, the following estimators are used:

$$\hat{P'}_1 = \frac{N_1}{N} \quad ; \quad \hat{P'}_2 = \frac{N_2}{N_1} \quad ; \quad \hat{P} = \hat{P'}_1 \cdot \hat{P'}_2 = \frac{N_2}{N} \tag{2}$$

3.2.1 Evaluation of $V(\hat{P})$

Assuming that $\hat{P'}_1$ and $\hat{P'}_2$ are independent random variables, the variance of $\hat{P}$ is:

$$V(\hat{P}) = V(\hat{P'}_1) \cdot V(\hat{P'}_2) + V(\hat{P'}_1) \cdot [E(\hat{P'}_2)]^2 + V(\hat{P'}_2) \cdot [E(\hat{P'}_1)]^2 \tag{3}$$

$V(\hat{P'}_1)$ and $V(\hat{P'}_2)$ can be derived as in [3] giving:

$$V(\hat{P'}_1) = \frac{K_1 \cdot P'_1}{a \cdot N} \quad (4a) \quad ; \quad V(\hat{P'}_2) = \frac{K_2 \cdot P'_2}{N \cdot P'_1} \quad (4b) \quad (4)$$

where:
$$K_1 = \sum_{\forall j} [P\{I_j \in C/I_i \in B\} - P_1] \quad (5)$$

$$K_2 = \sum_{\forall j} [P\{I_j \in A'/I_i \in A\} - P\{I_j \in A'/I_i \in B\}] \simeq \sum_{\forall j} [P\{I_j \in A/I_i \in A\} - P] \quad (6)$$

where A' means an event A belonging to the same interval $[B - D)$ as the instant i.

From formulas (3) and (4), and considering formulas (1), $V(\hat{P})$ can be written as:

$$V(\hat{P}) = \frac{P}{N}\left(K_1 \cdot P_2 \frac{K_2}{N \cdot P} + K_1 \cdot P_2 + K_2\right) \simeq \frac{P}{N} K_2 \quad (7)$$

3.2.2 Comparison between K_1 and K_2

From formula (5), it is clear that:

$$K_1 \leq \sum_{\forall j} [P\{I_j \in C/I_i \in C\} - P_1] \quad (8)$$

If the system behaves in an analogous way when the parameter S is near L and when it is near T, then:

$$\sum_{\forall j} [P\{I_j \in C/I_i \in C\} - P_1] \simeq \sum_{\forall j} [P\{I_j \in A/I_i \in A\} - P] \quad (9)$$

and, consequently, considering (6) and (8), it will be $K_2 \geq K_1$.

In general, K_2 can be greater, equal or smaller than K_1 depending on the behaviour of the system when S is near L compared to the behaviour when S is near T but, in most cases K_1 and K_2 will have the same order of magnitude.

Let us now consider how K_1 and K_2 are affected by the introduction of hysteresis, or, in general, by the value of Q. Let us call the factors K_1 and K_2 without hysteresis K_{1w} and K_{2w}. Without hysteresis, $Q = 1$, T' becomes T and B becomes F, thus:

$$K_{1w} = \sum_{\forall j} [P\{I_j \in C/I_i \in F\} - P_1] = \mu \sum_{\forall j} [P\{I_j \in F/I_i \in F\} - P\{I_j \in F\}] \quad (10)$$

As K_1 can be written as:

$$K_1 = \mu \sum_{\forall j} [P\{I_j \in F/I_i \in B\} - P\{I_j \in F\}] \quad (11)$$

we see that $K_1 \leq K_{1w}$. Based on the same reasoning, the larger Q is, the smaller K_1 will be.

K_2 is also affected by the introduction of hysteresis, but in a negligible manner. In formula (6), the hysteresis affects the definition of event A': A'(with hysteresis) $\supset A'$ (without hysteresis), thus $K_2 \geq K_{2w}$. In the same manner, the larger Q is, the greater K_2 will be. But these variations of K_2 are so negligible that they are not noted in the aproximation given in the second part of formula (6).

The above reasoning is confirmed by formula (7). Since the hysteresis does not affect $V(\hat{P})$ when RESTART is not used, it should be: $K_1 \cdot P_2 + K_2 =$ constant for any T', thus $\Delta K_2 = -P_2 \cdot \Delta K_1$.

3.3 ANALYSIS OF SIMULATIONS WITH RESTART

The estimators of P'_1, P'_2 and P are now:

$$\hat{P'}_1 = \frac{N_1}{N} \quad ; \quad \hat{P'}_2 = \frac{N_2}{R \cdot N_1} \quad ; \quad \hat{P} = \hat{P'}_1 \cdot \hat{P'}_2 = \frac{N_2}{R \cdot N} \qquad (12)$$

3.3.1 Variance of $V(\hat{P})$

Formulas (3) and (4a) also apply when RESTART is used. However, P'_2 is estimated in a different manner, then $V(\hat{P'}_2)$ is different. Let us define:

- Y_{ij} as the random variable which indicates the number of events A which occur in the jth retrial made starting in the event B_i, i. e., in the ith event B appearing in the simulation;
- $\overline{Y}_i$ as the random variable which indicates the expected value of Y_{ij} conditioned to the system state B_i.

Considering that the N_1 random variables $\overline{Y}_i$ are mutually independent as well as the R random variables Y_{ij} obtained for each B_i, $V(\hat{P'}_2)$ is derived as in [3], giving:

$$V(\hat{P'}_2) = \frac{K_2 \cdot P'_2}{N \cdot P'_1}\left(\frac{1}{R} + b \cdot P_2 \frac{R-1}{R}\right) \simeq \frac{K_2 \cdot P'_2}{N \cdot P'_1}\left(\frac{1}{R} + b \cdot P_2\right) \qquad (13)$$

where:
$$b = \frac{V_1}{K_2 \cdot P_2 \cdot P'_2} \quad ; \quad V_1 = E\left[\overline{Y}_i - P'_2\right]^2 \qquad (14)$$

On the basis of formulas (3), (4a) and (13), $V(\hat{P})$ can be written as:

$$V(\hat{P}) = \frac{P}{N}\left[K_1 \cdot P_2 + \left(K_2 + K_1 \cdot P_2 \frac{K_2}{N \cdot P}\right) \cdot \left(\frac{1}{R} + b \cdot P_2\right)\right] \simeq$$
$$\simeq \frac{P}{N}\left[K_1 \cdot P_2 + K_2\left(\frac{1}{R} + b \cdot P_2\right)\right] \qquad (15)$$

The comparison of (7) and (15) clearly indicates the advantage of RESTART, which is greater when b is smaller.

3.3.2 Bounds of b

Let us first compare the value of b with hysteresis and without it, under the assumptions considered above. Let us call variables $\overline{Y}_i$, P'_2, V_1 and b in the case without hysteresis $\overline{Y}_{iw}$, P'_{2w}, V_{1w} and b_w. The random variables $\overline{Y}_i$ have been considered mutually independent, regardless of the distance from T' to T, thus also when $T' = T$. Consequently, when $T' < T$, the random variables indicating the expected number of events A in the second or other subsequent interval $[F, G)$ of an interval $[B, D)$ must be considered independent from the system state in the starting event B of the interval $[B, D)$. Thus:

$$\overline{Y}_i = \overline{Y}_{iw} + (m_i - 1)P'_{2w} \qquad (16)$$

where m_i is the random variable indicating the expected number of intervals $[F, G)$ in an interval $[B, D)$ conditioned to the system state in the starting instant B_i. Thus, V_1 and b can be written as:

$$V_1 = E\left[\overline{Y}_{iw} - P'_{2w} + (m_i - m)P'_{2w}\right]^2 \simeq E\left[\overline{Y}_{iw} - P'_{2w}\right]^2 = V_{1w} \qquad (17)$$

$$b = \frac{V_{1w}}{K_2 \cdot P_2 \cdot m \cdot P'_{2w}} = \frac{b_w}{m} \tag{18}$$

Thus we see that b is smaller when there is hysteresis, and, in the same manner, the larger Q is, the smaller b will be.

Therefore, the upper bound of b derived in [3] for the case without hysteresis also applies for the case with hysteresis. The lower bound also applies since it is zero. Thus, the bounds of b are:

$$0 \leq b \leq \frac{\mu}{K_2} \left[\frac{P'_{2Mw}}{P'_{2w}} - 1 \right] \tag{19}$$

where P'_{2Mw} is the maximum of $\overline{Y}_{iw}$ for any i.

4 OPTIMIZATION OF RESTART PARAMETERS

The cost of simulation (in computer time) has been meausured taking as unit time the time to simulate an instant of interest. Thus, let us define r as the restoration cost, i.e, the cost of restoring the system state of event B, measured in that unit time. The total simulation cost, C, is given by:

$$C = N + N_1(l+r)(R-1) \simeq N(1 + P_1 \frac{l+r}{a} R) \tag{20}$$

Since the objective of RESTART is to decrease C for a same $V(\hat{P})$ or vice versa, and C and $V(\hat{P})$ are inversely proportional, the product $C \cdot V(\hat{P})$ is the function which should be minimized. The parameters to optimize are P_1, Q and R.

The optimization will be made assuming that K_1, K_2, r, μ, l and m are independent from P_1, and that K_1, K_2 and r (and, obviously, μ) are independent from Q.

In sections 3.2.2 and 3.3.2, we have seen that the introduction of hysteresis or, in general, the increase of Q produces an increase of K_1 and a decrease of b, which, in turn, as will be seen later, produce an increase of the gain obtained with RESTART. This fact is an additional advantage of the introduction of hysteresis which, however, due to the difficulty of its evaluation, is not considered in the optimization process.

From (15) and (20), it is derived that the optimization of $C \cdot V(\hat{P})$ should be made into two steps:

- A first step, in which a value of Q which minimizes the ratio $y = (l+r)/a$ is obtained.
- A second step which, based on the obtained value of y the optimum values of P_1 and R are obtained.

4.1 OPTIMIZATION OF THE VALUE OF Q

Let us define:

$$x = T - T' \quad ; \quad y = \frac{l(x)+r}{a(x)} \quad ; \quad \alpha = \frac{P\{S \geq T'\}}{P\{S \geq T'+1\}} \tag{21}$$

Assuming that α is independent from the value of x, we can write:

$$Q = \alpha^x \tag{22}$$

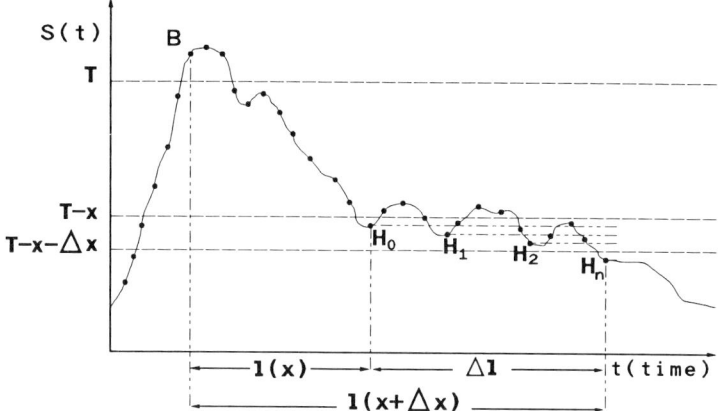

Figure 3: Example to illustrate how l depends on x.

Let us derive how l and a depend on x. Figure 3 shows an example of system evolution in which the instants of interest are represented by dots. In this figure, if the threshold T' were $T - x$, event D would be H_o and, if T' were $T - x - \Delta x$, D would be H_n. Thus, $\Delta l = l(x + \Delta x) - l(x)$ is the expected number of instants of interest in the interval $[H_o, H_n)$. Given an instant of interest H_i, let us define H_{i+1} as the closest subsequent instant in which $S(H_{i+1}) < S(H_i)$, $S(H_{i+1})$ and $S(H_i)$ being the values of the parameter S in the instants H_{i+1} and H_i respectively. The expected number of instants of interest in an interval $[H_i, H_{i+1})$ can be assumed independent from x and from Δx, and the expected number of instants H in the interval $[H_0, H_n)$ can be assumed independent from x and proportional to Δx. Therefore, we can write:

$$\Delta l = k \cdot \Delta x \tag{23}$$

where k is a system dependent constant. $\Delta a = a(x + \Delta x) - a(x)$ will be the part of Δl in which $S \geq T$. Since the only condition guaranteed in the interval $[H_o, H_n)$ is that $S \geq T - x - \Delta x$, it is possible to state that:

$$\Delta a = \Delta l \cdot P\{S \geq T / S \geq T - x - \Delta x\} = \frac{k \cdot \Delta x}{\alpha^{x + \Delta x}} \tag{24}$$

Using differential notation, (23) and (24) become:

$$dl = k \cdot dx \quad ; \quad da = \frac{k \cdot dx}{\alpha^x} \tag{25}$$

Since $l(o) = a(o) = \mu$, we have:

$$l(x) = \mu + k \cdot x \quad ; \quad a(x) = \mu + \frac{k}{\ln \alpha}\left(1 - \frac{1}{\alpha^x}\right) \tag{26}$$

From formulas (21), (22) and (26) and defining $\delta = k/\ln \alpha$, the following formula is derived for y:

$$y = \frac{1 + \frac{\tau}{\mu} + \frac{\delta}{\mu} \ln Q}{1 + \frac{\delta}{\mu}\left(1 - \frac{1}{Q}\right)} \tag{27}$$

The value Q_0 of Q which minimizes y is obtained from the following equation:

$$Q_o = 1 + \frac{\frac{r}{\mu} + \frac{\delta}{\mu} \ln Q_o}{1 + \frac{\delta}{\mu}} \tag{28}$$

Substituting this value in (27), the minimum value, y_o, of y is obtained:

$$y_o = 1 + \frac{\frac{r}{\mu} + \frac{\delta}{\mu} \ln Q_o}{1 + \frac{\delta}{\mu}} \tag{29}$$

Note that $y_o = Q_o$. This equality can be interpreted in economical terms: $l + r$ represents the total cost of producing a events C; thus $(l + r)/a = y$ represents the mean cost and $d(l + r)/da$, which, according to (22) and (25), is equal to Q, represents the marginal cost. The optimal production is reached when marginal cost equals mean cost, i.e., when $Q = y$.

In most practical cases δ is unknown and thus Q_o cannot be evaluated by (28). This problem has been solved by the following approximation of Q_o:

$$Q_o \simeq \sqrt{1 + \frac{r}{\mu}} \tag{30}$$

Although (30) is a rough approximation, the approximate value of y_o obtained from (30) and (27) is very close to the optimum one obtained from (28) and (29).

Thus, using (30) we assure that the simulation will closely reach the optimum value of y_0, but we do not know which is this value of y_o that the simulation reaches since it depends on δ. The value of the reached y_o is needed to optimize P_1 and R. Fortunately, an error in the optimization of P_1 and R due to a rough estimation of y_o has very small impact on the gain obtained with RESTART. Thus, the following rough estimation of y_o is enough for optimizing P_1 and R:

$$y_o \simeq \sqrt{1 + \frac{r}{\mu}} \tag{31}$$

Table 1 shows how the gain, G_a, obtained with this approximation is very close to the optimum one, G_o, for a very wide range of values of r/μ and δ/μ which cover most practical cases.

Nevertheless, formulas (30) and (31) require to know r and μ. r can be estimated by analyzing the computer program or comparing computer times in two test runs with different number R of retrials. Normally, r will only depend on the program itself, but not on the input data. Thus, only one evaluation may be enough for all the applications. However, μ may be different for each application and, in some cases, may be difficult to estimate. For these cases, a procedure in which the simulation adjusts by itself the value of T' may be useful. The procedure, which needs r as input data, may be as follows:

When the first event B occurs, the program makes several sets of retrials, first set with $T' = T$, and each of the next ones decreasing T' in a small constant quantity Δ. In each set of retrials, l and m are measured and $(l+r)/m$ evaluated. In the begining, $(l+r)/m$ decreases for each new T' until arriving to the optimum T', T'_o, moment in which $(l+r)/m$ starts to increase. The process finishes when this minimum of $(l+r)/m$, and thus T'_o, has been found. (Given that μ is independent from T', $y = (l+r)/(m \cdot \mu)$ will also have its minimum for T'_o). The optimum T'_o could be checked when the second event B occurs.

Since μ is measured in the first set of retrials ($\mu = l$ when $T' = T$), the procedure may also provide y_o, (the minimum y), which could be used to adjust the values of P_1 and R.

r/μ	0	1			10			100		
δ/μ	Any	1	10	100	1	10	100	1	10	100
G_i	7900	5250	5250	5250	1300	1300	1300	150	150	150
G_w	7900	5600	5600	5600	2400	2400	2400	800	800	800
G_o	7900	5950	6700	7400	3000	4650	6450	1100	2250	4500
G_a	7900	5800	6700	7150	2750	4650	5600	800	2250	3650

$P = 10^{-9}$; $K_1/K_2 + b = 1$; Parameters used: Q, $P_1 = \sqrt{P/y}$, $R = 1/\sqrt{P \cdot y}$
being: G_i : $Q = y = 1$; G_w : $Q = 1, y = 1 + r/\mu$; G_o : Formulas (28) and (29)
G_w : Formulas (30) and (31).

Table 1: Impact of the restoration cost and of the hysteresis on the gain provided by RESTART

4.2 OPTIMIZATION OF THE VALUES OF P_1 AND R

Once the optimum value of Q, Q_o, which minimizes $y = (l+r)/a$ has been obtained, and this minimum y, y_o, has been evaluated, the optimum values of P_1 and R, P_{1o} and R_o, which minimizes $C \cdot V(\hat{P})$, can be obtained from (15) and (20). They are given by:

$$P_{1o} = \sqrt{\left(\frac{K_1}{K_2} + b\right)\frac{1}{y_o} \cdot P} \quad ; \quad R_o = \frac{1}{\sqrt{\left(\frac{K_1}{K_2} + b\right) y_o \cdot P}} \quad (32)$$

5 GAIN OBTAINED WITH RESTART

The optimum values of P, Q_o, P_{1o} and R_o derived in section 4 lead to the following value of $C \cdot V(\hat{P})$:

$$C \cdot V(\hat{P}) = 4 \cdot K_2 \cdot P \cdot \sqrt{\left(\frac{K_1}{K_2} + b\right) \cdot y_o \cdot P} \quad (33)$$

Since the product $C \cdot V(\hat{P})$ when RESTART is not used is $K_2 \cdot P$, the gain, G_o, obtained is:

$$G_o = \frac{1}{4 \cdot \sqrt{\left(\frac{K_1}{K_2} + b\right) \cdot y_o \cdot P}} \quad (34)$$

Formula (34) can be difficult to evaluate in practice due to the difficulty of knowing K_1, K_2 and b. But, based on the bounds of b given by formula (19), we can write:

$$\frac{1}{4\sqrt{\left[\frac{K_1}{K_2} + \frac{\mu}{K_2}\left(\frac{P'_{2Mw}}{P'_{2w}} - 1\right)\right] \cdot y_o \cdot P}} \leq G_o \leq \frac{1}{4\sqrt{\frac{K_1}{K_2} \cdot y_o \cdot P}} \quad (35)$$

It most applications, according to section (3.2.2), it will be $K_1 \leq K_2$, $\mu \leq K_2$. Thus formula (36) will be normally a safe-side approximation of (35):

$$\frac{1}{4\sqrt{\frac{P'_{2Mw}}{P'_{2w}} y_o \cdot P}} \leq G_o \leq \frac{1}{4\sqrt{y_o \cdot P}} \quad (36)$$

In [3], tables giving the bounds of the gain were provided. Dramatic gains, able to reduce 115 days of computer time with normal simulation to few hours or even few minutes were reported. The robustness of RESTART for errors in the choice of the parameters was also shown: large gains were obtained even for values of the parameters equal to 1/10 or 10 times their optimum values. This robustness makes unnecessary a careful analysis to obtain a very important advantage from RESTART. The results shown in [3] apply to cases in which the restoration cost, r, is zero. Table 1 shows numerical results on the impact of the restoration cost and on how this impact can be reduced with the use of hysteresis. It can be seen that if hysteresis is not used and especially if r is ignored in the optimization of P_1 and R, the restoration cost may have a significant impact on the gain (see G_i). This impact is reduced considering the restoration cost in the optimization of P_1 and R, eventhough hysteresis is not used (see G_w). The use of hysteresis (see G_o or G_a) produces a further improvement of the gain, mainly if δ/μ is large, in which case the effect of the restoration cost is nearly compensated.

6 REDUCTION OF THE RESTORATION COST

The paper has shown up to now how the use of hysteresis and an appropriate adjustment of the parameters can reduce the impact of the restoration cost. Let see now how to reduce the restoration cost itself. Two complementary proposals are made for this purpose:

- To have a suitable event scheduling algorithm.
- To improve the memory management mechanism.

With respect to the first proposal: when an event B occurs, a number of future events have been previously scheduled. If the schedule of those events is maintained when the system state B is restored to initiate each new retrial, a similar system evolution can be expected in all the retrials starting in a same B; thus, b will be large and, consequently, the gain will be small. If, on the contrary, the pending events are rescheduled in each restoration, the restoration cost may be large. This problem may be solved taking into account that in traffic engineering simulations, most events are scheduled in a time which is deterministic or negative exponentially distributed or, in case of discrete time simulations, geometrically distributed. Some other often used distributions, as erlangian or hyperexponential, can be obtained as a combination of the previous ones.

In case of deterministic schedules, it is obvious that a reschedule should not be made. In case of events with negative exponential schedules, their memoryless property allows to have along the entire simulation a particular scheduling algorithm which proves to be very useful in the restorations. This algorithm consists of associating all the individual events of this type which are pending at any moment in only one global event with an occurrence rate (i.e. with an inverse of the mean scheduling time) equal to the sum of the occurrence rates of the individual events which it represents. Each time the global event occurs, the corresponding individual event can be determined by roulette. Using this technique, only one exponentially distributed event is presently scheduled in any moment, and thus, also when an event B occurs. Consequently, only one event has to be rescheduled in each retrial: the restoration cost is reduced without impacting on the value of b. A similar technique may be used for geometrically distributed schedules.

With respect to the memory management mechanism, the following procedure is proposed: Divide the RAM memory used by the program into three zones, which will be called X, Y and Z, X and Z having the same size. All the variables which have to be saved (or restored) when an event B (or D) occurs are stored in zone X. The other ones are stored in Y. The program performs its normal work with only zones X and Y, zone

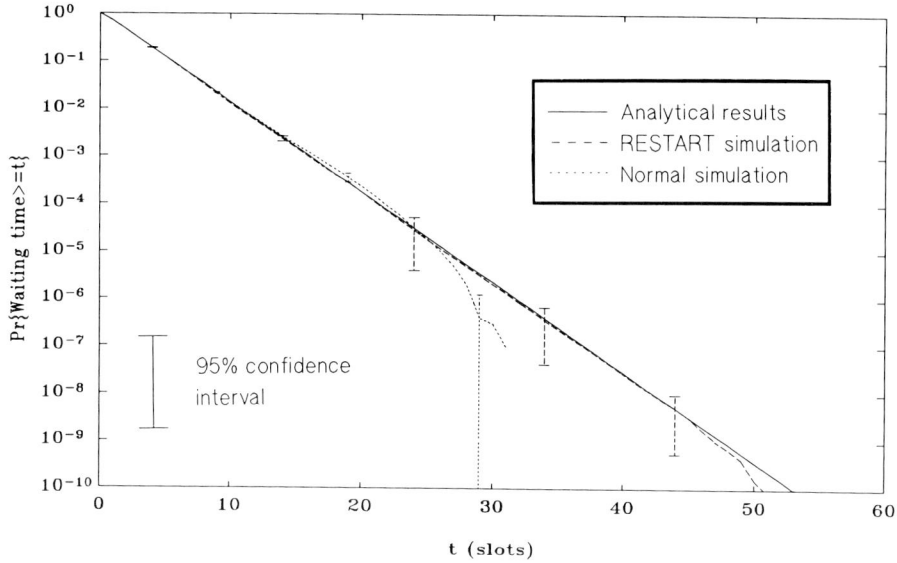

Figure 4: Waiting time distribution in an ATM multiplexer.

Z being only used to save the system state when an event B occurs. Thus, when an event B occurs, the zone X is dumped in zone Z (i.e, the information stored in X is copied as a whole in Z), and, when each retrial is initiated, the information stored in Z is dumped back in X. These memory dumps are quite faster than a copy made variable by variable. Unfortunately, these dumps cannot be made with many programming languages. We are presently investigating which language allows them.

7 APPLICATION EXAMPLE

RESTART has already been successfully used to simulate an ATM multiplexer with a large variety of source models, as shown in [4]. The source model chosen to be presented here is analytically tractable in order to allow comparison of results. This source model, the same which was presented in [3], is binomial: 80 Bernonilli sources, each with $p = 0.01$, giving a total load of 0.8 Erl. Events A and C were defined by the waiting time of the arriving cells. The restoration cost, which in the first version of the program was 180, was reduced to 15 by adopting the event scheduling algorithm proposed in section 6. The value of μ was estimated equal to 6.3. Thus, assuming $P = 10^{-9}$ and $K_1/K_2 + b = 1$, and following formulas (30), (31) and (32), the parameters $Q_o = 1.84$, $P_{1o} = 2.33 \cdot 10^{-5}$ and R=23321 have been chosen, corresponding to the thresholds $T = 24$ and $T' = 23$.

Figure 4 shows the results obtained by analytical formulas, by normal simulation of 10 million cells and by RESTART simulation of 6.2 million cells, both simulations taking up the same computer time. We can observe that the normal simulation gives accurate results up to 10^{-5}, while the results obtained with RESTART were accurate up to 10^{-9}.

8 CONCLUSIONS

The way of coping with the restoration cost in the RESTART method for accelerating rare event simulations has been presented. The paper, apart from giving recommendations to reduce this cost, proves how the use of hysteresis and an appropriate adjustment of the parameters can reduce its impact, leading to gains similar to those obtained in a hypothetical case without restoration cost.

Thus, the conclusions on the use of RESTART reported in [3] can be maintained. These conclusions can be summarized as follows:

- It allows a dramatic reduction of simulation time for an equal confidence of the results;
- It is general enough to be applied to most simulation models;
- It has a straightforward application to each particular case;
- It allows the application of simulation to evaluate future systems and networks, as e.g. ATM, where the required time for the simulation makes the application of classical methods impracticable.

ACKNOWLEDGMENTS

The authors wish to thank the members of the Traffic Performance Division of Telefónica I+D for their fruitful comments and helpful cooperation. In particular, we thank J. Andrade, who applied RESTART to a practical case and provided the results presented in section 7.

REFERENCES

[1] Frost, V. S., LaRue, W. W. and Shanmugan, K. S., *Efficient Techniques for the Simulation of Computer Communications Networks*, in: IEEE J. Select. Areas Commun., vol. 6, no. 1, 1988.

[2] Frater, M. R. et al, *Fast Simulation of Rare Events Using Reverse-Time Models*, in: Proc. of the ITC Specialist Seminar, Adelaide, 1989.

[3] Villén-Altamirano, M. and J., *RESTART: A Method for Accelerating Rare Event Simulations*, 13th ITC, Copenhagen, June 1991. Proc. in: Queueing Performance and Control in ATM, North-Holland.

[4] Andrade, J., Burakowski, W. and Villén-Altamirano, M., *Characterization of Cell Traffic Generated by an ATM Source*, 13th ITC, Copenhagen, June 1991. Proc. in: Tele and Data Traffic, North-Holland.

THE NEW CRITERION FOR THE COMPARISON OF QUEUEING SYSTEMS AND SYSTEMS WITH LOSSES

A. Kucherjavy

Central Research Institute of Telecommunications, Leningrad Branch (LONIIS), Warschawskaya 11, Leningrad, 196128, USSR

One of well known teletraffic theory principles is loss systems applicaition for high capacity multichannel groups (voice channel devices) with large average holding times and queueing systems application for groups having small number of channels (control devices)[1,2] with small average holding times.

Reasonable loss systems application for voice devices is based on wait and loss systems loss difference, which is more high with queueing systems. We can see it from the second Erlang formula:

$$P\{\gamma>0\}=E_v(y)/(1-y(1-E_v(y))/v) > E_v(y) \qquad (1)$$

where $P\{\gamma>0\}$ is waiting probability, $E_v(y)$ loss probability of a loss system.

Time loss for queueing systems is taken to be equal to waiting probability $P\{\gamma>0\}$. As waiting calls with large average holding time values have high average waiting times even for minimum loads, these waiting calls could be considered to be lost ones. However, in any case, there is a certain number of calls considered by subscribers to be waiting for a tolerable time period (e.g., the dial tone waiting during 1-2s).

Taking customers behaviour into consideration, there is no reason to consider such calls to be lost. Therefore, we propose the following criterion for loss systems and queueing systems comparing:

$$P\{\gamma>t^*\}=E_v(y) \qquad (2)$$

where t^* is specified waiting time providing no loses of waiting calls.

Criterion (2) evaluates the nature of subscriber and telephone network interaction. Solving equation $P\{\gamma > t\} = E_v(y)$, we can find out function $t_1(y,v)$; comparing its values with specified t^* we can determine reasonable application of service disciplines.

Figure 1. shows dependence of t_1 on channel group capacity v for different traffic values per line c. Dependency analisis shows that increase of channel group capacity for the same c values results in decrease of t_1 value. The same tendency takes place at traffic decrease per line with constant v. So, we conclude, that there is $\{y,v\}$ values domain, where a queueing system for voice channel is preferable to a loss system. In order to get an approximation value t_1 we can use the following formula derived from requirement (2):

$$t_1 = y(1-E_v(y))\bar{t}/v(v-y) , \qquad (3)$$

where $\bar{t}$ is mean holding time.

Proceeding from quality analysis (Figure 1) to quantity analsis, note that for channel groups with capacity v =10-20 a queueing system is preferable when c =0.2-0.3 Erl. Such traffic per line values take place in subscribers networks.

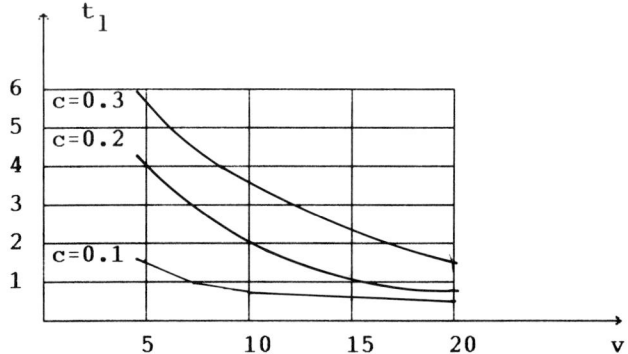

Fig.1. Dependence of t_1 value with different traffic per line c.

It should be noted that there are no hard constrained values for dial tone waiting. CCITT Recommendation Q.543 [3] specifies maximum possible delay. However, this value is normalized for total number of calls including waiting ones.In terms of practical considerations, accounting measured dial tone waiting times delayed due to equipment faults [4],allowed waiting time for dial tone waiting could be specified as 1-2s.

Example.Consider reasonable wait service discipline for out going traffic in DX-200 subscribers module.This module concentrates outgoing traffic from 64 subscribers to 30 channel group (equal to a single PCM channel).With traffic per subscribers line equal to 0.1Erl and average holding time 72s t_1 value is criterion, but CCITT Recommendation applied to total number of calls.Thus,in DX-200 subscribers module with outgoing connectiones,it is reasonable to use a queueing system.

The results derived and example considered show that at determined {y,v} values a queueing system application for voice channels migth be preferable to a loss system application.

REFERENCES:

1.B.S.Livshitz,A.P.Pshenichnikov,A.D.Harcevich."Teletraffic Theory",Moscow,Svjas,1979,224p.
2.M.A.Shneps."Information Distribution Systems",Moscow,Svjas, 1979,344p.
3.CCITT.Recommendation Q.543,v.6,FASC.6.5,Geneva,1989.
4.A.E.Kucherjavy."Statistical Approaches to Switching Equipment Monitoring",Proc.International Seminar on Teletraffic Theory and Computer Modelling.Sofia,Bulgaria,March 21-26,1988, p.261-272.

Author Index

AKIYAMA, M.	Japan	137
ANIDO, G.	Australia	305
ASGERSEN, Chr.	Denmark	61
BOLLA, R.	Italy	317
Le BON, A.	Canada	353
BORIK, M.J.	Czecho-Slovakia	575, 591
BOUMZEBRA, N.	Canada	539
BROCHIN, F.M.	Japan	269
BRUSILOVSKY, S.	USSR	607
BURAKOWSKI, W.	Poland	623
BURSZTYNOWSKI, D.	Poland	623
CARBONE, P.	Canada	97
CHLEBUS, E.	Poland	169
COTTON, M.	Canada	513
CRAIGNOU, B.	France	85
DAHMS, H.	Germany	439
DALLOS, G	Hungary	427
DAVOLI, F.	Italy	317
DUTKIEWICZ, E.	Australia	305
DZIONG, Z.	Poland	293
EHRIEL, I.	USSR	371
FILIPIAK, J.	Poland	11
GHASSEMI, M.	USA	663
GIRARD, A.	Canada	539
GOLDSHTEIN, B.	USSR	607
GRANEL, E.	Spain	489
GREAVES, D.J.	England	417
HÉBUTERNE, G.	France	365
HENDERSON, W.	South Australia	33
HORNE, G.A.	United Kingdom	149
HOSHIAI, Y.	Japan	581
HULICKI, Z.	Poland	473
JAROCIŃSKI, M.	Poland	207, 647

KANIYIL, J.	Japan	185
KARLSSON, J.M.	Sweden	281
KONVIT, M.	Czecho-Slovakia	575, 591
KOSITPAIBOON, R.	Canada	97
KROMOŁOWSKI, W.	Poland	453
KUCHERJAVY, A.	USSR	687
KUPRYANOVA, M.V.	USSR	461
KWIATKOWSKI, M.	Poland	525
LEBOURGES, M.	France	563
LIAO, K.-Q.	Canada	293
LIAU, B.	France	539
LINDGREN, G.	Sweden	635
LOMBARDO, A.	Italy	409
LUBACZ, J.	Poland	207, 647
MARIE, M.I.	Egypt	601
MARTÍN, F.E.	Spain	489
MASON, L.G.	Canada	293, 513
de MIGUEL, M.	Spain	73
MIZUSAVA, J.	Japan	137
MOUSTAFA, M.S.	Egypt	601
NAKATSUKA, S.	Japan	397
NOGUCHI, S.	Japan	185
NOWIKOW, P.	Poland	655
ONOZATO, Y.	Japan	185
PALAZZO, S.	Italy	409
PANNO, D.	Italy	409
PIGNATELLI, R.	Italy	409
PIÓRO, M.	Poland	73, 647
PITA, I.	Spain	73
POPESCU, A.	Sweden	381
RERLE, R.	USSR	607
RICHARDS, P.S.	Canada	49
ROBERTS, J.W.	France	221
ROOSMA, A.H.	Netherlands	257
ROSENBERG, C.	Canada	353, 365
RYDÉN, T.	Sweden	635
SAKSENA, V.R.	USA	111
SALZBORN, F.J.M.	South Australia	199
SATO, H.	Japan	397
SCHMITT, J.A.	USA	551
SCHNEPS-SCHNEPPE, M.A.	Latvia, USSR	123
SCHONFELD, T.J.	USA	111
SINGH, R.P.	USA	381

SKOOG, R.A.	USA	663
SOKOLOV, N.	USSR	617
SOTO, O.G.	Spain	207
SRIRAM, K.	USA	3
STERN, D.	France	163
SUSANNA, L.	Italy	409
SYSKI, W.	Poland	623
SZYMANOWSKI, M.	Poland	453
TOMASZEWSKI, A.	Poland	207, 647
UHL, T.	FRG	329
ULMER, J.	FRG	329
VEIRØ, B.	Denmark	343
VESELOVSKY, G.G.	USSR	461
VILLEN-ALTAMIRANO, J.	Spain	675
VILLEN-ALTAMIRANO, M.	Spain	675
WAJDA, K.	Poland	655
WERNIK, M.	Canada	97
WOŹNIAK, J.	Poland	501
YOKOTANI, T.	Japan	397
YOSHINO, H.	Japan	581
YUNUS, M.N.	Malaysia	245
ZHANG, Z.	USA	233
ZIELIŃSKI, K.	Poland	417